Springer Optimization and Its Applications

Volume 228

Aims and Scope

Optimization has continued to expand in all directions at an astonishing rate. New algorithmic and theoretical techniques are continually developing and the diffusion into other disciplines is proceeding at a rapid pace, with a spot light on machine learning, artificial intelligence, and quantum computing. Our knowledge of all aspects of the field has grown even more profound. At the same time, one of the most striking trends in optimization is the constantly increasing emphasis on the interdisciplinary nature of the field. Optimization has been a basic tool in areas not limited to applied mathematics, engineering, medicine, economics, computer science, operations research, and other sciences.

The series **Springer Optimization and Its Applications (SOIA)** aims to publish state-of-the-art expository works (monographs, contributed volumes, textbooks, handbooks) that focus on theory, methods, and applications of optimization. Topics covered include, but are not limited to, nonlinear optimization, combinatorial optimization, continuous optimization, stochastic optimization, Bayesian optimization, optimal control, discrete optimization, multi-objective optimization, and more. New to the series portfolio include Works at the intersection of optimization and machine learning, artificial intelligence, and quantum computing.

Volumes from this series are indexed by Web of Science, zbMATH, Mathematical Reviews, and SCOPUS.

Jein-Shan Chen

Complementarity Functions in Optimization

Springer

Jein-Shan Chen
Department of Mathematics
National Taiwan Normal University
Taipei, Taiwan

ISSN 1931-6828 ISSN 1931-6836 (electronic)
Springer Optimization and Its Applications
ISBN 978-3-032-06923-8 ISBN 978-3-032-06924-5 (eBook)
https://doi.org/10.1007/978-3-032-06924-5

This Springer imprint is published by the registered company Springer Nature Switzerland AG
The registered company address is: Gewerbestrasse 11, 6330 Cham, Switzerland

Preface

Complementarity functions occupy a central role in the field of optimization, as they enable the reformulation of the Karush-Kuhn-Tucker (KKT) conditions and the original optimization problem into either an equivalent system of equations or an unconstrained minimization problem. This reformulation opens the door to a wealth of novel approaches aimed at solving these equivalent forms, thereby generating solution candidates for the original problem. While two distinguished monographs have already addressed related areas of complementarity problems—focusing primarily on the existence of solutions, their stability, sensitivity analysis, and corresponding solution methods—this book adopts a distinct perspective. Here, we devote our primary attention to the study of complementarity functions themselves. Specifically, our goal is to present a comprehensive overview of their key properties and structural features, as well as to provide guiding principles for the construction of new complementarity functions. Moreover, we will illustrate how these functions can be effectively employed in algorithmic applications.

Chapter 1 offers an overview and essential background materials that lay the groundwork for the analyses presented in the subsequent chapters. In Chap. 2, we delve into the development of complementarity functions within the framework of nonlinear complementarity problems (NCPs), introducing several novel ideas and systematic techniques for constructing new NCP functions. Building upon these foundations, Chap. 3 extends the discussion to broader settings, encompassing second-order cones, positive semidefinite cones, and symmetric cones. These two chapters form the core of this monograph and are closely aligned with its central theme. Throughout Chaps. 2 and 3, the analysis is conducted within finite-dimensional spaces. Nevertheless, the development requires the careful handling of various inequalities and intricate technical arrangements. While the objectives in many instances may appear straightforward, the corresponding arguments can be laborious and subtle. Readers may, however, find valuable techniques and insights embedded within these detailed derivations.

In general, the unified analysis of certain C-functions presented in Sect. 3.3 encompasses and recovers the results discussed in Chap. 2, as well as Sects. 3.1 and 3.2, since symmetric cones naturally include the nonnegative orthant, the positive semidefinite cone, and the second-order cone as special cases. However, establishing certain properties within this unified framework may necessitate additional conditions. At first glance, one may observe parallel or analogous results across these various settings. This resemblance arises from the progression of C-function developments—originating in the classical NCP setting, advancing through the positive semidefinite and second-order cone contexts, and culminating in the broader symmetric cone framework. From a historical perspective, it is instructive to preserve this developmental pathway. Doing so not only allows readers to trace the evolution of these ideas but also helps them discern the subtle distinctions between settings and appreciate the diverse techniques required in each case.

In Chap. 4, we turn our attention to a selection of algorithms that employ complementarity functions. Broadly speaking, four distinct approaches are explored: the merit function approach, the nonsmooth function approach, the smoothing function approach, and the regularization approach. Within each framework, we present specially designed optimization algorithms that leverage the properties of complementarity functions, illustrating their practical utility and adaptability in algorithmic development.

In Chap. 5, we showcase the applications of complementarity functions within the framework of neural network methods, which differ fundamentally from traditional optimization techniques. For these dynamical systems, the primary concerns lie in the behavior of solution trajectories and the stability of the system, rather than the convergence rate or iteration count typically emphasized in conventional optimization algorithms. In particular, we focus on applying these methods to nonlinear complementarity problems and optimization problems involving second-order cones, providing illustrative examples to highlight their effectiveness in such contexts.

This book encapsulates my two decades of research on complementarity functions. Much like my earlier Springer monograph, "SOC Functions and Their Applications", it is dedicated, once more, to the cherished memory of my supervisor, Prof. Paul Tseng. I am profoundly grateful to have had the privilege of his mentorship. His unwavering encouragement and profound insight have continuously shaped and guided my academic journey. Though he tragically went missing in 2009, his exemplary dedication to research and his inspiring attitude remain etched in my heart. I wish to express my sincere appreciation to all my esteemed co-authors whose collaborative efforts have contributed to the material presented in this book: Prof. Shaohua Pan, Prof. Yu-Lin Chang, Prof. Chun-Hsu Ko, Prof. Xinhe Miao, Prof. Juhe Sun, Prof. Chu-Chin Hu, Dr. Ching-Yu Yang, Dr. Chien-Hao Huang, Dr. Thanh Chieu Nguyen, Dr. Harold Alcantara, among others. Working alongside them has been not only intellectually rewarding but also a source of great personal joy. My gratitude also extends to Xiaoni Chi and others who kindly assisted with proofreading, ensuring the clarity and precision of the text. Lastly, I owe my deepest thanks to my family-Vivian, Benjamin, and Ian along with Ian's beloved Doggy, Olah and Lil Bos.

Their unwavering support, love, and encouragement have been a constant source of strength, empowering me to pursue and persevere in my academic endeavors.

Taipei, Taiwan
June 2025

Jein-Shan Chen

Contents

Notations

- Throughout this book, an n-dimensional vector $x = (x_1, x_2, \cdots, x_n) \in \mathbb{R}^n$ means a *column* vector, i.e.,

$$x = \begin{bmatrix} x_1 \\ x_2 \\ \vdots \\ x_n \end{bmatrix}.$$

  In other words, without ambiguity, we also write the column vector as $x = (x_1, x_2, \cdots, x_n)$.
- $\mathbb{R}^n_+$ means $\{x = (x_1, x_2, \ldots, x_n) | x_i \geq 0, \text{ for all } i = 1, 2, \ldots, n\}$, whereas $\mathbb{R}^n_{++}$ denotes $\{x = (x_1, x_2, \ldots, x_n) | x_i > 0, \forall i = 1, 2, \ldots, n\}$.
- $\langle \cdot, \cdot \rangle$ denotes the Euclidean inner product.
- $\| \cdot \|$ is the Euclidean norm.
- ^{T} means transpose of a vector or a matrix.
- $B(x, \delta)$ denotes the neighborhood of x with radius $\delta > 0$.
- $\mathbb{R}^{n \times n}$ denotes the space of $n \times n$ real matrices.
- I represents an identity matrix of suitable dimension.
- For any symmetric matrices $A, B \in \mathbb{R}^{n \times n}$, we write $A \succeq B$ (respectively, $A \succ B$) to mean $A - B$ is positive semidefinite (respectively, positive definite).
- $\mathcal{S}^n$ denotes the space of $n \times n$ symmetric matrices; and $\mathcal{S}^n_+$ means the space of $n \times n$ symmetric positive semidefinite matrices.
- $\mathcal{O}$ denotes the set of $P \in \mathbb{R}^{n \times n}$ that are orthogonal, i.e., $P^{\mathrm{T}} = P^{-1}$.
- Given a set S, we denote $\bar{S}$, $\mathrm{int}(S)$ and $\mathrm{bd}(S)$ by the closure, the interior and the boundary of S, respectively.
- A function $f : \mathbb{R}^n \to (-\infty, \infty]$ is said to be proper if $f(\zeta) < \infty$ for at least one $\zeta \in \mathbb{R}^n$ and $f(\zeta) > -\infty$ for all $\zeta \in \mathbb{R}^n$.
- For a mapping $f : \mathbb{R}^n \to \mathbb{R}$, $\nabla f(x)$ denotes the gradient of f at x.
- For a closed proper convex function $f : \mathbb{R}^n \to (-\infty, \infty]$, we denote its domain by $\mathrm{dom} f := \{\zeta \in \mathbb{R}^n | f(\zeta) < \infty\}$.

- For a closed proper convex function $f : \mathbb{R}^n \to (-\infty, \infty]$, we denote the subdifferential of f at $\widehat{\zeta}$ by

$$\partial f(\widehat{\zeta}) := \left\{ w \in \mathbb{R}^n | f(\zeta) \geq f(\widehat{\zeta}) + \langle w, \zeta - \widehat{\zeta} \rangle, \forall \zeta \in \mathbb{R}^n \right\}.$$

- $C^{(i)}(J)$ denotes the family of functions which are defined on $J \subseteq \mathbb{R}^n$ to $\mathbb{R}$ and have continuous i-th derivative.
- For any differentiable mapping $F = (F_1, F_2, \cdots, F_m) : \mathbb{R}^n \to \mathbb{R}^m$, $\nabla F(x) = [\nabla F_1(x) \cdots \nabla F_m(x)]$ is a $n \times m$ matrix which denotes the transpose Jacobian of F at x.
- For a real valued function $f : J \to \mathbb{R}$, $f'(t)$ and $f''(t)$ denote the first derivative and second-order derivative of f at the differentiable point $t \in J$, respectively.
- For a mapping $F : S \subseteq \mathbb{R}^n \to \mathbb{R}^m$, $\partial F(x)$ denotes the subdifferential of F at x, while $\partial_B F(x)$ denotes the B-subdifferential of F at x.
- For nonnegative scalars α and β, we write $\alpha = O(\beta)$ to mean $\alpha \leq C\beta$, with C independent of α and β.
- We denote $K^* := \{y | \langle x, y \rangle \geq 0 \, \forall x \in K\}$ the dual cone of K, given any closed convex cone K.
- For any $x \in \mathbb{R}^n$, $(x)_+$ is used to denote the orthogonal projection of x onto $\mathcal{K}$, whereas $(x)_-$ means the orthogonal projection of x onto $-\mathcal{K}$.
- For any $x, y \in \mathbb{R}^n$, we write $x \succeq_{\mathcal{K}^n} y$ if $x - y \in \mathcal{K}^n$; and write $x \succ_{\mathcal{K}^n} y$ if $x - y \in \text{int}(\mathcal{K}^n)$.

List of Figures

List of Tables

Chapter 1
Backgrounds and Overviews

Abstract In this chapter, we provide a concise overview of complementarity problems frequently encountered in optimization, along with the various contexts from which complementarity functions naturally arise. Additionally, we revisit essential background material pertinent to the study of complementarity functions. Notably, the framework of *Euclidean Jordan algebra* offers a powerful and unifying approach for addressing a wide range of complementarity problems. To this end, we introduce the fundamental concepts of Euclidean Jordan algebra, beginning with the notion of *symmetric cones*, which play a central role in both complementarity problems and the construction of complementarity functions.

1.1 Symmetric Cones, Spectral Decomposition, and Löwner Function

Let $\mathbb{V}$ be a Euclidean space endowed with an inner product $\langle \cdot, \cdot \rangle$. A subset $K \subseteq \mathbb{V}$ is called a *cone* if

$$x \in K,\ \lambda \geq 0 \quad \Longrightarrow \quad \lambda x \in K.$$

A cone K which contains no line is said to be *pointed*, namely, $K \cap (-K) = \{0\}$. If K is also convex, then K is said to be a *convex cone*. Regarding convex cones, the following facts are well known.

(a) A set $K \subseteq \mathbb{V}$ is a convex cone if and only if

$$\lambda K \subseteq K \quad \forall \lambda \geq 0 \quad \text{and} \quad K + K \subseteq K.$$

(b) A set $K \subseteq \mathbb{V}$ is a convex cone if and only if it contains all nonnegative linear combinations of points in K.

(c) Let K be a convex set. The set

$$\{\lambda x \mid x \in K, \lambda \geq 0\}$$

J. Chen, *Complementarity Functions in Optimization*, Springer Optimization and Its Applications 228, https://doi.org/10.1007/978-3-032-06924-5_1

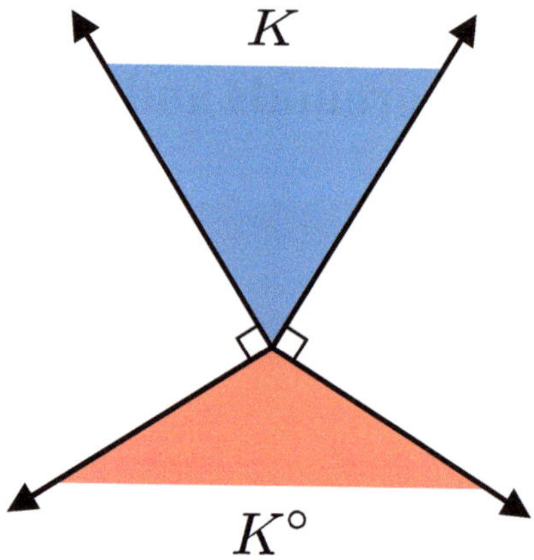

Fig. 1.1 The graph of K°

is the smallest convex cone containing K.

For any set $E \subseteq \mathbb{V}$, the set

$$E^\circ = \{y \in \mathbb{V} \mid \langle y, x\rangle \le 1, \quad \forall x \in E\}$$

is called the *polar set* of E. If E is a closed convex cone, $\lambda E \subseteq E$ for all $\lambda \ge 0$. Hence, the condition $\langle y, x\rangle \le 1, \forall x \in E$ is equivalent to $\langle y, x\rangle \le 0, \forall x \in E$. Therefore, the *polar cone* of a cone K is defined as

$$K^\circ = \{y \in \mathbb{V} \mid \langle y, x\rangle \le 0, \quad \forall x \in K\}.$$

To visualize the graph of K°, please see Fig. 1.1. Let K be a nonempty closed convex cone, it is also known that $K = K^{\circ\circ}$. For a closed cone $K \subseteq \mathbb{V}$, its *dual cone* K^* is given by

$$K^* := \{y \in \mathbb{V} \mid \langle x, y\rangle \ge 0, \quad \forall x \in K\}.$$

Let $\mathbb{V}$ be an n-dimensional vector space over the real field $\mathbb{R}$, endowed with a bilinear mapping $(x, y) \mapsto x \circ y$ from $\mathbb{V} \times \mathbb{V}$ into $\mathbb{V}$. The pair $(\mathbb{V}, \circ)$ is called a *Jordan algebra* [1–3] if the following two conditions are satisfied:

(i) $x \circ y = y \circ x$ for all $x, y \in \mathbb{V}$,
(ii) $x \circ (x^2 \circ y) = x^2 \circ (x \circ y)$ for all $x, y \in \mathbb{V}$.

Note that a Jordan algebra is not necessarily associative, i.e., $x \circ (y \circ z) = (x \circ y) \circ z$ may not hold for all $x, y, z \in \mathbb{V}$. We refer an element $e \in \mathbb{V}$ as the *identity* element if $x \circ e = e \circ x = x$ for all $x \in \mathbb{V}$. A Jordan algebra $(\mathbb{V}, \circ)$ with an identity element e is called a **Euclidean Jordan algebra** if there is an inner product, $\langle \cdot, \cdot\rangle$, such that

(iii) $\langle x \circ y, z\rangle = \langle y, x \circ z\rangle$ for all $x, y, z \in \mathbb{V}$.

Given a Euclidean Jordan algebra $(\mathbb{V}, \circ, \langle \cdot, \cdot \rangle)$, we denote the set of squares as

$$\mathcal{K} := \left\{ x^2 \mid x \in \mathbb{V} \right\}. \tag{1.1}$$

By [1, Theorem III.2.1], the set $\mathcal{K}$ described by (1.1) is called a **symmetric cone**, which means that $\mathcal{K}$ is a self-dual closed convex cone with nonempty interior and for any two elements $x, y \in \text{int}(\mathcal{K})$, there exists an invertible linear transformation $T : \mathbb{V} \to \mathbb{V}$ such that $T(\mathcal{K}) = \mathcal{K}$ and $T(x) = y$.

Example 1.1 The vector space $\mathbb{V} = \mathbb{R}^n$ with the usual inner product $\langle x, y \rangle = x^\top y$ can be made into a Euclidean Jordan algebra by defining

$$x \circ y = x \odot y,$$

where $\odot$ denotes the Hadamard product operator, i.e., $(x \odot y)_i = x_i y_i$ for $i = 1, \ldots, n$. Then, the set of squares $\mathcal{K}$ is precisely the nonnegative orthant $\mathbb{R}^n_+$, i.e., $\mathcal{K} = \mathbb{R}^n_+$.

Example 1.2 For $n > 1$, another bilinear mapping $\circ$ can be defined on $\mathbb{V} = \mathbb{R}^n$ if we write $x \in \mathbb{R}^n$ as $x = (x_1, \bar{x}_2) \in \mathbb{R} \times \mathbb{R}^{n-1}$. For any $x, y \in \mathbb{R}^n$, we define

$$x \circ y = \begin{bmatrix} x^\top y \\ x_1 \bar{y}_2 + y_1 \bar{x}_2 \end{bmatrix}. \tag{1.2}$$

The resulting Euclidean Jordan algebra is known as the *Jordan spin algebra* which we denote by $\mathbb{L}^n$. Its cone of squares is precisely the second-order cone (SOC for short), also called Lorentz cone and denoted by $\mathbb{L}^n_+$. In other words,

$$\mathcal{K} = \mathbb{L}^n_+ := \left\{ (x_1, \bar{x}_2) \in \mathbb{R} \times \mathbb{R}^{n-1} \mid \|\bar{x}_2\| \le x_1 \right\}. \tag{1.3}$$

Example 1.3 The vector space $\mathbb{V} = \mathbb{S}^n$ with inner product $\langle X, Y \rangle = \text{tr}(X^\top Y)$ and the bilinear map

$$X \circ Y = \frac{1}{2}(XY + YX)$$

forms a Euclidean Jordan algebra. Its cone of squares is precisely the set of all positive semidefinite matrices $\mathbb{S}^n_+$. In other words,

$$\mathcal{K} = \mathbb{S}^n_+ := \left\{ X \in \mathbb{S}^n \mid u^\top X u \ge 0, \ \ \forall u \in \mathbb{R}^n \right\}.$$

For any given $x \in \mathbb{V}$, let $\zeta(x)$ be the degree of the minimal polynomial of x, i.e.,

$$\zeta(x) := \min \left\{ k \,\middle|\, \{e, x, x^2, \ldots, x^k\} \text{ are linearly dependent} \right\}.$$

Then, the *rank* of $\mathbb{V}$ is defined as $\max\{\zeta(x) \mid x \in \mathbb{V}\}$. Here, we use r to denote the rank of the underlying Euclidean Jordan algebra. Recall that an element $c \in \mathbb{V}$ is *idempotent* if $c^2 = c$. Two idempotents c_i and c_j are said to be *orthogonal* if $c_i \circ c_j = 0$. One says that $\{c_1, c_2, \ldots, c_k\}$ is a *complete system of orthogonal idempotents* if

$$c_j^2 = c_j, \quad c_j \circ c_i = 0 \text{ if } j \neq i \text{ for all } j, i = 1, 2, \ldots, k, \quad \text{and} \sum_{j=1}^{k} c_j = e.$$

An idempotent is *primitive* if it is nonzero and cannot be written as the sum of two other nonzero idempotents. We call a complete system of orthogonal primitive idempotents a *Jordan frame*. The following Spectral Decomposition Theorem is very important in subsequent analysis under Euclidean Jordan algebra.

Theorem 1.1 ([1, Theorem III.1.2]) *Suppose that* $\mathbb{V}$ *is a Euclidean Jordan algebra with the rank* r. *Then for any* $x \in \mathbb{V}$, *there exists a Jordan frame* $\{c_1, \ldots, c_r\}$ *and real numbers* $\lambda_1(x), \ldots, \lambda_r(x)$, *arranged in the decreasing order* $\lambda_1(x) \geq \lambda_2(x) \geq \cdots \geq \lambda_r(x)$, *such that*

$$x = \lambda_1(x)c_1 + \lambda_2(x)c_2 + \cdots + \lambda_r(x)c_r. \tag{1.4}$$

The numbers $\lambda_j(x)$ *(counting multiplicities), which are uniquely determined by* x, *are called the eigenvalues and* $\operatorname{tr}(x) = \sum_{j=1}^{r} \lambda_j(x)$ *the trace of* x.

Since, by [1, Proposition III.1.5], a Jordan algebra $(\mathbb{V}, \circ)$ with an identity element $e \in \mathbb{V}$ is Euclidean if and only if the symmetric bilinear form $\operatorname{tr}(x \circ y)$ is positive definite, we may define another inner product on $\mathbb{V}$ by $\langle x, y \rangle := \operatorname{tr}(x \circ y)$ for any $x, y \in \mathbb{V}$. The inner product $\langle \cdot, \cdot \rangle$ is associative by [1, Proposition II. 4.3], i.e., $\langle x, y \circ z \rangle = \langle y, x \circ z \rangle$ for any $x, y, z \in \mathbb{V}$. Accordingly, we let $\| \cdot \|$ be the norm on $\mathbb{V}$ induced by the inner product, namely,

$$\|x\| := \sqrt{\langle x, x \rangle} = \left(\sum_{j=1}^{r} \lambda_j^2(x) \right)^{1/2}, \quad \forall x \in \mathbb{V}.$$

Then, by the Schwartz inequality, it is easy to verify that

$$\|x \circ y\| \leq \|x\| \cdot \|y\|, \quad \forall\, x, y \in \mathbb{V}. \tag{1.5}$$

For any given $x \in (\mathbb{V}, \circ, \langle \cdot, \cdot \rangle)$, let $\mathcal{L}(x)$ be the linear operator of $\mathbb{V}$ defined by

$$\mathcal{L}(x)y := x \circ y \quad \forall y \in \mathbb{V}. \tag{1.6}$$

It is noted that $\mathcal{L}(x)$ is symmetric with respect to the inner product $\langle \cdot, \cdot \rangle$ in the sense of

$$\langle \mathcal{L}(x)y, z \rangle = \langle y, \mathcal{L}(x)z \rangle \quad \forall y, z \in \mathbb{V}.$$

Suppose that $(\mathbb{V}, \circ, \langle\cdot,\cdot\rangle)$ is a simple Euclidean Jordan algebra of rank r and $\{c_1, c_2, \dots, c_r\}$ is a Jordan frame of $\mathbb{V}$. From [1, Lemma IV.], we know that the operators $\mathcal{L}(c_j)$, $j = 1, 2, \dots, r$ commute and admit a simultaneous diagonalization. In particular, for $i, j \in \{1, 2, \dots, r\}$, define the subspaces

$$\begin{aligned} \mathbb{V}_{ii} &:= \{x \in \mathbb{V} \mid x \circ c_i = x\} = \mathbb{R}c_i, \\ \mathbb{V}_{ij} &:= \left\{x \in \mathbb{V} \mid c_i \circ x = c_j \circ x = \frac{1}{2}x\right\} \text{ when } i \neq j. \end{aligned}$$

From [1, Corollary IV.2.6], it says that

$$\dim(\mathbb{V}_{ij}) = \dim(\mathbb{V}_{st}) \quad \text{for any } i \neq j \in \{1, 2, \dots, r\} \text{ and } s \neq t \in \{1, 2, \dots, r\},$$

and $n = r + \frac{d}{2}r(r-1)$, where d denotes this common dimension. Moreover, from [1, Theorem IV.2.1], we have the second version of decomposition.

Theorem 1.2 ([1, Theorem IV.2.1]) *The space $\mathbb{V}$ is the orthogonal direct sum of subspaces $\mathbb{V}_{ij}$ $(1 \le i \le j \le r)$, i.e., $\mathbb{V} = \oplus_{i\le j}\mathbb{V}_{ij}$. Furthermore,*

$$\begin{aligned} \mathbb{V}_{ij} \circ \mathbb{V}_{ij} &\subset \mathbb{V}_{ii} + \mathbb{V}_{ij}, \\ \mathbb{V}_{ij} \circ \mathbb{V}_{jk} &\subset \mathbb{V}_{ik}, \text{ if } i \neq k, \\ \mathbb{V}_{ij} \circ \mathbb{V}_{kl} &= \{0\}, \text{ if } \{i, j\} \cap \{k, l\} = \emptyset. \end{aligned}$$

Hence, given any fixed Jordan frame $\{c_1, c_2, \dots, c_r\}$, we can write any element $x \in \mathbb{V}$ as

$$x = \sum_{i=1}^{r} x_i c_i + \sum_{i<j} x_{ij},$$

where $x_i \in \mathbb{R}$ and $x_{ij} \in \mathbb{V}_{ij}$. The expression $\sum_{i=1}^{r} x_i c_i + \sum_{i<j} x_{ij}$ is called the Peirce decomposition of x.

The decomposition in Theorem 1.1 is called the *spectral decomposition*, whereas the decomposition in Theorem 1.2 is called the *Peirce decomposition*. For different elements x and y in $\mathbb{V}$, the Jordan frames in their spectral decompositions are different. To the contrast, x and y share the same Jordan frame in the Peirce decomposition. A Euclidean Jordan algebra is called *simple* if it cannot be written as a direct sum of the other two Euclidean Jordan algebras. It is known that every Euclidean Jordan algebra can be written as a direct sum of simple ones, which are not themselves direct sums in a nontrivial way. In finite dimensions, the simple Euclidean Jordan algebras come from the following five basic structures.

Theorem 1.3 ([1, Chapter V.3.7]) *Every simple Euclidean Jordan algebra is isomorphic to one of the followings.*

(i) The Jordan spin algebra $\mathbb{L}^n$.
(ii) The algebra $\mathbb{S}^n$ *of* $n \times n$ *real symmetric matrices.*
(iii) The algebra $\mathbb{H}^n$ *of all* $n \times n$ *complex Hermitian matrices.*
(iv) The algebra $\mathbb{Q}^n$ *of all* $n \times n$ *quaternion Hermitian matrices.*
(v) The algebra $\mathbb{O}^3$ *of all* 3×3 *octonion Hermitian matrices.*

Given an n-dimensional Euclidean Jordan algebra $(\mathbb{V}, \langle \cdot, \cdot \rangle, \circ)$ with $\mathcal{K}$ being its corresponding symmetric cone in $\mathbb{V}$. For any scalar function $f : \mathbb{R} \to \mathbb{R}$, we define a vector-valued function $f^{\mathrm{sc}}(x)$ (called Löwner function) on $\mathbb{V}$ as

$$f^{\mathrm{sc}}(x) = f(\lambda_1(x))c_1 + f(\lambda_2(x))c_2 + \cdots + f(\lambda_r(x))c_r, \tag{1.7}$$

where $x \in \mathbb{V}$ has the spectral decomposition as

$$x = \lambda_1(x)c_1 + \lambda_2(x)c_2 + \cdots + \lambda_r(x)c_r.$$

As mentioned earlier, when $\mathbb{V}$ represents the Jordan spin algebra $\mathbb{L}_n$, $\mathcal{K}$ corresponds to the second-order cone (SOC) given as in (1.3). For convenience, we also denote it by $\mathcal{K}^n$, which means a single SOC, that is,

$$\mathcal{K}^n := \{(x_1, x_2) \in \mathbb{R} \times \mathbb{R}^{n-1} \mid \|x_2\| \le x_1\}.$$

In particular, when $n = 1$, $\mathcal{K}^n$ reduces to the set of nonnegative real numbers $\mathbb{R}_+$. Under such case, the spectral decomposition (1.4) of $x = (x_1, x_2) \in \mathbb{R} \times \mathbb{R}^{n-1}$ appeared in Theorem 1.1 becomes

$$x = \lambda_1(x)u_x^{(1)} + \lambda_2(x)u_x^{(2)}, \tag{1.8}$$

where $\lambda_1(x)$, $\lambda_2(x)$, $u_x^{(1)}$ and $u_x^{(2)}$ with respect to $\mathcal{K}^n$ are given by

$$\lambda_i(x) = x_1 + (-1)^i \|x_2\|, \tag{1.9}$$

$$u_x^{(i)} = \begin{cases} \frac{1}{2}\left(1, (-1)^i \frac{x_2}{\|x_2\|}\right) & \text{if } x_2 \neq 0, \\ \frac{1}{2}\left(1, (-1)^i w\right) & \text{if } x_2 = 0, \end{cases} \tag{1.10}$$

for $i = 1, 2$, with w being any vector in $\mathbb{R}^{n-1}$ satisfying $\|w\| = 1$. If $x_2 \neq 0$, the decomposition (1.8) is unique. The determinant and trace of x are defined as $\det(x) := \lambda_1(x)\lambda_2(x)$ and $\mathrm{tr}(x) := \lambda_1(x) + \lambda_2(x)$, respectively.

Under the SOC setting, the Löwner function defined as in (1.7) reduces to so-called SOC-function f^{soc} studied in [4–8]. More specifically, with this spectral decomposition, for any scalar function $f : J \subseteq \mathbb{R} \to \mathbb{R}$, the Löwner function f^{sc} associated with $\mathcal{K}^n$ reduces to $f^{\mathrm{soc}} : S \subseteq \mathbb{R}^n \to \mathbb{R}^n$ as below:

$$f^{\mathrm{soc}}(x) = f(\lambda_1(x))u_x^{(1)} + f(\lambda_2(x))u_x^{(2)} \quad \forall x = (x_1, x_2) \in \mathbb{R} \times \mathbb{R}^{n-1}, \tag{1.11}$$

where J is an interval (finite or infinite, open or closed) of $\mathbb{R}$, and S is the domain of f^{soc} determined by f.

In the SOC setting, Chen, Chen, and Tseng [8] demonstrated that the Löwner function f^{soc} inherits several key properties from the underlying function f, including continuity, Lipschitz continuity, directional differentiability, Fréchet differentiability, continuous differentiability, and semismoothness. The Hölder continuity of both f^{soc} and f was further established in [4]. Sun and Sun [9] extended some of these foundational results to the broader context of symmetric cones, specifically regarding f^{sc}. Moreover, the SOC trace function associated with f can be defined as follows:

$$f^{\text{tr}}(x) := f(\lambda_1(x)) + f(\lambda_2(x)) = \text{tr}(f^{\text{soc}}(x)) \quad \forall x \in S. \tag{1.12}$$

Chen, Liao, and Pan [10] built up the following relation between f^{tr} and f^{soc}

$$\nabla f^{\text{tr}}(x) = (f')^{\text{soc}}(x) \quad \text{and} \quad \nabla^2 f^{\text{tr}}(x) = \nabla (f')^{\text{soc}}(x) \quad \forall x \in \text{int}S.$$

By employing the Schur Complement Theorem, they establish the convexity of SOC trace functions as well as compositions involving these functions. Several of these functions play a pivotal role in penalty and barrier function methods for second-order cone programs (SOCPs). Furthermore, certain fundamental inequalities related to second-order cones are instrumental in demonstrating the convexity properties of these functions. For a more comprehensive discussion on the roles and applications of f^{soc} and f^{tr}, defined in (1.11) and (1.12), respectively, the reader is referred to [7].

When $\mathbb{V}$ represents the algebra $\mathbb{S}^n$ of $n \times n$ real symmetric matrices, what do the spectral decomposition and the Löwner function look like? For any $X \in \mathbb{S}^n$, its (repeated) eigenvalues $\lambda_1, \ldots, \lambda_n$ are real and it admits a spectral decomposition of the form:

$$X = P \,\text{diag}\big[\lambda_1, \ldots, \lambda_n\big]\, P^{\mathsf{T}}, \tag{1.13}$$

for some orthogonal matrix P, where $\text{diag}[\lambda_1, \ldots, \lambda_n]$ denotes the $n \times n$ diagonal matrix with its ith diagonal entry λ_i. In fact, the spectral decomposition (1.13) corresponds to the spectral decomposition in Theorem 1.1. To see this, letting $P = \big[\mathbf{u}_1 \mid \mathbf{u}_2 \mid \cdots \mid \mathbf{u}_n\big]$, by taking the Jordan frame $\{c_1, c_2, \ldots, c_n\}$ as

$$\left\{\mathbf{u}_1\mathbf{u}_1^{\mathsf{T}}, \mathbf{u}_2\mathbf{u}_2^{\mathsf{T}}, \ldots, \mathbf{u}_n\mathbf{u}_n^{\mathsf{T}}\right\},$$

it can be verified that the spectral decomposition (1.4) of X, that is,

$$X = \lambda_1(X)c_1 + \lambda_2(X)c_2 + \cdots + \lambda_r(X)c_r,$$

reduces to the above matrix decomposition (1.13). Likewise, under the $\mathbb{S}^n_+$ setting, the Löwner function defined as in (1.7) reduces to a matrix-valued function. More specifically, for any function $f : \mathbb{R} \to \mathbb{R}$, we can define a corresponding function $f^{\text{mat}} : \mathbb{S}^n \to \mathbb{S}^n$ [11, 12] by

$$f^{\text{mat}}(X) := P \operatorname{diag}\left[f(\lambda_1), \ldots, f(\lambda_n)\right] P^{\mathsf{T}}. \tag{1.14}$$

It is known that $f^{\text{mat}}(X)$ is well-defined (independent of the ordering of $\lambda_1, \ldots, \lambda_n$ and the choice of P) and belongs to $\mathbb{S}^n$, see [11, Chapter V] and [12, Sect. 6.2]. Moreover, a result of Daleckii and Krein showed that if f is continuously differentiable, then f^{mat} is differentiable and its Jacobian $\nabla f^{\text{mat}}(X)$ has a simple formula, see [11, Theorem V.3.3]; also see [13, Proposition 4.3].

The function f^{mat} was used to develop non-interior continuation methods for solving semidefinite programs and semidefinite complementarity problems in [14]. Another related method was studied in [15]. Further studies of f^{mat} in the case of $f(\xi) = |\xi|$ and $f(\xi) = \max\{0, \xi\}$ were conducted in [16, 17], obtaining results such as strong semismoothness, formulas for directional derivatives, and necessary/sufficient conditions for strong stability of an isolated solution to semidefinite complementarity problem (SDCP).

The SOC function f^{soc} defined as in (1.11) has a connection to the matrix-valued f^{mat} given as in (1.14) via a special mapping. To see this, in light of the Löwner operator $\mathcal{L}(\cdot)$ given as in (1.6), for any $x = (x_1, x_2) \in \mathbb{R} \times \mathbb{R}^{n-1}$, we define a linear mapping from $\mathbb{R}^n$ to $\mathbb{R}^n$ as

$$\begin{aligned} L_x : \mathbb{R}^n &\longrightarrow \mathbb{R}^n \\ y &\longmapsto L_x y := \begin{bmatrix} x_1 & x_2^{\mathsf{T}} \\ x_2 & x_1 I \end{bmatrix} y. \end{aligned} \tag{1.15}$$

It can be easily verified that $x \circ y = L_x y$ for all $y \in \mathbb{R}^n$, and L_x is positive definite (and hence invertible) if and only if $x \in \operatorname{int}(\mathcal{K}^n)$. However, $L_x^{-1} y \neq x^{-1} \circ y$, for some $x \in \operatorname{int}(\mathcal{K}^n)$ and $y \in \mathbb{R}^n$, i.e., $L_x^{-1} \neq L_{x^{-1}}$. The mapping L_x defined as in (1.15) will be used to relate f^{soc} to f^{mat}; see relation (1.17) in Proposition 1.1. For convenience, in the subsequent contexts, we sometimes omit the variable notion x in $\lambda_i(x)$ and $u_x^{(i)}$ for $i = 1, 2$.

Proposition 1.1 *Let $x = (x_1, x_2) \in \mathbb{R} \times \mathbb{R}^{n-1}$ with spectral values $\lambda_1(x)$, $\lambda_2(x)$ given by (1.9) and spectral vectors $u_x^{(1)}$, $u_x^{(2)}$ given by (1.10). We denote $z := x_2$ if $x_2 \neq 0$; otherwise let z be any nonzero vector in $\mathbb{R}^{n-1}$. Then, the following results hold.*

(a) For any $t \in \mathbb{R}$, the matrix $L_x + tM_z$ has eigenvalues $\lambda_1(x)$, $\lambda_2(x)$, and $x_1 + t$ of multiplicity $n - 2$, where

$$M_z := \begin{bmatrix} 0 & 0 \\ 0 & I - \frac{zz^{\mathsf{T}}}{\|z\|^2} \end{bmatrix}. \tag{1.16}$$

(b) For any $f : \mathbb{R} \to \mathbb{R}$ and any $t \in \mathbb{R}$, we have

$$f^{\text{soc}}(x) = f^{\text{mat}}(L_x + tM_z)e. \tag{1.17}$$

Proof (a) It is straightforward to verify that, for any $x = (x_1, x_2) \in \mathbb{R} \times \mathbb{R}^{n-1}$, the eigenvalues of L_x are $\lambda_1(x), \lambda_2(x)$, as given by (1.9), and x_1 of multiplicity $n-2$. Its corresponding orthonormal set of eigenvectors is

$$\sqrt{2}u_x^{(1)}, \sqrt{2}u_x^{(2)}, u_x^{(i)} = (0, u_2^{(i)}), \quad i = 3, \ldots, n,$$

where $u_x^{(1)}, u_x^{(2)}$ are the spectral vectors with $w = \frac{z}{\|z\|}$ whenever $x_2 = 0$, and $u_2^{(3)}, \ldots, u_2^{(n)}$ is any orthonormal set of vectors that span the subspace of $\mathbb{R}^{n-2}$ orthogonal to z. Thus,

$$L_x = U\mathrm{diag}[\lambda_1(x), \lambda_2(x), x_1, \ldots, x_1]U^\top,$$

where $U := \left[\sqrt{2}u_x^{(1)}\ \sqrt{2}u_x^{(2)}\ u_x^{(3)} \cdots u_x^{(n)}\right]$. In addition, by using $u_x^{(i)} = \left(0, u_2^{(i)}\right)$, $i = 3, \ldots, n$, it is not hard to verify that

$$U\,\mathrm{diag}[0, 0, 1, \ldots, 1]\,U^\top = \begin{bmatrix} 0 & 0 \\ 0 & \sum_{i=3}^{n} u_2^{(i)}(u_2^{(i)})^\top \end{bmatrix}.$$

Since $Q := \left[\frac{z}{\|z\|}\ u_2^{(3)} \cdots u_2^{(n)}\right]$ is an orthogonal matrix, we have

$$I = QQ^\top = \frac{zz^\top}{\|z\|^2} + \sum_{i=3}^{n} u_2^{(i)}(u_2^{(i)})^\top$$

and hence $\sum_{i=3}^{n} u_2^{(i)}(u_2^{(i)})^\top = I - \frac{zz^\top}{\|z\|^2}$. This together with (1.16) shows that

$$U\mathrm{diag}[0, 0, 1, \ldots, 1]\,U^\top = M_z.$$

Thus, we obtain

$$L_x + tM_z = U\mathrm{diag}[\lambda_1(x), \lambda_2(x), x_1 + t, \ldots, x_1 + t]\,U^\top, \tag{1.18}$$

which is the desired result.

(b) Applying (1.18) yields

$$\begin{aligned} f^{\mathrm{mat}}(L_x + tM_z)e &= U\mathrm{diag}\left[f(\lambda_1(x)), f(\lambda_2(x)), f(x_1 + t), \ldots, f(x_1 + t)\right]U^\top e \\ &= f(\lambda_1(x))u_x^{(1)} + f(\lambda_2(x))u_x^{(2)} \\ &= f^{\mathrm{soc}}(x), \end{aligned}$$

where the second equality uses the special form of U. Then, the proof is complete. □

In light of (1.7), let $\varphi : \mathbb{R} \to \mathbb{R}$ be a scalar valued function. There exists a vector-valued function associated with the Euclidean Jordan algebra $(\mathbb{V}, \circ, \langle \cdot, \cdot \rangle)$ given by

$$\varphi_{\mathbb{V}}(x) := \varphi(\lambda_1(x))c_1 + \varphi(\lambda_2(x))c_2 + \cdots + \varphi(\lambda_r(x))c_r, \tag{1.19}$$

where $x \in \mathbb{V}$ has the spectral decomposition $x = \sum_{j=1}^r \lambda_j(x)c_j$. The function $\varphi_{\mathbb{V}}$ is also called the Löwner operator [9]. When $\varphi(t)$ is chosen as $\max\{0, t\}$ and $\min\{0, t\}$ for $t \in \mathbb{R}$, respectively, $\varphi_{\mathbb{V}}$ becomes the metric projection operator onto $\mathcal{K}$ and $-\mathcal{K}$:

$$(x)_+ := \sum_{j=1}^r \max\{0, \lambda_j(x)\}c_j \quad \text{and} \quad (x)_- := \sum_{j=1}^r \min\{0, \lambda_j(x)\}c_j.$$

Theorem 1.4 ([9, Theorem 13]) *For any $x = \sum_{j=1}^r \lambda_j(x)c_j$, let $\varphi_{\mathbb{V}}$ be given as in (1.19). Then, $\varphi_{\mathbb{V}}$ is (continuously) differentiable at x if and only if φ is (continuously) differentiable at each $\lambda_j(x)$, $j = 1, 2, \ldots, r$. The derivative of $\varphi_{\mathbb{V}}$ at x, for any $h \in \mathbb{V}$, is*

$$\varphi'_{\mathbb{V}}(x)h = \sum_{j=1}^r \left[\varphi^{[1]}(\lambda(x))\right]_{jj} \langle c_j, h\rangle c_j + \sum_{1\le j<l\le r} 4\left[\varphi^{[1]}(\lambda(x))\right]_{jl} c_j \circ (c_l \circ h),$$

where

$$\left[\varphi^{[1]}(\lambda(x))\right]_{ij} := \begin{cases} \dfrac{\varphi(\lambda_i(x)) - \varphi(\lambda_j(x))}{\lambda_i(x) - \lambda_j(x)} & \text{if } \lambda_i(x) \ne \lambda_j(x) \\ \varphi'(\lambda_i(x)) & \text{if } \lambda_i(x) = \lambda_j(x) \end{cases}, \quad i, j = 1, 2, \ldots, r.$$

In fact, the Jacobian $\varphi'_{\mathbb{V}}(\cdot)$ is a linear and symmetric operator, which can be written as

$$\varphi'_{\mathbb{V}}(x) = \sum_{j=1}^r \varphi'(\lambda_j(x))\mathcal{P}(c_j) + 2\sum_{i,j=1, i\ne j}^r \left[\varphi^{[1]}(\lambda(x))\right]_{ij} \mathcal{L}(c_j)\mathcal{L}(c_i) \tag{1.20}$$

where $\mathcal{P}(x) := 2\mathcal{L}^2(x) - \mathcal{L}(x^2)$ for any $x \in M$ is called the *quadratic representation* of $\mathbb{V}$. Consider $x \in \mathbb{V}$ with the spectral decomposition $x = \sum_{j=1}^r \lambda_j(x)c_j$. For $i, j \in \{1, 2, \ldots, r\}$, let $\mathcal{C}_{ij}(x)$ be the orthogonal projection operator onto $\mathbb{V}_{ij}$. Then, there hold

$$\mathcal{C}_{ij}(x) = \mathcal{C}^*_{ij}(x),\ \mathcal{C}^2_{ij}(x) = \mathcal{C}_{ij}(x),\ \mathcal{C}_{ij}(x)\mathcal{C}_{kl}(x) = 0 \text{ if } \{i, j\} \ne \{k, l\},\ i, j, k, l = 1, \ldots, r \tag{1.21}$$

and

$$\sum_{1\le i\le j\le r} \mathcal{C}_{ij}(x) = \mathcal{I},$$

where $\mathcal{C}_{ij}^*$ is the adjoint (operator) of $\mathcal{C}_{ij}$. Moreover, by using [1, Theorem IV. 2.1], it indicates

$$\mathcal{C}_{jj}(x) = \mathcal{P}(c_j) \text{ and } \mathcal{C}_{ij}(x) = 4\mathcal{L}(c_i)\mathcal{L}(c_j) = 4\mathcal{L}(c_j)\mathcal{L}(c_i) = \mathcal{C}_{ji}(x),\ i, j = 1, 2, \ldots, r.$$

Note that the original notation in [1] for orthogonal projection operator is P_{ij}. However, to avoid confusion with another orthogonal projector $P_i(c_j)$ onto $\mathbb{V}(c, \alpha)$ and orthogonal matrix P which will be used later, we adopt $\mathcal{C}_{ij}$ instead.

With the orthogonal projection operators $\{\mathcal{C}_{ij}(x) \mid i, j = 1, 2, \ldots, r\}$, we have the following spectral decomposition theorem for $\mathcal{L}(x)$ and $\mathcal{L}(x^2)$; see [2, Chaps. VI–V].

Theorem 1.5 *Let $x \in \mathbb{V}$ have the spectral decomposition $x = \sum_{j=1}^r \lambda_j(x)c_j$ and $\mathcal{L}(\cdot)$ be defined as in (1.6). Then, the operator $\mathcal{L}(x)$ has the spectral decomposition*

$$\mathcal{L}(x) = \sum_{j=1}^{r} \lambda_j(x)\mathcal{C}_{jj}(x) + \sum_{1\le j<l\le r} \frac{1}{2}\left(\lambda_j(x) + \lambda_l(x)\right)\mathcal{C}_{jl}(x)$$

with the spectrum $\sigma(\mathcal{L}(x))$ consisting of all distinct numbers in

$$\left\{\frac{1}{2}(\lambda_j(x) + \lambda_l(x)) \mid j, l = 1, 2, \ldots, r\right\},$$

and $\mathcal{L}(x^2)$ has the spectral decomposition

$$\mathcal{L}(x^2) = \sum_{j=1}^{r} \lambda_j^2(x)\mathcal{C}_{jj}(x) + \sum_{1\le j<l\le r} \frac{1}{2}\left(\lambda_j^2(x) + \lambda_l^2(x)\right)\mathcal{C}_{jl}(x)$$

with the spectrum $\sigma(\mathcal{L}(x^2))$ consisting of all distinct numbers in

$$\left\{\frac{1}{2}\left(\lambda_j^2(x) + \lambda_l^2(x)\right) \mid j, l = 1, 2, \ldots, r\right\}.$$

Proposition 1.2 *For any $x \in \mathbb{V}$, the operator $\mathcal{L}(x^2) - \mathcal{L}^2(x)$ is positive semidefinite.*

Proof By Theorem 1.5 and (1.21), we can verify that $\mathcal{L}^2(x)$ has the spectral decomposition:

$$\mathcal{L}^2(x) = \sum_{j=1}^{r} \lambda_j^2(x)\mathcal{C}_{jj}(x) + \sum_{1\le j<l\le r} \frac{1}{4}\left(\lambda_j(x) + \lambda_l(x)\right)^2 \mathcal{C}_{jl}(x).$$

This means that the operator $\mathcal{L}(x^2) - \mathcal{L}^2(x)$ has the spectral decomposition

$$\mathcal{L}(x^2) - \mathcal{L}^2(x) = \sum_{1\le j<l\le r} \left[\frac{1}{2}\left(\lambda_j^2(x) + \lambda_l^2(x)\right) - \frac{1}{4}\left(\lambda_j(x) + \lambda_l(x)\right)^2\right]\mathcal{C}_{jl}(x).$$

Noting that the orthogonal projection operator is positive semidefinite on $\mathcal{V}$ and

$$\frac{\lambda_j^2(x) + \lambda_l^2(x)}{2} \geq \frac{(\lambda_j(x) + \lambda_l(x))^2}{4} \quad \text{for all} \quad j, l = 1, 2, \ldots, r,$$

from which the conclusion follows, utilizing the spectral decomposition of $\mathcal{L}(x^2) - \mathcal{L}^2(x)$.
□

1.2 Complementarity Problems

Complementarity conditions lie at the heart of both the theoretical foundations and numerical analysis of numerous optimization algorithms. They frequently emerge, for example, in the formulation of the Karush-Kuhn-Tucker (KKT) conditions in mathematical programming, which underpin most - if not all - of the algorithms explored in the subsequent chapters. Beyond their pivotal role in optimization, complementarity problems also provide a well-established unified framework for addressing equilibrium models arising in various applied disciplines, including operations research, engineering, and economics [18, 19].

The standard setting of a complementarity problem is a Euclidean space $\mathbb{V}$ endowed with an inner product $\langle \cdot, \cdot \rangle$, in which we define the complementarity problem as below.

Definition 1.1 Let $F : \mathbb{V} \to \mathbb{V}$ and let $\mathcal{K}$ be a cone in $\mathbb{V}$. The problem of finding a point $x \in \mathbb{V}$ that satisfies

$$x \in \mathcal{K}, \quad F(x) \in \mathcal{K}^*, \quad \text{and} \quad \langle x, F(x) \rangle = 0 \tag{1.22}$$

is known as a **complementarity problem.**

Below are some well-known and classic examples of complementarity problems.

Example 1.4 (*Nonlinear Complementarity Problem*) Let $\mathbb{V} = \mathbb{R}^n$ and consider the usual inner product $\langle x, y \rangle = x^{\mathsf{T}} y$. Setting $\mathcal{K} = \mathbb{R}^n_+$, then $\mathcal{K}^* = \mathcal{K}$, i.e., $\mathcal{K}$ is self-dual, and the complementarity problem (1.22) reduces to finding $x \in \mathbb{R}^n$ such that

$$x \geq 0, \quad F(x) \geq 0, \quad \text{and} \quad \langle x, F(x) \rangle = 0, \tag{1.23}$$

which is known as nonlinear complementarity problem (NCP).

Example 1.5 (*Linear Complementarity Problem*) Let $\mathbb{V} = \mathbb{R}^n$ and consider the usual inner product $\langle x, y \rangle = x^{\mathsf{T}} y$. From the NCP (1.23), there corresponds to linear complementarity problem (LCP) when F reduces to affine function $Mx + q$ where M is an $n \times n$ matrix and $q \in \mathbb{R}^n$. It is usually denoted by LCP(M, q) with the mathematical format

$$x \geq 0, \quad Mx + q \geq 0, \quad \text{and} \quad \langle x, Mx + q \rangle = 0. \tag{1.24}$$

The Linear Complementarity Problem (LCP) (1.24) is not only equivalent to mixed linear $0 - 1$ optimization, but is also equivalent to the mixed integer feasibility problem. In addition, there exist several notable variants of the LCP, such as the horizontal LCP and vertical LCP. For a comprehensive treatment of these topics, the reader is referred to the monograph [18] and the Encyclopedia of Optimization [20].

Example 1.6 (*Second-Order Cone Complementarity Problem*) Let $\mathbb{V} = \mathbb{R}^n$ be endowed with the inner product $\langle x, y \rangle = x^\top y$. The *Lorentz cone*, also known as the *second-order cone*, is defined as

$$\mathcal{K} = \mathbb{L}_+^n := \left\{ (x_1, \bar{x}_2) \in \mathbb{R} \times \mathbb{R}^{n-1} \mid \|\bar{x}_2\| \leq x_1 \right\}.$$

This cone is also self-dual, and the corresponding complementarity problem is to find $x \in \mathbb{R}^n$ such that

$$x \succeq_{\mathbb{L}_+^n} 0, \quad F(x) \succeq_{\mathbb{L}_+^n} 0, \quad \text{and} \quad \langle x, F(x) \rangle = 0, \tag{1.25}$$

which is called the second-order cone complementarity problem (SOCCP).

Example 1.7 (*Positive Semidefinite Cone Complementarity Problem*) Let $\mathbb{V} = \mathbb{S}^n$ be the vector space of all $n \times n$ symmetric matrices endowed with the inner product $\langle X, Y \rangle = \text{tr}(X^\top Y)$. We consider the cone of positive semidefinite matrices $\mathcal{K} = \mathbb{S}_+^n$, which is again self-dual. The resulting complementarity problem is the search for a matrix $X \in \mathbb{R}^{n \times n}$ such that

$$X \succeq 0, \quad F(X) \succeq 0, \quad \text{and} \quad \langle X, F(X) \rangle = 0, \tag{1.26}$$

known as the positive semidefinite cone complementarity problem (SDCP).

Example 1.8 (*Symmetric Cone Complementarity Problem*) Let $\mathbb{V}$ be the general Euclidean Jordan algebra introduced in Sect. 1.1 and $\mathcal{K}$ be the symmetric cone defined as in (1.1). Then, the complementarity problem (1.22) becomes

$$x \in \mathcal{K}, \quad F(x) \in \mathcal{K}, \quad \langle x, F(x) \rangle = 0, \tag{1.27}$$

which is called the symmetric cone complementarity problem (SCCP).

As highlighted in Theorem 1.3 (see also [1, Chap. V.3.7]), the cones featured in the aforementioned examples belong to the class of symmetric cones. Consequently, the SCCP (1.27) serves as a unified framework encompassing the NCP (1.23), the SOCCP (1.25), and the SDCP (1.26). A recent study [21] further introduces systematic methodologies for constructing general non-symmetric cones.

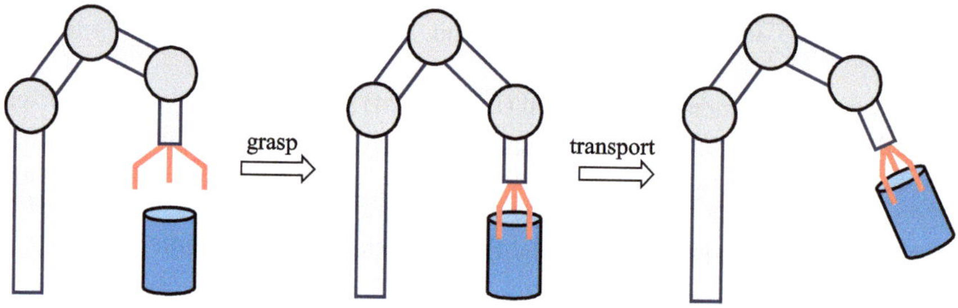

Fig. 1.2 Multifingered robot manipulation

In a Euclidean Jordan algebra, the orthogonality requirement in the complementarity problem (1.22) can also be expressed in terms of the Jordan product "$\circ$". In other words, from [22, Proposition 6], for $\mathcal{K}$ being a symmetric cone, there holds

$$\begin{aligned} & x \in \mathcal{K}, \;\; F(x) \in \mathcal{K}, \;\; \text{and } \langle x, F(x) \rangle = 0 \\ \Longleftrightarrow & x \in \mathcal{K}, \;\; F(x) \in \mathcal{K}, \;\; \text{and } x \circ F(x) = 0. \end{aligned} \tag{1.28}$$

Several nonlinear complementarity problems have already been introduced in well-known textbooks, such as [18, 19]. Therefore, we do not reiterate those examples here. Instead, we present a representative real-world problem arising from engineering applications, as discussed in [23]. Additional practical instances of nonlinear complementarity problems can be found in various domains, including multiuser power control in digital subscriber lines [24], three-dimensional frictional contact problems [25], and electric power markets [26, Sect. 5].

Figure 1.2 illustrates the multifingered robot grasping manipulation, where a multifingered robotic hand grasps and transports an object from an initial position to a final position. The dynamics of the object during this process can be described by the Newton-Euler equations, as outlined in [27, 28]. More precisely, the dynamic equation of the object is described by

$$\begin{aligned} \dot{y} &= v, \\ \dot{v} &= \frac{1}{m} R G_1 u + \begin{bmatrix} 0 \; 0 \; -g \end{bmatrix}^{\mathsf{T}}, \\ \dot{q} &= Q\omega, \\ \dot{\omega} &= I^{-1}\Big(RG_2 u - \omega \times (I\omega) \Big), \end{aligned} \tag{1.29}$$

where y is the position, v is the velocity, $q = [q_1 \; q_2 \; q_3]^{\mathsf{T}}$ is the quaternion, ω is the angular velocity, m is the object mass, I is the matrix of moment of initia, g is the gravity constant, u means the grasping forces which is represented by a matrix, $[G_1 \; G_2]$ is the contact matrix, R is the rotation matrix of the object, and Q can be expressed as

$$Q = \frac{1}{2}\begin{bmatrix} q_0 & q_3 & -q_2 \\ -q_3 & q_0 & q_1 \\ q_2 & -q_1 & q_0 \end{bmatrix} \quad \text{with} \quad q_0 = \sqrt{q_1^2 + q_2^2 + q_3^2}.$$

Moreover, the grasping forces are subject to the contact friction constraint, expressed as

$$\|(u_{i2}, u_{i3})\| \leq \mu \, u_{i1},$$

where u_{i1} is the normal force of the i-th finger, u_{i2} and u_{i3} are the friction forces of the i-th finger, $\|\cdot\|$ is the 2-norm, and μ is the friction coefficient.

In order to find the path that can be achieved with the minimum grasping forces, the optimal control problem is recast as

$$\begin{aligned} \min \quad & \int_0^T L dt \\ \text{s.t.} \quad & \dot{x} = f(x, u) \\ & x(0) = x_0 \\ & x(T) = x_T \\ & Du \in \mathcal{K}^d \times \mathcal{K}^d \times \cdots \times \mathcal{K}^d, \end{aligned} \tag{1.30}$$

where

$$L = \frac{u^T u}{2}, \quad x = \begin{bmatrix} y \\ v \\ q \\ \omega \end{bmatrix}.$$

In addition, f represents the right hand side of system (1.29), T is the control duration, x_0 and x_T are the initial and final states, respectively, D is the diagonal matrix with the friction coefficient, and $\mathcal{K}$ denotes the second-order cone, which is given by

$$\mathcal{K}^d := \left\{ \begin{bmatrix} z_1 \\ z_2 \end{bmatrix} \in \mathbb{R} \times \mathbb{R}^{d-1} \;\middle|\; \|z_2\| \leq z_1 \right\}.$$

The optimal control problem (1.30) can be addressed by applying Pontryagin's minimum principle; see, for example, [29–31]. In the language of optimization, this approach is equivalent to formulating the Karush-Kuhn-Tucker (KKT) conditions for problem (1.30), which consist of two key components. The first component involves a set of equalities pertaining to the Lagrange multipliers, while the second component captures the complementarity conditions. Specifically, by introducing the Hamiltonian function, the first part of the KKT conditions can be reformulated as follows:

$$
\begin{aligned}
\dot{x} - H_\lambda = \dot{x} - f(x,u) &= 0, \\
\dot{\lambda} + H_x = \dot{\lambda} + \lambda^\mathsf{T} f_x &= 0, \\
H_u = L_u + \lambda^\mathsf{T} f_u + \eta^\mathsf{T} D &= 0, \\
\phi(x(0), x(T)) &= 0, \\
\lambda(0) + \phi_{x(0)}^\mathsf{T} \sigma &= 0, \\
\lambda(T) + \phi_{x(T)}^\mathsf{T} \sigma &= 0,
\end{aligned}
$$

where λ, η, σ are the Lagrange multipliers, and $\phi(x(0), x(T)) = \begin{bmatrix} x(0) - x_0 \\ x(T) - x_T \end{bmatrix}$. The second part forms a second-order cone complementarity problem (SOCCP) as below:

$$
-\eta \in \mathcal{K}, \quad Du \in \mathcal{K}, \quad \eta^\mathsf{T} Du = 0, \tag{1.31}
$$

where $\mathcal{K} = \mathcal{K}^d \times \mathcal{K}^d \times \cdots \times \mathcal{K}^d$.

To conclude this section, we highlight the notion of "weighted complementarity problems" (WCP), which emerge in various equilibrium models in economics; see [32, 33]. The WCP can be viewed as a natural extension of the complementarity problem (1.27) and is characterized by the following mathematical formulation. Given a vector $w \in \mathcal{K}$, the goal of the weighted complementarity problem is to find $(x, s, y) \in \mathbb{V} \times \mathbb{V} \times \mathbb{R}^m$ such that

$$
x \in \mathcal{K}, \quad s \in \mathcal{K}, \quad F(x, s, y) = 0, \quad x \circ s = w, \tag{1.32}
$$

where $F : \mathbb{V} \times \mathbb{V} \times \mathbb{R}^m \to \mathbb{V} \times \mathbb{R}^m$ is a continuously differentiable nonlinear mapping. When the vector $w = 0$, the WCP (1.32) reduces to a mixed symmetric cone complementarity problem studied in [34]. When $w = 0$, $m = 0$, and $F(x, s, y) = f(x) - s$ with $f : \mathbb{V} \to \mathbb{V}$ being a continuously differentiable mapping, according to relation (1.28), the WCP (1.32) becomes the SCCP (1.27).

1.3 Complementarity Functions

The complementarity problem (1.22) essentially involves solving a system composed of inequalities defining the cones $\mathcal{K}$ and $\mathcal{K}^*$, along with an equation capturing the orthogonality condition. Rather than handling this system of inequalities and equation directly, we will demonstrate in the next section how it can be reformulated more conveniently and effectively through the use of complementarity functions. To this end, we present its definition as below.

Definition 1.2 A function $\phi : \mathbb{V} \times \mathbb{V} \to \mathbb{V}$ is called a **complementarity function** or a **C-function** if

$$
\phi(x, y) = 0 \iff x \in \mathcal{K}, \quad y \in \mathcal{K}^*, \quad \text{and} \quad \langle x, y \rangle = 0. \tag{1.33}
$$

In some cases, there exists a real-valued function $\psi : \mathbb{V} \times \mathbb{V} \to \mathbb{R}_+$, which also satisfies condition (1.33). Such a function is referred to as both a merit function and a C-function. This class of C-functions, along with their associated merit functions, plays a crucial role in the development of algorithms for solving the symmetric cone complementarity problem (SCCP) and symmetric cone programming (SCP). They have garnered significant attention in the contemporary optimization literature; see [9, 35–39] and references therein.

For self-dual closed convex cones $\mathcal{K}$ in a Euclidean space $\mathbb{V}$, we can always construct a complementarity function based on the projection mapping onto $\mathcal{K}$. We recall that given a set $\mathcal{K} \subseteq \mathbb{V}$, the *orthogonal projection* onto $\mathcal{K}$, denoted by $\Pi_{\mathcal{K}}$, is defined by

$$\Pi_{\mathcal{K}}(x) = \operatorname{argmin}_{y \in \mathcal{K}} \|y - x\|,$$

that is, $\Pi_{\mathcal{K}}(x)$ satisfies

$$\|\Pi_{\mathcal{K}}(x) - x\| \leq \|y - x\|, \quad \forall y \in \mathcal{K}.$$

A well-known result, the so-called *Projection Theorem*, is that for a nonempty closed and convex set $\mathcal{K}$, the projection $\Pi_{\mathcal{K}}(x)$ exists (which means the nearest-point) and is unique for each point $x \in \mathbb{V}$. Moreover, $\Pi_{\mathcal{K}}(x)$ is also the unique point satisfying the inequality

$$\langle x - \Pi_{\mathcal{K}}(x), z - \Pi_{\mathcal{K}}(x)\rangle \leq 0, \quad \forall z \in \mathcal{K}. \tag{1.34}$$

The proof of (1.34) is shown in Lemma 1.1(d) and other properties regarding projection mapping are summarized thereat, which are for subsequent needs. In addition, any point $x \in \mathbb{V}$ has a unique decomposition, known as the *Moreau decomposition*, given by

$$x = \Pi_{\mathcal{K}}(x) - \Pi_{\mathcal{K}^*}(-x),$$

where $\langle \Pi_{\mathcal{K}}(x), \Pi_{\mathcal{K}^*}(-x)\rangle = 0$. Very often, we also use $x_{\mathcal{K}}^+$ and $x_{\mathcal{K}}^-$ or $P_{\mathcal{K}}(x)$ and $P_{\mathcal{K}^*}(-x)$ to denote the projection of x onto $\mathcal{K}$ and $-\mathcal{K}^*$, respectively.

Lemma 1.1 *Let $\mathcal{K}$ be any closed convex cone in $\mathbb{R}^n$. For each $x \in \mathbb{R}^n$, let $x_{\mathcal{K}}^+$ and $x_{\mathcal{K}}^-$ denote the nearest-point (in the Euclidean norm) projection of x onto $\mathcal{K}$ and $-\mathcal{K}^*$, respectively. The following results hold.*

(a) For any $x \in \mathbb{R}^n$, we have $x = x_{\mathcal{K}}^+ + x_{\mathcal{K}}^-$ and $\|x\|^2 = \|x_{\mathcal{K}}^+\|^2 + \|x_{\mathcal{K}}^-\|^2$.
(b) For any $x \in \mathbb{R}^n$ and $y \in \mathcal{K}$, we have $\langle x, y\rangle \leq \langle x_{\mathcal{K}}^+, y\rangle$.
(c) If $\mathcal{K}$ is self-dual, then for any $x \in \mathbb{R}^n$ and $y \in \mathcal{K}$, we have $\left\|(x + y)_{\mathcal{K}}^+\right\| \geq \left\|x_{\mathcal{K}}^+\right\|$.
(d) For any $x, y \in \mathbb{R}^n$ and $z \in \mathcal{K}$, there hold

$$\left(x - x_{\mathcal{K}}^+\right)^{\mathsf{T}} \left(z - x_{\mathcal{K}}^+\right) \leq 0 \quad \text{and} \quad \left\|x_{\mathcal{K}}^+ - y_{\mathcal{K}}^+\right\| \leq \|x - y\|.$$

Proof (a) These are well-known results in convex geometry on representing x as the sum of its projection onto $\mathcal{K}$ and its polar $-\mathcal{K}^*$.

(b) Since $x_{\mathcal{K}}^- \in -\mathcal{K}^*$ and $y \in \mathcal{K}$, $\langle x_{\mathcal{K}}^-, y \rangle \leq 0$. By part(a), it is clear that $\langle x, y \rangle = \langle x_{\mathcal{K}}^+, y \rangle + \langle x_{\mathcal{K}}^-, y \rangle \leq \langle x_{\mathcal{K}}^+, y \rangle$.

(c) Since $\mathcal{K}$ is self-dual, we have $y \in \mathcal{K}^*$. Then, $(x + y)_{\mathcal{K}}^- - y \in -\mathcal{K}^*$. Since $x_{\mathcal{K}}^-$ is the nearest-point projection of x onto $-\mathcal{K}^*$, this implies

$$\|x_{\mathcal{K}}^- - x\| \leq \left\| \left((x + y)_{\mathcal{K}}^- - y \right) - x \right\|.$$

By part(a), this simplifies to $\|x_{\mathcal{K}}^+\| \leq \left\| (x + y)_{\mathcal{K}}^+ \right\|$.

(d) Consider the point $x_{\mathcal{K}}^+ + \alpha(z - x_{\mathcal{K}}^+) = \alpha z + (1 - \alpha)x_{\mathcal{K}}^+$ for $0 < \alpha \leq 1$. It belongs to $\mathcal{K}$ due to the convexity of $\mathcal{K}$. Since this point belongs to $\mathcal{K}$, we have

$$\left\| x - x_{\mathcal{K}}^+ - \alpha \left(z - x_{\mathcal{K}}^+ \right) \right\|^2 \geq \|x - x_{\mathcal{K}}^+\|^2 \quad \forall x \in \mathbb{R}^n.$$

Writing out the expression of the left hand side gives

$$\begin{aligned} \left\| x - x_{\mathcal{K}}^+ - \alpha \left(z - x_{\mathcal{K}}^+ \right) \right\|^2 &= \|x - x_{\mathcal{K}}^+\|^2 + \alpha^2 \|z - x_{\mathcal{K}}^+\|^2 - 2\alpha \left(x - x_{\mathcal{K}}^+ \right)^{\mathsf{T}} \left(z - x_{\mathcal{K}}^+ \right) \\ &\geq \qquad \left\| x - x_{\mathcal{K}}^+ \right\|^2. \end{aligned}$$

Then, we obtain

$$2\alpha (x - x_{\mathcal{K}}^+)^{\mathsf{T}} (z - x_{\mathcal{K}}^+) \leq \alpha^2 \|z - x_{\mathcal{K}}^+\|^2.$$

Dividing by α on both sides and letting $\alpha \to 0$ imply

$$\left(x - x_{\mathcal{K}}^+ \right)^{\mathsf{T}} \left(z - x_{\mathcal{K}}^+ \right) \leq 0, \tag{1.35}$$

which is the desired result.

For the second part, from the above inequality (1.35), we have $(w - x_{\mathcal{K}}^+)^{\mathsf{T}}(x - x_{\mathcal{K}}^+) \leq 0$ for all $w \in \mathcal{K}$. Noting $y_{\mathcal{K}}^+ \in \mathcal{K}$, it says that

$$\left(y_{\mathcal{K}}^+ - x_{\mathcal{K}}^+ \right)^{\mathsf{T}} \left(x - x_{\mathcal{K}}^+ \right) \leq 0$$

and similarly

$$\left(x_{\mathcal{K}}^+ - y_{\mathcal{K}}^+ \right)^{\mathsf{T}} \left(y - y_{\mathcal{K}}^+ \right) \leq 0.$$

Adding these two inequalities gives

$$\left(y_{\mathcal{K}}^+ - x_{\mathcal{K}}^+ \right)^{\mathsf{T}} \left(x - x_{\mathcal{K}}^+ - y + y_{\mathcal{K}}^+ \right) \leq 0,$$

which together with Schwartz inequality implies

$$\left\|y_{\mathcal{K}}^{+} - x_{\mathcal{K}}^{+}\right\|^2 \leq \left(y_{\mathcal{K}}^{+} - x_{\mathcal{K}}^{+}\right)^{\mathsf{T}} (y - x) \leq \left\|y_{\mathcal{K}}^{+} - x_{\mathcal{K}}^{+}\right\| \cdot \|y - x\|.$$

Then, the desired result follows and the proof is complete. □

Proposition 1.3 *Let $\mathcal{K} \subseteq \mathbb{V}$ be a self-dual closed convex cone. Define $\phi : \mathbb{V} \times \mathbb{V} \to \mathbb{V}$ as*

$$\phi_{\text{NR}}(x, y) = x - \Pi_{\mathcal{K}}(x - y). \tag{1.36}$$

Then, ϕ_{NR} is a C-function such that $\phi_{\text{NR}}(x, y) = \phi_{\text{NR}}(y, x)$ for all $x, y \in \mathbb{V}$.

Proof Using the Moreau decomposition of $x - y$ and noting that $\mathcal{K} = \mathcal{K}^*$ due to self-duality of $\mathcal{K}$, we have

$$x - y = \Pi_{\mathcal{K}}(x - y) - \Pi_{\mathcal{K}}(y - x).$$

This equation indicates $\phi_{\text{NR}}(x, y) = \phi_{\text{NR}}(y, x)$. Now, assume that $x \in \mathcal{K}$, $y \in \mathcal{K}$ and $\langle x, y\rangle = 0$. Using $\phi_{\text{NR}}(x, y) = \phi_{\text{NR}}(y, x)$, there hold

$$\begin{aligned} \|\phi_{\text{NR}}(x, y)\|^2 &= \langle \phi_{\text{NR}}(x, y), \phi_{\text{NR}}(y, x)\rangle \\ &= \langle x - \Pi_{\mathcal{K}}(x - y), y - \Pi_{\mathcal{K}}(y - x)\rangle \\ &= -\langle x, \Pi_{\mathcal{K}}(y - x)\rangle - \langle y, \Pi_{\mathcal{K}}(x - y)\rangle. \end{aligned} \tag{1.37}$$

Since $\mathcal{K}$ is self-dual and $x \in \mathcal{K}$, it is clear $\langle x, \Pi_{\mathcal{K}}(y - x)\rangle \geq 0$. Likewise, $\langle y, \Pi_{\mathcal{K}}(x - y)\rangle \geq 0$. From (1.37), we see that $\|\phi_{\text{NR}}(x, y)\|^2 \leq 0$, i.e., $\phi_{\text{NR}}(x, y) = 0$. Conversely, suppose that $\phi_{\text{NR}}(x, y) = 0$, that is $x = \Pi_{\mathcal{K}}(x - y)$. Then $x \in \mathcal{K}$. Since $\phi_{\text{NR}}(x, y) = \phi_{\text{NR}}(y, x)$, it also follows that $y \in \mathcal{K}$. It remains to show that $\langle x, y\rangle = 0$. By convexity of $\mathcal{K}$ and inequality (1.34), we have

$$0 \geq \langle (x - y) - x, z - x\rangle = \langle -y, z - x\rangle \tag{1.38}$$

for all $z \in \mathcal{K}$. Taking $z = 0 \in \mathcal{K}$, (1.38) gives $\langle x, y\rangle \leq 0$. Since $x \in \mathcal{K}$ and $\mathcal{K}$ is a cone, $z = 2x \in \mathcal{K}$. From (1.38) implies that $\langle x, y\rangle \geq 0$. Altogether, we achieve $\langle x, y\rangle = 0$. □

When $\mathcal{K}$ represents a symmetric cone, we denote by $\phi_{\text{NR}}^{\text{sc}}$ the function defined in (1.36), which has been shown to be strongly semismooth in [9]. More recently, the nonsingularity of Clarke's generalized Jacobian associated with the nonsmooth KKT system based on $\phi_{\text{NR}}^{\text{sc}}$ for linear SCP has been investigated in [40]. These contributions form the theoretical foundation for the development of nonsmooth Newton methods and smoothing Newton methods for solving the SCCPs and the SCPs. Another popular choice of ϕ satisfying Definition 1.2 is the Fischer-Burmeister (FB) complementarity function [22] defined as

$$\phi_{\text{FB}}^{\text{sc}}(x, y) := (x^2 + y^2)^{1/2} - (x + y) \quad \forall x, y \in \mathbb{V}, \tag{1.39}$$

where $x^2 = x \circ x$, and $x^{1/2}$ denotes the unique square root of $x \in \mathcal{K}$, i.e., $x^{1/2} \circ x^{1/2} = x$. Compared to the function ϕ^{sc}_{NR}, this function possesses a notable advantage: its squared norm induces a continuously differentiable merit function, which further enjoys a globally Lipschitz continuous gradient; see [39, 41] for details. This property significantly facilitates the globalization of nonsmooth Newton methods based on ϕ^{sc}_{FB}. Throughout this book, we frequently employ ϕ_{FB} and ϕ_{NR} directly in various settings associated with $\mathcal{K}$, whenever the context allows clear distinction without ambiguity.

For $\mathcal{K}$ representing symmetric cones, or certain classes of non-symmetric cones, there exist alternative approaches to constructing C-functions beyond the use of projections; see [42–46] for details. A more thorough discussion of these methods will be presented in Chap. 3. By employing a C-function ϕ satisfying Definition 1.2, the SOCCP (1.31) introduced in Sect. 1.2 can be further reformulated as a system of equations:

$$\phi(Du, -\eta) = 0.$$

In [47, 48], the complementarity function ϕ_{FB} is employed, which is a special case of (1.39) corresponding to SOC setting. In other words, by using the vector-valued function,

$$\phi_{FB}(\mathbf{a}, \mathbf{b}) := \left(\mathbf{a}^2 + \mathbf{b}^2\right)^{1/2} - (\mathbf{a} + \mathbf{b}) \tag{1.40}$$

for $\mathbf{a} = \begin{bmatrix} a_1 \\ a_2 \end{bmatrix} \in \mathbb{R} \times \mathbb{R}^{d-1}$, $\mathbf{b} = \begin{bmatrix} b_1 \\ b_2 \end{bmatrix} \in \mathbb{R} \times \mathbb{R}^{d-1}$, the SOCCP (1.31) is equivalent to

$$\phi_{FB}(Du, -\eta) = 0.$$

Here, the square term and square-root term in (1.40) are calculated via Jordan product

$$\mathbf{a} \circ \mathbf{b} = \begin{bmatrix} \mathbf{a}^T\mathbf{b} \\ a_1 b_2 + b_1 a_2 \end{bmatrix}.$$

In particular, the expressions for $\mathbf{a}^2$ and $\mathbf{a}^{1/2}$ are given by

$$\mathbf{a}^2 = \begin{bmatrix} \|\mathbf{a}\|^2 \\ 2a_1a_2 \end{bmatrix}$$

and

$$\mathbf{a}^{1/2} = \begin{bmatrix} s \\ \dfrac{a_2}{2s} \end{bmatrix} \quad \text{with} \quad s = \sqrt{\frac{1}{2}\left(a_1 + \sqrt{a_1^2 - \|a_2\|^2}\right)},$$

respectively.

1.4 Semismooth Functions, P-Functions, and P-Properties

To lay the groundwork for presenting the properties of existing NCP functions in the next chapter, we first recall some essential background concepts and materials that will play a crucial role in the subsequent analysis. To this end, we begin by briefly reviewing several notations. For a function $f : \mathbb{R}^n \to \mathbb{R}$, we denote by $\nabla f(x)$ and $\nabla^2 f(x)$ the gradient and Hessian of f, respectively. Besides, given a function $F : \mathbb{R}^n \to \mathbb{R}^m$, we denote by $JF(x)$ the Jacobian of F and we let $\nabla F(x) = JF(x)^{\mathrm{T}}$. Sometimes, to emphasize that the derivative is taken w.r.t. x, we write $J_x F(x)$ and $\nabla_x F(x)$, respectively.

We begin with the concept of semismoothness, originally introduced by Mifflin [49] for functionals and later extended to vector-valued functions by Qi and Sun [50]. As a preliminary, we first define the notion of strict continuity (also referred to as local Lipschitz continuity) at a point $x \in \mathbb{R}^n$; see [51, Chap. 9]. Specifically, a function F is said to be strictly continuous at x if there exist positive constants $\kappa > 0$ and $\delta > 0$ such that

$$\|F(y) - F(z)\| \le \kappa \|y - z\| \quad \forall y, z \in \mathbb{R}^n \ \text{ with } \ \|y - x\| \le \delta, \ \|z - x\| \le \delta;$$

and F is strictly continuous if F is strictly continuous at every $x \in \mathbb{R}^n$. If δ can be taken to be ∞, then F is Lipschitz continuous with Lipschitz constant κ. Define the function $\mathrm{lip}F : \mathbb{R}^k \to [0, \infty]$ by

$$\mathrm{lip}F(x) := \limsup_{\substack{y,z \to x \\ y \ne z}} \frac{\|F(y) - F(z)\|}{\|y - z\|}.$$

Then, F is strictly continuous at x if and only if $\mathrm{lip}F(x)$ is finite. We say F is directionally differentiable at $x \in \mathbb{R}^n$ if

$$F'(x; h) := \lim_{t \to 0^+} \frac{F(x + th) - F(x)}{t} \quad \text{exists} \quad \forall h \in \mathbb{R}^k;$$

and F is directionally differentiable if F is directionally differentiable at every $x \in \mathbb{R}^n$. F is differentiable (in the Fréchet sense) at $x \in \mathbb{R}^n$ if there exists a linear mapping $\nabla F(x) : \mathbb{R}^n \to \mathbb{R}^n$ such that

$$F(x + h) - F(x) - \nabla F(x)h = o(\|h\|).$$

We say that F is continuously differentiable if F is differentiable at every $x \in \mathbb{R}^n$ and ∇F is continuous.

If F is strictly continuous, then F is almost everywhere differentiable by Rademacher's Theorem–see [52] and [51, Sect. 9J]. In this case, the generalized Jacobian $\partial F(x)$ of F at x (in the Clarke sense) is defined as the convex hull of the generalized Jacobian $\partial_B F(x)$, where

$$\partial_B F(x) := \left\{ \lim_{x^j \to x} \nabla F(x^j) \;\middle|\; F \text{ is differentiable at } x^j \in \mathbb{R}^k \right\}.$$

The notation ∂_B is adopted from [53]. In [51, Chap. 9], the case of $n = 1$ is considered and the notations "$\bar{\nabla}$" and "$\bar{\partial}$" are used instead of, respectively, "∂_B" and "∂". In other words, $\partial F(x) = \mathrm{conv}\partial_B F(x)$. If $m = 1$, we also call $\partial F(x)$ the *generalized gradient* of F at x. The calculation of $\partial F(x)$ is usually difficult in practice, and Qi [54] proposed so-called *C-subdifferential* of F:

$$\partial_C F(x)^\mathsf{T} := \partial F_1(x) \times \cdots \times \partial F_m(x), \tag{1.41}$$

which is easier to compute than the generalized Jacobian $\partial F(x)$. Here, the right-hand side of (1.41) denotes the set of matrices in $\mathbb{R}^{n\times m}$ whose i-th column is given by the generalized gradient of the i-th component function F_i. In fact, by [52, Proposition 2.6.2], there holds

$$\partial F(x)^\mathsf{T} \subseteq \partial_C F(x)^\mathsf{T}.$$

Assume $F : \mathbb{R}^n \to \mathbb{R}^m$ is strictly continuous. We say F is semismooth at x if F is directionally differentiable at x and, for any $V \in \partial F(x + h)$, we have

$$F(x + h) - F(x) - Vh = o(\|h\|).$$

We say F is ρ-order semismooth at x ($0 < \rho < \infty$) if F is semismooth at x and, for any $V \in \partial F(x + h)$, we have

$$F(x + h) - F(x) - Vh = O(\|h\|^{1+\rho}). \tag{1.42}$$

A function F is said to be semismooth (respectively, ρ-order semismooth) if it possesses this property at every point $x \in \mathbb{R}^k$. In particular, F is referred to as strongly semismooth if it is 1-order semismooth. Notable examples of semismooth functions include convex functions and piecewise continuously differentiable functions. Furthermore, the composition of two semismooth (respectively, ρ-order semismooth) functions remains semismooth (respectively, ρ-order semismooth). The property of semismoothness plays a pivotal role in the design and analysis of nonsmooth Newton methods [50, 53], as well as in certain smoothing techniques discussed in the previous section. For more comprehensive treatments of semismooth functions, the reader is referred to [49, 50, 55].

Lemma 1.2 *Suppose $F : \mathbb{R}^n \to \mathbb{R}^n$ is strictly continuous and directionally differentiable in a neighborhood of $x \in \mathbb{R}^n$. Then, for any $0 < \rho < \infty$, the following two statements (where $O(\cdot)$ depends on F and x only) are equivalent:*

(a) *For any $h \in \mathbb{R}^n$ and any $V \in \partial F(x+h)$,*

$$F(x+h) - F(x) - Vh = o(\|h\|) \ \ (\textit{respectively}, \ O(\|h\|^{1+\rho})).$$

(b) *For any $h \in \mathbb{R}^n$ such that F is differentiable at $x+h$,*

$$F(x+h) - F(x) - \nabla F(x+h)h = o(\|h\|) \ \ (\textit{respectively}, \ O(\|h\|^{1+\rho})).$$

Proof Please see [17, Theorem 3.6]. □

The following lemmas, including a mean value theorem for vector-valued functions, will be essential for the subsequent analysis.

Lemma 1.3 *If $F : D \subseteq \mathbb{R}^n \to \mathbb{R}^m$ has a second derivative at each point of a convex set $D_0 \subseteq D$, then*

$$\|\nabla F(y) - \nabla F(x)\| \leq \sup_{0 \leq t \leq 1} \left\|\nabla^2 F(x + t(y-x))\right\| \cdot \|y - x\|.$$

Proof Please see [56, Theorem 3.3.5]. □

The Mean Value Theorem in Lemma 1.3 for a vector-valued function $F : \mathbb{R}^n \to \mathbb{R}^m$ is a bit different from the traditional one. More specifically, a vector-valued function $F : \mathbb{R}^n \to \mathbb{R}^m$ does not have a Mean Value Theorem in form of $F(y) = F(x) + \nabla F(z)^{\mathsf{T}}(y-x)$ where $z \in [x, y]$. To see a counterexample, define $F : \mathbb{R} \to \mathbb{R}^2$ as

$$F(t) = (t - t^2, t - t^3).$$

We compute that $F(0) = (0, 0)$, $F(1) = (0, 0)$, and $\nabla F(t) = \left[1 - 2t \ \ 1 - 3t^2\right]$. It can be seen that there does not exist $t \in [0, 1]$ satisfying $F(1) - F(0) = \nabla F(t)^{\mathsf{T}}(1 - 0)$.

Lemma 1.4 ([57, Lemma 1.3]) *Let $x = (x_1, x_2, \ldots, x_n) \in \mathbb{R}^n$ and $\|x\|_p := \left(\sum_{i=1}^{n} |x_i|^p\right)^{\frac{1}{p}}$. If $1 < p_1 < p_2$, then $\|x\|_{p_2} \leq \|x\|_{p_1} \leq n^{(\frac{1}{p_1} - \frac{1}{p_2})}\|x\|_{p_2}$.*

Proof We assume $x = (x_1, x_2, \ldots, x_n)$ is a nonzero vector since the inequality is trivial when $x = 0$. Thus, there exists at least one nonzero scalar component of x, say $x_{i_0} \neq 0$. Then, by noting $\frac{p_2}{p_1} > 1$, we obtain

$$
\begin{aligned}
\|x\|_{p_1}^{p_2} &= \left(\sum_{i=1}^{n} |x_i|^{p_1}\right)^{\frac{p_2}{p_1}} \\
&= \left(|x_{i_0}|^{p_1} + \sum_{i=1, i\neq i_0}^{n} |x_i|^{p_1}\right)^{\frac{p_2}{p_1}} \\
&= |x_{i_0}|^{p_2} \left(1 + \frac{\sum_{i=1, i\neq i_0}^{n} |x_i|^{p_1}}{|x_{i_0}|^{p_1}}\right)^{\frac{p_2}{p_1}} \\
&\geq |x_{i_0}|^{p_2} \left[1 + \left(\frac{\sum_{i=1, i\neq i_{i_0}}^{n} |x_i|^{p_1}}{|x_{i_0}|^{p_1}}\right)^{\frac{p_2}{p_1}}\right] \\
&= |x_{i_0}|^{p_2} + \left(\sum_{i=1, i\neq i_0}^{n} |x_i|^{p_1}\right)^{\frac{p_2}{p_1}} \\
&\geq |x_{i_0}|^{p_2} + \left(\sum_{i=1, i\neq i_0}^{n} |x_i|^{p_2}\right) \\
&= \|x\|_{p_2}^{p_2},
\end{aligned}
$$

where the first inequality uses the fact that $(1+t)^{\alpha} \geq 1 + t^{\alpha}$ for all $t > 0$ and $\alpha > 1$. This proves $\|x\|_{p_1} \geq \|x\|_{p_2}$.

To prove the reverse inequality, we will apply the Hölder Inequality,

$$
\left|x^{\mathsf{T}} y\right| \leq \|x\|_p \cdot \|y\|_q,
$$

where $\dfrac{1}{p} + \dfrac{1}{q} = 1$ and $1 \leq p \leq \infty$. This can be verified by the following:

$$
\begin{aligned}
\|x\|_{p_1} &= \left(\sum_{i=1}^{n} |x_i|^{p_1}\right)^{\frac{1}{p_1}} \\
&= \left(\sum_{i=1}^{n} 1 \cdot |x_i|^{p_1}\right)^{\frac{1}{p_1}} \\
&\leq \left[\left(\sum_{i=1}^{n} 1^{\frac{p_2}{p_2-p_1}}\right)^{\frac{p_2-p_1}{p_2}} \left(\sum_{i=1}^{n} (|x_i|^{p_1})^{\frac{p_2}{p_1}}\right)^{\frac{p_1}{p_2}}\right]^{\frac{1}{p_1}} \\
&= n^{\frac{p_2-p_1}{p_1 p_2}} \left(\sum_{i=1}^{n} |x_i|^{p_2}\right)^{\frac{1}{p_2}} \\
&= n^{(\frac{1}{p_1} - \frac{1}{p_2})} \|x\|_{p_2},
\end{aligned}
$$

where we set $\frac{1}{p} = \frac{p_2-p_1}{p_2}$ and $\frac{1}{q} = \frac{p_1}{p_2}$ in the Hölder Inequality. □

An important concept closely related to semismooth functions is that of SC^1 functions. We present its formal definition below.

Definition 1.3 A function $f : \mathbb{R}^n \to \mathbb{R}$ is said to be an SC^1 function if f is continuously differentiable and its gradient is semismooth.

The class of SC^1 functions can be regarded as lying between C^1 and C^2 functions. By introducing SC^1 functions, many results originally established for the minimization of C^2 functions can be extended to the minimization of SC^1 functions; see [58] and references therein. For further applications and a more comprehensive discussion on SC^1 functions, the reader is referred to the excellent book [19]. In addition to SC^1 functions, we also introduce the concept of LC^1 functions in this section.

Definition 1.4 A function $f : \mathbb{R}^n \to \mathbb{R}$ is called an LC^1 function if f is continuously differentiable and its gradient is locally Lipschitz continuous.

The class of LC^1 minimization problems was studied in [59], where the local superlinear convergence of an approximate Newton method was established under the assumption of semismoothness of the gradient function at a solution point. It is evident that any SC^1 function also qualifies as an LC^1 function. Additional concepts related to semismooth functions include piecewise smooth and almost smooth functions. It is well-known that piecewise smooth functions are prototypical examples of semismooth functions. However, recent studies have identified various semismooth functions that are not piecewise smooth; see [60] and references therein. Notable examples include the p-norm function with $1 < p < \infty$ defined on $\mathbb{R}^n$ for $n \geq 2$, the Euclidean norm function, pseudo-smooth NCP functions, and various smoothing functions.

Definition 1.5 The almost smooth (respectively, strongly almost smooth) functions are functions that are semismooth (respectively, strongly semismooth) on the whole space $\mathbb{R}^n$ and smooth everywhere except on sets with "dimension" less than $n - 1$ in the sense that the sets do not locally partition $\mathbb{R}^n$ into multiple connected components.

Now, we recall definitions of P-matrix, P-functions and P-property, along with several related concepts.

Definition 1.6 A matrix $M \in \mathbb{R}^{n\times n}$ is a

(a) P_0-matrix if every of its principal minors is nonnegative.
(b) P-matrix if every of its principal minors is positive.

It is clear that every P-matrix is also a P_0-matrix. Moreover, it is well-known that the Jacobian of any continuously differentiable P_0-function is itself a P_0-matrix. Below, we present one of the key characterizations of P_0-matrices, which will be utilized in subsequent analysis. For additional properties and a comprehensive discussion of P-matrices and P_0-matrices, the reader is referred to [18].

Lemma 1.5 ([18, Theorem 3.4.2]) *Let $M \in \mathbb{R}^{n\times n}$. The followings are equivalent:*

(a) *M is a P_0-matrix.*
(b) *For every nonzero vector x there exists an index i such that $x_i \neq 0$ and $x_i(Mx)_i \geq 0$.*
(c) *All real eigenvalues of M and its principal submatrices are nonnegative.*
(d) *For each $\varepsilon > 0$, $M + \varepsilon I$ is a P-matrix.*

Definition 1.7 Let $F : \mathbb{R}^n \to \mathbb{R}^n$, then

(a) F is monotone if $\langle x - y, F(x) - F(y)\rangle \geq 0$, for all $x, y \in \mathbb{R}^n$.
(b) F is strictly monotone if $\langle x - y, F(x) - F(y)\rangle > 0$, for all $x, y \in \mathbb{R}^n$ and $x \neq y$.
(c) F is strongly monotone with modulus $\mu > 0$ if $\langle x - y, F(x) - F(y)\rangle \geq \mu\|x - y\|^2$, for all $x, y \in \mathbb{R}^n$.
(d) F is a P_0-function if $\max\limits_{\substack{1\leq i\leq n \\ x_i \neq y_i}}(x_i - y_i)(F_i(x) - F_i(y)) > 0$, for all $x, y \in \mathbb{R}^n$ and $x \neq y$.
(e) F is a P-function if $\max\limits_{1\leq i\leq n}(x_i - y_i)(F_i(x) - F_i(y)) > 0$, for all $x, y \in \mathbb{R}^n$ and $x \neq y$.
(f) F is a uniform P-function with modulus $\mu > 0$ if $\max\limits_{1\leq i\leq n}(x_i - y_i)(F_i(x) - F_i(y)) \geq \mu\|x - y\|^2$, for all $x, y \in \mathbb{R}^n$.
(g) $\nabla F(x)$ is uniformly positive definite with modulus $\mu > 0$ if $d^{\mathsf{T}}\nabla F(x)d \geq \mu\|d\|^2$, for all $x \in \mathbb{R}^n$ and $d \in \mathbb{R}^n$.
(h) F is Lipschitz continuous if there exists a constant $L > 0$ such that $\|F(x) - F(y)\| \leq L\|x - y\|$, for all $x, y \in \mathbb{R}^n$.

From Definition 1.7, it is evident that strongly monotone functions are strictly monotone, and strictly monotone functions are, in turn, monotone. Furthermore, a function F is a P_0-function if it is monotone, and it is a uniform P-function with modulus $\mu > 0$ if F is strongly monotone with modulus $\mu > 0$. Additionally, when F is continuously differentiable, the following conclusions hold:

1. F is monotone if and only if $\nabla F(x)$ is positive semidefinite for all $x \in \mathbb{R}^n$.
2. F is strictly monotone if $\nabla F(x)$ is positive definite for all $x \in \mathbb{R}^n$.
3. F is strongly monotone if and only if $\nabla F(x)$ is uniformly positive definite.

Next, we introduce the definitions of Cartesian P-properties for a matrix $M \in \mathbb{R}^{n\times n}$, which can be viewed as special cases of the more general properties formulated by Chen and Qi [61] for linear transformations.

Definition 1.8 A matrix $M \in \mathbb{R}^{n\times n}$ is said to have

(a) the Cartesian P-property if for any $0 \neq x = (x_1, \ldots, x_m) \in \mathbb{R}^n$ with $x_i \in \mathbb{R}^{n_i}$, there exists an index $\nu \in \{1, 2, \ldots, m\}$ such that $\langle x_\nu, (Mx)_\nu\rangle > 0$;
(b) the Cartesian P_0-property if for any $0 \neq x = (x_1, \ldots, x_m) \in \mathbb{R}^n$ with $x_i \in \mathbb{R}^{n_i}$, there exists an index $\nu \in \{1, 2, \ldots, m\}$ such that $x_\nu \neq 0$ and $\langle x_\nu, (Mx)_\nu\rangle \geq 0$.

Clearly, when $m = n$ and $n_1 = \cdots = n_m = 1$, M having the Cartesian P-property (or P_0-property) coincides with M being a P-matrix (or P_0-matrix), which are introduced in [18]. Let M be an $n \times n$ matrix with elements m_{ij}. Then, M can be denoted by

$$M = \begin{bmatrix} M_{11} & M_{12} & \cdots & M_{1m} \\ M_{21} & M_{22} & \cdots & M_{2m} \\ \cdots & \cdots & \cdots & \cdots \\ M_{m1} & M_{m2} & \cdots & M_{mm} \end{bmatrix}, \tag{1.43}$$

where $M_{\nu l}$ for each $\nu = 1, \ldots, m$ and $l = 1, \ldots, m$ is an $n_\nu \times n_l$ matrix consisting of those elements m_{kj} with $k = n_{\nu-1} + 1, \ldots, n_\nu$, $j = n_{j-1} + 1, \ldots, n_j$ and $n_0 = 0$. Let S be a proper subset of $\{1, 2, \ldots, m\}$ and denote by $M(S)$ the matrix resulting from deleting the block matrix $M_{\nu l}$ with ν or l complementary to those indicated by S from M given as in (1.43). We call $M(S)$ *a principal block matrix* of M. By Definition 1.8, it is not hard to verify that every principal block matrix $M(S)$ must have the Cartesian P-property if the matrix M has the Cartesian P-property. When $m = n$ and $n_1 = \cdots = n_m = 1$, this reduces to the well-known fact that every principal submatrix of a P-matrix is again a P-matrix. Particularly, assume that the matrix M, by rearrangement, is written as

$$M = \begin{bmatrix} M_{\mathcal{J}\mathcal{J}} & M_{\mathcal{J}\mathcal{B}} \\ M_{\mathcal{B}\mathcal{J}} & M_{\mathcal{B}\mathcal{B}} \end{bmatrix}, \tag{1.44}$$

where $\mathcal{J}$ and $\mathcal{B}$ are index sets such that $\mathcal{J} \cup \mathcal{B} = \{1, 2, \ldots, m\}$ and $\mathcal{J} \cap \mathcal{B} = \emptyset$. Then, when M has the Cartesian P-property and $M_{\mathcal{J}\mathcal{J}}$ is nonsingular, we have the following result, which can be regarded as an extension of the fact that any Schur-complement of a P-matrix is also a P-matrix.

Proposition 1.4 *Suppose that M defined as in (1.44) has the Cartesian P-property and the matrix $M_{\mathcal{J}\mathcal{J}}$ is nonsingular. Then its Schur-complement in the matrix M, i.e.,*

$$\widehat{M}_{\mathcal{J}\mathcal{J}} = M_{\mathcal{B}\mathcal{B}} - M_{\mathcal{B}\mathcal{J}}(M_{\mathcal{J}\mathcal{J}})^{-1} M_{\mathcal{J}\mathcal{B}}$$

also has the Cartesian P-property.

Proof Let $y_{\mathcal{B}}$ be an arbitrary nonzero vector with the dimension same as $M_{\mathcal{B}\mathcal{B}}$. Let $x_{\mathcal{J}}$ be a vector with the dimension same as $M_{\mathcal{J}\mathcal{J}}$ such that

$$M_{\mathcal{J}\mathcal{J}} x_{\mathcal{J}} + M_{\mathcal{J}\mathcal{B}} y_{\mathcal{B}} = 0, \tag{1.45}$$

or equivalently,

$$x_{\mathcal{J}} = -(M_{\mathcal{J}\mathcal{J}})^{-1} M_{\mathcal{J}\mathcal{B}} y_{\mathcal{B}}. \tag{1.46}$$

Let $z = (x_{\mathcal{J}}, y_{\mathcal{B}}) \in \mathbb{R}^n$. Then, $z \neq 0$. From Definition 1.8(a) and the given assumption that M has the Cartesian P-property, there exists an index $i \in \{1, 2, \ldots, m\}$ such that

$$\langle z_i, (Mz)_i \rangle > 0. \tag{1.47}$$

Notice that the index i must belong to the set $\mathcal{B}$. If not, i.e., $i \in \mathcal{J}$, then from the definition of M we learn that the inequality (1.47) is equivalent to

$$\langle (x_{\mathcal{J}})_i, [M_{\mathcal{J}\mathcal{J}} x_{\mathcal{J}} + M_{\mathcal{J}\mathcal{B}} y_{\mathcal{B}}]_i \rangle > 0,$$

which obviously contradicts the equality (1.45). Now (1.47) is equivalent to

$$\langle (y_{\mathcal{B}})_i, [M_{\mathcal{B}\mathcal{J}} x_{\mathcal{J}} + M_{\mathcal{B}\mathcal{B}} y_{\mathcal{B}}]_i \rangle > 0.$$

Using the inequality and Eq. (1.46), we immediately have that

$$\begin{aligned}\langle (y_{\mathcal{B}})_i, [\widehat{M}_{\mathcal{J}\mathcal{J}} y_{\mathcal{B}}]_i \rangle &= \langle (y_{\mathcal{B}})_i, [M_{\mathcal{B}\mathcal{B}} y_{\mathcal{B}} - M_{\mathcal{B}\mathcal{J}} (M_{\mathcal{J}\mathcal{J}})^{-1} M_{\mathcal{J}\mathcal{B}} y_{\mathcal{B}}]_i \rangle \\ &= \langle (y_{\mathcal{B}})_i, [M_{\mathcal{B}\mathcal{B}} y_{\mathcal{B}} + M_{\mathcal{B}\mathcal{J}} x_{\mathcal{J}}]_i \rangle > 0.\end{aligned}$$

Thus, by Definition 1.8(a), the matrix $\widehat{M}_{\mathcal{J}\mathcal{J}}$ has the Cartesian P-property. □

Definition 1.9 ([22]) A matrix $M \in \mathbb{R}^{n\times n}$ is said to have

(a) the Jordan P-property (or the P_1-property) if $x \circ (Mx) \in -\mathcal{K} \Rightarrow x = 0$;
(b) the P-property if the condition that $L_{x_i} L_{(Mx)_i} = L_{(Mx)_i} L_{x_i}$, $i = 1, 2, \ldots, m$ and $x \circ (Mx) \in -\mathcal{K}$ necessarily implies $x = 0$;
(c) the P_0-property if $M + \varepsilon I$ for any $\varepsilon > 0$ has the P-property.

Proposition 1.5 *(a) If a matrix $M \in \mathbb{R}^{n\times n}$ has the Cartesian P-property, then it also has the Jordan P-property, and consequently the P-property.*
(b) If a matrix $M \in \mathbb{R}^{n\times n}$ has the Cartesian P_0-property, it has the P_0-property.

Proof (a) From Definition 1.9, it is not hard to see that the Jordan P-property implies the P-property. Therefore, we only need to prove the Cartesian P-property implies the Jordan P-property. Let $x = (x_1, \ldots, x_m) \in \mathbb{R}^n$ with $x_i \in \mathbb{R}^{n_i}$ be any vector such that $x \circ (Mx) \in -\mathcal{K}$. From the Cartesian structure of $\mathcal{K}$, we have

$$x_i \circ (Mx)_i \in -\mathcal{K}^{n_i} \quad \text{for } i = 1, 2, \ldots, m,$$

which, by the definition of Jordan product given by (1.2), means that

$$\langle x_i, (Mx)_i \rangle \le 0 \quad \text{for all } i = 1, 2, \ldots, m. \tag{1.48}$$

Now, suppose that $x \neq 0$. Then, from Definition 1.8(a), it follows that there exists an index $\nu \in \{1, 2, \ldots, m\}$ such that $\langle x_\nu, (Mx)_\nu \rangle > 0$, which clearly contradicts (1.48). Hence, M has the Jordan P-property.

(b) Observe that for any $\varepsilon > 0$, $M + \varepsilon I$ has the Cartesian P-property. By part (a) and Definition 1.9, M has the P_0-property. □

It should be noted that the Cartesian P_0-property does not necessarily entail the P-property. For example, let $m = 2$ and $n_1 = n_2 = 2$, and consider

$$M = \begin{bmatrix} 1\,1\,0\,0 \\ 1\,1\,0\,0 \\ 0\,0\,2\,2 \\ 0\,0\,2\,2 \end{bmatrix} \quad \text{and} \quad x = \begin{pmatrix} -2 \\ 2 \\ -1 \\ 1 \end{pmatrix}.$$

It is easy to verify that M has the Cartesian P_0-property, $x \circ (Mx) = (0, 0, 0, 0) \in -\mathcal{K} = -(\mathcal{K}^2 \times \mathcal{K}^2)$ and $L_x L_{Mx} = L_{Mx} L_x = 0$, but $x \neq 0$, i.e., M has not the P-property. Now, we are not clear whether the P-property implies the Cartesian P_0-property.

We now introduce the definitions of Cartesian P-properties for a nonlinear mapping $F : \mathbb{R}^n \to \mathbb{R}^n$ within the framework of second-order cones (SOCs). The foundational concepts of P-properties on Cartesian products in $\mathbb{R}^n$ were first formulated by Facchinei and Pang [19]. Subsequently, Chen and Qi [61], as well as Kong et al. [37], extended these notions to the settings of positive semidefinite cones and general Euclidean Jordan algebras, respectively. Building upon these developments, we present several nonlinear generalizations of the Cartesian P-properties in the context of $\mathcal{K}$, defined as follows.

Definition 1.10 A nonlinear mapping $F = (F_1, \ldots, F_m)$ with $F_i : \mathbb{R}^n \to \mathbb{R}^{n_i}$ is said to

(a) have the uniform Cartesian P-property if there exists a constant $\rho > 0$ such that, for any $x, y \in \mathbb{R}^n$, there is an index $\nu \in \{1, 2, \ldots, m\}$ such that

$$\langle x_\nu - y_\nu, F_\nu(x) - F_\nu(y) \rangle \geq \rho \|x - y\|^2;$$

(b) have the Cartesian P-property if for any $x, y \in \mathbb{R}^n$ with $x \neq y$, there exists an index $\nu \in \{1, 2, \ldots, m\}$ such that

$$x_\nu \neq y_\nu \quad \text{and} \quad \langle x_\nu - y_\nu, F_\nu(x) - F_\nu(y) \rangle > 0;$$

(c) have the Cartesian P_0-property if for any $x, y \in \mathbb{R}^n$ with $x \neq y$, there exists an index $\nu \in \{1, 2, \ldots, m\}$ such that

$$x_\nu \neq y_\nu \quad \text{and} \quad \langle x_\nu - y_\nu, F_\nu(x) - F_\nu(y) \rangle \geq 0;$$

(d) have the Cartesian R_{02}-property if for any sequence $\{x^k\}$ satisfying the condition

$$\|x^k\| \to +\infty, \quad \frac{[-x^k]_+}{\|x^k\|} \to 0, \quad \frac{[-F(x^k)]_+}{\|x^k\|} \to 0, \tag{1.49}$$

there exists an index $i \in \{1, 2, \ldots, m\}$ such that

$$\liminf_{k\to+\infty} \frac{\lambda_2\left[F_i(x^k)\circ x_i^k\right]}{\|x^k\|^2} > 0.$$

It is straightforward to verify the following one-way implications from Definition 1.10:

Uniform Cartesian P-property $\Longrightarrow$ Cartesian P-property $\Longrightarrow$ Cartesian P_0-property;
Uniform Cartesian P-property $\Longrightarrow$ Cartesian R_{02}-property.

Moreover, it is evident that when $m = 1$, the Cartesian P-property (or P_0-property) of the mapping F reduces to the strict monotonicity (or monotonicity) of F. If the mapping F is continuously differentiable and possesses the Cartesian P-property (or P_0-property), then its transposed Jacobian matrix $\nabla F(x)$ at any point $x \in \mathbb{R}^n$ inherits the corresponding Cartesian P-property. Furthermore, when F specializes to an affine function of the form $Mx + q$, the uniform Cartesian P-property of F is equivalent to the Cartesian P-property of the matrix M.

Proposition 1.6 *For any $\varepsilon > 0$, let $F_\varepsilon : \mathbb{R}^n \to \mathbb{R}^n$ be given by*

$$F_\varepsilon(x) \;:=\; F(x) + \varepsilon x. \tag{1.50}$$

(a) If F is a P_0-function, then the Jacobian matrices $F_\varepsilon'(x)$ for all $x \in \mathbb{R}^n$ are P-matrices. In particular, the function F_ε is a P-function.
(b) If F has the Cartesian P_0-property, then F_ε has the Cartesian P-property.

Proof Please see [62, Lemma 3.2] for part(a), whereas part(b) is clear by Definition 1.10 (b) and (c). □

It is worth noting that even if F possesses the Cartesian P-property, the perturbed function F_ε, as defined in (1.50), may fail to exhibit the uniform Cartesian P-property. A counterexample illustrating this phenomenon in the case $m = 1$ can be found in [62]. Lastly, in parallel with Definition 1.9, we introduce the notions of P-properties for nonlinear mappings within the framework of SOCs, which represent special instances of the broader concepts established in [63].

Definition 1.11 A nonlinear mapping $F = (F_1, \ldots, F_m) : \mathbb{R}^n \to \mathbb{R}^n$ is said to have

(a) the Jordan P-property if $(x - y) \circ (F(x) - F(y)) \in -\mathcal{K} \Rightarrow x = y$;
(b) the P-property if from the condition that $L_{x_i - y_i} L_{F_i(x)-F_i(y)} = L_{F_i(x)-F_i(y)} L_{x_i - y_i}$, $i = 1, 2, \ldots, m$ and $(x - y) \circ (F(x) - F(y)) \in -\mathcal{K}$ implies $x = y$;
(c) the P_0-property if $F(x) + \varepsilon x$ has the P-property for all $\varepsilon > 0$;
(d) the uniform Jordan P-property if there exists a constant $\varrho > 0$ such that, for any $\zeta, \xi \in \mathbb{R}^n$, there is an index $\nu \in \{1, 2, \ldots, m\}$ such that

$$\lambda_2[(\zeta_\nu - \xi_\nu) \circ (F_\nu(\zeta) - F_\nu(\xi))] \geq \varrho \|\zeta - \xi\|^2$$

where $\lambda_2[(\zeta_\nu - \xi_\nu) \circ (F_\nu(\zeta) - F_\nu(\xi))]$ means the second spectral value of $(\zeta_\nu - \xi_\nu) \circ (F_\nu(\zeta) - F_\nu(\xi))$;

(e) the linear growth if there is a constant $c > 0$ such that $\|F(\zeta)\| \leq \|F(0)\| + c\|\zeta\|$.

Proposition 1.7 *(a) If a mapping $F : \mathbb{R}^n \to \mathbb{R}^n$ has the Cartesian P-property, then it must have the Jordan P-property and the P-property.*

(b) If a mapping $F : \mathbb{R}^n \to \mathbb{R}^n$ has the Cartesian P_0-property, then it has the P_0-property.

Proof The proof is similar to that of Proposition 1.5, and we omit it. □

There are analogous concepts such as R_0-matrix, R_0-type function and R_0-type property, which play a crucial role in proving the existence of solutions and in establishing both local and global error bounds for the LCPs, the NCPs, and the SOCCPs, respectively. The precise definitions are provided below; for further details, we refer the reader to [18, 63, 64].

Definition 1.12 A matrix $M \in \mathbb{R}^{n\times n}$ is called an R_0-matrix if $\mathrm{SOL}(0, M) = \{0\}$, i.e., the linear complementarity problem

$$x \geq 0, \quad Mx \geq 0, \quad \langle x, Mx\rangle = 0$$

has 0 as its unique solution. Equivalently, M is an R_0-matrix if $x_i(Mx)_i = 0$ for all i and $x \geq 0$, and $Mx \geq 0$ implies $x = 0$.

It is known that P-matrix $\Rightarrow$ R_0-matrix. For defining the R_0-type function, we need the following notation. For any $x \in \mathbb{V}$, let $\lambda_i(x)$ for $i = 1, \ldots, r$ denote the spectral values of x and

$$\omega(x) := \max_{1\leq i\leq r} \lambda_i(x).$$

Definition 1.13 A function $F : \mathbb{V} \to \mathbb{V}$ is called

(a) an R_0^s-function if for any sequence $\{x_k\}$ that satisfies

$$\|x_k\| \to \infty, \quad \frac{(-x_k)_+}{\|x_k\|} \to 0, \quad \frac{(-F(x_k))_+}{\|x_k\|} \to 0,$$

we have

$$\liminf_{k\to\infty} \frac{\omega(\phi_{\mathrm{NR}}(x_k, F(x_k)))}{\|x_k\|} > 0;$$

(b) an R_{01}^s-function if for any sequence $\{x_k\}$ that satisfies

$$\|x_k\| \to \infty, \quad \frac{(-x_k)_+}{\|x_k\|} \to 0, \quad \frac{(-F(x_k))_+}{\|x_k\|} \to 0,$$

we have

$$\liminf_{k\to\infty} \frac{\langle x_k, F(x_k)\rangle}{\|x_k\|} > 0;$$

(c) an R^s_{02}-function if for any sequence $\{x_k\}$ that satisfies

$$\|x_k\| \to \infty, \quad \frac{(-x_k)_+}{\|x_k\|} \to 0, \quad \frac{(-F(x_k))_+}{\|x_k\|} \to 0,$$

we have

$$\liminf_{k\to\infty} \frac{\omega(x_k \circ F(x_k))}{\|x_k\|} > 0.$$

In the setting of SOC, i.e., $\mathbb{V} = \mathbb{R}^n$ and $\omega(x_k \circ F(x_k))$ reduces to $\lambda_2(x_k \circ F(x_k))$, there are R_{01}-function and R_{02}-function whose definitions are similar to R^s_{01}-function and R^s_{02}-function, respectively. The only distinction is that they incorporate $\|x_k\|^2$ in the denominator. In other words, a function $F : \mathbb{R}^n \to \mathbb{R}^n$ is called

(i) an R_{01}-function if for any sequence $\{x_k\}$ that satisfies

$$\|x_k\| \to \infty, \quad \frac{(-x_k)_+}{\|x_k\|} \to 0, \quad \frac{(-F(x_k))_+}{\|x_k\|} \to 0, \tag{1.51}$$

we have

$$\liminf_{k\to\infty} \frac{\langle x_k, F(x_k)\rangle}{\|x_k\|^2} > 0;$$

(ii) an R_{02}-function if for any sequence $\{x_k\}$ that satisfies

$$\|x_k\| \to \infty, \quad \frac{(-x_k)_+}{\|x_k\|} \to 0, \quad \frac{(-F(x_k))_+}{\|x_k\|} \to 0,$$

we have

$$\liminf_{k\to\infty} \frac{\lambda_2(x_k \circ F(x_k))}{\|x_k\|^2} > 0.$$

It is well known that every R_{01}-function is also an R_{02}-function, and that if F possesses the uniform Jordan P-property, then F is an R_{02}-function. Utilizing the inequality $\langle x, y\rangle \le \omega(x \circ y)\|e\|^2$ (see [63, Proposition 2.1(ii)]) together with Definition 1.13, it is straightforward to verify that $R^s_{01} \Longrightarrow R^s_{02}$. Moreover, by employing the Peirce Decomposition Theorem (Theorem 1.2), the following result establishes an additional implication: $R^s_0 \Longrightarrow R^s_{02}$.

Proposition 1.8 *If the function $F : \mathbb{V} \to \mathbb{V}$ is a R^s_0-function, then F is a R^s_{02}-function.*

Proof For the sake of simplicity, for any $x, y \in \mathbb{V}$, we let

$$x \sqcap y := x - (x - y)_+, \quad x \sqcup y := y + (x - y)_+.$$

It is easy to verify that $x \sqcup y := y + (x - y)_+ = x + (y - x)_+$. Moreover, these are commutative operations with

$$(x \sqcap y) \circ (x \sqcup y) = x \circ y, \quad x \sqcap y + x \sqcup y = x + y$$

and

$$x \sqcup y - x \sqcap y = |y - x| \in \mathcal{K}.$$

If we consider the element $x \sqcap y = x - (x - y)_+ \in \mathbb{V}$ and apply the Spectral Decomposition Theorem (Theorem 1.1), there exist a Jordan frame $\{e_1, e_2, \ldots, e_r\}$ and real numbers $\lambda_1, \lambda_2, \ldots, \lambda_r$ such that

$$x \sqcap y = \lambda_1 e_1 + \cdots + \lambda_r e_r.$$

On the other hand, considering the element $x \sqcup y = x + (y - x)_+ \in \mathbb{V}$ and applying the Peirce Decomposition Theorem (Theorem 1.2), we know

$$x \sqcup y = \sum_{i=1}^{r} x_i e_i + \sum_{i<j} x_{ij}$$

with $x_i \in \mathbb{R}$ and $x_{ij} \in \mathbb{V}_{ij}$. Without loss of generality, let $\lambda_1 = \omega(x \sqcap y)$. To proceed the arguments, we first establish an inequality:

$$x_1 \geq \lambda_1.$$

Note that

$$(x \sqcup y - x \sqcap y) = \sum_{i=1}^{r} (x_i - \lambda_i) e_i + \sum_{i<j} x_{ij} \in \mathcal{K}.$$

Thus, it follows that

$$\langle x \sqcup y - x \sqcap y, e_1 \rangle = (x_1 - \lambda_1) \|e_1\|^2 \geq 0,$$

which yields $x_1 \geq \lambda_1$. Now suppose R_0^s condition holds. Take a sequence $\{x_k\}$ satisfying the required condition in Definition 1.13(c), i.e.,

$$\|x_k\| \to \infty, \quad \frac{(-x_k)_+}{\|x_k\|} \to 0, \quad \frac{(-y_k)_+}{\|x_k\|} \to 0,$$

where $y_k := F(x_k)$. From R_0^s condition, we have

$$\liminf_{k \to \infty} \frac{\omega(x_k \sqcap y_k)}{\|x_k\|} = \liminf_{k \to \infty} \frac{\lambda_1}{\|x_k\|} > 0 \text{ and } \lambda_1 > 0. \tag{1.52}$$

For the element $x_k \circ y_k \in \mathbb{V}$, applying the Spectral Decomposition Theorem (Theorem 1.1) again, there exist a Jordan frame $\{f_1, f_2, \ldots, f_r\}$ and real numbers $\mu_1, \mu_2, \ldots, \mu_r$ with $\mu_1 \geq \mu_2 \geq \cdots \geq \mu_r$ such that

$$x_k \circ y_k = \mu_1 f_1 + \cdots + \mu_r f_r.$$

Then, we have $\omega(x_k \circ y_k) = \mu_1$. On the other hand,

$$\begin{aligned}
x_k \circ y_k &= (x_k \sqcap y_k) \circ (x_k \sqcup y_k) \\
&= (\lambda_1 e_1 + \cdots + \lambda_r e_r) \circ (\sum_{i=1}^{r} x_i e_i + \sum_{i<j} x_{ij}) \\
&= \sum_{i=1}^{r} \lambda_i x_i e_i + \sum_{i=1}^{r} \lambda_i e_i \circ (\sum_{i<j} x_{ij}) \\
&= \sum_{i=1}^{r} \lambda_i x_i e_i + \sum_{i=1}^{r} \frac{\lambda_i}{2} \sum_{i<j} x_{ij}.
\end{aligned}$$

Hence,

$$\begin{aligned}
\lambda_1 x_1 \langle e_1, e_1 \rangle &= \langle x_k \circ y_k, e_1 \rangle \\
&= \mu_1 \langle f_1, e_1 \rangle + \mu_2 \langle f_2, e_1 \rangle + \cdots + \mu_r \langle f_r, e_1 \rangle \\
&\leq \mu_1 \langle f_1, e_1 \rangle + \mu_1 \langle f_2, e_1 \rangle + \cdots + \mu_1 \langle f_r, e_1 \rangle \\
&\leq r \mu_1 \theta,
\end{aligned}$$

where $\theta = \max\{\langle f_1, e_1 \rangle, \ldots, \langle f_r, e_1 \rangle\}$. This leads to

$$\frac{\mu_1}{\|x_k\|} \geq \frac{\lambda_1 x_1 \langle e_1, e_1 \rangle}{r\theta \|x_k\|},$$

which combining with the formula (1.52) implies that

$$\liminf_{k\to\infty} \frac{\omega(x_k \circ y_k)}{\|x_k\|} = \liminf_{k\to\infty} \frac{\mu_1}{\|x_k\|} \geq \liminf_{k\to\infty} \frac{\lambda_1 x_1 \langle e_1, e_1 \rangle}{r\theta \|x_k\|} > 0,$$

where the second inequality holds due to $x_1 \geq \lambda_1 > 0$ and $\frac{\langle e_1, e_1 \rangle}{r\theta} > 0$. Therefore, the implication $R_0^s \Longrightarrow R_{02}^s$ holds. □

Next, we introduce the notion of weak R_0-type functions, which will be instrumental in establishing the boundedness of level sets for the SCCPs in Sect. 3.3.

Definition 1.14 A function $F : \mathbb{V} \to \mathbb{V}$ is called an R_0^w-function if for any sequence $\{x_k\}$ that satisfies

$$\|x_k\| \to \infty, \ \limsup_{k\to\infty} \omega((-x_k)_+) < \infty, \ \limsup_{k\to\infty} \omega((-F(x_k))_+) < \infty,$$

we have

$$\omega(x_k \sqcap F(x_k)) \to \infty.$$

When the mapping F is linear, specifically $F(x) = L(x) + q$ with $q \in \mathbb{V}$, the notions of R_0^s-function and R_0^w-function reduce to the classical R_0-property (or R_0-matrix) of L; that is, the associated SCLCP with $q = 0$ admits a unique zero solution. The proofs of these equivalences closely follow the arguments presented in [64, Proposition 2.2] and are therefore omitted here. Furthermore, by Definitions 1.13 and 1.14, we can readily establish the following relationship between R_0^s and R_0^w.

Proposition 1.9 *For the function $F : \mathbb{V} \to \mathbb{V}$, there holds*

$$R_0^s \Longrightarrow R_0^w.$$

Proof Suppose R_0^s condition holds. Take a sequence $\{x_k\}$ satisfying the required condition in Definition 1.14, i.e.,

$$\|x_k\| \to \infty, \quad \limsup_{k\to\infty} \omega((-x_k)_+) < \infty, \quad \limsup_{k\to\infty} \omega((-F(x_k))_+) < \infty.$$

It follows that

$$\|x_k\| \to \infty, \quad \frac{(-x_k)_+}{\|x_k\|} \to 0, \quad \frac{(-y_k)_+}{\|x_k\|} \to 0.$$

By the definition of R_0^s, we have

$$\liminf_{k\to\infty} \frac{\omega(x_k \sqcap y_k)}{\|x_k\|} > 0.$$

Combining with $\|x_k\| \to \infty$ implies that

$$\omega(x_k \sqcap y_k) \to \infty.$$

Therefore, the implication $R_0^s \Longrightarrow R_0^w$ holds. □

Definition 1.15 The mappings $G = (G_1, \ldots, G_m)$ and $F = (F_1, \ldots, F_m)$ are said to have the joint Cartesian R_{02}-property if for any sequence $\{\zeta^k\}$ satisfying the condition:

$$\|\zeta^k\| \to +\infty, \quad \frac{[-G(\zeta^k)]_+}{\|\zeta^k\|} \to 0, \quad \frac{[-F(\zeta^k)]_+}{\|\zeta^k\|} \to 0, \tag{1.53}$$

there exists an index $\nu \in \{1, 2, \ldots, m\}$ such that

$$\liminf_{k\to+\infty} \frac{\lambda_{\max}\left[G_\nu(\zeta^k) \circ F_\nu(\zeta^k)\right]}{\|\zeta^k\|} > 0.$$

Proposition 1.10 *Assume that $G(\zeta) \equiv \zeta$ for any $\zeta \in \mathbb{V}$ and F is a R_{02}-function. Then, G and F have the joint Cartesian R_{02}-property.*

Proof Suppose that F is an R_{02}-function. From Definition 3 of [38], for any sequence $\{\zeta^k\}$ satisfying the condition (1.53), there holds

$$\liminf_{k\to+\infty} \frac{\lambda_{\max}\left[\zeta^k \circ F(\zeta^k)\right]}{\|\zeta^k\|^2} > 0. \tag{1.54}$$

For each k, let $z^k = \zeta^k \circ F(\zeta^k)$ and suppose that it has the spectral decomposition $z^k = \sum_{j=1}^r \lambda_j(z^k)c_j^k$, where $\{c_1^k, \dots, c_r^k\} \subseteq \mathbb{V}$ is a Jordan frame. For convenience, we also denote $z^k = (z_1^k, \dots, z_m^k)$ with $z_i^k \in \mathbb{V}_i$. By the spectral decomposition of z^k, clearly,

$$z_i^k = \sum_{j=1}^r \lambda_j(z^k)(c_j^k)_i, \quad i = 1, 2, \dots, m, \tag{1.55}$$

with $c_j^k = \left((c_j^k)_1, \dots, (c_j^k)_m\right)$ for every $j \in \{1, 2, \dots, r\}$. Now, without loss of generality, we assume that $\lambda_{l_k}(z^k) = \lambda_{\max}(z^k)$ with $1 \le l_k \le r$. Then,

$$\lambda_{\max}(z^k) = \sum_{j=1}^r \lambda_j(z^k)\langle c_j^k, c_{l_k}^k\rangle = \sum_{i=1}^m \sum_{j=1}^r \lambda_j(z^k)\left\langle (c_j^k)_i, (c_{l_k}^k)_i\right\rangle.$$

Combining with (1.54) and (1.55), there exists an index $\nu \in \{1, 2, \dots, m\}$ such that

$$0 < \liminf_{k\to+\infty} \frac{\sum_{j=1}^r \lambda_j(z^k)\left\langle (c_j^k)_\nu, (c_{l_k}^k)_\nu\right\rangle}{\|\zeta^k\|^2} = \liminf_{k\to+\infty} \frac{\langle z_\nu^k, (c_{l_k}^k)_\nu\rangle}{\|\zeta^k\|^2}. \tag{1.56}$$

Suppose that z_ν^k as an element in the simple Euclidean Jordan algebra $(\mathbb{V}_\nu, \circ, \langle\cdot,\cdot\rangle)$ has the following spectral decomposition

$$z_\nu^k = \sum_{j=1}^{\bar{r}} \lambda_j(z_\nu^k)q_{\nu j}^k,$$

where $\{q_{\nu 1}^k, \dots, q_{\nu\bar{r}}^k\} \subseteq \mathbb{V}_\nu$ be the corresponding Jordan frame. Then,

$$\langle z_\nu^k, (c_{l_k}^k)_\nu\rangle \le \lambda_{\max}(z_\nu^k)\left\langle \sum_{j=1}^{\bar{r}} q_{\nu j}^k, (c_{l_k}^k)_\nu \right\rangle = \lambda_{\max}(z_\nu^k)\langle e_\nu, (c_{l_k}^k)_\nu\rangle, \tag{1.57}$$

where e_ν is the identity element in $\mathbb{V}_\nu$ and the inequality is since $(c_{l_k}^k)_\nu \in \mathcal{K}^\nu$ and $q_{\nu j}^k \in \mathcal{K}^\nu$ for every $j = 1, 2, \dots, \bar{r}$. From (1.56) and (1.57), it then follows that

$$0 < \liminf_{k\to+\infty} \frac{\langle z_\nu^k, (c_{l_k}^k)_\nu\rangle}{\|\zeta^k\|^2} \le \liminf_{k\to+\infty} \frac{\lambda_{\max}(z_\nu^k)}{\|\zeta^k\|} \cdot \frac{\langle e_\nu, (c_{l_k}^k)_\nu\rangle}{\|\zeta^k\|}.$$

Noting that $\langle e_\nu, (c^k_{l_k})_\nu \rangle$ is bounded for each k, we have that

$$\liminf_{k\to+\infty} \frac{\langle e_\nu, (c^k_{l_k})_\nu \rangle}{\|\zeta^k\|} = 0.$$

From the last two inequalities, it readily follows that

$$\liminf_{k\to+\infty} \frac{\lambda_{\max}(z^k_\nu)}{\|\zeta^k\|} > 0.$$

By Definition 1.15, the mappings G and F have the joint Cartesian R_{02}-property. □

Definition 1.16 The mappings $F, G : \mathbb{R}^n \to \mathbb{R}^n$ are said to have the joint $\tilde{R}_{01}$-property if for any sequence $\{\zeta^k\}$ with

$$\|\zeta^k\| \to +\infty, \quad \frac{[-G(\zeta^k)]_+}{\|\zeta^k\|} \to 0, \quad \frac{[-F(\zeta^k)]_+}{\|\zeta^k\|} \to 0, \tag{1.58}$$

there holds

$$\liminf_{k\to+\infty} \frac{\langle F(\zeta^k), G(\zeta^k) \rangle}{\|\zeta^k\|} > 0. \tag{1.59}$$

References

1. J. Faraut, A. Koranyi, *Analysis on Symmetric Cones* Oxford Mathematical Monographs. (Oxford University Press, New York, 1994)
2. M. Koecher, The Minnesota Notes on Jordan Algebras and Their Applications, in ed. by A. Brieg, S. Walcher (Springer, Berlin, 1999)
3. A. Korányi, Monotone functions on formally real Jordan algebras. Math. Ann. **269**, 73–76 (1984)
4. Y.-L. Chang, J.-S. Chen, The Hölder continuity of vector-valued functions associated with second-order cone. Pac. J. Optim. **8**, 135–141 (2012)
5. J.-S. Chen, Alternative proofs for some results of vector-valued functions associated with second-order cone. J. Nonlinear Convex Anal. **6**(2), 297–325 (2005)
6. J.-S. Chen, The convex and monotone functions associated with second-order cone. Optimization **55**, 363–385 (2006)
7. J.-S. Chen, *SOC Functions and Their Applications, Springer Optimization and Its Applications*, vol. 143 (Springer, Singapore, 2019)
8. J.-S. Chen, X. Chen, P. Tseng, Analysis of nonsmooth vector-valued functions associated with second-order cones. Math. Program. **101**, 95–117 (2004)
9. D. Sun, J. Sun, Löwner's operator and spectral functions in Euclidean Jordan algebras. Math. Oper. Res. **33**, 421–445 (2008)
10. J.-S. Chen, T.-K. Liao, S.-H. Pan, Using Schur complement theorem to prove convexity of some SOC-functions. J. Nonlinear Convex Anal. **13**(3), 421–431 (2012)
11. R. Bhatia, *Matrix Analysis* (Springer, New York, 1997)
12. R.A. Horn, C.R. Johnson, *Topics in Matrix Analysis* (Cambridge Press, Cambridge, United Kingdom, 1991)

13. X. Chen, H. Qi, P. Tseng, Analysis of nonsmooth symmetric-matrix-valued functions with applications to semidefinite complementarity problems. SIAM J. Optim. **13**, 960–985 (2003)
14. X. Chen, P. Tseng, Non-interior continuation methods for solving semidefinite complementarity problems. Math. Program. **95**, 431–474 (2003)
15. C. Kanzow, C. Nagel, Semidefinite programs: new search directions, smoothing-type methods, and numerical results. SIAM J. Optim. **13**(1), 1–23 (2002)
16. J.-S. Pang, D. Sun, J. Sun, Semismooth homeomorphisms and strong stability of semidefinite and Lorentz cone complementarity problems. Math. Oper. Res. **28**, 39–63 (2003)
17. D. Sun, J. Sun, Semismooth matrix valued functions. Math. Oper. Res. **27**, 150–169 (2002)
18. R.W. Cottle, J.-S. Pang, R.-E. Stone, *The Linear Complementarity Problem* (Academic Press, New York, 1992)
19. F. Facchinei, J.-S. Pang, *Finite-Dimensional Variational Inequalities and Complementarity Problems*, vol. I and II (Springer, New York, 2003)
20. C.A. Floudas, P.M. Pardalos, *Encyclopedia Optimization*, 2nd edn. (Springer, New York, 2009)
21. C.-Y. Yang, Y.-L. Chang, C.-C. Hu, J.-S. Chen, Novel constructions for closed convex cones through inequalities and support functions. J. Optim. Theory Appl. **205**(3), Article 57 (2025)
22. M.S. Gowda, R. Sznajder, J. Tao, Some P-properties for linear transformations on Euclidean Jordan algebras. Linear Algebr. Appl. **393**, 203–232 (2004)
23. C.-H. Ko, J.-S. Chen, Optimal grasping manipulation for multifingered robots using semismooth Newton method. Math. Probl. Eng. **2013**, Article ID 681710, 9 (2013)
24. N. Yamashita, Z.-Q. Luo, A nonlinear complementarity approach to multiuser power control for digital subscriber lines. Optim. Methods Softw. **19**, 633–652 (2004)
25. H.W. Zhang, J.Y. Li, S.-H. Pan, New second-order cone linear complementarity formulation and semi-smooth Newton algorithm for finite element analysis of 3D frictional contact problem. Comput. Methods Appl. Mech. Eng. **200**, 77–88 (2011)
26. P. Chen, G.-H. Lin, X. Zhu, F. Bai, Smoothing Newton method for nonsmooth second-order cone complementarity problems with application to electric power markets. J. Glob. Optim. **14**, 1168–1190 (1993)
27. W. Ham, H. Kwon, Sliding mode control for the hovering of helicopter by using quaternion dynamics, in *Proceedings of SICE Annual Conference, Sept. 13–18* (Waseda University, Tokyo, Japan, 2011), pp. 1024–1028
28. C.G. Mayhew, R.G. Sanfelice, A.R. Teel, On quaternion-based attitude control and the unwinding phenomenon, in *Proceedings of American Control Conference* (San Francisco, CA, USA, June 29–July 01, 2011), pp. 299–304
29. H.P. Geering, *Optimal Control with Engineering Applications* (Springer, Berlin, 2007)
30. D.G. Hull, *Optimal Control Theory for Applications* (Springer, Berlin, 2003)
31. F.L. Lewis, V.L. Syrmos, *Optimal Control* (Wiley, 1995)
32. F.A. Potra, Weighted complementarity problems—a new paradigm for computing equilibria. SIAM J. Optim. **22**, 1634–1654 (2012)
33. J. Tang, H. Zhang, A nonmonotone smoothing Newton algorithm for weighted complementarity problem. J. Optim. Theory Appl. **189**, 679–715 (2021)
34. A. Yoshise, Interior point trajectories and a homogenous model for nonlinear complementarity problems over symmetric cones. SIAM J. Optim. **17**, 1129–1153 (2006)
35. Z.-H. Huang, T. Ni, Smoothing algorithms for complementarity problems over symmetric cones. Comput. Optim. Appl. **45**, 557–579 (2010)
36. L.C. Kong, J. Sun, N.H. Xiu, A regularized smoothing Newton method for symmetric cone complementarity problems. SIAM J. Optim. **19**, 1028–1047 (2008)
37. L.C. Kong, L. Tuncel, N.H. Xiu, Vector-valued implicit Lagrangian for symmetric cone complementarity problems. Asia-Pac. J. Oper. Res. **26**, 199–233 (2009)
38. Y. Liu, L. Zhang, Y. Wang, Some properties of a class of merit functions for symmetric cone complementarity problems. Asia-Pac. J. Oper. Res. **23**, 473–496 (2006)
39. S.-H. Pan, J.-S. Chen, A one-parametric class of merit functions for the symmetric cone complementarity problem. J. Math. Anal. Appl. **355**, 195–215 (2009)

40. L.C. Kong, L. Tuncel, N.H. Xiu, Equivalent conditions for Jacobian nonsingularity in linear symmetric cone programming. J. Optim. Theory Appl. **148**, 364–389 (2011)
41. L.C. Kong, L. Tuncel, N.H. Xiu, The Fischer-Burmeister complementarity function on Euclidean Jordan algebras. Pac. J. Optim. **6**, 423–440 (2010)
42. Y. Lu, J.-S. Chen, On the self-duality and homogeneity of ellipsoidal cones. J. Nonlinear Convex Anal. **19**(8), 1355–1367 (2018)
43. P.-F. Ma, J.-S. Chen, C.-H. Huang, C.-H. Ko, Discovery of new complementarity functions for NCP and SOCCP. Comput. Appl. Math. **37**(5), 5727–5749 (2018)
44. X.-H. Miao, Y.-L. Chang, J.-S. Chen, On merit functions for p-order cone complementarity problem. Comput. Optim. Appl. **67**(1), 155–173 (2017)
45. X.-H. Miao, Y. Lu, J.-S. Chen, From symmetric cone optimization to nonsymmetric cone optimization: spectral decomposition, nonsmooth analysis, and projections onto nonsymmetric cones. Pac. J. Optim. **14**(3), 399–419 (2018)
46. X.-H. Miao, N. Qi, J.-S. Chen, Projection formula and one type of spectral factorization associated with p-order cone. J. Nonlinear Convex Anal. **18**(9), 1699–1705 (2017)
47. J.-S. Chen, P. Tseng, An unconstrained smooth minimization reformulation of the second-order cone complementarity problem. Math. Program. **104**, 293–327 (2005)
48. M. Fukushima, Z.-Q. Luo, P. Tseng, Smoothing functions for second-order cone complementarity problems. Comput. Optim. Appl. **17**, 159–182 (2000)
49. R. Mifflin, Semismooth and semiconvex functions in constrained optimization. SIAM J. Control Optim. **15**, 959–972 (1977)
50. L. Qi, J. Sun, A nonsmooth version of Newton's method. Math. Program. **58**, 353–367 (1993)
51. R.T. Rockafellar, R.J.-B. Wets, *Variational Analysis* (Springer, Berlin, 1998)
52. F.H. Clarke, *Optimization and Nonsmooth Analysis* (Wiley, New York, 1983)
53. L. Qi, Convergence analysis of some algorithms for solving nonsmooth equations. Math. Oper. Res. **18**, 227–244 (1993)
54. L. Qi, C-differentiability, C-Differential operators and generalized Newton methods, AMR 96/5, Applied Mathematics Report (University of New South Wales, Sydney, 1996)
55. A. Fischer, Solution of monotone complementarity problems with locally Lipschitzian functions. Math. Program. **76**, 513–532 (1997)
56. J.M. Ortega, W. Rheinboldt, *Iterative Solution of Nonlinear Equations in Several Variables* (SIAM, Philadelphia, 2000)
57. N.J. Higham, Estimating the matrix p-norm. Numer. Math. **62**, 539–555 (1992)
58. J.-S. Pang, L. Qi, A globally convergent Newton methods for convex SC^1 minimization problems. J. Optim. Theory Appl. **85**, 633–648 (1995)
59. L. Qi, Superlinearly convergent approximate Newton methods for LC^1 optimization problems. Math. Program. **64**, 277–294 (1994)
60. L. Qi, L.P. Tseng, On almost smooth functions and piecewise smooth functions. Nonlinear Anal.: Theory, Methods Appl. **67**(3), 773–794 (2007)
61. X. Chen, H. Qi, Cartesian P-property and its applications to the semidefinite linear complementarity problem. Math. Program. **106**, 177–201 (2006)
62. F. Facchinei, C. Kanzow, Beyond monotonicity in regularization methods for nonlinear complementarity problems. SIAM J. Control Optim. **37**, 1150–1161 (1999)
63. J. Tao, M.S. Gowda, Some P-properties for nonlinear transformations on Euclidean Jordan algebras. Math. Oper. Res. **4**, 985–1004 (2005)
64. B. Chen, Error bounds for R_0-type and monotone nonlinear complementarity problems. J. Optim. Theory Appl. **108**, 297–316 (2001)

Chapter 2
The Nonlinear Complementarity Functions

Abstract In this chapter, we delve into the development of complementarity functions within the framework of nonlinear complementarity problems (NCPs), introducing several novel ideas and systematic techniques for constructing new NCP functions.

Complementarity Problems (NCPs) constitute a fundamental class of variational inequalities, frequently emerging in the formulation of Karush-Kuhn-Tucker (KKT) conditions for optimization problems [1]. Beyond their role in optimization theory, the NCPs offer a powerful framework for analyzing equilibrium phenomena across a wide range of disciplines, including operations research, engineering, and economics [1–3].

Given a function $F : \mathbb{R}^n \to \mathbb{R}^n$, the problem of finding a point $x \in \mathbb{R}^n$ such that

$$x \geq 0, \quad F(x) \geq 0, \quad \text{and} \quad \langle x, F(x) \rangle = 0, \tag{2.1}$$

is precisely the nonlinear complementarity problem . Various approaches to solving this problem have been proposed, in which most of them utilize a so-called NCP function , that is, a function $\phi : \mathbb{R}^2 \to \mathbb{R}$ such that

$$\phi(a, b) = 0 \quad \Longleftrightarrow \quad a \geq 0, \quad b \geq 0, \quad \text{and} \quad ab = 0. \tag{2.2}$$

An NCP function is useful in solving the NCP (2.1) as it naturally exploits the structure of the problem. In particular, defining $\Phi_{\text{F}} : \mathbb{R}^n \to \mathbb{R}^n$ as

$$\Phi_{\text{F}}(x) = \begin{pmatrix} \phi(x_1, F_1(x)) \\ \vdots \\ \phi(x_n, F_n(x)) \end{pmatrix}, \tag{2.3}$$

it is clear to see that NCP (2.1) is equivalent to solving the system of equations $\Phi_{\text{F}}(x) = 0$. Based on the above discussion, there are roughly four main approaches to addressing the NCP (2.1), each utilizing an NCP function ϕ as defined in (2.2).

J. Chen, *Complementarity Functions in Optimization*, Springer Optimization and Its Applications 228, https://doi.org/10.1007/978-3-032-06924-5_2

(1) Merit function approach. The central idea of this approach is to reformulate the NCP as an unconstrained global minimization problem:

$$\min_{x \in \mathbb{R}^n} \Psi_{\mathrm{F}}(x) \quad \text{where} \quad \Psi_{\mathrm{F}}(x) := \frac{1}{2}\|\Phi_{\mathrm{F}}(x)\|^2. \tag{2.4}$$

Here, the objective function Ψ_{F} is also referred to as a merit function. It is evident that the global minimizers of problem (2.4) correspond precisely to the solutions of the NCP (2.1). As a result, attention is directed toward analyzing the structure and properties of Ψ_{F}, as well as developing effective solution methods for solving the minimization problem (2.4).

(2) Nonsmooth function approach. For this approach, it just employs the direct equivalent relation:

$$\text{NCP} \quad \Longleftrightarrow \quad \Phi_{\mathrm{F}}(x) = \begin{pmatrix} \phi(x_1, F_1(x)) \\ \vdots \\ \phi(x_n, F_n(x)) \end{pmatrix} = 0. \tag{2.5}$$

In other words, solving the NCP (2.1) is equivalent to finding solutions to the system of equations $\Phi_{\mathrm{F}}(x) = 0$. In general, the function Φ_{F} is nonsmooth, which gives rise to the term "nonsmooth function approach".

(3) Smoothing function approach. The functions Ψ_{F} and $\Phi\mathrm{F}$ used in the merit function and nonsmooth function approaches are often nondifferentiable. To address this, the smoothing approach introduces a family of smooth approximations. Specifically, one may construct a smooth function $\Psi\mathrm{F}^{\mu}$ with $\mu > 0$ such that

$$\Psi_{\mathrm{F}}^{\mu} \to \Psi_{\mathrm{F}} \quad \text{as} \quad \mu \to 0. \tag{2.6}$$

Alternatively, one may define a smooth approximation $\Phi\mathrm{F}^{\mu}$ satisfying

$$\Phi_{\mathrm{F}}^{\mu} \to \Phi_{\mathrm{F}} \quad \text{as} \quad \mu \to 0. \tag{2.7}$$

Since both $\Psi\mathrm{F}^{\mu}$ and $\Phi_{\mathrm{F}}{}^{\mu}$ are smooth, a wide range of well-established algorithms for smooth optimization or equation-solving can be employed to tackle problems (2.6) and (2.7), respectively. The subsequent analysis then focuses on identifying conditions under which solutions to the smoothed problems converge to those of the original nonsmooth formulations, namely, (2.4) or (2.5) as $\mu \to 0$.

(4) Regularization approach. Distinct from the previous three approaches, this method focuses on solving the original NCP (2.1) through a sequence of regularized complementarity problems, denoted as NCP(F_ε):

$$x \geq 0, \quad F_\varepsilon(x) \geq 0, \quad \langle x, F_\varepsilon(x)\rangle = 0, \tag{2.8}$$

where $\varepsilon > 0$ is a regularization parameter tending to zero, and F_ε is defined by

$$F_\varepsilon(x) \ := \ F(x) + \varepsilon x.$$

The central question in this approach is to determine under what conditions the solutions of the regularized problem NCP(F_ε), as defined in (2.8), converge to a solution of the original NCP (2.1) as $\varepsilon \to 0$.

Owing to their practical relevance, a wide variety of NCP functions have been proposed and thoroughly investigated in the literature [4]. Among these, one of the most widely used is the natural residual (NR) function [5], defined as

$$\phi_{\mathrm{NR}}(a, b) = \min\{a, b\} = a - [a - b]_+. \tag{2.9}$$

In contrast to the merit function defined in (2.4), the NR merit function $\Psi_{\mathrm{NR}} : \mathbb{R}^n \to \mathbb{R}_+$ is employed, and is given by

$$\Psi_{\mathrm{NR}}(x) := \frac{1}{2} \sum_{i=1}^{n} \phi_{\mathrm{NR}}^2(x_i \,, F_i(x)). \tag{2.10}$$

Analogously, for the nonsmooth function approach described in (2.5), the corresponding function Φ_{NR} is defined componentwise by replacing the generic NCP function ϕ with the natural residual ϕ_{NR} as given in (2.9), i.e.,

$$\Phi_{\mathrm{NR}}(x) = \begin{pmatrix} \phi_{\mathrm{NR}}(x_1, F_1(x)) \\ \vdots \\ \phi_{\mathrm{NR}}(x_n, F_n(x)) \end{pmatrix}.$$

Mangasarian and Solodov proposed another type of NCP function [6], which is defined by

$$\phi_{\mathrm{MS}}(a, b) = ab + \frac{1}{2\alpha}\Big(\max\{0, a - \alpha b\}^2 - a^2 + \max\{0, b - \alpha a\}^2 - b^2 \Big), \ \alpha > 1. \tag{2.11}$$

The NCP function ϕ_{MS} described above is differentiable, which is advantageous for the unconstrained minimization approach. However, it is important to note that an NCP function cannot, in general, be both convex and differentiable simultaneously; see [7, 8]. Consequently, the design of NCP functions that exhibit either convexity or differentiability, depending on the needs of a particular application, remains an important and ongoing area of research.

Another widely used NCP function is the Fischer-Burmeister (FB) function [9, 10], defined as

$$\phi_{\mathrm{FB}}(a, b) = \sqrt{a^2 + b^2} - (a + b). \tag{2.12}$$

The FB function has attracted considerable attention and has been extensively employed in numerous studies due to its favorable numerical properties. Several variants of ϕ_{FB} have also been explored in the literature [11]:

$$\phi_1(a,b) := \phi_{FB}(a,b) - \alpha a_+ b_+, \ \alpha > 0.$$
$$\phi_2(a,b) := \phi_{FB}(a,b) - \alpha (ab)_+, \ \alpha > 0.$$
$$\phi_3(a,b) := \sqrt{[\phi_{FB}(a,b)]^2 + \alpha (ab)^2}, \ \alpha > 0.$$
$$\phi_4(a,b) := \sqrt{[\phi_{FB}(a,b)]^2 + \alpha (a_+ b_+)^2}, \ \alpha > 0.$$
$$\phi_5(a,b) := \sqrt{[\phi_{FB}(a,b)]^2 + \alpha [(ab)_+]^2}, \ \alpha > 0.$$
$$\phi_6(a,b) := \sqrt{[\phi_{FB}(a,b)]^2 + \alpha [(ab)_+]^4}, \ \alpha > 0.$$
$$\phi_7(a,b) := \sqrt{[\phi_{FB}(a,b)_+]^2 + \alpha [(ab)_+]^2}, \ \alpha > 0.$$

In particular, it has been noted that the functions $\phi_2(a,b) = \phi_{FB}(a,b) - \alpha(ab)_+$ and $\phi_3(a,b) = \sqrt{[\phi_{FB}(a,b)]^2 + \alpha(ab)^2}$ are not recommended for practical use, as they lack certain desirable properties; see [11, p. 206].

A general framework for constructing NCP functions was first introduced by Mangasarian in [12]. The idea is to select a strictly increasing function $\theta : \mathbb{R} \to \mathbb{R}$ satisfying $\theta(0) = 0$, such that $a > b$ if and only if $\theta(a) > \theta(b)$. Under this setting, a vector z solves the complementarity problem (2.1) if and only if

$$\theta(|F_i(z) - z_i|) - \theta(F_i(z)) - \theta(z_i) = 0, \quad i = 1, \ldots, n.$$

An alternative construction was proposed by Luo and Tseng [13], which introduces a merit function $f_{LT} : \mathbb{R}^n \to \mathbb{R}$ defined by

$$f_{LT}(\zeta) := \psi_0(\langle \zeta, F(\zeta) \rangle) + \sum_{i=1}^{n} \psi_i(-\zeta_i, -F_i(\zeta)), \tag{2.13}$$

where $\psi_0 : \mathbb{R} \to [0,\infty)$ and $\psi_i : \mathbb{R}^2 \to [0,\infty)$ are continuous functions that vanish on the negative orthant only. The construction of the function f_{LT} is not derived from an NCP function. Nevertheless, it possesses several notable properties under certain assumptions [13]. In particular, f_{LT} is convex on $\mathbb{R}^n$ provided that the function $\langle x, F(x) \rangle$ and each component $-F_i(x)$, for $i = 1, \ldots, n$, are convex in x. Building upon the idea underlying the construction of f_{LT} as defined in (2.13), Kanzow, Yamashita, and Fukushima [14] introduced a class of NCP functions. Specifically, they considered the set of continuous functions $\Psi : \mathbb{R}^m \to [0,\infty)$ satisfying

$$\Psi(t) = 0 \iff t \le 0$$

which they denoted by Ψ^m. Then, for any $\Psi_0 \in \Psi^1$ and $\Psi_i \in \Psi^2$, each function $\phi_i : \mathbb{R}^2 \to \mathbb{R}$ defined by

$$\phi_i(a,b) = \Psi_0(a,b) + \Psi_i(-a,-b), \quad i = 1,\dots,n$$

is a nonnegative NCP function. A comprehensive survey of various merit functions can be found in [15]. More recently, a rigorous treatment of the construction of NCP functions was provided by Galántai in [4], which is notably the first work to compile a systematic list of existing NCP functions. In essence, most of these functions are extensions or variants of the previously discussed ϕ_{NR}, ϕ_{FB}, and ϕ_{MS}. Additional generalizations have also been proposed, building upon these foundational forms. In the following sections, we present a survey and review of recent developments in the design and analysis of new NCP functions.

2.1 Constructions of NCP Functions Based on ϕ_{FB}

2.1.1 Construction by Using p-Norm

There are several extensions of the FB function ϕ_{FB} given as in (2.12) in the literature. For example, Kanzow and Kleinmichel [16] extended ϕ_{FB} function to

$$\phi_\theta(a,b) := \sqrt{(a-b)^2 + \theta ab} - (a+b), \quad \theta \in (0,4).$$

Chen et al. [17] studied a penalized FB function

$$\phi_\lambda(a,b) := \lambda\phi_{\text{FB}}(a,b) + (1-\lambda)a_+b_+, \quad \lambda \in (0,1).$$

Additional forms of penalized Fischer-Burmeister (FB) functions have been explored by Sun and Qi [11]. In this section, we turn our attention to a notable extension of the classical FB function, denoted by ϕ_{FB}, which has recently garnered significant interest and has been the subject of extensive study. As observed in [4], it is particularly noteworthy that several nonlinear complementarity problem (NCP) functions bear a close resemblance to the FB function. Among them, the generalized FB function is of special interest:

$$\phi^p_{\text{FB}}(a,b) = \|(a,b)\|_p - (a+b), \quad p > 1 \tag{2.14}$$

This formulation represents a compelling generalization of ϕ_{FB} and has proven to be an effective tool for solving NCPs. Initially introduced by Tseng [18], the function ϕ^p_{FB} was established therein as a valid NCP function. Subsequent studies have further examined its properties and applications, as documented in [19–26]. Here, $|\cdot|_p$ denotes the l_p-norm, and the parameter p serves as a tunable variable that, as demonstrated in [21–24, 27], can potentially enhance the numerical performance of certain algorithms.

Accordingly, we define $\psi^p_{\text{FB}} : \mathbb{R}^2 \to \mathbb{R}_+$ by

$$\psi^p_{\text{FB}}(a, b) := \frac{1}{2}|\phi^p_{\text{FB}}(a, b)|^2. \tag{2.15}$$

For any given $p > 1$, the function ψ^p_{FB} is a nonnegative NCP function and smooth on $\mathbb{R}^2$ as will be seen later. Analogous to Φ_{F}, the function $\Phi^p_{\text{FB}} : \mathbb{R}^n \to \mathbb{R}^n$ given as

$$\Phi^p_{\text{FB}}(x) = \begin{pmatrix} \phi^p_{\text{FB}}(x_1, F_1(x)) \\ \cdot \\ \cdot \\ \cdot \\ \phi^p_{\text{FB}}(x_n, F_n(x)) \end{pmatrix} \tag{2.16}$$

yields a family of merit functions $\Psi^p_{\text{FB}} : \mathbb{R}^n \to \mathbb{R}$ for the NCP for which

$$\Psi^p_{\text{FB}}(x) := \frac{1}{2}\|\Phi^p_{\text{FB}}(x)\|^2 = \frac{1}{2}\sum_{i=1}^{n} \phi^p_{\text{FB}}(x_i, F_i(x))^2 = \sum_{i=1}^{n} \psi^p_{\text{FB}}(x_i, F_i(x)). \tag{2.17}$$

As will be demonstrated later, for any fixed $p > 1$ the function Ψ^p_{FB} serves as a continuously differentiable merit function for the NCP. This smoothness property makes it amenable to classical iterative approaches, such as the Newton method, which can be effectively employed to solve the NCP through unconstrained smooth optimization, namely:

$$\min_{x \in \mathbb{R}^n} \Psi^p_{\text{FB}}(x). \tag{2.18}$$

Proposition 2.1 *Let $\phi^p_{\text{FB}} : \mathbb{R}^2 \to \mathbb{R}$ be defined as in (2.14) where $p > 1$. Then, the following hold.*

(a) The function ϕ^p_{FB} is an NCP function, i.e., it satisfies (2.2).

(b) The function ϕ^p_{FB} is sub-additive, i.e., $\phi^p_{\text{FB}}(w + w') \le \phi^p_{\text{FB}}(w) + \phi^p_{\text{FB}}(w')$ for all $w, w' \in \mathbb{R}^2$.

(c) The function ϕ^p_{FB} is positive homogeneous, i.e., $\phi^p_{\text{FB}}(\alpha w) = \alpha\phi^p_{\text{FB}}(w)$ for all $w \in \mathbb{R}^2$ and $\alpha \ge 0$.

(d) The function ϕ^p_{FB} is convex, i.e., $\phi^p_{\text{FB}}(\alpha w + (1-\alpha)w') \le \alpha\phi^p_{\text{FB}}(w) + (1-\alpha)$ $\phi^p_{\text{FB}}(w')$ for all $w, w' \in \mathbb{R}^2$ and $\alpha \ge 0$.

(e) The function ϕ^p_{FB} is Lipschitz continuous with $L_1 = \sqrt{2} + 2^{(1/p-1/2)}$ when $1 < p < 2$, and with $L_2 = 1 + \sqrt{2}$ when $p \ge 2$. That is, $\left|\phi^p_{\text{FB}}(w) - \phi^p_{\text{FB}}(w')\right| \le L_2\|w - w'\|$ when $1 < p < 2$ and $\left|\phi^p_{\text{FB}}(w) - \phi^p_{\text{FB}}(w')\right| \le L_1\|w - w'\|$ when $p \ge 2$ for all $w, w' \in \mathbb{R}^2$.

(f) Given any point $(a, b) \in \mathbb{R}^2$, each element in the generalized gradient $\partial\phi^p_{\text{FB}}(a, b)$ has the representation $(\xi - 1, \zeta - 1)$ where, if $(a, b) \ne (0, 0)$,

$$(\xi, \zeta) = \left(\frac{\operatorname{sgn}(a) \cdot |a|^{p-1}}{\|(a,b)\|_p^{p-1}}, \frac{\operatorname{sgn}(b) \cdot |b|^{p-1}}{\|(a,b)\|_p^{p-1}} \right),$$

and otherwise (ξ, ζ) *is an arbitrary vector in* $\mathbb{R}^2$ *satisfying* $|\xi|^{\frac{p}{p-1}} + |\zeta|^{\frac{p}{p-1}} \leq 1$.

Proof (a) The proof can be seen in [18, p. 20]. For completeness, we here include it. Consider any $a \geq 0$ and $b \geq 0$ satisfying $ab = 0$. Then, we have either $a = 0$ or $b = 0$, which implies that $\phi_{\mathrm{FB}}^p(a,b) = \sqrt[p]{|a|^p} - a$ or $\phi_{\mathrm{FB}}^p(a,b) = \sqrt[p]{|b|^p} - b$. Considering $a \geq 0$ and $b \geq 0$, we thus have $\phi_{\mathrm{FB}}^p(a,b) = 0$. Conversely, consider any $(a,b) \in \mathbb{R}^2$ satisfying $\phi_{\mathrm{FB}}^p(a,b) = 0$. Then, there must hold $a \geq 0$ and $b \geq 0$. Otherwise, we have $\sqrt[p]{|a|^p + |b|^p} > (a+b)$ and hence contradicts the fact that $\phi_p(a,b) = 0$. Now we prove that one of a and b must be 0. Otherwise, $\|(a,b)\|_p < \|(a,b)\|_1 = a + b$. This obviously contradicts the fact that $\phi_{\mathrm{FB}}^p(a,b) = 0$. The two sides show that ϕ_{FB}^p is indeed an NCP function.

(b) Let $w = (a,b)$ and $w' = (c,d)$. Then, the desired result follows by

$$\begin{aligned}
\phi_{\mathrm{FB}}^p(w + w') &= \|(a,b) + (c,d)\|_p - (a+b+c+d) \\
&\leq \|(a,b)\|_p + \|(c,d)\|_p - (a+b) - (c+d) \\
&= \phi_{\mathrm{FB}}^p(a,b) + \phi_{\mathrm{FB}}^p(c,d) = \phi_{\mathrm{FB}}^p(w) + \phi_{\mathrm{FB}}^p(w'),
\end{aligned}$$

where the inequality is true since the triangle inequality holds for p-norm when $p > 1$.

(c) Let $w = (a,b) \in \mathbb{R}^2$ and $\alpha > 0$. Then the proof follows by

$$\phi_{\mathrm{FB}}^p(\alpha w) = \sqrt[p]{|\alpha a|^p + |\alpha b|^p} - (\alpha a + \alpha b) = \alpha \sqrt[p]{|a|^p + |b|^p} - \alpha(a+b) = \alpha \phi_{\mathrm{FB}}^p(w).$$

(d) This is true by part (b) and part (c).

(e) Let $w = (a,b)$ and $w' = (c,d)$, we have

$$\begin{aligned}
|\phi_{\mathrm{FB}}^p(w) - \phi_p(w')| &= \Big| \|(a,b)\|_p - (a+b) - \|(c,d)\|_p + (c+d) \Big| \\
&\leq \Big| \|(a,b)\|_p - \|(c,d)\|_p \Big| + |a-c| + |b-d| \\
&\leq \|(a,b) - (c,d)\|_p + \sqrt{2}\sqrt{|a-c|^2 + |b-d|^2} \\
&\leq \|(a,b) - (c,d)\|_p + \sqrt{2}\|(a,b) - (c,d)\| \\
&= \|w - w'\|_p + \sqrt{2}\|w - w'\|.
\end{aligned}$$

Then, by Lemma 1.4 (also see [28, Lemma 1.3]), i.e.,

$$\|x\|_{p_2} \leq \|x\|_{p_1} \leq n^{(1/p_1 - 1/p_2)} \|x\|_{p_2} \quad \text{for } x \in \mathbb{R}^n \quad \text{and } 1 < p_1 < p_2,$$

the desired results follow.

(f) This comes from direct computation. □

As shown below, the function ϕ^p_{FB} possesses several additional properties that are instrumental in establishing the results presented in the subsequent section.

Lemma 2.1 *Let $\phi^p_{\text{FB}} : \mathbb{R}^2 \to \mathbb{R}$ be defined as in (2.14) where $p > 1$. If $\{(a^k, b^k)\} \subseteq \mathbb{R}^2$ with ($a^k \to -\infty$) or ($b^k \to -\infty$) or ($a^k \to \infty$ and $b^k \to \infty$), then we have $|\phi^p_{\text{FB}}(a^k, b^k)| \to \infty$ for $k \to \infty$.*

Proof This result is also mentioned in [18, p. 20]. □

We now introduce another family of NCP functions that reformulate the nonlinear complementarity problem as an unconstrained minimization problem. In other words, these functions serve as a class of merit functions for the NCP. Given ϕ^p_{FB} as defined in (2.14), we define the function $\psi^p_{\text{FB}} : \mathbb{R}^2 \to \mathbb{R}_+$ for $p > 1$ by

$$\psi^p_{\text{FB}}(a, b) := \frac{1}{2}|\phi^p_{\text{FB}}(a, b)|^2. \tag{2.19}$$

This class of functions exhibits several desirable properties, as detailed below. Notably, for any fixed $p > 1$, the function ψ^p_{FB} is continuously differentiable everywhere, in contrast to ϕ^p_{FB}, which lacks differentiability at the origin.

Proposition 2.2 *Let ϕ^p_{FB}, ψ^p_{FB} be defined as in (2.14) and (2.19), respectively, where $p > 1$. Then, the following hold.*

(a) ψ^p_{FB} is an NCP function, i.e., it satisfies (2.2).
(b) $\psi^p_{\text{FB}}(a, b) \geq 0$ for all $(a, b) \in \mathbb{R}^2$.
(c) ψ^p_{FB} is continuously differentiable everywhere.
(d) $\nabla_a \psi^p_{\text{FB}}(a, b) \cdot \nabla_b \psi^p_{\text{FB}}(a, b) \geq 0$ for all $(a, b) \in \mathbb{R}^2$. The equality holds if and only if $\phi^p_{\text{FB}}(a, b) = 0$.
(e) $\nabla_a \psi^p_{\text{FB}}(a, b) = 0 \iff \nabla_b \psi^p_{\text{FB}}(a, b) = 0 \iff \phi^p_{\text{FB}}(a, b) = 0$.

Proof (a) Since $\psi^p_{\text{FB}}(a, b) = 0$ if and only if $\phi^p_{\text{FB}}(a, b) = 0$, the desired result is satisfied by Proposition 2.1(a).

(b) It is clear by definition of ψ^p_{FB}.

(c) From direct computation, we obtain $\nabla_a \psi^p_{\text{FB}}(0, 0) = \nabla_b \psi^p_{\text{FB}}(0, 0) = 0$. For $(a, b) \neq (0, 0)$, we have

$$\nabla_a \psi^p_{\text{FB}}(a, b) = \left(\frac{\operatorname{sgn}(a) \cdot |a|^{p-1}}{\|(a, b)\|_p^{p-1}} - 1 \right) \phi^p_{\text{FB}}(a, b) \tag{2.20}$$

$$\nabla_b \psi^p_{\text{FB}}(a, b) = \left(\frac{\operatorname{sgn}(b) \cdot |b|^{p-1}}{\|(a, b)\|_p^{p-1}} - 1 \right) \phi^p_{\text{FB}}(a, b). \tag{2.21}$$

where sgn$(\cdot)$ is the sign function. Clearly, $\left|\dfrac{\text{sgn}(a)\cdot|a|^{p-1}}{\|(a,b)\|_p^{p-1}}\right| \le 1$ and $\left|\dfrac{\text{sgn}(a)\cdot|a|^{p-1}}{\|(a,b)\|_p^{p-1}}\right|$ ≤ 1 (i.e., uniformly bounded) and moreover $\phi^p_{\rm FB}(a,b) \to 0$ as $(a,b) \to (0,0)$. Therefore, we have $\nabla_a \psi^p_{\rm FB}(a,b) \to 0$ and $\nabla_b \psi^p_{\rm FB}(a,b) \to 0$ as $(a,b) \to (0,0)$. This means that $\psi^p_{\rm FB}$ is continuously differentiable everywhere.

(d) From part(c), we know that if $(a,b) = (0,0)$, it is clear that $\nabla_a \psi^p_{\rm FB}(a,b) \cdot \nabla_b \psi^p_{\rm FB}(a,b) = 0$ and $\psi^p_{\rm FB}(a,b) = 0$. Now we assume that $(a,b) \neq (0,0)$. Then, $\nabla_a \psi^p_{\rm FB}(a,b) \cdot \nabla_b \psi^p_{\rm FB}(a,b)$ is

$$\left(\frac{\text{sgn}(a)\cdot|a|^{p-1}}{\|(a,b)\|_p^{p-1}} - 1\right)\left(\frac{\text{sgn}(b)\cdot|b|^{p-1}}{\|(a,b)\|_p^{p-1}} - 1\right)\phi^p_{\rm FB}(a,b)^2.$$

Again, from $\left|\dfrac{\text{sgn}(a)\cdot|a|^{p-1}}{\|(a,b)\|_p^{p-1}}\right| \le 1$ and $\left|\dfrac{\text{sgn}(b)\cdot|b|^{p-1}}{\|(a,b)\|_p^{p-1}}\right| \le 1$, it immediately yields that $\nabla_a \psi^p_{\rm FB}(a,b) \cdot \nabla_b \psi^p_{\rm FB}(a,b) \ge 0$ for all $(a,b) \in \mathbb{R}^2$. The equality holds if and only if $\phi^p_{\rm FB}(a,b) = 0$, $\dfrac{\text{sgn}(a)\cdot|a|^{p-1}}{\|(a,b)\|_p^{p-1}} = 1$ or $\dfrac{\text{sgn}(b)\cdot|b|^{p-1}}{\|(a,b)\|_p^{p-1}} = 1$. In fact, if $\dfrac{\text{sgn}(a)\cdot|a|^{p-1}}{\|(a,b)\|_p^{p-1}} = 1$, then we have $a > 0$ and $|a| = \|(a,b)\|_p$, which leads to $b = 0$ and hence $\phi^p_{\rm FB}(a,b) = \sqrt[p]{|a|^p} - a = a - a = 0$. Similarly, we have $\phi^p_{\rm FB}(a,b) = 0$ if $\dfrac{\text{sgn}(b)\cdot|b|^{p-1}}{\|(a,b)\|_p^{p-1}} = 1$. Thus, we conclude that the equality holds if and only if $\phi^p_{\rm FB}(a,b) = 0$.

(e) It is already seen in the last part of proof for part(d). □

It has been shown that if F is a monotone function [29] or a P_0-function [30], then any stationary point of $\Psi_{\rm F}$ is a global minimizer of the unconstrained optimization problem $\min_{x\in\mathbb{R}^n} \Psi_{\rm F}(x)$, and therefore constitutes a solution to the NCP. Furthermore, if F is strongly monotone [29] or a uniform P-function [30], the level sets of $\Psi_{\rm F}$ are guaranteed to be bounded. In what follows, we establish and prove analogous results for $\Psi^p_{\rm FB}$, assuming the same conditions as those in [29, 30]. The proofs of the subsequent propositions are inspired by the corresponding arguments found in these references.

Proposition 2.3 *Let $\Psi^p_{\rm FB} : \mathbb{R}^n \to \mathbb{R}$ be defined as (2.17) where $p > 1$. Then $\Psi^p_{\rm FB}(x) \ge 0$ for all $x \in \mathbb{R}^n$ and $\Psi^p_{\rm FB}(x) = 0$ if and only if x solves the NCP (2.1). Moreover, suppose that the NCP has at least one solution. Then, x is a global minimizer of $\Psi^p_{\rm FB}$ if and only if x solves the NCP.*

Proof The results follow from Proposition 2.2. □

Proposition 2.4 *Let $\Psi^p_{\text{FB}} : \mathbb{R}^n \to \mathbb{R}$ be defined as (2.17) where $p > 1$. Assume F is either monotone or P_0-function, then every stationary point of Ψ^p_{FB} is a global minima of (2.18), and therefore solves the original NCP.*

Proof (I) For the assumption of monotonicity of F, suppose that x^* is a stationary point of Ψ^p_{FB}. Then we have $\nabla\Psi^p_{\text{FB}}(x^*) = 0$ which implies that

$$\sum_{i=1}^{n}\left(\nabla_a\psi^p_{\text{FB}}(x_i^*, F_i(x^*))e_i + \nabla_b\psi^p_{\text{FB}}(x_i^*, F_i(x^*))\nabla F_i(x^*)\right) = 0, \tag{2.22}$$

where $e_i = (0, \ldots, 1, \ldots, 0)^T$. We denote $\nabla_a\psi^p_{\text{FB}}(x^*, F(x^*)) = (\ldots, \nabla_a\psi^p_{\text{FB}}(x_i^*, F_i(x^*)), \ldots)^{\mathsf{T}}$ and $\nabla_b\psi^p_{\text{FB}}(x^*, F(x^*)) = (\ldots, \nabla_b\psi^p_{\text{FB}}(x_i^*, F_i(x^*)), \ldots)^{\mathsf{T}}$, respectively. Then (2.22) can be abbreviated as

$$\nabla_a\psi^p_{\text{FB}}(x^*, F(x^*)) + \nabla F(x^*)\nabla_b\psi^p_{\text{FB}}(x^*, F(x^*)) = 0. \tag{2.23}$$

Now, multiplying (2.23) by $\nabla_b\psi^p_{\text{FB}}(x^*, F(x^*))^T$ leads to

$$\sum_{i=1}^{n}\left(\nabla_a\psi^p_{\text{FB}}(x_i^*, F_i(x^*)) \cdot \nabla_b\psi^p_{\text{FB}}(x_i^*, F_i(x^*))\right) + \nabla_b\psi^p_{\text{FB}}(x^*, F(x^*))^T\nabla F(x^*)\nabla_b\psi^p_{\text{FB}}(x^*, F(x^*)) = 0. \tag{2.24}$$

Since F is monotone, $\nabla F(x^*)$ is positive semidefinite, the second term of (2.24) is nonnegative. Moreover, each term in the first summation of (2.24) is nonnegative as well due to Proposition 2.2(d). Therefore, we have

$$\nabla_a\psi^p_{\text{FB}}(x_i^*, F_i(x^*)) \cdot \nabla_a\psi^p_{\text{FB}}(x_i^*, F_i(x^*)) = 0, \quad \forall i = 1, 2, \ldots, n,$$

which yields $\phi^p_{\text{FB}}(x_i^*, F_i(x^*)) = 0$ for all $i = 1, 2, \ldots, n$ by Proposition 2.2(e). Thus, $\Psi^p_{\text{FB}}(x^*) = 0$ which says x^* is a global minimizer of (2.18).

(II) If F is P_0-function and x^* is a stationary point of Ψ^p_{FB}, then $\Psi^p_{\text{FB}}(x^*) = 0$, which yields (2.23). Notice that $\nabla_a\psi^p_{\text{FB}}(a, b)$ and $\nabla_b\psi^p_{\text{FB}}(a, b)$ are given as forms of (2.20). If we denote $A(x^*)$ and $B(x^*)$ the possibly multi-valued $n \times n$ diagonal matrices whose diagonal elements are given by

$$A_{ii}(x^*) = \frac{\operatorname{sgn}(x_i^*) \cdot |x_i^*|^{p-1}}{\|(x_i^*, F_i(x^*))\|_p^{p-1}} \quad \text{if} \quad (x_i^*, F_i(x^*)) \neq (0, 0)$$

and

$$B_{ii}(x^*) = \frac{\operatorname{sgn}(F_i(x^*)) \cdot |F_i(x^*)|^{p-1}}{\|(x_i^*, F_i(x^*))\|_p^{p-1}} \quad \text{if} \quad (x_i^*, F_i(x^*)) \neq (0, 0).$$

If $(x_i^*, F_i(x^*)) = (0, 0)$, then we let $A(x^*) = B(x^*) = I$, i.e., the $n \times n$ identity matrix. With the notions of $A(x^*)$, $B(x^*)$ and (2.20), Eq. (2.23) can be rewritten as

$$\left[(A(x^*) - I) + \nabla F(x^*)(B(x^*) - I)\right]\Phi^p_{\text{FB}}(x^*) = 0. \tag{2.25}$$

We want to prove that $\Phi^p_{\text{FB}}(x^*) = 0$ (and hence $\Psi^p_{\text{FB}}(x^*) = 0$). Suppose not, i.e., $\Phi^p_{\text{FB}}(x^*) \neq 0$. Recall that $\Phi^p_{\text{FB}}(x^*) = 0$ if and only if (2.1) is satisfied and the i-th component of $\Phi^p_{\text{FB}}(x^*)$ is $\phi^p_{\text{FB}}(x_i^*, F_i(x^*))$. Thus, $\phi^p_{\text{FB}}(x_i, F_i(x^*)) \neq 0$ means one of the following occurs:

1. $x_i^* \neq 0$ and $F_i(x^*) \neq 0$.
2. $x_i^* = 0$ and $F_i(x^*) < 0$.
3. $x_i^* < 0$ and $F_i(x^*) = 0$.

In every case, we have $B_{ii}(x^*) \neq 1$ (since $B_{ii}(x^*) = 1$ if and only if $\phi^p_{\text{FB}}(x_i^*, F_i(x^*)) = 0$ by Proposition 2.2(d)(e)), so that $(B_{ii}(x^*) - 1) \cdot \phi^p_{\text{FB}}(x_i^*, F_i(x^*)) \neq 0$. Similar arguments apply for the vector $(A(x^*) - I)\Phi^p_{\text{FB}}(x^*)$. Thus, from the above, we can easily verify that if $\Phi^p_{\text{FB}}(x^*) \neq 0$ then $(B(x^*) - I)\Phi^p_{\text{FB}}(x^*)$ and $(A(x^*) - I)\Phi^p_{\text{FB}}(x^*)$ are both nonzero. Moreover, both of their nonzero elements are in the same positions, and such nonzero elements have the same sign. But, for Eq. (2.25) to hold, it would be necessary that $\nabla F(x^*)$ "revert the sign" of all the nonzero elements of $(B(x^*) - I)\Phi^p_{\text{FB}}(x^*)$, which contradicts the fact that $\nabla F(x^*)$ is a P_0-matrix by Lemma 1.5. □

Proposition 2.5 *Let $\Psi^p_{\text{FB}} : \mathbb{R}^n \to \mathbb{R}$ be defined as (2.17) where $p > 1$. Assume that F is either strongly monotone or uniform P-function, then the level sets*

$$\mathcal{L}(\Psi^p_{\text{FB}}, \gamma) := \left\{x \in \mathbb{R}^n \mid \Psi^p_{\text{FB}}(x) \leq \gamma\right\}$$

are bounded for all $\gamma \in \mathbb{R}$.

Proof (I) First, we consider the assumption of strong monotonicity of F. Suppose there exists an unbounded sequence $\{\|x^k\|\}_{k\in K} \to \infty$ with $\{x^k\}_{k\in K} \subseteq \mathcal{L}(\Psi^p_{\text{FB}}, \gamma)$ for some $\gamma \geq 0$, where K is a subset of N. We define the index set as

$$J := \left\{i \in \{1, 2, \ldots, n\} | \ \{x_i^k\} \text{ is unbounded}\right\}.$$

Since $\{x^k\}$ is unbounded, $J \neq \emptyset$. Let $\{z^k\}$ denote a bounded sequence defined by

$$z_i^k = \begin{cases} 0, & \text{if } i \in J, \\ x_i^k, & \text{if } i \notin J. \end{cases}$$

Then from the definition of $\{z^k\}$ and the strong monotonicity of F, we obtain

$$
\begin{aligned}
\mu \sum_{i\in J}(x_i^k)^2 &= \mu\|x^k - z^k\|^2 \\
&\le \langle x^k - z^k, F(x^k) - F(z^k)\rangle \\
&= \sum_{i=1}^{n}(x_i^k - z_i^k)(F_i(x^k) - F_i(z^k)) \\
&= \sum_{i\in J} x_i^k(F_i(x^k) - F_i(z^k)) \\
&\le \Big(\sum_{i\in J}(x_i^k)^2\Big)^{1/2} \sum_{i\in J} |(F_i(x^k) - F_i(z^k))|.
\end{aligned} \tag{2.26}
$$

Since $\sum_{i\in J}(x_i^k)^2 \neq 0$ for $k \in K$, then dividing by $\sum_{i\in J}(x_i^k)^2$ on both sides of (2.26) yields

$$
\mu\Big(\sum_{i\in J}(x_i^k)^2\Big)^{1/2} \le \sum_{i\in J} |(F_i(x^k) - F_i(z^k))|, \quad k \in K. \tag{2.27}
$$

On the other hand, we know $\{F_i(z^k)\}_{k\in K}$ is bounded ($i \in J$) due to $\{z^k\}_{k\in K}$ is bounded and F is continuous. Therefore from (2.27), we have

$$
\{|F_{i_0}(x^k)|\} \to \infty \quad \text{for some } i_0 \in J.
$$

Also, $\{\|x_{i_0}^k\|\} \to \infty$ by the definition of the index set J. Thus, Lemma 2.1 yields

$$
\phi_p(x_{i_0}^k, F_{i_0}(x^k)) \to \infty \quad \text{as} \quad k \to \infty.
$$

But this contradicts $\{x^k\} \subseteq \mathcal{L}(\Psi_{\mathrm{FB}}^p, \gamma)$.

(II) If F is uniform P-function, then the proof almost follows the same arguments as above. In particular, (2.26) is replaced by

$$
\begin{aligned}
\mu \sum_{i\in J}(x_i^k)^2 &= \mu\|x^k - z^k\|^2 \\
&\le \max_{1\le i\le n}(x_i^k - z_i^k)(F_i(x^k) - F_i(z^k)) \\
&= \max_{i\in J} x_i^k(F_i(x^k) - F_i(z^k)) \\
&= x_{j_0}^k(F_i(x^k) - F_i(z^k)) \\
&\le |x_{j_0}^k||(F_i(x^k) - F_i(z^k))|,
\end{aligned} \tag{2.28}
$$

where j_0 is one of the indices for which the max is attained. Then dividing by $|x_{j_0}^k|$ on both sides of (2.28) and the proof follows. □

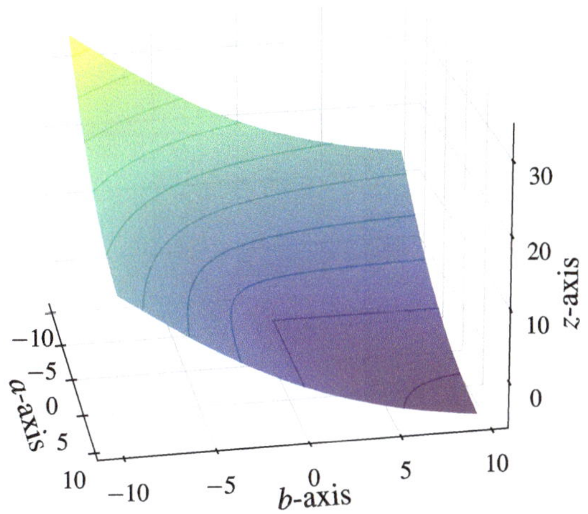

Fig. 2.1 The surface of $z = \phi^2_{\text{FB}}(a, b)$ with $(a, b) \in [-10, 10] \times [-10, 10]$

We now examine some geometric properties of the function ϕ^p_{FB} and offer interpretations of their significance. In particular, we present the family of surfaces defined by $\phi^p_{\text{FB}}(a, b)$ for various values of $p \in (1, +\infty)$; see Figs. 2.1 and 2.2. When the parameter p is fixed within this interval, Fig. 2.2 provides an intuitive visualization showing how the shape of the surface is influenced by the choice of p. From the definition of the p-norm, we recall that $\|(a, b)\|_1 := |a| + |b|$, and $\|(a, b)\|_\infty := \max\{|a|, |b|\}$. It follows trivially that $\phi^p_{\text{FB}}(a, b) \to \phi^1_{\text{FB}}(a, b) := |a| + |b| - (a + b)$ pointwise as $p \to 1$; see Fig. 2.2a and b. Conversely, as $p \to \infty$, we have $\phi^p_{\text{FB}}(a, b) \to \phi^\infty_{\text{FB}}(a, b) := \max\{|a|, |b|\} - (a + b)$, as illustrated in Fig. 2.2e and f. It is important to note that $\phi^1_{\text{FB}}(a, b)$ does not qualify as an NCP function, since $\phi^1_{\text{FB}}(a, b) = 0$ even when $a > 0$ and $b > 0$. In contrast, $\phi^\infty_{\text{FB}}(a, b)$ is indeed an NCP function, although it fails to be differentiable along the line $a = b$.

Lemma 2.2 ([31, Lemma 3.1]) *If $a > 0$ and $b > 0$, then $(a + b)^p > a^p + b^p$ for all $p \in (1, +\infty)$.*

Proof We present two different ways to prove this lemma.
(1) For any $p > 1$, $p = n + m$, where $n = [p]$ (the greatest integer less than or equal to p) and $m = p - n$, applying binomial theorem gives

$$\begin{aligned}(a + b)^p &= (a + b)^n (a + b)^m \\ &\geq (a^n + b^n)(a + b)^m \\ &= a^n (a + b)^m + b^n (a + b)^m \\ &\geq a^n a^m + b^n b^m \\ &= a^p + b^p.\end{aligned}$$

(2) Let $f(t) = (t + 1)^p - (t^p + 1)$. It is easy to verify that f is increasing on $[0, \infty)$ when $p > 1$. Hence, $f(a/b) \geq f(0) = 0$ which yields $(a + b)^p \geq a^p + b^p$. □

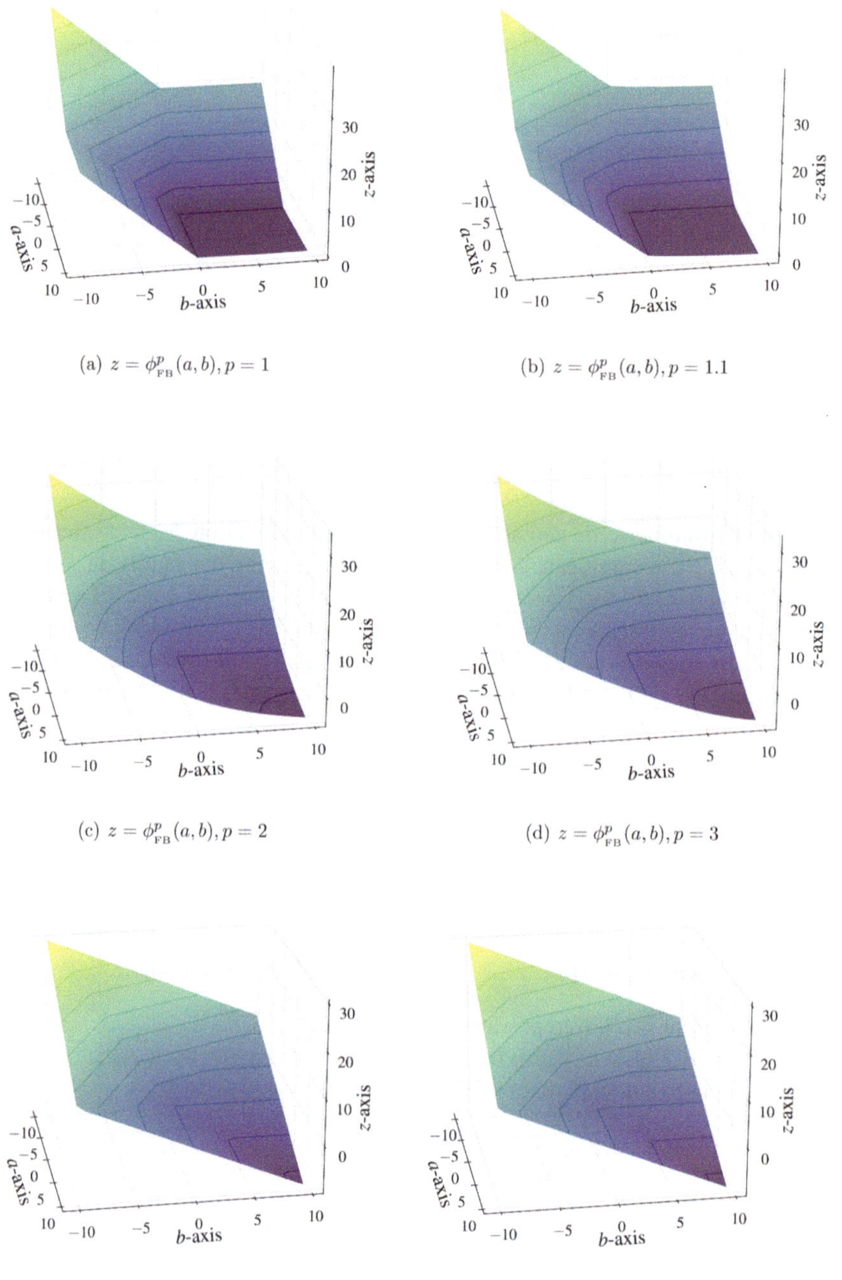

(a) $z = \phi^{p}_{\text{FB}}(a, b), p = 1$

(b) $z = \phi^{p}_{\text{FB}}(a, b), p = 1.1$

(c) $z = \phi^{p}_{\text{FB}}(a, b), p = 2$

(d) $z = \phi^{p}_{\text{FB}}(a, b), p = 3$

(e) $z = \phi^{p}_{\text{FB}}(a, b), p = 100$

(f) $z = \phi^{p}_{\text{FB}}(a, b), p = \infty$

Fig. 2.2 The surface of $z = \phi^{p}_{\text{FB}}(a, b)$ with different values of p

Lemma 2.3 ([21, Lemma 3.2]) *Let $\phi^p_{\rm FB} : \mathbb{R}^2 \to \mathbb{R}$ be given as in (2.14) where $p > 1$. Then, there holds*

$$\left(2 - 2^{\frac{1}{p}}\right) |\min\{a, b\}| \le \left|\phi^p_{\rm FB}(a, b)\right| \le \left(2 + 2^{\frac{1}{p}}\right) |\min\{a, b\}| .$$

Proof Without loss of generality, assume $a \ge b$. We will establish the desired results by examining the following two cases: (1) $a + b \le 0$ and (2) $a + b > 0$.

Case(1): $a + b \le 0$. In this case, we have

$$|\phi^p_{\rm FB}(a, b)| \ge \|(a, b)\|_p \ge |b| = |\min\{a, b\}| \ge (2 - 2^{\frac{1}{p}})|\min\{a, b\}|. \qquad (2.29)$$

On the other hand, since $a \ge b$ and $a + b \le 0$, we have $|b| \ge |a|$. Then

$$|\phi^p_{\rm FB}(a, b)| \le \|(a, b)\|_p - 2b = (2 + 2^{\frac{1}{p}})|b| = (2 + 2^{\frac{1}{p}})|\min\{a, b\}|.$$

Case(2): $a + b > 0$. If ab=0, then the desired inequality clearly holds. Thus, we discuss by two subcases:

(i) $ab < 0$. In this subcase, we have $a > 0$, $b < 0$, and $|a| > |b|$. Consequently,

$$\phi^p_{\rm FB}(a, b) \le |a| + |b| - (a + b) = -2b = 2|\min\{a, b\}| \le (2 + 2^{\frac{1}{p}})|\min\{a, b\}|,$$

and

$$\phi^p_{\rm FB}(a, b) \ge \|(a, b)\|_\infty - (a + b) = -b = |\min\{a, b\}| \ge (2 - 2^{\frac{1}{p}})|\min\{a, b\}|.$$

(ii) $ab > 0$. Now we have $a \ge b > 0$. Since for any $p > 1$ there holds that

$$0 \ge \phi^p_{\rm FB}(a, b) \ge \|(a, b)\|_\infty - (a + b) = a - (a + b) = -b = -\min\{a, b\},$$

we immediately obtain that

$$|\phi^p_{\rm FB}(a, b)| \le |\min\{a, b\}| \le (2 + 2^{\frac{1}{p}})|\min\{a, b\}|.$$

On the other hand, since $\phi^p_{\rm FB}(a, b) \le 0$, it follows that

$$|\phi^p_{\rm FB}(a, b)| = a + b - \|(a, b)\|_p = b\left[\left(\frac{a}{b} + 1\right) - \left(\left(\frac{a}{b}\right)^p + 1\right)^{1/p}\right].$$

Let $f(t) = t + 1 - (t^p + 1)^{1/p}$ for $t \ge 1$. Then

$$f'(t) = 1 - \left(\frac{t^p}{t^p + 1}\right)^{\frac{p-1}{p}}.$$

Notice that $f'(t) > 0$ for $t \geq 1$, and $f(1) = 2 - 2^{\frac{1}{p}}$, and hence we obtain that

$$|\phi^p_{\text{FB}}(a,b)| \geq (2 - 2^{\frac{1}{p}})b = (2 - 2^{\frac{1}{p}})|\min\{a,b\}| \quad \text{for any } p > 1. \tag{2.30}$$

All the aforementioned inequalities (2.29)–(2.30) imply that the desired inequality holds. □

Proposition 2.6 *Let* $\phi^p_{\text{FB}} : \mathbb{R}^2 \to \mathbb{R}$ *be given as in (2.14) where* $p \in (1, +\infty)$. *Then,*

(a) $(a > 0$ *and* $b > 0) \Longleftrightarrow \phi^p_{\text{FB}}(a,b) < 0$;
(b) $(a = 0$ *and* $b \geq 0)$ *or* $(b = 0$ *and* $a \geq 0) \Longleftrightarrow \phi^p_{\text{FB}}(a,b) = 0$;
(c) $b = 0$ *and* $a < 0 \Rightarrow \phi^p_{\text{FB}}(a,b) = -2a > 0$;
(d) $a = 0$ *and* $b < 0 \Rightarrow \phi^p_{\text{FB}}(a,b) = -2b > 0$.

Proof (a) If $a > 0$ and $b > 0$, it is easy to see $\phi^p_{\text{FB}}(a,b) < 0$ by Lemma 2.2. Conversely, using $\sqrt[p]{|a|^p + |b|^p} \geq |a|$ and $\sqrt[p]{|a|^p + |b|^p} \geq |b|$, we have $\sqrt[p]{|a|^p + |b|^p} \geq \max\{|a|, |b|\}$. Suppose $a \leq 0$ or $b \leq 0$, then we have $\max\{|a|, |b|\} \geq (a+b)$ which implies $\phi^p_{\text{FB}}(a,b) \geq 0$. This is a contradiction.

(b) By definition of $\phi^p_{\text{FB}}(a,b)$, we know

$$\phi^p_{\text{FB}}(a,0) = |a| - a = \begin{cases} 0 & a \geq 0, \\ -2a & a < 0, \end{cases} \qquad \phi_p(0,b) = |b| - b = \begin{cases} 0 & b \geq 0, \\ -2b & b < 0, \end{cases}$$

which say that $(a = 0$ and $b \geq 0)$ or $(b = 0$ and $a \geq 0) \Rightarrow \phi^p_{\text{FB}}(a,b) = 0$. Conversely, suppose $\phi^p_{\text{FB}}(a,b) = 0$. If $a < 0$ or $b < 0$, mimicking the arguments of part(a) yields

$$\sqrt[p]{|a|^p + |b|^p} > \max\{|a|, |b|\} > (a+b)$$

which implies $\phi^p_{\text{FB}}(a,b) > 0$. Thus, there must hold $a \geq 0$ and $b \geq 0$. Furthermore, one of a and b must be 0 from part(a).

The proof of (c) and (d) is direct from the proof of part(b). □

Proposition 2.6(a) demonstrates that $\phi^p_{\text{FB}}(a,b)$ is negative in the first quadrant of the $\mathbb{R}^2$-plane; see Fig. 2.3. Meanwhile, Proposition 2.6(b) establishes that $\phi^p_{\text{FB}}(a,b) = 0$ occurs only along the nonnegative coordinate axes, that is, when $a \geq 0, b = 0$ or $a = 0, b \geq 0$. In fact, this result is equivalent to asserting that $\phi^p_{\text{FB}}(a,b)$ satisfies the conditions of an NCP function. Furthermore, Proposition 2.6(b)–(d) collectively indicate that the parameter p has no influence on the value of $\phi^p_{\text{FB}}(a,b)$ along the a-axis and the b-axis.

Proposition 2.7 *Let* $\phi^p_{\text{FB}} : \mathbb{R}^2 \to \mathbb{R}$ *be given as in (2.14) where* $p \in (1, +\infty)$. *Then,*

(a) $\phi^p_{\text{FB}}(a,b) = \phi^p_{\text{FB}}(b,a)$;
(b) *if* $1 < p_1 < p_2$, *then* $\phi^{p_1}_{\text{FB}}(a,b) \geq \phi^{p_2}_{\text{FB}}(a,b)$.

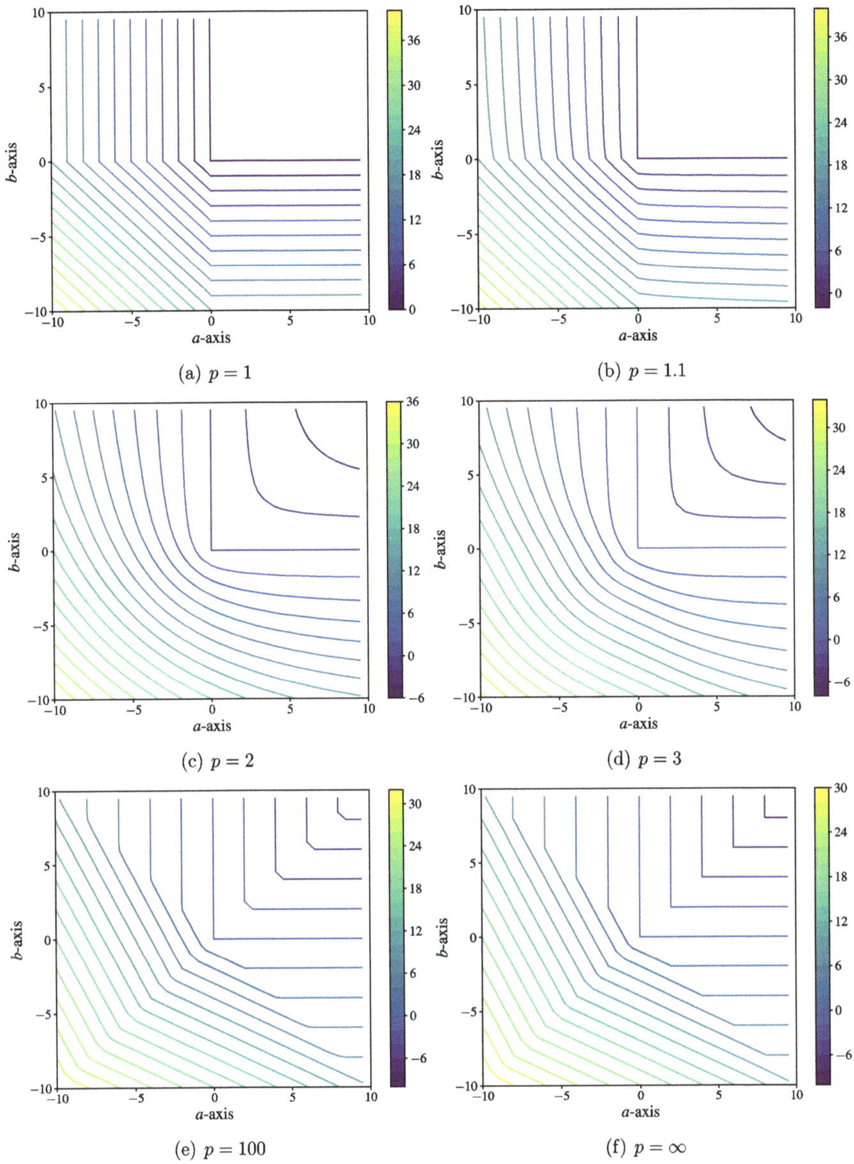

Fig. 2.3 Level curves of $z = \phi^p_{\rm FB}(a, b)$ with different values of p

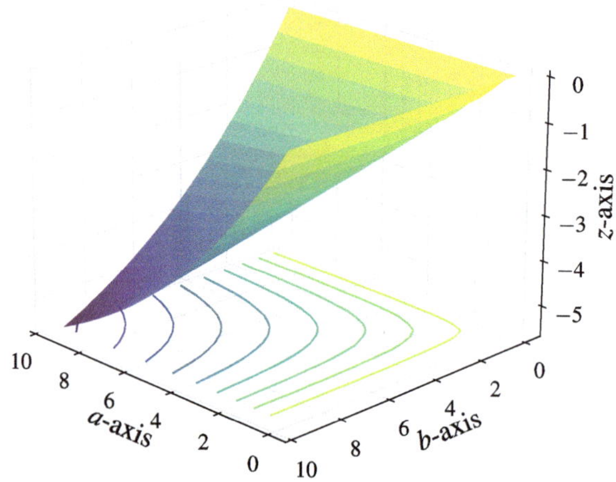

Fig. 2.4 The surface of $z = \phi_{FB}^2(a, b)$ with $(a, b) \in [0, 10] \times [0, 10]$

Proof Part(a) is trivial and part(b) is true by applying Lemma 1.4. □

Proposition 2.7(a) establishes the symmetry of $\phi_{FB}^p(a, b)$, indicating that there exist pairs of points symmetric about the line $a = b$ that share the same function value. In other words, the surface defined by $z = \phi_{FB}^p(a, b)$ exhibits identical geometric features in the second and fourth quadrants of the $\mathbb{R}^2$-plane; see Figs. 2.3, 2.4, and 2.5. Moreover, Proposition 2.1(d) shows that the surface is convex, as ϕ_{FB}^p itself is a convex function. Proposition 2.7(c) further reveals that the values of ϕ_{FB}^p decrease as the parameter p increases. In summary, the parameter p significantly influences the geometric structure of the surface.

Proposition 2.8 *If $\{a^k, b^k\} \subseteq \mathbb{R}^2$ with $(a^k \to -\infty)$ or $(b^k \to -\infty)$ or $(a^k \to +\infty$ and $b^k \to +\infty)$, then $|\phi_{FB}^p(a^k, b^k)| \to +\infty$ for $k \to +\infty$.*

Proof This can be found in [18, p. 20]. □

Proposition 2.8 highlights the increasing direction on the surface defined by $z = \phi_{FB}^p(a, b)$. This behavior is visually evident in the contour plot shown in Fig. 2.4, where darker shades correspond to lower surface heights. To gain a deeper understanding of the surface's structure, it is natural to examine certain characteristic curves lying on it. To this end, we consider a family of curves $\alpha_{r,p} : \mathbb{R} \to \mathbb{R}^3$ defined by:

$$\alpha_{r,p}(t) := \Big(r + t, r - t, \phi_{FB}^p(r + t, r - t)\Big) \tag{2.31}$$

where $r \in \mathbb{R}$ and $p \in (1, +\infty)$ are arbitrary but fixed. Geometrically, each curve $\alpha_{r,p}$ represents the intersection of the surface $z = \phi_{FB}^p(a, b)$ with the plane defined by $a + b = 2r$; see Fig. 2.5. In the following, we explore several key properties of these special curves.

Lemma 2.4 *Let $\phi_{FB}^p : \mathbb{R}^2 \to \mathbb{R}$ be given as in (2.14) where $p \in (1, +\infty)$. Fix any $r \in \mathbb{R}$, we define $f : \mathbb{R} \to \mathbb{R}$ as $f(t) := \phi_{FB}^p(r + t, r - t)$, then f is a convex function.*

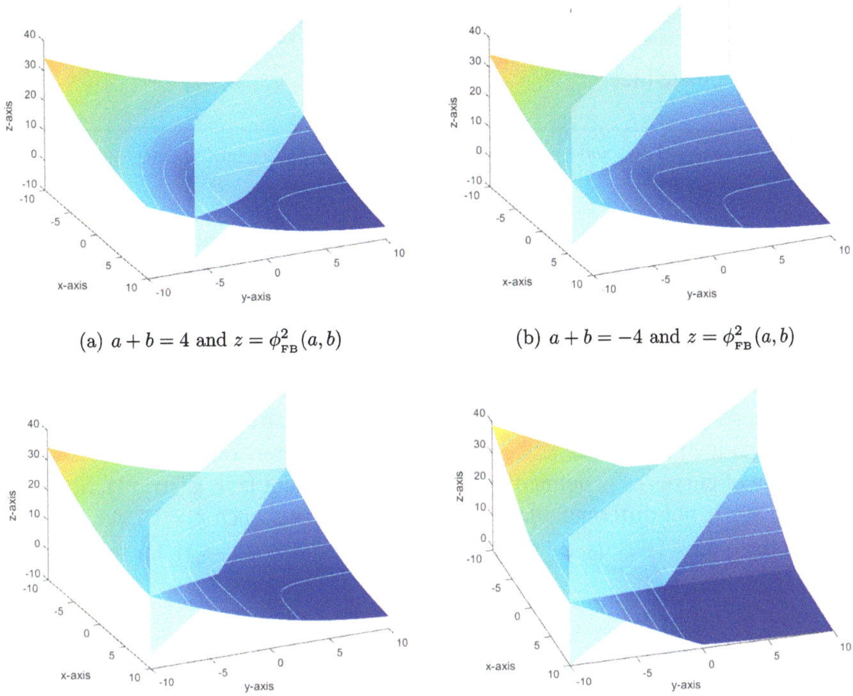

(a) $a+b=4$ and $z=\phi^2_{\rm FB}(a,b)$

(b) $a+b=-4$ and $z=\phi^2_{\rm FB}(a,b)$

(c) $a+b=0$ and $z=\phi^2_{\rm FB}(a,b)$

(d) $a+b=0$ and $z=\phi^{1.1}_{\rm FB}(a,b)$

Fig. 2.5 The curve intersected by surface $z=\phi^p_{\rm FB}(a,b)$ and plane $a+b=2r$

Proof By Proposition 2.7(b), we know that $\phi^p_{\rm FB}$ is a convex function. Observing that f is the composition of $\phi^p_{\rm FB}$ with an affine function, we conclude that f is also convex. Although the composition of two convex functions is not generally convex, convexity is preserved in this case due to the affine nature of one of the components. □

Proposition 2.9 *Let $\phi^p_{\rm FB}:\mathbb{R}^2\to\mathbb{R}$ be given as in (2.14) where $p>1$. Suppose a and b are constrained on the curve determined by $a+b=2r$ ($r\in\mathbb{R}$) and the surface. Then, $\phi^p_{\rm FB}(a,b)$ attains its minima $\phi^p_{\rm FB}(r,r)=2^{\frac{1}{p}}|r|-2r$ along this curve at $(a,b)=(r,r)$.*

Proof We know that $\phi^p_{\rm FB}(a,b)$ is differentiable everywhere except at the point $(0,0)$. Therefore, we consider two separate cases:

(i) Case (1): $r=0$. Since $a+b=0$, it follows that a and b have opposite signs, unless $a=b=0$. According to Proposition 2.6, this implies that $\phi^p_{\rm FB}(a,b)\geq 0$ in this scenario. In particular, $\phi^p_{\rm FB}(a,b)$ achieves its minimum value of zero at the origin, $(a,b)=(0,0)$.

(ii) Case (2): $r\neq 0$. Fix r and $p>1$. Let $f:\mathbb{R}\to\mathbb{R}$ and $g:\mathbb{R}\to\mathbb{R}$ be respectively defined as

$$f(t) := \phi^p_{\text{FB}}(r+t, r-t), \quad g(t) := |r+t|^p + |r-t|^p.$$

Then, we calculate that

$$f'(t) = \frac{g'(t)}{p(g(t))^{\frac{p-1}{p}}} \quad \text{and} \quad g'(t) = p\left[\text{sgn}(r+t)(r+t)^{p-1} - \text{sgn}(r-t)(r-t)^{p-1}\right].$$

We know $g(t) > 0$ for all $t \in \mathbb{R}$. It is clear $g'(0) = 0$, and hence $f'(0) = 0$. By Lemma 2.4, $f(t)$ is convex on $\mathbb{R}$. In addition, it is also continuous, therefore, $t = 0$ is a critical point of $f(t)$ which is also a global minimizer of $f(t)$. The proof is done since $a = b = r$ and $\phi^p_{\text{FB}}(r, r) = 2^{\frac{1}{p}}|r| - 2r$ when $t = 0$. □

Lemma 2.4 and Proposition 2.9 establish that the curve formed by the intersection of the plane $a + b = 2r$ and the surface $z = \phi^p_{\text{FB}}(a, b)$ is convex and attains its minimum at the point where $a = b$ (see Fig. 2.6). We now examine the curvature of the family of curves $\alpha_{r,p}$, defined as in (2.31), at the point $\left(r, r, \phi^p_{\text{FB}}(r, r)\right)$. Since the function ϕ^p_{FB} is not differentiable at $(a, b) = (0, 0)$ (i.e., when $r = 0$), we consider two nearby points, $\left(-t_0, t_0, \phi^p_{\text{FB}}(-t_0, t_0)\right)$ and $\left(t_0, -t_0, \phi^p_{\text{FB}}(t_0, -t_0)\right)$, for some $t_0 > 0$.

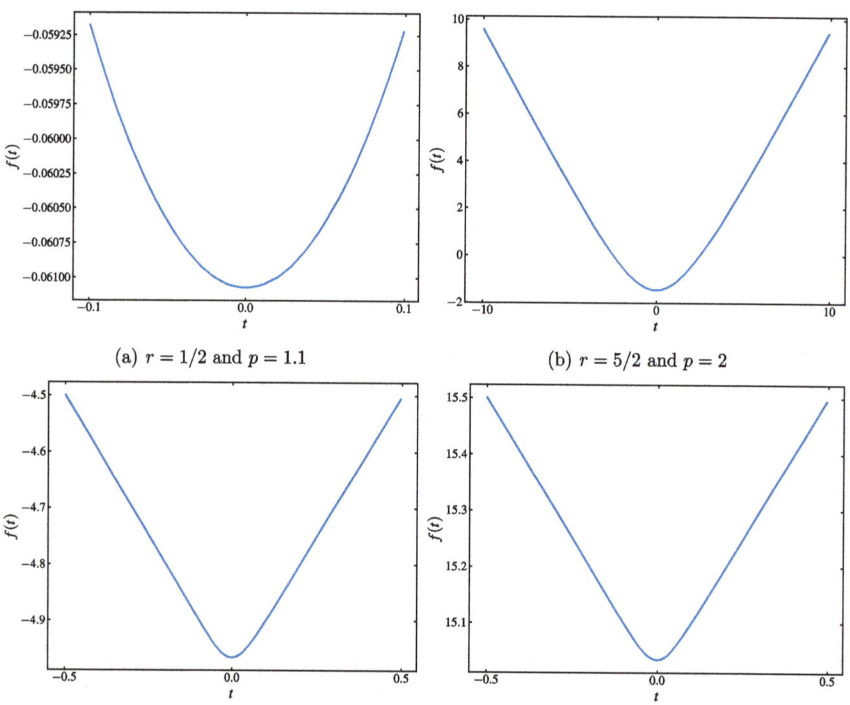

(a) $r = 1/2$ and $p = 1.1$ (b) $r = 5/2$ and $p = 2$

(c) $r = 5$ and $p = 100$ (d) $r = -5$ and $p = 100$

Fig. 2.6 The curve $f(t) = \phi^p_{\text{FB}}(r+t, r-t)$

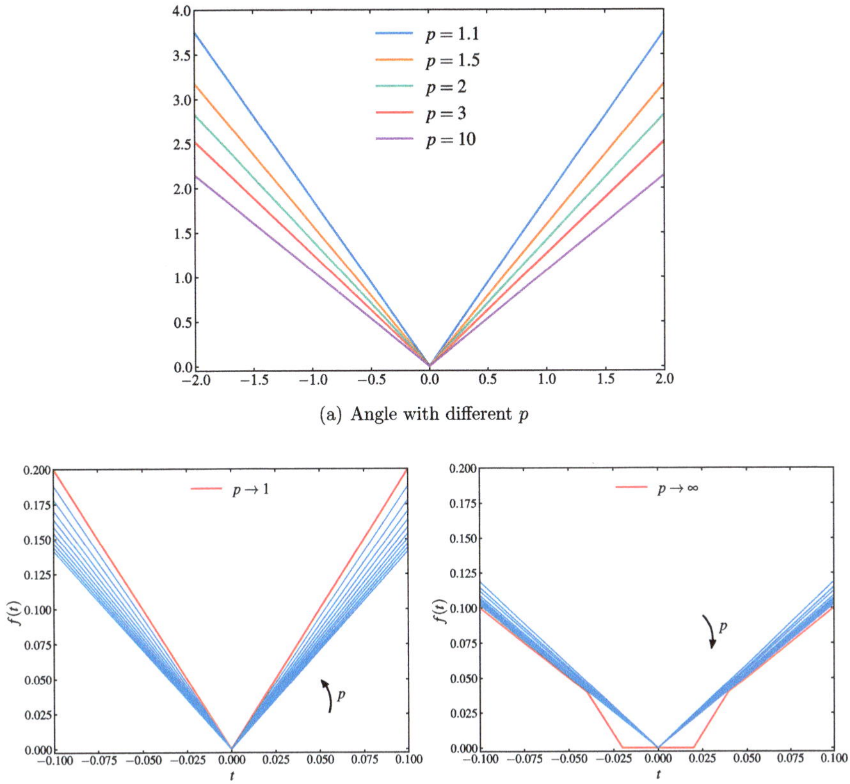

(a) Angle with different p

(b) The change of angle as $p \to 1$.

(c) The change of angle as $p \to +\infty$.

Fig. 2.7 Angle between vectors $\alpha_{0,p}(-t_0)$ and $\alpha_{0,p}(t_0)$

We then compute the cosine of the angle formed between $\alpha_{0,p}(-t_0)$ and $\alpha_{0,p}(t_0)$; see Fig. 2.7.

Proposition 2.10 *Let $\alpha_{r,p} : \mathbb{R} \to \mathbb{R}^3$ be defined as in (2.31), and $\cos_p(\theta)$ be cosine function of the angle between two vectors $\alpha_{0,p}(-t_0)$ and $\alpha_{0,p}(t_0)$ where $t_0 > 0$. Then,*

(a) $\cos_p(\theta) = \dfrac{2^{\frac{2}{p}} - 6}{\sqrt{\left(2^{\frac{2}{p}} - 2\right)^2 + 32}}$;

(b) $\cos_p(\theta) \to -\frac{1}{3}$ *as* $p \to 1$, *and* $\cos_p(\theta) \to -\frac{5}{33}$ *as* $p \to +\infty$;

(c) if $1 < p_1 < p_2$, *then* $\cos_{p_1}(\theta) < \cos_{p_2}(\theta)$.

Proof (a) By direct computation, we obtain

$$\begin{aligned}\cos_p(\theta) &= \frac{\alpha_{0,p}(-t_0)\cdot\alpha_{0,p}(t_0)}{\|\alpha_{0,p}(-t_0)\|\,\|\alpha_{0,p}(t_0)\|} \\ &= \frac{2^{\frac{2}{p}}-6}{\sqrt{(2^{\frac{2}{p}}+6)+2^{\frac{1}{p}+2}}\sqrt{(2^{\frac{2}{p}}+6)-2^{\frac{1}{p}+2}}} \\ &= \frac{2^{\frac{2}{p}}-6}{\sqrt{(2^{\frac{2}{p}}-2)^2+32}}.\end{aligned}$$

(b) From part(a), let $f:(1,+\infty)\to\mathbb{R}$ be $f(p):=\cos_p(\theta)$. Then $f(p)$ is continuous on $(1,+\infty)$. By taking the limit, we have $\cos_p(\theta)\to-\frac{1}{3}$ as $p\to 1$, and $\cos_p(\theta)\to-\frac{5}{33}$ as $p\to+\infty$.

(c) From part(b), we know $f'(p)=\frac{6-\left(1-\frac{\ln 2}{p}\right)2^{\frac{2}{p}}}{\sqrt{(2^{\frac{2}{p}}-2)^2+32}}$ which implies $f'(p)>0$ for all $p>1$. Therefore, $f(p)$ is a strictly increasing function on $(1,+\infty)$. □

Proposition 2.11 *Let $\alpha_{r,p}:\mathbb{R}\to\mathbb{R}^3$ be defined as in (2.31). Then, the following hold.*

(a) The curvature at point $\alpha_{r,p}(0)=\left(r,r,\phi^p_{\rm FB}(r,r)\right)$ is $\kappa_p(0)=\dfrac{(p-1)2^{\frac{1}{p}-1}}{|r|}$.
(b) $\kappa_p(0)\to 0$ as $p\to 1$ and $\kappa_p(0)\to+\infty$ as $p\to+\infty$.
(c) If $1<p_1<p_2$, then $\kappa_{p_1}(0)<\kappa_{p_2}(0)$.

Proof (a) Because $\alpha_{r,p}(t)=\left(r+t,r-t,\phi^p_{\rm FB}(r+t,r-t)\right)$, we know

$$\alpha'_{r,p}(0)=(1,-1,0)\quad\text{and}\quad\alpha''_{r,p}(0)=\left(0,0,\frac{(p-1)2^{\frac{1}{p}}}{|r|}\right).$$

Recall the formulation of curvature

$$\kappa_p(t)=\frac{|\alpha'_{r,p}(t)\wedge\alpha''_{r,p}(t)|}{|\alpha'_{r,p}(t)|^3},$$

where wage operator means the outer product of two vectors. Thus, we have

$$\kappa_p(0)=\frac{|\alpha'_{r,p}(0)\wedge\alpha''_{r,p}(0)|}{|\alpha'_{r,p}(0)|^3}=\frac{(p-1)2^{\frac{1}{p}-1}}{|r|}.$$

(b) Let $f : (1, +\infty) \to \mathbb{R}$ be defined as

$$f(p) := \kappa_p(0) = \frac{(p-1)2^{\frac{1}{p}-1}}{|r|},$$

then obviously $f(p)$ is continuous on $\mathbb{R}$. Thus, the desired result follows by taking the limit directly.

(c) From part(b), we compute

$$f'(p) = \frac{2^{\frac{1}{p}-1}}{|r|}\left(1 - \frac{\ln 2}{p} + \frac{\ln 2}{p^2}\right)$$

which implies $f'(p) > 0$ for all $p \in (1, +\infty)$. Thus, $f(p)$ is strictly increasing on $(1, +\infty)$. □

The preceding two propositions illustrate how the parameter p influences the geometric structure of the surface; see Fig. 2.8a and b. Proposition 2.11(b) states that as $p \to 1$, the curve approaches a straight line (Fig. 2.8c). Conversely, as $p \to +\infty$, the curve becomes increasingly sharp at the origin, and is no longer differentiable at $t = 0$ (Fig. 2.8d). In summary, the results presented in this section reveal that the parameter p significantly affects both the local and global geometric behavior of the surface defined by $z = \phi^p_{\rm FB}(a, b)$.

As previously discussed, the generalized FB function $\phi^p_{\rm FB}$ is convex and differentiable everywhere except at the point $(0, 0)$. In contrast, the function $\psi^p_{\rm FB}(a, b)$, defined in (2.19), is non-convex but remains continuously differentiable across its entire domain. Despite this key difference, $\phi^p_{\rm FB}$ and $\psi^p_{\rm FB}$ exhibit many similar geometric properties, as will be demonstrated. In what follows, we present several properties of $\psi^p_{\rm FB}$ and highlight the distinctions between $\psi^p_{\rm FB}$ and $\phi^p_{\rm FB}$.

Proposition 2.12 *Let $\psi^p_{\rm FB} : \mathbb{R}^2 \to \mathbb{R}$ be given as in (2.19) where $p \in (1, +\infty)$. Then,*

(a) $\psi^p_{\rm FB}(a, b) \geq 0$, $\forall (a, b) \in \mathbb{R}^2$;
(b) $\psi^p_{\rm FB}(a, b) = \psi^p_{\rm FB}(b, a)$, $\forall (a, b) \in \mathbb{R}^2$;
(c) $(a = 0$ and $b \geq 0)$ or $(b = 0$ and $a \geq 0) \Longleftrightarrow \psi^p_{\rm FB}(a, b) = 0$;
(d) $b = 0$ and $a < 0 \Rightarrow \psi^p_{\rm FB}(a, b) = 2a^2 > 0$;
(e) $a = 0$ and $b < 0 \Rightarrow \psi^p_{\rm FB}(a, b) = 2b^2 > 0$;
(f) $\psi^p_{\rm FB}$ is continuously differentiable everywhere.

Proof Parts (d) and (e) come from Proposition 2.6(c) and (d), please see [19, 20, 22] for the rest. □

Proposition 2.7(c) states that the value of $\phi^p_{\rm FB}$ decreases with respect to the parameter p. In contrast, $\psi^p_{\rm FB}$ does not exhibit this monotonicity in general. More precisely, this property holds for $\psi^p_{\rm FB}$ only within certain quadrants of the domain.

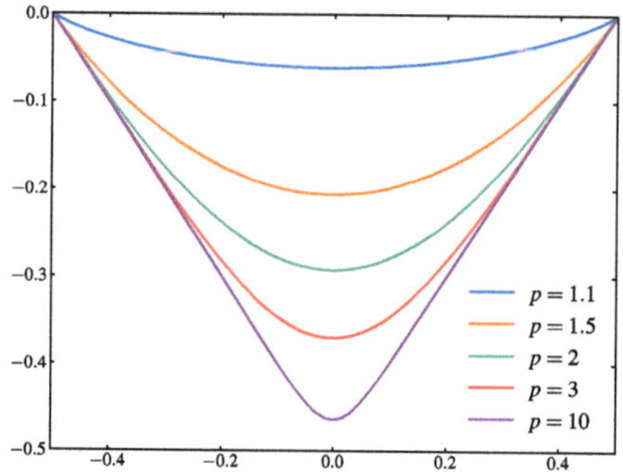

(a) The curvature with different p with $r = 1/2$

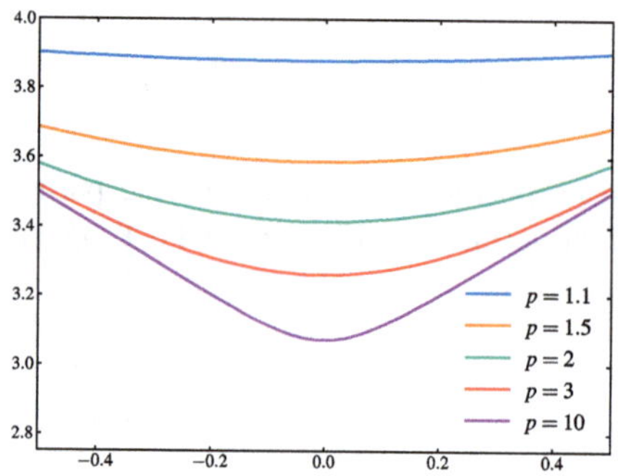

(b) The curvature with different p with $r = -1$

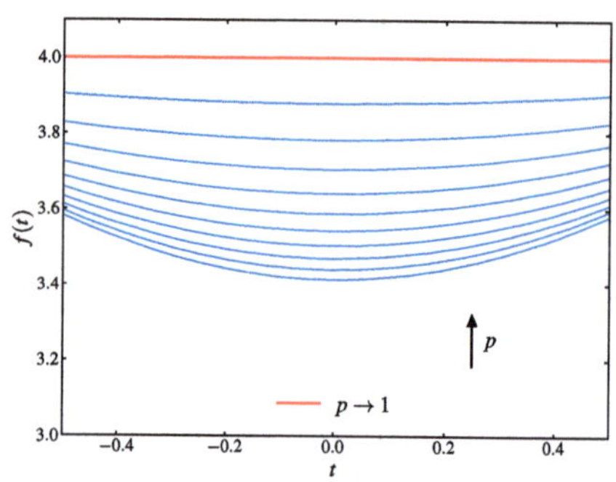

(c) The change of curvature as $p \to 1$.

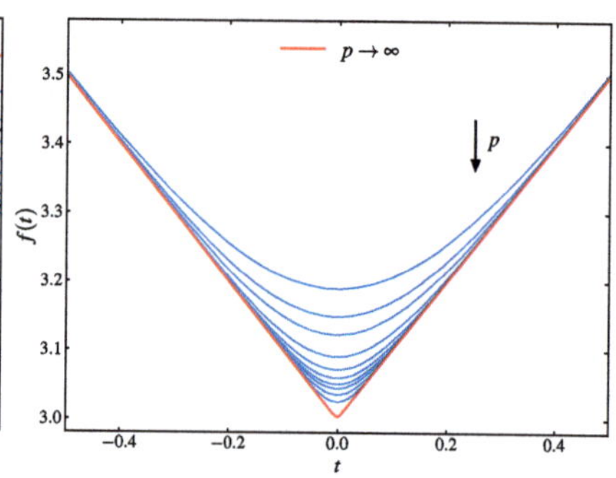

(d) The change of curvature as $p \to +\infty$.

Fig. 2.8 The curvature $\kappa_p(0)$ at point $\alpha_{r,p}(0)$

Proposition 2.13 *Suppose* $1 < p_1 < p_2$ *and* $(a, b) \in \mathbb{R}^2$. *Then,*

(a) if $a < 0$ *or* $b < 0$, *then* $\psi_{\text{FB}}^{p_1}(a, b) \geq \psi_{\text{FB}}^{p_2}(a, b)$;
(b) if $a > 0$ *and* $b > 0$, *then* $\psi_{\text{FB}}^{p_1}(a, b) \leq \psi_{\text{FB}}^{p_2}(a, b)$.

Proof (a) This is clear from Proposition 2.7(c).

(b) Suppose $a > 0$ and $b > 0$, from Proposition 2.6(a), we have $\phi^p_{\rm FB}(a,b) < 0$. Then Proposition 2.7(c) yields $\phi^{p_1}_{\rm FB}(a,b) \geq \phi^{p_2}_{\rm FB}(a,b)$, and hence $(\phi^{p_1}_{\rm FB})^2(a,b) \leq (\phi^{p_2}_{\rm FB})^2(a,b)$. □

Since $\psi^p_{\rm FB}$ is not convex in general, the counterpart to Proposition 2.9 is presented below.

Proposition 2.14 *Let $\psi^p_{\rm FB}(a,b)$ be defined as (2.19) with $a+b=2r$. Then, the following hold.*

(a) If $r \in \mathbb{R}^+$ and $a > 0, b > 0$, then $\psi^p_{\rm FB}(a,b)$ attains maxima $\left(2^{\frac{2}{p}-1} - 2^{\frac{1}{p}+1} + 2\right) r^2$ when $(a,b) = (r,r)$.

(b) If $r \in \mathbb{R}^- \cup \{0\}$, then $\psi^p_{\rm FB}(a,b)$ attains minima $\left(2^{\frac{2}{p}-1} + 2^{\frac{1}{p}+1} + 2\right) r^2$ when $(a,b) = (r,r)$.

Proof (a) When $a > 0$ and $b > 0$, Proposition 2.6(a) says that $\phi^p_{\rm FB}(a,b) < 0$. Since $\phi^2_{\rm FB}(a,b) > 0$, by Proposition 2.9, the minima of $\phi^p_{\rm FB}(a,b)$ becomes maxima of $\psi^p_{\rm FB}(a,b)$.

(b) This is a consequence of Proposition 2.9. □

The preceding results indicate that $\psi^p_{\rm FB}$ shares many geometric properties with $\phi^p_{\rm FB}$, as illustrated in Figs. 2.11 and 2.12. In particular, we define $\psi^1_{\rm FB}(a,b) := \frac{1}{2}|\phi^1_{\rm FB}(a,b)|^2$ and $\psi^\infty_{\rm FB}(a,b) := \frac{1}{2}|\phi^\infty_{\rm FB}(a,b)|^2$. Nonetheless, there remain key differences between $\phi^p_{\rm FB}$ and $\psi^p_{\rm FB}$. For instance, unlike $\phi^p_{\rm FB}$, the function $\psi^p_{\rm FB}$ is not convex. Figure 2.12 illustrates the direction of increase for $\psi^p_{\rm FB}$. It is also worth noting that $\psi^p_{\rm FB}(a,b)$ is nonnegative and exhibits distinct behaviors in the region where $a > 0$ and $b > 0$, as shown in Figs. 2.9, 2.10, and 2.11.

To further explore the geometric properties of $\psi^p_{\rm FB}$, we introduce a family of curves defined by

$$\beta_{r,p}(t) := \left(r+t, r-t, \psi^p_{\rm FB}(r+t, r-t)\right), \tag{2.32}$$

where r is a fixed real number and $t \in \mathbb{R}$. This family of curves represents the intersection between the plane $a+b=2r$ and the surface $z = \psi^p_{\rm FB}(a,b)$, as illustrated in Fig. 2.13.

Proposition 2.15 *Let $\beta_{r,p} : \mathbb{R} \to \mathbb{R}^3$ be defined as in (2.32). Then, the following hold.*

(a) The curvature at point $\beta_{r,p}(0) = \left(r, r, \psi^p_{\rm FB}(r,r)\right)$ is $\bar{\kappa}_p(0) = (p-1)2^{\frac{1}{p}} \left(1 - 2^{\frac{1}{p}-1}\right)$.

(b) $\bar{\kappa}_p(0) \to 0$ as $p \to 1$ and $\bar{\kappa}_p(0) \to +\infty$ as $p \to +\infty$.

(c) If $1 < p_1 < p_2$, then $\bar{\kappa}_{p_1}(0) < \bar{\kappa}_{p_2}(0)$.

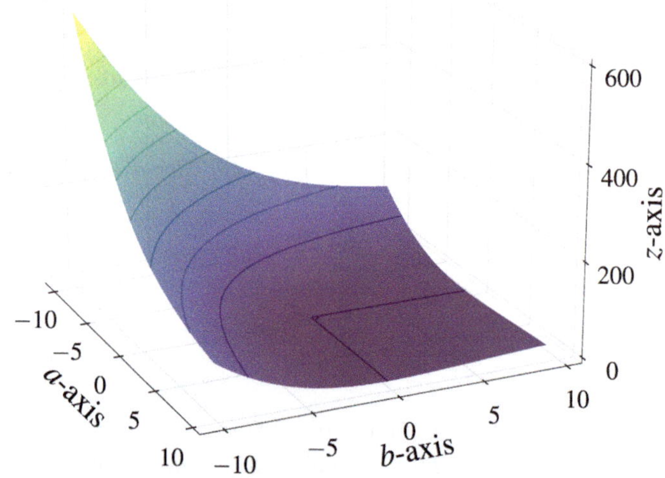

Fig. 2.9 The surface of $z = \psi_{\rm FB}^2(a, b)$ with $(a, b) \in [-10, 10] \times [-10, 10]$

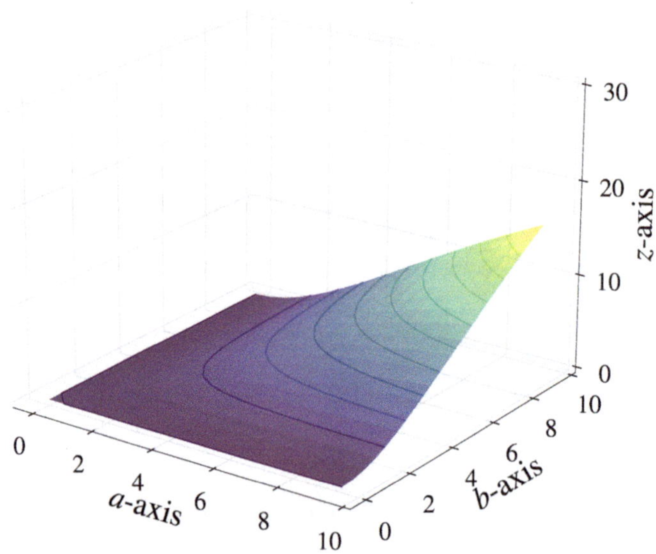

Fig. 2.10 The local surface of $z = \psi_{\rm FB}^2(a, b)$ with $(a, b) \in [0, 10] \times [0, 10]$

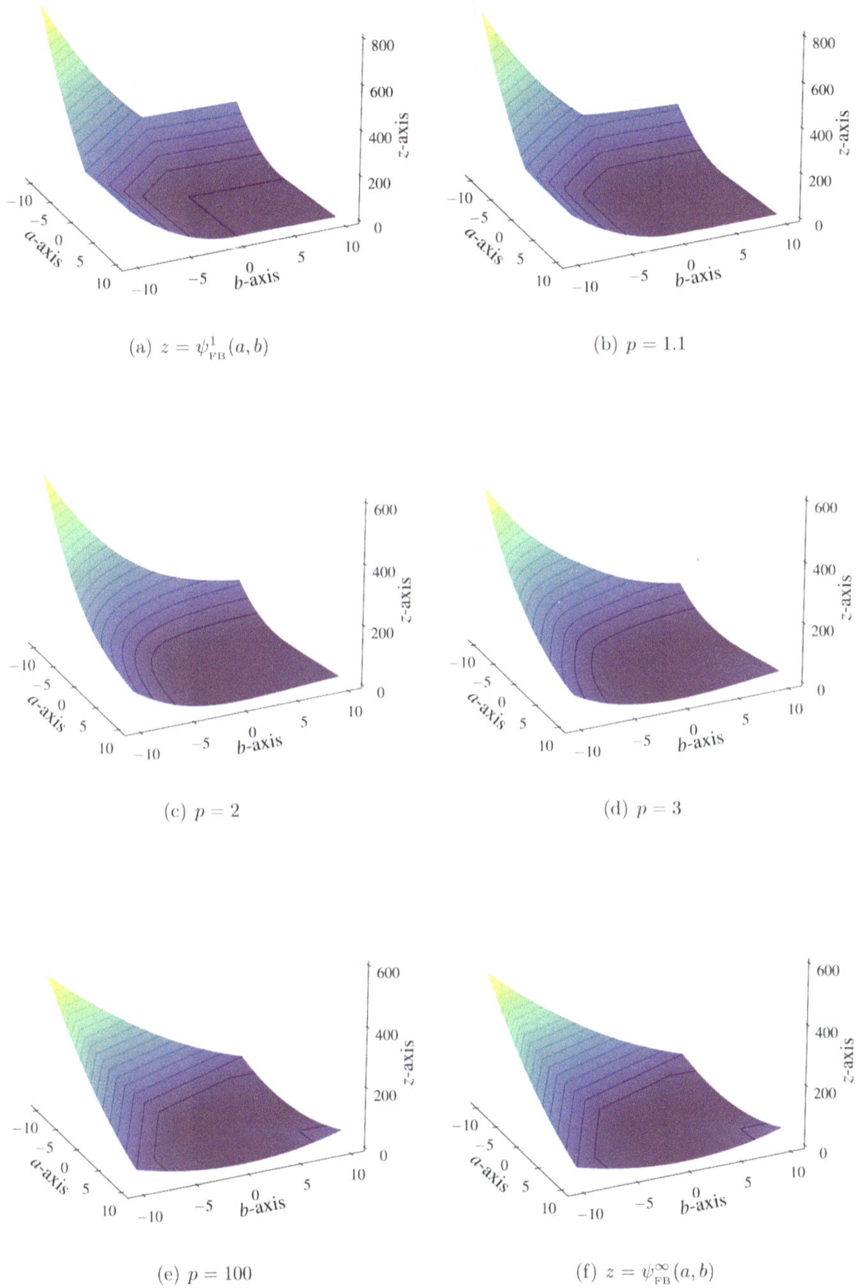

Fig. 2.11 The surface of $z = \psi^p_{\mathrm{FB}}(a, b)$ with different values of p

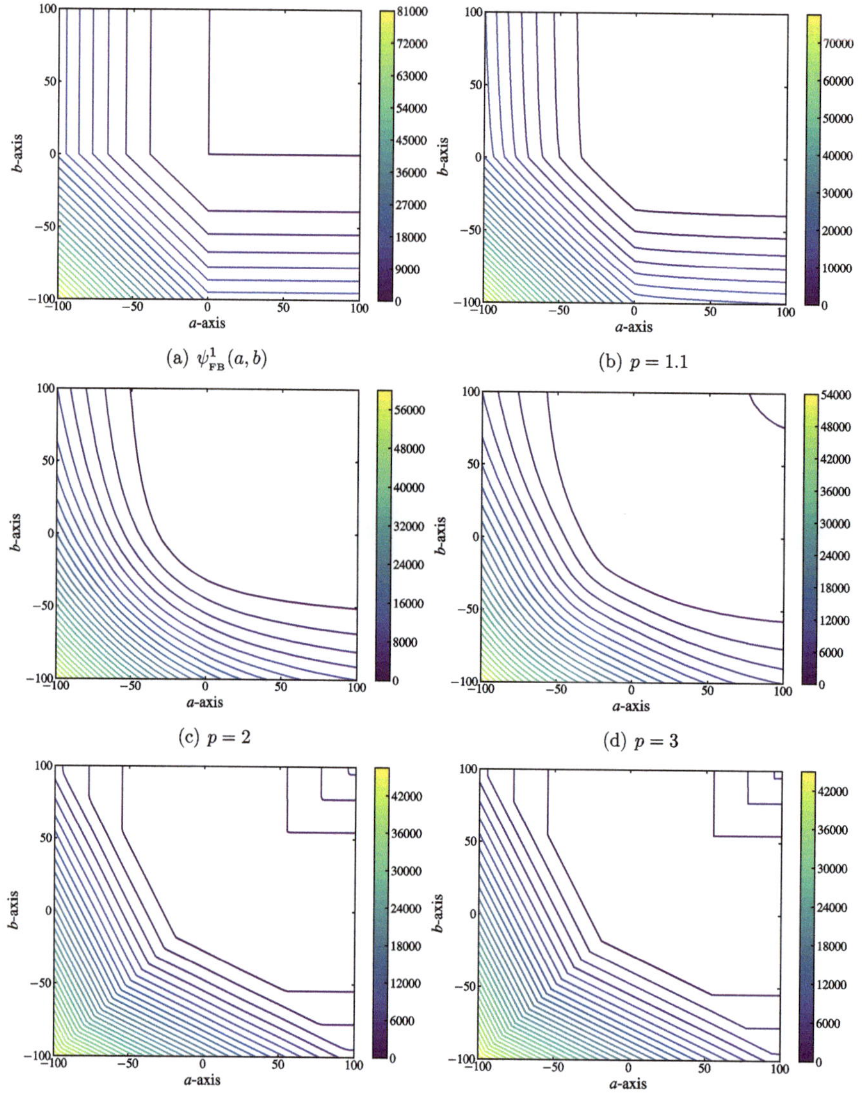

(a) $\psi^1_{\text{FB}}(a,b)$ (b) $p = 1.1$

(c) $p = 2$ (d) $p = 3$

(e) $p = 100$ (f) $\psi^\infty_{\text{FB}}(a,b)$

Fig. 2.12 Level curves of $z = \psi^p_{\text{FB}}(a,b)$ with different values of p

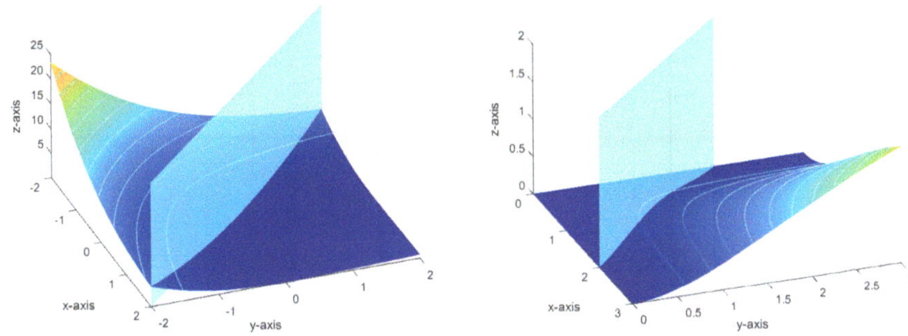

(a) $a+b=0$ and $z=\psi^2_{\rm FB}(a,b)$ (b) $a+b=2$ and $z=\psi^2_{\rm FB}(a,b)$

Fig. 2.13 The curve intersected by surface $z=\psi^p_{\rm FB}(a,b)$ and plane $a+b=2r$

Proof (a) From $\beta_{r,p}(t)=\big(r+t, r-t, \psi^p_{\rm FB}(r+t,r-t)\big)$, we know

$$\beta'_{r,p}(0)=(1,-1,0) \quad \text{and} \quad \beta''_{r,p}(0)=\left(0,0,(p-1)2^{\frac{2}{p}}-\mathrm{sgn}(r)(p-1)2^{\frac{1}{p}+1}\right)$$

which yields

$$\bar{\kappa}_p(r)=\frac{|\beta'_{r,p}(0)\wedge\beta''_{r,p}(0)|}{|\beta'_{r,p}(0)|^3}=(p-1)2^{\frac{1}{p}}(1-2^{\frac{1}{p}-1}).$$

(b) Let $f:(1,+\infty)\to\mathbb{R}$ be defined as $f(p):=\bar{\kappa}_p(0)=(p-1)2^{\frac{1}{p}}(1-2^{\frac{1}{p}-1})$. Then, the result follows by taking the limit directly.

(c) From part(b), it can be verified that $f'(p)>0$ for all $p\in(1,+\infty)$. Thus, $f(p)$ is strictly increasing on $(1,+\infty)$. □

Figure 2.14 illustrates how the shape of the curve evolves with varying values of p, particularly highlighting changes in curvature as p approaches 1 or tends toward infinity. As an addendum to part (a), we note that the curvature at two special points, $\beta_{r,p}(-r)=(0,2r,0)$ and $\beta_{r,p}(r)=(2r,0,0)$, is identical and given by $\bar{\kappa}_p(r)=\bar{\kappa}_p(-r)=\frac{1}{2}$. Although $\psi^p_{\rm FB}$ is differentiable everywhere, the mean curvature at the origin, $(0,0)$, does not exist. We summarize the similarities and differences between $\phi^p_{\rm FB}$ and $\psi^p_{\rm FB}$ below.

	$\phi^p_{\rm FB}(a,b)$	$\psi^p_{\rm FB}(a,b)$
Difference	Convex	Nonconvex
	Differentiable everywhere except $(0,0)$	Differentiable everywhere
	$\phi^p_{\rm FB}(a,b)<0$ when $a>0$ and $b>0$	$\psi^p_{\rm FB}(a,b)\geq 0, \forall(a,b)\in\mathbb{R}^2$

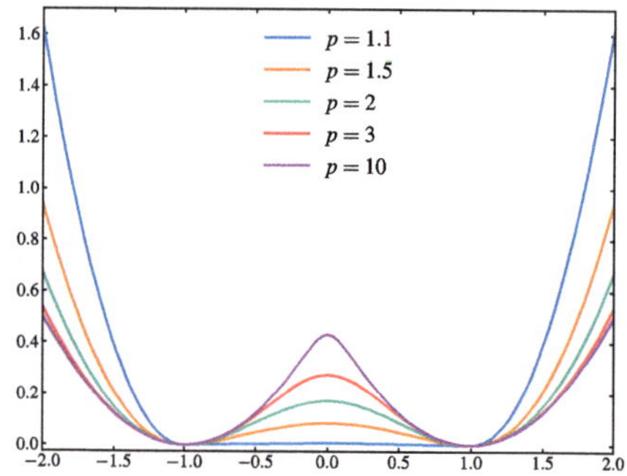

(a) The curvature with different p with $r = 1$

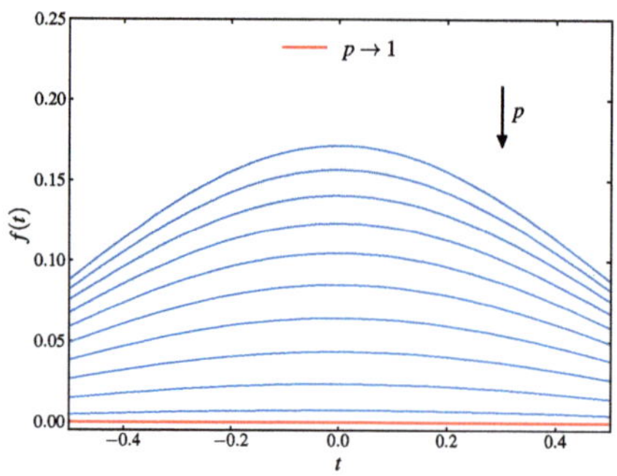

(b) The change of curvature as $p \to 1$ at $\beta_{1,p}(0)$.

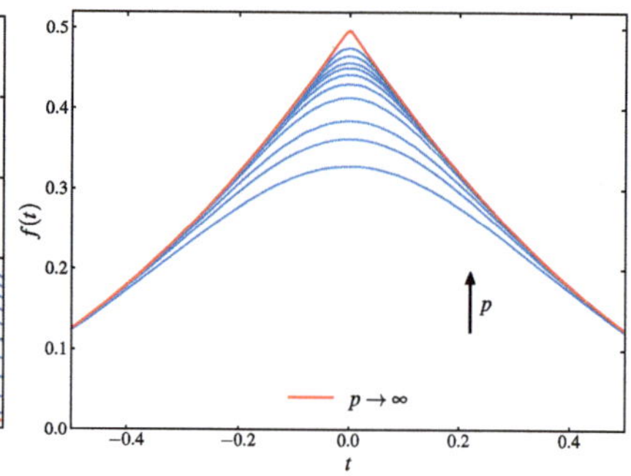

(c) The change of curvature as $p \to +\infty$ at $\beta_{1,p}(0)$.

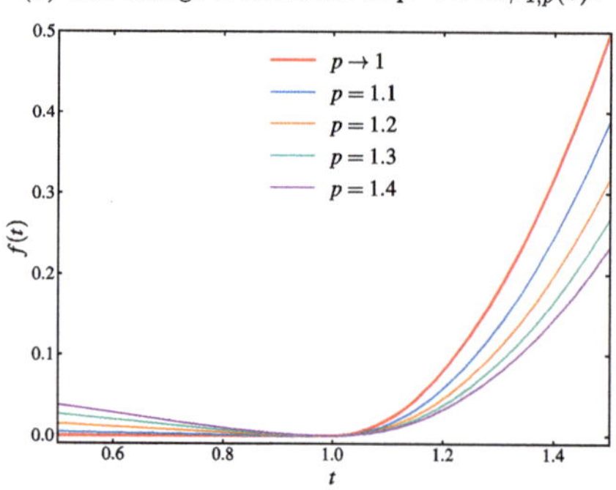

(d) The change of curvature as $p \to 1$ at $\beta_{1,p}(1)$.

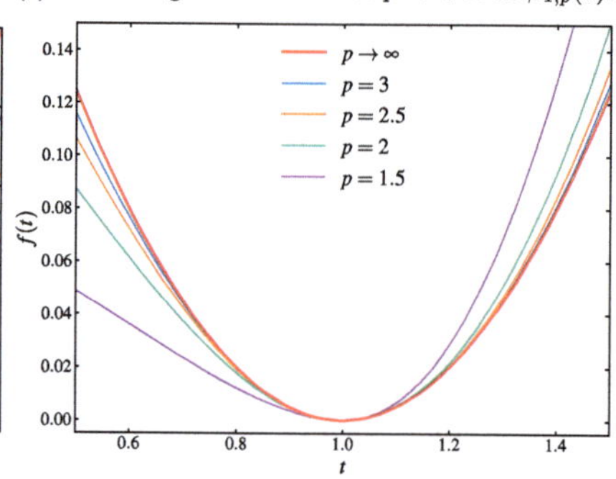

(e) The change of curvature as $p \to +\infty$ at $\beta_{1,p}(1)$.

Fig. 2.14 The curvature $\bar{\kappa}_p(0)$ at point $\beta_{r,p}(0)$

Similarity	(1) NCP function
	(2) Symmetry (i.e., $\phi^p_{\text{FB}}(a,b) = \phi^p_{\text{FB}}(b,a)$ and $\psi^p_{\text{FB}}(a,b) = \psi^p_{\text{FB}}(b,a)$)
	(3) The function is not affected by p on axes
	(4) When $(a^k \to -\infty)$ or $(b^k \to -\infty)$ or $(a^k, b^k \to +\infty)$, there have $\lvert\phi^p_{\text{FB}}(a^k, b^k)\rvert \to \infty$ and $\lvert\psi^p_{\text{FB}}(a^k, b^k)\rvert \to \infty$
	(5) non-coercive

Proposition 2.16 *The function $\Phi^p_{\text{FB}} : \mathbb{R}^n \to \mathbb{R}^n$ defined as in (2.16) is semismooth.*

Proof From Proposition 2.1(d), we know that ϕ^p_{FB} is convex and thus semismooth. Furthermore, each component of $\Phi^p_{\text{FB}}(x)$ is formed by composing the convex function $\phi^p_{\text{FB}} : \mathbb{R}^2 \to \mathbb{R}$ with the differentiable mapping $(x_i, F_i(x))^\top : \mathbb{R}^n \to \mathbb{R}^2$. Since both convex and differentiable functions are semismooth, and the composition of semismooth functions remains semismooth, it follows that Φ^p_{FB} itself is semismooth. □

Proposition 2.17 shows that ψ^p_{FB} is an SC^1 function. Consequently, if each F_i is also an SC^1 function, then Ψ^p_{FB} inherits this property. Before presenting the proof, we introduce a key technical lemma, which establishes that the gradient $\nabla\psi^p_{\text{FB}}$ is globally Lipschitz continuous, an essential result for our subsequent analysis.

Lemma 2.5 *The gradient of the function ψ^p_{FB} defined as (2.19) is Lipschitz continuous, that is, there exists $L > 0$ such that*

$$\|\nabla\psi^p_{\text{FB}}(a,b) - \nabla\psi^p_{\text{FB}}(c,d)\| \le L\|(a,b) - (c,d)\|,$$

for all $(a,b), (c,d) \in \mathbb{R}^2$.

Proof Based on the expressions for the gradient of ψ^p_{FB} given in (2.20) and (2.21), and by applying the chain rule and quotient rule (the computation, while routine, is somewhat tedious and thus omitted), we arrive at the following two cases.

(i) If p is even and $(a,b) \neq (0,0)$, then

$$\nabla^2_{aa}\psi^p_{\text{FB}}(a,b) = \left(\frac{a^{p-1}}{\|(a,b)\|_p^{p-1}} - 1\right)^2 + \frac{(p-1)a^{p-2}b^p}{\|(a,b)\|_p^{2p-1}}\Big(\|(a,b)\|_p - (a+b)\Big),$$

$$\begin{aligned}\nabla^2_{ab}\psi^p_{\text{FB}}(a,b) = \nabla^2_{ba}\psi^p_{\text{FB}}(a,b) &= \left(\frac{a^{p-1}}{\|(a,b)\|_p^{p-1}} - 1\right)\left(\frac{b^{p-1}}{\|(a,b)\|_p^{p-1}} - 1\right)\\ &\quad - \frac{(p-1)a^{p-1}b^{p-1}}{\|(a,b)\|_p^{2p-1}}\Big(\|(a,b)\|_p - (a+b)\Big),\end{aligned}$$

$$\nabla^2_{bb}\psi^p_{\text{FB}}(a,b) = \left(\frac{b^{p-1}}{\|(a,b)\|_p^{p-1}} - 1\right)^2 + \frac{(p-1)a^p b^{p-2}}{\|(a,b)\|_p^{2p-1}}\Big(\|(a,b)\|_p - (a+b)\Big).$$

It is clear that $\dfrac{|a|^{p-1}}{\|(a,b)\|_p^{p-1}} \leq 1$ and we also see that

$$|a|^{p-2} \cdot |b|^p \leq \Big(\max\{|a|,|b|\}\Big)^{2p-2} \leq \left(\sqrt[p]{|a|^p+|b|^p}\right)^{2p-2} \leq \|(a,b)\|_p^{2p-2},$$

which yields

$$\frac{|a|^{p-2}|b|^p}{\|(a,b)\|_p^{2p-2}} \leq 1. \tag{2.33}$$

Similarly, it can be verified that $\dfrac{|a|^p|b|^{p-2}}{\|(a,b)\|_p^{2p-2}} \leq 1$. On the other hand, by Lemma 1.4, we have

$$|a|+|b| \leq \sqrt{2}\sqrt{a^2+b^2} = \sqrt{2}\|(a,b)\|_2 \leq \sqrt{2}\cdot 2^{(1/2-1/p)}\|(a,b)\|_p = 2^{(1-1/p)}\|(a,b)\|_p.$$

Applying all the above, we can give an upper bound for $\nabla_{aa}^2\psi_{\text{FB}}^p(a,b)$ as below.

$$\begin{aligned}
&\left|\nabla_{aa}^2\psi_{\text{FB}}^p(a,b)\right| \\
\leq & \left(\frac{a^{p-1}}{\|(a,b)\|_p^{p-1}}+1\right)^2 + \frac{(p-1)|a|^{p-2}|b|^p}{\|(a,b)\|_p^{2p-2}} + \frac{(p-1)|a|^{p-2}|b|^p\cdot(|a|+|b|)}{\|(a,b)\|_p^{2p-1}} \\
\leq & 4+(p-1)+\frac{(p-1)|a|^{p-2}|b|^p\cdot 2^{(1-1/p)}\|(a,b)\|_p}{\|(a,b)\|_p^{2p-1}} \\
\leq & 4+(p-1)+(p-1)2^{(1-1/p)} \\
= & 4+(p-1)\left[1+2^{(1-1/p)}\right],
\end{aligned}$$

where the last inequality holds due to (2.33). By the same arguments, we also have

$$\left|\nabla_{bb}^2\psi_{\text{FB}}^p(a,b)\right| \leq 4+(p-1)\left[1+2^{(1-1/p)}\right].$$

Now, we estimate the upper bound for $\nabla_{ab}^2\psi_{\text{FB}}^p(a,b) = \nabla_{ba}^2\psi_{\text{FB}}^p(a,b)$ as below.

$$\begin{aligned}
&\left|\nabla_{ab}^2\psi_{\text{FB}}^p(a,b)\right| = \left|\nabla_{ba}^2\psi_{\text{FB}}^p(a,b)\right| \\
\leq & \left|\frac{a^{p-1}}{\|(a,b)\|_p^{p-1}}-1\right|\cdot\left|\frac{b^{p-1}}{\|(a,b)\|_p^{p-1}}-1\right|
\end{aligned}$$

$$
\begin{aligned}
&+\frac{(p-1)|a|^{p-1}|b|^{p-1}}{\|(a,b)\|_p^{2p-1}}\Big(\|(a,b)\|_p+(|a|+|b|)\Big)\\
\le&\left(\frac{|a|^{p-1}}{\|(a,b)\|_p^{p-1}}+1\right)\left(\frac{|b|^{p-1}}{\|(a,b)\|_p^{p-1}}+1\right)\\
&+\frac{(p-1)|a|^{p-1}|b|^{p-1}}{\|(a,b)\|_p^{2p-2}}+\frac{(p-1)|a|^{p-1}|b|^{p-1}\cdot(|a|+|b|)}{\|(a,b)\|_p^{2p-1}}\\
\le&\,4+(p-1)+\frac{(p-1)|a|^{p-1}|b|^{p-1}\cdot 2^{(1-1/p)}\|(a,b)\|_p}{\|(a,b)\|_p^{2p-1}}\\
\le&\,4+(p-1)+(p-1)2^{(1-1/p)}\\
=&\,4+(p-1)\Big[1+2^{(1-1/p)}\Big],
\end{aligned}
$$

where the third and fourth inequalities are true by the similar result as (2.33), that is, $\frac{|a|^{p-1}|b|^{p-1}}{\|(a,b)\|_p^{2p-2}} \le 1$.

(ii) If p is odd and $(a,b) \neq (0,0)$, then we obtain

$$
\begin{aligned}
\nabla^2_{aa}\psi^p_{\rm FB}(a,b) &= \left(\frac{\mathrm{sgn}(a)\cdot a^{p-1}}{\|(a,b)\|_p^{p-1}}-1\right)^2+\frac{\mathrm{sgn}(a)\mathrm{sgn}(b)\cdot(p-1)a^{p-2}b^p}{\|(a,b)\|_p^{2p-1}}\Big(\|(a,b)\|_p-(a+b)\Big),\\
\nabla^2_{ab}\psi^p_{\rm FB}(a,b) &= \nabla^2_{ba}\psi_p(a,b)=\left(\frac{\mathrm{sgn}(a)\cdot a^{p-1}}{\|(a,b)\|_p^{p-1}}-1\right)\left(\frac{\mathrm{sgn}(b)\cdot b^{p-1}}{\|(a,b)\|_p^{p-1}}-1\right),\\
&\quad -\frac{\mathrm{sgn}(a)\mathrm{sgn}(b)\cdot(p-1)a^{p-1}b^{p-1}}{\|(a,b)\|_p^{2p-1}}\Big(\|(a,b)\|_p-(a+b)\Big),\\
\nabla^2_{bb}\psi^p_{\rm FB}(a,b) &= \left(\frac{\mathrm{sgn}(b)\cdot b^{p-1}}{\|(a,b)\|_p^{p-1}}-1\right)^2+\frac{\mathrm{sgn}(a)\mathrm{sgn}(b)\cdot(p-1)a^{p}b^{p-2}}{\|(a,b)\|_p^{2p-1}}\Big(\|(a,b)\|_p-(a+b)\Big).
\end{aligned}
$$

In fact, the upper bounds for $\nabla^2_{aa}\psi^p_{\rm FB}(a,b)$, $\nabla^2_{ab}\psi^p_{\rm FB}(a,b)$, $\nabla^2_{bb}\psi^p_{\rm FB}(a,b)$ remain the same by following exactly the same steps as in the case where p is even. Thus, there exist a constant $L>0$ independent of (a,b) such that

$$\|\nabla^2\psi^p_{\rm FB}(a,b)\| \le L, \quad \forall\,(a,b)\neq(0,0)\in\mathbb{R}^2.$$

Then, by Lemma 1.3, we have

$$\|\nabla\psi_p(a,b)-\nabla\psi^p_{\rm FB}(c,d)\| \le L\|(a,b)-(c,d)\|, \tag{2.34}$$

for all $(a,b),(c,d)\in\mathbb{R}^2$ with $(0,0)\notin[(a,b),(c,d)]$. Moreover, (2.34) also holds in case $(a,b)=(c,d)=(0,0)$ since $\nabla_a\psi^p_{\rm FB}(a,b)=\nabla_b\psi^p_{\rm FB}(a,b)=0$. Therefore, we can assume $(a,b)\neq(0,0)$. From Proposition 2.2(c), $\psi^p_{\rm FB}$ is continuously differentiable for all $(a,b)\in\mathbb{R}^2$ with $\nabla\psi^p_{\rm FB}(0,0)=(0,0)$; then using a continuity argument, we obtain (2.34) remains true for all $(c, d) \in \mathbb{R}^2$. Thus, (2.34) holds for all $(a, b), (c, d) \in \mathbb{R}^2$ which says ψ_{FB}^p is globally Lipschitz continuous. □

Proposition 2.17 *The function ψ_{FB}^p defined as in (2.19) is an SC^1 function. Hence, if every F_i is an SC^1 function, then the function Ψ_{FB}^p given as in (2.17) is also an SC^1 function.*

Proof As established in Proposition 2.2(c), ψ_{FB}^p is continuously differentiable. It remains to verify that its gradient, $\nabla\psi_{\text{FB}}^p$, is semismooth. According to Lemma 2.5, $\nabla\psi_{\text{FB}}^p$ is Lipschitz continuous and, consequently, strictly continuous (i.e., locally Lipschitz continuous). Therefore, to prove the semismoothness of $\nabla\psi_{\text{FB}}^p$, it suffices to verify that it satisfies the condition in Lemma 1.2(b). More precisely, we only need to check semismoothness at the point (0, 0), since $\nabla\psi_{\text{FB}}^p$ is continuously differentiable, and hence semismooth, at all other points (as shown in the proof of Lemma 2.5). To this end, we have to verify that the equation in Lemma 1.2(b) is satisfied, i.e., for any $(h_1, h_2) \in \mathbb{R}^2$ such that $\nabla\psi_{\text{FB}}^p$ is differentiable at (h_1, h_2), we have

$$\nabla\psi_{\text{FB}}^p(h_1, h_2) - \nabla\psi_{\text{FB}}^p(0, 0) - \nabla^2\psi_{\text{FB}}^p(h_1, h_2) \cdot h = o(\|(h_1, h_2)\|). \tag{2.35}$$

In order to prove (2.35), we have two cases where p is even and p is odd.

For p is even, we denote (Ξ_1, Ξ_2) the left-hand side of (2.35). Then, we have

$$\begin{aligned} \begin{bmatrix} \Xi_1 \\ \Xi_2 \end{bmatrix} &:= \begin{bmatrix} k_1 \\ k_2 \end{bmatrix} \cdot \phi_{\text{FB}}^p(h_1, h_2) - \begin{bmatrix} 0 \\ 0 \end{bmatrix} \\ &- \begin{bmatrix} k_1^2 + \left(\frac{(p-1)h_1^{p-2}h_2^p}{\|(h_1,h_2)\|_p^{2p-1}}\right)\phi_{\text{FB}}^p(h_1, h_2) & k_1 \cdot k_2 - k_3\phi_{\text{FB}}^p(h_1, h_2) \\ k_1 \cdot k_2 - k_3\phi_{\text{FB}}^p(h_1, h_2) & k_2^2 + \left(\frac{(p-1)h_1^p h_2^{p-2}}{\|(h_1,h_2)\|_p^{2p-1}}\right)\phi_{\text{FB}}^p(h_1, h_2) \end{bmatrix} \cdot \begin{bmatrix} h_1 \\ h_2 \end{bmatrix}, \end{aligned} \tag{2.36}$$

where

$$\begin{aligned} k_1 &= \left(\frac{h_1^{p-1}}{\|(h_1, h_2)\|_p^{p-1}} - 1\right), \\ k_2 &= \left(\frac{h_2^{p-1}}{\|(h_1, h_2)\|_p^{p-1}} - 1\right), \\ k_3 &= \frac{(p-1)h_1^{p-1}h_2^{p-1}}{\|(h_1, h_2)\|_p^{2p-1}}. \end{aligned} \tag{2.37}$$

By plugging (2.37) into (2.36) and writing out Ξ_1 and Ξ_2, we obtain that $\Xi_1 = 0$ and $\Xi_2 = 0$. To see this, we compute Ξ_1 as below

$$
\begin{aligned}
\Xi_1 &= \left(\frac{h_1^{p-1}}{\|(h_1,h_2)\|_p^{p-1}} - 1\right)\phi_{\rm FB}^p(h_1,h_2) - \left(\frac{h_1^{p-1}}{\|(h_1,h_2)\|_p^{p-1}} - 1\right)^2 h_1 \\
&\quad - \frac{(p-1)h_1^{p-1}h_2^p}{\|(h_1,h_2)\|_p^{2p-1}} \cdot \phi_{\rm FB}^p(h_1,h_2) - \left(\frac{h_1^{p-1}}{\|(h_1,h_2)\|_p^{p-1}} - 1\right)\left(\frac{h_2^{p-1}}{\|(h_1,h_2)\|_p^{p-1}} - 1\right)h_2 \\
&\quad + \frac{(p-1)h_1^{p-1}h_2^p}{\|(h_1,h_2)\|_p^{2p-1}} \cdot \phi_{\rm FB}^p(h_1,h_2) \\
&= \phi_{\rm FB}^p(h_1,h_2)\left[\left(\frac{h_1^{p-1}}{\|(h_1,h_2)\|_p^{p-1}} - 1\right) - \frac{(p-1)h_1^{p-1}h_2^p}{\|(h_1,h_2)\|_p^{2p-1}} + \frac{(p-1)h_1^{p-1}h_2^p}{\|(h_1,h_2)\|_p^{2p-1}}\right] \\
&\quad - \left(\frac{h_1^{p-1}}{\|(h_1,h_2)\|_p^{p-1}} - 1\right)^2 h_1 - \left(\frac{h_1^{p-1}}{\|(h_1,h_2)\|_p^{p-1}} - 1\right)\left(\frac{h_2^{p-1}}{\|(h_1,h_2)\|_p^{p-1}} - 1\right)h_2 \\
&= \phi_{\rm FB}^p(h_1,h_2)\left(\frac{h_1^{p-1}}{\|(h_1,h_2)\|_p^{p-1}} - 1\right) - \left(\frac{h_1^{p-1}}{\|(h_1,h_2)\|_p^{p-1}} - 1\right)^2 h_1 \\
&\quad - \left(\frac{h_1^{p-1}}{\|(h_1,h_2)\|_p^{p-1}} - 1\right)\left(\frac{h_2^{p-1}}{\|(h_1,h_2)\|_p^{p-1}} - 1\right)h_2 \\
&= \left(\frac{h_1^{p-1}}{\|(h_1,h_2)\|_p^{p-1}} - 1\right)\left[\phi_{\rm FB}^p(h_1,h_2) - \left(\frac{h_1^{p-1}}{\|(h_1,h_2)\|_p^{p-1}} - 1\right)h_1 - \left(\frac{h_2^{p-1}}{\|(h_1,h_2)\|_p^{p-1}} - 1\right)h_2\right] \\
&= \left(\frac{h_1^{p-1}}{\|(h_1,h_2)\|_p^{p-1}} - 1\right)\left[\|(h_1,h_2)\|_p - \frac{h_1^p + h_2^p}{\|(h_1,h_2)\|_p^{p-1}}\right] \\
&= \left(\frac{h_1^{p-1}}{\|(h_1,h_2)\|_p^{p-1}} - 1\right)\cdot 0 \\
&= 0\,,
\end{aligned}
$$

where the second-to-last equality is true since $h_1^p + h_2^p = \|(h_1,h_2)\|_p^p$ when p is even. Similarly,

$$
\begin{aligned}
\Xi_2 &= \left(\frac{h_2^{p-1}}{\|(h_1,h_2)\|_p^{p-1}} - 1\right)\phi_{\rm FB}^p(h_1,h_2) - \left(\frac{h_2^{p-1}}{\|(h_1,h_2)\|_p^{p-1}} - 1\right)^2 h_2 \\
&\quad - \frac{(p-1)h_1^p h_2^{p-1}}{\|(h_1,h_2)\|_p^{2p-1}} \cdot \phi_{\rm FB}^p(h_1,h_2) - \left(\frac{h_1^{p-1}}{\|(h_1,h_2)\|_p^{p-1}} - 1\right)\left(\frac{h_2^{p-1}}{\|(h_1,h_2)\|_p^{p-1}} - 1\right)h_1 \\
&\quad + \frac{(p-1)h_1^p h_2^{p-1}}{\|(h_1,h_2)\|_p^{2p-1}} \cdot \phi_{\rm FB}^p(h_1,h_2) \\
&= \phi_{\rm FB}^p(h_1,h_2)\left[\left(\frac{h_2^{p-1}}{\|(h_1,h_2)\|_p^{p-1}} - 1\right) - \frac{(p-1)h_1^p h_2^{p-1}}{\|(h_1,h_2)\|_p^{2p-1}} + \frac{(p-1)h_1^p h_2^{p-1}}{\|(h_1,h_2)\|_p^{2p-1}}\right] \\
&\quad - \left(\frac{h_2^{p-1}}{\|(h_1,h_2)\|_p^{p-1}} - 1\right)^2 h_2 - \left(\frac{h_1^{p-1}}{\|(h_1,h_2)\|_p^{p-1}} - 1\right)\left(\frac{h_2^{p-1}}{\|(h_1,h_2)\|_p^{p-1}} - 1\right)h_1
\end{aligned}
$$

$$
\begin{aligned}
&= \phi_{\mathrm{FB}}^p(h_1,h_2)\left(\frac{h_2^{p-1}}{\|(h_1,h_2)\|_p^{p-1}}-1\right)-\left(\frac{h_2^{p-1}}{\|(h_1,h_2)\|_p^{p-1}}-1\right)^2 h_2 \\
&\quad -\left(\frac{h_1^{p-1}}{\|(h_1,h_2)\|_p^{p-1}}-1\right)\left(\frac{h_2^{p-1}}{\|(h_1,h_2)\|_p^{p-1}}-1\right)h_1 \\
&= \left(\frac{h_2^{p-1}}{\|(h_1,h_2)\|_p^{p-1}}-1\right)\left[\phi_{\mathrm{FB}}^p(h_1,h_2)-\left(\frac{h_1^{p-1}}{\|(h_1,h_2)\|_p^{p-1}}-1\right)h_1-\left(\frac{h_2^{p-1}}{\|(h_1,h_2)\|_p^{p-1}}-1\right)h_2\right] \\
&= \left(\frac{h_2^{p-1}}{\|(h_1,h_2)\|_p^{p-1}}-1\right)\left[\|(h_1,h_2)\|_p-\frac{h_1^p+h_2^p}{\|(h_1,h_2)\|_p^{p-1}}\right] \\
&= \left(\frac{h_1^{p-1}}{\|(h_1,h_2)\|_p^{p-1}}-1\right)\cdot 0 \\
&= 0\,,
\end{aligned}
$$

where the second-to-last equality is true since $h_1^p + h_2^p = \|(h_1,h_2)\|_p^p$ when p is even. From the above two expressions of Ξ_1 and Ξ_2, it implies that (2.35) is satisfied. Thus, $\nabla\psi_{\mathrm{FB}}^p$ is semismooth at (0, 0) for the case where p is even.

For odd values of p, the same line of reasoning applies, leading to analogous verifications. Thus, we conclude that $\nabla\psi_{\mathrm{FB}}^p$ is semismooth, and therefore ψ_{FB}^p is an SC^1 function. The second statement then follows directly from this result. □

We would like to highlight that for $p = 2$, the function ψ_{FB}^p was already shown to be an SC^1 function in [1, 30], with the first formal proof appearing in [30]. Proposition 2.17 extends this result to all $p \geq 2$, though its proof is considerably more intricate than in the quadratic case. With Lemma 2.5 and Proposition 2.17 established, we can now derive the following consequences.

Proposition 2.18 *If every F_i is an LC^1 function, then the function Φ_{FB}^p given as in (2.16) is strongly semsmooth.*

Proof It is known that ϕ_{FB}^p is semismooth, in fact, strongly semismooth. This follows from Proposition 2.1(d), Lemma 2.5, and [32, Theorem 7]. Moreover, since every LC^1 function is strongly semismooth, the result immediately follows. □

Proposition 2.19 *The function ψ_{FB}^p defined as in (2.19) is an LC^1 function. Hence, if every F_i is an LC^1 function, then the function Ψ_{FB}^p given as in (2.17) is also an LC^1 function.*

By applying Proposition 2.1(c), Proposition 2.16, and a result from [32], we immediately obtain an interesting property concerning the strong almost smoothness of Φ_{FB}^p. For further details on the notions of almost smooth and strongly almost smooth functions, we refer the reader to [32].

Proposition 2.20 *If every F_i is an LC^1 function, then the function Φ_{FB}^p defined as in (2.16) is strongly almost smooth function.*

Proof This result follows directly from Proposition 2.1(c), Proposition 2.16, and [32, Theorem 7]. □

2.2 Constructions of NCP Functions Based on ϕ_{NR}

2.2.1 Construction by Discrete Generalization

As discussed in Sect. 2.1, the generalized Fischer-Burmeister function ϕ^p_{FB}, defined in (2.14), encompasses the classical Fischer-Burmeister function as a special case and serves as a natural extension of the widely used ϕ_{FB} function. This extension replaces the Euclidean (2-norm) in $\phi_{\text{FB}}(a, b)$ with a general p-norm, providing what can be regarded as a "continuous generalization". A geometric perspective of ϕ^p_{FB} is presented in [26], while the impact of varying p on different algorithmic frameworks has been explored in [22, 24, 27, 31, 33]. In contrast, a natural question arises: "Is there a corresponding extension of the natural residual function?" The following diagram illustrates the core of this inquiry:

$$\phi_{\text{FB}}(a, b) = \|(a, b)\|_2 - (a + b) \longrightarrow \phi^p_{\text{FB}}(a, b) = \|(a, b)\|_p - (a + b)$$

$$\phi_{\text{NR}}(a, b) = \min\{a, b\} \longrightarrow \quad ???$$

While numerous NCP functions have been proposed as variants of the natural residual function ϕ_{NR}, no work in the literature has addressed a true extension of the natural residual function itself. The primary challenge lies in the absence of a continuous norm-based generalization, such as the one used for ϕ^p_{FB}. In this section, we provide an affirmative answer to this long-standing open question, as presented in [34]. Unlike the continuous generalization used for ϕ^p_{FB}, the approach here is based on a "discrete generalization". Specifically, we introduce the generalized natural residual function, denoted by ϕ^p_{NR}, defined as follows:

$$\phi^p_{\text{NR}}(a, b) = a^p - (a - b)^p_+ \quad \text{with } p > 1 \text{ being a positive odd integer,} \tag{2.38}$$

where $(a - b)^p_+ = [(a - b)_+]^p$ and $(a - b)_+ = \max\{a - b, 0\}$. Here, p being a positive odd integer is necessary, that is, we require that $p = 2k + 1$, where $k = 1, 2, 3, \ldots$. We will explain this later. Notice that when $p = 1$, ϕ^p_{NR} reduces to the natural residual function ϕ_{NR}, i.e., when $k = 0$, it corresponds to

$$\phi^1_{\text{NR}}(a, b) = a - (a - b)_+ = \min\{a, b\} = \phi_{\text{NR}}(a, b).$$

This is the motivation behind the term "generalized natural residual function". We emphasize once again that the proposed extension is based on a discrete generalization. For even values of p, the function ϕ^p_{NR} no longer qualifies as an NCP function in the traditional sense. However, a distinguishing feature of ϕ^p_{NR} is that it is twice continuously differentiable, as will be established in Proposition 2.23. In contrast, while the generalized Fischer-Burmeister function ϕ^p_{FB}, defined in (2.14), is not differentiable in general, the squared norm $\|\phi^p_{\text{FB}}(a, b)\|^2$ is differentiable everywhere.

As a result, the merit function approach typically employs $\|\phi_{\rm FB}^p(a,b)\|^2$, while the nonsmooth function approach makes direct use of $\phi_{\rm FB}^p(a,b)$. Unlike the nondifferentiability of $\phi_{\rm FB}^p$, the function $\phi_{\rm NR}^p$ with $p = 2k+1$ is twice continuously differentiable, making it especially attractive for algorithmic purposes. This smoothness enables the direct application of classical methods, such as Newton's method, to solve nonlinear complementarity problems (NCPs).

Proposition 2.21 *Let $\phi_{\rm NR}^p$ be defined as in (2.38). Then, $\phi_{\rm NR}^p$ is an NCP function.*

Proof First, we note that for any fixed real number $\xi \geq 0$ and odd integer p, the equation $t^p - \xi^p = 0$ has exactly one real solution $t = \xi$ because the function t^p is strictly monotone. Thus, we observe that

$$\begin{aligned} & \phi_{\rm NR}^p(a,b) = 0 \\ \Longleftrightarrow\; & a^p - (a-b)_+^p = 0 \\ \Longleftrightarrow\; & a - (a-b)_+ = 0 \\ \Longleftrightarrow\; & \min\{a,b\} = 0 \\ \Longleftrightarrow\; & a,\ b \geq 0,\ ab = 0. \end{aligned}$$

This shows that $\phi_{\rm NR}^p$ is an NCP-function. □

For p being an even integer, $\phi_{\rm NR}^p$ is not an NCP function. A counterexample is given as below

$$\phi_{\rm NR}^2(-2,-4) = (-2)^2 - (-2+4)_+^2 = 0.$$

Moreover, the function $\phi_{\rm NR}^p$ is neither convex nor concave function. To see this, taking $p = 3$ and using the following argument verify the assertion:

$$-1 = \phi_{\rm NR}^3(-1,-1) > \frac{1}{2}\phi_{\rm NR}^3(-2,-1) + \frac{1}{2}\phi_{\rm NR}^3(0,-1) = \frac{-8}{2} + \frac{-1}{2} = -\frac{9}{2}.$$

Proposition 2.22 *Let $p > 1$ be a positive odd integer. Then, we have*

$$[(a-b)_+]^p = [(a-b)^p]_+, \tag{2.39}$$

and hence

$$\phi_{\rm NR}^p(a,b) = a^p - [(a-b)_+]^p = a^p - [(a-b)^p]_+.$$

Proof For any $\alpha \in \mathbb{R}$, we know that $[\alpha]_+ = \frac{1}{2}(\alpha + |\alpha|)$. In addition, looking at the coefficients of the binomial $(1+x)^p$, we have

$$\sum_{j=0,\,\text{even}}^{p} C(p,j) = \sum_{j=0,\,\text{odd}}^{p} C(p,j) = \frac{1}{2}\sum_{j=0}^{p} C(p,j) = \frac{2^p}{2} = 2^{p-1}.$$

These two facts lead to

$$
\begin{aligned}
&[(a-b)_+]^p \\
&= \frac{1}{2^p}\,(a-b+|a-b|)^p \\
&= \frac{1}{2^p}\left(\sum_{j=0}^{p} C(p,j)|a-b|^j(a-b)^{p-j}\right) \\
&= \frac{1}{2^p}\left(\sum_{j=0,\,\text{even}}^{p} C(p,j)|a-b|^j(a-b)^{p-j} + \sum_{j=0,\,\text{odd}}^{p} C(p,j)|a-b|^j(a-b)^{p-j}\right) \\
&= \frac{1}{2^p}\left(\sum_{j=0,\,\text{even}}^{p} C(p,j)(a-b)^p + \sum_{j=0,\,\text{odd}}^{p} C(p,j)|a-b|(a-b)^{p-1}\right) \\
&= \frac{1}{2^p}\left(2^{p-1}(a-b)^p + 2^{p-1}|a-b|(a-b)^{p-1}\right) \\
&= \frac{1}{2}\left((a-b)^p + |a-b|(a-b)^{p-1}\right) \\
&= [(a-b)^p]_+,
\end{aligned}
$$

where the last equality holds because p is a positive odd integer. Thus, the proof is complete. □

In Proposition 2.22, observe that the equality in (2.39) holds exclusively when p is a positive odd integer. For even values of p, the identity $[(a-b)_+]^p = [(a-b)^p]_+$ no longer holds. This highlights the necessity of restricting p to positive odd integers in the definition of ϕ^p_{NR}. We now present an alternative formulation of ϕ^p_{NR} and establish its twice differentiability. To proceed, we first introduce a technical lemma.

Lemma 2.6 *Let $u(t) := |t|^p$ and $v(t) := t^p|t|$ where $p > 1$. Then,*

(a) the function $u(\cdot)$ is differentiable with $u'(t) = p\,\text{sgn}(t)|t|^{p-1}$;
(b) the function $v(\cdot)$ is differentiable with $v'(t) = (p+1)t^{p-1}|t|$.

Proof The arguments are straightforward which are omitted here. □

Proposition 2.23 *Let $p = 2k+1$ where $k = 1, 2, 3 \ldots$. Then, we have*

(a) $\phi^p_{\text{NR}}(a,b) = a^{2k+1} - \frac{1}{2}\left((a-b)^{2k+1} + (a-b)^{2k}|a-b|\right)$;
(b) ϕ^p_{NR} is continuously differentiable with

$$
\begin{aligned}
&\nabla\phi^p_{\text{NR}}(a,b) \\
&= p\begin{bmatrix} a^{p-1} - (a-b)^{p-2}(a-b)_+ \\ (a-b)^{p-2}(a-b)_+ \end{bmatrix};
\end{aligned}
$$

(c) ϕ^p_{NR} is twice continuously differentiable with

$$
\begin{aligned}
&\nabla^2\phi^p_{\text{NR}}(a,b) \\
&= p(p-1)\begin{bmatrix} a^{p-2} - (a-b)^{p-3}(a-b)_+ & (a-b)^{p-3}(a-b)_+ \\ (a-b)^{p-3}(a-b)_+ & -(a-b)^{p-3}(a-b)_+ \end{bmatrix}.
\end{aligned}
$$

Proof (a) The alternative expression is a direct consequence of Proposition 2.22.

(b) From Lemma 2.6, we compute

$$\begin{aligned}
&\frac{\partial \phi^p_{\mathrm{NR}}}{\partial a}(a,b)\\
&= \frac{\partial}{\partial a}\left(a^{2k+1} - \frac{1}{2}((a-b)^{2k+1} + (a-b)^{2k}|a-b|\right)\\
&= (2k+1)a^{2k} - \frac{(2k+1)}{2}(a-b)^{2k} - \frac{(2k+1)}{2}(a-b)^{2k-1}|a-b|
\end{aligned}$$

and

$$\begin{aligned}
&\frac{\partial \phi^p_{\mathrm{NR}}}{\partial b}(a,b)\\
&= \frac{\partial}{\partial b}\left(a^{2k+1} - \frac{1}{2}((a-b)^{2k+1} + (a-b)^{2k}|a-b|\right)\\
&= \frac{(2k+1)}{2}(a-b)^{2k} + \frac{(2k+1)}{2}(a-b)^{2k-1}|a-b|.
\end{aligned}$$

Hence, we obtain

$$\begin{aligned}
&\nabla \phi^p_{\mathrm{NR}}(a,b)\\
&= \frac{2k+1}{2}\begin{bmatrix} 2a^{2k} - (a-b)^{2k} - (a-b)^{2k-1}|a-b| \\ (a-b)^{2k} + (a-b)^{2k-1}|a-b| \end{bmatrix}\\
&= \frac{2k+1}{2}\begin{bmatrix} 2a^{2k} - 2(a-b)^{2k-1}(a-b)_+ \\ 2(a-b)^{2k-1}(a-b)_+ \end{bmatrix}\\
&= p\begin{bmatrix} a^{p-1} - (a-b)^{p-2}(a-b)_+ \\ (a-b)^{p-2}(a-b)_+ \end{bmatrix},
\end{aligned}$$

which proves part(b).

(c) Similarly, with Lemma 2.6 again, the Hessian matrix can be calculated as below.

$$\begin{aligned}
&\nabla^2 \phi^p_{\mathrm{NR}}(a,b)\\
&= k(2k+1)\begin{bmatrix} 2a^{2k-1} - (a-b)^{2k-1} - (a-b)^{2k-2}|a-b| & (a-b)^{2k-1} + (a-b)^{2k-2}|a-b| \\ (a-b)^{2k-1} + (a-b)^{2k-2}|a-b| & -(a-b)^{2k-1} - (a-b)^{2k-2}|a-b| \end{bmatrix}\\
&= p(p-1)\begin{bmatrix} a^{p-2} - (a-b)^{p-3}(a-b)_+ & (a-b)^{p-3}(a-b)_+ \\ (a-b)^{p-3}(a-b)_+ & -(a-b)^{p-3}(a-b)_+ \end{bmatrix}.
\end{aligned}$$

With this, it is clear that ϕ^p_{NR} is twice continuously differentiable. □

Proposition 2.24 *Let* ϕ^p_{NR} *be defined as in (2.38) with* $p > 1$ *being a positive odd integer. Then, the following hold.*

(a) $\phi^p_{\text{NR}}(a,b) > 0 \iff a > 0,\ b > 0.$

(b) ϕ^p_{NR} *is positive homogeneous of degree p, i.e.,* $\phi^p_{\text{NR}}(\alpha w) = \alpha^p \phi^p_{\text{NR}}(w)$ *for all* $w \in \mathbb{R}^2$ *and* $\alpha \geq 0$.

(c) ϕ^p_{NR} *is locally Lipschitz continuous, but not (globally) Lipschitz continuous.*

(d) ϕ^p_{NR} *is not* α*-Hölder continuous for any* $\alpha \in (0,1]$*, that is, the Hölder coefficient*

$$[\phi^p_{\text{NR}}]_{\alpha,\mathbb{R}^2} := \sup_{w \neq w'} \frac{|\phi^p_{\text{NR}}(w) - \phi^p_{\text{NR}}(w')|}{\|w - w'\|^\alpha}$$

is infinite.

(e) $$\nabla_a \phi^p_{\text{NR}}(a,b) \cdot \nabla_b \phi^p_{\text{NR}}(a,b) \begin{cases} > 0 \text{ on } \{(a,b) \mid a > b > 0 \text{ or } a > b > 2a\}, \\ = 0 \text{ on } \{(a,b) \mid a \leq b \text{ or } a > b = 2a \text{ or } a > b = 0\}, \\ < 0 \text{ otherwise.} \end{cases}$$

(f) $\nabla_a \phi^p_{\text{NR}}(a,b) \cdot \nabla_b \phi^p_{\text{NR}}(a,b) = 0$ *provided that* $\phi^p_{\text{NR}}(a,b) = 0$.

Proof (a) This result has been mentioned in [34, Lemma 2.2].

(b) It is clear by definition of ϕ^p_{NR}.

(c) Since continuously differentiability implies locally Lipschitz continuity, it remains to show that ϕ^p_{NR} is not Lipschitz continuous. Consider the restriction of ϕ^p_{NR} on the line $L := \{(a,b) \mid a = b > 0\}$. Note that for any $a > 0$, $\phi^p_{\text{NR}}(a,a) = a^p$, it suffices to show that $f(t) := t^p$ is not Lipschitz continuous. Indeed, for any $M > 0$, choosing $t = \max\{1, M\}$ and $t' = t + 1$ yields

$$\begin{aligned} \frac{|f(t) - f(t')|}{|t - t'|} &= (t+1)^p - t^p \\ &= (t+1)^{p-1} + (t+1)^{p-2}t + \cdots + t^{p-1} \\ &> p \cdot t^{p-1} \\ &> M. \end{aligned}$$

Hence, it follows that f is not Lipschitz continuous.

(d) As in the proof of part(c), we again restrict ϕ^p_{NR} on L and choose the same t. Hence, we also have

$$\frac{|f(t) - f(t')|}{|t - t'|^\alpha} > M$$

for any positive number M, that is, ϕ^p_{NR} is not α-Hölder continuous.

(e) According to Proposition 2.23, we know that

$$\begin{aligned} \nabla_a \phi^p_{\text{NR}}(a,b) \cdot \nabla_b \phi^p_{\text{NR}}(a,b) &= p^2 \cdot \left(a^{p-1} - (a-b)^{p-2}(a-b)_+\right)\left((a-b)^{p-2}(a-b)_+\right) \\ &= \begin{cases} p^2 \cdot \left(a^{p-1} - (a-b)^{p-1}\right)(a-b)^{p-1} & \text{if } a > b, \\ 0 & \text{if } a \leq b. \end{cases} \end{aligned}$$

When $a > b$, it is clear that $p^2 > 0$ and $(a-b)^{p-1} > 0$. Thus, we only consider the term $a^{p-1} - (a-b)^{p-1}$. Note that $p - 1$ is even, which implies

$$a^{p-1} = (a-b)^{p-1} \iff |a| = a-b \iff b = 0 \text{ or } b = 2a.$$

In addition to the case $a \le b$, there are two subcases $a > b = 0$ and $a > b = 2a$ such that $\nabla_a \phi^p_{\rm NR}(a,b) \cdot \nabla_b \phi^p_{\rm NR}(a,b) = 0$. On the other hand, we have

$$a^{p-1} > (a-b)^{p-1} \iff |a| > a-b \iff b > 0 \text{ or } b > 2a.$$

All the above says $\nabla_a \phi^p_{\rm NR}(a,b) \cdot \nabla_b \phi^p_{\rm NR}(a,b)$ is positive only when $a > b > 0$ or $a > b > 2a$. For the remainder case, it is not hard to verify $\nabla_a \phi^p_{\rm NR}(a,b) \cdot \nabla_b \phi^p_{\rm NR}(a,b) < 0$.

(f) It is clear from part(e). □

Finally, we present several variants of $\phi^p_{\rm NR}$. Analogous to the functions introduced in [11], these variants can also be verified as NCP functions.

$$\begin{aligned}
\varphi_1(a,b) &= \phi^p_{\rm NR}(a,b) + \alpha(a)_+(b)_+, \ \alpha > 0.\\
\varphi_2(a,b) &= \phi^p_{\rm NR}(a,b) + \alpha\left((a)_+(b)_+\right)^2, \ \alpha > 0.\\
\varphi_3(a,b) &= \left(\phi^p_{\rm NR}(a,b)\right)^2 + \alpha\left((ab)_+\right)^4, \ \alpha > 0.\\
\varphi_4(a,b) &= \left(\phi^p_{\rm NR}(a,b)\right)^2 + \alpha\left((ab)_+\right)^2, \ \alpha > 0.
\end{aligned}$$

Lemma 2.7 *The value of $\phi^p_{\rm NR}(a,b)$ is positive only in the first quadrant, i.e., $\phi^p_{\rm NR}(a,b) > 0$ if and only if $a > 0,\ b > 0$.*

Proof Since p is odd, the function $f(t) = t^p$ is strictly increasing. This observation allows us to verify that

$$\begin{aligned}
& a > 0,\ b > 0\\
\iff & a + b > |a-b|\\
\iff & a > \frac{a-b+|a-b|}{2}\\
\iff & a > (a-b)_+\\
\iff & a^p > (a-b)_+^p\\
\iff & \phi^p_{\rm NR}(a,b) > 0,
\end{aligned}$$

which is the desired result. □

Proposition 2.25 *All the above functions φ_i, $i \in \{1,2,3,4\}$ are NCP functions.*

Proof We will only show that φ_1 is an NCP-function and the same argument can be applied to the other cases. Let $\Omega := \{(a,b) \mid a > 0,\ b > 0\}$ and suppose $\varphi_1(a,b) = 0$. If $(a,b) \in \Omega$, then $\phi^p_{\rm NR}(a,b) > 0$ by Lemma 2.7, and hence, $\varphi_1(a,b) > 0$. This is a contradiction. Therefore, there must have $(a,b) \in \Omega^c$ which says $(a)_+(b)_+ = 0$. This further implies $\phi^p_{\rm NR}(a,b) = 0$ which is equivalent to $a, b \ge 0,\ ab = 0$. Then, one direction is proved. The converse direction is straightforward. □

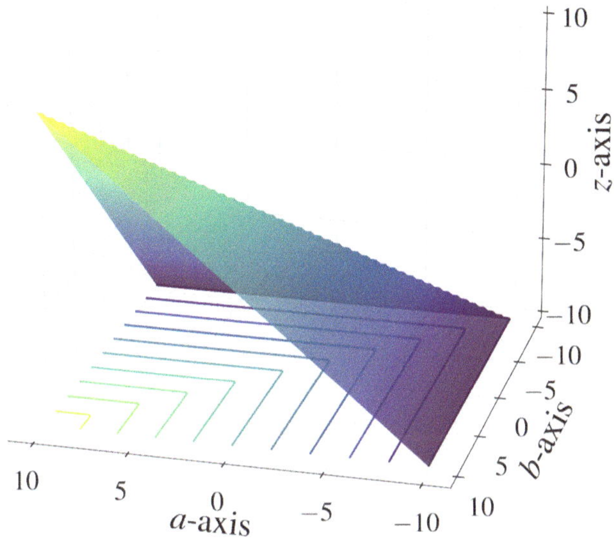

Fig. 2.15 The surface of $z = \phi^p_{\rm NR}(a, b)$ with $p = 1$ and $(a, b) \in [-10, 10] \times [-10, 10]$

We now illustrate the surfaces of $\phi^p_{\rm NR}$ for various values of p, providing further insight into this new family of NCP functions. Figure 2.15 shows the surface of $\phi_{\rm NR}(a, b)$, which is observed to be concave and increasing along the direction (t, t) in the first quadrant. In contrast, Fig. 2.16 displays the surface of $\phi^p_{\rm NR}(a, b)$, revealing that it is neither convex nor concave. Moreover, as noted in Lemma 2.7, $\phi^p_{\rm NR}(a, b)$ is positive only when both $a > 0$ and $b > 0$. The surfaces of $\phi^p_{\rm NR}$ for different values of p are depicted in Fig. 2.17.

2.2.2 *Construction by Symmetrizations*

In contrast to $\phi^p_{\rm FB}$ the function $\phi^p_{\rm NR}$ is derived through a "discrete generalization" and, quite remarkably, retains twice differentiability. This property allows for the direct application of various methods, such as Newton's method, to solve nonlinear complementarity problems (NCPs). However, unlike the graph of $\phi^p_{\rm FB}$, the graph of $\phi^p_{\rm NR}$ lacks symmetry, which may pose challenges in further analysis and in the development of solution algorithms. To address this, we aim to symmetrize $\phi^p_{\rm NR}$. Specifically, we propose two approaches to construct symmetric variants of this "generalized natural residual function", both of which continue to satisfy the conditions of NCP-functions. In doing so, we not only introduce new NCP functions and merit functions for the nonlinear complementarity problem, but also provide "symmetric counterparts" to the generalized Fischer-Burmeister function.

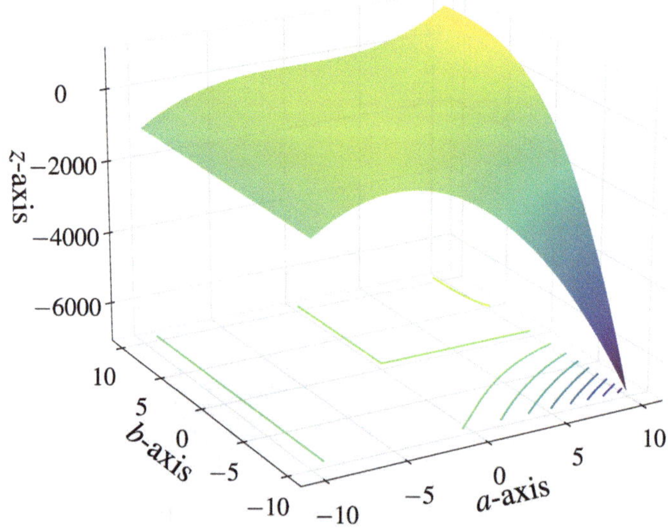

Fig. 2.16 The surface of $z = \phi^p_{\text{NR}}(a, b)$ with $p = 3$ and $(a, b) \in [-10, 10] \times [-10, 10]$

Next, we outline our approach to symmetrizing the "generalized natural residual function". The first step involves examining the graph of ϕ^p_{NR} as presented in [26]. To achieve symmetry in the graph of ϕ^p_{NR}, we consider the cases $a \geq b$ and $a \leq b$ separately. Motivated by the structure of ϕ^p_{NR}, we propose a first symmetrized version, denoted by $\phi^p_{\text{S-NR}} : \mathbb{R}^2 \to \mathbb{R}$, defined as follows:

$$\phi^p_{\text{S-NR}}(a, b) = \begin{cases} a^p - (a - b)^p & \text{if } a > b, \\ a^p = b^p & \text{if } a = b, \\ b^p - (b - a)^p & \text{if } a < b, \end{cases} \tag{2.40}$$

where $p > 1$ is a positive odd integer. As shown in Fig. 2.18, $\phi^p_{\text{S-NR}}$ is an NCP function whose graph is symmetric. However, $\phi^p_{\text{S-NR}}$ is not differentiable in general, prompting the natural question of whether a symmetric and differentiable variant can be constructed. To this end, we observe that the associated merit function $\|\phi^p_{\text{S-NR}}\|^2$ possesses the desired smoothness, although we seek a more direct and simpler differentiable formulation. Fortunately, we identify a second symmetrization of ϕ^p_{NR}, denoted by $\psi^p_{\text{S-NR}} : \mathbb{R}^2 \to \mathbb{R}_+$, defined as:

$$\psi^p_{\text{S-NR}}(a, b) = \begin{cases} a^p b^p - (a - b)^p b^p & \text{if } a > b, \\ a^p b^p = a^{2p} & \text{if } a = b, \\ a^p b^p - (b - a)^p a^p & \text{if } a < b, \end{cases} \tag{2.41}$$

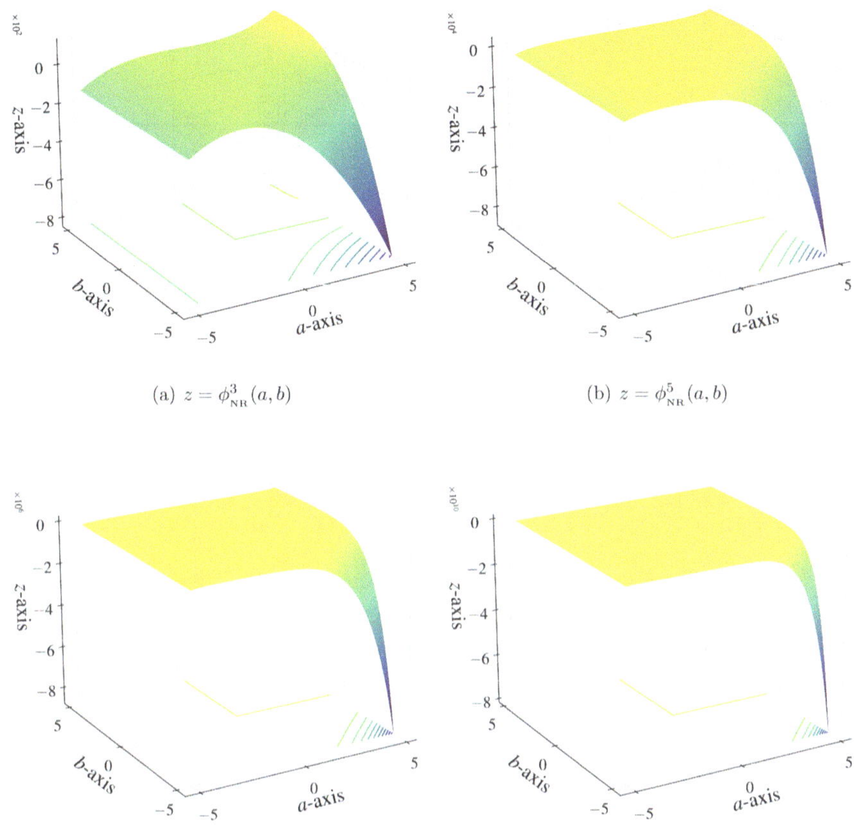

(a) $z = \phi^3_{\text{NR}}(a, b)$ (b) $z = \phi^5_{\text{NR}}(a, b)$

(c) $z = \phi^7_{\text{NR}}(a, b)$ (d) $z = \phi^{11}_{\text{NR}}(a, b)$

Fig. 2.17 The surface of $z = \phi^p_{\text{NR}}(a, b)$ with different values of p

where $p > 1$ is again a positive odd integer. The graph of $\psi^p_{\text{S-NR}}$ is shown in Fig. 2.19 and exhibits both symmetry and differentiability, fulfilling our desired properties for a well-behaved symmetrized NCP function.

Proposition 2.26 *Let $\phi^p_{\text{S-NR}}$ be defined in (2.40) with $p > 1$ being a positive odd integer. Then, $\phi^p_{\text{S-NR}}$ is an NCP function and is positive only on the first quadrant $\Omega = \{(a, b) \mid a > 0,\ b > 0\}$.*

Proof It is straightforward to verify that $\phi^p_{\text{S-NR}}$ is positive only in the first quadrant. Next, we demonstrate that $\phi^p_{\text{S-NR}}$ satisfies the properties of an NCP function. To this end, we proceed by analyzing three distinct cases. For $a > b$ and $\phi^p_{\text{S-NR}}(a, b) = 0$, it is clear to see $a^p - (a - b)^p = 0$, which implies that $a = a - b$. Thus, we conclude that $a > b = 0$. Similarly, for $a < b$ and $\phi^p_{\text{S-NR}}(a, b) = 0$, we have $0 = a < b$. For the third case, $a = b$ and $\phi^p_{\text{S-NR}}(a, b) = 0$, it is easy to see that $a = b = 0$. It is trivial

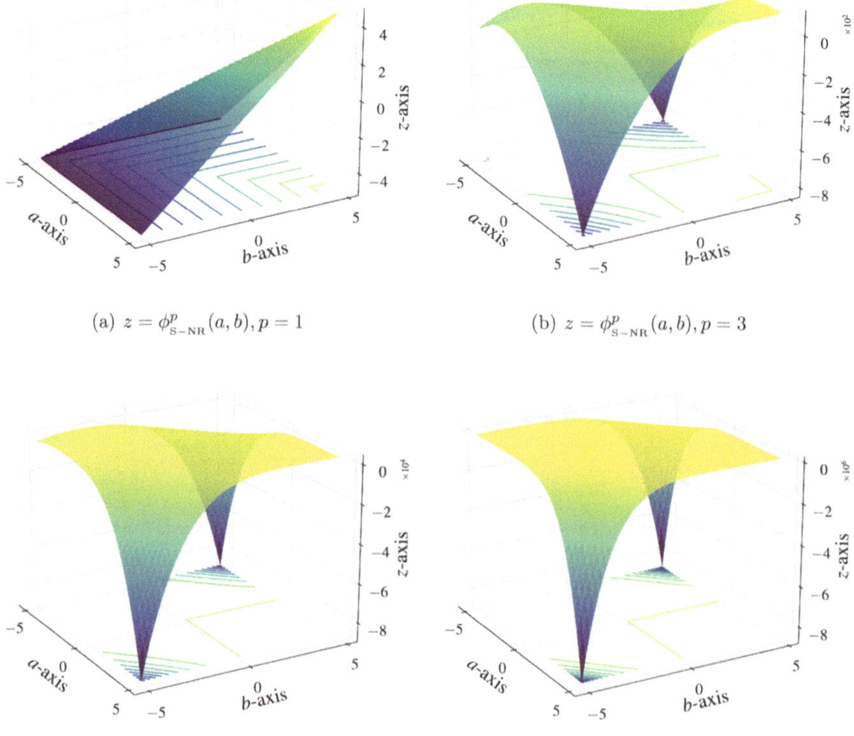

(a) $z = \phi^{p}_{_{\text{S-NR}}}(a, b), p = 1$ (b) $z = \phi^{p}_{_{\text{S-NR}}}(a, b), p = 3$

(c) $z = \phi^{p}_{_{\text{S-NR}}}(a, b), p = 5$ (d) $z = \phi^{p}_{_{\text{S-NR}}}(a, b), p = 7$

Fig. 2.18 The graphs of $\phi^{p}_{_{\text{S-NR}}}$ with different values of p

to check the converse way. In summary, $\phi^{p}_{_{\text{S-NR}}}$ satisfies that $\phi^{p}_{_{\text{S-NR}}}(a, b) = 0$ if and only if $a, b \geq 0$, $ab = 0$, and hence, it is an NCP function. □

For p being an even integer, $\phi^{p}_{_{\text{S-NR}}}$ is not an NCP function. A counterexample is given as below

$$\phi^{2}_{_{\text{S-NR}}}(-2, -4) = (-2)^2 - (-2 + 4)^2 = 0.$$

Moreover, the function $\phi^{p}_{_{\text{S-NR}}}$ is neither convex nor concave. To illustrate this, we set $p = 3$ and use the following arguments to verify the assertions:

$$1 = \phi^{3}_{_{\text{S-NR}}}(1, 1) < \frac{1}{2}\phi^{3}_{_{\text{S-NR}}}(0, 0) + \frac{1}{2}\phi^{3}_{_{\text{S-NR}}}(2, 2) = \frac{0}{2} + \frac{8}{2} = 4.$$

$$1 = \phi^{3}_{_{\text{S-NR}}}(1, 1) > \frac{1}{2}\phi^{3}_{_{\text{S-NR}}}(2, 0) + \frac{1}{2}\phi^{3}_{_{\text{S-NR}}}(0, 2) = \frac{0}{2} + \frac{0}{2} = 0.$$

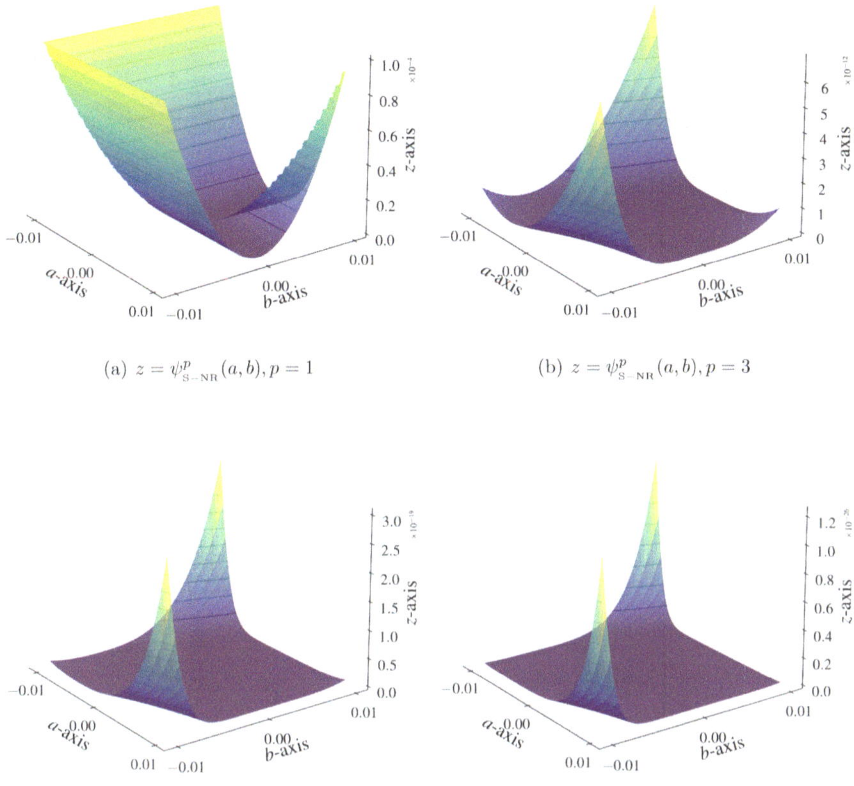

(a) $z = \psi^p_{\text{S-NR}}(a,b), p = 1$ (b) $z = \psi^p_{\text{S-NR}}(a,b), p = 3$

(c) $z = \psi^p_{\text{S-NR}}(a,b), p = 5$ (d) $z = \psi^p_{\text{S-NR}}(a,b), p = 7$

Fig. 2.19 The graphs of $\psi^p_{\text{S-NR}}$ with different values of p

Proposition 2.27 *Let $\phi^p_{\text{S-NR}}$ be defined in (2.40) with $p > 1$ being a positive odd integer. Then, the following hold.*

(a) An alternative expression of $\phi^p_{\text{S-NR}}$ is

$$\phi^p_{\text{S-NR}}(a,b) = \begin{cases} \phi^p_{\text{NR}}(a,b) & \text{if } a > b, \\ a^p = b^p & \text{if } a = b, \\ \phi^p_{\text{NR}}(b,a) & \text{if } a < b. \end{cases}$$

(b) The function $\phi^p_{\text{S-NR}}$ is not differentiable. However, $\phi^p_{\text{S-NR}}$ is continuously differentiable on the set $\Omega := \{(a,b) \mid a \neq b\}$ with

$$\nabla\phi^p_{\text{S-NR}}(a,b) = \begin{cases} p\,[\,a^{p-1} - (a-b)^{p-1},\ (a-b)^{p-1}\,]^T & \text{if } a > b, \\ p\,[\,(b-a)^{p-1},\ b^{p-1} - (b-a)^{p-1}\,]^T & \text{if } a < b. \end{cases}$$

In a more compact form,

$$\nabla \phi^p_{\text{S-NR}}(a,b) = \begin{cases} p\,[\,\phi^{p-1}_{\text{NR}}(a,b),\ (a-b)^{p-1}\,]^T & \text{if } a > b, \\ p\,[\,(b-a)^{p-1},\ \phi^{p-1}_{\text{NR}}(b,a)\,]^T & \text{if } a < b. \end{cases}$$

(c) *The function $\phi^p_{\text{S-NR}}$ is twice continuously differentiable on the set $\Omega = \{(a,b) \mid a \neq b\}$ with*

$$\nabla^2 \phi^p_{\text{S-NR}}(a,b) = \begin{cases} p(p-1)\begin{bmatrix} a^{p-2} - (a-b)^{p-2} & (a-b)^{p-2} \\ (a-b)^{p-2} & -(a-b)^{p-2} \end{bmatrix} & \text{if } a > b, \\ p(p-1)\begin{bmatrix} -(b-a)^{p-2} & (b-a)^{p-2} \\ (b-a)^{p-2} & b^{p-2} - (b-a)^{p-2} \end{bmatrix} & \text{if } a < b. \end{cases}$$

In a more compact form,

$$\nabla^2 \phi^p_{\text{S-NR}}(a,b) = \begin{cases} p(p-1)\begin{bmatrix} \phi^{p-2}_{\text{NR}}(a,b) & (a-b)^{p-2} \\ (a-b)^{p-2} & -(a-b)^{p-2} \end{bmatrix} & \text{if } a > b, \\ p(p-1)\begin{bmatrix} -(b-a)^{p-2} & (b-a)^{p-2} \\ (b-a)^{p-2} & \phi^{p-2}_{\text{NR}}(b,a) \end{bmatrix} & \text{if } a < b. \end{cases}$$

Proof The arguments are just direct computations, we omit them. □

Proposition 2.28 *Let $\phi^p_{\text{S-NR}}$ be defined as in (2.40) with $p > 1$ being a positive odd integer. Then, the following hold.*

(a) $\phi^p_{\text{S-NR}}(a,b) > 0 \iff a > 0,\ b > 0.$
(b) *$\phi^p_{\text{S-NR}}$ is positive homogeneous of degree p.*
(c) *$\phi^p_{\text{S-NR}}$ is not Lipschitz continuous.*
(d) *$\phi^p_{\text{S-NR}}$ is not α-Hölder continuous for any $\alpha \in (0,1]$.*
(e) $\nabla_a \phi^p_{\text{S-NR}}(a,b) \cdot \nabla_b \phi^p_{\text{S-NR}}(a,b) > 0$ *on* $\{(a,b) \mid a > b > 0\} \bigcup \{(a,b) \mid b > a > 0\}$.
(f) $\nabla_a \phi^p_{\text{S-NR}}(a,b) \cdot \nabla_b \phi^p_{\text{S-NR}}(a,b) = 0$ *provided that* $\phi^p_{\text{S-NR}}(a,b) = 0$ *and* $(a,b) \neq (0,0)$.

Proof (a) It is clear from Proposition 2.26 or [35, Proposition 2.1]).

(b) It follows from the definition of $\phi^p_{\text{S-NR}}$.

(c)–(d) The proof is similar to Proposition 2.24(c)–(d).

(e) It is enough to verify the case for $a > b > 0$ because for $b > a > 0$, the inequality will hold automatically due to $\phi^p_{\text{S-NR}}$ having a symmetric surface. To see this, according to Proposition 2.27(b), we have

$$\nabla_a \phi^p_{\text{S-NR}}(a,b) \cdot \nabla_b \phi^p_{\text{S-NR}}(a,b) = p^2 \cdot \left[a^{p-1} - (a-b)^{p-1}\right](a-b)^{p-1},$$

which yields the desired result by Proposition 2.24(e).

(f) This result also follows from the proof of Proposition 2.24(e). □

We now establish the semismoothness of $\phi^p_{\text{S-NR}}$. It is well known that any piecewise continuously differentiable function is semismooth. For completeness, we will verify this property step by step based on the definition. Through this process, we will not only confirm local Lipschitz continuity and characterize the generalized gradient, but also demonstrate that $\phi^p_{\text{S-NR}}$ is strongly semismooth. As a first step, we verify that the function is strictly continuous, that is, locally Lipschitz continuous. It is important to note, however, that $\phi^p_{\text{S-NR}}$ is not "globally" Lipschitz continuous, as established in Proposition 2.28(c).

Lemma 2.8 *Let $\phi^p_{\text{S-NR}}$ be defined as in (2.40) with $p > 1$ being a positive odd integer. Then, $\phi^p_{\text{S-NR}}$ is strictly continuous (locally Lipschitz continuous).*

Proof For any point $x = (a, b)$ with $a \neq b$, the continuous differentiability of $\phi^p_{\text{S-NR}}$ implies its locally Lipschitz continuity. It remains to show $\phi^p_{\text{S-NR}}$ is locally Lipschitz continuous on the line $L = \{(a, b) \mid a = b\}$.

To proceed the arguments, we present two inequalities that will be frequently used. Given any $x^0 = (a_0, a_0)$ and $\delta > 0$, let $N_\delta(x^0) := \{x \in \mathbb{R}^2 \mid \|x - x^0\| \leq \delta\}$. Then, for any $x = (x_1, x_2) \in N_\delta(x^0)$, we have two basic inequalities as follows:

$$|x_i| \leq \|x\| \leq \|x - x^0\| + \|x^0\| \leq \delta + \|x^0\| \quad \forall i = 1, 2. \tag{2.42}$$

$$|x_1 - x_2| \leq |x_1 - a_0| + |a_0 - x_2| \leq \|x - x^0\| + \|x^0 - x\| \leq 2\delta. \tag{2.43}$$

Now, for any $y, z \in N_\delta(x^0)$, we discuss four cases as below.

(i) For $y \in L$ and $z \in L$, we have

$$\begin{aligned}
\left|\phi^p_{\text{S-NR}}(y) - \phi^p_{\text{S-NR}}(z)\right| &= |y_1^p - z_1^p| \\
&= |y_1 - z_1| \cdot |y_1^{p-1} + y_1^{p-2} z_1 + \cdots + z_1^{p-1}| \\
&\leq \|y - z\| \cdot (|y_1|^{p-1} + |y_1|^{p-2} \cdot |z_1| + \cdots + |z_1|^{p-1}) \\
&\leq p(\delta + \|x^0\|)^{p-1} \|y - z\| \\
&= \kappa_1 \|y - z\|,
\end{aligned}$$

where $\kappa_1 := p(\delta + \|x^0\|)^{p-1}$ and the second inequality holds by (2.42).

(ii) For $y \notin L$ and $z \in L$ (or $y \in L$ and $z \notin L$), without loss of generality, we assume $y_1 > y_2$. Then, we have

$$\begin{aligned}
\left|\phi^p_{\text{S-NR}}(y) - \phi^p_{\text{S-NR}}(z)\right| &= |y_1^p - (y_1 - y_2)^p - z_1^p| \\
&\leq |y_1^p - z_1^p| + (y_1 - y_2)^p \\
&\leq \kappa_1 \|y - z\| + (y_1 - y_2)^{p-1} (|y_1 - z_1| + |z_1 - z_2| + |z_2 - y_2|) \\
&\leq \kappa_1 \|y - z\| + (2\delta)^{p-1} (\|y - z\| + \|z - y\|) \\
&= \kappa_2 \|y - z\|,
\end{aligned}$$

where $\kappa_2 := \kappa_1 + 2(2\delta)^{p-1}$ and the last inequality holds by (2.43).

(iii) For $y \notin L, z \notin L$ and y, z lie on the opposite side of L, i.e., $(y_1 - y_2)(z_1 - z_2) < 0$, without loss of generality, we assume $y_1 > y_2$ and $z_1 < z_2$. Since y, z lie on the opposite side of L, the line L and the segment $[y, z] := \{\lambda y + (1-\lambda)z \mid \lambda \in [0, 1]\}$ must intersect at a point $w \in [y, z] \cap L$. Thus, we have

$$\begin{aligned}
\left|\phi^p_{\text{S-NR}}(y) - \phi^p_{\text{S-NR}}(z)\right| &\le |\phi^p_{\text{S-NR}}(y) - \phi^p_{\text{S-NR}}(w)| + |\phi^p_{\text{S-NR}}(w) - \phi^p_{\text{S-NR}}(z)| \\
&\le \kappa_2\|y - w\| + \kappa_2\|w - z\| \\
&\le \kappa_2\|y - z\| + \kappa_2\|y - z\| \\
&= \kappa_3\|y - z\|,
\end{aligned}$$

where $\kappa_3 := 2\kappa_2$ and the third inequality holds because $w \in [y, z]$.

(iv) For $y \notin L, z \notin L$ and y, z lie on the same side of L, i.e., $(y_1 - y_2)(z_1 - z_2) > 0$, without loss of generality, we assume $y_1 > y_2$ and $z_1 > z_2$. Then, we have

$$\begin{aligned}
\left|\phi^p_{\text{S-NR}}(y) - \phi^p_{\text{S-NR}}(z)\right| &= |(y_1^p - (y_1 - y_2)^p) - (z_1^p - (z_1 - z_2)^p)| \\
&\le |y_1^p - z_1^p| + |(y_1 - y_2)^p - (z_1 - z_2)^p| \\
&\le \kappa_1\|y - z\| + 2p(2\delta)^{p-1}\|y - z\| \\
&= \kappa_4\|y - z\|
\end{aligned}$$

where $\kappa_4 = \kappa_1 + 2p(2\delta)^{p-1}$ and the second part is estimated as follows:

$$\begin{aligned}
&|(y_1 - y_2)^p - (z_1 - z_2)^p| \\
&= |(y_1 - y_2) - (z_1 - z_2)| \cdot |(y_1 - y_2)^{p-1} + \cdots + (z_1 - z_2)^{p-1}| \\
&\le (|y_1 - z_1| + |y_2 - z_2|)(|y_1 - y_2|^{p-1} + \cdots + |z_1 - z_2|^{p-1}) \\
&\le (\|y - z\| + \|y - z\|)p(2\delta)^{p-1} \\
&= 2p(2\delta)^{p-1}\|y - z\|.
\end{aligned}$$

From all the above, by choosing $\kappa = \max\{\kappa_1, \kappa_2, \kappa_3, \kappa_4\}$, we conclude that

$$\left|\phi^p_{\text{S-NR}}(y) - \phi^p_{\text{S-NR}}(z)\right| \le \kappa\|y - z\| \quad \text{for any } \; y, z \in N_\delta(x^0).$$

This means that $\phi^p_{\text{S-NR}}$ is locally Lipschitz continuous at x^0. Then, the proof is complete. □

Proposition 2.29 *Let $\phi^p_{\text{S-NR}}$ be defined as in (2.40) with $p > 1$ being a positive odd integer. Then, the generalized gradient of $\phi^p_{\text{S-NR}}$ is given by*

$$\partial\phi^p_{\text{S-NR}}(a, b) = \begin{cases} p\,[a^{p-1} - (a-b)^{p-1},\ (a-b)^{p-1}]^T & \text{if } a > b, \\ \{p\,[\alpha a^{p-1}, (1-\alpha)a^{p-1}]^T \mid \alpha \in [0, 1]\} & \text{if } a = b, \\ p\,[(b-a)^{p-1},\ b^{p-1} - (b-a)^{p-1}]^T & \text{if } a < b. \end{cases}$$

Proof We have already seen the $\partial\phi^p_{\rm S-NR}(a,b)$ when $a \neq b$ in [36]. For $a = b$, according to the definition of Clarke's generalized gradient, we claim that

$$\partial\phi^p_{\rm S-NR}(a,a) = \text{conv}\left\{ \lim_{(a_i,b_i)\to(a,a)} \nabla\phi^p_{\rm S-NR}(a_i,b_i) \mid \phi^p_{\rm S-NR} \text{ is differentiable at } (a_i,b_i) \in \mathbb{R}^2 \right\}.$$

To see this, we discuss three cases as below.

(i) If $a_i > b_i$, for any $i \geq n$ and sufficiently large n, then

$$\lim_{(a_i,b_i)\to(a,a)} \nabla\phi^p_{\rm S-NR}(a_i,b_i) = \lim_{(a_i,b_i)\to(a,a)} p \begin{bmatrix} a_i^{p-1} - (a_i - b_i)^{p-1} \\ (a_i - b_i)^{p-1} \end{bmatrix} = p \begin{bmatrix} a^{p-1} \\ 0 \end{bmatrix}.$$

(ii) If $a_i < b_i$, for any $i \geq n$ and sufficiently large n, then

$$\lim_{(a_i,b_i)\to(a,a)} \nabla\phi^p_{\rm S-NR}(a_i,b_i) = \lim_{(a_i,b_i)\to(a,a)} p \begin{bmatrix} (b_i - a_i)^{p-1} \\ b_i^{p-1} - (b_i - a_i)^{p-1} \end{bmatrix} = p \begin{bmatrix} 0 \\ a^{p-1} \end{bmatrix}.$$

(iii) For the remainder case, $\nabla\phi^p_{\rm S-NR}(a_i,b_i)$ has no limit as $(a_i,b_i) \to (a,a)$.

From all the above, we conclude that

$$\partial\phi^p_{\rm S-NR}(a,a) = \text{conv}\left\{ p \begin{bmatrix} a^{p-1} \\ 0 \end{bmatrix}, p \begin{bmatrix} 0 \\ a^{p-1} \end{bmatrix} \right\} = \left\{ p \begin{bmatrix} \alpha a^{p-1} \\ (1-\alpha)a^{p-1} \end{bmatrix} \middle| \alpha \in [0,1] \right\}.$$

Thus, the desired result follows. □

Lemma 2.9 *Let $\phi^p_{\rm S-NR}$ be defined as in (2.40) with $p > 1$ being a positive odd integer. Then, $\phi^p_{\rm S-NR}$ is a directional differentiable function.*

Proof For any point $x = (a,b)$ with $a \neq b$, the continuous differentiability of $\phi^p_{\rm S-NR}$ ensures its directional differentiability. Therefore, it remains verify directional differentiability along the line $L = \{(a,b) \mid a = b\}$. To this end, consider an arbitrary point $x = (a,a)$, a direction $h = (h_1,h_2)$, and $t > 0$. We proceed by examining the following three cases:

(i) If $h_1 = h_2$, then

$$\begin{aligned}
& \lim_{t\to 0^+} \frac{\phi^p_{\rm S-NR}(x+th) - \phi^p_{\rm S-NR}(x)}{t} \\
= & \lim_{t\to 0^+} \frac{(a+th_1)^p - a^p}{t} \\
= & \lim_{t\to 0^+} \frac{a^p + pa^{p-1}th_1 + \sum_{k=2}^p \binom{p}{k} a^{p-k} t^k h_1^k - a^p}{t} \\
= & \lim_{t\to 0^+} \left(pa^{p-1}h_1 + \sum_{k=2}^p \binom{p}{k} a^{p-k} t^{k-1} h_1^k \right) \\
= & pa^{p-1}h_1.
\end{aligned}$$

(ii) If $h_1 > h_2$, then

$$\begin{aligned}
&\lim_{t\to 0^+} \frac{\phi^p_{\text{S-NR}}(x+th) - \phi^p_{\text{S-NR}}(x)}{t} \\
&= \lim_{t\to 0^+} \frac{(a+th_1)^p - (th_1 - th_2)^p - a^p}{t} \\
&= \lim_{t\to 0^+} \frac{a^p + pa^{p-1}th_1 + \sum_{k=2}^{p} \binom{p}{k} a^{p-k} t^k h_1^k - t^p(h_1-h_2)^p - a^p}{t} \\
&= \lim_{t\to 0^+} \left(pa^{p-1}h_1 + \sum_{k=2}^{p} \binom{p}{k} a^{p-k} t^{k-1} h_1^k - t^{p-1}(h_1-h_2)^p \right) \\
&= pa^{p-1}h_1.
\end{aligned}$$

(iii) If $h_1 < h_2$, then

$$\begin{aligned}
&\lim_{t\to 0^+} \frac{\phi^p_{\text{S-NR}}(x+th) - \phi^p_{\text{S-NR}}(x)}{t} \\
&= \lim_{t\to 0^+} \frac{(a+th_2)^p - (th_2 - th_1)^p - a^p}{t} \\
&= \lim_{t\to 0^+} \frac{a^p + pa^{p-1}th_2 + \sum_{k=2}^{p} \binom{p}{k} a^{p-k} t^k h_2^k - t^p(h_2-h_1)^p - a^p}{t} \\
&= \lim_{t\to 0^+} \left(pa^{p-1}h_2 + \sum_{k=2}^{p} \binom{p}{k} a^{p-k} t^{k-1} h_2^k - t^{p-1}(h_2-h_1)^p \right) \\
&= pa^{p-1}h_2.
\end{aligned}$$

To sum up, we have verified the definition of directional differentiability for $\phi^p_{\text{S-NR}}$. This completes the proof. □

Proposition 2.30 *Let $\phi^p_{\text{S-NR}}$ be defined as in (2.40) with $p > 1$ being a positive odd integer. Then, $\phi^p_{\text{S-NR}}$ is a semismooth function. Moreover, $\phi^p_{\text{S-NR}}$ is strongly semismooth.*

Proof We shall proceed to directly establish the strong semismoothness of $\phi^p_{\text{S-NR}}$. Observe that $\phi^p_{\text{S-NR}}$ is twice continuously differentiable at any point $x = (a, b)$ with $a \neq b$, which immediately implies its strong semismoothness at such points. It thus remains to verify that $\phi^p_{\text{S-NR}}$ is strongly semismooth along the line $L = \{(a, b) \mid a = b\}$.

For any $x = (a, a)$, $h = (h_1, h_2)$, $V \in \partial\phi^p_{\text{S-NR}}(x + h)$ and $h \to 0$, we have the following inequality while $\|h\| \leq 1$:

$$\|h\|^p \leq \|h\|^2 \quad \text{for any } p \geq 2.$$

To prove the strong semismoothness of $\phi^p_{\text{S-NR}}$, we will apply this inequality and verify (1.42) by discussing three cases as below.

(i) If $h_1 = h_2$, then for any $\alpha \in [0, 1]$

$$\begin{aligned}
&\left|\phi^p_{\text{S-NR}}(x+h) - \phi^p_{\text{S-NR}}(x) - Vh\right| \\
=&\left|(a+h_1)^p - a^p - p\left[\alpha a^{p-1}, (1-\alpha)a^{p-1}\right]\begin{bmatrix} h_1 \\ h_1 \end{bmatrix}\right| \\
=&\left|a^p + pa^{p-1}h_1 + \sum_{k=2}^{p}\binom{p}{k}a^{p-k}h_1^k - a^p - pa^{p-1}h_1\right| \\
\leq& M_1(|h_1|^2 + \cdots + |h_1|^p) \\
\leq& M_1(\|h\|^2 + \cdots + \|h\|^p) \\
\leq& (p-1)M_1\|h\|^2,
\end{aligned}$$

where $M_1 = \max\left\{\binom{p}{k}|a|^{p-k} \mid k = 2, 3, \ldots, p\right\}$ and the last inequality holds when $\|h\| \leq 1$.

(ii) If $h_1 > h_2$, then

$$\begin{aligned}
&\left|\phi^p_{\text{S-NR}}(x+h) - \phi^p_{\text{S-NR}}(x) - Vh\right| \\
=&\left|(a+h_1)^p - (h_1-h_2)^p - a^p - p\left[(a+h_1)^{p-1} - (h_1-h_2)^{p-1}, (h_1-h_2)^{p-1}\right]\begin{bmatrix} h_1 \\ h_2 \end{bmatrix}\right| \\
=&\left|(a+h_1)^p - (h_1-h_2)^p - a^p - p(a+h_1)^{p-1}h_1 + p(h_1-h_2)^p\right| \\
=&\left|(a+h_1)^p - a^p - p\left(a^{p-1} + \sum_{k=1}^{p-1}\binom{p-1}{k}a^{p-1-k}h_1^k\right)h_1 + (p-1)(h_1-h_2)^p\right| \\
\leq&\underbrace{\left|(a+h_1)^p - a^p - pa^{p-1}h_1\right|}_{\Xi_1} + p\underbrace{\left|\sum_{k=1}^{p-1}\binom{p-1}{k}a^{p-1-k}h_1^{k+1}\right|}_{\Xi_2} + (p-1)\underbrace{\left|(h_1-h_2)^p\right|}_{\Xi_3}.
\end{aligned}$$

As $h \to 0$, we have the following estimations for each Ξ_i.

- $\Xi_1 \leq (p-1)M_1\|h\|^2$ by case (i).
- $\Xi_2 \leq \sum_{k=1}^{p-1}\binom{p-1}{k}|a|^{p-1-k}|h_1|^{k+1} \leq M_2(|h_1|^2 + \cdots + |h_1|^p) \leq (p-1)M_2\|h\|^2$, where $M_2 = \max\left\{\binom{p-1}{k}|a|^{p-1-k} \mid k = 1, 2, \ldots, p-1\right\}$.
- $\Xi_3 \leq \sum_{k=0}^{p}\binom{p}{k}|h_1|^{p-k}|h_2|^k \leq M_3(\|h\|^p + \cdots + \|h\|^p) \leq (p+1)M_3\|h\|^2$, where $M_3 = \max\left\{\binom{p}{k} \mid k = 0, 1, \ldots, p\right\}$.

Hence, we conclude that

$$\left|\phi^p_{\text{S-NR}}(x+h) - \phi^p_{\text{S-NR}}(x) - Vh\right| \leq M\|h\|^2,$$

where $M = (p-1)M_1 + p(p-1)M_2 + (p-1)(p+1)M_3$.

(iii) If $h_1 < h_2$, the argument is similar to the case (ii).

All the above together with Lemmas 2.8–2.9 prove that $\phi^p_{\text{S-NR}}$ is strongly semismooth. □

We now introduce additional variants of $\phi^p_{\text{S-NR}}$. As with the functions discussed in [11], the following variants can also be verified to satisfy the properties of NCP functions.

$$\widetilde{\phi}_1(a,b) = \phi^p_{\text{S-NR}}(a,b) + \alpha(a)_+(b)_+, \ \alpha > 0.$$
$$\widetilde{\phi}_2(a,b) = \phi^p_{\text{S-NR}}(a,b) + \alpha\left((a)_+(b)_+\right)^2, \ \alpha > 0.$$
$$\widetilde{\phi}_3(a,b) = \phi^p_{\text{S-NR}}(a,b) + \alpha\left((ab)_+\right)^4, \ \alpha > 0.$$
$$\widetilde{\phi}_4(a,b) = \phi^p_{\text{S-NR}}(a,b) + \alpha\left((ab)_+\right)^2, \ \alpha > 0.$$
$$\widetilde{\phi}_5(a,b) = \phi^p_{\text{S-NR}}(a,b) + \alpha\left((a)_+)^2((b)_+\right)^2, \ \alpha > 0.$$

Proposition 2.31 *All the above functions $\widetilde{\phi}_i(a,b)$ for $i \in \{1,2,3,4,5\}$ are NCP functions.*

Proof We only show that $\widetilde{\phi}_1(a,b)$ is an NCP function and the same argument can be applied to the other cases. First, we denote $\Omega := \{(a,b)\,|\,a > 0, b > 0\}$ the first quadrant and suppose that $\widetilde{\phi}_1(a,b) = 0$. If $(a,b) \in \Omega$, then $\phi^p_{\text{S-NR}}(a,b) > 0$ by Proposition 2.26, and hence, $\widetilde{\phi}_1(a,b) > 0$. This is a contradiction. Therefore, we must have $(a,b) \in \Omega^c$ which says $(a)_+(b)_+ = 0$. This further implies $\phi^p_{\text{S-NR}}(a,b) = 0$ which is equivalent to $a, b \geq 0,\ ab = 0$ by applying Proposition 2.26 again. Thus, $\widetilde{\phi}_1$ is an NCP function. □

Proposition 2.32 *Let $\psi^p_{\text{S-NR}}$ be defined in (2.41) with $p > 1$ being a positive odd integer. Then, $\psi^p_{\text{S-NR}}$ is an NCP function and is positive on the set*

$$\Omega = \{(a,b)\,|\,ab \neq 0\} \cup \{(a,b)\,|\,a < b = 0\} \cup \{(a,b)\,|\,0 = a > b\}.$$

Proof First of all, when $a < b = 0$, we have $\psi^p_{\text{S-NR}}(a,b) = a^{2p} > 0$. Similarly, when $0 = a > b$, we have $\psi^p_{\text{S-NR}}(a,b) = b^{2p} > 0$. For $0 \neq a > b \neq 0$, suppose that $b > 0$. Then, $a > (a-b)$ which implies $a^p > (a-b)^p$ and $b^p > 0$, and hence $a^p b^p - (a-b)^p b^p > 0$. On the other hand, suppose that $b < 0$. Then, $a < (a-b)$ which implies $a^p < (a-b)^p$ and $b^p < 0$. Thus, we also have $a^p b^p - (a-b)^p b^p > 0$. For $a = b \neq 0$, it is clear that $a^p b^p = a^{2p} > 0$. For the remaining case: $0 \neq a < b \neq 0$, the proof is similar to the case of $0 \neq a > b \neq 0$. From all the above, we prove that $\psi^p_{\text{S-NR}}$ is positive on the set Ω.

Next, we go on showing that $\psi^p_{\text{S-NR}}$ is an NCP function. Suppose that $a > b$ and $a^p b^p - (a-b)^p b^p = [a^p - (a-b)^p]b^b = 0$. If $b = 0$, then we have $a > b = 0$. Otherwise, we have $a = (a-b)$ which also yields that $a > b = 0$. Similarly, the condition $a < b$ and $a^p b^p - (b-a)^p a^p = 0$ implies that $b > a = 0$. The remaining case $a = b$ and $a^p b^p = 0$ gives that $a = b = 0$. Thus, from all the above, $\psi^p_{\text{S-NR}}$ is an NCP function. □

From Proposition 2.32, we conclude that $\psi^p_{\rm S-NR}$ is a merit function, as it is positive on Ω and vanishes precisely on the set $\{(a,b) \mid a \geq b = 0\} \cup \{(a,b) \mid 0 = a \leq b\}$. A bit further discussion of the function $\psi^p_{\rm S-NR}$ is provided as follows:

(i) For p being an even integer, $\psi^p_{\rm S-NR}$ is not an NCP function. A counterexample is given as below.

$$\psi^2_{\rm S-NR}(-2,-4) = (-2)^2(-4)^2 - (-2+4)^2(-4)^2 = 0.$$

(ii) The function $\psi^p_{\rm S-NR}$ is neither convex nor concave function. To see this, taking $p = 3$ and using the following argument verify the assertion.

$$1 = \psi^3_{\rm S-NR}(1,1) < \frac{1}{2}\psi^3_{\rm S-NR}(0,0) + \frac{1}{2}\psi^3_{\rm S-NR}(2,2) = \frac{0}{2} + \frac{64}{2} = 32.$$

$$1 = \psi^3_{\rm S-NR}(1,1) > \frac{1}{2}\psi^3_{\rm S-NR}(2,0) + \frac{1}{2}\psi^3_{\rm S-NR}(0,2) = \frac{0}{2} + \frac{0}{2} = 0.$$

Proposition 2.33 *Let $\psi^p_{\rm S-NR}$ be defined as in (2.41) with $p > 1$ being a positive odd integer. Then, the following hold.*

(a) An alternative expression of $\phi^p_{\rm S-NR}$ is

$$\psi^p_{\rm S-NR}(a,b) = \begin{cases} \phi^p_{\rm NR}(a,b)b^p & \text{if } a > b, \\ a^p b^p = a^{2p} & \text{if } a = b, \\ \phi^p_{\rm NR}(b,a)a^p & \text{if } a < b. \end{cases}$$

(b) The function $\psi^p_{\rm S-NR}$ is continuously differentiable with

$$\nabla\psi^p_{\rm S-NR}(a,b) = \begin{cases} p\,[\,a^{p-1}b^p - (a-b)^{p-1}b^p,\ a^p b^{p-1} - (a-b)^p b^{p-1} + (a-b)^{p-1}b^p\,]^{\rm T} & \text{if } a > b, \\ p\,[\,a^{p-1}b^p, a^p b^{p-1}\,]^{\rm T} = pa^{2p-1}[1\ ,1\,]^{\rm T} & \text{if } a = b, \\ p\,[\,a^{p-1}b^p - (b-a)^p a^{p-1} + (b-a)^{p-1}a^p,\ a^p b^{p-1} - (b-a)^{p-1}a^p\,]^{\rm T} & \text{if } a < b. \end{cases}$$

In a more compact form,

$$\nabla\psi^p_{\rm S-NR}(a,b) = \begin{cases} p\,[\,\phi^{p-1}_{\rm NR}(a,b)b^p,\ \phi^p_{\rm NR}(a,b)b^{p-1} + (a-b)^{p-1}b^p\,]^{\rm T} & \text{if } a > b, \\ p\,[\,a^{2p-1}, a^{2p-1}\,]^{\rm T} & \text{if } a = b, \\ p\,[\,\phi^p_{\rm NR}(b,a)a^{p-1} + (b-a)^{p-1}a^p,\ \phi^{p-1}_{\rm NR}(b,a)a^p\,]^{\rm T} & \text{if } a < b. \end{cases}$$

(c) *The function $\psi^p_{\text{S-NR}}$ is twice continuously differentiable with*

$$\nabla^2\psi^p_{\text{S-NR}}(a,b) = \begin{cases} p\begin{bmatrix} (p-1)[a^{p-2}-(a-b)^{p-2}]b^p & \begin{matrix}(p-1)(a-b)^{p-2}b^p \\ +p[a^{p-1}-(a-b)^{p-1}]b^{p-1}\end{matrix} \\ \begin{matrix}(p-1)(a-b)^{p-2}b^p \\ +p[a^{p-1}-(a-b)^{p-1}]b^{p-1}\end{matrix} & \begin{matrix}(p-1)[a^p-(a-b)^p]b^{p-2} \\ +2p(a-b)^{p-1}b^{p-1} \\ -(p-1)(a-b)^{p-2}b^p\end{matrix} \end{bmatrix} & \text{if } a>b, \\ p\begin{bmatrix} (p-1)a^{p-2}b^p & pa^{p-1}b^{p-1} \\ pa^{p-1}b^{p-1} & (p-1)a^pb^{p-2} \end{bmatrix} & \text{if } a=b, \\ p\begin{bmatrix} \begin{matrix}(p-1)[b^p-(b-a)^p]a^{p-2} \\ +2p(b-a)^{p-1}a^{p-1} \\ -(p-1)(b-a)^{p-2}a^p\end{matrix} & \begin{matrix}(p-1)(b-a)^{p-2}a^p \\ +p[b^{p-1}-(b-a)^{p-1}]a^{p-1}\end{matrix} \\ \begin{matrix}(p-1)(b-a)^{p-2}a^p \\ +p[b^{p-1}-(b-a)^{p-1}]a^{p-1}\end{matrix} & (p-1)[b^{p-2}-(b-a)^{p-2}]a^p \end{bmatrix} & \text{if } a<b. \end{cases}$$

Proof (a) It is clear to see this part.

(b) It is easy to verify the continuous differentiability of $\psi^p_{\text{S-NR}}(a,b)$ on the set $\{(a,b)\,|\,a>b \text{ or } a<b\}$. We only need to check the differentiability along the line $a=b$. Suppose that $h>k$, we observe that

$$\begin{aligned} &\psi^p_{\text{S-NR}}(a+h,a+k)-\psi^p_{\text{S-NR}}(a,a) \\ &= (a+h)^p(a+k)^p-(h-k)^pb^p-a^{2p} \\ &= \langle pa^{2p-1}(1,1),(h,k)\rangle + R(a,h,k). \end{aligned}$$

Here the remainder $R(a,h,k)$ is $o(h,k)$ function of h and k, since the degree of h and k of $R(a,h,k)$ is at least 2. Similarly, from the other two cases $h=k$ and $h<k$, we can conclude that $\nabla\psi^p_{\text{S-NR}}(a,a)=pa^{2p-1}(1,1)^{\mathsf{T}}$. In addition, the continuity of $\nabla\psi^p_{\text{S-NR}}(a,b)$ along the line $a=b$ is easy to verify.

(c) The arguments for this part are similar to those for part(b). We omit them. □

Likewise, we present some other variants of $\psi^p_{\text{S-NR}}$. Indeed, analogous to those functions in [11], the variants of $\psi^p_{\text{S-NR}}$ as below can be verified being NCP functions.

$$\begin{aligned} \widetilde{\psi}_1(a,b) &= \psi^p_{\text{S-NR}}(a,b)+\alpha(a)_+(b)_+,\ \alpha>0. \\ \widetilde{\psi}_2(a,b) &= \psi^p_{\text{S-NR}}(a,b)+\alpha\left((a)_+(b)_+\right)^2,\ \alpha>0. \\ \widetilde{\psi}_3(a,b) &= \psi^p_{\text{S-NR}}(a,b)+\alpha\left((ab)_+\right)^4,\ \alpha>0. \\ \widetilde{\psi}_4(a,b) &= \psi^p_{\text{S-NR}}(a,b)+\alpha\left((ab)_+\right)^2,\ \alpha>0. \\ \widetilde{\psi}_5(a,b) &= \psi^p_{\text{S-NR}}(a,b)+\alpha\left(((a)_+)^2((b)_+)\right)^2,\ \alpha>0. \end{aligned}$$

Proposition 2.34 *All the above functions $\widetilde{\psi}_i(a,b)$ for $i\in\{1,2,3,4,5\}$ are NCP functions.*

Proof We only show that $\widetilde{\psi}_1$ is a NCP-function and the same argument can be applied to the other cases. Let $\Omega := \{(a,b) \mid ab \neq 0\}$ and suppose that $\widetilde{\psi}_1(a,b) = 0$. If $(a,b) \in \Omega$, then $\psi^p_{\text{S-NR}}(a,b) > 0$ by Proposition 2.32; and hence, $\widetilde{\psi}_1(a,b) > 0$. This is a contradiction. Therefore, we must have $(a,b) \in \Omega^c$ which says $(a)_+(b)_+ = 0$. This further implies $\psi^p_{\text{S-NR}}(a,b) = 0$ which is equivalent to $a, b \geq 0,\ ab = 0$ by applying Proposition 2.32 again. Thus, $\widetilde{\psi}_1$ is an NCP function. □

Based on Figs. 2.18 and 2.19, we offer several observations regarding the surfaces of $\phi^p_{\text{S-NR}}$ and $\psi^p_{\text{S-NR}}$, as well as their algebraic properties. First, it is evident that $\phi^p_{\text{S-NR}}(a,b) = \phi^p_{\text{S-NR}}(b,a)$ and $\psi^p_{\text{S-NR}}(a,b) = \psi^p_{\text{S-NR}}(b,a)$, indicating that both surfaces are symmetric with respect to the line $a = b$. Regarding their algebraic structure, it can be verified that

$$\psi^p_{\text{S-NR}}(a,b) = \big[\min(a,b)\big]^2 \quad \text{for} \quad p = 1.$$

To see this, for example if $a > b$, we check that $a^1b^1 - (a-b)^1b^1 = b^2 = \min(a,b)^2$. On the other hand, for large $p = 3, 5, 7, \ldots$, the function $\psi^p_{\text{S-NR}}$ does not coincide with $\big[\min(a,b)\big]^{2p}$. Nonetheless, when we restrict $\psi^p_{\text{S-NR}}(a,b)$ on the line $a = b$ and two axes $a = 0$ and $b = 0$, we really have that

$$\psi^p_{\text{S-NR}}(a,b) = \big[\min(a,b)\big]^{2p}.$$

In summary, $\psi^p_{\text{S-NR}}$ can be regarded as a merit function associated with the original natural residual NCP function $\phi_{\text{NR}}(a,b) = \min(a,b)$. Notably, $\psi^p_{\text{S-NR}}$ is twice continuously differentiable, making it well-suited for the development of various algorithmic frameworks that exploit smoothness properties. However, it is important to point out that $\psi^p_{\text{S-NR}}$ does not satisfy the condition:

$$\nabla_a \psi^p_{\text{S-NR}}(a,b) \cdot \nabla_b \psi^p_{\text{S-NR}}(a,b) \geq 0. \quad \text{(cf. Property 2.2(d) in [27])}$$

For instance, setting $p = 3$ and evaluating at $(a,b) = (0,-1)$, we obtain $\nabla_a \psi^3_{\text{S-NR}}(0,-1) = 3$ and $\nabla_b \psi^3_{\text{S-NR}}(0,-1) = -6$, leading to a negative product. This may present challenges when analyzing convergence rates in certain optimization algorithms. Visually, the surface of $\phi^p_{\text{S-NR}}$ resembles The graph of $\phi^p_{\text{S-NR}}$ is neither convex nor concave. On the other hand, the surface of $\psi^p_{\text{S-NR}}$ is smooth yet also lacks convexity and concavity.

Proposition 2.35 *Let $\psi^p_{\text{S-NR}}$ be defined as in (2.41) with $p > 1$ being a positive odd integer. Then, the following hold.*

(a) $\psi^p_{\text{S-NR}}(a,b) \geq 0$ *for all* $(a,b) \in \mathbb{R}^2$.
(b) $\psi^p_{\text{S-NR}}$ *is positive homogeneous of degree* $2p$.
(c) $\psi^p_{\text{S-NR}}$ *is locally Lipschitz continuous, but not Lipschitz continuous.*
(d) $\psi^p_{\text{S-NR}}$ *is not* α*-Hölder continuous for any* $\alpha \in (0,1]$.
(e) $\nabla_a \psi^p_{\text{S-NR}}(a,b) \cdot \nabla_b \psi^p_{\text{S-NR}}(a,b) > 0$ *on the first quadrant* $\mathbb{R}^2_{++}$.

(f) $\psi^p_{\text{S-NR}}(a,b) = 0 \iff \nabla\psi^p_{\text{S-NR}}(a,b) = 0$. *In particular, we have* $\nabla_a\psi^p_{\text{S-NR}}(a,b) \cdot \nabla_b\psi^p_{\text{S-NR}}(a,b) = 0$ *provided that* $\psi^p_{\text{S-NR}}(a,b) = 0$.

Proof (a) This inequality follows from Proposition 2.32 or [35, Proposition 3.1].

(b) It is clear by the definition of $\psi^p_{\text{S-NR}}$.

(c)–(d) The proof is similar to Proposition 2.24(c)–(d).

(e) For convenience, we denote $\Lambda := \nabla_a\psi^p_{\text{S-NR}}(a,b) \cdot \nabla_b\psi^p_{\text{S-NR}}(a,b)$. Then, we proceed the proof by discussing three cases. For $a > b > 0$, we have

$$\Lambda = p^2 b^{2p-1} \cdot \left(a^{p-1} - (a-b)^{p-1}\right)\left(a^p - (a-b)^p + (a-b)^{p-1}b\right).$$

Note that $a > a - b > 0$ and $b > 0$, therefore we prove $\Lambda > 0$. Similarly, when $b > a > 0$, we also have $\Lambda > 0$. For the third case $a = b > 0$, it is clear that $\Lambda = p^2 a^{4p-2} > 0$.

(f) Note that $\psi^p_{\text{S-NR}}$ is a NCP-function, i.e.,

$$\psi^p_{\text{S-NR}}(a,b) = 0 \iff a \geq 0,\ b \geq 0,\ ab = 0.$$

From Proposition 2.33(b), we know $\nabla\psi^p_{\text{S-NR}}(a,b) = 0$ either when $a \geq b = 0$ or $b \geq a = 0$. Conversely, we suppose $\nabla\psi^p_{\text{S-NR}}(a,b) = 0$. For $a = b$,

$$\nabla\psi^p_{\text{S-NR}}(a,b) = pa^{2p-1}[1\,1]^{\mathsf{T}} = 0 \implies a = b = 0,$$

this proves that $\psi^p_{\text{S-NR}}(a,b) = 0$. For $a > b$, we know from Proposition 2.33(b) that

$$\begin{cases} a^{p-1}b^p - (a-b)^{p-1}b^p = 0, \\ a^p b^{p-1} - (a-b)^p b^{p-1} + (a-b)^{p-1}b^p = 0. \end{cases} \tag{2.44}$$

Note that it is clear to see that $b = 0$ satisfies the system (2.44). Assume $b \neq 0$, the system (2.44) becomes

$$a^{p-1} - (a-b)^{p-1} = 0, \tag{2.45}$$

$$a^p - (a-b)^p + (a-b)^{p-1}b = 0. \tag{2.46}$$

From (2.45), we obtain $(a-b)^{p-1} = a^{p-1}$. Then, substituting it into Eq. (2.46) yields

$$a^p - (a-b)a^{p-1} + a^{p-1}b = 0.$$

This implies $a^{p-1}b = 0$. Thus, we obtain $a = 0$. Again, by (2.45), we obtain $(-b)^{p-1} = 0$. This leads to a contradiction since we assume $b \neq 0$. Therefore, for $a > b$, we obtain that b must be zero, and hence $\psi^p_{\text{S-NR}}(a,b) = 0$. Similarly, when $a < b$, we also have $\psi^p_{\text{S-NR}}(a,b) = 0$. In summary, we conclude that $\psi^p_{\text{S-NR}}(a,b) = 0$ if and only if $\nabla\psi^p_{\text{S-NR}}(a,b) = 0$. □

Lastly, we investigate the growth behavior of ϕ^p_{NR}, $\phi^p_{\text{S-NR}}$, and $\psi^p_{\text{S-NR}}$ in Proposition 2.36. To this end, we first introduce a key lemma that will serve as the foundation for our analysis.

Lemma 2.10 *For any $x \in [0, 1]$ and any $k > 0$, we have*

$$(1-x)^k \leq \frac{1}{1+kx}.$$

Proof First, we define $f : [0, 1] \to \mathbb{R}$ by $f(x) = (1-x)^k(1+kx)$. A simple calculation yields $f'(x) = -k(k+1)x(1-x)^{k-1}$. Then, f monotonically decreases on $[0, 1]$ from $f(0) = 1$ to $f(1) = 0$. Consequently, $0 \leq f(x) \leq 1$, which completes the proof. □

Proposition 2.36 *Let $\{(a^k, b^k)\}_{k=1}^{\infty} \subseteq \mathbb{R}^2$ such that $|a^k| \to \infty$ and $|b^k| \to \infty$ as $k \to \infty$. Then, $|\phi^p_{\text{NR}}(a^k, b^k)| \to \infty$, $|\phi^p_{\text{S-NR}}(a^k, b^k)| \to \infty$, and $|\psi^p_{\text{S-NR}}(a^k, b^k)| \to \infty$.*

Proof (a) First, we verify that $|\phi^p_{\text{S-NR}}(a^k, b^k)| \to \infty$. To proceed, we consider three cases.

(i) Suppose $a^k \to \infty$ and $b^k \to \infty$. Note that for all $x \in [-1, 0]$ and $n \in \mathbb{N}$, there holds

$$(1+x)^n \leq (1-nx)^{-1}$$

which is due to Lemma 2.10. Thus, when $a > b > 0$, we have

$$\begin{aligned}
\phi^p_{\text{S-NR}}(a, b) =& a^p - (a-b)^p = a^p - a^p\left(1 - \frac{b}{a}\right)^p \\
\geq& a^p - a^p\left(1 - p\left(-\frac{b}{a}\right)\right)^{-1} \\
=& a^p - a^p\left(\frac{a}{a+pb}\right) \\
=& \frac{pa^p b}{a+pb} \\
\geq& \frac{pa^{p-1}b}{1+p} \\
\geq& \frac{pb^p}{1+p}.
\end{aligned}$$

Similarly, $\phi^p_{\text{S-NR}}(a, b) \geq \frac{pa^p}{1+p}$ for $b > a > 0$. Thus, $\phi^p_{\text{S-NR}}(a^k, b^k) \to \infty$ as $k \to \infty$.

(ii) Suppose $a^k \to -\infty$ and $b^k \to -\infty$. Observe that $\phi^p_{\text{S-NR}}(a, b) \leq a^p$ when $a > b$, and $\phi^p_{\text{S-NR}}(a, b) \leq b^p$ when $a < b$. Thus, $\phi^p_{\text{S-NR}}(a^k, b^k) \to -\infty$ as $k \to \infty$.

(iii) Suppose $a^k \to \infty$ and $b^k \to -\infty$. For $a > 0$ and $b < 0$, we have

$$(a-b)^p \geq a^p + (-b)^p = a^p - b^p.$$

Thus, $\phi^p_{\text{S-NR}}(a,b) = a^p - (a-b)^p \leq b^p$ and we conclude that $\phi^p_{\text{S-NR}}(a^k,b^k) \to -\infty$ as $k \to \infty$. In the case that $a^k \to -\infty$ and $b^k \to \infty$, we also have $\phi^p_{\text{S-NR}}(a^k,b^k) \to -\infty$ as $k \to \infty$ by symmetry of $\phi^p_{\text{S-NR}}$.

(b) Next, we show that $|\phi^p_{\text{NR}}(a^k,b^k)| \to \infty$. Again, we examine three cases.

(i) Suppose that $a^k \to -\infty$. Since $\phi^p_{\text{NR}}(a,b) = a^p - (a-b)^p_+ \leq a^p$ for all $(a,b) \in \mathbb{R}^2$, it is trivial to see that $\phi^p_{\text{NR}}(a^k,b^k) \to -\infty$.

(ii) Suppose that $a^k \to \infty$ and $b^k \to \infty$. For $a > b > 0$, then we have

$$\phi^p_{\text{NR}}(a,b) = \phi^p_{\text{S-NR}}(a,b) \geq \frac{pb^p}{1+p}.$$

For $0 \leq a < b$, it is clear that $\phi^p_{\text{NR}}(a,b) = a^p$. Then, we conclude that $\phi^p_{\text{NR}}(a^k,b^k) \to \infty$.

(iii) Suppose that $a^k \to \infty$ and $b^k \to -\infty$. For $a > 0$ and $b < 0$, we have

$$\phi^p_{\text{NR}}(a,b) = \phi^p_{\text{S-NR}}(a,b) \leq b^p$$

and so $\phi^p_{\text{NR}}(a^k,b^k) \to -\infty$. Thus, we have proved that $|\phi^p_{\text{NR}}(a^k,b^k)| \to \infty$.

(c) The last limit, $|\psi^p_{\text{S-NR}}(a^k,b^k)| \to \infty$, follows from the fact that

$$\psi^p_{\text{S-NR}}(a,b) = \begin{cases} \phi^p_{\text{S-NR}}(a,b)b^p & \text{if } a > b, \\ a^p b^p = a^{2p} & \text{if } a = b, \\ \phi^p_{\text{S-NR}}(b,a)a^p & \text{if } a < b. \end{cases}$$

and the inequalities obtained above for $\phi^p_{\text{S-NR}}$. □

2.2.3 *Construction by Continuous Generalization*

From a numerical perspective, one may naturally ask whether a "continuous generalization" of the natural residual (NR) function ϕ_{NR} exists, namely, a generalization parametrized by p taking values over a continuous interval. In this section, we propose such a continuous-type generalization of the NR function. The proposed function, while lacking a symmetric surface, admits two symmetrized forms that also depend continuously on the parameter p. Our generalization is defined as follows:

$$\widetilde{\phi}^p_{\text{NR}}(a,b) = \text{sgn}(a)|a|^p - [(a-b)_+]^p, \tag{2.47}$$

where $p \in (0,\infty)$, and the sign function is defined by

$$\mathrm{sgn}(t) := \begin{cases} 1, \text{ if } t > 0, \\ 0, \text{ if } t = 0, \\ -1, \text{ if } t < 0. \end{cases}$$

It is not hard to verify that $\widetilde{\phi}^p_{\rm NR}$ is an NCP function. Indeed, observe that

$$\widetilde{\phi}^p_{\rm NR}(a,b) = f(a) - f((a-b)_+),$$

where $f(t) = \mathrm{sgn}(t)|t|^p$, a bijective function. Consequently,

$$\widetilde{\phi}^p_{\rm NR}(a,b) = 0 \Longleftrightarrow f(a) = f((a-b)_+) \Longleftrightarrow a = (a-b)_+ \Longleftrightarrow \phi_{\rm NR}(a,b) = 0.$$

Notably, when p is an odd integer, $\widetilde{\phi}^p_{\rm NR}$ coincides with $\phi^p_{\rm NR}$, meaning this continuous generalization subsumes the discrete-type extension defined in (2.38). We further remark that this type of transformation, applying a monotonic bijective function f to $\phi_{\rm NR}$, can be applied to any NCP function of the form $\phi = \phi_1 - \phi_2$, a fact also noted in [4].

It is clear to see that the function (2.47) does not have a symmetric surface. Employing the same strategy as in [35], we propose two symmetrizations of $\widetilde{\phi}^p_{\rm NR}$ as

$$\widetilde{\phi}^p_{\rm S-NR}(a,b) = \begin{cases} \mathrm{sgn}(a)|a|^p - (a-b)^p \text{ if } a \geq b, \\ \mathrm{sgn}(b)|b|^p - (b-a)^p \text{ if } a < b, \end{cases} \tag{2.48}$$

and

$$\widetilde{\psi}^p_{\rm S-NR}(a,b) = \begin{cases} \mathrm{sgn}(a)\mathrm{sgn}(b)|a|^p|b|^p - \mathrm{sgn}(b)(a-b)^p|b|^p \text{ if } a \geq b, \\ \mathrm{sgn}(a)\mathrm{sgn}(b)|a|^p|b|^p - \mathrm{sgn}(a)(b-a)^p|a|^p \text{ if } a < b, \end{cases} \tag{2.49}$$

where $p > 0$. Notice that $\widetilde{\phi}^p_{\rm S-NR} = \phi^p_{\rm S-NR}$ and $\widetilde{\psi}^p_{\rm S-NR} = \psi^p_{\rm S-NR}$ whenever p is odd.

Proposition 2.37 *Let $\widetilde{\phi}^p_{\rm NR}$, $\widetilde{\phi}^p_{\rm S-NR}$, and $\widetilde{\psi}^p_{\rm S-NR}$ be defined as in (2.47), (2.48), and (2.49), respectively. For any $p > 0$, the functions $\widetilde{\phi}^p_{\rm NR}$, $\widetilde{\phi}^p_{\rm S-NR}$, and $\widetilde{\psi}^p_{\rm S-NR}$ are NCP functions. Moreover, $\widetilde{\phi}^p_{\rm NR}(a,b) > 0$ ($\widetilde{\phi}^p_{\rm S-NR}(a,b) > 0$) if and only if $a > 0$ and $b > 0$, while $\widetilde{\psi}^p_{\rm S-NR}(a,b) \geq 0$ for all $(a,b) \in \mathbb{R}^2$.*

Proof That $\widetilde{\phi}^p_{\rm NR}$ is an NCP function follows from the above discussion. Moreover, note that $a > 0$ and $b > 0$ if and only if $a > (a-b)_+$. Since $f(t) = \mathrm{sgn}(t)|t|^p$ is strictly increasing, we see that $a > 0$ and $b > 0$ if and only if $\mathrm{sgn}(a)|a|^p > \mathrm{sgn}((a-b)_+)|(a-b)_+|^p$, i.e. $\widetilde{\phi}^p_{\rm NR}(a,b) > 0$. On the other hand, observe that

$$\widetilde{\phi}^p_{\rm S-NR}(a,b) = \begin{cases} \widetilde{\phi}^p_{\rm NR}(a,b) \text{ if } a \geq b, \\ \widetilde{\phi}^p_{\rm NR}(b,a) \text{ if } a < b, \end{cases} \tag{2.50}$$

and

$$\widetilde{\psi}^p_{\text{S-NR}}(a, b) = \begin{cases} \text{sgn}(b)|b|^p \widetilde{\phi}^p_{\text{NR}}(a, b) \text{ if } a \geq b, \\ \text{sgn}(a)|a|^p \widetilde{\phi}^p_{\text{NR}}(b, a) \text{ if } a < b. \end{cases} \tag{2.51}$$

Using above identities and the fact that $\widetilde{\phi}^p_{\text{NR}}$ is an NCP function, then $\widetilde{\phi}^p_{\text{S-NR}}$ and $\widetilde{\psi}^p_{\text{S-NR}}$ are also NCP functions with algebraic signs as specified in the proposition. □

In light of the preceding proposition, the functions $\widetilde{\phi}^p_{\text{NR}}$, $\widetilde{\phi}^p_{\text{S-NR}}$, and $\widetilde{\psi}^p_{\text{S-NR}}$ may be regarded as continuous generalizations of ϕ^p_{NR}, $\phi^p_{\text{S-NR}}$, and $\psi^p_{\text{S-NR}}$, respectively. We now proceed to establish several fundamental properties of these functions, which will later play a key role in the development of a neural network-based approach. We begin by examining their smoothness properties. Throughout this discussion, $C^1(\Omega)$ and $C^2(\Omega)$ denote the spaces of continuously differentiable and twice continuously differentiable functions on a domain $\Omega \subset \mathbb{R}^n$, respectively.

Proposition 2.38 *The following result holds:*

(a) If $p > 1$, the function $\widetilde{\phi}^p_{\text{NR}} \in C^1(\mathbb{R}^2)$ and its gradient is given by

$$\nabla \widetilde{\phi}^p_{\text{NR}}(a, b) = p \begin{bmatrix} |a|^{p-1} - (a-b)^{p-1}\text{sgn}((a-b)_+) \\ (a-b)^{p-1}\text{sgn}((a-b)_+) \end{bmatrix}.$$

If $p > 2$, then $\widetilde{\phi}^p_{\text{NR}} \in C^2(\mathbb{R}^2)$ and its Hessian is given by

$$\nabla^2 \widetilde{\phi}^p_{\text{NR}}(a, b) = p(p-1) \begin{bmatrix} \text{sgn}(a)|a|^{p-2} - (a-b)^{p-2}\text{sgn}((a-b)_+) & (a-b)^{p-2}\text{sgn}((a-b)_+) \\ (a-b)^{p-2}\text{sgn}((a-b)_+) & -(a-b)^{p-2}\text{sgn}((a-b)_+) \end{bmatrix}.$$

(b) If $p > 1$, the function $\widetilde{\phi}^p_{\text{S-NR}} \in C^1(\Omega)$ where $\Omega := \{(a, b) \mid a \neq b\}$. In this case, the gradient of $\widetilde{\phi}^p_{\text{S-NR}}$ is given by

$$\nabla \widetilde{\phi}^p_{\text{S-NR}}(a, b) = \begin{cases} p\,[\,|a|^{p-1} - (a-b)^{p-1},\ (a-b)^{p-1}\,]^T \text{ if } a > b, \\ p\,[\,(b-a)^{p-1},\ |b|^{p-1} - (b-a)^{p-1}\,]^T \text{ if } a < b. \end{cases}$$

Further, $\widetilde{\phi}^p_{\text{S-NR}}$ is differentiable at $(0, 0)$ with $\nabla \widetilde{\phi}^p_{\text{S-NR}}(0, 0) = [0, 0]^T$. If $p > 2$, then $\widetilde{\phi}^p_{\text{S-NR}} \in C^2(\Omega)$ with Hessian given by

$$\nabla^2 \widetilde{\phi}^p_{\text{S-NR}}(a, b) = \begin{cases} p(p-1) \begin{bmatrix} \text{sgn}(a)|a|^{p-2} - (a-b)^{p-2} & (a-b)^{p-2} \\ (a-b)^{p-2} & -(a-b)^{p-2} \end{bmatrix} \text{ if } a > b, \\ p(p-1) \begin{bmatrix} -(b-a)^{p-2} & (b-a)^{p-2} \\ (b-a)^{p-2} & \text{sgn}(b)|b|^{p-2} - (b-a)^{p-2} \end{bmatrix} \text{ if } a < b. \end{cases}$$

(c) If $p > 1$, then $\widetilde{\psi}^p_{\text{S-NR}} \in C^1(\mathbb{R}^2)$ whose gradient is given by

$$\nabla \widetilde{\psi}^p_{\text{S-NR}}(a, b) = \begin{cases} p \begin{bmatrix} \text{sgn}(b)|b|^p(|a|^{p-1} - (a-b)^{p-1}) \\ \text{sgn}(a)|a|^p|b|^{p-1} - (a-b)^p|b|^{p-1} + \text{sgn}(b)(a-b)^{p-1}|b|^p \end{bmatrix} & \text{if } a > b, \\ p|a|^{2p-1} \begin{bmatrix} 1 \\ 1 \end{bmatrix} & \text{if } a = b, \\ p \begin{bmatrix} \text{sgn}(b)|a|^{p-1}|b|^p - (b-a)^p|a|^{p-1} + \text{sgn}(a)(b-a)^{p-1}|a|^p \\ \text{sgn}(a)|a|^p(|b|^{p-1} - (b-a)^{p-1}) \end{bmatrix} & \text{if } a < b, \end{cases}$$

If $p > 2$, then $\widetilde{\psi}^p_{\rm S-NR} \in C^2(\mathbb{R}^2)$ whose Hessian is given by

$$
\nabla^2 \widetilde{\psi}^p_{\rm S-NR}(a,b) = \begin{cases}
p \begin{bmatrix}
\begin{array}{c} (p-1)[\mathrm{sgn}(a)\mathrm{sgn}(b)|a|^{p-2}|b|^p] \\ -(p-1)(a-b)^{p-2}\mathrm{sgn}(b)|b|^p \end{array} & \begin{array}{c} (p-1)(a-b)^{p-2}\mathrm{sgn}(b)|b|^p \\ +p[|a|^{p-1}-(a-b)^{p-1}]|b|^{p-1} \end{array} \\[2ex]
\begin{array}{c} (p-1)(a-b)^{p-2}\mathrm{sgn}(b)|b|^p \\ +p[|a|^{p-1}-(a-b)^{p-1}]|b|^{p-1} \end{array} & \begin{array}{c} (p-1)[\mathrm{sgn}(a)\mathrm{sgn}(b)|a|^p|b|^{p-2}] \\ -(p-1)(a-b)^p\mathrm{sgn}(b)|b|^{p-2} \\ +2p(a-b)^{p-1}|b|^{p-1} \\ -(p-1)(a-b)^{p-2}\mathrm{sgn}(b)|b|^p \end{array}
\end{bmatrix} & \text{if } a > b, \\[4ex]
p \begin{bmatrix}
(p-1)\mathrm{sgn}(a)\mathrm{sgn}(b)|a|^{p-2}|b|^p & p|a|^{p-1}|b|^{p-1} \\
p|a|^{p-1}|b|^{p-1} & (p-1)\mathrm{sgn}(a)\mathrm{sgn}(b)|a|^p|b|^{p-2}
\end{bmatrix} & \text{if } a = b, \\[4ex]
p \begin{bmatrix}
\begin{array}{c} (p-1)[\mathrm{sgn}(a)\mathrm{sgn}(b)|a|^{p-2}|b|^p] \\ -(p-1)(b-a)^p\mathrm{sgn}(a)|a|^{p-2} \\ +2p(b-a)^{p-1}|a|^{p-1} \\ -(p-1)(b-a)^{p-2}\mathrm{sgn}(a)|a|^p \end{array} & \begin{array}{c} (p-1)(b-a)^{p-2}\mathrm{sgn}(a)|a|^p \\ +p[|b|^{p-1}-(b-a)^{p-1}]|a|^{p-1} \end{array} \\[2ex]
\begin{array}{c} (p-1)(b-a)^{p-2}\mathrm{sgn}(a)|a|^p \\ +p[|b|^{p-1}-(b-a)^{p-1}]|a|^{p-1} \end{array} & \begin{array}{c} (p-1)[\mathrm{sgn}(a)\mathrm{sgn}(b)|a|^p|b|^{p-2}] \\ -(p-1)(b-a)^{p-2}\mathrm{sgn}(a)|a|^p \end{array}
\end{bmatrix} & \text{if } a < b.
\end{cases}
$$

Proof Note that $f(t) = \mathrm{sgn}(t)|t|^p$ is continuously differentiable when $p > 1$ with $f'(t) = p|t|^{p-1}$. Moreover, f is twice continuously differentiable when $p > 2$ with $f''(t) = p(p-1)\mathrm{sgn}(t)|t|^{p-2}$. Using these and the alternative formulas given in (2.50) and (2.51), the gradients and Hessians can be easily obtained. The calculations are omitted. □

The above proposition serves as a generalization of results found in [34, Proposition 2.2], [35, Propositions 2.2 and 3.2], and [37, Proposition 4.3]. In a similar vein, the following result extends [37, Propositions 3.4, 4.5 and 5.4].

Proposition 2.39 *Let $p > 1$. Then, the following hold:*

(a) $$\nabla_a \widetilde{\phi}^p_{\rm NR}(a,b) \cdot \nabla_b \widetilde{\phi}^p_{\rm NR}(a,b) \begin{cases} > 0 \text{ on } \{(a,b) \mid a > b > 0 \text{ or } a > b > 2a\}, \\ = 0 \text{ on } \{(a,b) \mid a \le b \text{ or } a > b = 2a \text{ or } a > b = 0\}, \\ < 0 \text{ otherwise.} \end{cases}$$

(b) $$\nabla_a \widetilde{\phi}^p_{\rm S-NR}(a,b) \cdot \nabla_b \widetilde{\phi}^p_{\rm S-NR}(a,b) \begin{cases} > 0 \text{ on } \{(a,b) \mid a > b > 0 \text{ or } a > b > 2a\} \\ \quad \text{and on } \{(a,b) \mid b > a > 0 \text{ or } b > a > 2b\}, \\ = 0 \text{ on } \{(a,b) \mid \widetilde{\phi}^p_{\rm S-NR}(a,b) = 0 \text{ or } a > b = 2a \text{ or } b > a = 2b\}, \\ < 0 \text{ otherwise.} \end{cases}$$

(c) $\nabla_a \widetilde{\psi}^p_{\rm S-NR}(a,b) \cdot \nabla_b \widetilde{\psi}^p_{\rm S-NR}(a,b) > 0$ *on the first quadrant* $\mathbb{R}^2_{++}$, *and* $\widetilde{\psi}^p_{\rm S-NR}(a,b) = 0 \Longleftrightarrow \nabla \widetilde{\psi}^p_{\rm S-NR}(a,b) = 0$.

Proof Using Proposition 2.38(a),

$$\begin{aligned}
&\nabla_a\widetilde{\phi}^p_{\text{NR}}(a,b)\cdot\nabla_b\widetilde{\phi}^p_{\text{NR}}(a,b)\\
&= p^2[|a|^{p-1}-(a-b)^{p-1}\text{sgn}((a-b)_+)](a-b)^{p-1}\text{sgn}((a-b)_+)\\
&= \begin{cases} p^2[|a|^{p-1}-(a-b)^{p-1}](a-b)^{p-1} & \text{if } a>b\\ 0 & \text{if } a\le b\end{cases}.
\end{aligned}$$

Suppose now that $a > b$. Since $g(t) := t^{p-1}$ is a strictly increasing function on $[0,\infty)$, $|a|^{p-1}-(a-b)^{p-1}>0$ if and only if $|a|>a-b$, which happens if and only if $b>0$ or $b>2a$. This establishes Proposition 2.39(a). Statement (b) easily follows from part(a), while part(c) can be easily verified using the result of Proposition 2.38(c). □

Like what we did earlier, we now analyze the growth behavior of the proposed families of functions in Proposition 2.40, which serves as a continuous counterpart to Proposition 2.36.

Proposition 2.40 *Let* $\phi\in\{\widetilde{\phi}^p_{\text{NR}},\widetilde{\phi}^p_{\text{S-NR}},\widetilde{\psi}^p_{\text{S-NR}}\}$. *Then,* $|\phi(a^k,b^k)|\to\infty$ *for any sequence* $\{(a^k,b^k)\}_{k=1}^{\infty}$ *in* $\mathbb{R}^2$ *such that* $|a^k|\to\infty$ *and* $|b^k|\to\infty$.

Proof The proposition follows from Lemma 2.10 and analogous arguments for Proposition 2.36. □

2.3 Constructions of NCP Functions Based on ϕ^p_{FB}

In this section, we explore several extensions based on the Fischer–Burmeister function ϕ^p_{FB}. Recall that the FB function is defined by

$$\phi_{\text{FB}}(a,b)=\sqrt{a^2+b^2}-a-b,\quad \forall(a,b)\in\mathbb{R}^2,$$

and one of its generalizations, proposed by Kanzow and Kleinmichel [16], is given by

$$\phi_\theta(a,b) := \sqrt{(a-b)^2+\theta ab}-a-b,\quad \theta\in(0,4),\quad \forall(a,b)\in\mathbb{R}^2. \tag{2.52}$$

It has been shown in [16, 19, 20, 22, 23, 38] that both ϕ_θ defined in (2.52) and ϕ^p_{FB} given in (2.14) enjoy several desirable properties, including strong semismoothness, Lipschitz continuity, and continuous differentiability. Furthermore, the corresponding merit functions associated with ϕ_θ and ϕ^p_{FB} have been shown to possess the SC^1 property (i.e., they are continuously differentiable with semismooth gradients) and the LC^1 property (i.e., they are continuously differentiable with Lipschitz continuous gradients), under appropriate assumptions.

2.3.1 Construction by Using Parameter

The above idea of introducing a parameter can likewise be applied to ϕ_{FB}^p,

$$\phi_{\text{FB}}^p(a,b) = \sqrt[p]{|a|^p + |b|^p} - a - b, \quad p \in (1, \infty).$$

In fact, motivated by those functions studied in [16, 22], we consider the following class of functions [25]:

$$\phi_{\theta,p}(a,b) := \sqrt[p]{\theta(|a|^p + |b|^p) + (1-\theta)|a-b|^p} - a - b, \quad p > 1, \quad \theta \in (0,1]. \tag{2.53}$$

Accordingly, there is an associated unconstrained minimization:

$$\Psi_{\theta,p}(x) := \frac{1}{2}\sum_{i=1}^{n} \phi_{\theta,p}^2(x_i, F_i(x)). \tag{2.54}$$

An important question arises: is the function $\phi_{\theta,p}$ an NCP function? If so, do the functions defined in (2.53) and (2.54) inherit the same desirable properties, such as strong semismoothness, Lipschitz continuity, and smooth merit function characteristics, as the previously studied functions mentioned above? Furthermore, how do the merit function methods based on (2.53) and (2.54) perform in terms of numerical behavior and practical efficiency?

In this section, we provide a partial answer to the questions raised above. Specifically, we demonstrate that the function $\phi_{\theta,p}$, defined in (2.53), is indeed an NCP function. We also examine several desirable properties of $\phi_{\theta,p}$ and its associated nonnegative merit function, including strong semismoothness, Lipschitz continuity, and continuous differentiability. For convenience, we define

$$\eta_{\theta,p}(a,b) := \sqrt[p]{\theta(|a|^p + |b|^p) + (1-\theta)|a-b|^p}, \quad p > 1, \quad \theta \in (0,1]. \tag{2.55}$$

Proposition 2.41 *The function $\phi_{\theta,p}$ defined by (2.53) is an NCP function.*

Proposition 2.42 *The function $\eta_{\theta,p}$ defined by (2.55) is a norm on $\mathbb{R}^2$ for all $p > 1, \theta \in (0,1]$.*

Proposition 2.43 *Let $\phi_{\theta,p}$ be defined by (2.53), then for all $\theta \in (0,1]$ and $p > 1$,*

(i) $\phi_{\theta,p}$ is sub-additive, i.e., $\phi_{\theta,p}((a,b)+(c,d)) \le \phi_{\theta,p}(a,b) + \phi_{\theta,p}(c,d)$ for all $(a,b), (c,d) \in \mathbb{R}^2$;

(ii) $\phi_{\theta,p}$ is positive homogenous, i.e., $\phi_{\theta,p}(\alpha(a,b)) = \alpha\phi_{\theta,p}(a,b)$ for all $(a,b) \in \mathbb{R}^2$ and $\alpha > 0$;

(iii) $\phi_{\theta,p}$ is a convex function on $\mathbb{R}^2$, i.e., $\phi_{\theta,p}(\alpha(a,b)+(1-\alpha)(c,d)) \le \alpha\phi_{\theta,p}(a,b) + (1-\alpha)\phi_{\theta,p}(c,d)$ for all $(a,b), (c,d) \in \mathbb{R}^2$ and $\alpha \in [0,1]$;

(iv) $\phi_{\theta,p}$ *is Lipschitz continuous on* $\mathbb{R}^2$*;*
(v) $\phi_{\theta,p}$ *is continuously differentiable on* $\mathbb{R}^2\backslash\{(0,0)\}$*;*
(vi) $\phi_{\theta,p}$ *is strongly semismooth on* $\mathbb{R}^2$*.*

Proof By using $\phi_{\theta,p}((a,b)) = \eta_{\theta,p}(a,b) - (a+b)$ and Proposition 2.42, we can obtain that the results (i), (ii), and (iii) hold.

(iv) Since $\eta_{\theta,p}$ is a norm on $\mathbb{R}^2$ from Proposition 2.42 and any two norms in finite dimensional space are equivalent, it follows that there exists a positive constant κ such that

$$\eta_{\theta,p}(a,b) \le \kappa\|(a,b)\|, \quad \forall (a,b) \in \mathbb{R}^2,$$

where $\|\cdot\|$ represents the Euclidean norm on $\mathbb{R}^2$. Hence, for all $(a,b), (c,d) \in \mathbb{R}^2$, there holds

$$\begin{aligned}
\left|\phi_{\theta,p}(a,b) - \phi_{\theta,p}(c,d)\right| &= |\eta_{\theta,p}(a,b) - (a+b) - \eta_{\theta,p}(c,d) + (c+d)| \\
&\le |\eta_{\theta,p}(a,b) - \eta_{\theta,p}(c,d)| + |a-c| + |b-d| \\
&\le \eta_{\theta,p}(a-c, b-d) + \sqrt{2}\|(a-c, b-d)\| \\
&\le \kappa\|(a-c, b-d)\| + \sqrt{2}\|(a-c, b-d)\| \\
&= (\kappa + \sqrt{2})\|(a-c, b-d)\|.
\end{aligned}$$

This says that $\phi_{\theta,p}$ is Lipschitz continuous with Lipschitz constant $\kappa + \sqrt{2}$, i.e., the result (iv) holds.

(v) If $(a,b) \ne (0,0)$, then $\eta_{\theta,p}(a,b) \ne 0$ by Proposition 2.42. By direct calculations, we obtain

$$\frac{\partial \phi_{\theta,p}(a,b)}{\partial a} = \frac{\theta \mathrm{sgn}(a)|a|^{p-1} + (1-\theta)\mathrm{sgn}(a-b)|a-b|^{p-1}}{\eta_{\theta,p}(a,b)^{p-1}} - 1; \tag{2.56}$$

$$\frac{\partial \phi_{\theta,p}(a,b)}{\partial b} = \frac{\theta \mathrm{sgn}(b)|b|^{p-1} - (1-\theta)\mathrm{sgn}(a-b)|a-b|^{p-1}}{\eta_{\theta,p}(a,b)^{p-1}} - 1, \tag{2.57}$$

where $\mathrm{sgn}(\cdot)$ is the symbol function. Then, it is easy to see from (2.56) and (2.57) that the result (v) holds.

(vi) Since $\phi_{\theta,p}$ is a convex function by the result (iii), we know that it is a semismooth function. Noticing that $\phi_{\theta,p}$ is continuously differentiable except (0, 0), it is sufficient to prove that it is strongly semismooth at (0, 0). For any $(h,k) \in \mathbb{R}^2\backslash\{(0,0)\}$, $\phi_{\theta,p}$ is differentiable at (h,k), and hence, $\nabla\phi_{\theta,p}(h,k) = \left(\frac{\partial\phi_{\theta,p}(h,k)}{\partial a}, \frac{\partial\phi_{\theta,p}(h,k)}{\partial b}\right)^T$. Thus, we have

$$
\begin{aligned}
&\phi_{\theta,p}((0,0)+(h,k)) - \phi_{\theta,p}(0,0) - \left(\frac{\partial \phi_{\theta,p}(h,k)}{\partial a}, \frac{\partial \phi_{\theta,p}(h,k)}{\partial b}\right)\begin{pmatrix} h \\ k \end{pmatrix} \\
&= \sqrt[p]{\theta(|h|^p+|k|^p)+(1-\theta)|h-k|^p} - (h+k) \\
&\quad - \left(\frac{\operatorname{sgn}(h)|h|^{p-1} + \operatorname{sgn}(h-k)|h-k|^{p-1}}{\eta_{\theta,p}(h,k)^{p-1}} - 1\right)h \\
&\quad - \left(\frac{\operatorname{sgn}(k)|k|^{p-1} - \operatorname{sgn}(h-k)|h-k|^{p-1}}{\eta_{\theta,p}(h,k)^{p-1}} - 1\right)k \\
&= \sqrt[p]{\theta(|h|^p+|k|^p)+(1-\theta)|h-k|^p} \\
&\quad - \frac{\operatorname{sgn}(h)|h|^{p-1}h + \operatorname{sgn}(k)|k|^{p-1}k + \operatorname{sgn}(h-k)|h-k|^{p-1}(h-k)}{\eta_{\theta,p}(h,k)^{p-1}} \\
&= \sqrt[p]{\theta(|h|^p+|k|^p)+(1-\theta)|h-k|^p} - \frac{|h|^p+|k|^p+|h-k|^p}{\eta_{\theta,p}(h,k)^{p-1}} \\
&= \eta_{\theta,p}(h,k) - \frac{|h|^p+|k|^p+|h-k|^p}{\eta_{\theta,p}(h,k)^{p-1}} \\
&= \frac{\eta_{\theta,p}(h,k)^p - (|h|^p+|k|^p+|h-k|^p)}{\eta_{\theta,p}(h,k)^{p-1}} \\
&= 0 \\
&= O(\|(h,k)\|^2).
\end{aligned}
$$

Then, we obtain that $\phi_{\theta,p}$ is strongly semismooth. □

Proposition 2.44 *Let $\phi_{\theta,p}$ be defined by (2.53) and $\{(a^k, b^k)\} \subseteq \mathbb{R}^2$. Then,*

$$|\phi_{\theta,p}(a^k, b^k)| \to \infty$$

if one of the following conditions is satisfied. (i) $a^k \to -\infty$; (ii) $b^k \to -\infty$; (iii) $a^k \to \infty$ and $b^k \to \infty$.

Proof (i) Suppose that $a^k \to -\infty$. If $\{b^k\}$ is bounded from above, then the result holds trivially. When $b^k \to \infty$, we have $-a^k > 0$ and $b^k > 0$ for all k sufficiently large, and hence,

$$\sqrt[p]{\theta(|a^k|^p+|b^k|^p)+(1-\theta)|a^k-b^k|^p} - b^k \geq \sqrt[p]{\theta|b^k|^p+(1-\theta)|b^k|^p} - b^k = 0.$$

This, together with $-a^k \to \infty$ and the definition of $\phi_{\theta,p}$, implies that the result holds.

(ii) For the case of $b^k \to -\infty$, a similar analysis yields the result of the proposition.

(iii) Suppose that $a^k \to \infty$ and $b^k \to \infty$. Since $p > 1$ and $\theta \in (0, 1]$, we have

$$(1-\theta)|a^k - b^k|^p \leq (1-\theta)(|a^k|^p + |b^k|^p)$$

for all sufficiently large k. Thus, for all sufficiently large k,

$$\sqrt[p]{\theta(|a^k|^p + |b^k|^p) + (1-\theta)|a^k - b^k|^p} \le \sqrt[p]{|a^k|^p + |b^k|^p},$$

and hence,

$$(a^k + b^k) - \sqrt[p]{\theta(|a^k|^p + |b^k|^p) + (1-\theta)|a^k - b^k|^p} \ge (a^k + b^k) - \sqrt[p]{|a^k|^p + |b^k|^p}.$$

By [22, Lemma 3.1] we know that $(a^k + b^k) - \sqrt[p]{|a^k|^p + |b^k|^p} \to \infty$ as $k \to \infty$ when the condition (iii) is satisfied. Thus, we obtain that

$$|\phi_{\theta p}(a^k, b^k)| = (a^k + b^k) - \sqrt[p]{\theta(|a^k|^p + |b^k|^p) + (1-\theta)|a^k - b^k|^p} \to \infty$$

as $k \to \infty$, which completes the proof. □

We now define a nonnegative merit function, associated with the function $\phi_{\theta,p}$, as follows:

$$\psi_{\theta,p}(a,b) := \frac{1}{2}\phi_{\theta,p}^2(a,b), \quad p > 1, \quad \theta \in (0,1], \ (a,b) \in \mathbb{R}^2. \tag{2.58}$$

Proposition 2.45 *Let $\psi_{\theta,p}$ be defined by* (2.58), *then for all $\theta \in (0,1]$ and $p > 1$,*

(i) $\psi_{\theta,p}$ *is an NCP function;*
(ii) $\psi_{\theta,p}(a,b) \ge 0$ *for all* $(a,b) \in \mathbb{R}^2$*;*
(iii) $\psi_{\theta,p}$ *is continuously differentiable on* $\mathbb{R}^2$*;*
(vi) $\psi_{\theta,p}$ *is strongly semismooth on* $\mathbb{R}^2$*;*
(v) $\frac{\partial \psi_{\theta,p}(a,b)}{\partial a} \cdot \frac{\partial \psi_{\theta,p}(a,b)}{\partial b} \ge 0$ *for all* $(a,b) \in \mathbb{R}^2$, *where the equality holds if and only if* $\phi_{\theta,p}(a,b) = 0$*;*
(vi) $\frac{\partial \psi_{\theta,p}(a,b)}{\partial a} = 0 \iff \frac{\partial \psi_{\theta,p}(a,b)}{\partial b} = 0 \iff \phi_{\theta,p}(a,b) = 0.$

Proof By the definition of $\psi_{\theta,p}$, it is easy to see that the results (i) and (ii) hold.

(iii). By using Proposition 2.43 and the definition of $\psi_{\theta,p}$, it is sufficient to prove that $\psi_{\theta,p}$ is differentiable at (0, 0) and the gradient is continuous at (0, 0). In fact, for all $(a,b) \in \mathbb{R}^2 \backslash \{(0,0)\}$, we have

$$\begin{aligned}
|\phi_{\theta p}(a,b)| &= \left|\sqrt[p]{\theta(|a|^p + |b|^p) + (1-\theta)|a-b|^p} - a - b\right| \\
&\le \left|\sqrt[p]{\theta|a|^p} + \sqrt[p]{\theta|b|^p} + \sqrt[p]{(1-\theta)|a-b|^p}\right| + |a| + |b| \\
&\le |a| + |b| + |a-b| + |a| + |b| \\
&\le 3(|a| + |b|),
\end{aligned}$$

where the second inequality follows from $p > 1$ and the third inequality follows from $\theta \in (0,1]$. Hence,

$$\psi_{\theta,p}(a,b) - \psi_{\theta,p}(0,0) = \frac{1}{2}\phi_{\theta,p}^2(a,b) \le \frac{1}{2}(3(|a| + |b|))^2 \le O(|a|^2 + |b|^2).$$

Thus, similar to that of [39, Proposition 1], we can achieve that $\psi_{\theta,p}$ is differentiable at $(0, 0)$ with $\nabla\psi_{\theta,p}(0, 0) = (0, 0)^T$. Now, we prove that for all $(a, b) \in \mathbb{R}^2\backslash\{(0, 0)\}$,

$$\left|\frac{\theta \text{sgn}(a)|a|^{p-1} + (1-\theta)\text{sgn}(a-b)|a-b|^{p-1}}{\eta_{\theta,p}(a,b)^{p-1}}\right| \leq 1, \tag{2.59}$$

$$\left|\frac{\theta \text{sgn}(b)|b|^{p-1} - (1-\theta)\text{sgn}(a-b)|a-b|^{p-1}}{\eta_{\theta,p}(a,b)^{p-1}}\right| \leq 1. \tag{2.60}$$

In fact,

$$\begin{aligned}
&\left|\frac{\theta \text{sgn}(a)|a|^{p-1} + (1-\theta)\text{sgn}(a-b)|a-b|^{p-1}}{\eta_{\theta p}(a,b)^{p-1}}\right| \\
\leq &\frac{\theta|a|^{p-1} + (1-\theta)|a-b|^{p-1}}{\eta_{\theta,p}(a,b)^{p-1}} \\
= &\frac{\theta^{\frac{1}{p}}|\theta^{\frac{1}{p}}a|^{p-1} + (1-\theta)^{\frac{1}{p}}|(1-\theta)^{\frac{1}{p}}(a-b)|^{p-1}}{\eta_{\theta p}(a,b)^{p-1}} \\
\leq &\frac{((\theta^{\frac{1}{p}})^p + ((1-\theta)^{\frac{1}{p}})^p)^{\frac{1}{p}}((|\theta^{\frac{1}{p}}a|^{p-1})^{\frac{p}{p-1}} + (|(1-\theta)^{\frac{1}{p}}(a-b)|^{p-1})^{\frac{p}{p-1}})^{\frac{p-1}{p}}}{\eta_{\theta,p}(a,b)^{p-1}} \\
= &\frac{(\theta + (1-\theta))(x^p + z^p)^{\frac{p-1}{p}}}{\eta_{\theta,p}(a,b)^{p-1}} \\
= &\frac{(x^p + z^p)^{\frac{p-1}{p}}}{(x^p + y^p + z^p)^{\frac{p-1}{p}}} \\
= &\left(\frac{x^p + z^p}{x^p + y^p + z^p}\right)^{\frac{p-1}{p}} \\
\leq &1,
\end{aligned}$$

where $x := |\theta^{\frac{1}{p}}a|^p$, $y := |\theta^{\frac{1}{p}}b|^p$, $z := |(1-\theta)^{\frac{1}{p}}(a-b)|^p$; the first inequality follows from the triangle inequality; the second inequality follows from the well known Hölder inequality; the second equality follows from the definitions of x and z; the third equality follows from the definitions of $\eta_{\theta,p}(a, b)$, x, y, and z; and the third inequality follows from the fact that x, y, and z are all nonnegative. Therefore, (2.59) holds. Similar analysis will derive that (2.60) holds.

Thus, it follows from (2.59) and (2.60) that both $\frac{\partial\phi_{\theta,p}(a,b)}{\partial a}$ and $\frac{\partial\phi_{\theta,p}(a,b)}{\partial b}$ are uniformly bounded. Since $\phi_{\theta,p}(a, b) \to 0$ as $(a, b) \to (0, 0)$, we obtain the desired result.

(iv) Since the composition of strongly semismooth function is also strongly semismooth (see [10, Theorem 19]), by Proposition 2.43(vi) and the definition of $\psi_{\theta,p}$ we obtain that the desired result holds.

(v) It is obvious that $\frac{\partial \psi_{\theta,p}(a,b)}{\partial a} = 0$ when $(a, b) = (0, 0)$. Now, suppose that $(a, b) \neq (0, 0)$. Since

$$\frac{\partial \psi_{\theta,p}(a,b)}{\partial a} \cdot \frac{\partial \psi_{\theta,p}(a,b)}{\partial b} = \frac{\partial \phi_{\theta,p}(a,b)}{\partial a} \cdot \frac{\partial \phi_{\theta,p}(a,b)}{\partial b} \cdot \phi_{\theta,p}(a,b)^2, \quad (2.61)$$

by (2.56), (2.57), (2.59), and (2.60), we obtain that $\frac{\partial \phi_{\theta,p}(a,b)}{\partial a} \leq 0$ and $\frac{\partial \phi_{\theta,p}(a,b)}{\partial b} \leq 0$ for all $(a, b) \in \mathbb{R}^2$, that is, the first result of (v) holds. In addition, from (2.61) it is obvious that the sufficient condition of the second result of (v) holds. Now, we suppose that $\frac{\partial \psi_{\theta,p}(a,b)}{\partial a} \cdot \frac{\partial \psi_{\theta,p}(a,b)}{\partial b} = 0$. Then, it is sufficient to prove that $\phi_{\theta,p}(a, b) = 0$ when $\frac{\partial \phi_{\theta,p}(a,b)}{\partial a} \cdot \frac{\partial \phi_{\theta,p}(a,b)}{\partial b} = 0$. Suppose that $\frac{\partial \phi_{\theta,p}(a,b)}{\partial a} = 0$, without loss of generality. From the proof of (iii) in this proposition, it is easy to see that it must be $y = 0$, and hence, $b = 0$. After a simple symbol discussion for (2.56), we may get $a \geq 0$. Hence $\phi_{\theta,p}(a, b) = 0$ by Proposition 2.41. So, the result (v) holds.

(vi) Since

$$\frac{\partial \psi_{\theta,p}(a,b)}{\partial a} = \frac{\partial \phi_{\theta,p}(a,b)}{\partial a}\phi_{\theta,p}(a,b), \qquad \frac{\partial \psi_{\theta,p}(a,b)}{\partial b} = \frac{\partial \phi_{\theta,p}(a,b)}{\partial b}\phi_{\theta,p}(a,b),$$

the result (vi) is immediately satisfied from the above analysis. □

Proposition 2.46 *The gradient function of the function $\psi_{\theta,p}$, defined by* (2.58) *with $p \geq 2$ and $\theta \in (0, 1]$, is Lipschitz continuous, that is, there exists a positive constant L such that*

$$\|\nabla \psi_{\theta,p}(a,b) - \nabla \psi_{\theta,p}(c,d)\| \leq L\|(a,b) - (c,d)\| \quad (2.62)$$

holds for all $(a, b), (c, d) \in \mathbb{R}^2$.

Proof It follows from the definition of $\psi_{\theta,p}$ and the proof of Proposition 2.45(iii) that $\nabla \psi_{\theta,p}(a,b) = \nabla \phi_{\theta,p}(a,b)\phi_{\theta p}(a,b)$ when $(a, b) \neq (0, 0)$, and $\nabla \psi_{\theta,p}(0, 0) = (0, 0)^{\mathrm{T}}$. From Proposition 2.45(iii), we know that $\psi_{\theta,p}$ is continuous differentiable. The proof is divided into three cases.

Case 1: If $(a, b) = (c, d) = (0, 0)$, it follows from Proposition 2.45 that $\nabla \psi_{\theta,p}(0, 0) = (0, 0)$, and hence, (2.62) holds for all positive number L.
Case 2: Consider the case that one of (a, b) and (c, d) is $(0, 0)$, but not all. We assume that $(a, b) \neq (0, 0)$ and $(c, d) = (0, 0)$, without loss of generality. Then,

$$\begin{aligned}
\|\nabla\psi_{\theta,p}(a,b)-\nabla\psi_{\theta,p}(c,d)\| &= \|\nabla\psi_{\theta,p}(a,b)-(0,0)\| \\
&= \|\nabla\phi_{\theta,p}(a,b)\phi_{\theta,p}(a,b)-(0,0)\| \\
&= \|\nabla\phi_{\theta,p}(a,b)\|\phi_{\theta,p}(a,b) \\
&= \|\nabla\phi_{\theta,p}(a,b)\||\phi_{\theta,p}(a,b)-\phi_{\theta,p}(0,0)| \\
&\le L\|(a,b)-(0,0)\|,
\end{aligned}$$

where the inequality follows from the fact that $\{\|\nabla\phi_{\theta,p}(a,b)\|\}$ is uniformly bounded on $\mathbb{R}^2$ (which can be obtained from the proof of Proposition 2.45(iii)) and $\phi_{\theta,p}$ is Lipschitz continuous on $\mathbb{R}^2$ given in Proposition 2.43(iv). Hence, (2.62) holds for some positive constant L.

Case 3: If both (a,b) and (c,d) are not $(0,0)$, we will apply Lemma 1.3 to prove that (2.62) holds for this case. For simplicity, we denote

$$\begin{aligned}
\hat{h}_1 &:= \frac{\theta\,\text{sgn}(a)|a|^{p-1}+(1-\theta)\text{sgn}(a-b)|a-b|^{p-1}}{\eta_{\theta,p}^{p-1}(a,b)}; \\
\hat{h}_2 &:= \frac{\theta\,\text{sgn}(b)|b|^{p-1}-(1-\theta)\text{sgn}(a-b)|a-b|^{p-1}}{\eta_{\theta,p}^{p-1}(a,b)}; \\
\hat{a}_1 &:= (\theta|a|^{p-2}+(1-\theta)|a-b|^{p-2})\eta_{\theta,p}^{p}(a,b); \\
\hat{a}_2 &:= -\hat{h}_1^2\eta_{\theta,p}^{2p-2}(a,b); \\
\hat{b}_1 &:= -(1-\theta)|a-b|^{p-2}\eta_{\theta,p}^{p}(a,b); \\
\hat{b}_2 &:= -\hat{h}_1\hat{h}_2\eta_{\theta p}^{2p-2}(a,b); \\
\hat{c}_1 &:= (\theta|b|^{p-2}+(1-\theta)|a-b|^{p-2})\eta_{\theta,p}^{p}(a,b); \\
\hat{c}_2 &:= -\hat{h}_2^2\eta_{\theta p}^{2p-2}(a,b).
\end{aligned}$$

When $(a,b)\neq(0,0)$, by direct calculations, we have

$$\begin{aligned}
\frac{\partial^2\psi_{\theta,p}(a,b)}{\partial a^2} &= (\hat{h}_1-1)^2+(p-1)\frac{\hat{a}_1+\hat{a}_2}{\eta_{\theta,p}^{2p-1}(a,b)}(\eta_{\theta,p}(a,b)-(a+b)); \\
\frac{\partial^2\psi_{\theta,p}(a,b)}{\partial a\partial b} &= (\hat{h}_1-1)(\hat{h}_2-1)+(p-1)\frac{\hat{b}_1+\hat{b}_2}{\eta_{\theta,p}^{2p-1}(a,b)}(\eta_{\theta,p}(a,b)-(a+b)); \\
\frac{\partial^2\psi_{\theta,p}(a,b)}{\partial b^2} &= (\hat{h}_2-1)^2+(p-1)\frac{\hat{c}_1+\hat{c}_2}{\eta_{\theta p}^{2p-1}(a,b)}(\eta_{\theta,p}(a,b)-(a+b)); \\
\frac{\partial^2\psi_{\theta,p}(a,b)}{\partial b\partial a} &= \frac{\partial^2\psi_{\theta,p}(a,b)}{\partial a\partial b},
\end{aligned}$$

where the last equality follows from the fact that $\frac{\partial^2\psi_{\theta,p}(a,b)}{\partial a\partial b}$ and $\frac{\partial^2\psi_{\theta,p}(a,b)}{\partial b\partial a}$ are continuous when $(a,b)\neq(0,0)$. Since for any $p\geq 2$, $\eta_{\theta,p}(\cdot,\cdot)$ is a norm on $\mathbb{R}^2$ by Proposition 2.42, it is easy to verify that

$$|a+b| \le |a| + |b| \le \sqrt[p]{|a|^p + |b|^p} + \sqrt[p]{|a|^p + |b|^p} = 2\|(a,b)\|_p \le 2\kappa^* \eta_{\theta p}(a,b),$$

where $\kappa^* > 0$ is a constant depending on θ and p. Thus, we have

$$\begin{aligned}
\frac{\hat{a}_1}{\eta_{\theta p}^{2p-2}(a,b)} &= \frac{\theta|a|^{p-2} + (1-\theta)|a-b|^{p-2}}{\eta_{\theta p}^{p-2}(a,b)} \\
&= \frac{\theta|a|^{p-2}}{\eta_{\theta p}^{p-2}(a,b)} + \frac{(1-\theta)|a-b|^{p-2}}{\eta_{\theta p}^{p-2}(a,b)} \\
&\le \theta^{\frac{2}{p}} + (1-\theta)^{\frac{2}{p}} \\
&\le 2.
\end{aligned}$$

Similarly, we have

$$\frac{|\hat{b}_1|}{\eta_{\theta p}^{2p-2}(a,b)} \le 1; \quad \frac{\hat{c}_1}{\eta_{\theta p}^{2p-2}(a,b)} \le 2.$$

These, together with the results $|\hat{h}_1| \le 1$ and $|\hat{h}_2| \le 1$ given in Proposition 2.45, yield

$$\frac{|\hat{a}_2|}{\eta_{\theta,p}^{2p-2}(a,b)} \le 1; \quad \frac{|\hat{b}_2|}{\eta_{\theta,p}^{2p-2}(a,b)} \le 1; \quad \frac{|\hat{c}_2|}{\eta_{\theta,p}^{2p-2}(a,b)} \le 1.$$

Thus, we compute that

$$\begin{aligned}
\left|\frac{\partial^2 \psi_{\theta,p}(a,b)}{\partial a^2}\right| &= \left|(\hat{h}_1 - 1)^2 + (p-1)\frac{\hat{a}_1 + \hat{a}_2}{\eta_{\theta,p}^{2p-1}(a,b)}(\eta_{\theta,p}(a,b) - (a+b))\right| \\
&\le |(\hat{h}_1 - 1)^2| + (p-1)\left(\left|\frac{\hat{a}_1 + \hat{a}_2}{\eta_{\theta p}^{2p-1}(a,b)}\eta_{\theta,p}(a,b)\right| + \left|\frac{\hat{a}_1 + \hat{a}_2}{\eta_{\theta,p}^{2p-1}(a,b)}(a+b)\right|\right) \\
&\le 4 + (1+2\kappa^*)(p-1)\left(\frac{\hat{a}_1}{\eta_{\theta,p}^{2p-2}(a,b)} + \frac{|\hat{a}_2|}{\eta_{\theta p}^{2p-2}(a,b)}\right) \\
&\le 4 + 3(1+2\kappa^*)(p-1); \\
\left|\frac{\partial^2 \psi_{\theta,p}(a,b)}{\partial a \partial b}\right| &= \left|(\hat{h}_1 - 1)(\hat{h}_2 - 1) + (p-1)\frac{\hat{a}_1 + \hat{a}_2}{\eta_{\theta,p}^{2p-1}(a,b)}(\eta_{\theta,p}(a,b) - (a+b))\right| \\
&\le |(\hat{h}_1 - 1)(\hat{h}_2 - 1)| \\
&\quad + (p-1)\left(\left|\frac{\hat{b}_1 + \hat{b}_2}{\eta_{\theta,p}^{2p-1}(a,b)}\eta_{\theta,p}(a,b)\right| + \left|\frac{\hat{b}_1 + \hat{b}_2}{\eta_{\theta,p}^{2p-1}(a,b)}(a+b)\right|\right) \\
&\le 4 + (1+2\kappa^*)(p-1)\left(\frac{|\hat{b}_1|}{\eta_{\theta,p}^{2p-2}(a,b)} + \frac{|\hat{b}_2|}{\eta_{\theta,p}^{2p-2}(a,b)}\right) \\
&\le 4 + 2(1+2\kappa^*)(p-1); \\
\left|\frac{\partial^2 \psi_{\theta,p}(a,b)}{\partial b^2}\right| &= \left|(\hat{h}_2 - 1)^2 + (p-1)\frac{\hat{c}_1 + \hat{c}_2}{\eta_{\theta,p}^{2p-1}(a,b)}(\eta_{\theta,p}(a,b) - (a+b))\right|
\end{aligned}$$

$$
\begin{aligned}
&\le |(\hat{h}_2 - 1)^2| + (p-1)\left(\left|\frac{\hat{c}_1 + \hat{c}_2}{\eta_{\theta,p}^{2p-1}(a,b)}\eta_{\theta,p}(a,b)\right| + \left|\frac{\hat{c}_1 + \hat{c}_2}{\eta_{\theta,p}^{2p-1}(a,b)}(a+b)\right|\right)\\
&\le 4 + (1+2\kappa^*)(p-1)\left(\frac{\hat{c}_1}{\eta_{\theta,p}^{2p-2}(a,b)} + \frac{|\hat{c}_2|}{\eta_{\theta,p}^{2p-2}(a,b)}\right)\\
&\le 4 + 3(1+2\kappa^*)(p-1).
\end{aligned}
$$

Hence, there exists a positive constant L such that (2.62) holds by Lemma 1.3. □

It should be noted that $\nabla\psi_{\theta,p}$ is not Lipschitz continuous for all $\theta \in (0,1]$ when $p \in (1,2)$. In fact, if we fixed $\theta = 1$, for $(a,b) \neq (0,0)$ and $(c,d) \neq (0,0)$, we have

$$
\begin{aligned}
&\|\nabla\psi_{1p}(a,b) - \nabla\psi_{1p}(c,d)\|\\
=&\|\nabla\phi_{1p}(a,b)\phi_{1p}(a,b) - \nabla\phi_{1p}(c,d)\phi_{1p}(c,d)\|\\
\ge&\left|\frac{\mathrm{sgn}(a)|a|^{p-1}}{\|(a,b)\|_p^{p-1}}\phi_{1p}(a,b) - \frac{\mathrm{sgn}(c)|c|^{p-1}}{\|(c,d)\|_p^{p-1}}\phi_{1p}(c,d) + \phi_{1p}(c,d) - \phi_{1p}(a,b)\right|\\
\ge&\left|\frac{\mathrm{sgn}(a)|a|^{p-1}}{\|(a,b)\|_p^{p-1}}\phi_{1p}(a,b) - \frac{\mathrm{sgn}(c)|c|^{p-1}}{\|(c,d)\|_p^{p-1}}\phi_{1p}(c,d)\right| - |\phi_{1p}(c,d) - \phi_{1p}(a,b)|\\
\ge&\left|\frac{\mathrm{sgn}(a)|a|^{p-1}}{\|(a,b)\|_p^{p-1}}\phi_{1p}(a,b) - \frac{\mathrm{sgn}(c)|c|^{p-1}}{\|(c,d)\|_p^{p-1}}\phi_{1p}(c,d)\right| - (\kappa + \sqrt{2})\|(c,d) - (a,b)\|,
\end{aligned}
$$

where $\kappa + \sqrt{2}$ is given in Proposition 2.43(iv). If we let $(a,b) = (1,-n)$, $(c,d) = (-1,-n)$ with $n \in (1,\infty)$, we have

$$
\begin{aligned}
&\left|\frac{\mathrm{sgn}(a)|a|^{p-1}}{\|(a,b)\|_p^{p-1}}\phi_{1p}(a,b) - \frac{\mathrm{sgn}(c)|c|^{p-1}}{\|(c,d)\|_p^{p-1}}\phi_{1p}(c,d)\right|\\
=&\frac{\sqrt[p]{1+n^p} + (n-1)}{(1+n^p)^{(p-1)/p}} + \frac{\sqrt[p]{1+n^p} + (n+1)}{(1+n^p)^{(p-1)/p}}\\
=&2\frac{\sqrt[p]{1+n^p} + n}{(1+n^p)^{(p-1)/p}}\\
\ge&\frac{4n}{(1+n^p)^{(p-1)/p}}\\
=&\frac{4n^{2-p}n^{p-1}}{(1+n^p)^{(p-1)/p}}\\
=&\frac{4n^{2-p}}{(1+(1/n)^p)^{(p-1)/p}}\\
\ge&n^{2-p},
\end{aligned}
$$

where the first and the second inequalities follow from $2 > p > 1$ and $n > 1$. Since $\|(a, b) - (c, d)\| = 2$ and $n \in (1, \infty)$, from the above inequalities it is easy to verify that $\nabla\psi_{1p}$ is not Lipschitz continuous.

We now turn our attention to the merit function for the NCP as defined in (2.54), and proceed to examine several of its key properties. These properties serve as the theoretical foundation for the algorithms presented in Chap. 5. Furthermore, we explore the semismoothness related characteristics of the merit function. To this end, we define

$$\Phi_{\theta,p}(x) := \begin{pmatrix} \phi_{\theta,p}(x_1, F_1(x)) \\ \cdots \\ \phi_{\theta,p}(x_n, F_n(x)) \end{pmatrix}. \tag{2.63}$$

Consequently, the merit function defined in (2.54) can be expressed as

$$\Psi_{\theta,p}(x) = \frac{1}{2}\|\Phi_{\theta,p}(x)\|^2 = \sum_{i=1}^{n} \psi_{\theta,p}(x_i, F_i(x)). \tag{2.64}$$

Proposition 2.47 *(i) The function $\psi_{\theta,p}$ defined by (2.58) with $p \geq 2$ and $\theta \in (0, 1]$ is an SC^1 function. Hence, if every F_i is an SC^1 function, then the function $\Psi_{\theta,p}$ defined by (2.64) with $p \geq 2$ and $\theta \in (0, 1]$ is also an SC^1 function.*

(ii) If every F_i is an LC^1 function, then the function $\Phi_{\theta p}$ defined by (2.63) with $p > 1$ and $\theta \in (0, 1]$ is strongly semismooth.

(iii) The function $\psi_{\theta,p}$ defined by (2.58) with $p \geq 2$ and $\theta \in (0, 1]$ is an LC^1 function. Hence, if every F_i is an LC^1 function, then the function $\Psi_{\theta,p}$ defined by (2.64) with $p \geq 2$ and $\theta \in (0, 1]$ is also an LC^1 function.

Proof (i) By Proposition 2.45, it suffices to establish the semismoothness of $\nabla\psi_{\theta,p}$. From the proof of Proposition 2.46, it is evident that $\nabla\psi_{\theta,p}(a, b)$ is continuously differentiable for all $(a, b) \neq (0, 0)$. Thus, our focus narrows to demonstrating the semismoothness of $\nabla\psi_{\theta,p}(a, b)$ at the point $(0, 0)$.

For any direction $(h_1, h_2) \in \mathbb{R}^2 \setminus \{(0, 0)\}$, it is known that $\nabla\psi_{\theta,p}$ is differentiable at (h_1, h_2). Consequently, it remains to show that

$$\nabla\psi_{\theta,p}(h_1, h_2) - \nabla\psi_{\theta,p}(0, 0) - \nabla^2\psi_{\theta,p}(h_1, h_2) \cdot (h_1, h_2)^{\mathsf{T}} = o(\|(h_1, h_2)\|).$$

In fact, let $\hat{a}_1, \hat{a}_2, \hat{b}_1, \hat{b}_2, \hat{c}_1, \hat{c}_2$ be similarly defined as those in Proposition 2.46 with (a, b) being replaced by (h_1, h_2). Denote

$$\hat{h}_3 := (p-1)\frac{\hat{a}_1+\hat{a}_2}{\eta_{\theta,p}^{2p-1}(h_1,h_2)}\phi_{\theta,p}(h_1,h_2);$$
$$\hat{h}_4 := (p-1)\frac{\hat{b}_1+\hat{b}_2}{\eta_{\theta,p}^{2p-1}(h_1,h_2)}\phi_{\theta,p}(h_1,h_2);$$
$$\hat{h}_5 := (p-1)\frac{\hat{c}_1+\hat{c}_2}{\eta_{\theta,p}^{2p-1}(h_1,h_2)}\phi_{\theta,p}(h_1,h_2),$$

and

$$\begin{aligned}
m_1 &:= (\theta|h_1|^{p-2}+(1-\theta)|h_1-h_2|^{p-2})\eta_{\theta,p}^p(h_1,h_2)h_1 - \hat{h}_1^2\eta_{\theta,p}^{2p-2}(h_1,h_2)h_1;\\
m_2 &:= (1-\theta)|h_1-h_2|^{p-2}\eta_{\theta,p}^p(h_1,h_2)h_2 + \hat{h}_1\hat{h}_2\eta_{\theta,p}^{2p-2}(h_1,h_2)h_2;\\
m_3 &:= (\theta|h_1|^{p-2}+(1-\theta)|h_1-h_2|^{p-2})\eta_{\theta,p}^p(h_1,h_2)h_1\\
&\quad -(1-\theta)|h_1-h_2|^{p-2}\eta_{\theta,p}^p(h_1,h_2)h_2;\\
m_4 &:= \hat{h}_1\hat{h}_2\eta_{\theta,p}^{2p-2}(h_1,h_2)h_2 + \hat{h}_1^2\eta_{\theta,p}^{2p-2}(h_1,h_2)h_1;\\
m_5 &:= (\theta\mathrm{sgn}(h_1)|h_1|^{p-1}+(1-\theta)\mathrm{sgn}(h_1-h_2)|h_1-h_2|^{p-1})\eta_{\theta,p}^p(h_1,h_2);\\
m_6 &:= \hat{h}_1\hat{h}_2\eta_{\theta,p}^{2p-2}(h_1,h_2)h_2 + \hat{h}_1^2\eta_{\theta,p}^{2p-2}(h_1,h_2)h_1.
\end{aligned}$$

Then,

$$\begin{aligned}
\begin{pmatrix} H_1 \\ H_2 \end{pmatrix} := \begin{pmatrix} \hat{h}_1-1 \\ \hat{h}_2-1 \end{pmatrix}\cdot\phi_{\theta,p}(h_1,h_2) - \begin{pmatrix} 0 \\ 0 \end{pmatrix}\\
-\begin{pmatrix} (\hat{h}_1-1)^2+\hat{h}_3 & (\hat{h}_1-1)(\hat{h}_2-1)+\hat{h}_4 \\ (\hat{h}_1-1)(\hat{h}_2-1)+\hat{h}_4 & (\hat{h}_2-1)^2+\hat{h}_5 \end{pmatrix}\cdot\begin{pmatrix} h_1 \\ h_2 \end{pmatrix}.
\end{aligned}$$

and hence,

$$\begin{aligned}
H_1 &= (\hat{h}_1-1)\phi_{\theta,p}(h_1,h_2) - ((\hat{h}_1-1)^2+\hat{h}_3)h_1 - ((\hat{h}_1-1)(\hat{h}_2-1)+\hat{h}_4)h_2\\
&= (\hat{h}_1-1)\phi_{\theta,p}(h_1,h_2) - \hat{h}_3h_1 - \hat{h}_4h_2 - (\hat{h}_1-1)((\hat{h}_1-1)h_1+(\hat{h}_2-1)h_2)\\
&= (\hat{h}_1-1)\phi_{\theta,p}(h_1,h_2) - \hat{h}_3h_1 - \hat{h}_4h_2 - (\hat{h}_1-1)\phi_{\theta,p}(h_1,h_2)\\
&= -(p-1)\left(\frac{\hat{a}_1+\hat{a}_2}{\eta_{\theta,p}^{2p-1}(h_1,h_2)}h_1 + \frac{\hat{b}_1+\hat{b}_2}{\eta_{\theta,p}^{2p-1}(h_1,h_2)}h_2\right)\phi_{\theta,p}(h_1,h_2)\\
&= -(p-1)\phi_{\theta,p}(h_1,h_2)\left(\frac{m_1-m_2}{\eta_{\theta,p}^{2p-1}(h_1,h_2)}\right)\\
&= -(p-1)\phi_{\theta,p}(h_1,h_2)\left(\frac{m_3-m_4}{\eta_{\theta,p}^{2p-1}(h_1,h_2)}\right)
\end{aligned}$$

$$
\begin{aligned}
&= -(p-1)\phi_{\theta,p}(h_1,h_2)\left(\frac{m_5 - m_6}{\eta_{\theta,p}^{2p-1}(h_1,h_2)}\right)\\
&= -(p-1)\phi_{\theta,p}(h_1,h_2)\left(\hat{h}_1 - \hat{h}_1\frac{\hat{h}_1 h_1 + \hat{h}_2 h_2}{\eta_{\theta,p}(h_1,h_2)}\right)\\
&= -(p-1)\phi_{\theta,p}(h_1,h_2)(\hat{h}_1 - \hat{h}_1)\\
&= 0,
\end{aligned}
$$

where the third equality follows from $\hat{h}_1 h_1 + \hat{h}_2 h_2 = \eta_{\theta,p}$ given in the proof of Proposition 2.43 and the definition of $\phi_{\theta,p}$, the fourth equality follows from the definitions of $\hat{h}_3$, $\hat{h}_4$, the fifth equality follows from the definitions of $\hat{a}_1$, $\hat{a}_2$, $\hat{b}_1$, $\hat{b}_2$, and the eighth equality follows from $\hat{h}_1 h_1 + \hat{h}_2 h_2 = \eta_{\theta,p}$ given in the proof of Proposition 2.43.

Similar analysis yields $H_2 = 0$. Thus, $\nabla\psi_{\theta,p}$ is semismooth. Furthermore, $\psi_{\theta,p}$ is a SC^1 function.

(ii) Since an LC^1 function is strongly semismooth, and the composition of strongly semismooth functions preserves strong semismoothness, it follows from Proposition 2.43(vi) that the desired result holds.

(iii) Utilizing the results established above, it is straightforward to verify that assertion (iii) holds. □

The conclusions of Proposition 2.47(i) and (iii) no longer hold when $p \in (1,2)$ for all $\theta \in (0,1]$, due to the fact that $\nabla\psi_{\theta,p}$ is, in general, not locally Lipschitz continuous. For instance, consider the points $(a,b) = (\frac{1}{n}, -1)$ and $(c,d) = (-\frac{1}{n}, -1)$; it can be shown that $\nabla\psi_{\theta,p}$ fails to be Lipschitz continuous in any neighborhood of $(0,-1)$.

Proposition 2.48 *Let $\Psi_{\theta,p} : \mathbb{R}^n \to \mathbb{R}$ be defined by* (2.64) *with $p > 1$ and $\theta \in (0,1]$. Then, $\Psi_{\theta,p}(x) \geq 0$ for all $x \in \mathbb{R}^n$ and $\Psi_{\theta,p}(x) = 0$ if and only if x solves the NCP* (2.1). *Moreover, suppose that the solution set of the NCP* (2.1) *is nonempty, then x is a global minimizer of $\Psi_{\theta,p}$ if and only if x solves the NCP* (2.1).

Proof The result follows from Proposition 2.45 immediately. □

Proposition 2.49 *Let $\Psi_{\theta,p} : \mathbb{R}^n \to \mathbb{R}$ be defined by* (2.64) *with $p > 1, \theta \in (0,1]$. Suppose that F is either a monotone function or a P_0-function, then every stationary point of $\Psi_{\theta,p}$ is a global minima of $\min_{x\in\mathbb{R}^n} \Psi_{\theta,p}(x)$, and therefore solves the NCP* (2.1).

Proof Relying on Proposition 2.45 and [22, Lemma 2.1], the proof of this proposition closely follows the argument presented in [22, Proposition 3.4], and is thus omitted. □

Proposition 2.50 *Let $\Psi_{\theta,p}$ be defined by* (2.64) *with $\theta \in (0,1]$ and $p > 1$. Suppose that F is either a strongly monotone function or a uniform P-function. Then the level sets*

$$\mathcal{L}(\Psi_{\theta,p}, \gamma) := \{x \in \mathbb{R}^n \,|\, \Psi_{\theta,p}(x) \le \gamma\}$$

are bounded for all $\gamma \in \mathbb{R}$.

Proof By invoking Proposition 2.44, the proof proceeds in a manner analogous to that of [22, Proposition 3.5], and is therefore omitted here. □

2.3.2 Construction by Using Penalized Term

An alternative approach involves introducing a penalization term in place of a parameter. In this section, we explore this technique by analyzing the following merit function $\Psi_{\alpha,p} : \mathbb{R}^n \to \mathbb{R}$, defined as

$$\Psi_{\alpha,p}(x) := \sum_{i=1}^{n} \psi_{\alpha,p}(x_i \, , \, F_i(x)), \tag{2.65}$$

where $\psi_{\alpha,p} : \mathbb{R}^2 \to \mathbb{R}_+$ is an NCP function given by

$$\psi_{\alpha,p}(a,b) := \frac{\alpha}{2}(\max\{0, ab\})^2 + \psi_p(a,b) = \frac{\alpha}{2}(ab)_+^2 + \frac{1}{2}(\|(a,b)\|_p - (a+b))^2 \tag{2.66}$$

with $\alpha \ge 0$ being a real parameter. When $\alpha = 0$, the function $\psi_{\alpha,p}$ reduces to ψ_p, making $\psi_{\alpha,p}$ an extension of ψ_{FB}^p. Moreover, $\psi_{\alpha,p}$ generalizes the function ψ_α investigated by Yamada, Yamashita, and Fukushima in [40], which corresponds to the case $p = 2$. In what follows, we examine several advantageous properties of the merit function $\psi_{\alpha,p}$ that are instrumental in the subsequent analysis. We also establish mild conditions under which the merit function $\Psi_{\alpha,p}$ possesses bounded level sets and provides a global error bound.

The next lemma demonstrates that $\psi_{\alpha,p}$ shares many of the favorable properties of ψ_p. Additionally, when $\alpha > 0$, it enjoys a significant property that ψ_{FB}^p lacks; see Lemma 2.11(f).

Lemma 2.11 *The function* $\psi_{\alpha,p}$ *defined by (2.66) has the following favorable properties:*

(a) $\psi_{\alpha,p}$ *is an NCP function and* $\psi_{\alpha,p} \ge 0$ *for all* $(a,b) \in \mathbb{R}^2$.
(b) $\psi_{\alpha,p}$ *is continuously differentiable everywhere, and moreover, if* $(a,b) \ne (0,0)$,

$$\begin{aligned}
\nabla_a \psi_{\alpha,p}(a,b) &= \alpha b (ab)_+ + \left(\frac{\operatorname{sgn}(a) \cdot |a|^{p-1}}{\|(a,b)\|_p^{p-1}} - 1 \right) \phi_p(a,b), \\
\nabla_b \psi_{\alpha,p}(a,b) &= \alpha a (ab)_+ + \left(\frac{\operatorname{sgn}(b) \cdot |b|^{p-1}}{\|(a,b)\|_p^{p-1}} - 1 \right) \phi_p(a,b);
\end{aligned} \tag{2.67}$$

and otherwise $\nabla_a \psi_{\alpha,p}(0,0) = \nabla_b \psi_{\alpha,p}(0,0) = 0$.

(c) For $p \geq 2$, the gradient of $\psi_{\alpha,p}$ is Lipschitz continuous on any nonempty bounded set S, i.e., there exists $L > 0$ such that for any $(a, b), (c, d) \in S$,

$$\|\nabla\psi_{\alpha,p}(a,b) - \nabla\psi_{\alpha,p}(c,d)\| \leq L\|(a,b) - (c,d)\|.$$

(d) $\nabla_a\psi_{\alpha,p}(a,b) \cdot \nabla_b\psi_{\alpha,p}(a,b) \geq 0$ for any $(a,b) \in \mathbb{R}^2$, and furthermore, the equality holds if and only if $\psi_{\alpha,p}(a,b) = 0$.
(e) $\nabla_a\psi_{\alpha,p}(a,b) = 0 \Longleftrightarrow \nabla_b\psi_{\alpha,p}(a,b) = 0 \Longleftrightarrow \psi_{\alpha,p}(a,b) = 0$.
(f) Suppose that $\alpha > 0$. If $a \to -\infty$ or $b \to -\infty$ or $ab \to \infty$, then $\psi_{\alpha,p}(a,b) \to \infty$.

Proof Parts (a), (b), and (f) directly follow from the definition of $\psi_{\alpha,p}$ and Proposition 3.2(a) and (c) and [22, Lemma 3.1]. It remains to show parts (c)–(e).

(c) Notice that the functions $a(ab)_+$ and $b(ab)_+$ for any $a, b \in \mathbb{R}$ are Lipschitz continuous on any nonempty bounded set S, whereas $\phi_p(a,b)$ is Lipschitz continuous on $\mathbb{R}^2$ by [22, Proposition 3.1(e)]. Therefore, by the expression of $\nabla\psi_{\alpha,p}(a,b)$ and the boundedness of

$$\left(\frac{\operatorname{sgn}(a)\cdot|a|^{p-1}}{\|(a,b)\|_p^{p-1}} - 1\right) \quad \text{and} \quad \left(\frac{\operatorname{sgn}(b)\cdot|b|^{p-1}}{\|(a,b)\|_p^{p-1}} - 1\right),$$

it is not hard to verify that the gradient $\nabla\psi_{\alpha,p}(a,b)$ is Lipschitz continuous on S for $p \geq 2$.

(d) If $(a,b) = (0,0)$, part (d) clearly holds. Now suppose that $(a,b) \neq (0,0)$. Then,

$$\begin{aligned}
&\nabla_a\psi_{\alpha,p}(a,b)\cdot\nabla_b\psi_{\alpha,p}(a,b) \\
&= \left(\frac{\operatorname{sgn}(a)\cdot|a|^{p-1}}{\|(a,b)\|_p^{p-1}} - 1\right)\left(\frac{\operatorname{sgn}(b)\cdot|b|^{p-1}}{\|(a,b)\|_p^{p-1}} - 1\right)(\phi_{\text{FB}}^p)^2(a,b) \qquad (2.68)\\
&\quad + \alpha^2 ab(ab)_+{}^2 + \alpha a(ab)_+\left(\frac{\operatorname{sgn}(a)\cdot|a|^{p-1}}{\|(a,b)\|_p^{p-1}} - 1\right)\phi_{\text{FB}}^p(a,b) \\
&\quad + \alpha b(ab)_+\left(\frac{\operatorname{sgn}(b)\cdot|b|^{p-1}}{\|(a,b)\|_p^{p-1}} - 1\right)\phi_{\text{FB}}^p(a,b).
\end{aligned}$$

Since

$$ab(ab)_+{}^2 \geq 0, \quad \frac{\operatorname{sgn}(a)\cdot|a|^{p-1}}{\|(a,b)\|_p^{p-1}} - 1 \leq 0, \quad \text{and} \quad \frac{\operatorname{sgn}(b)\cdot|b|^{p-1}}{\|(a,b)\|_p^{p-1}} - 1 \leq 0, \qquad (2.69)$$

it suffices to show that the last two terms of (2.68) are nonnegative. We next claim that

$$\alpha a(ab)_+ \left(\frac{\operatorname{sgn}(a) \cdot |a|^{p-1}}{\|(a,b)\|_p^{p-1}} - 1 \right) \phi_p(a,b) \geq 0, \quad \forall\, (a,b) \neq (0,0). \tag{2.70}$$

If $a \leq 0$, then $\phi_{\text{FB}}^p(a,b) \geq 0$, which together with the second inequality in (2.69) implies that (2.70) holds. If $a > 0$ and $b > 0$, then $\phi_{\text{FB}}^p(a,b) < 0$, which implies (2.70) by a similar reason. If $a > 0$ and $b \leq 0$, then $(ab)_+ = 0$, and hence (2.70) holds. Similarly, we have that

$$\alpha b(ab)_+ \left(\frac{\operatorname{sgn}(b) \cdot |b|^{p-1}}{\|(a,b)\|_p^{p-1}} - 1 \right) \phi_{\text{FB}}^p(a,b) \geq 0, \quad \forall\, (a,b) \neq (0,0).$$

Consequently, $\nabla_a \psi_{\alpha,p}(a,b) \cdot \nabla_b \psi_{\alpha,p}(a,b) \geq 0$. From (2.68), $\nabla_a \psi_{\alpha,p}(a,b) \cdot \nabla_b \psi_{\alpha,p}(a,b) = 0$ if and only if $\{a = 0$ or $(a \geq 0$ and $b = 0)$ or $\phi_p(a,b)=0\}$ and $\{b = 0$ or $(b \geq 0$ and $a = 0)$ or $\phi_{\text{FB}}^p(a,b) = 0\}$ and $\{ab = 0\}$. Thus, $\nabla_a \psi_\alpha(a,b) \cdot \nabla_b \psi_{\alpha,p}(a,b) = 0$ if and only if $\psi_{\alpha,p}(a,b) = 0$.

(e) If $\psi_{\alpha,p}(a,b) = 0$, then $ab = 0$ and $\phi_{\text{FB}}^p(a,b) = 0$ by part (a), which in turn implies that $\nabla_a \psi_{\alpha,p}(a,b) = 0$ and $\nabla_b \psi_{\alpha,p}(a,b) = 0$. Next, we claim that $\nabla_a \psi_{\alpha,p}(a,b) = 0$ implies $\psi_{\alpha,p}(a,b) = 0$. Suppose that $\nabla_a \psi_{\alpha,p}(a,b) = 0$. Then,

$$\alpha b(ab)_+ = -\left(\frac{\operatorname{sgn}(a) \cdot |a|^{p-1}}{\|(a,b)\|_p^{p-1}} - 1 \right) \phi_{\text{FB}}^p(a,b). \tag{2.71}$$

We can verify that the equality (2.71) implies $b = 0, a \geq 0$ or $b > 0, a = 0$. Under the two cases, we achieve $\psi_{\alpha,p}(a,b) = 0$. Similarly, $\nabla_b \psi_{\alpha,p}(a,b) = 0$ also implies $\psi_{\alpha,p}(a,b) = 0$. □

Note that $ab \to \infty$ does not necessarily imply $\psi_p(a,b) \to \infty$, which means ψ_{FB}^p does not share Lemma 2.11(f). In fact, for $\alpha = 0$, the lemma needs to be modified as "if $(a \to \infty)$ or $(b \to \infty)$ or $(a \to \infty$ and $b \to \infty)$, then $\psi_{\alpha,p}(a,b) \to \infty$". As we will see later, Lemma 2.11(f) is useful for proving that the level sets of $\Psi_{\alpha,p}$ are bounded. Besides, by Lemma 2.11(a), we immediately have the following result.

Proposition 2.51 *Let $\Psi_{\alpha,p}$ be defined as in (2.65). Then, $\Psi_{\alpha,p}(x) \geq 0$ for all $x \in \mathbb{R}^n$ and $\Psi_{\alpha,p}(x) = 0$ if and only if x solves the NCP. Moreover, if the NCP has at least one solution, then x is a global minimizer of $\Psi_{\alpha,p}$ if and only if x solves the NCP.*

Proposition 2.51 indicates that the NCP can be recast as the unconstrained minimization:

$$\min_{x \in \mathbb{R}^n} \Psi_{\alpha,p}(x). \tag{2.72}$$

In general, finding a global minimizer of $\Psi_{\alpha,p}$ is a challenging task. Hence, it is crucial to identify conditions under which a stationary point of $\Psi_{\alpha,p}$ is guaranteed to be a global minimum. By applying Lemma 2.11(d) along with the proof techniques used in [29, Theorem 3.5], one can establish that each stationary point of $\Psi_{\alpha,p}$ is a global minimizer if and only if the mapping F is a P_0-function.

Proposition 2.52 *Let F be a P_0-function. Then $x^* \in \mathbb{R}^n$ is a global minimum of the unconstrained optimization problem (2.72) if and only if x^* is a stationary point of $\Psi_{\alpha,p}$.*

The following proposition demonstrates that the unconstrained minimization (2.72) admits a stationary point under fairly mild assumptions on the mapping F. Given that similar results and related analyses can be found in [20, Proposition 4.1], [29, Theorem 3.8], and [14, Theorem 4.1], we omit the proof here.

Proposition 2.53 *The function $\Psi_{\alpha,p}$ has bounded level sets $\mathcal{L}(\Psi_{\alpha,p}, \gamma)$ for all $\gamma \in \mathbb{R}$, if F is monotone and the NCP is strictly feasible (i.e., there exists $\hat{x} > 0$ such that $F(\hat{x}) > 0$) when $\alpha > 0$, or F is a uniform P-function when $\alpha \geq 0$.*

In what follows, we show that the merit functions $\Psi^p_{\rm FB}$, $\Psi_{\rm NR}$, and $\Psi_{\alpha,p}$ exhibit the same order of growth on any bounded set.

Proposition 2.54 *Let $\Psi_{\rm NR}$, $\Psi^p_{\rm FB}$, and $\Psi_{\alpha,p}$ be defined as in (2.10), (2.17), and (2.65), respectively. Let S be an arbitrary bounded set. Then, for any $p > 1$, we have*

$$\left(2 - 2^{\frac{1}{p}}\right)^2 \Psi_{\rm NR}(x) \leq \Psi^p_{\rm FB}(x) \leq \left(2 + 2^{\frac{1}{p}}\right)^2 \Psi_{\rm NR}(x) \quad \text{for all } x \in \mathbb{R}^n \tag{2.73}$$

and

$$\left(2 - 2^{\frac{1}{p}}\right)^2 \Psi_{\rm NR}(x) \leq \Psi_{\alpha,p}(x) \leq \left(\alpha B^2 + (2 + 2^{\frac{1}{p}})^2\right) \Psi_{\rm NR}(x) \quad \text{for all } x \in S, \tag{2.74}$$

where B is a constant defined by $B = \max\limits_{1 \leq i \leq n} \left\{ \sup\limits_{x \in S} \{\max\{|x_i|, |F_i(x)|\}\} \right\} < \infty$.

Proof The inequality in (2.73) is direct by Lemma 2.3 and the definitions of Ψ_p and $\Psi_{\rm NR}$. In addition, from Lemma 2.3 and the definition of $\Psi_{\alpha,p}$, it follows that

$$\Psi_{\alpha,p}(x) \geq \left(2 - 2^{\frac{1}{p}}\right)^2 \Psi_{\rm NR}(x) \quad \text{for all } x \in \mathbb{R}^n.$$

We next prove the inequality on the right-hand side of (2.74). We claim that, for each i,

$$(x_i F_i(x))_+ \leq B|\min\{x_i, F_i(x)\}| \quad \text{for all } x \in S. \tag{2.75}$$

Without loss of generality, suppose $F_i(x) \geq x_i$. If $F_i(x) \geq x_i \geq 0$, it follows that

$$(x_i F_i(x))_+ = x_i F_i(x) = F_i(x)|\min\{x_i, F_i(x)\}| \leq B|\min\{x_i, F_i(x)\}|.$$

If $F_i(x) \geq 0 \geq x_i$, then $(x_i F_i(x))_+ = 0$. If $0 \geq F_i(x) \geq x_i$, it follows that

$$(x_i F_i(x))_+ = |x_i F_i(x)| \leq |x_i|^2 \leq B|\min\{x_i, F_i(x)\}|.$$

Thus, (2.75) holds for all $x \in S$. By Lemma 2.3 and (2.75), for all $i = 1, \ldots, n$ and $x \in S$,

$$\psi_{\alpha,p}(x_i, F_i(x)) \leq \left\{\alpha B^2 + (2 + 2^{\frac{1}{p}})^2\right\} \min\{x_i, F_i(x)\}^2$$

holds for any $p > 1$. The proof is then complete by the definition of $\Psi_{\alpha,p}$ and Ψ_{NR}. □

Proposition 2.55 *Let Ψ^p_{FB} and $\Psi_{\alpha,p}$ be defined by (2.17) and (2.65) respectively, and S be any bounded set. Then, for any $p > 1$ and all $x \in S$, we have the following inequalities:*

$$\frac{(2 - 2^{\frac{1}{p}})^2}{\left(\alpha B^2 + (2 + 2^{\frac{1}{p}})^2\right)} \Psi_{\alpha,p}(x) \leq \Psi^p_{\text{FB}}(x) \leq \frac{(2 + 2^{\frac{1}{p}})^2}{(2 - 2^{\frac{1}{p}})^2} \Psi_{\alpha,p}(x)$$

where B is the constant defined as in Proposition 2.54.

Proof This is an immediate consequence of Proposition 2.54. □

Since Ψ^p_{FB}, Ψ_{NR} and $\Psi_{\alpha,p}$ exhibit the same order of growth on bounded sets, it follows that any one of them yields a global error bound for the NCP if the others do as well. In what follows, we establish that $\Psi_{\alpha,p}$ provides a global error bound for the NCP when $\alpha > 0$, even in the absence of Lipschitz continuity of the mapping F.

Proposition 2.56 *Let $\Psi_{\alpha,p}$ be defined as in (2.65). Suppose that F is a uniform P-function with modulus $\mu > 0$. If $\alpha > 0$, then there exists a constant $\kappa_1 > 0$ such that*

$$\|x - x^*\| \leq \kappa_1 \Psi_{\alpha,p}(x)^{\frac{1}{4}} \quad \text{for all } x \in \mathbb{R}^n;$$

if $\alpha = 0$ and S is any bounded set, there exists a constant $\kappa_2 > 0$ such that

$$\|x - x^*\| \leq \kappa_2 \left(\max\left\{\Psi_{\alpha,p}(x), \sqrt{\Psi_{\alpha,p}(x)}\right\}\right)^{\frac{1}{2}} \quad \text{for all } x \in S;$$

where $x^* = (x_1^*, \ldots, x_n^*)$ is the unique solution for the NCP (2.1).

Proof Since F is a uniform P-function, the NCP has the unique solution, and moreover,

$$\begin{aligned}
\mu\|x - x^*\|^2 &\leq \max_{1\leq i\leq n} (x - x^*)(F_i(x) - F_i(x^*)) \\
&= \max_{1\leq i\leq n} \{x_i F_i(x) - x_i^* F_i(x) - x_i F_i(x^*) + x_i^* F_i(x^*)\} \\
&= \max_{1\leq i\leq n} \{x_i F_i(x) - x_i^* F_i(x) - x_i F_i(x^*)\} \\
&\leq \max_{1\leq i\leq n} \tau_i\{(x_i F_i(x))_+ + (-F_i(x))_+ + (-x_i)_+\},
\end{aligned} \tag{2.76}$$

where $\tau_i := \max\{1, x_i^*, F_i(x^*)\}$. We next prove that for all $(a, b) \in \mathbb{R}^2$,

$$(-a)_+{}^2 + (-b)_+{}^2 \le \left[\|(a, b)\|_p - (a + b)\right]^2. \tag{2.77}$$

Without loss of generality, suppose $a \ge b$. If $a \ge b \ge 0$, then (2.77) holds obviously. If $a \ge 0 \ge b$, then $\|(a, b)\|_p - (a + b) \ge -b \ge 0$, which in turn implies that

$$(-a)_+{}^2 + (-b)_+{}^2 = b^2 \le \left[\|(a, b)\|_p - (a + b)\right]^2.$$

If $0 \ge a \ge b$, then $(-a)_+{}^2 + (-b)_+{}^2 = a^2 + b^2 \le \left[\|(a, b)\|_p - (a + b)\right]^2$. Hence, (2.77) follows.

Suppose that $\alpha > 0$. Using the inequality (2.77), we then obtain that

$$\begin{aligned}
\left[(ab)_+ + (-a)_+ + (-b)_+\right]^2 =& (ab)_+^2 + (-b)_+^2 + (-a)_+^2 + 2(ab)_+(-a)_+ \\
&+ 2(-a)_+(-b)_+ + 2(ab)_+(-b)_+ \\
\le& (ab)_+^2 + (-b)_+^2 + (-a)_+^2 + (ab)_+^2 + (-a)_+^2 \\
&+ (-a)_+^2 + (-b)_+^2 + (ab)_+^2 + (-b)_+^2 \\
\le& 3\left[(ab)_+^2 + \left(\|(a, b)\|_p - (a + b)\right)^2\right] \\
\le& \tau\left[\frac{\alpha}{2}(ab)_+^2 + \frac{1}{2}\left(\|(a, b)\|_p - (a + b)\right)^2\right] \\
=& \tau\psi_{\alpha,p}(a, b) \quad \text{for all } (a, b) \in \mathbb{R}^2,
\end{aligned} \tag{2.78}$$

where $\tau := \max\left\{\frac{6}{\alpha}, 6\right\} > 0$. Combining (2.78) with (2.76) and letting $\hat{\tau} = \max_{1 \le i \le n} \tau_i$, we get

$$\begin{aligned}
\mu\|x - x^*\|^2 \le& \max_{1 \le i \le n} \tau_i \left\{\tau\psi_{\alpha,p}(x_i, F_i(x))\right\}^{1/2} \\
\le& \hat{\tau}\tau^{1/2} \max_{1 \le i \le n} \psi_{\alpha,p}(x_i, F(x))^{1/2} \\
\le& \hat{\tau}\tau^{1/2} \left\{\sum_{i=1}^{n} \{\psi_{\alpha,p}(x_i, F_i(x))\right\}^{1/2} \\
=& \hat{\tau}\tau^{1/2}\Psi_{\alpha,p}(x, F(x))^{1/2}.
\end{aligned}$$

From this, the first desired result follows immediately by setting $\kappa_1 := \left[\hat{\tau}\tau^{1/2}/\mu\right]^{1/2}$.

Suppose that $\alpha = 0$. From the proof of Proposition 2.54, the inequality (2.75) holds. Combining with equations (2.76)–(2.77), it then follows that for all $x \in S$,

$$
\begin{aligned}
\mu\|x - x^*\|^2 \leq & \max_{1\leq i\leq n} \tau_i \left[B|\min\{x_i, F_i(x)\}| + (\psi^p_{\rm FB}(x_i, F_i(x)))^{1/2}\right] \\
\leq & \hat{\tau} \max_{1\leq i\leq n} \left[\sqrt{2}\hat{B}\psi^p_{\rm FB}(x_i, F_i(x)) + (\psi^p_{\rm FB}(x_i, F_i(x)))^{1/2}\right] \\
\leq & \sqrt{2}\hat{\tau}\hat{B}\left(\Psi^p_{\rm FB}(x) + \sqrt{\Psi^p_{\rm FB}(x)}\right) \\
\leq & 4\hat{\tau}\hat{B}\max\left\{\Psi^p_{\rm FB}(x), \sqrt{\Psi^p_{\rm FB}(x)}\right\} \\
= & 4\hat{\tau}\hat{B}\max\left\{\Psi_{\alpha,p}(x), \sqrt{\Psi_{\alpha,p}(x)}\right\},
\end{aligned}
$$

where $\hat{B} = B/(2 - 2^{\frac{1}{p}})$ and the second inequality is due to Lemma 2.3. Letting $\kappa_2 := 2\left[\hat{\tau}\hat{B}/\mu\right]^{1/2}$, we obtain the desired result from the above inequality. □

The following lemma is crucial in the proof of Proposition 2.57, which is pivotal in establishing the convergence rate of the algorithm.

Lemma 2.12 *For all $(a, b) \neq (0, 0)$ and $p > 1$, we have the following inequality:*

$$
\left(\frac{\operatorname{sgn}(a)\cdot|a|^{p-1} + \operatorname{sgn}(b)\cdot|b|^{p-1}}{\|(a,b)\|_p^{p-1}} - 2\right)^2 \geq \left(2 - 2^{\frac{1}{p}}\right)^2.
$$

Proof If $a = 0$ or $b = 0$, the inequality holds obviously. Then we complete the proof by considering three cases: (i) $a > 0$ and $b > 0$, (ii) $a < 0$ and $b < 0$, and (iii) $ab < 0$.

Case (i): Without loss of generality, we suppose $a \geq b > 0$. Then, we have

$$
\frac{|a|^{p-1} + |b|^{p-1}}{\|(a,b)\|_p^{p-1}} = \frac{\left(|\frac{a}{b}|\right)^{p-1} + 1}{\left(\left(|\frac{a}{b}|\right)^p + 1\right)^{1-\frac{1}{p}}}.
$$

Let $f(t) := \dfrac{t^{p-1} + 1}{(t^p + 1)^{1-\frac{1}{p}}}$ for any $t > 0$. By computation, we obtain that

$$
f'(t) = \frac{t^{p-2}(p-1)(1-t)}{(t^p+1)^2}, \quad \forall t > 0.
$$

Since $f'(t) < 0$ for $t \geq 1$ and $f(1) = 2^{\frac{1}{p}}$, it follows that $f(t) \leq 2^{\frac{1}{p}}$ for $t \geq 1$. Therefore,

$$
\frac{|a|^{p-1} + |b|^{p-1}}{\|(a,b)\|_p^{p-1}} \leq 2^{\frac{1}{p}} \quad \text{for } p > 1,
$$

which in turn implies that $2 - \dfrac{|a|^{p-1} + |b|^{p-1}}{\|(a,b)\|_p^{p-1}} \geq 2 - 2^{\frac{1}{p}}$ for $p > 1$. Squaring both sides then leads to the desired inequality.

Case (ii): By similar arguments as in case (i), we obtain

$$2 - 2^{\frac{1}{p}} \leq 2 - \frac{|a|^{p-1} + |b|^{p-1}}{\|(a,b)\|_p^{p-1}} \leq 2 + \frac{|a|^{p-1} + |b|^{p-1}}{\|(a,b)\|_p^{p-1}} \quad \text{for } p > 1,$$

from which the result follows immediately.

Case (iii): Again, we suppose $|a| \geq |b|$ and therefore have

$$2^{\frac{1}{p}} \geq \frac{|a|^{p-1} + |b|^{p-1}}{\|(a,b)\|_p^{p-1}} \geq \frac{|a|^{p-1} - |b|^{p-1}}{\|(a,b)\|_p^{p-1}} \quad \text{for } p > 1.$$

Thus $2 - 2^{\frac{1}{p}} \leq 2 - \dfrac{|a|^{p-1} - |b|^{p-1}}{\|(a,b)\|_p^{p-1}}$ for $p > 1$ and the desired result is also satisfied. □

Proposition 2.57 *Let $\psi_{\alpha,p}$ be given as in (2.66). Then, for all $x \in \mathbb{R}^n$ and $p > 1$,*

$$\|\nabla_a \psi_{\alpha,p}(x, F(x)) + \nabla_b \psi_{\alpha,p}(x, F(x))\|^2 \geq 2\left(2 - 2^{\frac{1}{p}}\right)^2 \Psi^p_{\text{FB}}(x),$$

and particularly, for all x belonging to any bounded set S and $p > 1$,

$$\|\nabla_a \psi_{\alpha,p}(x, F(x)) + \nabla_b \psi_{\alpha,p}(x, F(x))\|^2 \geq \frac{2(2 - 2^{\frac{1}{p}})^4}{\left(\alpha B^2 + (2 + 2^{\frac{1}{p}})^2\right)} \Psi_{\alpha,p}(x)$$

where B is defined as in Proposition 2.54 and

$$\nabla_a \psi_{\alpha,p}(x, F(x)) := \left(\nabla_a \psi_{\alpha,p}(x_1, F_1(x)), \ldots, \nabla_a \psi_{\alpha,p}(x_n, F_n(x))\right)^{\mathsf{T}},$$

$$\nabla_b \psi_{\alpha,p}(x, F(x)) := \left(\nabla_b \psi_{\alpha,p}(x_1, F_1(x)), \ldots, \nabla_b \psi_{\alpha,p}(x_n, F_n(x))\right)^{\mathsf{T}}.$$

Proof The second part of the conclusions is direct by Proposition 2.55 and the first part. From the definition of $\nabla_a \psi_{\alpha,p}(x, F(x))$, $\nabla_b \psi_{\alpha,p}(x, F(x))$ and $\Psi^p_{\text{FB}}(x)$, the first part of the conclusions is equivalent to proving that the following inequality

$$\left(\nabla_a \psi_{\alpha,p}(a,b) + \nabla_b \psi_{\alpha,p}(a,b)\right)^2 \geq 2\left(2 - 2^{\frac{1}{p}}\right)^2 \psi^p_{\text{FB}}(a,b) \tag{2.79}$$

holds for all $(a,b) \in \mathbb{R}^2$. When $(a,b) = (0,0)$, the inequality (2.79) clearly holds. Suppose $(a,b) \neq (0,0)$. Then, it follows from Eq. (2.67) that

$$
\begin{aligned}
&\left(\nabla_a \psi_{\alpha,p}(a,b) + \nabla_b \psi_{\alpha,p}(a,b)\right)^2 \\
&= \left\{ \alpha(a+b)(ab)_+ + (\phi_p(a,b)) \left(\frac{\text{sgn}(a)\cdot|a|^{p-1} + \text{sgn}(b)\cdot|b|^{p-1}}{\|(a,b)\|_p^{p-1}} - 2 \right) \right\}^2 \\
&= \alpha^2(a+b)^2(ab)_+^2 + (\phi_p(a,b))^2 \left(\frac{\text{sgn}(a)\cdot|a|^{p-1} + \text{sgn}(b)\cdot|b|^{p-1}}{\|(a,b)\|_p^{p-1}} - 2 \right)^2 \\
&\quad +2\alpha(a+b)(ab)_+(\phi_p(a,b)) \left(\frac{\text{sgn}(a)\cdot|a|^{p-1} + \text{sgn}(b)\cdot|b|^{p-1}}{\|(a,b)\|_p^{p-1}} - 2 \right).
\end{aligned}
$$

Now, we claim that for all $(a,b) \neq (0,0) \in \mathbb{R}^2$,

$$
2\alpha(a+b)(ab)_+(\phi_p(a,b)) \left(\frac{\text{sgn}(a)\cdot|a|^{p-1} + \text{sgn}(b)\cdot|b|^{p-1}}{\|(a,b)\|_p^{p-1}} - 2 \right) \geq 0. \quad (2.80)
$$

If $ab \leq 0$, then $(ab)_+ = 0$ and the inequality (2.79) is clear. If $a, b > 0$, then by noting that

$$
\left(\frac{\text{sgn}(a)\cdot|a|^{p-1} + \text{sgn}(b)\cdot|b|^{p-1}}{\|(a,b)\|_p^{p-1}} - 2 \right) \leq 0, \quad \forall (a,b) \neq (0,0) \in \mathbb{R}^2 \quad (2.81)
$$

and $\phi^p_{\text{FB}}(a,b) \leq 0$, the inequality (2.80) also holds. If $a, b < 0$, then $\phi^p_{\text{FB}}(a,b) \geq 0$, which together with (2.81) then yields the inequality (2.80). Thus, we prove that the inequality (2.80) holds for all $(a,b) \neq (0,0)$. Using Lemma 2.12 and equations (2.80)–(2.81), we readily obtain the inequality (2.79) holds for all $(a,b) \neq (0,0)$. Then, the proof is complete. □

2.3.3 Construction by Using Parameter and Penalized Term

We combine both previously discussed ideas, parameterization and penalization, to construct a new class of NCP functions. In contrast to the function $\phi_{\theta,p}$ introduced in (2.53), defined by

$$
\phi_{\theta,p}(a,b) = \sqrt[p]{\theta(|a|^p + |b|^p) + (1-\theta)(|a-b|^p)} - (a+b),
$$

we now propose an alternative extension. Specifically, we define the function $\psi_{\alpha,\theta,p} : \mathbb{R}^2 \to \mathbb{R}_+$ as

$$
\psi_{\alpha,\theta,p}(a,b) := \frac{\alpha}{2}(\max\{0, ab\})^2 + \psi_{\theta,p}(a,b) \quad (2.82)
$$

where $\alpha \geq 0$ is a real parameter. The corresponding merit function $\Psi_{\alpha,\theta,p} : \mathbb{R}^n \to \mathbb{R}_+$ is defined by

$$\Psi_{\alpha,\theta,p}(x) := \sum_{i=1}^{n} \psi_{\alpha,\theta,p}(x_i\ ,\ F_i(x)). \tag{2.83}$$

It is worth noting that $\psi_{\alpha,\theta,p}$ encompasses several well-known functions, such as ψ_{FB}, ψ_{FB}^p, ψ_θ, $\psi_{\theta,p}$, and the function ψ_7 from [11], as special cases. Although $\psi_{\alpha,\theta,p}$ is constructed by penalizing the function $\psi_{\theta,p}$ studied in [25], we explore additional and more favorable properties of $\psi_{\alpha,\theta,p}$ here. In particular, we show that the merit function $\Psi_{\alpha,\theta,p}$ possesses bounded level sets and provides a global error bound for the NCP under mild assumptions, properties that were not addressed in [25]. Moreover, as highlighted in [17], the penalized Fischer-Burmeister function not only exhibits stronger theoretical properties than the classical FB function, but also demonstrates superior numerical performance. This further motivates our consideration of this generalized class of NCP functions. Indeed, a unified investigation into the properties of various penalized NCP functions and their variants has been conducted; for further details, we refer the reader to [41].

Lemma 2.13 *The function $\psi_{\alpha,\theta,p}$ defined by (2.82) has the following favorable properties:*

(a) $\psi_{\alpha,\theta,p}$ is an NCP function and $\psi_{\alpha,\theta,p} \geq 0$ for all $(a, b) \in \mathbb{R}^2$.
(b) $\psi_{\alpha,\theta,p}$ is continuously differentiable everywhere. Moreover, if $(a, b) \neq (0, 0)$,

$$\begin{aligned}
&\nabla_a \psi_{\alpha,\theta,p}(a,b) \\
&= \alpha b(ab)_+ + \left(\frac{\theta \text{sgn}(a) \cdot |a|^{p-1} + (1-\theta)\text{sgn}(a-b)|a-b|^{p-1}}{[\theta(|a|^p + |b|^p) + (1-\theta)|a-b|^p)]^{(p-1)/p}} - 1 \right) \phi_{\theta,p}(a,b), \\
&\nabla_b \psi_{\alpha,\theta,p}(a,b) \\
&= \alpha a(ab)_+ + \left(\frac{\theta \text{sgn}(b) \cdot |b|^{p-1} - (1-\theta)\text{sgn}(a-b)|a-b|^{p-1}}{[\theta(|a|^p + |b|^p) + (1-\theta)|a-b|^p)]^{(p-1)/p}} - 1 \right) \phi_{\theta,p}(a,b),
\end{aligned} \tag{2.84}$$

and otherwise, $\nabla_a \psi_{\alpha,\theta,p}(0,0) = \nabla_b \psi_{\alpha,\theta,p}(0,0) = 0$.
(c) For $p \geq 2$, the gradient of $\psi_{\alpha,\theta,p}$ is Lipschitz continuous on any nonempty bounded set S, i.e., there exists $L > 0$ such that for any $(a, b), (c, d) \in S$,

$$\|\nabla \psi_{\alpha,\theta,p}(a,b) - \nabla \psi_{\alpha,\theta,p}(c,d)\| \leq L\|(a,b) - (c,d)\|.$$

(d) $\nabla_a \psi_{\alpha,\theta,p}(a,b) \cdot \nabla_b \psi_{\alpha,\theta,p}(a,b) \geq 0$ for any $(a, b) \in \mathbb{R}^2$, and the equality holds if and only if $\psi_{\alpha,\theta,p}(a,b) = 0$.
(e) $\nabla_a \psi_{\alpha,\theta,p}(a,b) = 0 \Longleftrightarrow \nabla_b \psi_{\alpha,\theta,p}(a,b) = 0 \Longleftrightarrow \psi_{\alpha,\theta,p}(a,b) = 0$.
(f) Suppose that $\alpha > 0$. If $a \to -\infty$ or $b \to -\infty$ or $ab \to \infty$, then $\psi_{\alpha,\theta,p}(a,b) \to \infty$.

Proof (a) It is clear that $\psi_{\alpha,\theta,p}(a,b) \geq 0$ for all $(a, b) \in \mathbb{R}^2$ from the definition of $\psi_{\alpha,\theta,p}$. Then by [25, Proposition 2.1], we have

$$\psi_{\alpha,\theta,p}(a,b) = 0 \Longleftrightarrow \frac{\alpha}{2}(\max\{0, ab\})^2 = 0 \text{ and } \psi_{\theta,p}(a,b) = 0 \Longleftrightarrow a \geq 0,\ \ b \geq 0,\ \ ab = 0.$$

Hence, $\psi_{\alpha,\theta,p}$ is an NCP function.

(b) First, direct calculations give the partial derivatives of $\psi_{\alpha,\theta,p}$. Then, using $\alpha b(ab)_+ \to (0,0)$ and $\alpha a(ab)_+ \to (0,0)$ as $(a,b) \to (0,0)$, we have $\frac{\alpha}{2}(\max\{0,ab\})^2$ is continuously differentiable everywhere. By [25, Proposition 2.5], it is known that $\psi_{\theta,p}$ is continuously differentiable everywhere. In view of the expression of $\nabla\psi_{\alpha,\theta,p}(a,b)$, $\psi_{\alpha,\theta,p}$ is also continuously differentiable everywhere.

(c) First, we claim that $a(ab)_+$ for any $a,b \in \mathbb{R}$ is Lipschitz continuous on any nonempty bounded set S. For any $(a,b) \in S$ and $(c,d) \in S$, without loss of generality, we may assume that $a^2+b^2 \le k$ and $c^2+d^2 \le k$ which imply $|a| \le k+1$, $|b| \le k+1$, $|c| \le k+1$ and $|d| \le k+1$. Then,

$$\begin{aligned}
&\left|a(ab)_+ - c(cd)_+\right| \\
=&\frac{1}{2}\left|a^2b + a|ab| - c^2d - c|cd|\right| \\
=&\frac{1}{2}\left|a^2b - a^2d + a^2d - c^2d + a|ab| - c|ab| + c|ab| - c|cd|\right| \\
\le&\frac{1}{2}\left(|a^2b - a^2d| + |a^2d - c^2d| + \big|a|ab| - c|ab|\big| + \big|c|ab| - c|cd|\big|\right) \\
=&\frac{1}{2}\left(a^2|b-d| + |a+c||d||a-c| + |ab||a-c| + |c||ab-cd|\right) \\
\le&\frac{1}{2}\Big[k|b-d| + (|a|+|c|)|d||a-c| + k|a-c| + (k+1)|ab-ad+ad-cd|\Big] \\
\le&\frac{1}{2}\Big[k|b-d| + 2(k+1)^2|a-c| + k|a-c| + (k+1)^2(|b-d|+|a-c|)\Big] \\
=&\frac{1}{2}\Big\{\left[2(k+1)^2 + k + (k+1)^2\right]|a-c| + \left[k+(k+1)^2\right]|b-d|\Big\} \\
\le& l\big(|a-c| + |b-d|\big) \\
\le&\sqrt{2}l\|(a,b)-(c,d)\|,
\end{aligned}$$

where $l = 2(k+1)^2 + k + (k+1)^2$. Hence, the mapping $a(ab)_+$ is Lipschitz continuous on any nonempty bounded set S and so is $\alpha a(ab)_+$. Similarly, $\alpha b(ab)_+$ is Lipschitz continuous on any nonempty bounded set S. All of these imply the gradient function of the function $\frac{\alpha}{2}(\max\{0,ab\})^2$ is Lipschitz continuous on any bounded set S. On the other hand, by [25, Theorem 2.1], the gradient function of the function $\psi_{\theta,p}$ with $p \ge 2$, $\theta \in (0,1]$ is Lipschitz continuous. Thus, the gradient of $\psi_{\alpha,\theta,p}$ is Lipschitz continuous on any nonempty bounded set S.

(d) If $(a,b) = (0,0)$, part(d) clearly holds. Now we assume that $(a,b) \ne (0,0)$. Then,

$$\begin{aligned}
&\nabla_a\psi_{\alpha,\theta,p}(a,b)\cdot\nabla_b\psi_{\alpha,\theta,p}(a,b) \qquad (2.85)\\
=&\ cd\phi_{\theta,p}^2(a,b) + \alpha^2ab{(ab)_+}^2 + \alpha a(ab)_+c\phi_{\theta,p}(a,b) + \alpha b(ab)_+d\phi_{\theta,p}(a,b),
\end{aligned}$$

where

$$c = \left(\frac{\theta \mathrm{sgn}(a) \cdot |a|^{p-1} + (1-\theta)\mathrm{sgn}(a-b)|a-b|^{p-1}}{[\theta(|a|^p + |b|^p) + (1-\theta)|a-b|^p)]^{(p-1)/p}} - 1 \right),$$

$$d = \left(\frac{\theta \mathrm{sgn}(b) \cdot |b|^{p-1} - (1-\theta)\mathrm{sgn}(a-b)|a-b|^{p-1}}{[\theta(|a|^p + |b|^p) + (1-\theta)|a-b|^p)]^{(p-1)/p}} - 1 \right).$$

From the proof of [25, Proposition 2.5], we know $ab(ab)_+^2 \geq 0$ and

$$\left(\frac{\theta \mathrm{sgn}(a) \cdot |a|^{p-1} + (1-\theta)\mathrm{sgn}(a-b)|a-b|^{p-1}}{[\theta(|a|^p + |b|^p) + (1-\theta)|a-b|^p)]^{(p-1)/p}} - 1 \right) \leq 0,$$

$$\left(\frac{\theta \mathrm{sgn}(b) \cdot |b|^{p-1} - (1-\theta)\mathrm{sgn}(a-b)|a-b|^{p-1}}{[\theta(|a|^p + |b|^p) + (1-\theta)|a-b|^p)]^{(p-1)/p}} - 1 \right) \leq 0, \tag{2.86}$$

it suffices to show that the last two terms of (2.85) are nonnegative. For this purpose, we claim that

$$\alpha a(ab)_+ \left(\frac{\theta \mathrm{sgn}(a) \cdot |a|^{p-1} + (1-\theta)\mathrm{sgn}(a-b)|a-b|^{p-1}}{[\theta(|a|^p + |b|^p) + (1-\theta)|a-b|^p)]^{(p-1)/p}} - 1 \right) \phi_{\theta,p}(a,b) \geq 0 \tag{2.87}$$

for all $(a, b) \neq (0, 0)$. If $a \leq 0$ and $b \leq 0$, then $\phi_{\theta,p}(a, b) \geq 0$, which together with the second inequality in (2.86) implies that (2.87) holds. If $a \leq 0$ and $b \geq 0$, then $(ab)_+ = 0$, which says that (2.87) holds. If $a > 0$ and $b > 0$, then $|a|^p + |b|^p \geq |a - b|^p$. Thus, $\phi_{\theta,p}(a, b) \leq \phi_p(a, b) \leq 0$, which together with the second inequality in (2.86) yields (2.87). If $a > 0$ and $b \leq 0$, then $(ab)_+ = 0$, and hence (2.87) holds. Similarly, we also have

$$\alpha b(ab)_+ \left(\frac{\theta \mathrm{sgn}(b) \cdot |b|^{p-1} - (1-\theta)\mathrm{sgn}(a-b)|a-b|^{p-1}}{[\theta(|a|^p + |b|^p) + (1-\theta)|a-b|^p)]^{(p-1)/p}} - 1 \right) \phi_{\theta,p}(a,b) \geq 0$$

for all $(a, b) \neq (0, 0)$. Consequently, $\nabla_a \psi_{\alpha,\theta,p}(a, b) \cdot \nabla_b \psi_{\alpha,\theta,p}(a, b) \geq 0$. Besides, by the proof of [25, Proposition 2.5], we know $c = 0$ if and only if $b = 0$ and $a > 0$; $d = 0$ if and only if $a = 0$ and $b > 0$. This together with (2.85) says $\nabla_a \psi_{\alpha,\theta,p}(a, b) \cdot \nabla_b \psi_{\alpha,\theta,p}(a, b) = 0$ if and only if $\{\psi_{\theta,p}(a, b) = 0 \text{ and } \alpha^2 ab(ab)_+{}^2 = 0\}$ or $\{c = 0\}$ or $\{d = 0\}$ if and only if $\{\psi_{\theta,p}(a, b) = 0 \text{ and } ab \leq 0\}$ or $\{c = 0\}$ or $\{d = 0\}$ if and only if $\psi_{\theta,p}(a, b) = 0$ and $\frac{\alpha}{2}(\max\{0, ab\})^2 = 0$ if and only if $\psi_{\alpha,\theta,p}(a, b) = 0$.

(e) If $\psi_{\alpha,\theta,p}(a, b) = 0$, then $\frac{\alpha}{2}(\max\{0, ab\})^2 = 0$ and $\psi_{\theta,p}(a, b) = 0$, which imply $ab \leq 0$ and $\phi_{\theta,p}(a, b) = 0$. Hence, $\nabla_a \psi_{\alpha,\theta,p}(a, b) = 0$ and $\nabla_b \psi_{\alpha,\theta,p}(a, b) = 0$. Now, it remains to show that $\nabla_a \psi_{\alpha,\theta,p}(a, b) = 0$ implying $\psi_{\alpha,\theta,p}(a, b) = 0$. Suppose that $\nabla_a \psi_{\alpha,\theta,p}(a, b) = 0$, which yields

$$\alpha b(ab)_+ = -\left(\frac{\theta \mathrm{sgn}(a) \cdot |a|^{p-1} + (1-\theta)\mathrm{sgn}(a-b)|a-b|^{p-1}}{[\theta(|a|^p + |b|^p) + (1-\theta)|a-b|^p)]^{(p-1)/p}} - 1 \right) \phi_{\theta,p}(a,b). \tag{2.88}$$

We will argue that the equality (2.88) implies $\big(b = 0, a \geq 0\big)$ or $\big(b > 0, a = 0\big)$. To see this, we let

$$
\begin{aligned}
c &= \alpha b(ab)_+,\\
d &= -\left(\frac{\theta \operatorname{sgn}(a)\cdot |a|^{p-1} + (1-\theta)\operatorname{sgn}(a-b)|a-b|^{p-1}}{[\theta(|a|^p + |b|^p) + (1-\theta)|a-b|^p)]^{(p-1)/p}} - 1\right)\phi_{\theta,p}(a,b),\\
e &= \left(\frac{\theta \operatorname{sgn}(a)\cdot |a|^{p-1} + (1-\theta)\operatorname{sgn}(a-b)|a-b|^{p-1}}{[\theta(|a|^p + |b|^p) + (1-\theta)|a-b|^p)]^{(p-1)/p}} - 1\right).
\end{aligned}
$$

It is not hard to observe that $\big(e \leq 0\big)$ and $\big(e = 0$ implies $b = 0\big)$ which are helpful for the following discussions.

Case 1: $b = 0$ and $a < 0$. Then, it leads to $c = 0$ but $d \neq 0$, which violates (2.88).

Case 2: $b < 0$ and $a \geq 0$. Then, we have $e < 0$, and hence $c = 0$ but $d \neq 0$, which violates (2.88).

Case 3: $b < 0$ and $a < 0$. Then, we have $e < 0$ and $\phi_{\theta,p}(a,b) > 0$, which yield $c \leq 0$ but $d > 0$. This contradicts to (2.88) too.

Case 4: $b > 0$ and $a > 0$. Then, we have $e < 0$ and $\phi_{\theta,p}(a,b) < 0$, which imply $c \geq 0$ but $d < 0$. This contradicts to (2.88) too.

Case 5: $b > 0$ and $a < 0$. Similar arguments as in Case 2 cause a contradiction.

Thus, (2.88) implies $\big(b = 0, a \geq 0\big)$ or $\big(b > 0, a = 0\big)$, and each of which always yields $\psi_{\alpha,\theta,p}(a,b) = 0$. By symmetry, $\nabla_b \psi_{\alpha,\theta,p}(a,b) = 0$ also implies $\psi_{\alpha,\theta,p}(a,b) = 0$.

(f) If $a \to -\infty$ or $b \to -\infty$, from [25, Proposition 2.4], we know $|\phi_{\theta,p}(a,b)| \to \infty$. In addition, the fact $\frac{\alpha}{2}(\max\{0, ab\})^2 \geq 0$ gives $\psi_{\alpha,\theta,p}(a,b) \to \infty$. If $ab \to \infty$, since $\alpha > 0$, we have $\frac{\alpha}{2}(\max\{0, ab\})^2 \to \infty$. This together with $\psi_{\theta,p}(a,b) \geq 0$ says $\psi_{\alpha,\theta,p}(a,b) \to \infty$. □

Proposition 2.58 *Let $\Psi_{\alpha,\theta,p}$ be defined as in (2.83). Then $\Psi_{\alpha,\theta,p}(x) \geq 0$ for all $x \in \mathbb{R}^n$ and $\Psi_{\alpha,\theta,p}(x) = 0$ if and only if x solves the NCP. Moreover, if the NCP has at least one solution, then x is a global minimizer of $\Psi_{\alpha,\theta,p}$ if and only if x solves the NCP.*

Proof Since $\psi_{\theta,p}$ is a NCP function, from [25, Proposition 2.5], we have that x solving the NCP $\Longleftrightarrow x \geq 0,\ F(x) \geq 0,\ \langle x, F(x)\rangle = 0 \Longleftrightarrow x \geq 0,\ F(x) \geq 0,\ x_i F_i(x) = 0$ for all $i \in \{1, 2, \ldots, n\} \Longleftrightarrow \Psi_{\alpha,\theta,p}(x) = 0$. Besides, $\Psi_{\alpha,\theta,p}(x)$ is nonnegative. Thus, if x solves the NCP, then x is a global minimizer of $\Psi_{\alpha,\theta,p}$. Next, we claim that if the NCP has at least one solution, then x is a global minimizer of $\Psi_{\alpha,\theta,p} \Longrightarrow x$ solves the NCP. Suppose x does not solve the NCP. From x solves the NCP $\Longleftrightarrow$ $\Psi_{\alpha,\theta,p}(x) = 0$ and $\Psi_{\alpha,\theta,p}(x)$ is nonnegative, it is clear $\Psi_{\alpha,\theta,p}(x) > 0$. However, by assumption, the NCP has a solution, say y, which makes that $\Psi_{\alpha,\theta,p}(y) = 0$. Then, we reach a contradiction that $\Psi_{\alpha,\theta,p}(x) > 0 = \Psi_{\alpha,\theta,p}(y)$ and x is a global minimizer of $\Psi_{\alpha,\theta,p}$. Thus, the proof is complete. □

Like what we have done in previous sections, Proposition 2.58 indicates that the NCP can be recast as the unconstrained minimization:

$$\min_{x \in \mathbb{R}^n} \Psi_{\alpha,\theta,p}(x). \tag{2.89}$$

Then, we establish analogous results based on this reformulation.

Proposition 2.59 *Let F be a P_0-function. Then $x^* \in \mathbb{R}^n$ is a global minimum of the unconstrained optimization problem (2.89) if and only if x^* is a stationary point of $\Psi_{\alpha,\theta,p}$.*

Proposition 2.60 *The function $\Psi_{\alpha,\theta,p}$ has bounded level sets $\mathcal{L}(\Psi_{\alpha,\theta,p}, c)$ for all $c \in \mathbb{R}$, if F is monotone and the NCP is strictly feasible (i.e., there exists $\hat{x} > 0$ such that $F(\hat{x}) > 0$) when $\alpha > 0$, or F is a uniform P-function when $\alpha \geq 0$.*

Proof From [17], if F is a monotone function with a strictly feasible point, then the following condition holds: for every sequence $\{x^k\}$ such that $\|x^k\| \to \infty$, $(-x^k)_+ < \infty$, and $(-F(x^k))_+ < \infty$, we have $\max\limits_{1\leq i\leq n} \left\{(x_i^k)_+ F_i(x^k)_+\right\} \to \infty$. Suppose that there exists an unbounded sequence $x^k \subseteq \mathcal{L}(\Psi_{\alpha,\theta,p}, c)$ for some $c \in \mathbb{R}$. Since $\Psi_{\alpha,\theta,p}(x^k) \leq c$, there is no index i such that $x_i^k \to -\infty$ or $F_i(x^k) \to -\infty$ by Lemma 2.13(f). Hence, $\max\limits_{1\leq i\leq n} \left\{(x_i^k)_+ F_i(x^k)_+\right\} \to \infty$. Also, there is an index j, and at least a subsequence x_j^k such that $\left\{(x_j^k)_+ F_j(x^k)_+\right\} \to \infty$. However, this implies that $\Psi_{\alpha,\theta,p}(x^k)$ is unbounded by Lemma 2.13(f), contracting to the assumption on level sets. Another part of the proof is similar to the proof of [22, Proposition 3.5]. □

Lemma 2.14 *Let $\phi_{\theta,p} : \mathbb{R}^2 \to \mathbb{R}$ be defined as in (2.53). Then, for any $p > 1$ and all $\theta \in (0, 1]$, there holds*

$$(2 - 2^{\frac{1}{p}})|\min\{a, b\}| \leq |\phi_{\theta,p}(a, b)| \leq (2 + 2^{\frac{1}{p}})|\min\{a, b\}|. \tag{2.90}$$

Proof Without loss of generality, we assume $a \geq b$. We will prove the desired results by considering the following two cases: (1) $a + b \leq 0$ and (2) $a + b > 0$.
Case(1): $a + b \leq 0$. In this case, we need to discuss two subcases:
(i) $|a|^p + |b|^p \geq |a - b|^p$. In this subcase, we have

$$\begin{aligned}
|\phi_{\theta,p}(a, b)| &\geq |\sqrt[p]{\theta(|a - b|^p) + (1 - \theta)(|a - b|^p)} - (a + b)| \\
&= |\sqrt[p]{(|a - b|^p)} - (a + b)| \\
&= |(|a - b| - (a + b)| \\
&= |a - b - (a + b)| \\
&= |2b| \\
&= 2|\min\{a, b\}| \\
&\geq (2 - 2^{\frac{1}{p}})|\min\{a, b\}|
\end{aligned} \tag{2.91}$$

On the other hand, since $|a|^p + |b|^p \geq |a-b|^p$ and by [21, Lemma 3.2], we have

$$|\phi_{\theta,p}(a,b)| \leq |\phi^p_{\mathrm{FB}}(a,b)| \leq (2+2^{\frac{1}{p}})|\min\{a,b\}|.$$

(ii) $|a|^p + |b|^p < |a-b|^p$. Since $|a|^p + |b|^p < |a-b|^p$ and by [21, Lemma 3.2], we have

$$|\phi_{\theta,p}(a,b)| > |\phi^p_{\mathrm{FB}}(a,b)| \geq (2-2^{\frac{1}{p}})|\min\{a,b\}|.$$

On the other hand, by the discussion of Case(1),

$$|\phi_{\theta,p}(a,b)| < 2|b| \leq (2+2^{\frac{1}{p}})|\min\{a,b\}|.$$

Case(2): $a+b>0$. If ab=0, then (2.90) clearly holds. Thus, we proceed with the arguments by discussing two subcases:
(i) $ab<0$. In this subcases, we have $a>0, b<0, |a|>|b|$. By Lemma 2.2, $|a|^p + |b|^p \leq |a-b|^p$. Then,

$$\phi_{\theta,p}(a,b) \geq \phi^p_{\mathrm{FB}}(a,b) \geq |a|-a-b \geq -b = |\min\{a,b\}| \geq (2-2^{\frac{1}{p}})|\min\{a,b\}|.$$

On the other hand,

$$\phi_{\theta,p}(a,b) \leq |a-b|-(a+b) = -2b = 2|\min\{a,b\}| \leq (2+2^{\frac{1}{p}})|\min\{a,b\}|.$$

(ii) $ab>0$. In this subcases, we have $a \geq b > 0$, $|a|^p+|b|^p \geq |a-b|^p$. By Lemma 2.2, $\phi_{\theta,p}(a,b) \leq \phi_p(a,b) \leq 0$. Notice that $\phi_{\theta,p}(a,b) \geq |a-b|-(a+b) = -2b$ $= -2\min\{a,b\}$, and hence we obtain that

$$|\phi_{\theta,p}(a,b)| \leq 2|\min\{a,b\}| \leq (2+2^{\frac{1}{p}})|\min\{a,b\}|.$$

On the other hand, since $\phi_{\theta,p}(a,b) \leq \phi_p(a,b) \leq 0$, and by [21, Lemma 3.2], and hence we obtain that

$$|\phi_{\theta,p}(a,b)| \geq |\phi_p(a,b)| \geq (2-2^{\frac{1}{p}})|\min\{a,b\}|. \tag{2.92}$$

All the aforementioned inequalities (2.91)–(2.92) imply that (2.90) holds. □

Proposition 2.61 *Let $\Psi_{\theta,p}$, Ψ_{NR} and $\Psi_{\alpha,\theta,p}$ be defined as in (2.54), (2.10) and (2.83), respectively. Let S be an arbitrary bounded set. Then, for any $p>1$, we have*

$$(2-2^{\frac{1}{p}})^2\Psi_{\mathrm{NR}}(x) \leq \Psi_{\theta,p}(x) \leq (2+2^{\frac{1}{p}})^2\Psi_{\mathrm{NR}}(x) \quad \text{for all } x \in \mathbb{R}^n \tag{2.93}$$

and

$$(2-2^{\frac{1}{p}})^2\Psi_{\rm NR}(x) \le \Psi_{\alpha,\theta,p}(x) \le (\alpha B^2 + (2+2^{\frac{1}{p}})^2)\Psi_{\rm NR}(x) \quad \text{for all } x \in S, \tag{2.94}$$

where B is a constant defined by $B = \max_{1\le i\le n}\left\{\sup_{x\in S}\{\max\{|x_i|, |F_i(x)|\}\}\right\} < \infty.$

Proof The inequality in (2.93) is direct by Lemma 2.14 and the definitions of $\Psi_{\theta,p}$ and $\Psi_{\rm NR}$. In addition, from Lemma 2.14 and the definition of $\Psi_{\alpha,\theta,p}$, it follows that

$$\Psi_{\alpha,\theta,p}(x) \ge \left(2-2^{\frac{1}{p}}\right)^2 \Psi_{\rm NR}(x) \quad \text{for all } x \in \mathbb{R}^n.$$

It remains to prove the inequality on the right-hand side of (2.94). From the proof of [21, Proposition 3.1], we know for each i,

$$(x_i F_i(x))_+ \le B|\min\{x_i, F_i(x)\}| \quad \text{for all } x \in S. \tag{2.95}$$

By Lemma 2.14 and (2.95), for all $i = 1, \ldots, n$ and $x \in S$,

$$\psi_{\alpha,\theta,p}(x_i, F_i(x)) \le \frac{1}{2}\left\{\alpha B^2 + (2+2^{\frac{1}{p}})^2\right\}\min\{x_i, F_i(x)\}^2$$

holds for any $p > 1$. The proof is then complete by the definitions of $\Psi_{\alpha,\theta,p}$ and $\Psi_{\rm NR}$. □

Proposition 2.62 *Let $\Psi_{\theta,p}$ and $\Psi_{\alpha,\theta,p}$ be defined by (2.54) and (2.83), respectively; and S be any bounded set. Then, for any $p > 1$ and all $x \in S$, we have the following inequalities:*

$$\frac{(2-2^{\frac{1}{p}})^2}{\left(\alpha B^2 + (2+2^{\frac{1}{p}})^2\right)}\Psi_{\alpha,\theta,p}(x) \le \Psi_{\theta,p}(x) \le \frac{(2+2^{\frac{1}{p}})^2}{(2-2^{\frac{1}{p}})^2}\Psi_{\alpha,\theta,p}(x)$$

where B is the constant defined as in Proposition 2.61.

Proof It follows from Proposition 2.61 directly. □

Proposition 2.63 *Let $\Psi_{\alpha,\theta,p}$ be defined as in (2.83). Suppose that F is a uniform P-function with modulus $\mu > 0$. If $\alpha > 0$, then there exists a constant $\kappa_1 > 0$ such that*

$$\|x - x^*\| \le \kappa_1 \Psi_{\alpha,\theta,p}(x)^{\frac{1}{4}} \quad \text{for all } x \in \mathbb{R}^n;$$

if $\alpha = 0$ and S is any bounded set, there exists a constant $\kappa_2 > 0$ such that

$$\|x - x^*\| \le \kappa_2 \left(\max\left\{\Psi_{\alpha,\theta,p}(x), \sqrt{\Psi_{\alpha,\theta,p}(x)}\right\}\right)^{\frac{1}{2}} \quad \text{for all } x \in S;$$

where $x^* = (x_1^*, \ldots, x_n^*)$ is the unique solution for the NCP.

Proof By the proof of [21, Theorem 3.4], we have

$$\mu\|x - x^*\|^2 \le \max_{1\le i\le n} \tau_i\{(x_i F_i(x))_+ + (-F_i(x))_+ + (-x_i)_+\}, \tag{2.96}$$

where $\tau_i := \max\{1, x_i^*, F_i(x^*)\}$. We next prove that for all $(a, b) \in \mathbb{R}^2$,

$$(-a)_+{}^2 + (-b)_+{}^2 \le [\phi_{\theta,p}(a, b)]^2. \tag{2.97}$$

To see this, without loss of generality, we assume $a \ge b$ and discuss three cases:
(i) If $a \ge b \ge 0$, then (2.97) holds obviously.
(ii) If $a \ge 0 \ge b$, then $|a|^p + |b|^p \le |a - b|^p$ by Lemma 2.2, which implies $\phi_{\theta,p}(a, b) \ge \|(a, b)\|_p - (a + b) \ge -b \ge 0$. Hence, $(-a)_+{}^2 + (-b)_+{}^2 = b^2 \le [\phi_{\theta,p}(a, b)]^2$.
(iii) If $0 \ge a \ge b$, then $(-a)_+{}^2 + (-b)_+{}^2 = a^2 + b^2 \le [\phi_{\theta,p}(a, b)]^2$. Hence, (2.97) follows.

Suppose that $\alpha > 0$. Using the inequality (2.97), we then obtain that

$$\begin{aligned}
\left[(ab)_+ + (-a)_+ + (-b)_+\right]^2 &= (ab)_+^2 + (-b)_+^2 + (-a)_+^2 + 2(ab)_+(-a)_+ \\
&\quad + 2(-a)_+(-b)_+ + 2(ab)_+(-b)_+ \\
&\le (ab)_+^2 + (-b)_+^2 + (-a)_+^2 + (ab)_+^2 + (-a)_+^2 \\
&\quad + (-a)_+^2 + (-b)_+^2 + (ab)_+^2 + (-b)_+^2 \\
&\le 3\left[(ab)_+^2 + [\phi_{\theta,p}(a, b)]^2\right] \\
&\le \tau\left[\frac{\alpha}{2}(ab)_+^2 + \frac{1}{2}[\phi_{\theta,p}(a, b)]^2\right] \\
&= \tau\psi_{\alpha,\theta,p}(a, b),
\end{aligned} \tag{2.98}$$

where $\tau := \max\left\{\frac{6}{\alpha}, 6\right\} > 0$. Combining (2.98) with (2.96) and letting $\hat{\tau} = \max_{1\le i\le n} \tau_i$, we get

$$\begin{aligned}
\mu\|x - x^*\|^2 &\le \max_{1\le i\le n} \tau_i \left\{\tau\psi_{\alpha,\theta,p}(x_i, F_i(x))\right\}^{1/2} \\
&\le \hat{\tau}\tau^{1/2} \max_{1\le i\le n} \psi_{\alpha,\theta,p}(x_i, F_i(x))^{1/2} \\
&\le \hat{\tau}\tau^{1/2}\left\{\sum_{i=1}^n \{\psi_{\alpha,\theta,p}(x_i, F_i(x))\right\}^{1/2} \\
&= \hat{\tau}\tau^{1/2}\Psi_{\alpha,\theta,p}(x)^{1/2}.
\end{aligned}$$

From this, the first desired result follows immediately by setting $\kappa_1 := \left[\hat{\tau}\tau^{1/2}/\mu\right]^{1/2}$.

Suppose that $\alpha = 0$. From the proof of Proposition 2.61, the inequality (2.95) holds. Combining with equations (2.96)–(2.97), it then follows that for all $x \in S$,

$$
\begin{aligned}
\mu\|x-x^*\|^2 &\le \max_{1\le i\le n} \tau_i \left[B|\min\{x_i, F_i(x)\}| + 2(\psi_{\theta,p}(x_i, F_i(x)))^{1/2}\right] \\
&\le \hat{\tau} \max_{1\le i\le n} \left[\sqrt{2}\hat{B}(\psi_{\theta,p}(x_i, F_i(x)))^{1/2} + 2(\psi_{\theta,p}(x_i, F_i(x)))^{1/2}\right] \\
&\le (\sqrt{2}\hat{B}+2)\hat{\tau}(\Psi_{\theta,p}(x))^{1/2} \\
&= (\sqrt{2}\hat{B}+2)\hat{\tau}(\Psi_{\alpha,\theta,p}(x))^{1/2} \\
&\le (\sqrt{2}\hat{B}+2)\hat{\tau}(\max\left\{\Psi_{\alpha,\theta,p}(x), \sqrt{\Psi_{\alpha,\theta,p}(x)}\right\})
\end{aligned}
$$

where $\hat{B} = B/(2-2^{\frac{1}{p}})$, $\hat{\tau} = \max_{1\le i\le n} \tau_i$ and the second inequality is from Lemma 2.14. Letting $\kappa_2 := \left[(\sqrt{2}\hat{B}+2)\hat{\tau}/\mu\right]^{1/2}$, we obtain the desired result from the above inequality. □

Lemma 2.15 *For all $(a, b) \neq (0, 0)$ and $p > 1$, we have the following inequality:*

$$
\left(\frac{\theta[\operatorname{sgn}(a)\cdot|a|^{p-1} + \operatorname{sgn}(b)\cdot|b|^{p-1}]}{[\theta(|a|^p+|b|^p)+(1-\theta)|a-b|^p)]^{(p-1)/p}} - 2\right)^2 \ge \left(2-2^{\frac{1}{p}}\right)^2, \quad \forall\theta\in(0,1].
$$

Proof If $a = 0$ or $b = 0$, the inequality holds obviously. Then we complete the proof by considering three cases: (i) $a > 0$ and $b > 0$, (ii) $a < 0$ and $b < 0$, and (iii) $ab < 0$. Case (i): Since $\theta \in (0, 1]$ and $p > 1$, it follows that $\theta^{1/p} \le 1$. Now, by the proof of [21, Lemma 3.3], we have

$$
\begin{aligned}
&\frac{\theta[\operatorname{sgn}(a)\cdot|a|^{p-1} + \operatorname{sgn}(b)\cdot|b|^{p-1}]}{[\theta(|a|^p+|b|^p)+(1-\theta)|a-b|^p)]^{(p-1)/p}} \\
&= \frac{\theta[|a|^{p-1}+|b|^{p-1}]}{[\theta(|a|^p+|b|^p)+(1-\theta)|a-b|^p)]^{(p-1)/p}} \\
&\le \frac{\theta[|a|^{p-1}+|b|^{p-1}]}{[\theta(|a|^p+|b|^p)]^{(p-1)/p}} \\
&= \frac{\theta^{1/p}[|a|^{p-1}+|b|^{p-1}]}{[(|a|^p+|b|^p)]^{(p-1)/p}} \\
&\le 2^{1/p} \quad \text{for } p > 1.
\end{aligned}
$$

Therefore, it yields

$$
2 - \frac{\theta[|a|^{p-1}+|b|^{p-1}]}{[\theta(|a|^p+|b|^p)+(1-\theta)|a-b|^p)]^{(p-1)/p}} \ge 2 - 2^{\frac{1}{p}}
$$

for $p > 1$. Squaring both sides then leads to the desired inequality.

Case (ii): By similar arguments as in case (i), we obtain

$$\begin{aligned}
& 2 - 2^{\frac{1}{p}} \\
\le\, & 2 - \frac{\theta[|a|^{p-1} + |b|^{p-1}]}{[\theta(|a|^p + |b|^p) + (1-\theta)|a-b|^p)]^{(p-1)/p}} \\
\le\, & 2 + \frac{\theta[|a|^{p-1} + |b|^{p-1}]}{[\theta(|a|^p + |b|^p) + (1-\theta)|a-b|^p)]^{(p-1)/p}} \quad \text{for } p > 1,
\end{aligned}$$

from which the result follows immediately.

Case (iii): Again, we suppose $|a| \ge |b|$ and therefore have

$$\begin{aligned}
& 2^{\frac{1}{p}} \\
\ge\, & \frac{\theta[|a|^{p-1} + |b|^{p-1}]}{[\theta(|a|^p + |b|^p) + (1-\theta)|a-b|^p)]^{(p-1)/p}} \\
\ge\, & \frac{\theta[|a|^{p-1} - |b|^{p-1}]}{[\theta(|a|^p + |b|^p) + (1-\theta)|a-b|^p)]^{(p-1)/p}} \quad \text{for } p > 1.
\end{aligned}$$

Thus, it gives

$$2 - 2^{\frac{1}{p}} \le 2 - \frac{\theta[|a|^{p-1} - |b|^{p-1}]}{[\theta(|a|^p + |b|^p) + (1-\theta)|a-b|^p)]^{(p-1)/p}}$$

for $p > 1$ and the desired result is also satisfied. □

Proposition 2.64 *Let $\psi_{\alpha,\theta,p}$ be given as in (2.82). Then, for all $x \in \mathbb{R}^n$ and $p > 1$,*

$$\left\| \nabla_a \psi_{\alpha,\theta,p}(x, F(x)) + \nabla_b \psi_{\alpha,\theta,p}(x, F(x)) \right\|^2 \ge 2\left(2 - 2^{\frac{1}{p}}\right)^2 \Psi_{\theta,p}(x) \quad \forall \theta \in (0, 1].$$

In particular, for all x belonging to any bounded set S and $p > 1$,

$$\left\| \nabla_a \psi_{\alpha,\theta,p}(x, F(x)) + \nabla_b \psi_{\alpha,\theta,p}(x, F(x)) \right\|^2 \ge \frac{2(2 - 2^{\frac{1}{p}})^4}{\left(\alpha B^2 + (2 + 2^{\frac{1}{p}})^2\right)} \Psi_{\alpha,\theta,p}(x) \quad \forall \theta \in (0, 1],$$

where B is defined as in Proposition 2.61 and

$$\begin{aligned}
\nabla_a \psi_{\alpha,\theta,p}(x, F(x)) &:= \Big(\nabla_a \psi_{\alpha,\theta,p}(x_1, F_1(x)), \ \ldots, \nabla_a \psi_{\alpha,\theta,p}(x_n, F_n(x)) \Big)^{\mathsf{T}}, \\
\nabla_b \psi_{\alpha,\theta,p}(x, F(x)) &:= \Big(\nabla_b \psi_{\alpha,\theta,p}(x_1, F_1(x)), \ \ldots, \nabla_b \psi_{\alpha,\theta,p}(x_n, F_n(x)) \Big)^{\mathsf{T}}.
\end{aligned}$$

Proof The second part of the conclusions is direct by Corollary 2.62 and the first part. Thus, it remains to show the first part. From the definitions of $\nabla_a \psi_{\alpha,\theta,p}(x, F(x))$, $\nabla_b \psi_{\alpha,\theta,p}(x, F(x))$ and $\Psi_{\theta,p}(x)$, showing the first part is equivalent to proving that the following inequality

$$\left(\nabla_a \psi_{\alpha,\theta,p}(a,b) + \nabla_b \psi_{\alpha,\theta,p}(a,b)\right)^2 \geq 2\left(2 - 2^{\frac{1}{p}}\right)^2 \psi_{\theta,p}(a,b) \tag{2.99}$$

holds for all $(a, b) \in \mathbb{R}^2$. When $(a, b) = (0, 0)$, the inequality (2.99) clearly holds. Suppose $(a, b) \neq (0, 0)$. Then, it follows from Eq. (2.84) that

$$\begin{aligned}
&\left(\nabla_a \psi_{\alpha,\theta,p}(a,b) + \nabla_b \psi_{\alpha,\theta,p}(a,b)\right)^2 \\
&= \left\{\alpha(a+b)(ab)_+ + (\phi_{\theta,p}(a,b))\left(\frac{\theta[\mathrm{sgn}(a)\cdot|a|^{p-1} + \mathrm{sgn}(b)\cdot|b|^{p-1}]}{[\theta(|a|^p+|b|^p)+(1-\theta)|a-b|^p)]^{(p-1)/p}} - 2\right)\right\}^2 \\
&= \alpha^2(a+b)^2(ab)_+^2 + (\phi_{\theta,p}(a,b))^2\left(\frac{\theta[\mathrm{sgn}(a)\cdot|a|^{p-1} + \mathrm{sgn}(b)\cdot|b|^{p-1}]}{[\theta(|a|^p+|b|^p)+(1-\theta)|a-b|^p)]^{(p-1)/p}} - 2\right)^2 \\
&\quad +2\alpha(a+b)(ab)_+(\phi_{\theta,p}(a,b))\left(\frac{\theta[\mathrm{sgn}(a)\cdot|a|^{p-1} + \mathrm{sgn}(b)\cdot|b|^{p-1}]}{[\theta(|a|^p+|b|^p)+(1-\theta)|a-b|^p)]^{(p-1)/p}} - 2\right).
\end{aligned} \tag{2.100}$$

Now, we claim that for all $(a, b) \neq (0, 0) \in \mathbb{R}^2$,

$$2\alpha(a+b)(ab)_+(\phi_{\theta,p}(a,b))\left(\frac{\theta[\mathrm{sgn}(a)\cdot|a|^{p-1} + \mathrm{sgn}(b)\cdot|b|^{p-1}]}{[\theta(|a|^p+|b|^p)+(1-\theta)|a-b|^p)]^{(p-1)/p}} - 2\right) \geq 0. \tag{2.101}$$

If $ab \leq 0$, then $(ab)_+ = 0$ and the inequality (2.101) is clear. If $a, b > 0$, then by the proof of Lemma 2.15, we have

$$\left(\frac{\theta[\mathrm{sgn}(a)\cdot|a|^{p-1} + \mathrm{sgn}(b)\cdot|b|^{p-1}]}{[\theta(|a|^p+|b|^p)+(1-\theta)|a-b|^p)]^{(p-1)/p}} - 2\right) \leq 0, \quad \forall (a,b) \neq (0,0) \in \mathbb{R}^2 \tag{2.102}$$

and $\phi_{\theta,p}(a, b) \leq 0$, which imply the inequality (2.101) also holds. If $a, b < 0$, then $\phi_{\theta,p}(a, b) \geq 0$, which together with (2.102) yields the inequality (2.101). Thus, we obtain that the inequality (2.101) holds for all $(a, b) \neq (0, 0)$. Now using Lemma 2.15 and equations (2.100)–(2.101), we readily obtain the inequality (2.99) holds for all $(a, b) \neq (0, 0)$. The proof is thus complete. □

2.3.4 Construction by Discrete Generalization

We may also extend the concept of "discrete generalization", as introduced in Sect. 2.2, to the Fischer-Burmeister (FB) function. This leads to the definition of a new function, denoted by $\phi^p_{\text{D-FB}}$, given by

$$\phi^p_{\text{D-FB}}(a,b) = \left(\sqrt{a^2+b^2}\right)^p - (a+b)^p, \tag{2.103}$$

where $p > 1$ is a positive odd integer and $(a, b) \in \mathbb{R}^2$. Observe that when $p = 1$, $\phi^p_{\text{D-FB}}$ reduces to the standard Fischer-Burmeister function. We will show that $\phi^p_{\text{D-FB}}$ is an NCP function and, notably, is twice continuously differentiable without requiring the squaring of its norm. However, it is important to note that if p is even, the function $\phi^p_{\text{D-FB}}$ no longer satisfies the properties necessary to qualify as an NCP function. Although differentiability of $\phi^p_{\text{D-FB}}$ is advantageous, it does not imply that Newton's method can be directly applied in all cases, since the Jacobian at a degenerate solution to the NCP may be singular (see [42, 43]). Nonetheless, this differentiability feature opens the door to applying various methods, such as derivative-free algorithms, for solving the NCP directly and effectively.

Lemma 2.16 *Suppose that $p = 2k + 1$ where $k = 1, 2, 3, \ldots$. Then, for any $u, v \in \mathbb{R}$, we have $u^p = v^p$ if and only if $u = v$.*

Proof The proof is straightforward and can be found in [44, Theorem 1.12]. Here, we provide an alternative proof.
"$\Leftarrow$" It is trivial.
"$\Rightarrow$" For $v = 0$, since $u^p = v^p$, we have $u = v = 0$. For $v \neq 0$, from $f(t) = t^p - 1$ being a strictly monotone increasing function for any $t \in \mathbb{R}$, we have $\left(\frac{u}{v}\right)^p - 1 = 0$ if and only if $\frac{u}{v} = 1$, which implies $u = v$. Thus, the proof is complete. □

Lemma 2.17 *Let $\phi^p_{\text{D-FB}}$ be defined as in* (2.103) *where p is a positive odd integer. Then, the value of $\phi^p_{\text{D-FB}}(a, b)$ is negative only in the first quadrant, i.e., $\phi^p_{\text{D-FB}}(a, b) < 0$ if and only if $a > 0$, $b > 0$.*

Proof We know that $f(t) = t^p$ is a strictly increasing function when p is odd. Using this fact yields

$$\begin{aligned}
& a > 0,\ b > 0 \\
\Longleftrightarrow\ & a + b > 0 \quad \text{and} \quad ab > 0 \\
\Longleftrightarrow\ & \sqrt{a^2+b^2} < a + b \\
\Longleftrightarrow\ & \left(\sqrt{a^2+b^2}\right)^p < (a+b)^p \\
\Longleftrightarrow\ & \phi^p_{\text{D-FB}}(a,b) < 0,
\end{aligned}$$

which proves the desired result. □

Proposition 2.65 *Let $\phi^p_{\text{D-FB}}$ be defined as in* (2.103) *where p is a positive odd integer. Then, the function $\phi^p_{\text{D-FB}}$ is an NCP function.*

Proof Suppose $\phi^p_{\text{D-FB}}(a, b) = 0$, which says $\left(\sqrt{a^2+b^2}\right)^p = (a+b)^p$. Using p being a positive odd integer and applying Lemma 2.16, we have

$$\left(\sqrt{a^2+b^2}\right)^p = (a+b)^p \quad \Longleftrightarrow \quad \sqrt{a^2+b^2} = a+b.$$

It is well known that $\sqrt{a^2+b^2} = a+b$ is equivalent to $a, b \geq 0, ab = 0$ because ϕ_{FB} is an NCP-function. This shows that $\phi^p_{\text{D-FB}}(a,b) = 0$ implies $a, b \geq 0, ab = 0$. The converse direction is trivial. Thus, we prove that $\phi^p_{\text{D-FB}}$ is an NCP-function. □

We now provide a more detailed discussion of the newly introduced NCP function $\phi^p_{\text{D-FB}}$.

(a) For p being an even integer, $\phi^p_{\text{D-FB}}$ is not an NCP function. A counterexample is given as below.

$$\phi^p_{\text{D-FB}}(-5, 0) = (-5)^2 - (-5)^2 = 0.$$

(b) The surface of $\phi^p_{\text{D-FB}}$ is symmetric, that is, $\phi^p_{\text{D-FB}}(a,b) = \phi^p_{\text{D-FB}}(b,a)$.

(c) The function $\phi^p_{\text{D-FB}}(a,b)$ is positive homogenous of degree p, i.e., $\phi^p_{\text{D-FB}}(\alpha(a,b)) = \alpha^p \phi^p_{\text{D-FB}}(a,b)$.

(d) The function $\phi^p_{\text{D-FB}}$ is neither convex nor concave function. To see this, taking $p = 3$ and using the following argument verify the assertion.

$$0 = \phi^3_{\text{D-FB}}(1,1) > \frac{1}{2}\phi^3_{\text{D-FB}}(0,0) + \frac{1}{2}\phi^3_{\text{D-FB}}(2,2) = \frac{1}{2} \times 0 + \frac{1}{2}\left(2^{\frac{9}{2}} - 2^6\right) = \frac{1}{2}\left(2^{\frac{9}{2}} - 2^6\right)$$

and

$$2^{-\frac{3}{2}} = \phi^3_{\text{D-FB}}\left(-\frac{1}{2}, \frac{1}{2}\right) < \frac{1}{2}\phi^3_{\text{D-FB}}(-1,0) + \frac{1}{2}\phi^3_{\text{D-FB}}(0,1) = \frac{1}{2} \times 2 + \frac{1}{2} \times 0 = 1.$$

Proposition 2.66 *Let $\phi^p_{\text{D-FB}}$ be defined as in* (2.103) *where p is a positive odd integer. Then, the following hold.*

(a) For $p > 1$, $\phi^p_{\text{D-FB}}$ is continuously differentiable with

$$\nabla \phi^p_{\text{D-FB}}(a,b) = p \begin{bmatrix} a(\sqrt{a^2+b^2})^{p-2} - (a+b)^{p-1} \\ b(\sqrt{a^2+b^2})^{p-2} - (a+b)^{p-1} \end{bmatrix}.$$

(b) For $p > 1$, $\phi^p_{\text{D-FB}}$ is twice continuously differentiable with $\nabla^2 \phi^p_{\text{D-FB}}(0,0) = \begin{bmatrix} 0 & 0 \\ 0 & 0 \end{bmatrix}$, and for $(a,b) \neq (0,0)$,

$$\nabla^2 \phi^p_{\text{D-FB}}(a,b) = \begin{bmatrix} \dfrac{\partial^2 \phi^p_{\text{D-FB}}}{\partial a^2} & \dfrac{\partial^2 \phi^p_{\text{D-FB}}}{\partial a \partial b} \\ \dfrac{\partial^2 \phi^p_{\text{D-FB}}}{\partial b \partial a} & \dfrac{\partial^2 \phi^p_{\text{D-FB}}}{\partial b^2} \end{bmatrix}, \tag{2.104}$$

where

$$\frac{\partial^2 \phi^p_{\mathrm{D\text{-}FB}}}{\partial a^2} = p\left\{[(p-1)a^2+b^2](\sqrt{a^2+b^2})^{p-4} - (p-1)(a+b)^{p-2}\right\},$$
$$\frac{\partial^2 \phi^p_{\mathrm{D\text{-}FB}}}{\partial a \partial b} = p[(p-2)ab(\sqrt{a^2+b^2})^{p-4} - (p-1)(a+b)^{p-2}] = \frac{\partial^2 \phi^p_{\mathrm{D\text{-}FB}}}{\partial b \partial a},$$
$$\frac{\partial^2 \phi^p_{\mathrm{D\text{-}FB}}}{\partial b^2} = p\left\{[a^2+(p-1)b^2](\sqrt{a^2+b^2})^{p-4} - (p-1)(a+b)^{p-2}\right\}.$$

Proof (a) The differentiability of $\phi^p_{\mathrm{D\text{-}FB}}$, along with the computations of its first and second derivatives, follows directly from standard calculus and is therefore omitted here for brevity.

(b) For $p > 3$, the verifications are straightforward. The trick part is the case of $p = 3$. Thus, we only show that $\phi^p_{\mathrm{D\text{-}FB}}$ is twice continuously differentiable whenever $p = 3$. In fact, for $(a, b) \neq (0, 0)$, $\phi^3_{\mathrm{D\text{-}FB}}$ is twice continuously differentiable with $\nabla^2 \phi^3_{\mathrm{D\text{-}FB}}$ satisfying (2.104). It remains to claim $\nabla^2 \phi^3_{\mathrm{D\text{-}FB}}(0,0) = \begin{bmatrix} 0 & 0 \\ 0 & 0 \end{bmatrix}$, and $\nabla^2 \phi^3_{\mathrm{D\text{-}FB}}$ is continuous at $(0, 0)$. First, we note that

$$\nabla \phi^3_{\mathrm{D\text{-}FB}}(a,b) - \nabla \phi^3_{\mathrm{D\text{-}FB}}(0,0) = 3\begin{bmatrix} a(\sqrt{a^2+b^2}) - (a+b)^2 \\ b(\sqrt{a^2+b^2}) - (a+b)^2 \end{bmatrix},$$

and

$$\begin{aligned}\left\|\begin{bmatrix} a(\sqrt{a^2+b^2}) - (a+b)^2 \\ b(\sqrt{a^2+b^2}) - (a+b)^2 \end{bmatrix}\right\| &\le \left\|(\sqrt{a^2+b^2})\begin{bmatrix} a \\ b \end{bmatrix}\right\| + \left\|(a+b)^2\begin{bmatrix} 1 \\ 1 \end{bmatrix}\right\| \\ &= a^2+b^2+\sqrt{2}(a+b)^2 \\ &= (1+\sqrt{2})\underbrace{(a^2+b^2)}_{\text{(i)}} + 2\underbrace{\sqrt{2}ab}_{\text{(ii)}}.\end{aligned}$$

(i) $\dfrac{(a^2+b^2)}{\sqrt{a^2+b^2}} = \sqrt{a^2+b^2} \to 0$ as $(a,b) \to (0,0)$.

(ii) $\dfrac{|\sqrt{2}ab|}{\sqrt{a^2+b^2}} \le \dfrac{\sqrt{2}|ab|}{\sqrt{2|ab|}} = \sqrt{|ab|} \to 0$ as $(a,b) \to (0,0)$, where the inequality holds by arithmetic-geometric mean inequality.

Hence, we have

$$\lim_{(a,b)\to(0,0)} \frac{\|\nabla \phi^3_{\mathrm{D\text{-}FB}}(a,b) - \nabla \phi^3_{\mathrm{D\text{-}FB}}(0,0)\|}{\sqrt{a^2+b^2}} = 0,$$

i.e., $\phi^3_{\mathrm{D\text{-}FB}}$ is twice differentiable and $\nabla^2 \phi^3_{\mathrm{D\text{-}FB}}(0,0) = \begin{bmatrix} 0 & 0 \\ 0 & 0 \end{bmatrix}$. Secondly, we claim each second partial derivative is continuous at $(0, 0)$. For

$$\frac{\partial^2 \phi^3_{\text{D-FB}}}{\partial a^2} = 3\left(\frac{2a^2+b^2}{\sqrt{a^2+b^2}} - 2(a+b)\right) = 3\left(\sqrt{a^2+b^2} + \frac{a^2}{\sqrt{a^2+b^2}} - 2(a+b)\right),$$

it is clear that $\sqrt{a^2+b^2} \to 0$, $a+b \to 0$ as $(a,b) \to (0,0)$. And the second term $\frac{a^2}{\sqrt{a^2+b^2}}$ also tends to zero because

$$\frac{a^2}{\sqrt{a^2+b^2}} = |a| \cdot \frac{|a|}{\sqrt{a^2+b^2}} \leq |a| \to 0.$$

Hence, $\frac{\partial^2 \phi^3_{\text{D-FB}}}{\partial a^2}$ is continuous at $(0,0)$. For $\frac{\partial^2 \phi^3_{\text{D-FB}}}{\partial b^2}$, the proof is similar. For

$$\frac{\partial^2 \phi^p_{\text{D-FB}}}{\partial a \partial b} = 3\left(\frac{ab}{\sqrt{a^2+b^2}} - 2(a+b)\right) = \frac{\partial^2 \phi^p_{\text{D-FB}}}{\partial b \partial a},$$

it is obvious that $\frac{\partial^2 \phi^p_{\text{D-FB}}}{\partial a \partial b}$ tends to zero, where the first term tends to zero by (ii). Therefore, we obtain $\phi^3_{\text{D-FB}}$ is twice continuously differentiable at $(0,0)$, which is the desired result. □

Proposition 2.67 *Let $\phi^p_{\text{D-FB}}$ be defined as in (2.103) where $p > 1$ being a positive odd integer. Then, the following hold.*

(a) $\phi^p_{\text{D-FB}}(a,b) < 0 \iff a > 0,\ b > 0.$
(b) $\phi^p_{\text{D-FB}}$ *is locally Lipschitz continuous, but not Lipschitz continuous.*
(c) $\phi^p_{\text{D-FB}}$ *is not α-Hölder continuous for any $\alpha \in (0,1]$.*
(d) $\nabla_a \phi^p_{\text{D-FB}}(a,b) \cdot \nabla_b \phi^p_{\text{D-FB}}(a,b) > 0$ *on the first quadrant* $\mathbb{R}^2_{++}$.
(e) $\nabla_a \phi^p_{\text{D-FB}}(a,b) \cdot \nabla_b \phi^p_{\text{D-FB}}(a,b) = 0$ *provided that* $\phi^p_{\text{D-FB}}(a,b) = 0$.

Proof (a) It follows from Lemma 2.17 immediately.

(b)–(c) The arguments are similar to Proposition 2.24(c)–(d).

(d) According to Proposition 2.66, we have

$$\begin{aligned}
&\nabla_a \phi^p_{\text{D-FB}}(a,b) \cdot \nabla_b \phi^p_{\text{D-FB}}(a,b) \\
&= p\left[a(\sqrt{a^2+b^2})^{p-2} - (a+b)^{p-1}\right] \cdot p\left[b(\sqrt{a^2+b^2})^{p-2} - (a+b)^{p-1}\right] \\
&= p^2\left[ab(a^2+b^2)^{p-2} + (a+b)^{2p-2} - (a+b)^{p-1}(\sqrt{a^2+b^2})^{p-2} \cdot (a+b)\right] \\
&= p^2\left[ab(a^2+b^2)^{p-2} + (a+b)^{2p-2} - (a+b)^{p}(\sqrt{a^2+b^2})^{p-2}\right] \\
&= p^2\left[ab(a^2+b^2)^{p-2} + (a+b)^{p}\left((a+b)^{p-2} - (\sqrt{a^2+b^2})^{p-2}\right)\right].
\end{aligned}$$

Since $a > 0$, $b > 0$ and $p-2$ is also an odd number, the term $(a+b)^{p-2} - (\sqrt{a^2+b^2})^{p-2}$ is always positive by part(a). This clearly implies the desired result.

(e) From Proposition 2.65, we know $\phi^p_{\text{D-FB}}$ is an NCP function, which implies

$$\phi^p_{\text{D-FB}}(a,b) = 0 \Longleftrightarrow a \geq 0,\ b \geq 0,\ ab = 0.$$

When $a \geq 0$ and $b = 0$, we have $\nabla_a \phi^p_{\text{D-FB}}(a,0) = a(\sqrt{a^2})^{p-2} - a^{p-1} = a^{p-1} - a^{p-1} = 0$. Similarly, when $b \geq 0$ and $a = 0$, we have $\nabla_b \phi^p_{\text{D-FB}}(0,b) = 0$. In summary, we conclude $\nabla_a \phi^p_{\text{D-FB}}(a,b) \cdot \nabla_b \phi^p_{\text{D-FB}}(a,b) = 0$ provided that $\phi^p_{\text{D-FB}} = 0$. □

We next present several variants of $\phi^p_{\text{D-FB}}$. In fact, similar to the functions proposed in [11], these variants can be verified to satisfy the defining properties of NCP functions.

$$\begin{aligned}
\phi_1(a,b) &= \phi^p_{\text{D-FB}}(a,b) - \alpha(a)_+(b)_+,\ \alpha > 0.\\
\phi_2(a,b) &= \phi^p_{\text{D-FB}}(a,b) - \alpha\left((a)_+(b)_+\right)^2,\ \alpha > 0.\\
\phi_3(a,b) &= [\phi^p_{\text{D-FB}}(a,b)]^2 + \alpha\left((ab)_+\right)^4,\ \alpha > 0.\\
\phi_4(a,b) &= [\phi^p_{\text{D-FB}}(a,b)]^2 + \alpha\left((ab)_+\right)^2,\ \alpha > 0.
\end{aligned}$$

Proposition 2.68 *All the above functions ϕ_i for $i \in \{1,2,3,4\}$ are NCP functions.*

Proof Applying Lemma 2.17, the arguments are similar to those in [34, Proposition 2.4], which are omitted here. □

Indeed, in light of Lemma 2.16, we can construct additional variants of $\phi^p_{\text{D-FB}}$, each of which constitutes a novel NCP function. More specifically, let k and m be positive integers, and let $f : \mathbb{R} \times \mathbb{R} \to \mathbb{R}$, and $g : \mathbb{R} \times \mathbb{R} \to \mathbb{R}$ be functions such that $g(a,b) \neq 0$ for all $a, b \in \mathbb{R}$. Then, the following constructions yield new variants of $\phi^p_{\text{D-FB}}$ that satisfy the properties of NCP functions.

$$\phi_5(a,b) = \left[g(a,b)\left(\sqrt{a^2+b^2} + f(a,b)\right)\right]^{\frac{2k+1}{2m+1}} - \left[g(a,b)\left(a+b+f(a,b)\right)\right]^{\frac{2k+1}{2m+1}}.$$

$$\phi_6(a,b) = \left[g(a,b)(\sqrt{a^2+b^2} - a - b)\right]^{\frac{2k+1}{2m+1}}.$$

$$\phi_7(a,b) = \left[g(a,b)(\sqrt{a^2+b^2} - a + f(a,b))\right]^{\frac{2k+1}{2m+1}} - [g(a,b)(b+f(a,b))]^{\frac{2k+1}{2m+1}}.$$

$$\phi_8(a,b) = \left[g(a,b)(\sqrt{a^2+b^2} - a + f(a,b))\right]^{\frac{2k+1}{2m+1}} - [g(a,b)(b+f(a,b))]^{\frac{2k+1}{2m+1}}.$$

$$\phi_9(a,b) = e^{\phi_i(a,b)} - 1 \text{ where } i = 5,6,7,8.$$

$$\phi_{10}(a,b) = \ln(|\phi_i(a,b)| + 1) \text{ where } i = 5,6,7,8.$$

Proposition 2.69 *All the above functions ϕ_i for $i \in \{5,6,7,8,9,10\}$ are NCP functions.*

Proof This is an immediate consequence of Propositions 2.65–2.68. In particular, by Lemma 2.16 and $g(a,b) \neq 0$ for $a, b \in \mathbb{R}$, we have

$$\phi_5(a,b)=0$$

$$
\begin{aligned}
&\Longleftrightarrow\left[g(a,b)\left(\sqrt{a^2+b^2}+f(a,b)\right)\right]^{\frac{2k+1}{2m+1}}=\left[g(a,b)(a+b+f(a,b))\right]^{\frac{2k+1}{2m+1}}\\
&\Longleftrightarrow\left\{\left[g(a,b)\left(\sqrt{a^2+b^2}+f(a,b)\right)\right]^{\frac{2k+1}{2m+1}}\right\}^{2m+1}=\left\{\left[g(a,b)(a+b+f(a,b))\right]^{\frac{2k+1}{2m+1}}\right\}^{2m+1}\\
&\Longleftrightarrow\left[g(a,b)\left(\sqrt{a^2+b^2}+f(a,b)\right)\right]^{2k+1}=\left[g(a,b)(a+b+f(a,b))\right]^{2k+1}\\
&\Longleftrightarrow g(a,b)\left(\sqrt{a^2+b^2}+f(a,b)\right)=g(a,b)(a+b+f(a,b))\\
&\Longleftrightarrow\left(\sqrt{a^2+b^2}+f(a,b)\right)=(a+b+f(a,b))\\
&\Longleftrightarrow\sqrt{a^2+b^2}=a+b.
\end{aligned}
$$

The other functions ϕ_i for $i \in \{6, 7, 8, 9, 10\}$ is similar to ϕ_5. □

Proposition 2.70 *Suppose that $\phi(a,b) = \varphi_1(a,b) - \varphi_2(a,b)$ is an NCP function on $\mathbb{R} \times \mathbb{R}$ and k and m are positive integers. Then, $\left[\phi(a,b)\right]^{\frac{2k+1}{2m+1}}$ and $\left[\varphi_1(a,b)\right]^{\frac{2k+1}{2m+1}} - \left[\varphi_2(a,b)\right]^{\frac{2k+1}{2m+1}}$ are NCP functions.*

Proof Using k and m being positive integers and applying Lemma 2.16, we have

$$
\begin{aligned}
&\left[\phi(a,b)\right]^{\frac{2k+1}{2m+1}}=0\\
\Longleftrightarrow&\left\{\left[\phi(a,b)\right]^{\frac{2k+1}{2m+1}}\right\}^{2m+1}=0\\
\Longleftrightarrow&\left[\phi(a,b)\right]^{2k+1}=0\\
\Longleftrightarrow&\phi(a,b)=0.
\end{aligned}
$$

Similarly, we have

$$
\begin{aligned}
&\left[\varphi_1(a,b)\right]^{\frac{2k+1}{2m+1}}-\left[\varphi_2(a,b)\right]^{\frac{2k+1}{2m+1}}=0\\
\Longleftrightarrow&\left[\varphi_1(a,b)\right]^{\frac{2k+1}{2m+1}}=\left[\varphi_2(a,b)\right]^{\frac{2k+1}{2m+1}}\\
\Longleftrightarrow&\left\{\left[\varphi_1(a,b)\right]^{\frac{2k+1}{2m+1}}\right\}^{2m+1}=\left\{\left[\varphi_2(a,b)\right]^{\frac{2k+1}{2m+1}}\right\}^{2m+1}\\
\Longleftrightarrow&\left[\varphi_1(a,b)\right]^{2k+1}=\left[\varphi_2(a,b)\right]^{2k+1}\\
\Longleftrightarrow&\varphi_1(a,b)=\varphi_2(a,b)\\
\Longleftrightarrow&\phi(a,b)=0.
\end{aligned}
$$

The above arguments together with the assumption of $\phi(a,b)$ being an NCP function yield the desired result. □

Couple remarks regarding Proposition 2.70 are pointed out as below

(a) When k is a positive odd integer and m is a positive integer, $\left[\phi(a,b)\right]^k$ is an NCP function. Whenever perturbing the parameter k, we obtain new NCP functions. For example, if $\phi(a,b)$ is an NCP-function, then $\left[\phi(a,b)\right]^{k+\frac{1}{2m+1}}$ is an

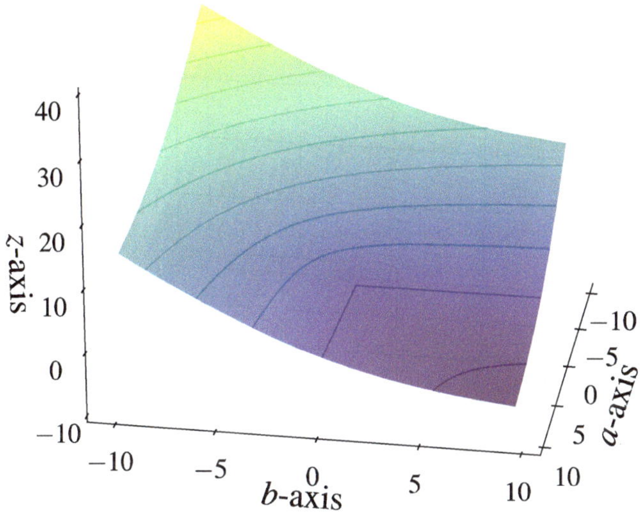

Fig. 2.20 The surface of $z = \phi_{\text{D−FB}}(a, b)$ and $(a, b) \in [-10, 10] \times [-10, 10]$

NCP function. We can determine suitable and nice NCP functions among these functions according to their numerical performance.

(b) When k is a positive even integer and m is a positive integer, $\left[\phi(a, b)\right]^k$ cannot be an NCP function. However, $\left[\phi(a, b)\right]^{k+\frac{1}{2m+1}}$ is an NCP function, which offers an way to construct new NCP functions for even k. This approach opens an entirely new avenue for constructing NCP functions, offering a flexible and systematic framework for generating novel formulations with desirable analytical and computational properties.

To conclude this section, we illustrate the surfaces of $\phi^p_{\text{D−FB}}$ for various values of p, providing a visual perspective that offers deeper insight into the structure and behavior of this new family of NCP functions. Figure 2.20 is the surface if $\phi_{\text{D−FB}}(a, b)$ from which we see that it is convex. Figure 2.21 presents the surface of $\phi^3_{\text{D−FB}}(a, b)$ in which we see that it is neither convex nor concave as mentioned earlier. In addition, the value of $\phi^p_{\text{D−FB}}(a, b)$ is negative only when $a > 0$ and $b > 0$ as mentioned in Lemma 2.17. The surfaces of $\phi^p_{\text{D−FB}}$ with various values of p are shown in Fig. 2.22.

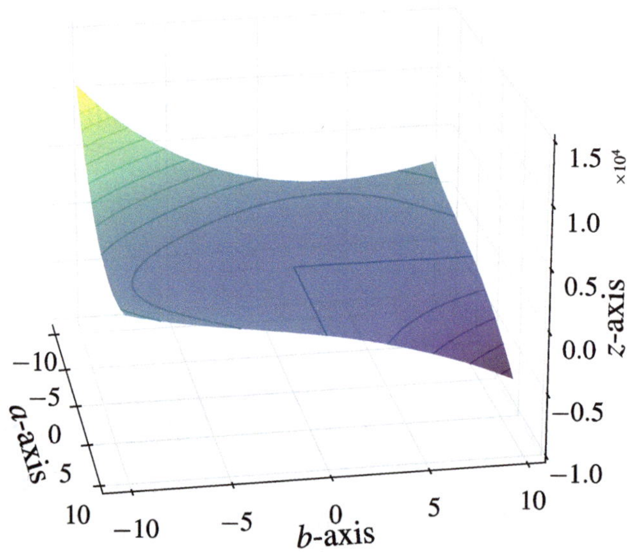

Fig. 2.21 The surface of $z = \phi^3_{\text{D-FB}}(a, b)$ and $(a, b) \in [-10, 10] \times [-10, 10]$

2.4 Constructions of NCP Functions Involving Certain Functions

To motivate this section, we note that there exists an alternative approach to deriving the functions ϕ^p_{NR} and $\phi^p_{\text{D-FB}}$, as discussed in Sects. 2.2 and 2.3, respectively. In [4], a method is proposed that constructs new NCP functions from existing ones through monotone transformations. This approach is, in fact, inspired by a key lemma presented in [4], which serves as the foundation for the transformation-based construction.

Lemma 2.18 *Assume that ϕ is continuous and $\phi(a, b) = f_1(a, b) - f_2(a, b)$. Let $\theta : \mathbb{R} \to \mathbb{R}$ be a strictly monotone increasing and continuous function. Then, the function ϕ is an NCP function if and only if $\psi_\theta(a, b) = \theta(f_1(a, b)) - \theta(f_2(a, b))$ is an NCP function.*

Proof Please see [4, Lemma 15]. □

In light of Lemma 2.18, we define the function $\theta = \theta_p$ as $\theta_p(t) = \text{sgn}(t)|t|^p$, where "sgn($t$)" denotes the sign function and $p \geq 1$. To illustrate, consider the Fischer-Burmeister function, for which we take $f_1(a, b) = \sqrt{a^2 + b^2}$ and $f_2(a, b) = a + b$. For the natural residual function, we set $f_1(a, b) = a$ and $f_2(a, b) = (a - b)_+$. With these choices, it can be verified that both $\phi^p_{\text{D-FB}}$ and ϕ^p_{NR} (with p restricted to odd integers) can be derived from the more general formulation ψ_{θ_p}. In this sense, ψ_{θ_p} encompasses both functions as special cases, and may thus be viewed as a form of

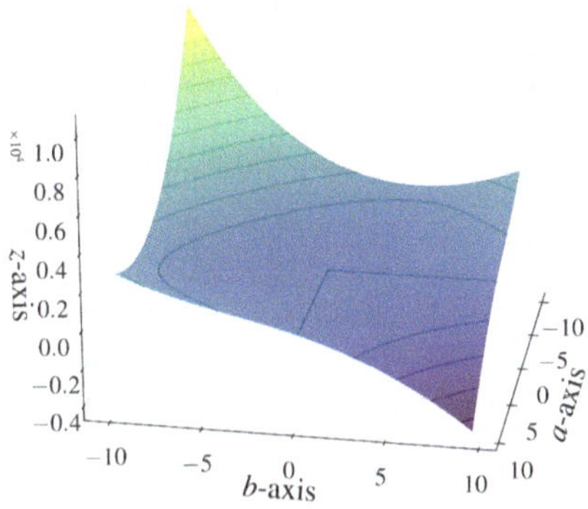

(a) $z = \phi^3_{\mathrm{D-FB}}(a, b)$

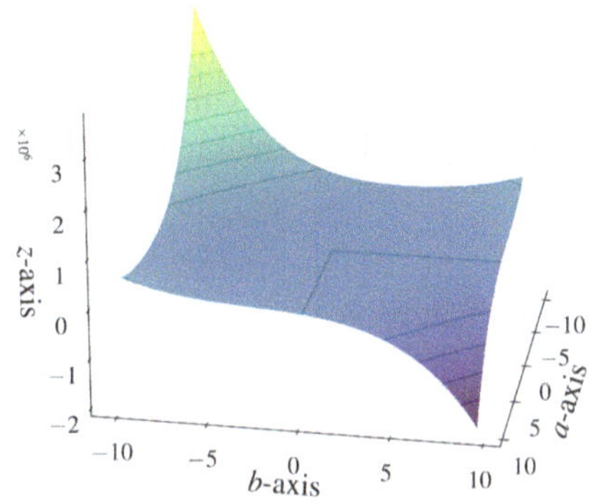

(b) $z = \phi^5_{\mathrm{D-FB}}(a, b)$

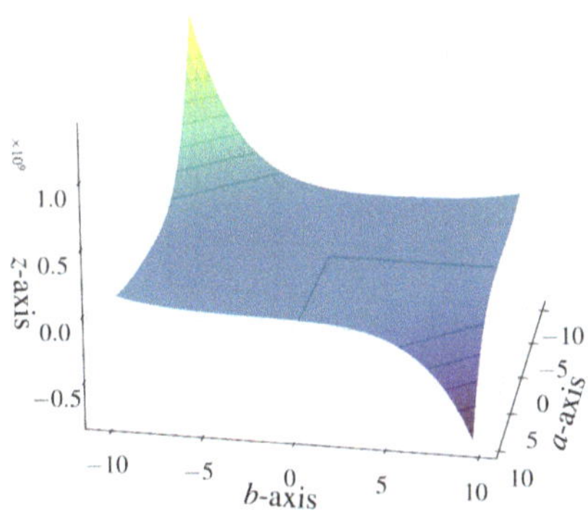

(c) $z = \phi^7_{\mathrm{D-FB}}(a, b)$

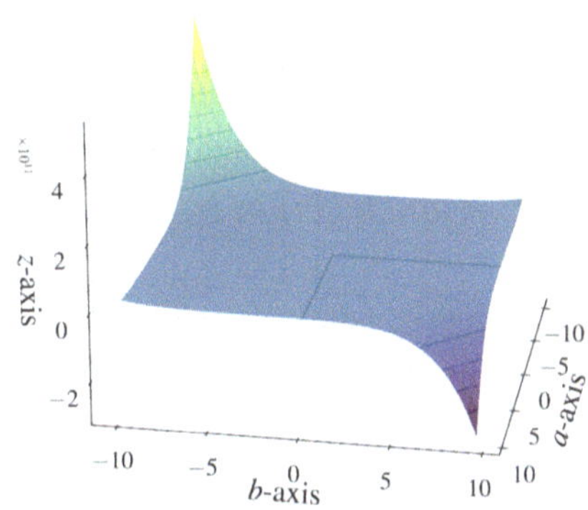

(d) $z = \phi^9_{\mathrm{D-FB}}(a, b)$

Fig. 2.22 The surface of $z = \phi^p_{\mathrm{D-FB}}(a, b)$ with different values of p

"continuous generalization". However, our preference is to interpret these constructions through the lens of "discrete generalization", which more accurately reflects the nature of our approach. Specifically, for the function $\phi^p_{\mathrm{NR}}(a, b) = a^p - (a - b)^p_+$, it is essential that p be an odd integer to ensure that the resulting function retains the defining properties of an NCP function. This condition underscores the discrete nature of the generalization: the validity of the function fundamentally depends on specific, discrete values of the parameter p. Therefore, the central idea behind our new families of NCP functions is rooted in discrete generalization, rather than in a smooth or continuous extension. This distinction forms the basis for our terminology and conceptual framework. On the other hand, if we consider the FB function $\phi_{\mathrm{FB}}(a, b) = \sqrt{a^2 + b^2} - (a + b)$. When plugging $p = 2$ into θ_p, we obtain a corresponding NCP function

$$\psi_{\theta_2}(a,b) = a^2 + b^2 - \text{sgn}(a+b)(a+b)^2,$$

which doesn't coincide with the form

$$\phi^p_{\text{D-FB}}(a,b) = \left(\sqrt{a^2+b^2}\right)^2 - (a+b)^2.$$

Thus, the functions $\phi^p_{\text{D-FB}}$ and ϕ^p_{NR} only with positive odd integer p can be retrieved from the way proposed in [4]. Again, it requires p to be a positive odd integer to guarantee that both $\phi^p_{\text{D-FB}}$ and ϕ^p_{NR} are NCP functions. In view of all the above, we still call them discrete-type families of NCP functions.

The aforementioned construction in Lemma 2.18 or [4, Lemma 15] relies on specific functions $\theta(\cdot)$ that satisfy particular conditions. Motivated by this concept, we propose novel construction ways that incorporate alternative classes of functions, thereby expanding the toolkit for generating new complementarity functions. Another noteworthy point is that both $\phi^p_{\text{D-FB}}$ and ϕ^p_{NR} serve not only as NCP functions, but also as complementarity functions for second-order cone complementarity problems (SOCCPs); further details can be found in Chap. 3.

2.4.1 Construction by Using Certain Functions

As previously indicated, we introduce a novel approach for constructing continuous NCP functions through the use of specific auxiliary functions. Let $\theta : \mathbb{R} \to \mathbb{R}$ be a continuous function, and define $\phi^p_\theta : \mathbb{R}^2 \to \mathbb{R}$ by

$$\phi^p_\theta(a,b) = \|(a,b)\|_p - (\theta(b)a + \theta(a)b), \quad p \geq 1. \tag{2.105}$$

It is evident that ϕ^p_θ is a continuous and symmetric function, meaning that $\phi^p_\theta(a,b) = \phi^p_\theta(b,a)$. With an appropriate choice of θ, this construction gives rise to an NCP function. Our analysis proceeds by considering two distinct cases, determined by the value of p.

I. The case of $p = 1$.

We first consider the case of $p = 1$, that is, $\phi^p_\theta : \mathbb{R}^2 \to \mathbb{R}$ is given by

$$\phi^1_\theta(a,b) = |a| + |b| - \big(\theta(b)a + \theta(a)b\big). \tag{2.106}$$

Proposition 2.71 *Let $\theta : \mathbb{R} \to \mathbb{R}$ such that $\theta(0) = 1$, $\theta(t) > 1$ for all $t > 0$, and $-1 < \theta(t) < 1$ for all $t < 0$. Then, the function ϕ^1_θ defined by* (2.106) *is an NCP function. Moreover, $\phi^1_\theta(a,b) \leq 0$ if and only if $(a,b) \in \mathbb{R}^2_+$.*

Proof Observe that we may rewrite ϕ^1_θ as

$$\phi^1_\theta(a,b) = a\big(\text{sgn}(a) - \theta(b)\big) + b\big(\text{sgn}(b) - \theta(a)\big),$$

where

$$\operatorname{sgn}(t) := \begin{cases} 1 \text{ if } t > 0, \\ 0 \text{ if } t = 0, \\ -1 \text{ if } t < 0. \end{cases}$$

Then, it is easy to verify that

$$\phi_\theta^1(a,b) = \begin{cases} 0 & \text{if } a, b \geq 0 \,\&\, ab = 0, \\ a(1-\theta(b)) + b(1-\theta(a)) & \text{if } a > 0 \,\&\, b > 0, \\ -a(1+\theta(b)) + b(1-\theta(a)) & \text{if } a < 0 \,\&\, b \geq 0, \\ -a(1+\theta(b)) - b(1+\theta(a)) & \text{if } a < 0 \,\&\, b < 0. \end{cases} \tag{2.107}$$

By our hypotheses on θ, we see that $\phi_\theta^1(a,b) < 0$ for the second case, and $\phi_\theta^1(a,b) > 0$ for the third and last cases. Finally, by symmetry of ϕ_θ^1, we have $\phi_\theta^1(a,b) > 0$ when $a > 0$ and $b < 0$ as in the third case. In other words, $\phi_\theta^1(a,b) = 0$ if and only if $a, b \geq 0$ and $ab = 0$. This says that ϕ_θ^1 is an NCP function. □

An important implication of Proposition 2.73 is encapsulated in the following result, which characterizes the growth behavior of the NCP function ϕ_θ^1. This consequence plays a key role in establishing the coerciveness of Φ_{F}, as defined in (2.3) (see [30]), which in turn facilitates the convergence analysis of relevant algorithms. We omit the proof, as it follows directly from the explicit formula of ϕ_θ^1 given in (2.107). It is worth emphasizing, however, that the strict inequality conditions on the limits of θ as $x \to \pm\infty$ are essential to preclude the emergence of indeterminate products.

Proposition 2.72 *Let θ satisfy the hypothesis of Proposition 2.73 such that $\lim_{t\to\infty} \theta(t) > 1$ and $-1 < \lim_{t\to-\infty} \theta(t) < 1$. Then, $|\phi_\theta^1(a^k, b^k)| \to \infty$ as $k \to \infty$ for any sequence $\{(a^k, b^k)\} \subseteq \mathbb{R}^2$ with $|a^k| \to \infty$ and $|b^k| \to \infty$.*

For the remainder of this section, we assume that $\theta(\cdot)$ satisfies the conditions specified in Proposition 2.73 when $p = 1$. A straightforward choice for θ is any monotonically increasing function whose range lies within $(-1, \infty)$, passes through the point (0, 1), and is strictly monotonic in a neighborhood of zero.

Example 2.1 The functions

$$\theta_1(t) = e^t, \quad \theta_2(t) = \frac{\sqrt{t^2+4}+t}{2}, \quad \text{and} \quad \theta_3(t) = \frac{2}{1+e^-t}$$

clearly satisfy the conditions of Proposition 2.73 and Corollary 2.72. The graphs of $\phi_{\theta_i}^1(a,b)$ for $i = 1, 2, 3$ are shown in Figs. 2.23a, 2.24a, and 2.25a. For each i, it is evident that the function $\phi_{\theta_i}^1$ is non-positive on $\mathbb{R}_+^2$ and has the growth behavior as described in Corollary 2.72. In addition, $\phi_{\theta_i}^1$ is a nonsmooth nonconvex function for all i. In particular, the function has sharp trace curves corresponding to $a = 0$ and $b = 0$, which are the points of non-differentiability of ϕ_θ^1.

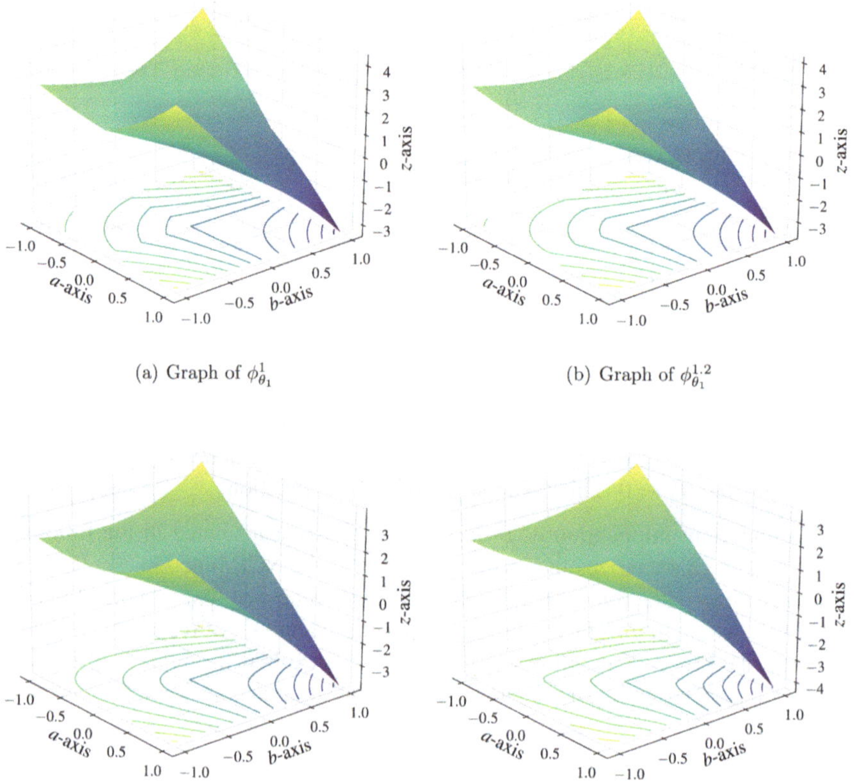

(a) Graph of $\phi_{\theta_1}^{1}$

(b) Graph of $\phi_{\theta_1}^{1.2}$

(c) Graph of $\phi_{\theta_1}^{2}$

(d) Graph of $\phi_{\theta_1}^{10}$

Fig. 2.23 Graphs of $\phi_{\theta_1}^{p}$ for different values of p where $\theta_1(t) = e^t$

II. The case of $p > 1$.

We now turn our attention to the case $p > 1$ and investigate the conditions under which ϕ_θ^p constitutes an NCP function. These conditions closely resemble those presented in Proposition 2.73, with a few key distinctions. Specifically, strict inequality at $t = 1$ is not required; however, a stronger lower bound on $\theta(t)$ is necessary for $t < 0$.

Proposition 2.73 *Let* $p > 1$. *Suppose* $\theta : \mathbb{R} \to \mathbb{R}$ *such that* $\theta(0) = 1$, $\theta(t) \geq 1$ *for all* $t > 0$, *and* $-2^{\frac{1-p}{p}} \leq \theta(t) \leq 1$ *for all* $t < 0$. *Then, the function* ϕ_θ^p *defined by* (2.105) *is an NCP function. Moreover,* $\phi_\theta^p(a, b) \leq 0$ *if and only if* $(a, b) \in \mathbb{R}_+^2$.

Proof Since ϕ_θ^p is symmetric w.r.t. the line $a = b$, it suffices to check the values of ϕ_θ^p on the region $a \leq b$. We carefully consider four cases.

(i) If $a = 0$ and $b > 0$, then $\phi_\theta^p(a, b) = |b| - \theta(0)b = 0$ since $\theta(0) = 1$.

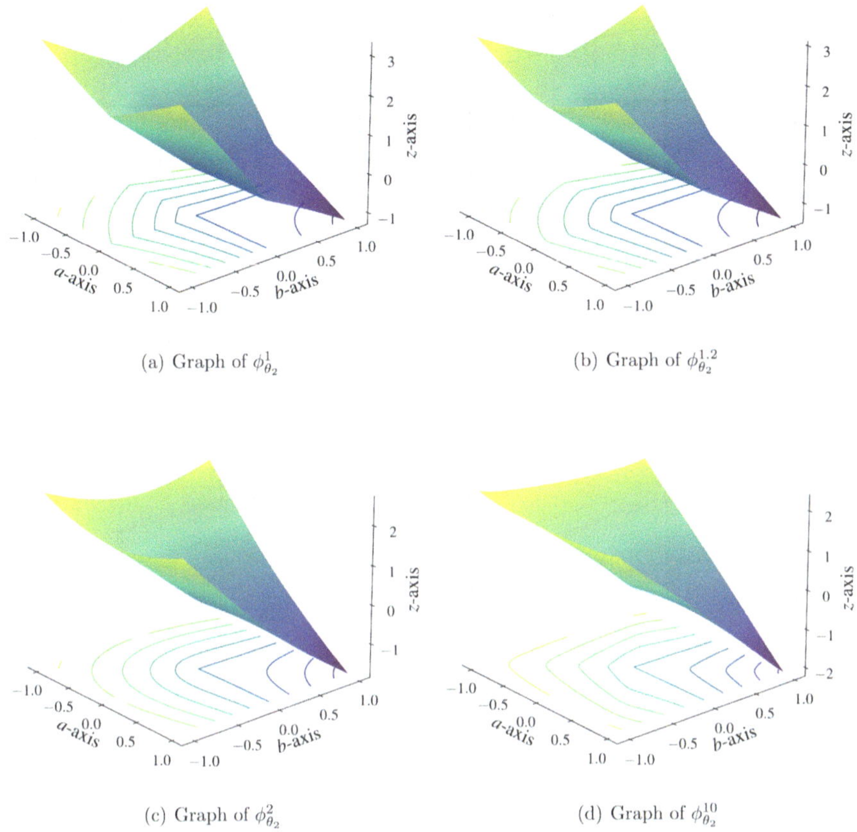

(a) Graph of $\phi^1_{\theta_2}$

(b) Graph of $\phi^{1.2}_{\theta_2}$

(c) Graph of $\phi^2_{\theta_2}$

(d) Graph of $\phi^{10}_{\theta_2}$

Fig. 2.24 Graphs of $\phi^p_{\theta_2}$ for different values of p where $\theta_2(t) = \frac{\sqrt{t^2+4}+t}{2}$

(ii) Suppose $a > 0$ and $b > 0$. Due to $p > 1$, we have $\|(a,b)\|_p = (a^p + b^p)^{\frac{1}{p}} < a + b$ which in turn yields

$$\phi^p_\theta(a,b) < a + b - (\theta(b)a + \theta(a)b) = a(1-\theta(b)) + b(1-\theta(a)).$$

Because $\theta(t) \geq 1$ for any $t > 0$ it follows that $\phi^p_\theta(a,b) < 0$.

(iii) Suppose $a < 0$ and $b \geq 0$. In this case, we have that $\|(a,b)\|_p > a + b$. Thus,

$$\phi^p_\theta(a,b) > a + b - (\theta(b)a + \theta(a)b) = a(1-\theta(b)) + b(1-\theta(a)).$$

Since $b \geq 0$, we have $1 - \theta(b) \leq 0$ and so the term $a(1-\theta(b))$ is nonnegative. On the other hand, $1 - \theta(a) > 0$ since $a < 0$ which means that the term $b(1-\theta(a))$ is likewise nonnegative. Hence, $\phi^p_\theta(a,b) > 0$.

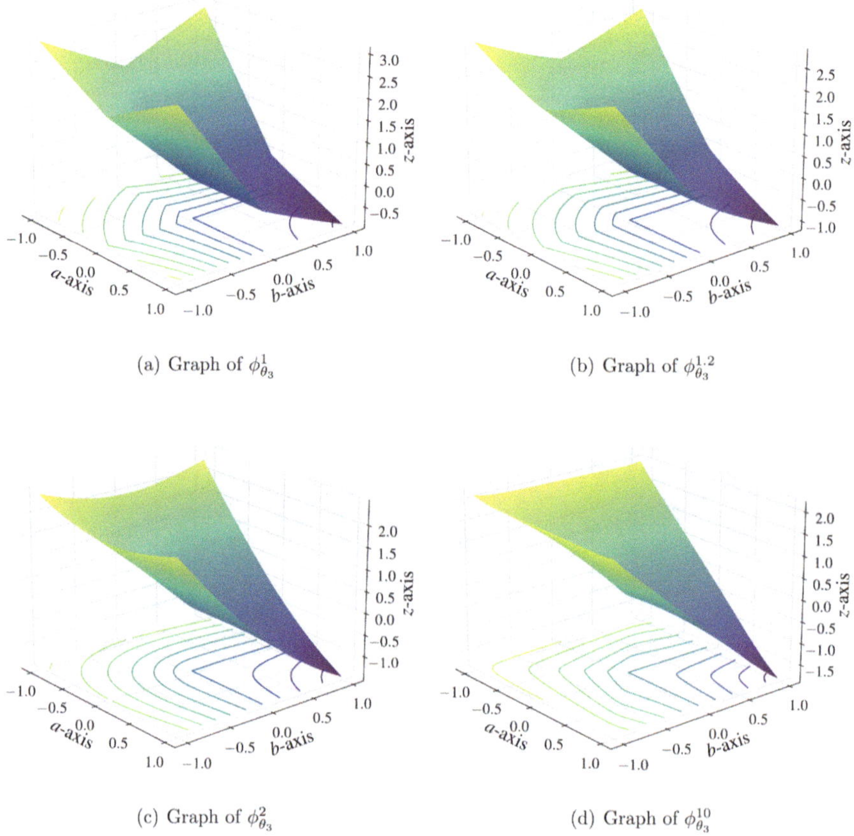

(a) Graph of $\phi^1_{\theta_3}$ (b) Graph of $\phi^{1.2}_{\theta_3}$

(c) Graph of $\phi^2_{\theta_3}$ (d) Graph of $\phi^{10}_{\theta_3}$

Fig. 2.25 Graphs of $\phi^p_{\theta_3}$ for different values of p where $\theta_3(t) = \frac{2}{1+e^{-t}}$

(iv) Finally, suppose that $a < 0$ and $b < 0$. The function $t \mapsto t^p$ is strictly convex on $[0, \infty)$ since $p > 1$. Thus,

$$\|(a, b)\|_p^p = |a|^p + |b|^p > 2^{1-p}(|a| + |b|)^p,$$

which implies that $\|(a, b)\|_p > 2^{\frac{1-p}{p}}(|a| + |b|) = -2^{\frac{1-p}{p}}(a + b)$. Consequently,

$$\begin{aligned}
\phi^p_\theta(a, b) >& -2^{\frac{1-p}{p}}(a + b) - (\theta(b)a + \theta(a)b) \\
=& -a(2^{\frac{1-p}{p}} + \theta(b)) - b(2^{\frac{1-p}{p}} + \theta(a)) \\
\geq& 0
\end{aligned}$$

where the last inequality follows from the assumption that $\theta(t) \geq -2^{\frac{1-p}{p}}$ for all $t \leq 0$.

From the above four cases, it is clear that $\phi_\theta^p(a,b) \le 0$ only on $\mathbb{R}_+^2$. This completes the proof. □

Proposition 2.74 *Let θ satisfy the hypothesis of Proposition 2.73 such that $\lim_{t\to\infty}\theta(t) > 1$ and $-2^{\frac{1-p}{p}} < \lim_{t\to-\infty}\theta(t) < 1$. Then, $|\phi_\theta^p(a^k,b^k)| \to \infty$ as $k \to \infty$ for any sequence $\{(a^k,b^k)\} \subseteq \mathbb{R}^2$ with $|a^k| \to \infty$ and $|b^k| \to \infty$.*

Proof The result follows from the inequalities obtained from cases (ii), (iii) and (iv) in the proof of Proposition 2.73. □

For all cases where $p > 1$, we henceforth assume that $\theta(\cdot)$ satisfies the conditions outlined in Proposition 2.73. We now proceed to illustrate this with a few examples.

Example 2.2 Observe that by taking $\theta(t) \equiv 1$, we obtain the generalized FB function (2.14). Hence, the family of NCP functions given by (2.105) subsumes the class of generalized FB functions.

Example 2.3 As in Example 2.3, consider θ_i for $i = 1, 2, 3$. Then, for any $p > 1$, the function $\phi_{\theta_i}^p$ is an NCP function by Proposition 2.73. Notice from Figs. 1–3 (subfigures (b) to (d)) that the graphs of $\phi_{\theta_i}^p$ $(p > 1)$ look "smoother" than that of $\phi_{\theta_i}^1$. In particular, ϕ_θ^p is not differentiable only at the origin. Finally, $\phi_{\theta_i}^p$ is also nonconvex similar to $\phi_{\theta_i}^1$ in Example 2.3.

It is well established that no complementarity function can simultaneously possess both differentiability and convexity [7, 8]. In fact, such a function may lack both properties. The following two propositions demonstrate that this is indeed true for ϕ_θ^p. As observed in Examples 2.3 and 2.3, ϕ_θ^p fails to be convex. We now assert that this non-convexity holds more generally.

Proposition 2.75 *Suppose that θ is strictly increasing on some interval $I = [0, t_0)$. Then, ϕ_θ^p is not convex.*

Proof Suppose that ϕ_θ^p is convex, due to $\phi_\theta^p(0,0) = 0$, it must be the case that $\phi_\theta^p(\lambda a, \lambda b) \le \lambda\phi_\theta^p(a,b)$ for any $\lambda \in [0,1]$ and any $u, v \in \mathbb{R}$. Taking any $a, b \in I$ yields

$$\begin{aligned}
&\phi_\theta^p(\lambda a, \lambda b) - \lambda\phi_\theta^p(a,b)\\
&= \|(\lambda a, \lambda b)\|_p - (\lambda\theta(\lambda b)a + \lambda\theta(\lambda a)b) - \lambda(\|(a,b)\|_p\\
&\quad -(\theta(b)a + \theta(a)b))\\
&= \lambda a(\theta(b) - \theta(\lambda b)) + \lambda b(\theta(a) - \theta(\lambda a)).
\end{aligned}$$

Since $\lambda \in [0,1]$, we have that $\lambda a, \lambda b \in I$. By the strict monotonicity assumption on θ in I, there has $\phi_\theta^p(\lambda a, \lambda b) - \lambda\phi_\theta^p(a,b) > 0$. Hence, ϕ_θ^p is not convex. □

Proposition 2.76 *Suppose that θ is continuously differentiable and satisfies the conditions of Proposition 2.73 if $p = 1$ or Proposition 2.73 if $p > 1$. Then, ϕ_θ^p is semismooth. Moreover, the generalized gradient of ϕ_θ^1 is described by*

$$\partial\phi_\theta^1(a,b) = \begin{cases} \{[\mathrm{sgn}(a) - \theta'(a)b - \theta(b),\ \mathrm{sgn}(b) - \theta'(b)a - \theta(a)]^\mathsf{T}\} & \text{if } a \neq 0 \,\&\, b \neq 0 \\ \{[0,\ 2\lambda - 1 - a\theta'(0) - \theta(a)]^\mathsf{T} \mid \lambda \in [0,1]\} & \text{if } a > 0 \,\&\, b = 0 \\ \{[2\lambda - 1 - b\theta'(0) - \theta(b),\ 0]^\mathsf{T} \mid \lambda \in [0,1]\} & \text{if } a = 0 \,\&\, b > 0 \\ \{[-2,\ 2\lambda - 1 - a\theta'(0) - \theta(a)]^\mathsf{T} \mid \lambda \in [0,1]\} & \text{if } a < 0 \,\&\, b = 0 \\ \{[-2,\ 2\lambda - 1 - b\theta'(0) - \theta(b)]^\mathsf{T} \mid \lambda \in [0,1]\} & \text{if } a = 0 \,\&\, b < 0 \\ \{[\xi,\zeta]^\mathsf{T} \mid \xi,\zeta \in [-2,0]\,\} & \text{if } a = b = 0 \end{cases}$$

and for $p > 1$, we have

$$\partial\phi_\theta^p(a,b) = \begin{cases} \left\{\left[\frac{\mathrm{sgn}(a)|a|^{p-1}}{\|(a,b)\|_p^{1-p}} - \theta(b) - b\theta'(a),\ \frac{\mathrm{sgn}(b)|b|^{p-1}}{\|(a,b)\|_p^{1-p}} - \theta(a) - a\theta'(b)\right]^\mathsf{T}\right\} & \text{if } (a,b) \neq (0,0) \\ \left\{[\xi - 1, \zeta - 1]^\mathsf{T} \mid |\xi|^{\frac{p}{p-1}} + |\zeta|^{\frac{p}{p-1}} \le 1\right\} & \text{if } a = b = 0. \end{cases}$$

Proof Note that the mapping $f : (a,b) \mapsto \|(a,b)\|_p$ is a convex map and is therefore semismooth. Because $g : (a,b) \mapsto -(\theta(b)a + \theta(a)b$ is smooth (and hence semismooth), their sum $f + g = \phi_\theta^p$ is semismooth. Now, we compute the generalized gradient of ϕ_θ^1. It is clear that ϕ_θ^1 is differentiable only on $D := \{(a,b) : a \neq 0 \text{ and } b \neq 0\}$. Then, its gradient

$$\nabla\phi_\theta^1(a,b) = \begin{bmatrix} \mathrm{sgn}(a) - \theta'(a)b - \theta(b) \\ \mathrm{sgn}(b) - \theta'(b)a - \theta(a) \end{bmatrix} \quad \forall (a,b) \in D,$$

coincides with the generalized gradient on D. Suppose then that $(a,b) \notin D$. First, we consider the case when $a > 0$ and $b = 0$. By definition of Clarke's generalized gradient $\partial\phi_\theta^1(a,b) = \mathrm{conv}\left(\partial_B\phi_\theta^1(a,b)\right)$, i.e., the convex hull of the B-subdifferential

$$\partial_B\phi_\theta^1(a,b) = \big\{g \in \mathbb{R}^2 \mid \exists\{(a_k,b_k)\}_{k=1}^\infty \subseteq D \text{ s.t. } (a_k,b_k) \to (a,b) \text{ and } \nabla\phi_\theta^1(a_k,b_k) \to g\big\}.$$

Let $\{(a_k,b_k)\}_{k=1}^\infty \subseteq D$ such that $(a_k,b_k) \to (a,0)$. For all sufficiently large k, we have $a_k > 0$. If $b_k > 0$ for all k sufficiently large, then

$$\begin{aligned} \lim_{k\to\infty} \nabla\phi_\theta^1(a_k,b_k) &= \lim_{k\to\infty} \begin{bmatrix} \mathrm{sgn}(a_k) - \theta'(a_k)b_k - \theta(b_k) \\ \mathrm{sgn}(b_k) - \theta'(b_k)a_k - \theta(a_k) \end{bmatrix} \\ &= \begin{bmatrix} 1 - \theta'(a)\cdot 0 - \theta(0) \\ 1 - \theta'(0)\cdot a - \theta(a) \end{bmatrix} \\ &= \begin{bmatrix} 0 \\ 1 - a\theta'(0) - \theta(a) \end{bmatrix}, \end{aligned}$$

where we used the fact that θ is continuously differentiable and that $\theta(0) = 1$. If $b_k < 0$ for all k sufficiently large, then

$$\lim_{k\to\infty} \nabla\phi_\theta^1(a_k, b_k) = \begin{bmatrix} 1 - \theta'(a) \cdot 0 - \theta(0) \\ -1 - \theta'(0) \cdot a - \theta(a) \end{bmatrix} = \begin{bmatrix} 0 \\ -1 - a\theta'(0) - \theta(a) \end{bmatrix}.$$

In other cases, $\nabla\phi_\theta^1(a_k, b_k)$ has no limit. Hence,

$$\partial_B \phi_\theta^1(a, 0) = \left\{[0, 1 - a\theta'(0) - \theta(a)]^\mathsf{T}, [0, -1 - a\theta'(0) - \theta(a)]^\mathsf{T}\right\}$$

and the result for the case $a > 0$ and $b = 0$ follows by taking the convex hull. We omit the proof of the other cases as the arguments are similar. Finally, note that ϕ_θ^p is differentiable on $\mathbb{R}^2$ except at $(0, 0)$. The computation of the generalized gradient $\phi_\theta^p(0, 0)$ is similar to the computation of $\partial\phi_{\text{FB}}^p(0, 0)$ shown as in [20]. This completes the proof. □

Finally, we explore several variants and generalizations of ϕ_θ^p. In addition, we propose specific functions that can be employed to construct new NCP functions from existing ones. To facilitate this discussion, we denote by $(t)_+$ the projection of t onto the nonnegative real line, that is,

$$(t)_+ := \begin{cases} t & \text{if } t \geq 0, \\ 0 & \text{if } t < 0. \end{cases}$$

For convenience, we define $\hat{\phi}_\theta^{p,i}$ for $i = 1, 2, 3$ as follows:

$$\begin{aligned} \hat{\phi}_\theta^{p,1}(a, b) &= \phi_\theta^p(a, b) - \alpha(a)_+(b)_+ \\ \hat{\phi}_\theta^{p,2}(a, b) &= \phi_\theta^p(a, b) - \alpha(ab)_+^2 \\ \hat{\phi}_\theta^{p,3}(a, b) &= \phi_\theta^p(a, b) - \alpha(a)_+^2(b)_+^2 \end{aligned}$$

where $\alpha > 0$. For any $p \geq 1$ and $(a, b) \in \mathbb{R}_{++}^2$, we know from Proposition 2.73 and Proposition 2.73 that $\hat{\phi}_\theta^{p,i}(a, b) < 0$. Moreover, $\hat{\phi}_\theta^{p,i}(a, b) = \phi_\theta^p(a, b) > 0$ for all $(a, b) \notin \mathbb{R}_+^2$. Consequently, these three variants are easily to be seen as NCP functions as well.

Proposition 2.77 *The functions $\hat{\phi}_\theta^{p,i}$ are all NCP functions for any $\alpha > 0$ and $i = 1, 2, 3$.*

In recent years, both "continuous" and "discrete" generalizations of NCP functions have attracted considerable interest; see [20, 22, 34, 35]. These generalizations typically involve a tunable parameter q, which has been shown to significantly enhance the numerical performance of certain algorithms based on NCP functions [22, 27, 45]. Moreover, such extensions can yield NCP functions with distinct analytical properties [34, 35]. For example, the generalized Fischer–Burmeister (FB) function (2.14) serves as a continuous generalization of the classical FB function (2.12), with $p \in (1, \infty)$. The standard FB function is recovered by setting $p = 2$. In parallel, discrete generalizations have also been proposed. One notable instance is the natural residual (NR) function,

$$\phi_{\text{NR}}(a, b) = \min\{a, b\} = a - (a - b)_+$$

which remains a widely used NCP function alongside the FB function. A discrete generalization of the NR function, proposed in [34], is given by

$$\phi_{\text{NR}}^q(a, b) = a^q - [(a - b)_+]^q \tag{2.108}$$

where q is a positive odd integer. When $q = 1$, the original NR function is recovered. The term "discrete" reflects the fact that q is restricted to positive odd integers. An intriguing feature of the generalized NR function (2.108) is its twice differentiability for $q > 3$, a property not shared by the original NR function. This enhanced smoothness renders ϕ_{NR}^q particularly suitable for algorithms that require differentiable NCP functions.

We wish to highlight that the technique underlying the second type of generalization discussed above, namely, the discrete generalization, can be systematically applied to NCP functions of the form

$$\phi(a, b) = \bar{\phi}_1(a, b) - \bar{\phi}_2(a, b). \tag{2.109}$$

In other words, the function

$$\phi^q(a, b) := [\bar{\phi}_1(a, b)]^q - [\bar{\phi}_2(a, b)]^q$$

is always a discrete generalization of ϕ given in (2.109), where q is a positive odd integer. As a matter of fact, we can further extend such technique by considering any family of injective functions $\{f_q\}$. More precisely, consider the function

$$\phi_{f_q}(a, b) := f_q(\bar{\phi}_1(a, b)) - f_q(\bar{\phi}_2(a, b)) \tag{2.110}$$

which can be readily shown to be an NCP function whenever f_q is injective and ϕ is an NCP function of the form given in (2.109). This transformation, as expressed in (2.110), has also been observed in [4]. For example, the discrete generalized NR function (2.108) can be obtained by applying this transformation to the standard NR function using the map $f_q(t) = t^q$, where $q > 0$ is an odd integer. Applying the same transformation to our NCP function ϕ_θ^p, we arrive at the discrete generalization

$$(\phi_\theta^p)^q := \|(a, b)\|_p^q - (\theta(b)a + \theta(a)b)^q,$$

where q is again a positive odd integer. As noted earlier, such generalizations may yield NCP functions with distinct analytical properties. In particular, it is straightforward to verify that $(\phi_\theta^p)^q$ is continuously differentiable on $\mathbb{R}^2$ whenever $q \geq p > 1$, in contrast to the original function ϕ_θ^p, which is not differentiable at the origin.

Another discrete generalization of ϕ_θ^p can be obtained by applying the same map $f_q(t) = t^q$ to the equivalent form of ϕ_θ^p given by

$$\phi_\theta^p(a,b) = \phi_{\text{FB}}^p(a,b) - \big[a(\theta(b)-1) + b(\theta(a)-1)\big]. \tag{2.111}$$

This yields another symmetric generalization

$$(\phi_\theta^p)_{\text{FB}}^q(a,b) = [\phi_{\text{FB}}^p(a,b)]^q - \big[(a(\theta(b)-1) + b(\theta(a)-1)\big]^q.$$

For $q = 1$, Proposition 2.76 ensures the semismoothness of ϕ_θ^p. Interestingly, the discrete generalization introduced above yields smooth NCP functions for any $p > 1$ and odd integers $q \geq 3$. This fact is straightforward to verify, and we omit the proof for brevity. These results are consolidated in Proposition 2.78. It is also worth noting that all of the aforementioned generalizations preserve symmetry. More generally, the transformation in (2.110) produces symmetric NCP functions when applied to our proposed function ϕ_θ^p as well as to its alternative representation given in (2.111).

Proposition 2.78 *Suppose θ is continuously differentiable and satisfies the conditions of Proposition 2.73 if $p = 1$, or Proposition 2.73 if $p > 1$. Let $q \geq 1$ be an odd integer. Then,*

$$(\phi_\theta^p)^q(a,b) := \|(a,b)\|_p^q - \big(\theta(b)a + \theta(a)b\big)^q$$

is a discrete generalization of ϕ_θ^p, which is smooth if $q \geq p > 1$. Additionally,

$$(\phi_\theta^p)_{\text{FB}}^q(a,b) := [\phi_{\text{FB}}^p(a,b)]^q - [(a(\theta(b)-1) + b(\theta(a)-1)]^q$$

is also a discrete generalizations of ϕ_θ^p, which is smooth if $q \geq 3$ and $p > 1$

It is worth noting that the function $f_q(t) = t^q$, with $q \geq 1$ an odd integer, is commonly used to enhance the numerical performance of algorithms. In the context of neural network-based approaches to optimization, such functions are often referred to as activation functions. Their primary purpose is to improve convergence rates, and several alternative activation functions have been proposed in the literature. A few notable examples include:

1. Bipolar Sigmoid Function [46, 47]:

$$f_q(t) = \frac{1 - e^{-qt}}{1 + e^{-qt}}, \quad q > 0.$$

2. Power-Sigmoid Function [46, 47]:

$$f_q(t) = \begin{cases} \frac{1+e^{-q_1}}{1-e^{-q_1}} \cdot \frac{1-e^{-q_1 t}}{1+e^{-q_1 t}} & \text{if } |t| < 1 \\ t^{q_2} & \text{if } |t| \geq 1 \end{cases}$$

where $q = (q_1, q_2)$, $q_1 > 2$ and $q_2 \geq 3$ is an odd integer.

3. Smooth Power-Sigmoid Function [46, 47]:

$$f_q(t) = \frac{1}{2} \cdot \frac{1+e^{-q_1}}{1-e^{-q_1}} \cdot \frac{1-e^{-q_1 t}}{1+e^{-q_1 t}} + \frac{1}{2} t^{q_2}$$

where $q = (q_1, q_2)$, $q_1 > 2$ and $q_2 \geq 3$.

4. Sign-Bi-Power Function [48]:

$$f_q(t) = \begin{cases} |t|^q + |t|^{\frac{1}{q}} \text{ if } t > 0 \\ 0 \text{ if } t = 0 \\ -|t|^q - |t|^{\frac{1}{q}} \text{ if } t < 0 \end{cases}, \quad q > 0.$$

All of the functions mentioned above are injective mappings that can be employed to transform an NCP function of the form (2.109). However, these transformations do not constitute generalizations in the sense described earlier. A true generalization, as illustrated, is only possible if there exists a parameter $\bar{q}$ such that $f_{\bar{q}}(t) \equiv t$. Nonetheless, we observe that the function $\frac{f_q}{2}$ yields a continuous generalization via the transformation (2.110) when f_q is chosen to be the "sign-bi-power function". In any case, a promising direction for future research lies in exploring how these injective functions might enhance the numerical efficiency of NCP function, based solution methods, much in the same way they are used to improve performance in neural network approaches. For the power function $f_q(t) = t^q$, and the resulting generalized NR function, some encouraging numerical results have already been reported in [45]. Finally, we remark that the composite map $f_q \circ \phi_\theta^p$ is also a valid NCP function, provided that f_q is injective and satisfies $f_q(0) = 0$, as is the case for the aforementioned activation functions. This approach is also worth consideration in numerical implementations.

2.4.2 Construction by Using Invertible Functions

In this section, we introduce a novel approach to constructing NCP functions, an idea that, to the best of our knowledge, is new to the literature. Specifically, we identify conditions under which the class of invertible functions can be effectively utilized to generate new NCP functions. This development is motivated by the discovery and structural analysis of three particular NCP functions, which reveal a unifying pattern. Illustrative examples of the resulting NCP functions, accompanied by their graphical representations, are also provided.

We begin by presenting the following three NCP functions, which serve as the inspiration for our proposed construction:

$$\phi_{\ln-\max}(a,b) = \ln(e^{|a|} + e^{|b|} - 1) - \max(a,b); \tag{2.112}$$
$$\phi_{\ln-\text{sum}}(a,b) = \ln(e^{|a|} + e^{|b|} - 1) - (a+b); \tag{2.113}$$
$$\phi_{\text{abs}-\exp}(a,b) = |a| + |b| - e^{a}b - e^{b}a. \tag{2.114}$$

The first two functions, $\phi_{\ln-\max}$ and $\phi_{\ln-\text{sum}}$, are constructed using the exponential and logarithmic functions, with additional terms designed to ensure that the function evaluates to zero along the nonnegative axes. These NCP functions were discovered through an examination of the construction techniques in [49, 50]. It is important to note that all three functions involve absolute value terms, rendering them nondifferentiable. Consequently, for the purposes of subsequent analysis, we must work with their subdifferentials in the sense of Clarke [51]. To that end, we make use of the sign function:

$$\text{sgn}(t) := \begin{cases} 1 & \text{if } t > 0, \\ 0 & \text{if } t = 0, \\ -1 & \text{if } t < 0. \end{cases}$$

along with the definition of the convex hull of all limit points of Jacobian sequences.

Proposition 2.79 *Let $\phi_{\ln-\max} : \mathbb{R}^2 \to \mathbb{R}$ be defined in (2.112), that is,*

$$\phi_{\ln-\max}(a,b) = \ln(e^{|a|} + e^{|b|} - 1) - \max(a,b).$$

Then, the following hold.

(a) The function $\phi_{\ln-\max}$ is an NCP function.
(b) The subdifferential of $\phi_{\ln-\max}$ is described by

$$
\partial\phi_{\ln-\max}(a,b) = \begin{cases}
\left\{\left(\frac{e^a}{e^a+e^b-1}-1,\ \frac{e^b}{e^a+e^b-1}\right)\right\} & \text{if } (a,b)\in \mathrm{I}_1=\{(a,b)\,|\,a,b>0 \text{ and } a>b\}.\\
\left\{\left(\frac{e^a}{e^a+e^b-1},\ \frac{e^b}{e^a+e^b-1}-1\right)\right\} & \text{if } (a,b)\in \mathrm{I}_2=\{(a,b)\,|\,a,b>0 \text{ and } b>a\}.\\
\left\{\left(\frac{-e^{-a}}{e^{-a}+e^b-1},\ \frac{e^b}{e^{-a}+e^b-1}-1\right)\right\} & \text{if } (a,b)\in \mathrm{I}_3=\{(a,b)\,|\,a<0,\ b>0\}.\\
\left\{\left(\frac{-e^{-a}}{e^{-a}+e^{-b}-1},\ \frac{-e^{-b}}{e^{-a}+e^{-b}-1}-1\right)\right\} & \text{if } (a,b)\in \mathrm{I}_4=\{(a,b)\,|\,a,b<0 \text{ and } b>a\}.\\
\left\{\left(\frac{-e^{-a}}{e^{-a}+e^{-b}-1}-1,\ \frac{-e^{-b}}{e^{-a}+e^{-b}-1}\right)\right\} & \text{if } (a,b)\in \mathrm{I}_5=\{(a,b)\,|\,a,b<0, \text{ and } a>b\}\\
\left\{\left(\frac{e^a}{e^a+e^{-b}-1}-1,\ \frac{-e^{-b}}{e^a+e^{-b}-1}\right)\right\} & \text{if } (a,b)\in \mathrm{I}_6=\{(a,b)\,|\,a>0,\ b<0\}\\
\left\{(0,\rho)\,|\,\frac{-1}{e^a}\le\rho\le\frac{1}{e^a}\right\} & \text{if } (a,b)\in \mathrm{L}_1=\{(a,b)\,|\,a>0,\ b=0\}\\
\{(\rho,0)\,|\,-1\le\rho\le 1\} & \text{if } (a,b)\in \mathrm{L}_3=\{(a,b)\,|\,a=0,\ b>0\}\\
\mathrm{conv}\left\{\begin{pmatrix}\frac{e^a}{2e^a-1}-1\\ \frac{e^a}{2e^a-1}\end{pmatrix},\begin{pmatrix}\frac{e^a}{2e^a-1}\\ \frac{e^a}{2e^a-1}-1\end{pmatrix}\right\} & \text{if } (a,b)\in \mathrm{L}_2=\{(a,b)\,|\,a,b>0, \text{ and } a=b\}.\\
\{(-1,\rho-1)\,|\,-e^a\le\rho\le e^a\} & \text{if } (a,b)\in \mathrm{L}_4=\{(a,b)\,|\,a<0,\ b=0\}.\\
\left\{(\rho-1,-1)\,|\,-e^b\le\rho\le e^b\right\} & \text{if } (a,b)\in \mathrm{L}_6=\{(a,b)\,|\,a=0,\ b<0\}.\\
\mathrm{conv}\left\{\begin{pmatrix}\frac{-e^{-a}}{2e^{-a}-1}\\ \frac{-e^{-a}}{2e^{-a}-1}-1\end{pmatrix},\begin{pmatrix}\frac{-e^{-a}}{2e^{-a}-1}-1\\ \frac{-e^{-a}}{2e^{-a}-1}\end{pmatrix}\right\} & \text{if } (a,b)\in \mathrm{L}_5=\{(a,b)\,|\,a,b<0, \text{ and } a=b\}.\\
\mathrm{conv}\left\{\begin{matrix}(0,1),(1,0),\\(-1,-2),(-2,-1)\end{matrix}\right\} & \text{if } (a,b)=(0,0).
\end{cases}
$$

where conv(S) *denotes the convex hull of the set* S.

Proof (a) "$\Rightarrow$" Suppose $\phi_{\ln-\max}(a,b)=0$, we need to show $a\ge 0,\ b\ge 0,\ ab=0$. To proceed, we discuss two cases.

(i) If $a\ge b$, then $e^{|a|}+e^{|b|}-1=e^a$. The left-hand side of this equality is greater than or equal to 1 since the absolute value is always nonnegative. Hence, e^a must be greater than or equal to 1. This leads to a being greater than or equal to 0. Thus, $a\ge 0$ and $e^{|b|}-1=0$, which says $a\ge 0$ and $b=0$.

(ii) If $b\ge a$, by the symmetric form of the function $\phi_{\ln-\max}$, we see that $a=0,\ b\ge 0$. Therefore, $a\ge 0,\ b=0$ or $a=0,\ b\ge 0$. This is equivalent to $a\ge 0,\ b\ge 0,\ ab=0$.

"$\Leftarrow$" Conversely, if $a\ge 0,\ b\ge 0,\ ab=0$, then $a\ge 0,\ b=0$ or $a=0,\ b\ge 0$. For $a\ge 0,\ \ b=0$, then $\phi_{\ln-\max}(a,b)=\ln(e^{|a|}+e^{|b|}-1)-\max(a,b)=|a|-a=0$. For $a=0,\ b\ge 0$, then $\phi_{\ln-\max}(a,b)=\ln(e^{|a|}+e^{|b|}-1)-\max(a,b)=|b|-b=0$. Thus, the proof is done.

(b) Note that $\phi_{\ln-\max}$ is differentiable at $(a,b)\in \mathrm{I}_1\sim \mathrm{I}_6$, whereas it is not differentiable in other cases. Hence, we need to calculate each case separately.

Case (1): $(a,b)\ \in\ \mathrm{I}_1=\{(a,b)\,|\,a,b>0 \text{ and } a>b\}$.

$$
\begin{aligned}
&\nabla\phi_{\ln-\max}(a,b)\\
&=\left(\frac{e^{|a|}a}{(e^{|a|}+e^{|b|}-1)|a|}-1,\ \frac{e^{|b|}b}{(e^{|a|}+e^{|b|}-1)|b|}\right)\\
&=\left(\frac{e^a}{e^a+e^b-1}-1,\ \frac{e^b}{e^a+e^b-1}\right)
\end{aligned}
$$

Case (2): $(a,b)\ \in\ \mathrm{I}_2=\{(a,b)\,|\,a,b>0 \text{ and } b>a\}$.

$$\begin{aligned}
&\nabla\phi_{\ln-\max}(a,b) \\
&= \left(\frac{e^{|a|}a}{(e^{|a|}+e^{|b|}-1)|a|},\ \frac{e^{|b|}b}{(e^{|a|}+e^{|b|}-1)|b|}-1\right) \\
&= \left(\frac{e^{a}}{e^{a}+e^{b}-1},\ \frac{e^{b}}{e^{a}+e^{b}-1}-1\right)
\end{aligned}$$

Case (3): $(a,b) \in \mathrm{I}_3 = \{(a,b)\,|\,a<0,\ b>0\}$.

$$\begin{aligned}
&\nabla\phi_{\ln-\max}(a,b) \\
&= \left(\frac{e^{|a|}a}{(e^{|a|}+e^{|b|}-1)|a|},\ \frac{e^{|b|}b}{(e^{|a|}+e^{|b|}-1)|b|}-1\right) \\
&= \left(\frac{-e^{-a}}{e^{-a}+e^{b}-1},\ \frac{e^{b}}{e^{-a}+e^{b}-1}-1\right)
\end{aligned}$$

Case (4): $(a,b) \in \mathrm{I}_4 = \{(a,b)\,|\,a,b<0,\ \text{and } b>a\}$.

$$\begin{aligned}
&\nabla\phi_{\ln-\max}(a,b) \\
&= \left(\frac{e^{|a|}a}{(e^{|a|}+e^{|b|}-1)|a|},\ \frac{e^{|b|}b}{(e^{|a|}+e^{|b|}-1)|b|}-1\right) \\
&= \left(\frac{-e^{-a}}{e^{-a}+e^{-b}-1},\ \frac{-e^{-b}}{e^{-a}+e^{-b}-1}-1\right)
\end{aligned}$$

Case (5): $(a,b) \in \mathrm{I}_5 = \{(a,b)\,|\,a,b<0,\ \text{and } a>b\}$.

$$\begin{aligned}
&\nabla\phi_{\ln-\max}(a,b) \\
&= \left(\frac{e^{|a|}a}{(e^{|a|}+e^{|b|}-1)|a|}-1,\ \frac{e^{|b|}b}{(e^{|a|}+e^{|b|}-1)|b|}\right) \\
&= \left(\frac{-e^{-a}}{e^{-a}+e^{-b}-1}-1,\ \frac{-e^{-b}}{e^{-a}+e^{-b}-1}\right)
\end{aligned}$$

Case (6): $(a,b) \in \mathrm{I}_6 = \{(a,b)\,|\,a>0,\ b<0\}$.

$$\begin{aligned}
&\nabla\phi_{\ln-\max}(a,b) \\
&= \left(\frac{e^{|a|}a}{(e^{|a|}+e^{|b|}-1)|a|}-1,\ \frac{e^{|b|}b}{(e^{|a|}+e^{|b|}-1)|b|}\right) \\
&= \left(\frac{e^{a}}{e^{a}+e^{-b}-1}-1,\ \frac{-e^{-b}}{e^{a}+e^{-b}-1}\right)
\end{aligned}$$

Case (7): $(a, b) \in \mathrm{L}_1 = \{(a, b) \mid a > 0,\ b = 0\}$.
Since the point $(a, 0)$ is adjacent to the region I_1 and I_6, let $\{(a_k, b_k)\}$ be the sequence such that $\lim_{k\to\infty}(a_k, b_k) = (a, 0)$.
If $\{(a_k, b_k)\} \subseteq \mathrm{I}_1 = \{(a, b) | a, b > 0 \text{ and } a > b\}$, then

$$\begin{aligned}
&\lim_{k\to\infty} \nabla\phi_{\ln-\max}(a_k, b_k)\\
&= \lim_{k\to\infty}\left(\frac{e^{a_k}}{e^{a_k}+e^{b_k}-1} - 1,\ \frac{e^{b_k}}{e^{a_k}+e^{b_k}-1}\right)\\
&= \left(\frac{e^{a}}{e^{a}+e^{0}-1} - 1,\ \frac{e^{0}}{e^{a}+e^{0}-1}\right)\\
&= \left(0, \frac{1}{e^a}\right).
\end{aligned}$$

If $\{(a_k, b_k)\} \subseteq \mathrm{I}_6 = \{(a, b) \mid a > 0,\ b < 0\}$, then

$$\begin{aligned}
&\lim_{k\to\infty} \nabla\phi_{\ln-\max}(a_k, b_k)\\
&= \lim_{k\to\infty}(\frac{e^{a_k}}{e^{a_k}+e^{-b_k}-1} - 1,\ \frac{-e^{-b_k}}{e^{a_k}+e^{-b_k}-1})\\
&= (\frac{e^{a}}{e^{a}+e^{0}-1} - 1,\ \frac{-e^{0}}{e^{a}+e^{0}-1})\\
&= (0, \frac{-1}{e^a})
\end{aligned}$$

Thus, by definition of subdifferential, we have

$$\partial\phi_{\ln-\max}(a, b) = \mathrm{co}\left\{\begin{pmatrix}0\\ \frac{1}{e^a}\end{pmatrix}, \begin{pmatrix}0\\ \frac{-1}{e^a}\end{pmatrix}\right\} = \left\{(0, \rho) \mid \frac{-1}{e^a} \le \rho \le \frac{1}{e^a}\right\},$$

Case (8) $\sim$ (12): For the region L_i, $i = 2, \ldots, 6$ and the origin point $(0, 0)$, the way to calculate the subdifferential is similar to L_1, so we omit them here. □

The next proposition concerns the second NCP function, $\phi_{\ln-\mathrm{sum}}$. In fact, computing the subdifferential of $\phi_{\ln-\max}$ is significantly more intricate than that of $\phi_{\ln-\mathrm{sum}}$, due to the presence of the $\max(a, b)$ term. When combined with the absolute value expressions, this term induces a more complex partitioning of the domain into regions where the function is differentiable. Moreover, some of the results pertaining to $\phi_{\ln-\mathrm{sum}}$ can be derived using elements from Proposition 2.79. Before proceeding to the proof, we highlight two notable observations regarding $\phi_{\ln-\mathrm{sum}}$. First, observe that $\max(a, b) = a + b$ when $a \ge 0$, $b \ge 0$, $ab = 0$. Consequently, under these conditions, $\phi_{\ln-\max}(a, b) = \phi_{\ln-\mathrm{sum}}(a, b) = 0$. These observations lead us to conjecture that $\phi_{\ln-\mathrm{sum}}$ also qualifies as an NCP function. To confirm this, it is necessary to evaluate the behavior of $\phi_{\ln-\mathrm{sum}}$ across the remaining regions of the domain.

Proposition 2.80 *Let $\phi_{\ln -\text{sum}} : \mathbb{R}^2 \to \mathbb{R}$ be defined in (2.113), that is,*

$$\phi_{\ln -\text{sum}}(a,b) = \ln(e^{|a|} + e^{|b|} - 1) - (a+b).$$

Then, the following hold.

(a) The function $\phi_{\ln -\text{sum}}$ is an NCP function.
(b) The subdifferential of $\phi_{\ln -\text{sum}}$ is described by

$$\partial\phi_{\ln -\text{sum}}(a,b) = \begin{cases} \left\{\left(\frac{e^{|a|}a}{(e^{|a|}+e^{|b|}-1)|a|} - 1,\ \frac{e^{|b|}b}{(e^{|a|}+e^{|b|}-1)|b|} - 1\right)\right\} & \text{if } a \neq 0 \text{ and } b \neq 0. \\ \left\{(\rho - 1, 0) \mid \le \frac{-1}{e^{|b|}} \le \rho \le \frac{1}{e^{|b|}}\right\} & \text{if } a = 0, b > 0. \\ \left\{(\rho - 1, -2) \mid \frac{-1}{e^{|b|}} \le \rho \le \frac{1}{e^{|b|}}\right\} & \text{if } a = 0, b < 0. \\ \left\{(0, \rho - 1) \mid \frac{-1}{e^{|a|}} \le \rho \le \frac{1}{e^{|a|}}\right\} & \text{if } a > 0, b = 0. \\ \left\{(-2, \rho - 1) \mid \frac{-1}{e^{|a|}} \le \rho \le \frac{1}{e^{|a|}}\right\} & \text{if } a < 0, b = 0. \\ \{(\xi, \eta) \mid -2 \le \xi, \eta \le 0\} & \text{if } a = b = 0. \end{cases}$$

Proof To prove part (a), we must verify that $\phi_{\ln -\text{sum}}$ satisfies condition (2.2). From Proposition 2.79, it is evident that $\phi_{\ln -\text{sum}}(a,b) = 0$ on the nonnegative portions of the a, b-axes, and that the function is strictly positive on their negative sides. Therefore, it suffices to examine the behavior of $\phi_{\ln -\text{sum}}$ within the four quadrants of the ab-plane. Assume that $\phi_{\ln -\text{sum}}(a,b) = 0$. To analyze this scenario, we proceed by considering the following four cases corresponding to the quadrants of the plane.

Case (i): If $a > 0$ and $b > 0$, then $e^a + e^b - 1 = e^a e^b$. Then, we have

$$0 = e^a(e^b - 1) + (e^b - 1) = (e^a - 1)(e^b - 1).$$

Since $a > 0$ and $b > 0$, we should have $(e^a - 1)(e^b - 1) > 0$ which leads to a contradiction. Thus, $\phi_{\ln -\text{sum}}(a,b) \neq 0$ in case (i).

Case (ii): If $a < 0$ and $b > 0$, then $e^{-a} + e^b - 1 = e^a e^b$ and hence $\frac{1}{e^a} + e^b - 1 = e^a e^b$, which gives $(e^a)^2 e^b - 1 - e^a e^b + e^a = 0$. It follows that

$$0 = e^a e^b(e^a - 1) + (e^a - 1) = (e^a e^b + 1)(e^a - 1).$$

But, $e^a e^b + 1 > 0$ and $e^a - 1 < 0$, it says $(e^a e^b + 1)(e^a - 1) < 0$, which contradicts the above equation. Thus, $\phi_{\ln -\text{sum}}(a,b) \neq 0$ in case (ii).

Case (iii): Suppose $a < 0$ and $b < 0$. Obviously, $e^{|a|} + e^{|b|} - 1$ is greater than 1, and $e^a e^b = e^{a+b}$ is less than 1. Hence, $\phi_{\ln -\text{sum}}(a,b) \neq 0$ in case (iii).

Case (iv): If $a > 0$ and $b < 0$, then $\phi_{\ln -\text{sum}} \neq 0$ by noting the symmetry of $\phi_{\ln -\text{sum}}(a,b)$ and the arguments in case (ii).

To sum up, from all the above, we prove that $\phi_{\ln -\text{sum}}(a,b) = 0 \Longleftrightarrow a \geq 0, b = 0$ or $a = 0,\ b \geq 0 \Longleftrightarrow a \geq 0,\ b \geq 0,\ ab = 0$.

(b) Again, by using the definition of subdifferential, we calculate each case separately.

Case (1): If $a \neq 0$ and $b \neq 0$, then $\phi_{\ln-\text{sum}}$ is differentiable. Then, we have

$$\nabla\phi_{\ln-\text{sum}}(a,b) = \left(\frac{e^{|a|}a}{(e^{|a|}+e^{|b|}-1)|a|} - 1, \frac{e^{|b|}b}{(e^{|a|}+e^{|b|}-1)|b|} - 1\right). \quad (2.115)$$

Case (2): Suppose $a = 0$ and $b > 0$. we compute subdifferential by the definition of convex hull of all limits points of Jacobian sequence. Let $(a_k, b_k) \to (0, b)$ as $k \to \infty$. Applying (2.115) yields

$$\lim_{k\to\infty} \nabla\phi_{\ln-\text{sum}}(a_k,b_k) = \begin{cases} \left(\frac{1}{e^b} - 1, 0\right) & \text{if } \{(a_k,b_k)\} \subseteq \{(a,b) \mid a > 0, b > 0\}. \\ \left(\frac{-1}{e^b} - 1, 0\right) & \text{if } \{(a_k,b_k)\} \subseteq \{(a,b) \mid a < 0, b > 0\}. \end{cases}$$

This concludes $\partial\phi_{\ln-\text{sum}}(a,b) = \{(\rho - 1, 0) \mid \rho \in [\frac{-1}{e^b}, \frac{1}{e^b}]\}$. The other cases exclude the case $a = b = 0$, which are similar to the above cases. Therefore, it remains to prove the case when $a = b = 0$.

For the case $a = b = 0$, let $(a_k, b_k) \to (0, 0)$. Compute the limit of (2.115) as $(a_k, b_k) \to (0, 0)$, we have

$$\lim_{k\to\infty} \nabla\phi_{\ln-\text{sum}}(a_k,b_k) = \begin{cases} (0,0) & \text{if } \{(a_k,b_k)\} \subseteq \{(a,b) \mid a > 0, b > 0\}. \\ (-2,0) & \text{if } \{(a_k,b_k)\} \subseteq \{(a,b) \mid a < 0, b > 0\}. \\ (-2,-2) & \text{if } \{(a_k,b_k)\} \subseteq \{(a,b) \mid a < 0, b < 0\}. \\ (0,-2) & \text{if } \{(a_k,b_k)\} \subseteq \{(a,b) \mid a > 0, b < 0\}. \end{cases}$$

$$\text{Hence } \partial\phi_{\ln-\text{sum}}(0,0) = \text{co}\left\{\begin{pmatrix}0\\0\end{pmatrix}, \begin{pmatrix}0\\-2\end{pmatrix}, \begin{pmatrix}-2\\0\end{pmatrix}, \begin{pmatrix}-2\\-2\end{pmatrix}\right\} = \{(\xi,\eta) \mid -2 \leq \xi, \eta \leq 0\}.$$

□

Proposition 2.81 *Let $\phi_{\text{abs-exp}} : \mathbb{R}^2 \to \mathbb{R}$ be defined as in (2.114), that is,*

$$\phi_{\text{abs-exp}}(a,b) = |a| + |b| - e^a b - e^b a.$$

Then, the following hold.

(a) The function $\phi_{\text{abs-exp}}$ is an NCP function.
(b) The subdifferential of $\phi_{\text{abs-exp}}$ is described as

$$\partial\phi_{\text{abs-exp}}(a,b) = \begin{cases} \{(\text{sgn}(a) - e^a b - e^b, \text{sgn}(b) - e^b a - e^a)\} & \text{if } a \neq 0,\ b \neq 0. \\ \{(0, \rho - a - e^a) \mid -1 \leq \rho \leq 1\} & \text{if } a > 0,\ b = 0. \\ \{(\rho - b - e^b, 0) \mid -1 \leq \rho \leq 1\} & \text{if } a = 0,\ b > 0. \\ \{(-2, \rho - a - e^a) \mid -1 \leq \rho \leq 1\} & \text{if } a < 0,\ b = 0. \\ \{(\rho - b - e^b, -2) \mid -1 \leq \rho \leq 1\} & \text{if } a = 0,\ b < 0. \\ \{(\xi,\eta) \mid -2 \leq \xi, \eta \leq 0\} & \text{if } a = b = 0. \end{cases}$$

Proof (a) First, we rewrite the function $\phi_{\text{abs-exp}}$ as

$$\phi_{\text{abs-exp}}(a,b) = a(\text{sgn}(a) - e^b) + b(\text{sgn}(b) - e^a),$$

which possesses the below piecewise expression:

$$\phi_{\text{abs-exp}}(a,b) = \begin{cases} 0 & \text{if } a \geq 0,\ b \geq 0, \text{ and } ab = 0. \\ a(1-e^b) + b(1-e^a) & \text{if } a > 0,\ b > 0. \\ -a(1+e^b) + b(1-e^a) & \text{if } a < 0,\ b \geq 0. \\ -a(1+e^b) - b(1+e^a) & \text{if } a < 0,\ b < 0. \end{cases}$$

It is noted that $\phi_{\text{abs-exp}}(a,b)$ is negative in the second case, and positive in the third and last cases. In light of the symmetry of $\phi_{\text{abs-exp}}(a,b)$, we obtain that $\phi_{\text{abs-exp}}(a,b)$ is also positive on $a \geq 0$, $b < 0$. Therefore, $\phi_{\text{abs-exp}}$ is an NCP function.

(b) We discuss a few cases in order to calculate the subdifferential of $\phi_{\text{abs-exp}}$.

Case (1): If $a \neq 0$ and $b \neq 0$, we have

$$\nabla\phi_{\text{abs-exp}}(a,b) = \left(\text{sgn}(a) - e^a b - e^b,\ \text{sgn}(b) - e^b a - e^a\right). \tag{2.116}$$

Case (2): Suppose $a > 0,\ b = 0$ and $(a_k, b_k) \to (a, 0)$ as $k \to \infty$. From expression (2.116), we know

$$\lim_{k\to\infty} \nabla\phi_{\text{abs-exp}}(a_k,b_k) = \begin{cases} (0, 1-a-e^a) & \text{if } \{(a_k,b_k)\} \subseteq \{(a,b)\,|\,a>0, b>0\}. \\ (0, -1-a-e^a) & \text{if } \{(a_k,b_k)\} \subseteq \{(a,b)\,|\,a>0, b<0\}. \end{cases}$$

Then, it follows that $\partial\phi_{\text{abs-exp}}(a,b) = \{(0, \rho - a - e^a)\,|\,-1 \leq \rho \leq 1\}$.

For the other cases except for the case when $a = b = 0$, the calculation is similar to case (2), so we omit them here.

Case (3): Suppose $a = b = 0$ and $(a_k, b_k) \to (0,0)$ as $k \to \infty$. From expression (2.116), we compute that

$$\lim_{k\to\infty} \nabla\phi_{\text{abs-exp}}(a_k,b_k) = \begin{cases} (0,0) & \text{if } \{(a_k,b_k)\} \subseteq \{(a,b)\,|\,a>0, b>0\}. \\ (-2,0) & \text{if } \{(a_k,b_k)\} \subseteq \{(a,b)\,|\,a<0, b>0\}. \\ (-2,-2) & \text{if } \{(a_k,b_k)\} \subseteq \{(a,b)\,|\,a<0, b<0\}. \\ (0,-2) & \text{if } \{(a_k,b_k)\} \subseteq \{(a,b)\,|\,a>0, b<0\}. \end{cases}$$

This concludes $\partial\phi_{\text{abs-exp}}(0,0) = \{(\xi,\eta)\,|\,-2 \leq \xi, \eta \leq 0\}$. □

It is worth noting that the function $\phi_{\text{abs-exp}}$ was discovered through a different line of reasoning than the previous two NCP functions. The key distinction lies in the interpretation of the term $|a| + |b|$, which can be seen as the l_1-norm of the vector (a,b), while the exponential terms e^a, e^b serve as monotone cofactors applied to b and a, respectively. By examining the structural patterns underlying these three newly introduced NCP functions, we identified a broader framework for generating NCP functions through the use of invertible functions. In particular, the first two functions, $\phi_{\ln-\max}(a,b)$ and $\phi_{\ln-\text{sum}}(a,b)$, both feature the term $\ln(e^{|a|}+e^{|b|}-1)$, which prominently involves invertible functions. Motivated by this observation, we propose the following generalization of the common term:

$$f(f^{-1}(|a|)+f^{-1}(|b|)-f^{-1}(0)),$$

where f is a real-valued function defined on a suitable domain and subject to certain structural assumptions. Accordingly, their natural extended formats become

$$\phi(a,b)=f(f^{-1}(|a|)+f^{-1}(|b|)-1)-\max(a,b); \tag{2.117}$$
$$\phi(a,b)=f(f^{-1}(|a|)+f^{-1}(|b|)-1)-(a+b); \tag{2.118}$$
$$\phi(a,b)=f(f^{-1}(|a|)+f^{-1}(|b|)-1)-g(a)b-g(b)a. \tag{2.119}$$

Clearly, if $f(t)=\ln t$, then the functions (2.117) and (2.118) reduce to those functions (2.112) and (2.113). If $f^{-1}(t)=t+1$ and $g(t)=e^t$, then the function (2.119) reduces to the function (2.114). In this Section, we provide a complete discussion on under what conditions of f, $(f^{-1})'$ and g, the above functions defined as in (2.117), (2.118) and (2.119) will be NCP functions.

Proposition 2.82 *Suppose f is a real valued function defined on $\mathbb{R}$ with $f(1)=0$ and $f|_I$ denotes the restricted function of f on $I\subseteq\mathbb{R}$. If $f|_I$ satisfies one of the following conditions:*

(a) $f|_I:[1,\infty)\to[0,\infty)$ is invertible, or
(b) $f|_I:(-\infty,1]\to[0,\infty)$ is invertible,

then $\phi_f(a,b)=f(f^{-1}(|a|)+f^{-1}(|b|)-f^{-1}(0))-\max(a,b)$ is an NCP function.

Proof (a) Without loss of ambiguity, we still use f instead of $f|_I$ in our analysis. Since $f:[1,\infty)\to[0,\infty)$ is invertible and $f(1)=0$, f is strictly monotone increasing on $[1,\infty)$. In addition, f^{-1} is also strictly monotone increasing. To verify that ϕ_f is an NCP function, we need to show that ϕ_f satisfies condition (2.2).

"$\Rightarrow$" Suppose $\phi_f(a,b)=0$, we consider the two regions on the a,b-plane which are $a\geq b$ and $b\geq a$.

Case (1): If $a\geq b$, then $f(f^{-1}(|a|)+f^{-1}(|b|)-f^{-1}(0))=\max(a,b)=a$. Since $f^{-1}(|a|)+f^{-1}(|b|)-f^{-1}(0)\geq 1$ and f is strictly monotone increasing on $[1,\infty)$, so $a=f(f^{-1}(|a|)+f^{-1}(|b|)-f^{-1}(0))\geq f(1)=0$. Then, $f^{-1}(|b|)-1=f^{-1}(a)-f^{-1}(|a|)=0$ since a is nonnegative. This says $b=0,\ a\geq 0$.

Case (2): If $b\geq a$, by the symmetric form of the function, we obtain $a=0,\ b\geq 0$. Therefore, $a\geq 0,\ b=0$ or $b\geq 0,\ a=0$. This is equivalent to $a\geq 0, b\geq 0, ab=0$.

"$\Leftarrow$" Conversely, suppose that $a\geq 0$, $b\geq 0$, $ab=0$. Then, we have $a\geq 0,\ b=0$ or $b=0,\ a\geq 0$. If $a\geq 0,\ b=0$, it is trivial that $\phi_f(a,b)=|a|-a=0$. If $b\geq 0,\ a=0$, it is also clear that $\phi_f(a,b)=|b|-b=0$.

(b) Since $f : (-\infty, 1] \to [0, \infty)$ is invertible and $f(1) = 0$, f is strictly monotone decreasing. In addition, f^{-1} is also strictly monotone decreasing. Next, we show that ϕ_f is an NCP function.

"$\Rightarrow$" Suppose $\phi_f(a, b) = 0$, we consider the two regions on the a, b-plane which are $a \geq b$ and $b \geq a$.

Case (1): If $a \geq b$, then $f(f^{-1}(|a|) + f^{-1}(|b|) - f^{-1}(0)) = a$. Since $f^{-1}(|a|) + f^{-1}(|b|) - f^{-1}(0) \leq 1$ and f is strictly monotone decreasing on $(-\infty, 1]$, $a = f(f^{-1}(|a|) + f^{-1}(|b|) - f^{-1}(0)) \geq f(1) = 0$. Then, $f^{-1}(|b|) - 1 = f^{-1}(a) - f^{-1}(|a|) = 0$ since a is nonnegative. Hence, $b = 0,\ a \geq 0$.

Case (2): If $b \geq a$, by the symmetric form of the function, we obtain that $a = 0,\ b \geq 0$. Therefore, $a \geq 0,\ b = 0$ or $b \geq 0,\ a = 0$. This is equivalent to $a \geq 0$, $b \geq 0$, $ab = 0$.

"$\Leftarrow$" Conversely, suppose that $a \geq 0$, $b \geq 0$, $ab = 0$. Then, we have $a \geq 0,\ b = 0$ or $b \geq 0,\ a = 0$. For $a \geq 0,\ b = 0$, it is clear that $\phi_f(a, b) = |a| - a = 0$. For $b \geq 0,\ a = 0$, it is also trivial that $\phi_f(a, b) = |b| - b = 0$. □

Example 2.4 Here are examples of f satisfying condition in Proposition 2.82(a).

1. $f_1(t) = (t-1)\big|_{[1,\infty)}$.
2. $f_2(t) = \ln(t)\big|_{[1,\infty)}$.
3. $f_3(t) = (t-1)^{1/2}\big|_{[1,\infty)}$.
4. $f_4(t) = (t-1)^{1/5}\big|_{[1,\infty)}$.

Then, their corresponding NCP functions are shown as below and their graphs are depicted in Fig. 2.26.

1. $\phi_{f_1}(a, b) = |a| + |b| - \max(a, b)$.
2. $\phi_{f_2}(a, b) = \ln(e^{|a|} + e^{|b|} - 1) - \max(a, b)$.
3. $\phi_{f_3}(a, b) = \|(a, b)\|_2 - \max(a, b)$.
4. $\phi_{f_4}(a, b) = \|(a, b)\|_5 - \max(a, b)$.

Proposition 2.83 *Suppose f is a continuously differentiable real valued function with $f(1) = 0$. If f satisfies the following conditions:*

(i) $f : [1, \infty) \to [0, \infty)$ is invertible, and
(ii) $(f^{-1})'$ is strictly monotone on $[0, \infty)$,

then $\phi_f(a, b) = f(f^{-1}(|a|) + f^{-1}(|b|) - f^{-1}(0)) - (a + b)$ is an NCP function.

Proof To verify that ϕ_f qualifies as an NCP function, we must demonstrate that $\phi_f(a, b) = 0$ if and only if $a, b \geq 0$ and $ab = 0$; in other words, the function vanishes exclusively along the nonnegative sides of the a, b-axes. To this end, we examine the behavior of ϕ_f in each of the four quadrants of the ab-plane. For the second, third, and fourth quadrants, the analysis relies solely on the monotonicity of f^{-1}. However, when analyzing the first quadrant, where both $a > 0$ and $b > 0$, we additionally

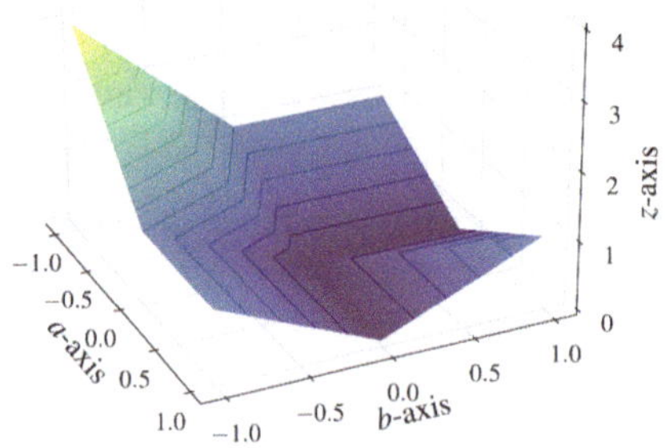

(a) Graph of ϕ_{f_1} in Example 2.4

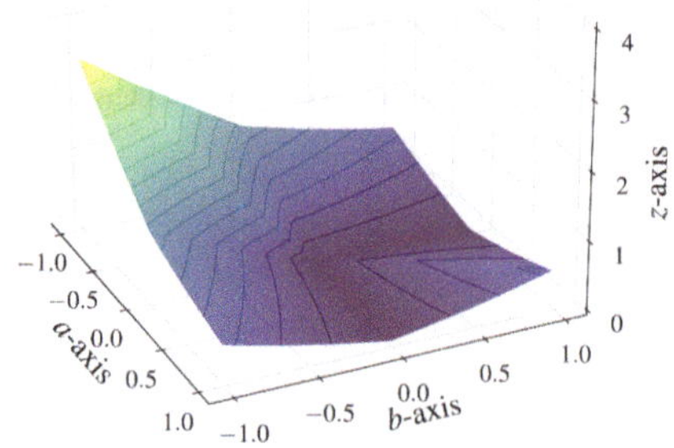

(b) Graph of ϕ_{f_2} in Example 2.4

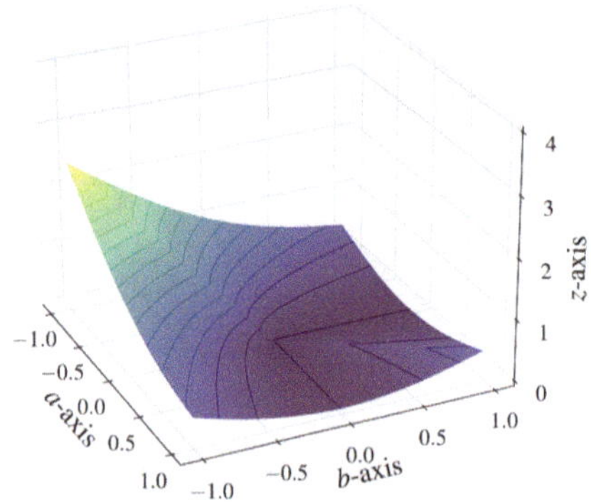

(c) Graph of ϕ_{f_3} in Example 2.4

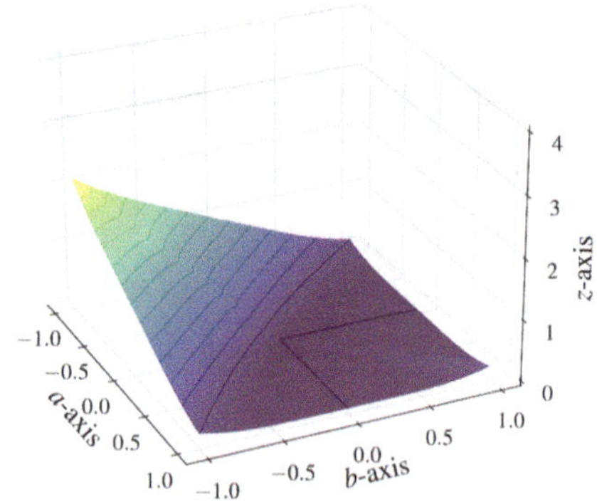

(d) Graph of ϕ_{f_4} in Example 2.4

Fig. 2.26 Graphs of NCP functions shown in Example 2.4

require the monotonicity of the derivative $(f^{-1})'$ to establish the necessary properties of ϕ_f.

Case (1): Suppose $a > 0$ and $b > 0$. If $(f^{-1})'$ is strictly monotone increasing on $[0, \infty)$, then we have

$$\begin{aligned}
& f^{-1}(a+b) - f^{-1}(b) \\
= & \int_b^{a+b} (f^{-1})'(x)dx \\
> & \int_0^{a} (f^{-1})'(x)dx \\
= & f^{-1}(a) - f^{-1}(0) \\
= & f^{-1}(a) - 1.
\end{aligned}$$

Thus, $1 < f^{-1}(a) + f^{-1}(b) - 1 < f^{-1}(a+b)$. Since f is strictly monotone increasing on $[1, \infty)$, so

$$f(f^{-1}(a) + f^{-1}(b) - 1) < f(f^{-1}(a+b)) = a + b.$$

Thus,

$$\phi_f(a, b) = f(f^{-1}(a) + f^{-1}(b) - 1) - (a + b) < 0.$$

If $(f^{-1})'$ is strictly monotone decreasing on $[0, \infty)$, then we have

$$\begin{aligned} & f^{-1}(a+b) - f^{-1}(b) \\ = & \int_b^{a+b} (f^{-1})'(x)dx \\ < & \int_0^a (f^{-1})'(x)dx \\ = & f^{-1}(a) - f^{-1}(0) \\ = & f^{-1}(a) - 1. \end{aligned}$$

Thus, $1 < f^{-1}(a+b) < f^{-1}(a) + f^{-1}(b) - 1$. Since f is strictly monotone increasing on $[1, \infty)$, we see

$$f(f^{-1}(a) + f^{-1}(b) - 1) > f(f^{-1}(a+b)) = a + b.$$

Then, it is clear that

$$\phi_f(a, b) = f(f^{-1}(a) + f^{-1}(b) - 1) - (a + b) > 0.$$

Case (2): Suppose $a < 0$ and $b > 0$. Under this case, if $a + b > 0$, since f^{-1} is strictly monotone increasing on $[0, \infty)$, so we have $f^{-1}(|b|) - f^{-1}(0) > f^{-1}(a + b) - f^{-1}(|a|)$. Thus,

$$f^{-1}(|a|) + f^{-1}(|b|) - 1 > f^{-1}(a+b) > 1.$$

Since f is strictly monotone increasing on $[1, \infty)$, we have

$$f(f^{-1}(|a|) + f^{-1}(|b|) - 1) > f(f^{-1}(a+b)) = a + b.$$

If $a + b \leq 0$, we still have $f(f^{-1}(|a|) + f^{-1}(|b|) - 1) > a + b$ because f is positive on $(1, \infty)$ and $f^{-1}(|a|) + f^{-1}(|b|) - 1 > 1$. Thus, there holds

$$\phi_f(a, b) = f(f^{-1}(|a|) + f^{-1}(|b|) - 1) - (a + b) > 0.$$

Case (3): Suppose $a < 0$ and $b < 0$. Since $f^{-1}(|a|) + f^{-1}(|b|) - 1 > 1$, $f(1) = 0$, and f is strictly monotone increasing on $[1, \infty)$, $f(f^{-1}(|a|) + f^{-1}(|b|) - 1) > 0$. Thus, we have

$$\phi_f(a, b) = f(f^{-1}(|a|) + f^{-1}(|b|) - 1) - (a + b) > 0.$$

Case (4): Suppose $a > 0$ and $b < 0$. This case is the symmetric case of $a < 0, b > 0$. Thus, there holds

$$\phi_f(a, b) = f(f^{-1}(|a|) + f^{-1}(|b|) - 1) - a + b > 0.$$

Case (5): Suppose $a \geq 0,\ b = 0$ or $a = 0,\ b \geq 0$. In this case, ϕ_f is zero.

Case (6): Suppose $a < 0,\ b = 0$ or $a = 0,\ b < 0$. In this case, ϕ_f is positive.

In summary, ϕ_f is zero only on the nonnegative sides of a, b-axes. □

Proposition 2.84 *Suppose f is a continuously differentiable real valued function with $f(1) = 0$. If f satisfies the following conditions:*

(i) $f : (-\infty, 1] \to [0, \infty)$ is invertible, and
(ii) $(f^{-1})'$ is strictly monotone on $[0, \infty)$,

then $\phi_f(a, b) = f(f^{-1}(|a|) + f^{-1}(|b|) - f^{-1}(0)) - (a + b)$ is an NCP function.

Proof The proof is similar to that in Proposition 2.83. □

Example 2.5 Here are examples of f satisfying the conditions in Proposition 2.83.

1. $f_1(t) = \ln(t)\big|_{[1,\infty)}$.
2. $f_2(t) = (t - 1)^{1/2}\big|_{[1,\infty)}$.
3. $f_3(t) = (t - 1)^{1/5}\big|_{[1,\infty)}$.

Then, their corresponding NCP functions are as below and their graphs are depicted in Fig. 2.27.

1. $\phi_{f_1}(a, b) = \ln(e^{|a|} + e^{|b|} - 1) - (a + b)$.
2. $\phi_{f_2}(a, b) = \|(a, b)\|_2 - (a + b)$.
3. $\phi_{f_3}(a, b) = \|(a, b)\|_5 - (a + b)$.

Proposition 2.83 establishes a sufficient condition on the function f and the derivative $(f^{-1})'$ to ensure that ϕ_f is an NCP function. However, this condition is not necessary. In fact, there exist functions f for which $(f^{-1})'$ is neither strictly increasing nor strictly decreasing, yet the resulting ϕ_f still satisfies the defining properties of an NCP function. To illustrate this, we present two counterexamples where f does not satisfy the strict monotonicity requirement on $(f^{-1})'$, but ϕ_f nonetheless remains a valid NCP function.

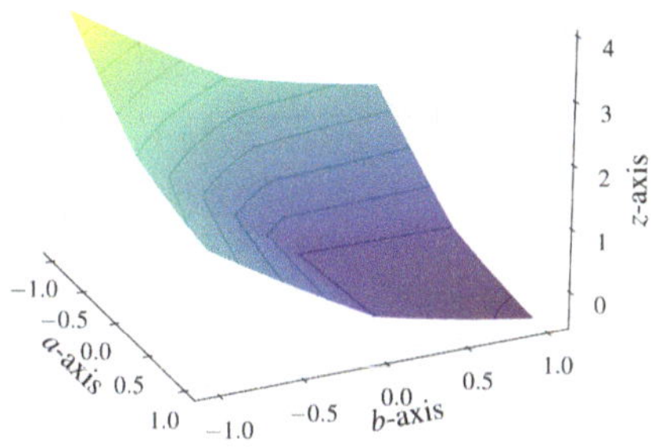

(a) Graph of ϕ_{f_1} in Example 2.5

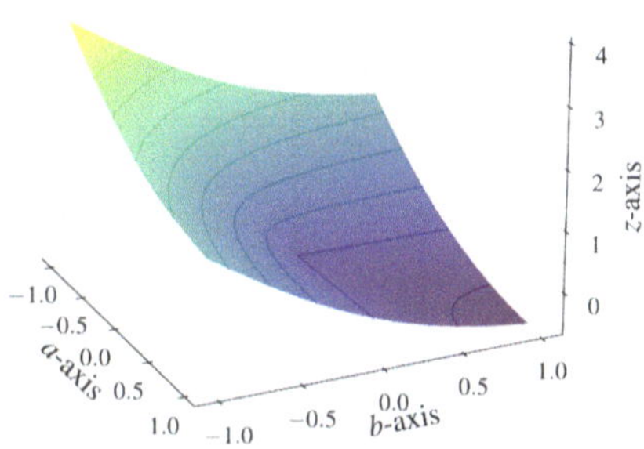

(b) Graph of ϕ_{f_2} in Example 2.5

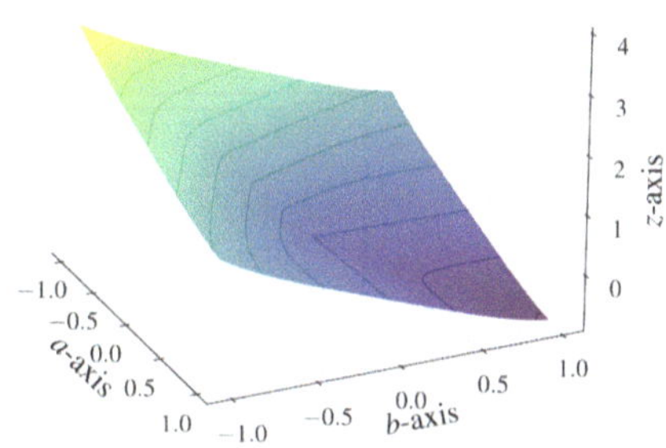

(c) Graph of ϕ_{f_3} in Example 2.5

Fig. 2.27 Graphs of NCP functions shown in Example 2.5

Example 2.6 Let f be a real valued function defined by

$$f(t) = \begin{cases} -\sqrt{38-2t}+6, & \text{if } 1 \le t \le 18.5. \\ \sqrt{2t-36}+4, & \text{if } 18.5 \le t \le 20. \\ \frac{t}{2}-4, & \text{if } 20 \le t. \end{cases}$$

Then, we compute that

$$f^{-1}(t) = \begin{cases} -\frac{t^2}{2}+6t+1, & \text{if } 0 \le t \le 5. \\ \frac{t^2}{2}-4t+26, & \text{if } 5 \le t \le 6. \\ 2t+8, & \text{if } 6 \le t. \end{cases}$$

and

$$(f^{-1})'(t) = \begin{cases} -t+6, & \text{if } 0 \le t \le 5. \\ t-4, & \text{if } 5 \le t \le 6. \\ 2, & \text{if } 6 \le t. \end{cases}$$

The graphs of f, f^{-1} and $(f^{-1})'$ are given as in Fig. 2.28. Consider $\phi_f(a, b) = f(f^{-1}(|a|) + f^{-1}(|b|) - f^{-1}(0)) - (a + b)$, we see that ϕ_f is zero on the nonnegative sides of the a, b-axes and positive on the negative sides of the a, b-axes. In addition, by using the monotonicity of f^{-1}, ϕ is positive on the second, third, and fourth quadrant due to Proposition 2.83. Thus, we only have to check the value of ϕ_f on the first quadrant. From the expression of $(f^{-1})'$, we can draw a diagram of

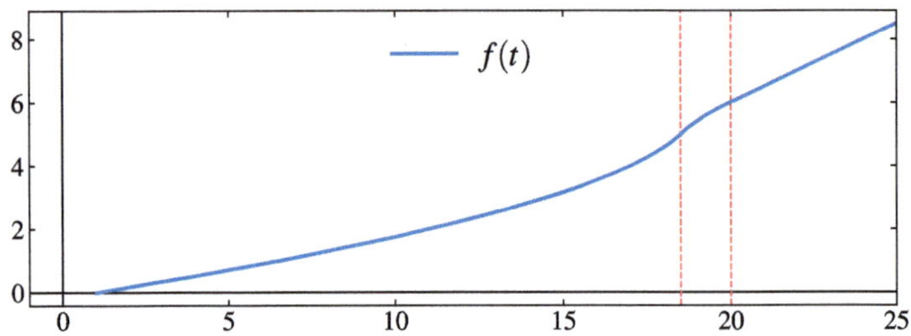

(a) Graph of f in Example 2.6

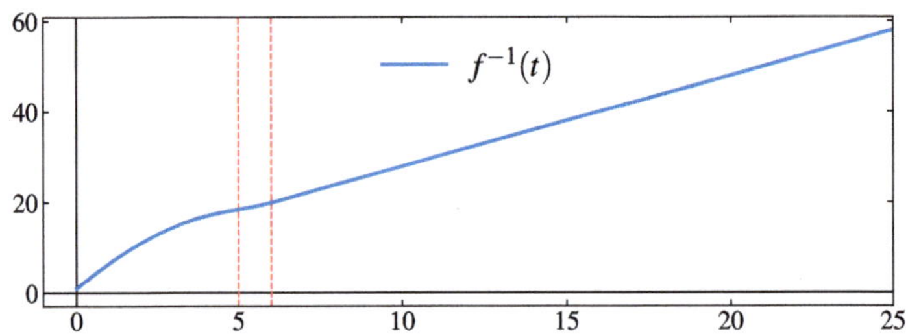

(b) Graph of f^{-1} in Example 2.6

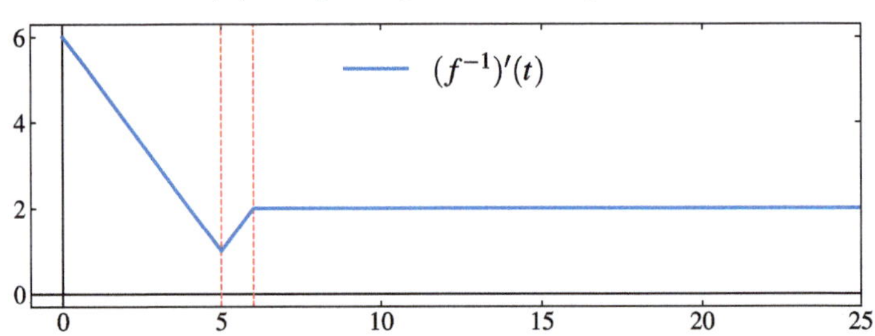

(c) Graph of $(f^{-1})'$ in Example 2.6

Fig. 2.28 Graphs of f, f^{-1} and $(f^{-1})'$ in Example 2.6

the function and easily find that $\int_0^a (f^{-1})'(t)dt > \int_b^{a+b} (f^{-1})'(t)dt$ for all $a, b > 0$. This implies

$$\begin{aligned}
& f^{-1}(a) - 1 > f^{-1}(a+b) - f^{-1}(b) \\
\Longrightarrow & f^{-1}(a) + f^{-1}(b) - 1 > f^{-1}(a+b) > 1 \\
\Longrightarrow & f(f^{-1}(a) + f^{-1}(b) - 1) > f(f^{-1}(a+b)) = a+b \\
\Longrightarrow & f(f^{-1}(a) + f^{-1}(b) - 1) - (a+b) > 0,
\end{aligned}$$

which says $\phi_f(a,b) > 0$ on the first quadrant. Hence, ϕ_f is an NCP function, whose graph is shown in Fig. 2.30a.

Example 2.7 Let f be a real valued function defined by

$$f(t) = \begin{cases} \sqrt{2t-1} - 1, & \text{if } 1 \le t \le 18.5. \\ -\sqrt{-2t+73} + 11, & \text{if } 18.5 \le t \le 24. \\ \sqrt{2t-23} + 1, & \text{if } 24 \le t. \end{cases}$$

Then, we compute that

$$f^{-1}(t) = \begin{cases} \frac{t^2}{2} + t + 1, & \text{if } 0 \le t \le 5. \\ -\frac{t^2}{2} + 11t - 24, & \text{if } 5 \le t \le 6. \\ \frac{t^2}{2} - t + 12, & \text{if } 6 \le t. \end{cases}$$

and

$$(f^{-1})'(t) = \begin{cases} t + 1, & \text{if } 0 \le t \le 5. \\ -t + 11, & \text{if } 5 \le t \le 6. \\ t - 1, & \text{if } 6 \le t. \end{cases}$$

The graphs of f, f^{-1} and $(f^{-1})'$ are given as in Fig. 2.29. Consider $\phi_f(a,b) = f(f^{-1}(|a|) + f^{-1}(|b|) - f^{-1}(0)) - (a+b)$, we see that ϕ_f is zero on the nonnegative sides of the a, b-axes and positive on the negative sides of the a, b-axes. In addition, by using the monotonicity of f^{-1}, ϕ_f is positive on the second, third, and fourth quadrant due to Proposition 2.83. Thus, we only have to check the value of ϕ_f on the first quadrant. From the expression of $(f^{-1})'$, we can draw a diagram of the function and easily find that $\int_0^a (f^{-1})'(x)dx < \int_b^{a+b} (f^{-1})'(x)dx \ \forall a, b > 0$. This implies

$$\begin{aligned}
& f^{-1}(a) - 1 < f^{-1}(a+b) - f^{-1}(b) \\
\Longrightarrow & 1 < f^{-1}(a) + f^{-1}(b) - 1 < f^{-1}(a+b) \\
\Longrightarrow & f(f^{-1}(a) + f^{-1}(b) - 1) < f(f^{-1}(a+b)) = a+b \\
\Longrightarrow & f(f^{-1}(a) + f^{-1}(b) - 1) - (a+b) < 0,
\end{aligned}$$

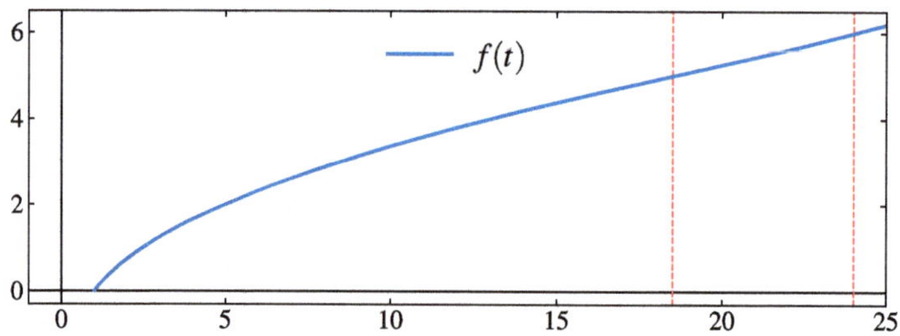

(a) Graph of f in Example 2.7

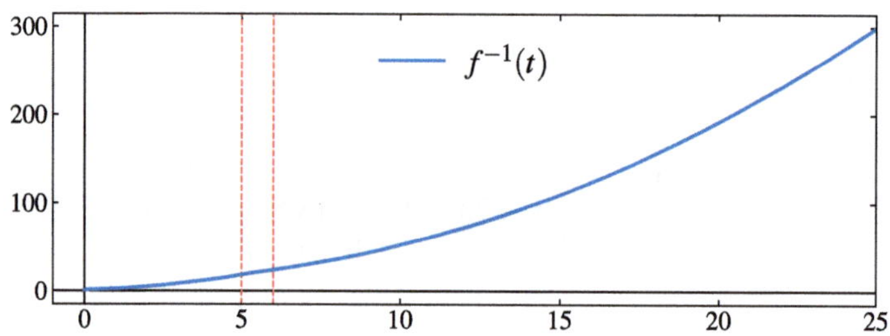

(b) Graph of f^{-1} in Example 2.7

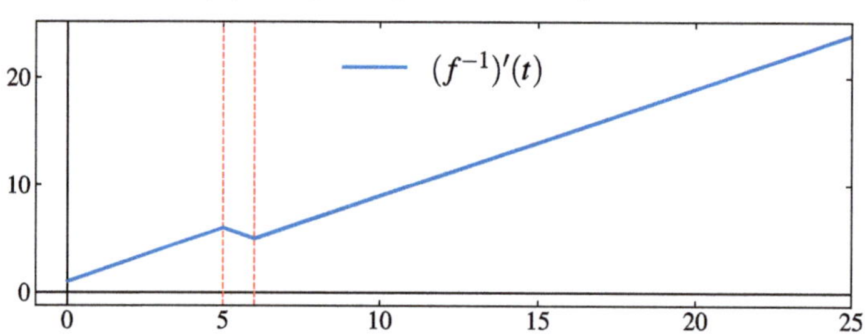

(c) Graph of $(f^{-1})'$ in Example 2.7

Fig. 2.29 Graphs of f, f^{-1} and $(f^{-1})'$ in Example 2.7

which says $\phi_f(a, b) < 0$ on the first quadrant. Hence, ϕ_f is an NCP function, whose graph is shown in Fig. 2.30b.

There exists a promising avenue for further extending the class of NCP functions described in Proposition 2.83. Specifically, we observe that by incorporating additional functions, subject to suitable conditions, applied to the negative parts of a and b, one can formulate an entirely new class of NCP functions. This extension leads to greater flexibility in construction while preserving the key properties of NCP functions. We formalize this idea in the next proposition.

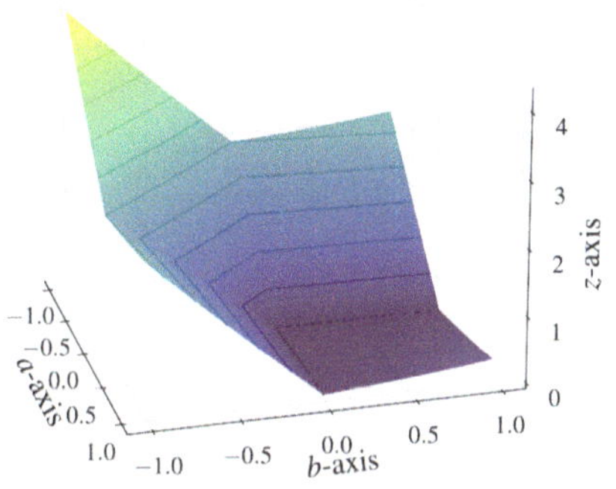

(a) Graph of ϕ_{f_2} in Example 2.6

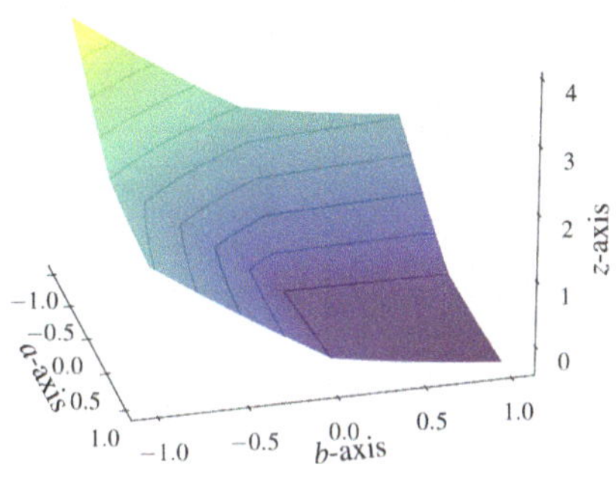

(b) Graph of ϕ_{f_3} in Example 2.7

Fig. 2.30 Graphs of two ϕ_f functions in Examples 2.6 and 2.7

Proposition 2.85 *Suppose f is a continuously differentiable real valued function with $f(1) = 0$ and g is a real valued function with $g(0) = 1$. If f and g satisfy the following conditions:*

(i) $f : [1, \infty) \to [0, \infty)$ *is invertible;*
(ii) $(f^{-1})'$ *is strictly monotone increasing;*
(iii) $g(0) = 1$, $g(t) \geq 1\ \forall t > 0$, *and* $1 \geq g(t) > \frac{-1}{2}\ \forall t < 0$.

Then, $\phi_{f,g}(a, b) = f(f^{-1}(|a|) + f^{-1}(|b|) - f^{-1}(0)) - (g(b)a + g(a)b)$ *is an NCP function.*

Proof To show that ϕ_f is an NCP function, we have to verify that ϕ_f is zero only on the nonnegative sides of the a, b-axes. To this end, we check all the regions of the a, b-plane.

Case (1): Suppose $a > 0$ and $b > 0$. By Proposition 2.83, we have

$$f(f^{-1}(|a|) + f^{-1}(|b|) - 1) < a + b \leq g(b)a + g(a)b,$$

which yields

$$f(f^{-1}(|a|) + f^{-1}(|b|) - 1) - (g(b)a + g(a)b) < 0.$$

Case (2): Suppose $a < 0$ and $b > 0$. By Proposition 2.83, we have

$$f(f^{-1}(|a|) + f^{-1}(|b|) - 1) > a + b \geq g(b)a + g(a)b,$$

which implies

$$f(f^{-1}(|a|) + f^{-1}(|b|) - 1) - (g(b)a + g(a)b) > 0.$$

Case (3): Suppose $a < 0$ and $b < 0$. Since f^{-1} and $(f^{-1})'$ are both strictly monotone increasing on $[0, \infty)$, we know that f^{-1} is strictly convex on $[0, \infty)$. This indicates that

$$f^{-1}(|a|) + f^{-1}(|b|) - 1 > 2f^{-1}\left(\frac{|a| + |b|}{2}\right) - 1 > f^{-1}\left(\frac{|a| + |b|}{2}\right) > 1.$$

In addition, using f being strictly monotone increasing on $[1, \infty)$, it gives

$$f(f^{-1}(|a|) + f^{-1}(|b|) - 1) > f\left(2f^{-1}\left(\frac{|a| + |b|}{2}\right) - 1\right) > f(f^{-1}\left(\frac{|a| + |b|}{2})\right) = \frac{|a| + |b|}{2}.$$

Thus, we obtain

$$\begin{aligned}
& f(f^{-1}(|a|) + f^{-1}(|b|) - 1) - g(b)a - g(a)b \\
> & \frac{|a| + |b|}{2} - (g(b)a + g(a)b) \\
= & \frac{a}{2}(\text{sgn}(a) - 2g(b)) + \frac{b}{2}(\text{sgn}(b) - 2g(a)) \\
> & 0.
\end{aligned}$$

Case (4): Suppose $a > 0$ and $b < 0$. This case is the symmetric case of $a < 0,\ b > 0$. Due to

$$f(f^{-1}(|a|) + f^{-1}(|b|) - 1) > a + b \geq g(b)a + g(a)b,$$

it is clear to see

$$f(f^{-1}(|a|) + f^{-1}(|b|) - 1) - (g(b)a + g(a)b) > 0.$$

Case (5): Suppose $a \geq 0, b = 0$ or $a = 0, b \geq 0$. In this case, $\phi_{f,g}$ is zero.

Case (6): Suppose $a < 0, b = 0$ or $a = 0, b < 0$. In this case, $\phi_{f,g}$ is positive.

From all the above, $\phi_{f,g}$ is zero only on the nonnegative sides of the a, b-axes. Hence, $\phi_{f,g}$ is an NCP function. □

Example 2.8 Here are examples of f and g satisfying those conditions in Proposition 2.85.

(a) Three examples for function f:

a. $f_1(t) = \ln(t)\big|_{[1,\infty)}$.
b. $f_2(t) = (t-1)^{1/2}\big|_{[1,\infty)}$.
c. $f_3(t) = (t-1)^{1/5}\big|_{[1,\infty)}$.

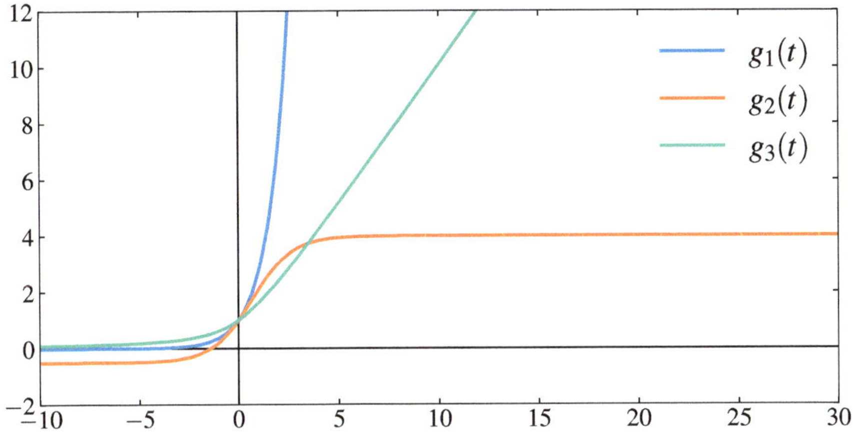

Fig. 2.31 Graphs of g functions given in Example 2.8

(b) Three examples for function g, see Fig. 2.31:

a. $g_1(t) = e^t$.

b. $g_2(t) = \dfrac{4 - e^{-t}}{1 + 2e^{-t}}$.

c. $g_3(t) = \dfrac{\sqrt{t^2 + 4} + t}{2}$.

Then, applying those f and g functions in Example 2.8, we generate the following nine NCP functions, see Fig. 2.32.

$$\phi_1(a, b) = \ln(e^{|a|} + e^{|b|} - 1) - e^b a - e^a b,$$

$$\phi_2(a, b) = \ln(e^{|a|} + e^{|b|} - 1) - \frac{4 - e^{-b}}{1 + 2e^{-b}} a - \frac{4 - e^{-a}}{1 + 2e^{-a}} b,$$

$$\phi_3(a, b) = \ln(e^{|a|} + e^{|b|} - 1) - \frac{\sqrt{b^2 + 4} + b}{2} a - \frac{\sqrt{a^2 + 4} + a}{2} b,$$

$$\phi_4(a, b) = \|(a, b)\|_2 - e^b a - e^a b,$$

$$\phi_5(a, b) = \|(a, b)\|_2 - \frac{4 - e^{-b}}{1 + 2e^{-b}} a - \frac{4 - e^{-a}}{1 + 2e^{-a}} b,$$

$$\phi_6(a, b) = \|(a, b)\|_2 - \frac{\sqrt{b^2 + 4} + b}{2} a - \frac{\sqrt{a^2 + 4} + a}{2} b,$$

$$\phi_7(a, b) = \|(a, b)\|_5 - e^b a - e^a b,$$

$$\phi_8(a, b) = \|(a, b)\|_5 - \frac{4 - e^{-b}}{1 + 2e^{-b}} a - \frac{4 - e^{-a}}{1 + 2e^{-a}} b,$$

$$\phi_9(a, b) = \|(a, b)\|_5 - \frac{\sqrt{b^2 + 4} + b}{2} a - \frac{\sqrt{a^2 + 4} + a}{2} b.$$

Fig. 2.32 Graphs of generated NCP functions by Proposition 2.85

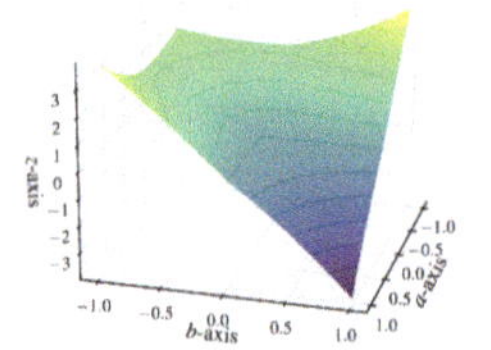

(a) Graph of ϕ_1 from Example 2.8

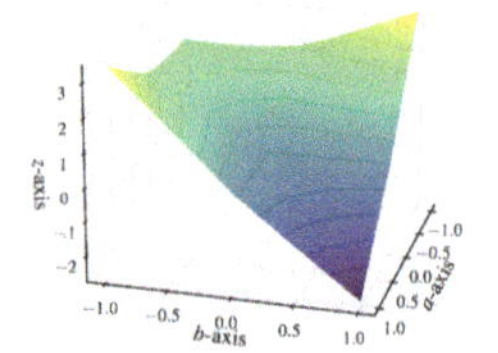

(b) Graph of ϕ_2 from Example 2.8

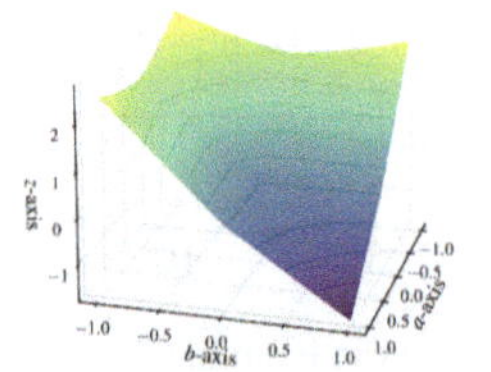

(c) Graph of ϕ_3 from Example 2.8

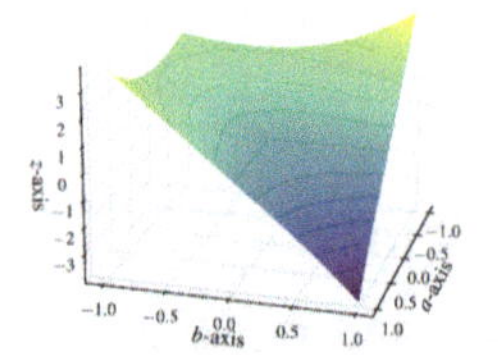

(d) Graph of ϕ_4 from Example 2.8

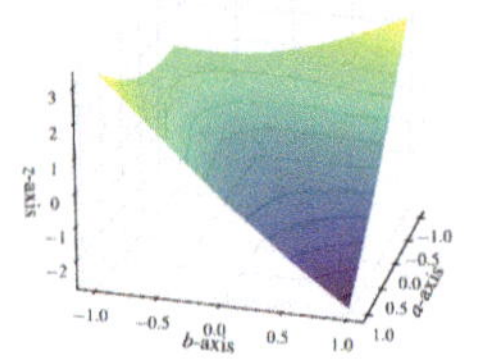

(e) Graph of ϕ_5 from Example 2.8

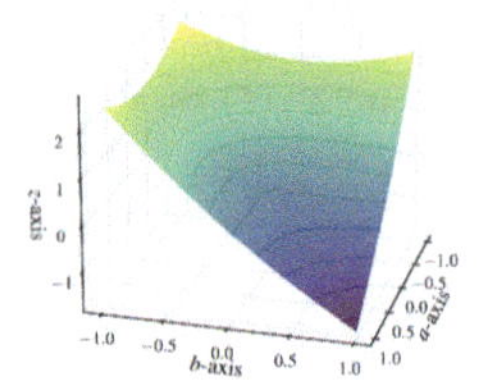

(f) Graph of ϕ_6 from Example 2.8

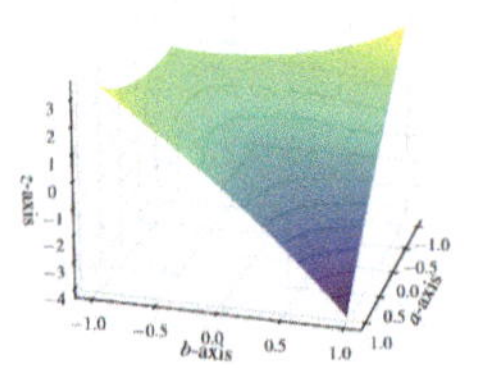

(g) Graph of ϕ_7 from Example 2.8

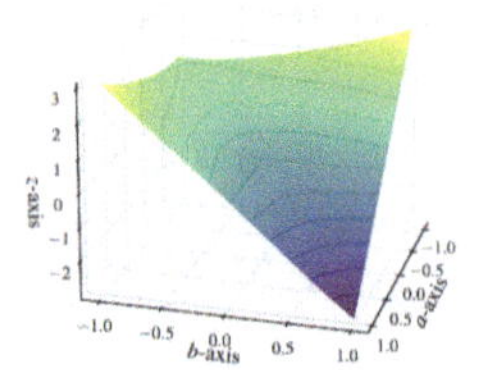

(h) Graph of ϕ_8 from Example 2.8

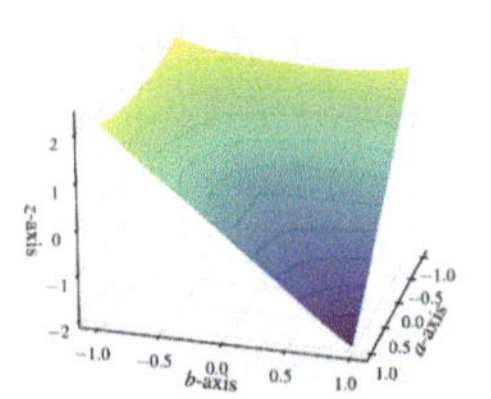

(i) Graph of ϕ_9 from Example 2.8

The following proposition serves as a counterpart to Proposition 2.84, with arguments closely paralleling those used in the proof of Proposition 2.85. The result relies on the monotonicity and concavity of f^{-1} over the interval $[0, \infty)$. For brevity, we omit the proof.

Proposition 2.86 *Suppose f is a continuously differentiable real valued function with $f(1) = 0$ and g is a real valued function with $g(0) = 1$. If f and g satisfy the following conditions:*

(i) $f : (-\infty, 1] \to [0, \infty)$ *is invertible;*
(ii) $(f^{-1})'$ *is strictly monotone decreasing;*
(iii) $g(0) = 1$, $g(t) \geq 1\ \forall t > 0$, *and* $1 \geq g(t) > \frac{-1}{2}\ \forall t < 0$.

Then, $\phi_{_{f,g}}(a, b) = f(f^{-1}(|a|) + f^{-1}(|b|) - f^{-1}(0)) - (g(b)a + g(a)b)$ *is an NCP function.*

In Proposition 2.85, consider the specific choice $f(t) = t - 1$ In this case, we have $f(f^{-1}(|a|) + f^{-1}(|b|) - 1) = |a| + |b|$, and notably, $(f^{-1})'$ is not strictly monotone increasing. Nevertheless, we observe that if an auxiliary function g satisfies a strict inequality condition, rather than an equality condition, on the interval $(0, \infty)$, then the resulting function $\phi_{_{f,g}}$ remains a valid NCP function. This observation motivates an extended case of Proposition 2.85. A related class of NCP functions of this type was previously introduced in [50], and can be viewed as a specific instance connected to the framework established in Proposition 2.85.

Example 2.9 Suppose that $\phi(a, b) = |a| + |b| - (g(a)b + g(b)a)$, where $g : \mathbb{R} \to \mathbb{R}$ satisfying

$$g(0) = 1,\ \ g(t) > 1 \quad \forall t > 0, \quad \text{and} \quad 1 \geq g(t) > -1 \quad \forall t < 0.$$

Then, ϕ is an NCP function. Note that the condition $1 \geq g(t) > -1$ for all $t < 0$ is a bit weaker than $1 \geq g(t) > -\frac{1}{2}$ for all $t < 0$, used in Proposition 2.85 and Proposition 2.86. This distinction arises because our analysis involves both f and g, whereas the approach in [50] considers only g.

For instance, consider the same examples of g presented in Example 2.8, which also satisfy the conditions mentioned above.

1. $g_1(t) = e^t$.
2. $g_2(t) = \dfrac{4 - e^{-t}}{1 + 2e^{-t}}$.
3. $g_3(t) = \dfrac{\sqrt{t^2 + 4} + t}{2}$.

The corresponding NCP functions are then constructed as follows; see Fig. 2.33 for their graphical representations.

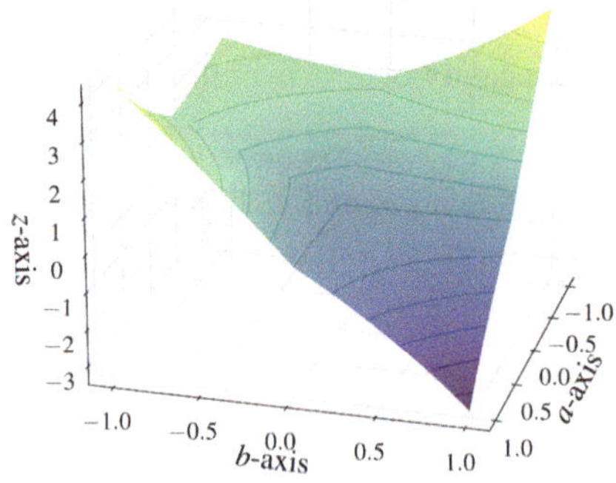

(a) Graph of ϕ_{g_1} Example 2.9

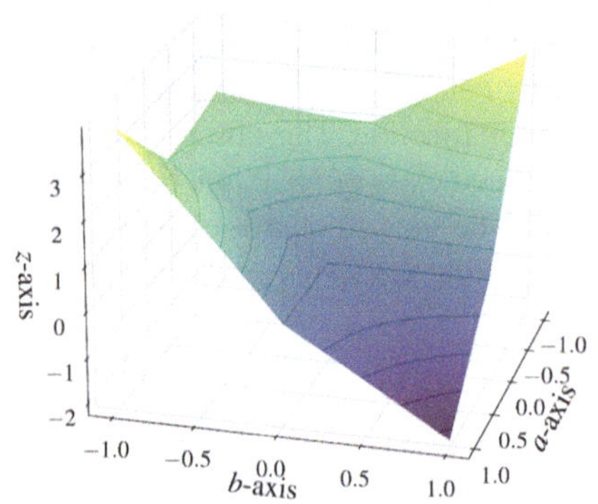

(b) Graph of ϕ_{g_2} Example 2.9

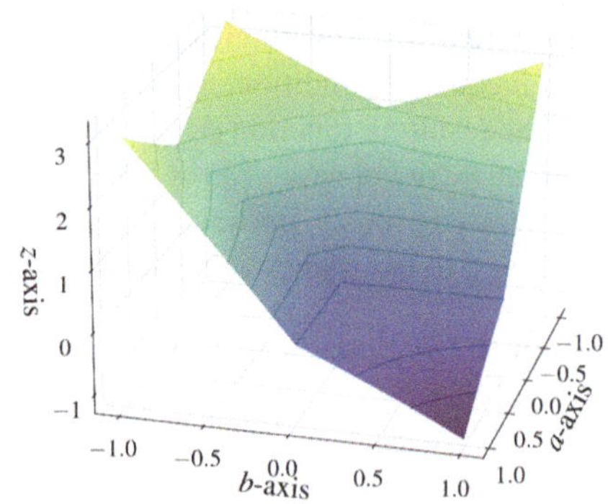

(c) Graph of ϕ_{g_3} Example 2.9

Fig. 2.33 Graphs of NCP functions generated in Example 2.9

1. $\phi_{g_1}(a, b) = |a| + |b| - e^b a - e^a b$.
2. $\phi_{g_2}(a, b) = |a| + |b| - \dfrac{4 - e^{-b}}{1 + 2e^{-b}} a - \dfrac{4 - e^{-a}}{1 + 2e^{-a}} b$.
3. $\phi_{g_3}(a, b) = |a| + |b| - \dfrac{\sqrt{b^2 + 4} + b}{2} a - \dfrac{\sqrt{a^2 + 4} + a}{2} b$.

References

1. F. Facchinei, J.-S. Pang, *Finite-Dimensional Variational Inequalities and Complementarity Problems*, vol. I and II (Springer, New York, 2003)
2. M.C. Ferris, O.L. Mangasarian, J.-S. Pang (eds.), *Complementarity: Applications* Algorithms and Extensions. (Kluwer Academic Publishers, Dordrecht, 2001)

3. M.C. Ferris, J.-S. Pang, Engineering and economic applications of complementarity problems. SIAM Rev. **39**, 669–713 (1997)
4. A. Galantai, Properties and construction of NCP functions. Comput. Optim. Appl. **52**, 805–824 (2012)
5. J.-S. Pang, A B-differentiable equation based, globally and locally quadratically convergent algorithm for nonlinear programs, complementarity and variational inequality problems. Math. Program. **51**, 101–131 (1991)
6. O.L. Mangasarian, M.V. Solodov, Nonlinear complementarity as unconstrained and constrained minimization. Math. Program. **62**, 2777–297 (1993)
7. S.M. Miri, S. Effati, On generalized convexity of nonlinear complementarity functions. J. Optim. Theory Appl. **164**, 723–730 (2015)
8. C.-H. Huang, J.-S. Chen, J.E. Martinez-Legaz, Differentiability v.s. convexity of complementarity functions. Optim. Lett. **11**(1), 209–216 (2017)
9. A. Fischer, A special Newton-type optimization method. Optimization **24**, 269–284 (1992)
10. A. Fischer, Solution of monotone complementarity problems with locally Lipschitzian functions. Math. Program. **76**, 513–532 (1997)
11. D. Sun, L. Qi, On NCP functions. Comput. Optim. Appl. **13**, 201–220 (1999)
12. O.L. Mangasarian, Equivalence of the complementarity problem to a system of nonlinear equations. SIAM J. Appl. Math. **31**, 89–92 (1976)
13. Z.Q. Luo, P. Tseng, A new class of merit functions for the nonlinear complementarity problem, in *Complementarity and Variational Problems: State of the Art*. ed. by M.C. Ferris, J.S. Pang (SIAM, Philadelphia, 1997), pp.204–225
14. C. Kanzow, Y. Yamashita, M. Fukushima, New NCP functions and their properties. J. Optim. Theory Appl. **97**, 115–135 (1997)
15. A. Fischer, H. Jiang, Merit functions for complementarity and related problems: a survey. Math. Program. **76**, 513–532 (1997)
16. C. Kanzow, H. Kleinmichel, A new class of semismooth Newton method for nonlinear complementarity problems. Comput. Optim. Appl. **11**, 227–251 (1998)
17. B. Chen, X. Chen, C. Kanzow, A penalized Fischer-Burmeister NCP-function: theoretical investigation and numerical results. Math. Program. **88**, 211–216 (2000)
18. P. Tseng, Global behaviour of a class of merit functions for the nonlinear complementarity problem. J. Optim. Theory Appl. **89**, 17–37 (1996)
19. J.-S. Chen, The Semismooth-related properties of a merit function and a descent method for the nonlinear complementarity problem. J. Global Optim. **36**, 565–580 (2006)
20. J.-S. Chen, On some NCP-functions based on the generalized Fischer-Burmeister function. Asia-Pac. J. Oper. Res. **24**(3), 401–420 (2007)
21. J.-S. Chen, H.-T. Gao, S.-H. Pan, An R-linearly convergent derivative-free algorithm for the NCPs based on the generalized Fischer-Burmeister merit function. J. Comput. Appl. Math. **232**(2), 455–471 (2009)
22. J.-S. Chen, S.-H. Pan, A family of NCP functions and a descent method for the nonlinear complementarity problem. Comput. Optim. Appl. **40**, 389–404 (2008)
23. J.-S. Chen, S.-H. Pan, A regularization semismooth Newton method based on the generalized Fischer-Burmeister function for P_0-NCPs. J. Comput. Appl. Math. **220**(1–2), 464–479 (2008)
24. J.-S. Chen, S.-H. Pan, T.-C. Lin, A smoothing Newton method based on the generalized Fischer-Burmeister function for MCPs. Nonlinear Anal. Theory Methods Appl. **72**(9–10), 3739–3758 (2010)
25. S.-L. Hu, Z.-H. Huang, J.-S. Chen, Properties of a family of generalized NCP-functions and a derivative free algorithm for complementarity problems. J. Comput. Appl. Math. **230**, 69–82 (2009)
26. H.-Y. Tsai, J.-S. Chen, Geometric views of the generalized Fischer-Burmeister function and its induced merit function. Appl. Math. Comput. **237**, 31–59 (2014)
27. J.-S. Chen, C.-H. Ko, S.-H. Pan, A neural network based on generalized Fischer-Burmeister function for nonlinear complementarity problems. Inf. Sci. **180**, 697–711 (2010)
28. N.J. Higham, Estimating the matrix p-norm. Numer. Math. **62**, 539–555 (1992)

29. C. Geiger, C. Kanzow, On the resolution of monotone complementarity problems. Comput. Optim. Appl. **5**, 155–173
30. F. Facchinei, J. Soares, A new merit function for nonlinear complementarity problems and a related algorithm. SIAM J. Optim. **7**, 225–247 (1997)
31. J.-S. Chen, Z.-H. Huang, C.-Y. She, A new class of penalized NCP-functions and its properties. Comput. Optim. Appl. **50**(1), 49–73 (2011)
32. L. Qi, L.P. Tseng, On almost smooth functions and piecewise smooth functions. Nonlinear Anal.: Theory, Methods Appl. **67**(3), 773–794 (2007)
33. J.-S. Chen, S.-H. Pan, C.-Y. Yang, Numerical comparison of two effective methods for mixed complementarity problems. J. Comput. Appl. Math. **234**(3), 667–683 (2010)
34. J.-S. Chen, C.-H. Ko, X.-R. Wu, What is the generalization of natural residual function for NCP. Pac. J. Optim. **12**(1), 19–27 (2016)
35. Y.-L. Chang, J.-S. Chen, C.-Y. Yang, Symmetrization of generalized natural residual function for NCP. Oper. Res. Lett. **43**, 354–358 (2015)
36. P.-F. Ma, J.-S. Chen, C.-H. Huang, C.-H. Ko, Discovery of new complementarity functions for NCP and SOCCP. Comput. Appl. Math. **37**(5), 5727–5749 (2018)
37. C.-H. Huang, K.-J. Weng, J.-S. Chen, H.-W. Chu, M.-Y. Li, On four discrete-type families of NCP-functions. J. Nonlinear Convex Anal. **20**(2), 283–306 (2019)
38. S.-H. Pan, J.-S. Chen, A semismooth Newton method for SOCCPs based on a one-parametric class of complementarity functions. Comput. Optim. Appl. **45**(1), 59–88 (2010)
39. J.-S. Chen, P. Tseng, An unconstrained smooth minimization reformulation of the second-order cone complementarity problem. Math. Program. **104**, 293–327 (2005)
40. K. Yamada, N. Yamashita, M. Fukushima, A new derivative-free descent method for the nonlinear complementarity problems, in *Nonlinear Optimization and Related Topics*. ed. by G.D. Pillo, F. Giannessi (Kluwer Academic Publishers, Netherlands, 2000), pp.463–489
41. J. Wang, C. Gu, G. Wang, Penalized NCP-functions for nonlinear complementarity problems and a scaling algorithm. J. Ind. Manag. Optim. (2021). https://doi.org/10.3934/jimo.2021171
42. C. Kanzow, Nonlinear complementarity as unconstrained optimization. J. Optim. Theory Appl. **88**, 139–155 (1996)
43. C. Kanzow, H. Kleinmichel, A class of Newton-Type methods for equality and inequality constrained optimization. Optim. Methods Softw. **5**, 173–198 (1995)
44. L. Baggett, *Analysis of Functions of a Single Variable* (University of Colorado, 2012). http://cnx.org/content/col11249/1.1/
45. J.H. Alcantara, J.-S. Chen, Neural networks based on three classes of NCP-functions for solving nonlinear complementarity problems. Neurocomputing **359**, 102–113 (2019)
46. Y. Zhang, Z. Fan, Z. Li, Zhang neural network for online solution of time-varying Sylvester equation, in *Proceedings of the 2nd International Conference on Advances in Computation and Intelligence, ISICA'07* (Springer, Berlin), pp. 276–285
47. Y. Zhang, S.S. Ge, Design and analysis of a general recurrent neural network model for time-varying matrix inversion. IEEE Trans. Neural Netw. **16**, 1477–1490 (2005)
48. S. Li, S. Chen, B. Liu, Accelerating a recurrent neural network to finite-time convergence for solving time-varying Sylvester equation by using a sign-bi-power activation function. Neural Process. Lett. **37**, 189–205 (2013)
49. J.H. Alcantara, J.-S. Chen, A novel generalization of the natural residual function and a neural network approach for the NCP. Neurocomputing **413**, 368–382 (2020)
50. J.H. Alcantara, C.-H. Lee, C.T. Nguyen, Y.-L. Chang, J.-S. Chen, On construction of new NCP functions. Oper. Res. Lett. **48**, 115–121 (2020)
51. F.H. Clarke, *Optimization and Nonsmooth Analysis* (Wiley, New York, 1983)

Chapter 3
General Complementarity Functions

Abstract In this chapter, we broaden our scope to explore more general classes of complementarity functions linked to other types of symmetric cones, including C-functions associated with second-order cones (SOC), the cone of positive semidefinite matrices, and the unified symmetric cone.

The NCP functions introduced in Chap. 1 are, in fact, C-functions associated with the nonnegative orthant $\mathbb{R}^n_+$, a special instance of symmetric cones. In this chapter, we broaden our scope to explore more general classes of complementarity functions linked to other types of symmetric cones [1], including C-functions associated with second-order cones (SOC), the cone of positive semidefinite matrices, and the unified symmetric cone.

3.1 Complementarity Functions Associated with SOC

In this section, we examine the complementarity problem within the framework of second-order cones, known as the second-order cone complementarity problem (SOCCP). The objective is to find a vector $\zeta \in \mathbb{R}^n$ that satisfies

$$\langle F(\zeta), \zeta \rangle = 0, \quad F(\zeta) \in \mathcal{K}, \quad \zeta \in \mathcal{K}, \tag{3.1}$$

where $\langle \cdot, \cdot \rangle$ denotes the Euclidean inner product, $F : \mathbb{R}^n \to \mathbb{R}^n$ is a smooth (i.e., continuously differentiable) mapping, and $\mathcal{K}$ is the Cartesian product of second-order cones. In other words,

$$\mathcal{K} = \mathcal{K}^{n_1} \times \cdots \times \mathcal{K}^{n_m}, \tag{3.2}$$

where $m, n_1, \ldots, n_m \geq 1$, $n_1 + \cdots + n_m = n$, and

$$\mathcal{K}^{n_i} := \{(x_1, x_2) \in \mathbb{R} \times \mathbb{R}^{n_i - 1} \mid \|x_2\| \leq x_1\}, \tag{3.3}$$

with $\|\cdot\|$ denoting the Euclidean norm and $\mathcal{K}^1$ denoting the set of nonnegative reals $\mathbb{R}_+$.

J. Chen, *Complementarity Functions in Optimization*, Springer Optimization and Its Applications 228, https://doi.org/10.1007/978-3-032-06924-5_3

A special case of (3.2)–(3.3) arises when $\mathcal{K} = \mathbb{R}^n_+$, the nonnegative orthant in $\mathbb{R}^n$, corresponding to the setting where $m = n$ and $n_1 = \cdots = n_m = 1$. In this case, the SOCCP (3.1) reduces to the classical NCP, a cornerstone in optimization theory with wide-ranging applications in engineering and economics; see, for example, [2–5]. A broader formulation of the SOCCP seeks a vector $\zeta \in \mathbb{R}^n$ satisfying

$$\langle F(\zeta), G(\zeta)\rangle = 0, \quad F(\zeta) \in \mathcal{K}, \quad G(\zeta) \in \mathcal{K}, \tag{3.4}$$

where $\langle \cdot, \cdot \rangle$ denotes the Euclidean inner product, and $F : \mathbb{R}^n \to \mathbb{R}^n$, $G : \mathbb{R}^n \to \mathbb{R}^n$ are smooth (i.e., continuously differentiable) mappings. Unless stated otherwise, throughout this chapter we assume $\mathcal{K} = \mathcal{K}^n$ for simplicity, that is, $\mathcal{K}$ is a single second-order cone. However, the ensuing analysis readily extends to the general case where $\mathcal{K}$ takes the form of a Cartesian product as in (3.2).

Recall that when $\mathcal{K}$ is the second-order cone $\mathcal{K}^n$, a function $\phi : \mathbb{R}^n \times \mathbb{R}^n \to \mathbb{R}^n$ is called a C-function associated with the SOC if

$$\begin{aligned} \phi(x, y) = 0 &\Longleftrightarrow x, y \in \mathcal{K}^n, \; x \circ y = 0, \\ &\Longleftrightarrow x, y \in \mathcal{K}^n, \; \langle x, y\rangle = 0. \end{aligned} \tag{3.5}$$

Such C-functions are particularly valuable in addressing the SOCCP, as they enable reformulation into nonsmooth equations. Over the years, various methods have been proposed to solve the SOCCPs. These include interior-point methods [6–10] and non-interior smoothing Newton methods [11–13]. More recently, an alternative approach was introduced in [14], where the SOCCP is reformulated as an unconstrained smooth minimization problem. The idea is to construct a smooth function $\psi : \mathbb{R}^n \times \mathbb{R}^n \to \mathbb{R}_+$ such that

$$\psi(x, y) = 0 \quad \Longleftrightarrow \quad x \in \mathcal{K}^n, \quad y \in \mathcal{K}^n, \quad \langle x, y\rangle = 0. \tag{3.6}$$

Such a function ψ is referred to as a merit function, and it also qualifies as a C-function. With this formulation, the SOCCP can be rewritten as the unconstrained smooth (global) minimization problem:

$$\min_{\zeta \in \mathbb{R}^n} \; \psi(F(\zeta), \zeta). \tag{3.7}$$

This reformulation allows the application of standard gradient-based optimization techniques, such as conjugate gradient and quasi-Newton methods [15, 16]. As discussed in [14], this approach offers several advantages. However, its effectiveness critically depends on the appropriate choice of the merit function ψ.

We begin with the NCP function introduced by Mangasarian and Solodov, as presented in (2.11). Under the second-order cone (SOC) setting, this function admits two potential extensions. The first is the implicit Lagrangian function $\psi_{\text{MS}} : \mathbb{R}^n \times \mathbb{R}^n \to \mathbb{R}_+$, which is parameterized by $\alpha > 1$ and defined as follows:

$$\psi_{\rm MS}(x,y) := \max_{u,v\in\mathcal{K}^n}\left\{\langle x, y-v\rangle - \langle y,u\rangle - \frac{1}{2\alpha}(\|x-u\|^2+\|y-v\|^2)\right\}$$
$$=\langle x,y\rangle + \frac{1}{2\alpha}\left(\|(x-\alpha y)_+\|^2 - \|x\|^2 + \|(y-\alpha x)_+\|^2 - \|y\|^2\right). \quad (3.8)$$

This function is also extended to semidefinite complementarity problems (SDCPs) by Tseng [17] and general symmetric cone complementarity problems (SCCPs) by Kong et al. [18]. The second extension is the vector-valued implicit Lagrangian function, $\phi_{\rm MS} : \mathbb{R}^n \times \mathbb{R}^n \to \mathbb{R}^n$, defined by

$$\phi_{\rm MS}(x,y) := x\circ y + \frac{1}{2\alpha}\left[(x-\alpha y)_+^2 - x^2 + (y-\alpha x)_+^2 - y^2\right] \quad \forall x,y\in\mathbb{R}^n, \alpha>1. \quad (3.9)$$

Both $\psi_{\rm MS}$ and $\phi_{\rm MS}$ are C-functions associated with the SOC; see Proposition 3.1 below. Furthermore, according to [18, Theorem 3.2(b)] (also refer to Sect. 3.3), the function $\psi_{\rm MS}$ serves as a merit function derived from the trace of the $\phi_{\rm MS}$ function. It is worth noting that the C-function $\phi_{\rm MS}$ is primarily constructed via the projection onto the second-order cone, providing a natural extension of the original $\phi_{\rm MS}$ defined in (2.11). The results that follow generalize several classical findings, particularly those in [19–21], from the NCP framework to the SOC setting.

Proposition 3.1 *For any fixed* $\alpha > 1$ *and all* $x, y \in \mathbb{R}^n$, *we have the following results.*

(a) $\psi_{\rm MS}(x,y) = 0 \iff x\in\mathcal{K}^n,\ y\in\mathcal{K}^n,\ \langle x,y\rangle = 0 \iff \phi_{\rm MS}(x,y)=0.$
(b) $\phi_{\rm MS}$ *and* $\psi_{\rm MS}$ *are continuously differentiable everywhere, with*

$$\nabla_x\psi_{\rm MS}(x,y) = y + \alpha^{-1}\left((x-\alpha y)_+ - x\right) - (y-\alpha x)_+,$$
$$\nabla_y\psi_{\rm MS}(x,y) = x + \alpha^{-1}\left((y-\alpha x)_+ - y\right) - (x-\alpha y)_+.$$

(c) *The gradient function* $\nabla\psi_{\rm MS}$ *is globally Lipschitz continuous.*
(d) $\langle x, \nabla_x\psi_{\rm MS}(x,y)\rangle + \langle y, \nabla_y\psi_{\rm MS}(x,y)\rangle = 2\psi_{\rm MS}(x,y).$
(e) $\langle \nabla_x\psi_{\rm MS}(x,y), \nabla_y\psi_{\rm MS}(x,y)\rangle \geq 0.$
(f) $\psi_{\rm MS}(x,y) = 0$ *if and only if* $\nabla_x\psi_{\rm MS}(x,y) = 0$ *and* $\nabla_y\psi_{\rm MS}(x,y)=0.$
(g) $(\alpha-1)\|\phi_{\rm NR}(x,y)\|^2 \geq \psi_{\rm MS}(x,y) \geq (1-\alpha^{-1})\|\phi_{\rm NR}(x,y)\|^2.$
(h) $\alpha^{-1}(\alpha-1)^2\psi_{\rm MS}(x,y) \leq \|\nabla_x\psi_{\rm MS}(x,y)+\nabla_y\psi_{\rm MS}(x,y)\|^2 \leq 2\alpha(\alpha-1)$ $\psi_{\rm MS}(x,y).$

Proof The proofs of parts (a)–(b) and (e)–(f) can be found in [18]. Parts (c) and (d) follow directly from the explicit forms of $\psi_{\rm MS}$ and its gradient $\nabla\psi_{\rm MS}$. Part (g) is a straightforward application of [19, Proposition 2.2], taking $\widetilde{\pi} = -\psi_{\rm MS}$. Finally, part (h) follows easily from [22, Theorem 4.2], together with results from parts (b) and (g). □

In the setting of SOC, as mentioned in Chap. 1, for any $x=(x_1,x_2)\in\mathbb{R}\times\mathbb{R}^{n-1}$ and $y=(y_1,y_2)\in\mathbb{R}\times\mathbb{R}^{n-1}$, their *Jordan product* associated with $\mathcal{K}^n$ is

$$x \circ y = (\langle x, y \rangle,\ y_1 x_2 + x_1 y_2).$$

The identity element under this product is $e := (1, 0, \cdots, 0)^{\mathsf{T}} \in \mathbb{R}^n$. We use the notation x^2 to denote the Jordan product $x \circ x$, and $x + y$ to represent the usual componentwise addition of vectors. It is well known that $x^2 \in \mathcal{K}^n$ for any $x \in \mathbb{R}^n$. Moreover, for any $x \in \mathcal{K}^n$, there exists a unique vector $x^{1/2} \in \mathcal{K}^n$ such that $(x^{1/2})^2 = x^{1/2} \circ x^{1/2} = x$. Based on this, the Fischer–Burmeister function associated with the SOC is defined by

$$\phi_{\text{FB}}(x, y) := (x^2 + y^2)^{1/2} - x - y, \tag{3.10}$$

for all $(x, y) \in \mathbb{R}^n \times \mathbb{R}^n$; it is a map from $\mathbb{R}^n \times \mathbb{R}^n$ to $\mathbb{R}^n$. It was shown in [12] that $\phi_{\text{FB}}(x, y) = 0$ if and only if $x \in \mathcal{K}^n,\ y \in \mathcal{K}^n,\ \langle x, y \rangle = 0$. Hence, $\psi_{\text{FB}} : \mathbb{R}^n \times \mathbb{R}^n \to \mathbb{R}_+$ induced from

$$\psi_{\text{FB}}(x, y) := \frac{1}{2}\|\phi_{\text{FB}}(x, y)\|^2 \tag{3.11}$$

is a merit function for the SOCCP.

It is known that SOCCP can be reduced to an SDCP by observing that, for any $x = (x_1, x_2) \in \mathbb{R} \times \mathbb{R}^{n-1}$, we have $x \in \mathcal{K}^n$ if and only if

$$L_x := \begin{bmatrix} x_1 & x_2^{\mathsf{T}} \\ x_2 & x_1 I \end{bmatrix}$$

is positive semidefinite (also see [12, p. 437] and [23]). However, this reduction increases the problem dimension from n to $n(n + 1)/2$ and it is not known whether this increase can be mitigated by exploiting the special "arrow" structure of L_x.

Lemma 3.1 *Suppose that* $x = (x_1, x_2) \in \mathbb{R} \times \mathbb{R}^{n-1}$ *has the spectral decomposition (1.8) with spectral values* λ_1, λ_2 *and spectral vectors* $u_x^{(1)}, u_x^{(2)}$. *Then, the following results hold.*

(a) $x^2 = \lambda_1^2 u^{(1)} + \lambda_2^2 u^{(2)} \in \mathcal{K}^n$.
(b) If $x \in \mathcal{K}^n$, *then* $0 \le \lambda_1 \le \lambda_2$ *and* $x^{1/2} = \sqrt{\lambda_1}\, u^{(1)} + \sqrt{\lambda_2}\, u^{(2)}$.
(c) If $x \in \text{int}(\mathcal{K}^n)$, *then* $0 < \lambda_1 \le \lambda_2$, $\det(x) = \lambda_1 \lambda_2$, *and* L_x *is invertible with*

$$L_x^{-1} = \frac{1}{\det(x)} \begin{bmatrix} x_1 & -x_2^{\mathsf{T}} \\ -x_2 & \dfrac{\det(x)}{x_1} I + \dfrac{1}{x_1} x_2 x_2^{\mathsf{T}} \end{bmatrix}.$$

(d) $x \circ y = L_x y$ *for all* $y \in \mathbb{R}^n$, *and* $L_x \succ 0$ *if and only if* $x \in \text{int}(\mathcal{K}^n)$.

Proof Please refer to [12] for detailed proof. □

Since $x^2, y^2 \in \mathcal{K}^n$ for any $x, y \in \mathbb{R}^n$, we have $x^2 + y^2 = (\|x\|^2 + \|y\|^2, 2x_1 x_2 + 2y_1 y_2) \in \mathcal{K}^n$, which implies

$$x^2 + y^2 \notin \text{int}(\mathcal{K}^n) \iff \|x\|^2 + \|y\|^2 = 2\|x_1 x_2 + y_1 y_2\|. \tag{3.12}$$

The spectral values of $x^2 + y^2$ are as below:

$$\begin{aligned} \lambda_1 &:= \|x\|^2 + \|y\|^2 - 2\|x_1x_2 + y_1y_2\|, \\ \lambda_2 &:= \|x\|^2 + \|y\|^2 + 2\|x_1x_2 + y_1y_2\|. \end{aligned} \tag{3.13}$$

For convenience, we introduce some notational conventions that will be used throughout the remainder of this section. For any vectors $x = (x_1, x_2) \in \mathbb{R} \times \mathbb{R}^{n-1}$ and $y = (y_1, y_2) \in \mathbb{R} \times \mathbb{R}^{n-1}$, we define the mappings $w, z : \mathbb{R}^n \times \mathbb{R}^n \to \mathbb{R}^n$ as follows:

$$\begin{aligned} w &= (w_1, w_2) = (w_1(x, y), w_2(x, y)) = w(x, y) := x^2 + y^2, \\ z &= (z_1, z_2) = (z_1(x, y), z_2(x, y)) = z(x, y) := (x^2 + y^2)^{1/2}. \end{aligned} \tag{3.14}$$

Clearly, $w \in \mathcal{K}^n$ with $w_1 = \|x\|^2 + \|y\|^2$ and $w_2 = 2(x_1x_2 + y_1y_2)$. By denoting

$$\bar{w}_2 = \begin{cases} \dfrac{w_2}{\|w_2\|} = \dfrac{x_1x_2 + y_1y_2}{\|x_1x_2 + y_1y_2\|} & \text{if } w_2 \neq 0 \\ \text{any vector in } \mathbb{R}^{n-1} \text{ satisfying} \|\bar{w}_2\| = 1 & \text{if } w_2 = 0 \end{cases}$$

and using Lemma 3.1 (b) and (c), we can express z as

$$z = \left(\frac{\sqrt{\lambda_2(w)} + \sqrt{\lambda_1(w)}}{2}, \frac{\sqrt{\lambda_2(w)} - \sqrt{\lambda_1(w)}}{2} \bar{w}_2 \right) \in \mathcal{K}^n. \tag{3.15}$$

We now present several key technical lemmas essential for the analysis that follows. The first lemma characterizes specific properties of vectors x and y when the sum $x^2 + y^2$ lies on the boundary of $\mathcal{K}^n$. The second establishes an upper bound for two squared terms in terms of a quantity that reflects the proximity of $x^2 + y^2$ to the boundary of $\mathcal{K}^n$. The remaining lemmas provide additional useful relationships. Collectively, these results form the foundation for the subsequent analysis in this chapter.

Lemma 3.2 *For any $x = (x_1, x_2), y = (y_1, y_2) \in \mathbb{R} \times \mathbb{R}^{n-1}$ with $x^2 + y^2 \notin \text{int}(\mathcal{K}^n)$, we have*

$$\begin{aligned} x_1^2 &= \|x_2\|^2, \\ y_1^2 &= \|y_2\|^2, \\ x_1y_1 &= x_2^\top y_2, \\ x_1y_2 &= y_1x_2. \end{aligned}$$

Proof By (3.12), $\|x\|^2 + \|y\|^2 = 2\|x_1x_2 + y_1y_2\|$. Thus $\left(\|x\|^2 + \|y\|^2\right)^2 = 4\|x_1x_2 + y_1y_2\|^2$, so that

$$\|x\|^4 + 2\|x\|^2\|y\|^2 + \|y\|^4 = 4(x_1x_2 + y_1y_2)^\top (x_1x_2 + y_1y_2).$$

Notice that $\|x\|^2 = x_1^2 + \|x_2\|^2$ and $\|y\|^2 = y_1^2 + \|y_2\|^2$. Thus,

$$\left(x_1^2 + \|x_2\|^2\right)^2 + 2\|x\|^2\|y\|^2 + \left(y_1^2 + \|y_2\|^2\right)^2 = 4x_1^2\|x_2\|^2 + 8x_1y_1x_2^{\mathsf{T}}y_2 + 4y_1^2\|y_2\|^2.$$

Simplifying the above expression yields

$$\left(x_1^2 - \|x_2\|^2\right)^2 + \left(y_1^2 - \|y_2\|^2\right)^2 + \left(2\|x\|^2\|y\|^2 - 8x_1y_1x_2^{\mathsf{T}}y_2\right) = 0.$$

The first two terms are nonnegative. The third term is also nonnegative because

$$\begin{aligned}\|x\|^2\|y\|^2 &= \left(x_1^2 + \|x_2\|^2\right)\left(y_1^2 + \|y_2\|^2\right) \\ &\geq \left(2|x_1|\|x_2\|\right)\left(2|y_1|\|y_2\|\right) \\ &= 4|x_1||y_1|\|x_2\|\|y_2\| \\ &\geq 4x_1y_1x_2^{\mathsf{T}}y_2.\end{aligned}$$

Hence

$$x_1^2 = \|x_2\|^2, \quad y_1^2 = \|y_2\|^2, \quad 2\|x\|^2\|y\|^2 - 8x_1y_1x_2^{\mathsf{T}}y_2 = 0.$$

Substituting $x_1^2 = \|x_2\|^2$ and $y_1^2 = \|y_2\|^2$ into the last equation, the resulting three equations imply $x_1y_1 = x_2^{\mathsf{T}}y_2$.

It remains to prove that $x_1y_2 = y_1x_2$. If $x_1 = 0$, then $\|x_2\| = |x_1| = 0$ so this relation is true. Symmetrically, if $y_1 = 0$, then this relation is also true. Suppose that $x_1 \neq 0$ and $y_1 \neq 0$. Then, $x_2 \neq 0$, $y_2 \neq 0$, and

$$x_1y_1 = x_2^{\mathsf{T}}y_2 = \|x_2\|\|y_2\|\cos\theta = |x_1||y_1|\cos\theta,$$

where θ is the angle between x_2 and y_2. Hence, $\cos\theta \in \{-1, 1\}$, i.e., $y_2 = \alpha x_2$ for some $\alpha \neq 0$. This yields

$$x_1y_1 = x_2^{\mathsf{T}}y_2 = \alpha\|x_2\|^2 = \alpha x_1^2,$$

so that $y_1/x_1 = \alpha$. Thus, $y_2 = x_2y_1/x_1$. □

Lemma 3.3 *For any* $x = (x_1, x_2)$, $y = (y_1, y_2) \in \mathbb{R} \times \mathbb{R}^{n-1}$ *with* $x_1x_2 + y_1y_2 \neq 0$, *we have*

$$\begin{aligned}\left(x_1 - \frac{(x_1x_2 + y_1y_2)^{\mathsf{T}}x_2}{\|x_1x_2 + y_1y_2\|}\right)^2 &\leq \left\|x_2 - x_1\frac{x_1x_2 + y_1y_2}{\|x_1x_2 + y_1y_2\|}\right\|^2 \\ &\leq \|x\|^2 + \|y\|^2 - 2\|x_1x_2 + y_1y_2\|.\end{aligned}$$

In other words,

$$\left(x_1 + (-1)^i x_2^{\mathsf{T}}\bar{w}_2\right)^2 \leq \left\|x_2 + (-1)^i x_1\bar{w}_2\right\|^2 \leq \lambda_i(w) \quad \text{for } i = 1, 2.$$

Proof The first inequality can be seen by expanding the square on both sides and using the Cauchy–Schwarz inequality. It remains to prove the second inequality. Let us multiply both sides of this inequality by

$$\|x_1x_2 + y_1y_2\|^2 = x_1^2\|x_2\|^2 + 2x_1y_1x_2^{\mathsf{T}}y_2 + y_1^2\|y_2\|^2$$

and let L and R denote, respectively, the left-hand side and the right-hand side. Since $x_1x_2 + y_1y_2 \neq 0$, the second inequality is equivalent to $R - L \geq 0$. We have

$$\begin{aligned}
L &= \left(\|x_2\|^2 - 2x_1\frac{(x_1x_2 + y_1y_2)^{\mathsf{T}}x_2}{\|x_1x_2 + y_1y_2\|} + x_1^2\right)\|x_1x_2 + y_1y_2\|^2 \\
&= \|x_2\|^2\left(x_1^2\|x_2\|^2 + 2x_1y_1x_2^{\mathsf{T}}y_2 + y_1^2\|y_2\|^2\right) \\
&\quad -2x_1\left(x_1\|x_2\|^2 + y_1x_2^{\mathsf{T}}y_2\right)\|x_1x_2 + y_1y_2\| \\
&\quad +x_1^2\left(x_1^2\|x_2\|^2 + 2x_1y_1x_2^{\mathsf{T}}y_2 + y_1^2\|y_2\|^2\right) \\
&= x_1^2\|x_2\|^4 + 2x_1y_1x_2^{\mathsf{T}}y_2\|x_2\|^2 + y_1^2\|x_2\|^2\|y_2\|^2 \\
&\quad -2x_1^2\|x_2\|^2\|x_1x_2 + y_1y_2\| - 2x_1y_1x_2^{\mathsf{T}}y_2\|x_1x_2 + y_1y_2\| \\
&\quad +x_1^4\|x_2\|^2 + 2x_1^3y_1x_2^{\mathsf{T}}y_2 + x_1^2y_1^2\|y_2\|^2,
\end{aligned}$$

and

$$\begin{aligned}
R &= \left(\|x\|^2 + \|y\|^2 - 2\|x_1x_2 + y_1y_2\|\right)\|x_1x_2 + y_1y_2\|^2 \\
&= \left(x_1^2 + \|x_2\|^2 - 2\|x_1x_2 + y_1y_2\|\right)\|x_1x_2 + y_1y_2\|^2 + \|y\|^2\|x_1x_2 + y_1y_2\|^2 \\
&= \left(x_1^2 + \|x_2\|^2 - 2\|x_1x_2 + y_1y_2\|\right)\left(x_1^2\|x_2\|^2 + 2x_1y_1x_2^{\mathsf{T}}y_2 + y_1^2\|y_2\|^2\right) \\
&\quad +\|y\|^2\|x_1x_2 + y_1y_2\|^2 \\
&= x_1^4\|x_2\|^2 + 2x_1^3y_1x_2^{\mathsf{T}}y_2 + x_1^2y_1^2\|y_2\|^2 + x_1^2\|x_2\|^4 + 2x_1y_1x_2^{\mathsf{T}}y_2\|x_2\|^2 \\
&\quad +y_1^2\|x_2\|^2\|y_2\|^2 - 2x_1^2\|x_2\|^2\|x_1x_2 + y_1y_2\| - 4x_1y_1x_2^{\mathsf{T}}y_2\|x_1x_2 + y_1y_2\| \\
&\quad -2y_1^2\|y_2\|^2\|x_1x_2 + y_1y_2\| + \|y\|^2\|x_1x_2 + y_1y_2\|^2.
\end{aligned}$$

Thus, taking the difference and using the Cauchy–Schwarz inequality yields

$$\begin{aligned}
&R - L \\
&= \|y\|^2\|x_1x_2 + y_1y_2\|^2 - 2x_1y_1x_2^{\mathsf{T}}y_2\|x_1x_2 + y_1y_2\| - 2y_1^2\|y_2\|^2\|x_1x_2 + y_1y_2\| \\
&= y_1^2\|x_1x_2 + y_1y_2\|^2 + \|y_2\|^2\|x_1x_2 + y_1y_2\|^2 - 2y_1y_2^{\mathsf{T}}(x_1x_2 + y_1y_2)\|x_1x_2 + y_1y_2\|
\end{aligned}$$

$$
\begin{aligned}
&\geq y_1^2\|x_1x_2 + y_1y_2\|^2 + \|y_2\|^2\|x_1x_2 + y_1y_2\|^2 - 2|y_1|\|y_2\|\ \|x_1x_2 + y_1y_2\|^2 \\
&= (|y_1| - \|y_2\|)^2\ \|x_1x_2 + y_1y_2\|^2 \\
&\geq 0.
\end{aligned}
$$

Then, the proof is complete. □

Lemma 3.4 *There exists a scalar constant $C > 0$ such that*

$$
\|L_x L^{-1}_{(x^2+y^2)^{1/2}}\|_F \leq C, \quad \|L_y L^{-1}_{(x^2+y^2)^{1/2}}\|_F \leq C
$$

for all $(x, y) \neq (0, 0)$ satisfying $x^2 + y^2 \in \operatorname{int}(\mathcal{K}^n)$. ($\|A\|_F$ denotes the Frobenius norm of $A \in \mathbb{R}^{n\times n}$.)

Proof Consider any $(x, y) \neq (0, 0)$ satisfying $x^2 + y^2 \in \operatorname{int}(\mathcal{K}^n)$. Let λ_1, λ_2 be the spectral values of $x^2 + y^2$ and let $z := (x^2 + y^2)^{1/2}$. Then, z is given by (3.15), i.e.,

$$
z_1 = \frac{\sqrt{\lambda_1} + \sqrt{\lambda_2}}{2}, \qquad z_2 = \frac{\sqrt{\lambda_2} - \sqrt{\lambda_1}}{2} w_2,
$$

with λ_1, λ_2 given by (3.13), and $w_2 := \dfrac{x_1x_2 + y_1y_2}{\|x_1x_2 + y_1y_2\|}$ if $x_1x_2 + y_1y_2 \neq 0$; otherwise w_2 is any vector satisfying $\|w_2\| = 1$. Using Lemma 3.1(c), we have that

$$
\begin{aligned}
&L_x L_z^{-1} \\
&= \frac{1}{\det(z)} \begin{bmatrix} x_1z_1 - x_2^{\mathsf T}z_2 & -x_1z_2^T + \frac{\det(z)}{z_1}x_2^{\mathsf T} + \frac{x_2^{\mathsf T}z_2}{z_1}z_2^{\mathsf T} \\ x_2z_1 - x_1z_2 & -x_2z_2^{\mathsf T} + \frac{x_1\det(z)}{z_1}I + \frac{x_1}{z_1}z_2z_2^{\mathsf T} \end{bmatrix} \qquad (3.16) \\
&= \frac{1}{\sqrt{\lambda_1}\sqrt{\lambda_2}} \begin{bmatrix} \frac{\sqrt{\lambda_1}+\sqrt{\lambda_2}}{2}x_1 + \frac{\sqrt{\lambda_1}-\sqrt{\lambda_2}}{2}x_2^{\mathsf T}w_2 & \frac{\sqrt{\lambda_1}-\sqrt{\lambda_2}}{2}x_1w_2^{\mathsf T} + \frac{2\sqrt{\lambda_1}\sqrt{\lambda_2}}{\sqrt{\lambda_1}+\sqrt{\lambda_2}}x_2^{\mathsf T} \\ & \quad + \frac{(\sqrt{\lambda_1}-\sqrt{\lambda_2})^2}{2(\sqrt{\lambda_1}+\sqrt{\lambda_2})}x_2^{\mathsf T}w_2w_2^{\mathsf T} \\ \frac{\sqrt{\lambda_1}+\sqrt{\lambda_2}}{2}x_2 + \frac{\sqrt{\lambda_1}-\sqrt{\lambda_2}}{2}x_1w_2 & \frac{\sqrt{\lambda_1}-\sqrt{\lambda_2}}{2}x_2w_2^T + \frac{2\sqrt{\lambda_1}\sqrt{\lambda_2}}{\sqrt{\lambda_1}+\sqrt{\lambda_2}}x_1I \\ & \quad + \frac{(\sqrt{\lambda_1}-\sqrt{\lambda_2})^2}{2(\sqrt{\lambda_1}+\sqrt{\lambda_2})}x_1w_2w_2^T \end{bmatrix} \\
&= \begin{bmatrix} \frac{(x_1 + x_2^{\mathsf T}w_2)}{2\sqrt{\lambda_2}} + \frac{(x_1 - x_2^{\mathsf T}w_2)}{2\sqrt{\lambda_1}} & \left(\frac{x_1w_2^{\mathsf T}}{2\sqrt{\lambda_2}} - \frac{x_1w_2^{\mathsf T}}{2\sqrt{\lambda_1}}\right) + \frac{2x_2^{\mathsf T}}{\sqrt{\lambda_1} + \sqrt{\lambda_2}} \\ & + \frac{\frac{\sqrt{\lambda_2}}{\sqrt{\lambda_1}} - 2 + \frac{\sqrt{\lambda_1}}{\sqrt{\lambda_2}}}{2(\sqrt{\lambda_1}+\sqrt{\lambda_2})}x_2^{\mathsf T}w_2w_2^{\mathsf T} \\ \frac{(x_2 + x_1w_2)}{2\sqrt{\lambda_2}} + \frac{(x_2 - x_1w_2)}{2\sqrt{\lambda_1}} & \left(\frac{x_2w_2^{\mathsf T}}{2\sqrt{\lambda_2}} - \frac{x_2w_2^{\mathsf T}}{2\sqrt{\lambda_1}}\right) + \frac{2x_1I}{\sqrt{\lambda_1} + \sqrt{\lambda_2}} \\ & + \frac{\frac{\sqrt{\lambda_2}}{\sqrt{\lambda_1}} - 2 + \frac{\sqrt{\lambda_1}}{\sqrt{\lambda_2}}}{2(\sqrt{\lambda_1}+\sqrt{\lambda_2})}x_1w_2w_2^{\mathsf T} \end{bmatrix}.
\end{aligned}
$$

Since $\lambda_2 \geq \|x\|^2$, we see that $\sqrt{\lambda_2} \geq |x_1|$ and $\sqrt{\lambda_2} \geq \|x_2\|$. Also, $\|w_2\| = 1$. Thus, terms that involve dividing x_1 or x_2 or x_1w_2 or $x_2^{\mathsf T}w_2$ or $x_1w_2w_2^{\mathsf T}$ or $x_2^{\mathsf T}w_2w_2^{\mathsf T}$ by $\sqrt{\lambda_2}$ or $\sqrt{\lambda_1} + \sqrt{\lambda_2}$ are uniformly bounded. Also, $\sqrt{\lambda_1}/\sqrt{\lambda_2} \leq 1$. Thus

$$
\begin{aligned}
& L_x L_z^{-1} \\
&= \begin{bmatrix} O(1) + \dfrac{(x_1 - x_2^\mathsf{T} w_2)}{2\sqrt{\lambda_1}} & O(1) - \dfrac{x_1 w_2^\mathsf{T}}{2\sqrt{\lambda_1}} + \dfrac{\frac{\sqrt{\lambda_2}}{\sqrt{\lambda_1}}}{2(\sqrt{\lambda_1} + \sqrt{\lambda_2})} x_2^\mathsf{T} w_2 w_2^\mathsf{T} \\ O(1) + \dfrac{(x_2 - x_1 w_2)}{2\sqrt{\lambda_1}} & O(1) - \dfrac{x_2 w_2^\mathsf{T}}{2\sqrt{\lambda_1}} + \dfrac{\frac{\sqrt{\lambda_2}}{\sqrt{\lambda_1}}}{2(\sqrt{\lambda_1} + \sqrt{\lambda_2})} x_1 w_2 w_2^\mathsf{T} \end{bmatrix} \\
&= \begin{bmatrix} O(1) + \dfrac{(x_1 - x_2^\mathsf{T} w_2)}{2\sqrt{\lambda_1}} & O(1) - \dfrac{x_1 w_2^\mathsf{T}}{2(\sqrt{\lambda_1} + \sqrt{\lambda_2})} - \dfrac{\sqrt{\lambda_2}(x_1 - x_2^\mathsf{T} w_2)}{2(\sqrt{\lambda_1} + \sqrt{\lambda_2})\sqrt{\lambda_1}} w_2^\mathsf{T} \\ O(1) + \dfrac{(x_2 - x_1 w_2)}{2\sqrt{\lambda_1}} & O(1) - \dfrac{x_2 w_2^\mathsf{T}}{2(\sqrt{\lambda_1} + \sqrt{\lambda_2})} - \dfrac{\sqrt{\lambda_2}(x_2 - x_1 w_2)}{2(\sqrt{\lambda_1} + \sqrt{\lambda_2})\sqrt{\lambda_1}} w_2^\mathsf{T} \end{bmatrix},
\end{aligned}
$$

where $O(1)$ denote terms that are uniformly bounded, with bound independent of (x, y). By Lemma 3.3, if $x_1 x_2 + y_1 y_2 \neq 0$, then $|x_1 - x_2^\mathsf{T} w_2| \le \|x_2 - x_1 w_2\| \le \sqrt{\lambda_1}$. If $x_1 x_2 + y_1 y_2 = 0$, then $\lambda_1 = \|x\|^2 + \|y\|^2$ so that, by choosing w_2 to further satisfy $x_2^\mathsf{T} w_2 = 0$ (in addition to $\|w_2\| = 1$), we obtain

$$|x_1 - x_2^\mathsf{T} w_2| \le \|x_2 - x_1 w_2\| = \|x\| \le \sqrt{\lambda_1}.$$

Thus, all terms in $L_x L_z^{-1}$ are uniformly bounded. □

Lemma 3.5 *(a) For any $x \in \mathcal{K}^n$, $y \in \mathbb{R}^n$ with $x^2 - y^2 \in \mathcal{K}^n$, we have $x - y \in \mathcal{K}^n$.*
(b) For any $x, y \in \mathbb{R}^n$ and $w \in \mathcal{K}^n$ such that $w^2 - x^2 - y^2 \in \mathcal{K}^n$, we have $L_w^2 \succeq L_x^2 + L_y^2$.

Proof Please see [12, Proposition 3.4]. □

Lemma 3.6 *(a) For any $x \in \mathbb{R}^n$, $\langle x, (x)_- \rangle = \|(x)_-\|^2$ and $\langle x, (x)_+ \rangle = \|(x)_+\|^2$.*
(b) For any $x \in \mathbb{R}^n$ and $y \in \mathbb{R}^n$, we have

$$x \in \mathcal{K}^n \iff \langle x, y \rangle \ge 0 \quad \forall y \in \mathcal{K}^n. \tag{3.17}$$

(c) Let $x = (x_1, x_2) \in \mathbb{R} \times \mathbb{R}^{n-1}$ and $y = (y_1, y_2) \in \mathbb{R} \times \mathbb{R}^{n-1}$. Then, we have

$$\langle x, y \rangle \le \sqrt{2}\, \|(x \circ y)_+\|. \tag{3.18}$$

Proof (a) By definition of trace, we know that $tr(x \circ y) = 2\langle x, y \rangle$. Thus,

$$
\begin{aligned}
\langle x, (x)_- \rangle &= \frac{1}{2}\mathrm{tr}\big(x \circ (x)_-\big) \\
&= \frac{1}{2}\mathrm{tr}\big([(x)_+ + (x)_-] \circ (x)_-\big) \\
&= \frac{1}{2}\mathrm{tr}\big((x)_-^2\big) \\
&= \|(x)_-\|^2,
\end{aligned}
$$

where the last inequality is from the definition of trace again. Similar arguments applied for $\langle x, (x)_+ \rangle = \|(x)_+\|^2$.

(b) Since $\mathcal{K}^n$ is self-dual, that is $\mathcal{K}^n = (\mathcal{K}^n)^*$. Hence, the desired result follows.

(c) First, we observe the fact that

$$\begin{aligned} x \in \mathcal{K}^n &\Longleftrightarrow (x)_+ = x, \\ x \in -\mathcal{K}^n &\Longleftrightarrow (x)_+ = 0, \\ x \notin \mathcal{K}^n \cup -\mathcal{K}^n &\Longleftrightarrow (x)_+ = \lambda_2 u^{(2)}, \end{aligned}$$

where λ_2 is the bigger spectral value of x with the corresponding spectral vector $u^{(2)}$. Hence, we have three cases.

Case (1): If $x \circ y \in \mathcal{K}^n$, then $(x \circ y)_+ = x \circ y$. By definition of Jordan product of x and y as (1.2), i.e., $x \circ y = (\langle x, y \rangle\ ,\ x_1 y_2 + y_1 x_2)$. It is clear that $\|(x \circ y)_+\| \geq \langle x, y \rangle$ and hence (3.18) holds.

Case (2): If $x \circ y \in -\mathcal{K}^n$, then $(x \circ y)_+ = 0$. Since $x \circ y \in -\mathcal{K}^n$, by definition of Jordan product again, we have $\langle x, y \rangle \leq 0$. Hence, it is true that $\sqrt{2}\|(x \circ y)_+\| \geq \langle x, y \rangle$.

Case (3): If $x \circ y \notin \mathcal{K}^n \cup -\mathcal{K}^n$, then $(x \circ y)_+ = \lambda_2 u^{(2)}$ where

$$\begin{aligned} \lambda_2 &= \langle x, y \rangle + \|x_1 y_2 + y_1 x_2\|, \\ u^{(2)} &= \frac{1}{2}\left(1\ ,\ \frac{x_1 y_2 + y_1 x_2}{\|x_1 y_2 + y_1 x_2\|}\right). \end{aligned}$$

If $\langle x, y \rangle \leq 0$, then (3.18) is trivial. Thus, we can assume $\langle x, y \rangle > 0$. In fact, the desired inequality (3.18) follows from the below.

$$\begin{aligned} \|(x \circ y)_+\|^2 &= \frac{1}{2}\lambda_2^2 \\ &= \frac{1}{2}\left(\langle x, y \rangle^2 + 2\langle x, y \rangle \cdot \|x_1 y_2 + y_1 x_2\| + \|x_1 y_2 + y_1 x_2\|^2\right) \\ &\geq \frac{1}{2}\langle x, y \rangle^2, \end{aligned}$$

where the first equality is by $\|u^{(2)}\| = 1/\sqrt{2}$. □

Lemma 3.7 *Let ϕ_{FB} and ψ_{FB} be given by* (3.10) *and* (3.11), *respectively. For any $(x, y) \in \mathbb{R}^n \times \mathbb{R}^n$, we have*

$$4\psi_{\text{FB}}(x, y) \geq 2\left\|\phi_{\text{FB}}(x, y)_+\right\|^2 \geq \left\|(-x)_+\right\|^2 + \left\|(-y)_+\right\|^2.$$

Proof The first inequality follows from Lemma 1.1(a). It remains to show the second inequality. By Lemma 1.1(d), $(x^2+y^2)^{1/2}-x \in \mathcal{K}^n$. Since $\mathcal{K}^n$ is self-dual, then Lemma 1.1(c) yields

$$\left\| \left((x^2+y^2)^{1/2}-x-y\right)_+ \right\|^2 \geq \left\| (-y)_+ \right\|^2.$$

By a symmetric argument,

$$\left\| \left((x^2+y^2)^{1/2}-x-y\right)_+ \right\|^2 \geq \left\| (-x)_+ \right\|^2.$$

Adding the above two inequalities yields the desired second inequality. □

Lemma 3.8 *Let ϕ_{FB} and ψ_{FB} be given by (3.10) and (3.11), respectively. For any $\{(x^k, y^k)\}_{k=1}^{\infty} \subseteq \mathbb{R}^n \times \mathbb{R}^n$, let $\lambda_1^k \leq \lambda_2^k$ and $\mu_1^k \leq \mu_2^k$ denote the spectral values of x^k and y^k, respectively. Then the following results hold.*

(a) If $\lambda_1^k \to -\infty$ or $\mu_1^k \to -\infty$, then $\psi_{\text{FB}}(x^k, y^k) \to \infty$.
(b) Suppose that $\{\lambda_1^k\}$ and $\{\mu_1^k\}$ are bounded below. If $\lambda_2^k \to \infty$ or $\mu_2^k \to \infty$, then $\langle x, x^k \rangle + \langle y, y^k \rangle \to \infty$ for any $x, y \in \text{int}(\mathcal{K}^n)$.

Proof (a) This follows from Lemma 3.7 and the fact that

$$2\left\|(-x^k)_+\right\|^2 = \sum_{i=1}^{2} \left(\max\{0, -\lambda_i^k\}\right)^2$$

and similarly for $\|(-y^k)_+\|^2$; see [12, Property 2.2 and Proposition 3.3].

(b) Fix any $x = (x_1, x_2), y = (y_1, y_2) \in \mathbb{R} \times \mathbb{R}^{n-1}$ with $\|x_2\| < x_1$, $\|y_2\| < y_1$. Using the spectral decomposition

$$x^k = \left(\frac{\lambda_1^k + \lambda_2^k}{2}, \frac{\lambda_2^k - \lambda_1^k}{2} w_2^k\right) \quad \text{with } \|w_2^k\| = 1,$$

we have

$$\langle x, x^k \rangle = \left(\frac{\lambda_1^k + \lambda_2^k}{2}\right) x_1 + \left(\frac{\lambda_2^k - \lambda_1^k}{2}\right) x_2^{\mathrm{T}} w_2^k = \frac{\lambda_1^k}{2}(x_1 - x_2^{\mathrm{T}} w_2^k) + \frac{\lambda_2^k}{2}(x_1 + x_2^{\mathrm{T}} w_2^k). \tag{3.19}$$

Since $\|w_2^k\| = 1$, we have $x_1 - x_2^{\mathrm{T}} w_2^k \geq x_1 - \|x_2\| > 0$ and $x_1 + x_2^{\mathrm{T}} w_2^k \geq x_1 - \|x_2\| > 0$. Since $\{\lambda_1^k\}$ is bounded below, the first term on the right-hand side of (3.19) is bounded below. If $\{\lambda_2^k\} \to \infty$, then the second term on the right-hand side of (3.19) tends to infinity. Hence, $\langle x, x^k \rangle \to \infty$. A similar argument shows that $\langle y, y^k \rangle$ is bounded below. Thus, $\langle x, x^k \rangle + \langle y, y^k \rangle \to \infty$. If $\{\mu_2^k\} \to \infty$, the argument is symmetric to the one above. □

3.1.1 The Functions ϕ_{FB} and ψ_{FB} in SOC Setting

A. The functions ϕ_{FB} and ψ_{FB}

Proposition 3.2 *Let $\phi_{\text{FB}} : \mathbb{R}^n \times \mathbb{R}^n \to \mathbb{R}^n$ be given by* (3.10). *Then, the function ϕ_{FB} is a C-function associated with SOC, that is, it satisfies* (3.5). *In other words, it holds*

$$\begin{aligned}\phi_{\text{FB}}(x, y) = 0 &\Longleftrightarrow x, y \in \mathcal{K}^n, \ x \circ y = 0, \\ &\Longleftrightarrow x, y \in \mathcal{K}^n, \ \langle x, y \rangle = 0.\end{aligned}$$

Proof Please see [12, Proposition 2.1]. □

Proposition 3.3 *Let $\phi_{\text{FB}} : \mathbb{R}^n \times \mathbb{R}^n \to \mathbb{R}^n$ be given by* (3.10). *Then, the function ϕ_{FB} is strongly semismooth.*

Proof The proof relies on the relationship between the singular value decomposition (SVD) of a nonsymmetric matrix and the spectral decomposition of a symmetric matrix; see [24, Corollary 3.3] for details. □

Proposition 3.4 *Let ϕ_{FB} be given by* (3.10). *Then, the function ψ_{FB} given by* (3.11) *has the following properties.*

(a) $\psi_{\text{FB}} : \mathbb{R}^n \times \mathbb{R}^n \to \mathbb{R}_+$ *is a C-function associated with SOC, which satisfies* (3.6).

(b) ψ_{FB} *is differentiable at every* $(x, y) \in \mathbb{R}^n \times \mathbb{R}^n$. *Moreover,* $\nabla_x \psi_{\text{FB}}(0, 0) = \nabla_y \psi_{\text{FB}}(0, 0) = 0$. *If* $(x, y) \neq (0, 0)$ *and* $x^2 + y^2 \in \text{int}(\mathcal{K}^n)$, *then*

$$\begin{aligned}\nabla_x \psi_{\text{FB}}(x, y) &= \left(L_x L^{-1}_{(x^2+y^2)^{1/2}} - I\right) \phi_{\text{FB}}(x, y), \\ \nabla_y \psi_{\text{FB}}(x, y) &= \left(L_y L^{-1}_{(x^2+y^2)^{1/2}} - I\right) \phi_{\text{FB}}(x, y).\end{aligned}$$

If $(x, y) \neq (0, 0)$ *and* $x^2 + y^2 \notin \text{int}(\mathcal{K}^n)$, *then* $x_1^2 + y_1^2 \neq 0$ *and*

$$\nabla_x \psi_{\text{FB}}(x, y) = \left(\frac{x_1}{\sqrt{x_1^2 + y_1^2}} - 1\right) \phi_{\text{FB}}(x, y), \tag{3.20}$$

$$\nabla_y \psi_{\text{FB}}(x, y) = \left(\frac{y_1}{\sqrt{x_1^2 + y_1^2}} - 1\right) \phi_{\text{FB}}(x, y). \tag{3.21}$$

Proof (a) This result follows from Proposition 3.2 directly.

(b) In light of the behavior of $x^2 + y^2$, we proceed with the analysis by considering three distinct cases.

Case (1): $x = y = 0$.
For any $h, k \in \mathbb{R}^n$, let $\mu_1 \le \mu_2$ be the spectral values and let $v^{(1)}$, $v^{(2)}$ be the corresponding spectral vectors of $h^2 + k^2$. Then, by Lemma 3.1(b),

$$\begin{aligned}\left\|(h^2+k^2)^{1/2}-h-k\right\| &= \|\sqrt{\mu_1}v^{(1)}+\sqrt{\mu_2}v^{(2)}-h-k\| \\ &\le \sqrt{\mu_1}\,\|v^{(1)}\|+\sqrt{\mu_2}\,\|v^{(2)}\|+\|h\|+\|k\| \\ &= \frac{1}{\sqrt{2}}(\sqrt{\mu_1}+\sqrt{\mu_2})+\|h\|+\|k\|.\end{aligned}$$

Also

$$\begin{aligned}\mu_1 \;\le\; \mu_2 &= \|h\|^2+\|k\|^2+2\|h_1h_2+k_1k_2\| \\ &\le \|h\|^2+\|k\|^2+2|h_1|\|h_2\|+2|k_1|\|k_2\| \\ &\le 2\|h\|^2+2\|k\|^2.\end{aligned}$$

Combining the above two inequalities yields

$$\begin{aligned}\psi_{\text{FB}}(h,k)-\psi_{\text{FB}}(0,0) &= \left\|(h^2+k^2)^{1/2}-h-k\right\|^2 \\ &\le \left(\frac{1}{\sqrt{2}}(\sqrt{\mu_1}+\sqrt{\mu_2})+\|h\|+\|k\|\right)^2 \\ &\le \left(\frac{2}{\sqrt{2}}\sqrt{2\|h\|^2+2\|k\|^2}+\|h\|+\|k\|\right)^2 \\ &= O(\|h\|^2+\|k\|^2).\end{aligned}$$

This shows that ψ_{FB} is differentiable at $(0, 0)$ with

$$\nabla_x \psi_{\text{FB}}(0,0) = \nabla_y \psi_{\text{FB}}(0,0) = 0.$$

Case (2): $(x, y) \ne (0, 0)$ and $x^2 + y^2 \in \operatorname{int}(\mathcal{K}^n)$.
Since $x^2 + y^2 \in \operatorname{int}(\mathcal{K}^n)$, Proposition 5.2 of [12] implies that ϕ_{FB} is continuously differentiable at (x, y). Since ψ_{FB} is the composition of ϕ with $x \mapsto \frac{1}{2}\|x\|^2$, then ψ_{FB} is continuously differentiable at (x, y). The expressions (3.4) for $\nabla_x \psi_{\text{FB}}(x, y)$ and $\nabla_y \psi_{\text{FB}}(x, y)$ follow from the chain rule for differentiation and the expression for the Jacobian of ϕ_{FB} given in [12, Proposition 5.2] (also see [12, Corollary 5.4]).

Case (3): $(x, y) \ne (0, 0)$ and $x^2 + y^2 \notin \operatorname{int}(\mathcal{K}^n)$.
By (3.12), $\|x\|^2 + \|y\|^2 = 2\|x_1x_2 + y_1y_2\|$. Since $(x, y) \ne (0, 0)$, this also implies $x_1x_2 + y_1y_2 \ne 0$, so Lemmas 3.2 and 3.3 are applicable. By (3.15),

$$(x^2+y^2)^{1/2} = \left(\frac{\sqrt{\lambda_1}+\sqrt{\lambda_2}}{2},\; \frac{\sqrt{\lambda_2}-\sqrt{\lambda_1}}{2}w_2\right),$$

where λ_1, λ_2 are given by (3.13) and $w_2 := \dfrac{x_1x_2 + y_1y_2}{\|x_1x_2 + y_1y_2\|}$. Thus, $\lambda_1 = 0$ and $\lambda_2 > 0$. Since $x_1x_2 + y_1y_2 \neq 0$, we have $x_1'x_2' + y_1'y_2' \neq 0$ for all $(x', y') \in \mathbb{R}^n \times \mathbb{R}^n$ sufficiently near to (x, y). Moreover,

$$
\begin{aligned}
2\psi_{\text{FB}}(x', y') &= \left\|(x'^2 + y'^2)^{1/2} - x' - y'\right\|^2 \\
&= \left\|\left(x'^2 + y'^2\right)^{1/2}\right\|^2 + \|x' + y'\|^2 - 2\left\langle\left(x'^2 + y'^2\right)^{1/2}, x' + y'\right\rangle \\
&= \|x'\|^2 + \|y'\|^2 + \|x' + y'\|^2 - 2\left\langle\left(x'^2 + y'^2\right)^{1/2}, x' + y'\right\rangle,
\end{aligned}
$$

where the third equality uses the observation that $\|z\|^2 = \langle z^2, e\rangle$ for any $z \in \mathbb{R}^n$. Since $\|x'\|^2 + \|y'\|^2 + \|x' + y'\|^2$ is clearly differentiable in (x', y'), it suffices to show that

$$
\begin{aligned}
&2\left\langle\left(x'^2 + y'^2\right)^{1/2}, x' + y'\right\rangle \\
&= (\sqrt{\mu_2} + \sqrt{\mu_1})(x_1' + y_1') + (\sqrt{\mu_2} - \sqrt{\mu_1})\frac{(x_1'x_2' + y_1'y_2')^\top(x_2' + y_2')}{\|x_1'x_2' + y_1'y_2'\|} \\
&= \sqrt{\mu_2}\left(x_1' + y_1' + \frac{(x_1'x_2' + y_1'y_2')^\top(x_2' + y_2')}{\|x_1'x_2' + y_1'y_2'\|}\right) \\
&\quad + \sqrt{\mu_1}\left(x_1' + y_1' - \frac{(x_1'x_2' + y_1'y_2')^\top(x_2' + y_2')}{\|x_1'x_2' + y_1'y_2'\|}\right)
\end{aligned}
\tag{3.22}
$$

is differentiable at $(x', y') = (x, y)$, where μ_1, μ_2 are the spectral values of $x'^2 + y'^2$, i.e., $\mu_i = \|x'\|^2 + \|y'\|^2 + 2(-1)^i\|x_1'x_2' + y_1'y_2'\|$. Since $\lambda_2 > 0$, we see that the first term on the right-hand side of (3.22) is differentiable at $(x', y') = (x, y)$. We claim that the second term on the right-hand side of (3.22) is $o(\|h\| + \|k\|)$ with $h := x' - x, k := y' - y$, i.e., it is differentiable with zero gradient. To see this, notice that $x_1x_2 + y_1y_2 \neq 0$, so that $\mu_1 = \|x'\|^2 + \|y'\|^2 - 2\|x_1'x_2' + y_1'y_2'\|$, viewed as a function of (x', y'), is differentiable at $(x', y') = (x, y)$. Moreover, $\mu_1 = \lambda_1 = 0$ when $(x', y') = (x, y)$. Thus, first-order Taylor's expansion of μ_1 at (x, y) yields

$$
\mu_1 \;=\; O(\|x' - x\| + \|y' - y\|) \;=\; O(\|h\| + \|k\|).
$$

Also, since $x_1x_2 + y_1y_2 \neq 0$, by the product and quotient rules for differentiation, the function

$$
x_1' + y_1' - \frac{(x_1'x_2' + y_1'y_2')^\top(x_2' + y_2')}{\|x_1'x_2' + y_1'y_2'\|}
\tag{3.23}
$$

is differentiable at $(x', y') = (x, y)$. Moreover, the function (3.23) has value 0 at $(x', y') = (x, y)$. This is because

$$x_1 + y_1 - \frac{(x_1x_2 + y_1y_2)^{\mathsf{T}}(x_2 + y_2)}{\|x_1x_2 + y_1y_2\|} = x_1 - w_2^{\mathsf{T}}x_2 \; + \; y_1 - w_2^{\mathsf{T}}y_2 \; = \; 0 + 0,$$

where $w_2 := (x_1x_2 + y_1y_2)/\|x_1x_2 + y_1y_2\|$ and the last equality uses the fact that, by Lemma 3.3 and $\|x\|^2 + \|y\|^2 = 2\|x_1x_2 + y_1y_2\|$, we have $w_2^{\mathsf{T}}x_2 = x_1$, $w_2^{\mathsf{T}}y_2 = y_1$. (By symmetry, Lemma 3.3 still holds when x and y are switched.) Thus, the function (3.23) is $O(\|h\| + \|k\|)$ in magnitude. This together with $\mu_1 = O(\|h\| + \|k\|)$ shows that the second term on the right of (3.22) is $O((\|h\| + \|k\|)^{3/2}) = o(\|h\| + \|k\|)$.

Thus, we have shown that ψ_{FB} is differentiable at (x, y). Moreover, the preceding argument shows that $2\nabla\psi_{\text{FB}}(x, y)$ is the sum of the gradient of $\|x'\|^2 + \|y'\|^2 + \|x' + y'\|^2$ and the gradient of the first term on the right of (3.22), evaluated at $(x', y') = (x, y)$. The gradient of $\|x'\|^2 + \|y'\|^2 + \|x' + y'\|^2$ with respect to x', evaluated at $(x', y') = (x, y)$, is $4x + 2y$. Using the product and quotient rules for differentiation, the gradient of the first term on the right of (3.22) with respect to x_1', evaluated at $(x', y') = (x, y)$, works out to be

$$\begin{aligned} &\frac{x_1 + w_2^{\mathsf{T}}x_2}{\sqrt{\lambda_2}}\left(x_1 + y_1 + w_2^{\mathsf{T}}(x_2 + y_2)\right) \\ &+\sqrt{\lambda_2}\left(1 + \frac{x_2^{\mathsf{T}}(x_2 + y_2)}{\|x_1x_2 + y_1y_2\|} - \frac{w_2^{\mathsf{T}}(x_2 + y_2)}{\|x_1x_2 + y_1y_2\|}w_2^{\mathsf{T}}x_2\right) \\ =\; &\frac{2x_1(x_1 + y_1)}{\sqrt{x_1^2 + y_1^2}} + 2\sqrt{x_1^2 + y_1^2}, \end{aligned}$$

where $w_2 := (x_1x_2 + y_1y_2)/\|x_1x_2 + y_1y_2\|$ and the equality uses Lemma 3.2 and the fact that, by Lemma 3.3 and $\|x\|^2 + \|y\|^2 = 2\|x_1x_2 + y_1y_2\|$, we have $w_2^{\mathsf{T}}x_2 = x_1$, $w_2^{\mathsf{T}}y_2 = y_1$. Similarly, the gradient of the first term on the right of (3.22) with respect to x_2', evaluated at $(x', y') = (x, y)$, works out to be

$$\begin{aligned} &\frac{x_2 + w_2x_1}{\sqrt{\lambda_2}}\left(x_1 + y_1 + w_2^{\mathsf{T}}(x_2 + y_2)\right) \\ &+\sqrt{\lambda_2}\left(\frac{2x_1x_2 + (x_1 + y_1)y_2}{\|x_1x_2 + y_1y_2\|} - \frac{w_2^{\mathsf{T}}(x_2 + y_2)}{\|x_1x_2 + y_1y_2\|}w_2x_1\right) \\ =\; &2\,\frac{2x_1x_2 + (x_1 + y_1)y_2}{\sqrt{x_1^2 + y_1^2}}. \end{aligned}$$

In particular, the equality uses the fact that, by Lemma 3.2, we have $x_1y_2 = y_1x_2$ and $\|x_1x_2 + y_1y_2\| = x_1^2 + y_1^2$, so that $w_2x_1 = x_2$ and $\lambda_2 = 4(x_1^2 + y_1^2)$. Thus, combining the preceding gradient expressions yields

$$2\nabla_x\psi_{\text{FB}}(x, y) = 4x + 2y - \begin{bmatrix} 2\sqrt{x_1^2 + y_1^2} \\ 0 \end{bmatrix} - \frac{2}{\sqrt{x_1^2 + y_1^2}}\begin{bmatrix} x_1(x_1 + y_1) \\ 2x_1x_2 + (x_1 + y_1)y_2 \end{bmatrix}.$$

Using $\|x_1x_2 + y_1y_2\| = x_1^2 + y_1^2$ and $\lambda_2 = 4(x_1^2 + y_1^2)$, we can also write

$$(x^2 + y^2)^{1/2} = \left(\sqrt{x_1^2 + y_1^2}, \ \frac{x_1x_2 + y_1y_2}{\sqrt{x_1^2 + y_1^2}} \right),$$

so that

$$\phi_{\text{FB}}(x, y) = \left(\sqrt{x_1^2 + y_1^2} - (x_1 + y_1), \ \frac{x_1x_2 + y_1y_2}{\sqrt{x_1^2 + y_1^2}} - (x_2 + y_2) \right). \tag{3.24}$$

Using the fact that $x_1y_2 = y_1x_2$, we can rewrite the above expression for $\nabla_x \psi_{\text{FB}}(x, y)$ in the form of (3.20). By symmetry, (3.21) also holds. □

Proposition 3.5 *Let* ϕ_{FB} *be given by* (3.10)*. Then, the function* ψ_{FB} *given by* (3.11) *is smooth everywhere on* $\mathbb{R}^n \times \mathbb{R}^n$.

Proof By Proposition 3.4, ψ_{FB} is differentiable everywhere on $\mathbb{R}^n \times \mathbb{R}^n$. We will show that $\nabla \psi_{\text{FB}}$ is continuous at every $(a, b) \in \mathbb{R}^n \times \mathbb{R}^n$. By the symmetry between x and y in $\nabla \psi_{\text{FB}}$, it suffices to verify that $\nabla_x \psi_{\text{FB}}$ is continuous at every $(a, b) \in \mathbb{R}^n \times \mathbb{R}^n$.

Case (1): $a = b = 0$.
By Proposition 3.4, $\nabla_x \psi_{\text{FB}}(0, 0) = 0$. Thus, we need to show that $\nabla_x \psi_{\text{FB}}(x, y) \to 0$ as $(x, y) \to (0, 0)$. We consider two subcases: (i) $(x, y) \neq (0, 0)$ and $x^2 + y^2 \in \text{int}(\mathcal{K}^n)$ and (ii) $(x, y) \neq (0, 0)$ and $x^2 + y^2 \notin \text{int}(\mathcal{K}^n)$. In subcase (i), we have from Proposition 3.4 that $\nabla_x \psi_{\text{FB}}(x, y)$ is given by the expression (3.4). By Lemma 3.4, $L_x L^{-1}_{(x^2+y^2)^{1/2}}$ is uniformly bounded, with bound independent of (x, y). Also, ϕ_{FB} given by (3.10) is continuous at $(0, 0)$ so that $\phi_{\text{FB}}(x, y) \to 0$ as $(x, y) \to (a, b)$. It follows from (3.4) that $\nabla_x \psi_{\text{FB}}(x, y) \to 0$ as $(x, y) \to (a, b)$ in subcase (i). In subcase (ii), we have from Proposition 3.4 that $\nabla_x \psi_{\text{FB}}(x, y)$ is given by the expression (3.20). Clearly $x_1/\sqrt{x_1^2 + y_1^2}$ is uniformly bounded, with bound independent of (x, y). Also, $\phi_{\text{FB}}(x, y) \to 0$ as $(x, y) \to (a, b)$. It follows from (3.20) that $\nabla_x \psi_{\text{FB}}(x, y) \to 0$ as $(x, y) \to (a, b)$ in subcase (ii).

Case (2): $(a, b) \neq (0, 0)$ and $a^2 + b^2 \in \text{int}(\mathcal{K}^n)$.
It was already shown in the proof of Proposition 3.4 that ψ_{FB} is continuously differentiable at (a, b).

Case (3): $(a, b) \neq (0, 0)$ and $a^2 + b^2 \notin \text{int}(\mathcal{K}^n)$.
By (3.12), $\|a\|^2 + \|b\|^2 = 2\|a_1a_2 + b_1b_2\|$. By Proposition 3.4, we have $a_1^2 + b_1^2 > 0$ and

$$\nabla_x \psi_{\text{FB}}(a, b) = \left(\frac{a_1}{\sqrt{a_1^2 + b_1^2}} - 1 \right) \phi(a, b).$$

We need to show that $\nabla_x \psi_{\text{FB}}(x, y) \to \nabla_x \psi_{\text{FB}}(a, b)$. We consider two cases: (i) $(x, y) \neq (0, 0)$ and $x^2 + y^2 \in \text{int}(\mathcal{K}^n)$ and (ii) $(x, y) \neq (0, 0)$ and $x^2 + y^2 \notin \text{int}(\mathcal{K}^n)$. In subcase (ii), we have from Proposition 3.4 that $\nabla_x \psi_{\text{FB}}(x, y)$ is given by the expression (3.20). This expression is continuous at (a, b). Thus, $\nabla_x \psi_{\text{FB}}(x, y) \to \nabla_x \psi_{\text{FB}}(a, b)$ as $(x, y) \to (a, b)$ in subcase (ii). The remainder of our proof treats subcase (i). In subcase (i), we have from Proposition 3.4 that $\nabla_x \psi_{\text{FB}}(x, y)$ is given by the expression (3.4), i.e.,

$$\begin{aligned}
\nabla_x \psi_{\text{FB}}(x, y) &= \left(L_x L^{-1}_{(x^2+y^2)^{1/2}} - I \right) \phi_{\text{FB}}(x, y) \\
&= L_x L^{-1}_{(x^2+y^2)^{1/2}} (x^2 + y^2)^{1/2} - L_x L^{-1}_{(x^2+y^2)^{1/2}} (x + y) - \phi_{\text{FB}}(x, y) \\
&= x - L_x L^{-1}_{(x^2+y^2)^{1/2}} (x + y) - \phi_{\text{FB}}(x, y).
\end{aligned}$$

Also, by Lemma 3.2, we have $\|a_1 a_2 + b_1 b_2\| = \frac{1}{2}\|a\|^2 + \frac{1}{2}\|b\|^2 = a_1^2 + b_1^2$ and $a_1 b_2 = b_1 a_2$, implying that (see (3.13), (3.15))

$$\begin{aligned}
\frac{a_1}{\sqrt{a_1^2 + b_1^2}} (a^2 + b^2)^{1/2} &= \frac{a_1}{\sqrt{a_1^2 + b_1^2}} \left(\sqrt{a_1^2 + b_1^2}, \frac{a_1 a_2 + b_1 b_2}{\sqrt{a_1^2 + b_1^2}} \right) \\
&= \left(a_1, \frac{a_1^2 a_2 + a_1 b_1 b_2}{a_1^2 + b_1^2} \right) \\
&= \left(a_1, \frac{a_1^2 a_2 + b_1^2 a_2}{a_1^2 + b_1^2} \right) \\
&= (a_1, a_2) \\
&= a.
\end{aligned}$$

This together with (3.20) yields

$$\begin{aligned}
\nabla_x \psi_{\text{FB}}(a, b) &= \left(\frac{a_1}{\sqrt{a_1^2 + b_1^2}} - 1 \right) \phi(a, b) \\
&= \frac{a_1}{\sqrt{a_1^2 + b_1^2}} \left[(a^2 + b^2)^{1/2} - (a + b) \right] - \phi_{\text{FB}}(a, b) \\
&= \frac{a_1}{\sqrt{a_1^2 + b_1^2}} (a^2 + b^2)^{1/2} - \frac{a_1}{\sqrt{a_1^2 + b_1^2}} (a + b) - \phi_{\text{FB}}(a, b) \\
&= a - \frac{a_1}{\sqrt{a_1^2 + b_1^2}} (a + b) - \phi_{\text{FB}}(a, b).
\end{aligned}$$

Since ϕ_{FB} is continuous, to prove $\nabla_x \psi_{\text{FB}}(x, y) \to \nabla_x \psi_{\text{FB}}(a, b)$ as $(x, y) \to (a, b)$, it suffices to show that

$$L_x L^{-1}_{(x^2+y^2)^{1/2}} x \to \frac{a_1}{\sqrt{a_1^2+b_1^2}} a \quad \text{as } (x,y) \to (a,b), \tag{3.25}$$

$$L_x L^{-1}_{(x^2+y^2)^{1/2}} y \to \frac{a_1}{\sqrt{a_1^2+b_1^2}} b \quad \text{as } (x,y) \to (a,b). \tag{3.26}$$

Note $\|a\|^2 + \|b\|^2 = 2\|a_1a_2 + b_1b_2\|$ and $(a,b) \neq (0,0)$, there has $a_1a_2 + b_1b_2 \neq 0$. Thus, by taking (x,y) sufficiently near to (a,b), we can assume that $x_1x_2 + y_1y_2 \neq 0$. Let $z := (x^2+y^2)^{1/2}$. Then z is given by (3.15) with λ_1, λ_2 given by (3.13) and $w_2 := \dfrac{x_1x_2+y_1y_2}{\|x_1x_2+y_1y_2\|}$. In addition, $\det(z) = z_1^2 - \|z_2\|^2 = \sqrt{\lambda_1\lambda_2}$. Let $(\zeta_1, \zeta_2) := L_x L_z^{-1} x$. Then, (3.25) reduces to

$$\zeta_1 \to \frac{a_1^2}{\sqrt{a_1^2+b_1^2}} \quad \text{and} \quad \zeta_2 \to \frac{a_1}{\sqrt{a_1^2+b_1^2}} a_2 \quad \text{as } (x,y) \to (a,b). \tag{3.27}$$

We prove (3.27) below. By Lemma 3.2, as $(x,y) \to (a,b)$,

$$\lambda_1 \to 0, \quad \lambda_2 \to \|a\|^2 + \|b\|^2 + 2\|a_1a_2 + b_1b_2\| = 4(a_1^2+b_1^2), \quad z_1 \to \sqrt{a_1^2+b_1^2}. \tag{3.28}$$

Using (3.16), we calculate the first component of $L_x L_z^{-1} x$ to be

$$\begin{aligned}
\zeta_1 &:= \frac{1}{\det(z)} \left(x_1^2 z_1 - x_2^T z_2 x_1 - x_1 z_2^T x_2 + \frac{\det(z)}{z_1} \|x_2\|^2 + \frac{(x_2^T z_2)^2}{z_1} \right), \\
&= \frac{\|x_2\|^2}{z_1} + \frac{1}{z_1 \det(z)} \left(x_1^2 z_1^2 - 2 x_2^T z_2 x_1 z_1 + (x_2^T z_2)^2 \right) \\
&= \frac{\|x_2\|^2}{z_1} + \frac{(x_1 z_1 - x_2^T z_2)^2}{z_1 \det(z)}.
\end{aligned}$$

In addition, applying Lemma 3.2 and (3.28) yields

$$\frac{\|x_2\|^2}{z_1} \to \frac{\|a_2\|^2}{\sqrt{a_1^2+b_1^2}} = \frac{a_1^2}{\sqrt{a_1^2+b_1^2}}.$$

Thus, to prove the first relation in (3.27), it suffices to show that

$$\frac{(x_1 z_1 - x_2^T z_2)^2}{z_1 \det(z)} \to 0 \quad \text{as } (x,y) \to (a,b).$$

Note that

$$\frac{(x_1 z_1 - x_2^T z_2)^2}{z_1 \det(z)} = \frac{1}{z_1\sqrt{\lambda_1\lambda_2}} \left(x_1 \frac{\sqrt{\lambda_1}+\sqrt{\lambda_2}}{2} + \frac{\sqrt{\lambda_1}-\sqrt{\lambda_2}}{2} x_2^T w_2 \right)^2$$

$$
\begin{aligned}
&= \frac{1}{z_1\sqrt{\lambda_1\lambda_2}}\left(x_1\sqrt{\lambda_1} + \frac{\sqrt{\lambda_2}-\sqrt{\lambda_1}}{2}\left(x_1 - x_2^{\mathsf T} w_2\right)\right)^2 \\
&= \frac{1}{z_1\sqrt{\lambda_2}}\Bigg(x_1^2\sqrt{\lambda_1} + x_1\left(\sqrt{\lambda_2}-\sqrt{\lambda_1}\right)\left(x_1 - x_2^{\mathsf T} w_2\right) \\
&\quad + \frac{(\sqrt{\lambda_2}-\sqrt{\lambda_1})^2}{4\sqrt{\lambda_1}}\left(x_1 - x_2^{\mathsf T} w_2\right)^2\Bigg).
\end{aligned} \tag{3.29}
$$

We also have from (3.28) that $\lambda_1 \to 0$, $\sqrt{\lambda_2} \to 2\sqrt{a_1^2+b_1^2} > 0$, and $z_1 \to \sqrt{a_1^2+b_1^2} > 0$. Moreover, by Lemma 3.3 and $w_2 = \dfrac{x_1x_2 + y_1y_2}{\|x_1x_2 + y_1y_2\|}$,

$$
\frac{\left(x_1 - x_2^{\mathsf T} w_2\right)^2}{\sqrt{\lambda_1}} \to 0 \quad \text{as } (x, y) \to (a, b).
$$

Thus, the right-hand side of (3.29) tends to zero as $(x, y) \to (a, b)$. This proves the first relation in (3.27). Using (3.16), we calculate the last $n-1$ components of $L_x L_z^{-1} x$ to be

$$
\begin{aligned}
\zeta_2 &:= \frac{1}{\det(z)}\left(x_1x_2z_1 - x_1^2 z_2 - x_2^{\mathsf T} z_2 x_2 + \frac{x_1\det(z)}{z_1}x_2 + \frac{x_1}{z_1} z_2 z_2^{\mathsf T} x_2\right) \\
&= \frac{x_1}{z_1}x_2 + \frac{1}{\det(z)}\left[(x_1z_1 - x_2^{\mathsf T} z_2)x_2 + x_1\left(\frac{x_2^T z_2}{z_1} - x_1\right)z_2\right] \\
&= \frac{x_1}{z_1}x_2 + \frac{(x_1z_1 - x_2^{\mathsf T} z_2)}{\det(z)}\left(x_2 - \frac{x_1}{z_1}z_2\right).
\end{aligned}
$$

Also, by (3.28), we obtain

$$
\frac{x_1}{z_1}x_2 \;\to\; \frac{a_1}{\sqrt{a_1^2+b_1^2}}\, a_2.
$$

Therefore, to prove the second relation in (3.27), it suffices to show that

$$
\frac{(x_1z_1 - x_2^{\mathsf T} z_2)}{\det(z)}\left(x_2 - \frac{x_1}{z_1}z_2\right) \to 0 \quad \text{as } (x, y) \to (a, b).
$$

First, $\dfrac{(x_1z_1 - x_2^{\mathsf T} z_2)}{\det(z)}$ is bounded as $(x, y) \to (a, b)$ because, by (3.15),

$$
\frac{(x_1z_1 - x_2^{\mathsf T} z_2)}{\det(z)} = \frac{1}{\sqrt{\lambda_1\lambda_2}}\left(x_1\frac{\sqrt{\lambda_1}+\sqrt{\lambda_2}}{2} - \frac{\sqrt{\lambda_2}-\sqrt{\lambda_1}}{2}x_2^{\mathsf T} w_2\right)
$$

$$= \frac{1}{\sqrt{\lambda_1\lambda_2}}\left(x_1\sqrt{\lambda_1} + \frac{\sqrt{\lambda_2}-\sqrt{\lambda_1}}{2}\left(x_1 - x_2^{\mathrm{T}}w_2\right)\right)$$
$$= \frac{x_1}{\sqrt{\lambda_2}} + \frac{\sqrt{\lambda_2}-\sqrt{\lambda_1}}{2\sqrt{\lambda_1\lambda_2}}\left(x_1 - x_2^{\mathrm{T}}w_2\right)$$
$$= \frac{x_1}{\sqrt{\lambda_2}} + \frac{1-\sqrt{\lambda_1}/\sqrt{\lambda_2}}{2}\frac{\left(x_1 - x_2^{\mathrm{T}}w_2\right)}{\sqrt{\lambda_1}},$$

and the first term on the right-hand side converges to $a_1/\sqrt{4(a_1^2+b_1^2)}$ (see (3.28)) while the second term is bounded by (3.13) and Lemma 3.3. Second, $x_2 - \frac{x_1}{z_1}z_2 \to 0$ as $(x, y) \to (a, b)$ because, by (3.15) and (3.28),

$$\begin{aligned} x_2 - \frac{x_1}{z_1}z_2 \to & a_2 - \frac{a_1}{\sqrt{a_1^2+b_1^2}}\frac{\sqrt{4(a_1^2+b_1^2)}}{2}\frac{a_1a_2+b_1b_2}{\|a_1a_2+b_1b_2\|} \\ =& a_2 - \frac{a_1^2a_2+a_1b_1b_2}{\|a_1a_2+b_1b_2\|} \\ =& a_2 - \frac{a_1^2a_2+b_1^2a_2}{a_1^2+b_1^2} \\ =& a_2 - a_2 \\ =& 0, \end{aligned}$$

where the second equality is due to Lemma 3.2, so that $a_1b_2 = b_1a_2$ and $\|a_1a_2 + b_1b_2\| = a_1^2 + b_1^2$. This proves the second relation in (3.27).

Thus, we have proven (3.25). An analogous argument can be used to prove (3.26), which we omit for simplicity. This shows that $\nabla_x\psi_{\mathrm{FB}}(x, y) \to \nabla_x\psi_{\mathrm{FB}}(a, b)$ as $(x, y) \to (a, b)$ in subcase (i). □

Proposition 3.6 *Let ϕ_{FB} and ψ_{FB} be given by* (3.10) *and* (3.11)*, respectively. For any $(x, y) \in \mathbb{R}^n \times \mathbb{R}^n$, we have the following results.*

(a)

$$\langle x, \nabla_x\psi_{\mathrm{FB}}(x, y)\rangle + \langle y, \nabla_y\psi_{\mathrm{FB}}(x, y)\rangle = \left\|\phi_{\mathrm{FB}}(x, y)\right\|^2. \tag{3.30}$$

(b)

$$\langle \nabla_x\psi_{\mathrm{FB}}(x, y), \nabla_y\psi_{\mathrm{FB}}(x, y)\rangle \geq 0, \tag{3.31}$$

with equality holding if and only if $\phi_{\mathrm{FB}}(x, y) = 0$.

Proof Case (1): $x = y = 0$. From Proposition 3.4, $\nabla_x\psi_{\mathrm{FB}}(x, y) = \nabla_y\psi_{\mathrm{FB}}(x, y) = 0$, so the proposition is true.

Case (2): $(x, y) \neq (0, 0)$ and $x^2 + y^2 \in \mathrm{int}(\mathcal{K}^n)$. By Proposition 3.4, we have

$$\nabla_x \psi_{\text{FB}}(x, y) = \left(L_x L_z^{-1} - I \right) \phi_{\text{FB}}(x, y),$$
$$\nabla_y \psi_{\text{FB}}(x, y) = \left(L_y L_z^{-1} - I \right) \phi_{\text{FB}}(x, y),$$

where we let $z := (x^2 + y^2)^{1/2}$. For simplicity, we will write $\phi_{\text{FB}}(x, y)$ as ϕ_{FB}. Thus,

$$\begin{aligned}
&\langle x, \nabla_x \psi_{\text{FB}}(x, y)\rangle + \langle y, \nabla_y \psi_{\text{FB}}(x, y)\rangle \\
&= \langle x, (L_x L_z^{-1} - I)\phi_{\text{FB}}\rangle + \langle y, (L_y L_z^{-1} - I)\phi_{\text{FB}}\rangle \\
&= \langle (L_z^{-1} L_x - I)x, \phi_{\text{FB}}\rangle + \langle (L_z^{-1} L_y - I)y, \phi_{\text{FB}}\rangle \\
&= \langle L_z^{-1} L_x x + L_z^{-1} L_y y - x - y, \phi_{\text{FB}}\rangle \\
&= \langle L_z^{-1}(x^2 + y^2) - x - y, \phi_{\text{FB}}\rangle \\
&= \langle L_z^{-1} z^2 - x - y, \phi_{\text{FB}}\rangle \\
&= \langle z - x - y, \phi_{\text{FB}}\rangle \\
&= \|\phi_{\text{FB}}\|^2,
\end{aligned}$$

where the next-to-last equality follows from $L_z z = z^2$, so that $L_z^{-1} z^2 = z$. This proves (3.30). Similarly,

$$\begin{aligned}
\langle \nabla_x \psi_{\text{FB}}(x, y), \nabla_y \psi_{\text{FB}}(x, y)\rangle &= \langle (L_x L_z^{-1} - I)\phi_{\text{FB}}, (L_y L_z^{-1} - I)\phi_{\text{FB}}\rangle \\
&= \langle (L_x - L_z) L_z^{-1}\phi_{\text{FB}}, (L_y - L_z) L_z^{-1}\phi_{\text{FB}}\rangle \qquad (3.32) \\
&= \langle (L_y - L_z)(L_x - L_z) L_z^{-1}\phi_{\text{FB}}, L_z^{-1}\phi_{\text{FB}}\rangle.
\end{aligned}$$

Let S be the symmetric part of $(L_y - L_z)(L_x - L_z)$. Then

$$\begin{aligned}
S &= \frac{1}{2}\Big((L_y - L_z)(L_x - L_z) + (L_x - L_z)(L_y - L_z) \Big) \\
&= \frac{1}{2}\Big(L_x L_y + L_y L_x - L_z(L_x + L_y) - (L_x + L_y)L_z + 2L_z^2 \Big) \\
&= \frac{1}{2}(L_z - L_x - L_y)^2 + \frac{1}{2}(L_z^2 - L_x^2 - L_y^2).
\end{aligned}$$

Since $z \in \mathcal{K}^n$ and $z^2 = x^2 + y^2$, Lemma 3.5 yields $L_z^2 - L_x^2 - L_y^2 \succeq O$. Then, (3.32) gives

$$\begin{aligned}
&\langle \nabla_x \psi_{\text{FB}}(x, y), \nabla_y \psi_{\text{FB}}(x, y)\rangle \\
=&\langle S L_z^{-1}\phi_{\text{FB}}, L_z^{-1}\phi_{\text{FB}}\rangle \\
=&\frac{1}{2}\langle (L_z - L_x - L_y)^2 L_z^{-1}\phi_{\text{FB}}, L_z^{-1}\phi_{\text{FB}}\rangle + \frac{1}{2}\langle (L_z^2 - L_x^2 - L_y^2) L_z^{-1}\phi_{\text{FB}}, L_z^{-1}\phi_{\text{FB}}\rangle \\
\geq&\frac{1}{2}\langle (L_z - L_x - L_y)^2 L_z^{-1}\phi_{\text{FB}}, L_z^{-1}\phi_{\text{FB}}\rangle \\
=&\frac{1}{2}\|L_{\phi_{\text{FB}}} L_z^{-1}\phi_{\text{FB}}\|^2,
\end{aligned}$$

where the last equality uses $L_z - L_x - L_y = L_{z-x-y} = L_{\phi_{\text{FB}}}$. This proves (3.31). If the inequality in (3.31) holds with equality, then the above relation yields $\|L_{\phi_{\text{FB}}} L_z^{-1} \phi_{\text{FB}}\|^2 = 0$ and, by Lemma 3.1(d),

$$\phi_{\text{FB}} \circ (L_z^{-1} \phi_{\text{FB}}) = L_{\phi_{\text{FB}}} L_z^{-1} \phi_{\text{FB}} = 0.$$

Then, the definition of Jordan product yields

$$\langle \phi_{\text{FB}},\ L_z^{-1} \phi_{\text{FB}} \rangle = 0.$$

Since $z = (x^2 + y^2)^{1/2} \in \text{int}(\mathcal{K}^n)$ so that $L_z^{-1} \succ O$ (see Lemma 3.1(d)), this implies $\phi_{\text{FB}} = 0$. Conversely, if $\phi_{\text{FB}} = 0$, then it follows from (3.4) that

$$\langle \nabla_x \psi_{\text{FB}}(x, y), \nabla_y \psi_{\text{FB}}(x, y) \rangle = 0.$$

Case (3): $(x, y) \neq (0, 0)$ and $x^2 + y^2 \notin \text{int}(\mathcal{K}^n)$. By Proposition 3.4, we have

$$\nabla_x \psi_{\text{FB}}(x, y) = \left(\frac{x_1}{\sqrt{x_1^2 + y_1^2}} - 1 \right) \phi_{\text{FB}}(x, y),$$

$$\nabla_y \psi_{\text{FB}}(x, y) = \left(\frac{y_1}{\sqrt{x_1^2 + y_1^2}} - 1 \right) \phi_{\text{FB}}(x, y).$$

Thus,

$$\begin{aligned}
&\langle x, \nabla_x \psi_{\text{FB}}(x, y) \rangle + \langle y, \nabla_y \psi_{\text{FB}}(x, y) \rangle \\
&= \left(\frac{x_1}{\sqrt{x_1^2 + y_1^2}} - 1 \right) \langle x, \phi_{\text{FB}}(x, y) \rangle + \left(\frac{y_1}{\sqrt{x_1^2 + y_1^2}} - 1 \right) \langle y, \phi_{\text{FB}}(x, y) \rangle \\
&= \left\langle \left(\frac{x_1}{\sqrt{x_1^2 + y_1^2}} - 1 \right) x + \left(\frac{y_1}{\sqrt{x_1^2 + y_1^2}} - 1 \right) y, \phi_{\text{FB}}(x, y) \right\rangle \\
&= \left\langle \frac{x_1 x + y_1 y}{\sqrt{x_1^2 + y_1^2}} - x - y, \phi_{\text{FB}}(x, y) \right\rangle \\
&= \langle \phi_{\text{FB}}(x, y), \phi_{\text{FB}}(x, y) \rangle,
\end{aligned}$$

where the last equality uses (3.24). This proves (3.30). Similarly,

$$\begin{aligned}
&\langle \nabla_x \psi_{\text{FB}}(x, y), \ \nabla_y \psi_{\text{FB}}(x, y) \rangle \\
&= \left(\frac{x_1}{\sqrt{x_1^2 + y_1^2}} - 1 \right) \left(\frac{y_1}{\sqrt{x_1^2 + y_1^2}} - 1 \right) \|\phi_{\text{FB}}(x, y)\|^2 \\
&\geq 0.
\end{aligned}$$

This proves (3.31). If the inequality in (3.31) holds with equality, then either $\phi_{\text{FB}}(x, y) = 0$ or $\frac{x_1}{\sqrt{x_1^2+y_1^2}} = 1$ or $\frac{y_1}{\sqrt{x_1^2+y_1^2}} = 1$. In the second case, we have $y_1 = 0$ and $x_1 \geq 0$, so that Lemma 3.2 yields $y_2 = 0$ and $x_1 = \|x_2\|$. In the third case, we have $x_1 = 0$ and $y_1 \geq 0$, so that Lemma 3.2 yields $x_2 = 0$ and $y_1 = \|y_2\|$. Thus, in these two cases, we achieve $x \circ y = 0, \ x \in \mathcal{K}^n, \ y \in \mathcal{K}^n$. Then, by Proposition 3.2, $\phi_{\text{FB}}(x, y) = 0$. □

Lemma 3.9 *Let $\omega : \mathbb{R}^n \times \mathbb{R}^n \to \mathbb{R}^m$ be defined by $\omega(x, y) := u(x, y) \circ v(x, y)$, where $u, v : \mathbb{R}^n \times \mathbb{R}^n \to \mathbb{R}^m$ are differentiable mappings. Then, ω is differentiable and*

$$\begin{aligned}
\nabla_x \omega(x, y) &= \nabla_x u(x, y) L_{v(x,y)} + \nabla_x v(x, y) L_{u(x,y)}, \\
\nabla_y \omega(x, y) &= \nabla_y u(x, y) L_{v(x,y)} + \nabla_y v(x, y) L_{u(x,y)}.
\end{aligned} \tag{3.33}$$

Proof This is the product rule corresponding to the Jordan product. As its proof is straightforward, we omit the details. □

Lemma 3.10 *For any $x, y \in \mathbb{R}^n$, let $w(x, y) = (w_1, w_2)$ and $z(x, y) = (z_1, z_2)$ be defined as in (3.14). Suppose that $F(x, y) := L_x L_{z(x,y)}^{-1}(x + y)$ and $G(x, y) := L_y L_{z(x,y)}^{-1}(x + y)$. Then, we have*

(a) z is differentiable at $(x, y) \neq (0, 0) \in \mathbb{R}^n \times \mathbb{R}^n$ with $x^2 + y^2 \in \text{int}(\mathcal{K}^n)$. Moreover

$$\nabla_x z(x, y) = L_x L_{z(x,y)}^{-1}, \quad \nabla_y z(x, y) = L_y L_{z(x,y)}^{-1},$$

where

$$L_z^{-1} = \begin{cases} \begin{bmatrix} b & c\bar{w}_2^T \\ c\bar{w}_2 & aI + (b-a)\bar{w}_2\bar{w}_2^T \end{bmatrix} & \text{if } w_2 \neq 0; \\ \left(1/\sqrt{w_1}\right) I & \text{if } w_2 = 0, \end{cases} \tag{3.34}$$

with

$$\begin{aligned}
a &= \frac{2}{\sqrt{\lambda_2(w)} + \sqrt{\lambda_1(w)}}, \\
b &= \frac{1}{2}\left(\frac{1}{\sqrt{\lambda_2(w)}} + \frac{1}{\sqrt{\lambda_1(w)}} \right), \\
c &= \frac{1}{2}\left(\frac{1}{\sqrt{\lambda_2(w)}} - \frac{1}{\sqrt{\lambda_1(w)}} \right)
\end{aligned}$$

and $\bar{w}_2 = \frac{w_2}{\|w_2\|}$.

(b) *F, G are differentiable at $(x, y) \neq (0, 0) \in \mathbb{R}^n \times \mathbb{R}^n$ with $x^2 + y^2 \in \text{int}(\mathcal{K}^n)$. Moreover, $\|\nabla F(x, y)\|$, $\|\nabla G(x, y)\|$ are uniformly bounded at such points.*

Proof (a) That the function z is differentiable is an immediate consequence of [25]. See also [26, Proposition 4]. Since, $z^2(x, y) = x^2 + y^2$, applying the product rule (3.33) in Lemma 3.9 yields

$$2\nabla_x z(x, y) L_{z(x,y)} = 2L_x, \quad 2\nabla_y z(x, y) L_{z(x,y)} = 2L_y.$$

Hence, the desired results follow.

(b) For symmetry, it is enough to show that F is differentiable at $(x, y) \neq (0, 0)$ with $x^2 + y^2 \in \text{int}(\mathcal{K}^n)$ and that $\|\nabla_x F(x, y)\|$, $\|\nabla_y F(x, y)\|$ are uniformly bounded. It is clear that F is differentiable at such points. The key part is to show the uniform boundedness of $\|\nabla_x F(x, y)\|$, $\|\nabla_y F(x, y)\|$. Let λ_1, λ_2 be the spectral values of $x^2 + y^2$, then

$$\begin{aligned}
\lambda_1 &:= \|x\|^2 + \|y\|^2 - 2\|x_1 x_2 + y_1 y_2\|, \\
\lambda_2 &:= \|x\|^2 + \|y\|^2 + 2\|x_1 x_2 + y_1 y_2\|.
\end{aligned}$$

Thus, $z(x, y) := (x^2 + y^2)^{1/2}$ has the spectral values $\sqrt{\lambda_1}$, $\sqrt{\lambda_2}$ and

$$z(x, y) = (z_1, z_2) = \left(\frac{\sqrt{\lambda_1} + \sqrt{\lambda_2}}{2}, \frac{\sqrt{\lambda_2} - \sqrt{\lambda_1}}{2} w_2 \right),$$

where $w_2 := \dfrac{x_1 x_2 + y_1 y_2}{\|x_1 x_2 + y_1 y_2\|}$ if $x_1 x_2 + y_1 y_2 \neq 0$ and otherwise w_2 is any vector in $\mathbb{R}^{n-1}$ satisfying $\|w_2\| = 1$.

Now, let $u := L_{z(x,y)}^{-1}(x + y)$. By applying Lemma 3.1, we compute u as below.

$$\begin{aligned}
u &= L_{z(x,y)}^{-1}(x + y) \\
&= \frac{1}{\det(z(x, y))} \begin{bmatrix} z_1 & -z_2^T \\ -z_2 & \frac{\det(z(x, y))}{z_1} I + \frac{1}{z_1} z_2 z_2^T \end{bmatrix} \begin{bmatrix} x_1 + y_1 \\ x_2 + y_2 \end{bmatrix} \\
&= \frac{1}{\det(z(x, y))} \begin{bmatrix} (x_1 + y_1) z_1 - (x_2 + y_2)^T z_2 \\ -(x_1 + y_1) z_2 + \frac{\det(z)}{z_1} (x_2 + y_2) + \frac{(x_2 + y_2)^T z_2}{z_1} z_2 \end{bmatrix} \\
&:= \begin{bmatrix} u_1 \\ u_2 \end{bmatrix}.
\end{aligned}$$

We notice that $F(x, y) = L_x L_{z(x,y)}^{-1}(x + y) = L_x u = x \circ u$. Then by applying Lemma 3.9, we obtain

$$\nabla_x F(x, y) = L_u + \nabla_x u(x, y) L_x \quad \text{and} \quad \nabla_y F(x, y) = \nabla_y u(x, y) L_x.$$

To show that $\|\nabla_x F(x, y)\|$ is uniformly bounded, we shall verify that both $\|L_u\|$ and $\|\nabla_x u(x, y) L_x\|$ are uniformly bounded. We prove them as follows.

(i) To see $\|L_u\|$ is uniformly bounded, it is sufficient to argue that $|u_1|$, $\|u_2\|$ are both uniformly bounded. First, we argue that $|u_1|$ is uniformly bounded. From the above expression of u, we have

$$u_1 = \frac{1}{\det(z(x, y))}(x_1 z_1 - x_2^T z_2) + \frac{1}{\det(z(x, y))}(y_1 z_1 - y_2^T z_2).$$

Following the similar arguments as in Lemma 3.4 yields

$$\begin{aligned} u_1 &= \frac{1}{\det(z(x, y))}(x_1 z_1 - x_2^T z_2) + \frac{1}{\det(z(x, y))}(y_1 z_1 - y_2^T z_2) \\ &= \left[O(1) + \frac{(x_1 - x_2^T w_2)}{2\sqrt{\lambda_1}} \right] + \left[O(1) + \frac{(y_1 - y_2^T w_2)}{2\sqrt{\lambda_1}} \right], \end{aligned}$$

where $O(1)$ denotes terms that are uniformly bounded with bound independent of (x, y). Moreover, by Lemma 3.3, if $x_1 x_2 + y_1 y_2 \neq 0$ then $|x_1 - x_2^T w_2| \leq \|x_2 - x_1 w_2\| \leq \sqrt{\lambda_1}$. If $x_1 x_2 + y_1 y_2 = 0$ then $\lambda_1 = \|x\|^2 + \|y\|^2$ so that by choosing w_2 to further satisfy $x_2^T w_2 = 0$ we obtain $|x_1 - x_2^T w_2| \leq \|x_2 - x_1 w_2\| \leq \|x\| \leq \sqrt{\lambda_1}$. Similarly, it can be verified that $|y_1 - y_2^T w_2| \leq \sqrt{\lambda_1}$. Thus, $|u_1|$ is uniformly bounded.

Secondly, we argue that $\|u_2\|$ is also uniformly bounded. Again, using the expression of u and following the similar arguments as in Lemma 3.4, we obtain

$$\begin{aligned} u_2 &= \frac{1}{\det(z(x, y))}\left[-x_1 z_2 + \frac{\det(z(x, y))}{z_1} x_2 + \frac{x_2^T z_2}{z_1} z_2 \right] \\ &\quad + \frac{1}{\det(z(x, y))}\left[-y_1 z_2 + \frac{\det(z(x, y))}{z_1} y_2 + \frac{y_2^T z_2}{z_1} z_2 \right] \\ &= \left[O(1) - \frac{x_1 w_2}{2\sqrt{\lambda_1}} + \frac{\frac{\sqrt{\lambda_2}}{\sqrt{\lambda_1}}(x_2^T w_2)}{2(\sqrt{\lambda_1} + \sqrt{\lambda_2})} w_2 \right] + \left[O(1) - \frac{y_1 w_2}{2\sqrt{\lambda_1}} + \frac{\frac{\sqrt{\lambda_2}}{\sqrt{\lambda_1}}(y_2^T w_2)}{2(\sqrt{\lambda_1} + \sqrt{\lambda_2})} w_2 \right] \\ &= \left[O(1) - \frac{x_1 w_2}{2(\sqrt{\lambda_1} + \sqrt{\lambda_2})} - \frac{\sqrt{\lambda_2}(x_1 - x_2^T w_2)}{2(\sqrt{\lambda_1} + \sqrt{\lambda_2})\sqrt{\lambda_1}} w_2 \right] \\ &\quad + \left[O(1) - \frac{y_1 w_2}{2(\sqrt{\lambda_1} + \sqrt{\lambda_2})} - \frac{\sqrt{\lambda_2}(y_1 - y_2^T w_2)}{2(\sqrt{\lambda_1} + \sqrt{\lambda_2})\sqrt{\lambda_1}} w_2 \right]. \end{aligned}$$

Using the same explanations as above for u_1 yields that each term is uniformly bounded. Thus, $\|u_2\|$ is uniformly bounded. This together with $|u_1|$ being uniformly bounded implies that $\|\nabla_x F(x, y)\| = \|L_u\| = \left\| \begin{bmatrix} u_1 & u_2^T \\ u_2 & u_1 I \end{bmatrix} \right\|$ is also uniformly bounded.

(ii) Now, it comes to show that $\|\nabla_x u(x,y)L_x\|$ is uniformly bounded. From the definition of $u := L_{z(x,y)}^{-1}(x+y)$, we know that $z(x,y)\circ u = x+y$. Applying Lemma 3.9 gives

$$\nabla_x z(x,y)L_u + \nabla_x u(x,y)L_{z(x,y)} = I,$$

which leads to

$$\begin{aligned}
&\nabla_x u(x,y)L_{z(x,y)} = I - \nabla_x z(x,y)L_u = I - (L_x L_{z(x,y)}^{-1})L_u \\
\Longrightarrow &\nabla_x u(x,y) = \left(I - L_x L_{z(x,y)}^{-1} L_u\right) L_{z(x,y)}^{-1} \\
\Longrightarrow &\nabla_x u(x,y)L_x = \left(I - L_x L_{z(x,y)}^{-1} L_u\right) L_{z(x,y)}^{-1} L_x \\
\Longrightarrow &\nabla_x u(x,y)L_x = L_{z(x,y)}^{-1} L_x - L_x L_{z(x,y)}^{-1} L_u L_{z(x,y)}^{-1} L_x \\
\Longrightarrow &\nabla_x u(x,y)L_x = (L_x L_{z(x,y)}^{-1})^{\mathsf{T}} - (L_x L_{z(x,y)}^{-1})L_u(L_x L_{z(x,y)}^{-1})^{\mathsf{T}}.
\end{aligned}$$

Therefore,

$$\|\nabla_x u(x,y)L_x\| \le \left\|(L_x L_{z(x,y)}^{-1})^{\mathsf{T}}\right\| + \left\|L_x L_{z(x,y)}^{-1}\right\| \cdot \|L_u\| \cdot \left\|(L_x L_{z(x,y)}^{-1})^{\mathsf{T}}\right\|.$$

By Lemma 3.4, $\left\|L_x L_{z(x,y)}^{-1}\right\|$ is uniformly bounded, so is $\left\|(L_x L_{z(x,y)}^{-1})^{\mathsf{T}}\right\|$. This together with $\|L_u\|$ being uniformly bounded shown as above yields $\|\nabla_x u(x,y)L_x\|$ is uniformly bounded.

From (i) and (ii), it follows that $\|\nabla_x F(x,y)\|$ is uniformly bounded. A similar line of reasoning applies to $\left\|\nabla_y F(x,y)\right\|$; and therefore, $\|\nabla F(x,y)\|$ is uniformly bounded as well. This concludes the proof. □

Lemma 3.11 *Let $\psi_{\rm FB}$ be defined as in (3.11). Then, $\nabla\psi_{\rm FB}$ is continuously differentiable everywhere except for $(x,y)=(0,0)$. Moreover, $\left\|\nabla^2\psi_{\rm FB}(x,y)\right\|$ is uniformly bounded for all $(x,y)\neq(0,0)$.*

Proof For any $(x,y)\in\mathbb{R}^n\times\mathbb{R}^n$, let $z := (x^2+y^2)^{1/2}$. We prove this lemma by considering the following two cases.

(i) Consider all points $(x,y)\neq(0,0)$ with $x^2+y^2\in\text{int}(\mathcal{K}^n)$. Since

$$\begin{aligned}
\nabla_x\psi_{\rm FB}(x,y) &= \left(L_x L_z^{-1} - I\right)\phi_{\rm FB}(x,y) = x - L_x L_z^{-1}(x+y) - \phi_{\rm FB}(x,y), \\
\nabla_y\psi_{\rm FB}(x,y) &= \left(L_y L_z^{-1} - I\right)\phi_{\rm FB}(x,y) = y - L_y L_z^{-1}(x+y) - \phi_{\rm FB}(x,y),
\end{aligned}$$

we compute $\nabla^2\psi_{\rm FB}(x,y)$ as follows:

$$\nabla^2_{xx}\psi_{\rm FB}(x,y) = I - \nabla_x\left(L_x L_z^{-1}(x+y)\right) - \left(L_x L_z^{-1} - I\right), \tag{3.35}$$

$$\nabla^2_{xy}\psi_{\text{FB}}(x,y) = -\nabla_y\left(L_x L_z^{-1}(x+y)\right) - \left(L_y L_z^{-1} - I\right),$$
$$\nabla^2_{yx}\psi_{\text{FB}}(x,y) = -\nabla_x\left(L_y L_z^{-1}(x+y)\right) - \left(L_x L_z^{-1} - I\right),$$
$$\nabla^2_{yy}\psi_{\text{FB}}(x,y) = I - \nabla_y\left(L_y L_z^{-1}(x+y)\right) - \left(L_y L_z^{-1} - I\right).$$

The continuity of $\nabla^2\psi_{\text{FB}}$ at (x, y) thus follows. It is easy to see that $\|L_x L_z^{-1}\|$, $\|L_y L_z^{-1}\|$ are uniformly bounded by [14, Lemma 4] ($\|\cdot\|$ and $\|\cdot\|_F$ are equivalent in $\mathbb{R}^{n\times n}$). Let $F(x, y) := L_x L_z^{-1}(x + y)$ and $G(x, y) := L_y L_z^{-1}(x + y)$. By Lemma 3.10, we know that $\left\|\nabla_x\left(L_x L_z^{-1}(x+y)\right)\right\| = \|\nabla_x F(x, y)\|$ is uniformly bounded. Likewise, we have that $\left\|\nabla_y\left(L_x L_z^{-1}(x+y)\right)\right\|$, $\left\|\nabla_x\left(L_y L_z^{-1}(x+y)\right)\right\|$, $\left\|\nabla_y\left(L_y L_z^{-1}(x+y)\right)\right\|$ are all uniformly bounded. Thus, we can conclude that $\|\nabla^2_{xx}\psi_{\text{FB}}(x, y)\|$, $\|\nabla^2_{xy}\psi_{\text{FB}}(x, y)\|$, $\|\nabla^2_{yx}\psi_{\text{FB}}(x, y)\|$, $\|\nabla^2_{yy}\psi_{\text{FB}}(x, y)\|$ are all uniformly bounded which implies that $\|\nabla^2\psi_{\text{FB}}(x, y)\|$ is also uniformly bounded.

(ii) Consider all points $(x, y) \neq (0, 0)$ with $x^2 + y^2 \notin \text{int}(\mathcal{K}^n)$. Since

$$\nabla_x\psi_{\text{FB}}(x,y) = \left(\frac{x_1}{\sqrt{x_1^2+y_1^2}} - 1\right)\phi_{\text{FB}}(x,y) = x - \frac{x_1}{\sqrt{x_1^2+y_1^2}}(x+y) - \phi_{\text{FB}}(x,y),$$
$$\nabla_y\psi_{\text{FB}}(x,y) = \left(\frac{y_1}{\sqrt{x_1^2+y_1^2}} - 1\right)\phi_{\text{FB}}(x,y) = y - \frac{y_1}{\sqrt{x_1^2+y_1^2}}(x+y) - \phi_{\text{FB}}(x,y),$$

we compute $\nabla^2\psi_{\text{FB}}(x, y)$ as follows:

$$\nabla^2_{xx}\psi_{\text{FB}}(x,y) = I - \left(\frac{x_1}{\sqrt{x_1^2+y_1^2}}I + \frac{x_1y_1^2+y_1^3}{(x_1^2+y_1^2)^{3/2}}\begin{bmatrix}1 & 0\\ 0 & \mathbf{0}\end{bmatrix}\right) - \left(\frac{x_1}{\sqrt{x_1^2+y_1^2}} - 1\right)I, \quad (3.36)$$
$$\nabla^2_{xy}\psi_{\text{FB}}(x,y) = -\left(\frac{x_1}{\sqrt{x_1^2+y_1^2}}I - \frac{x_1^2y_1+x_1y_1^2}{(x_1^2+y_1^2)^{3/2}}\begin{bmatrix}1 & 0\\ 0 & \mathbf{0}\end{bmatrix}\right) - \left(\frac{y_1}{\sqrt{x_1^2+y_1^2}} - 1\right)I,$$
$$\nabla^2_{yx}\psi_{\text{FB}}(x,y) = -\left(\frac{y_1}{\sqrt{x_1^2+y_1^2}}I - \frac{x_1^2y_1+x_1y_1^2}{(x_1^2+y_1^2)^{3/2}}\begin{bmatrix}1 & 0\\ 0 & \mathbf{0}\end{bmatrix}\right) - \left(\frac{x_1}{\sqrt{x_1^2+y_1^2}} - 1\right)I,$$
$$\nabla^2_{yy}\psi_{\text{FB}}(x,y) = I - \left(\frac{y_1}{\sqrt{x_1^2+y_1^2}}I + \frac{x_1^3+x_1^2y_1}{(x_1^2+y_1^2)^{3/2}}\begin{bmatrix}1 & 0\\ 0 & \mathbf{0}\end{bmatrix}\right) - \left(\frac{y_1}{\sqrt{x_1^2+y_1^2}} - 1\right)I,$$

where $\mathbf{0}$ denotes the $(n-1)\times(n-1)$ zero matrix. Now we provide a sketch proof to verify $\nabla_{xx}\psi_{\text{FB}}$ is continuous. Let $(a, b) \neq (0, 0)$ and $a^2 + b^2 \notin \text{int}(\mathcal{K}^n)$. We want to prove that

$$\nabla_{xx}\psi_{\rm FB}(x,y) \to \nabla_{xx}\psi_{\rm FB}(a,b), \quad \text{as} \quad (x,y) \to (a,b). \tag{3.37}$$

Due to the neighborhood of such (a,b), we have to consider two subcases: (1) $(x,y) \neq (0,0)$ with $x^2+y^2 \in \text{int}(\mathcal{K}^n)$ and (2) $(x,y) \neq (0,0)$ with $x^2+y^2 \notin \text{int}(\mathcal{K}^n)$. It is clear that (3.37) holds in subcase (2) because the formula given in (3.36) is continuous. In subcase (1), we have

$$\begin{aligned}\nabla_{xx}\psi_{\rm FB}(x,y) &= I - \nabla_x\left(L_xL_z^{-1}(x+y)\right) - \left(L_xL_z^{-1} - I\right) \\ &= I - \left[L_u + \left(L_xL_z^{-1}\right)^T - \left(L_xL_z^{-1}\right)(L_u)\left(L_xL_z^{-1}\right)^{\rm T}\right] - \left(L_xL_z^{-1} - I\right).\end{aligned} \tag{3.38}$$

In view of (3.35), (3.36) and (3.38), it suffices to show the following three statements for (3.37) to be held in this subcase (1):

(a) $L_xL_z^{-1} \to \dfrac{a_1}{\sqrt{a_1^2+b_1^2}}I$, as $(x,y) \to (a,b)$.

(b) $L_u \to \dfrac{a_1+b_1}{\sqrt{a_1^2+b_1^2}}I$, as $(x,y) \to (a,b)$.

(c) $L_u - (L_xL_z^{-1})(L_u)(L_xL_z^{-1})^T \to \dfrac{a_1^2(a_1+b_1)}{(a_1^2+b_1^2)^{3/2}}I$, as $(x,y) \to (a,b)$.

First, we know from [14, Proposition 2] that there holds

$$L_xL_z^{-1}(x+y) \to \frac{a_1}{\sqrt{a_1^2+b_1^2}}(a+b) \quad \text{as} \quad (x,y) \to (a,b),$$

which implies $L_xL_z^{-1} \to \dfrac{a_1}{\sqrt{a_1^2+b_1^2}}I$, as $(x,y) \to (a,b)$ since both $(x+y)$ and $L_xL_z^{-1}$ are continuous and $(x+y) \to (a+b)$ when $(x,y) \to (a,b)$. Secondly, if we look into the entries of L_u and compare them with the entries of $L_xL_z^{-1}$ (see [14, Eq. (27)]), then it is clear that $L_u \to \dfrac{a_1+b_1}{\sqrt{a_1^2+b_1^2}}I$, as $(x,y) \to (a,b)$. Finally, part(c) follows immediately from parts (a) and (b). Thus, we complete the verifications of (3.37). The other cases can be argued similarly for $\nabla_{xy}\psi_{\rm FB}$, $\nabla_{yx}\psi_{\rm FB}$, and $\nabla_{yy}\psi_{\rm FB}$. In addition, it is also clear that each term in the above expressions (3.36) is uniformly bounded. Thus, we obtain that $\nabla^2\psi_{\rm FB}$ is continuously differentiable near (x,y) and $\|\nabla^2\psi_{\rm FB}(x,y)\|$ is uniformly bounded. □

Proposition 3.7 *Let $\psi_{\rm FB}$ be defined as (3.11). Then, $\nabla\psi_{\rm FB}$ is globally Lipschitz continuous, i.e., there exists a constant C such that for all (x,y), $(a,b) \in \mathbb{R}^n \times \mathbb{R}^n$,*

$$\left\|\nabla_x\psi_{\rm FB}(x,y) - \nabla_x\psi_{\rm FB}(a,b)\right\| \leq C\|(x,y)-(a,b)\|, \tag{3.39}$$

$$\left\| \nabla_y \psi_{\text{FB}}(x, y) - \nabla_y \psi_{\text{FB}}(a, b) \right\| \leq C \|(x, y) - (a, b)\|$$

and is semismooth everywhere.

Proof Due to symmetry, it suffices to establish the first part of (3.39). For any $(x, y) \in \mathbb{R}^n \times \mathbb{R}^n$, let $z := (x^2 + y^2)^{1/2}$.

(i) First, we prove that $\nabla_x \psi_{\text{FB}}$ is Lipschitz continuous at $(0, 0)$. We have to discuss three subcases for completing the proof of this part.
If $(x, y) = (0, 0)$, it is obvious that (3.39) is satisfied.
If $(x, y) \neq (0, 0)$ with $x^2 + y^2 \in \text{int}(\mathcal{K}^n)$, then

$$\left\| \nabla_x \psi_{\text{FB}}(x, y) - \nabla_x \psi_{\text{FB}}(0, 0) \right\| = \left\| \nabla_x \psi_{\text{FB}}(x, y) \right\| = \left\| x - L_x L_z^{-1}(x + y) - \phi_{\text{FB}}(x, y) \right\|.$$

It is already known that x and $\phi_{\text{FB}}(x, y)$ are Lipschitz continuous (see [24, Corollary 3.3]). In addition, Theorem 3.2.4 of [27, pp. 70] says that the uniform boundedness of $\nabla\left(L_x L_z^{-1}(x + y) \right)$ (by Lemma 3.10) yields the Lipschitz continuity. Thus, (3.39) is satisfied for this subcase.

If $(x, y) \neq (0, 0)$ with $x^2 + y^2 \notin \text{int}(\mathcal{K}^n)$, then

$$\left\| \nabla_x \psi_{\text{FB}}(x, y) - \nabla_x \psi_{\text{FB}}(0, 0) \right\| = \left\| \nabla_x \psi_{\text{FB}}(x, y) \right\| = \left\| x - \frac{x_1}{\sqrt{x_1^2 + y_1^2}}(x + y) - \phi_{\text{FB}}(x, y) \right\|.$$

Since $\left| \dfrac{x_1}{\sqrt{x_1^2 + y_1^2}} \right| \leq 1$ and both $(x + y)$, $\phi_{\text{FB}}(x, y)$ are known Lipschitz continuous, the desired result follows.

(ii) Secondly, we prove that $\nabla_x \psi_{\text{FB}}$ is Lipschitz continuous at $(a, b) \neq (0, 0)$. Let $(x, y) \in \mathbb{R}^n \times \mathbb{R}^n$, we wish to show that (3.39) is satisfied. In fact, if the line segment $[(a, b), (x, y)]$ does not contain the origin, then we can write

$$\left\| \nabla_x \psi_{\text{FB}}(x, y) - \nabla_x \psi_{\text{FB}}(a, b) \right\| \leq \left\| \int_0^1 \nabla^2 \psi_{\text{FB}}[(a, b) + t((x, y) - (a, b))]dt \right\| \leq C\|(x, y) - (a, b)\|,$$

where the first inequality is from the Mean-Value Theorem (see [27, Theorem 3.2.3]), and the second inequality is by Lemma 3.11. On the other hand, if the line segment $[(a, b), (x, y)]$ contains the origin, we can construct a sequence $\{(x^k, y^k)\}$ converging to (x, y) but for each k, the line segment $[(a, b), (x^k, y^k)]$ does not contain the origin and apply the above inequalities to obtain

$$\left\| \nabla_x \psi_{\text{FB}}(x^k, y^k) - \nabla_x \psi_{\text{FB}}(a, b) \right\| \leq C \left\| (x^k, y^k) - (a, b) \right\|,$$

which, by the continuity, implies

$$\left\| \nabla_x \psi_{\text{FB}}(x, y) - \nabla_x \psi_{\text{FB}}(a, b) \right\| \le C \|(x, y) - (a, b)\|.$$

Thus, (3.39) is satisfied.

To complete the proof of this theorem, we only need to verify that $\nabla \psi_{\text{FB}}$ is semismooth at the origin as, by Lemma 3.11, $\nabla \psi_{\text{FB}}$ is continuously differentiable near any $(0, 0) \neq (x, y) \in \mathbb{R}^n \times \mathbb{R}^n$. From Proposition 3.4(b)–(c), we know that for any $t \in \mathbb{R}_+$ and $(x, y) \in \mathbb{R}^n \times \mathbb{R}^n$, we have

$$\nabla \psi_{\text{FB}}(tx, ty) = t \nabla \psi_{\text{FB}}(x, y).$$

Thus, $\nabla \psi_{\text{FB}}$ is directionally differentiable at the origin and for any $(0, 0) \neq (x, y) \in \mathbb{R}^n \times \mathbb{R}^n$

$$\nabla^2 \psi_{\text{FB}}(x, y)(x, y) = (\nabla \psi_{\text{FB}})'((x, y); (x, y)) = \nabla \psi_{\text{FB}}(x, y).$$

This means that for any $(0, 0) \neq (x, y) \in \mathbb{R}^n \times \mathbb{R}^n$ converging to $(0, 0)$,

$$\nabla \psi_{\text{FB}}(x, y) - \nabla \psi_{\text{FB}}(0, 0) - \nabla^2 \psi_{\text{FB}}(x, y)(x, y) = \nabla \psi_{\text{FB}}(x, y) - 0 - \nabla \psi_{\text{FB}}(x, y) = 0,$$

which, together with the Lipschitz continuity of $\nabla \psi_{\text{FB}}$ and the directional differentiability of $\nabla \psi_{\text{FB}}$ at the origin ($\nabla \psi_{\text{FB}}$ is, however, not differentiable at the origin), shows that $\nabla \psi_{\text{FB}}(x, y)$ is (strongly) semismooth at the origin. □

Proposition 3.8 *Let ψ_{FB} be defined as in (3.11). Then, ψ_{FB} is an SC^1 function as well as an LC^1 function.*

Proof The results follow from Proposition 3.7 immediately. □

Returning to ϕ_{FB}, we now present the representation of the elements of the B-subdifferential $\partial_B \phi_{\text{FB}}(x, y)$ at a general point $(x, y) \in \mathbb{R}^n \times \mathbb{R}^n$.

Proposition 3.9 *Let ϕ_{FB} be defined as in (3.10). Given a general point $(x, y) \in \mathbb{R}^n \times \mathbb{R}^n$, each element in $\partial_B \phi_{\text{FB}}(x, y)$ is given by $[V_x - I \;\; V_y - I]$ with V_x and V_y having the following representation:*

(a) If $x^2 + y^2 \in \text{int}(\mathcal{K}^n)$, then $V_x = L_z^{-1} L_x$ and $V_y = L_z^{-1} L_y$.
(b) If $x^2 + y^2 \in \text{bd}(\mathcal{K}^n)$ and $(x, y) \neq (0, 0)$, then

$$\begin{aligned} V_x &\in \left\{ \frac{1}{2\sqrt{2w_1}} \begin{bmatrix} 1 & \bar{w}_2^{\mathsf{T}} \\ \bar{w}_2 & 4I - 3\bar{w}_2 \bar{w}_2^{\mathsf{T}} \end{bmatrix} L_x + \frac{1}{2} \begin{bmatrix} 1 \\ -\bar{w}_2 \end{bmatrix} u^{\mathsf{T}} \right\} \\ V_y &\in \left\{ \frac{1}{2\sqrt{2w_1}} \begin{bmatrix} 1 & \bar{w}_2^{\mathsf{T}} \\ \bar{w}_2 & 4I - 3\bar{w}_2 \bar{w}_2^{\mathsf{T}} \end{bmatrix} L_y + \frac{1}{2} \begin{bmatrix} 1 \\ -\bar{w}_2 \end{bmatrix} v^{\mathsf{T}} \right\} \end{aligned} \tag{3.40}$$

for some $u = (u_1, u_2), v = (v_1, v_2) \in \mathbb{R} \times \mathbb{R}^{n-1}$ satisfying $|u_1| \le \|u_2\| \le 1$ and $|v_1| \le \|v_2\| \le 1$, where $\bar{w}_2 = w_2 / \|w_2\|$.

(c) If $(x, y) = (0, 0)$*, then* $V_x \in \{L_{\hat{x}}\},\ V_y \in \{L_{\hat{y}}\}$ *for some* $\hat{x}, \hat{y}$ *with* $\|\hat{x}\|^2 + \|\hat{y}\|^2 = 1$*, or*

$$V_x \in \left\{ \frac{1}{2}\begin{bmatrix} 1 \\ \bar{w}_2 \end{bmatrix} \xi^{\mathsf{T}} + \frac{1}{2}\begin{bmatrix} 1 \\ -\bar{w}_2 \end{bmatrix} u^{\mathsf{T}} + 2\begin{bmatrix} 0 & 0 \\ (I - \bar{w}_2\bar{w}_2^{\mathsf{T}})s_2 & (I - \bar{w}_2\bar{w}_2^{\mathsf{T}})s_1 \end{bmatrix} \right\}$$
$$V_y \in \left\{ \frac{1}{2}\begin{bmatrix} 1 \\ \bar{w}_2 \end{bmatrix} \eta^{\mathsf{T}} + \frac{1}{2}\begin{bmatrix} 1 \\ -\bar{w}_2 \end{bmatrix} v^{\mathsf{T}} + 2\begin{bmatrix} 0 & 0 \\ (I - \bar{w}_2\bar{w}_2^{\mathsf{T}})\omega_2 & (I - \bar{w}_2\bar{w}_2^{\mathsf{T}})\omega_1 \end{bmatrix} \right\} \quad (3.41)$$

for some $u = (u_1, u_2), v = (v_1, v_2), \xi = (\xi_1, \xi_2), \eta = (\eta_1, \eta_2) \in \mathbb{R} \times \mathbb{R}^{n-1}$ *such that* $|u_1| \le \|u_2\| \le 1$, $|v_1| \le \|v_2\| \le 1$, $|\xi_1| \le \|\xi_2\| \le 1$, $|\eta_1| \le \|\eta_2\| \le 1$, $\bar{w}_2 \in \mathbb{R}^{n-1}$ *satisfying* $\|\bar{w}_2\| = 1$*, and* $s = (s_1, s_2)$*,* $\omega = (\omega_1, \omega_2) \in \mathbb{R} \times \mathbb{R}^{n-1}$ *satisfying* $\|s\|^2 + \|\omega\|^2 \le 1/2$.

Proof Let $D_{\phi_{\rm FB}}$ denote the set of points where $\phi_{\rm FB}$ is differentiable. From Lemma 3.10 and $\phi_{\rm FB}(x, y) = z(x, y) - (x + y)$, we know

$$(\phi_{\rm FB})'_x(x, y) = L_z^{-1}L_x - I, \quad (\phi_{\rm FB})'_y(x, y) = L_z^{-1}L_y - I \quad \forall (x, y) \in D_{\phi_{\rm FB}}.$$

(a) In this case, $\phi_{\rm FB}$ is continuously differentiable at (x, y) by Lemma 3.10. Hence, $\partial_B\phi_{\rm FB}(x, y)$ consists of a single element, i.e., $\phi'_{\rm FB}(x, y) = [L_z^{-1}L_x - I \ \ L_z^{-1}L_y - I]$, and the result is clear.

(b) Assume that $(x, y) \neq (0, 0)$ satisfies $x^2 + y^2 \in \mathrm{bd}(\mathcal{K}^n)$. Then $w \in \mathrm{bd}(\mathcal{K}^n)$ and $w_1 > 0$, which means $\|w_2\| = w_1 > 0$ and $\lambda_2(w) > \lambda_1(w) = 0$. Observe that, when $w_2 \neq 0$, the matrix L_z^{-1} in (3.34) can be decomposed as the sum of

$$L_1(w) := \frac{1}{2\sqrt{\lambda_1(w)}}\begin{bmatrix} 1 & -\bar{w}_2^{\mathsf{T}} \\ -\bar{w}_2 & \bar{w}_2\bar{w}_2^{\mathsf{T}} \end{bmatrix}$$

and

$$L_2(w) := \frac{1}{2\sqrt{\lambda_2(w)}}\begin{bmatrix} 1 & \bar{w}_2^{\mathsf{T}} \\ \bar{w}_2 & \dfrac{4\sqrt{\lambda_2(w)}}{\sqrt{\lambda_2(w)} + \sqrt{\lambda_1(w)}}(I - \bar{w}_2\bar{w}_2^{\mathsf{T}}) + \bar{w}_2\bar{w}_2^{\mathsf{T}} \end{bmatrix} \quad (3.42)$$

with $\bar{w}_2 = w_2/\|w_2\|$. Consequently, $(\phi_{\rm FB})'_x$ and $(\phi_{\rm FB})'_y$ can be rewritten as

$$(\phi_{\rm FB})'_x(x, y) = (L_1(w) + L_2(w))\,L_x - I, \quad (\phi_{\rm FB})'_y(x, y) = (L_1(w) + L_2(w))\,L_y - I. \quad (3.43)$$

Let $\{(x^k, y^k)\} \subseteq D_{\phi_{\rm FB}}$ be an arbitrary sequence converging to (x, y). Let $w^k = (w_1^k, w_2^k) = w(x^k, y^k)$ and $z^k = z(x^k, y^k)$ for each k, where $w(x, y)$ and $z(x, y)$ are given as in (3.14). Since $w_2 \neq 0$, we without loss of generality assume $\|w_2^k\| \neq 0$ for each k. Let $\bar{w}_2^k = w_2^k/\|w_2^k\|$ for each k. From (3.43), it follows that

$$(\phi_{\rm FB})'_x(x^k, y^k) = \big(L_1(w^k) + L_2(w^k)\big)\,L_{x^k} - I,$$

$$(\phi_{\rm FB})'_y(x^k, y^k) = \big(L_1(w^k) + L_2(w^k)\big)\, L_{y^k} - I.$$

Since $\lim_{k\to\infty} \lambda_1(w^k) = 0$, $\lim_{k\to\infty} \lambda_2(w^k) = 2w_1 > 0$ and $\lim_{k\to\infty} \bar{w}_2^k = \bar{w}_2$, we have

$$\lim_{k\to\infty} L_2(w^k)L_{x^k} = C(w)L_x \quad \text{and} \quad \lim_{k\to\infty} L_2(w^k)L_{y^k} = C(w)L_y$$

where

$$C(w) = \frac{1}{2\sqrt{2w_1}} \begin{bmatrix} 1 & \bar{w}_2^{\mathsf T} \\ \bar{w}_2 & 4I - 3\bar{w}_2\bar{w}_2^{\mathsf T} \end{bmatrix}.$$

Next we focus on the limit of $L_1(w^k)L_{x^k}$ and $L_1(w^k)L_{y^k}$ as $k \to \infty$. By computing,

$$\begin{aligned} L_1(w^k)L_{x^k} &= \frac{1}{2} \begin{bmatrix} u_1^k & (u_2^k)^{\mathsf T} \\ -u_1^k\bar{w}_2^k & -\bar{w}_2^k(u_2^k)^{\mathsf T} \end{bmatrix}, \\ L_1(w^k)L_{y^k} &= \frac{1}{2} \begin{bmatrix} v_1^k & (v_2^k)^{\mathsf T} \\ -v_1^k\bar{w}_2^k & -\bar{w}_2^k(v_2^k)^{\mathsf T} \end{bmatrix}, \end{aligned}$$

where

$$u_1^k = \frac{x_1^k - (x_2^k)^{\mathsf T}\bar{w}_2^k}{\sqrt{\lambda_1(w^k)}}, \quad u_2^k = \frac{x_2^k - x_1^k\bar{w}_2^k}{\sqrt{\lambda_1(w^k)}}, \quad v_1^k = \frac{y_1^k - (y_2^k)^{\mathsf T}\bar{w}_2^k}{\sqrt{\lambda_1(w^k)}}, \quad v_2^k = \frac{y_2^k - y_1^k\bar{w}_2^k}{\sqrt{\lambda_1(w^k)}}. \tag{3.44}$$

By Lemma 3.3, $|u_1^k| \le \|u_2^k\| \le 1$ and $|v_1^k| \le \|v_2^k\| \le 1$. So, taking the limit (possibly on a subsequence) on $L_1(w^k)L_{x^k}$ and $L_1(w^k)L_{y^k}$, respectively, gives

$$\begin{aligned} L_1(w^k)L_{x^k} &\to \frac{1}{2} \begin{bmatrix} u_1 & u_2^{\mathsf T} \\ -u_1\bar{w}_2 & -\bar{w}_2u_2^{\mathsf T} \end{bmatrix} = \frac{1}{2} \begin{bmatrix} 1 \\ -\bar{w}_2 \end{bmatrix} u^{\mathsf T} \\ L_1(w^k)L_{y^k} &\to \frac{1}{2} \begin{bmatrix} v_1 & v_2^{\mathsf T} \\ -v_1\bar{w}_2 & -\bar{w}_2v_2^{\mathsf T} \end{bmatrix} = \frac{1}{2} \begin{bmatrix} 1 \\ -\bar{w}_2 \end{bmatrix} v^{\mathsf T} \end{aligned} \tag{3.45}$$

for some $u = (u_1, u_2)$, $v = (v_1, v_2) \in \mathbb{R} \times \mathbb{R}^{n-1}$ satisfying $|u_1| \le \|u_2\| \le 1$ and $|v_1| \le \|v_2\| \le 1$. In fact, u and v are some accumulation point of the sequences $\{u^k\}$ and $\{v^k\}$, respectively. From Eqs. (3.44)–(3.45), we immediately obtain

$$\begin{aligned} (\phi_{\rm FB})'_x(x^k, y^k) &\to C(w)L_x + \frac{1}{2} \begin{bmatrix} 1 \\ -\bar{w}_2 \end{bmatrix} u^{\mathsf T} - I, \\ (\phi_{\rm FB})'_y(x^k, y^k) &\to C(w)L_y + \frac{1}{2} \begin{bmatrix} 1 \\ -\bar{w}_2 \end{bmatrix} v^{\mathsf T} - I. \end{aligned}$$

This shows that $\phi'_{\rm FB}(x^k, y^k) \to [V_x - I \;\; V_y - I]$ as $k \to \infty$ with V_x, V_y satisfying (3.40).

(c) Assume that $(x, y) = (0, 0)$. Let $\{(x^k, y^k)\} \subseteq D_{\phi_{\rm FB}}$ be an arbitrary sequence converging to (x, y). Let $w^k = (w_1^k, w_2^k) = w(x^k, y^k)$ and $z^k = z(x^k, y^k)$ for each k. Since $w = 0$, we without any loss of generality assume that $w_2^k = 0$ for all k, or $w_2^k \neq 0$ for all k.

Case (1): $w_2^k = 0$ for all k. From Lemma 3.10, it follows that $L_{z^k}^{-1} = (1/\sqrt{w_1^k})I$. Therefore,

$$(\phi_{\rm FB})'_x(x^k, y^k) = \frac{1}{\sqrt{w_1^k}} L_{x^k} - I \quad \text{and} \quad (\phi_{\rm FB})'_y(x^k, y^k) = \frac{1}{\sqrt{w_1^k}} L_{y^k} - I.$$

Since $w_1^k = \|x^k\|^2 + \|y^k\|^2$, every element in $(\phi_{\rm FB})'_x(x^k, y^k)$ and $(\phi_{\rm FB})'_y(x^k, y^k)$ is bounded. Taking limit (possibly on a subsequence) on $(\phi_{\rm FB})'_x(x^k, y^k)$ and $(\phi_{\rm FB})'_y(x^k, y^k)$, we obtain

$$(\phi_{\rm FB})'_x(x^k, y^k) \rightarrow L_{\hat{x}} - I \quad \text{and} \quad (\phi_{\rm FB})'_y(x^k, y^k) \rightarrow L_{\hat{y}} - I$$

for some vectors $\hat{x}, \hat{y} \in \mathbb{R}^n$ satisfying $\|\hat{x}\|^2 + \|\hat{y}\|^2 = 1$, where $\hat{x}$ and $\hat{y}$ are some accumulation point of the sequences $\left\{\frac{x^k}{\sqrt{w_1^k}}\right\}$ and $\left\{\frac{y^k}{\sqrt{w_1^k}}\right\}$, respectively. Thus, we prove that $\phi'_{\rm FB}(x^k, y^k) \rightarrow [V_x - I \;\; V_y - I]$ as $k \rightarrow \infty$ with $V_x \in \{L_{\hat{x}}\}$ and $V_y \in \{L_{\hat{y}}\}$.

Case (2): $w_2^k \neq 0$ for all k. Now $(\phi_{\rm FB})'_x(x^k, y^k)$ and $(\phi_{\rm FB})'_y(x^k, y^k)$ are given as in (3.43). Using the same arguments as part(b) and noting the boundedness of $\{\bar{w}_2^k\}$, we have

$$L_1(w^k)L_{x^k} \rightarrow \frac{1}{2}\begin{bmatrix} 1 \\ -\bar{w}_2 \end{bmatrix} u^{\mathsf{T}}, \quad L_1(w^k)L_{y^k} \rightarrow \frac{1}{2}\begin{bmatrix} 1 \\ -\bar{w}_2 \end{bmatrix} v^{\mathsf{T}} \tag{3.46}$$

for some $u = (u_1, u_2), v = (v_1, v_2) \in \mathbb{R} \times \mathbb{R}^{n-1}$ satisfying $|u_1| \leq \|u_2\| \leq 1$ and $|v_1| \leq \|v_2\| \leq 1$, and $\bar{w}_2 \in \mathbb{R}^{n-1}$ satisfying $\|\bar{w}_2\| = 1$. We next compute the limit of $L_2(w^k)L_{x^k}$ and $L_2(w^k)L_{y^k}$ as $k \rightarrow \infty$. By the definition of $L_2(w)$ in (3.42),

$$L_2(w^k)L_{x^k} = \frac{1}{2}\begin{bmatrix} \xi_1^k & (\xi_2^k)^{\mathsf{T}} \\ \xi_1^k \bar{w}_2^k + 4\left(I - \bar{w}_2^k(\bar{w}_2^k)^{\mathsf{T}}\right) s_2^k & \bar{w}_2^k(\xi_2^k)^{\mathsf{T}} + 4\left(I - \bar{w}_2^k(\bar{w}_2^k)^{\mathsf{T}}\right) s_1^k \end{bmatrix},$$

$$L_2(w^k)L_{y^k} = \frac{1}{2}\begin{bmatrix} \eta_1^k & (\eta_2^k)^{\mathsf{T}} \\ \eta_1^k \bar{w}_2^k + 4\left(I - \bar{w}_2^k(\bar{w}_2^k)^{\mathsf{T}}\right) \omega_2^k & \bar{w}_2^k(\eta_2^k)^{\mathsf{T}} + 4\left(I - \bar{w}_2^k(\bar{w}_2^k)^{\mathsf{T}}\right) \omega_1^k \end{bmatrix},$$

where

$$\xi_1^k = \frac{x_1^k + (x_2^k)^{\mathsf{T}}\bar{w}_2^k}{\sqrt{\lambda_2(w^k)}}, \quad \xi_2^k = \frac{x_2^k + x_1^k\bar{w}_2^k}{\sqrt{\lambda_2(w^k)}}, \quad \eta_1^k = \frac{y_1^k + (y_2^k)^{\mathsf{T}}\bar{w}_2^k}{\sqrt{\lambda_2(w^k)}}, \quad \eta_2^k = \frac{y_2^k + y_1^k\bar{w}_2^k}{\sqrt{\lambda_2(w^k)}},$$

and

$$s_1^k = \frac{x_1^k}{\sqrt{\lambda_2(w^k)} + \sqrt{\lambda_1(w^k)}}, \quad s_2^k = \frac{x_2^k}{\sqrt{\lambda_2(w^k)} + \sqrt{\lambda_1(w^k)}},$$
$$\omega_1^k = \frac{y_1^k}{\sqrt{\lambda_2(w^k)} + \sqrt{\lambda_1(w^k)}}, \quad \omega_2^k = \frac{y_2^k}{\sqrt{\lambda_2(w^k)} + \sqrt{\lambda_1(w^k)}}.$$

By Lemma 3.3, $|\xi_1^k| \le \|\xi_2^k\| \le 1$ and $|\eta_1^k| \le \|\eta_2^k\| \le 1$. In addition,

$$\|s^k\|^2 + \|\omega^k\|^2 = \frac{\|x^k\|^2 + \|y^k\|^2}{2(\|x^k\|^2 + \|y^k\|^2) + 2\sqrt{\lambda_1(w^k)}\sqrt{\lambda_2(w^k)}} \le \frac{1}{2}.$$

Hence, taking limit (possibly on a subsequence) on $L_2(w^k)L_{x^k}$ and $L_2(w^k)L_{y^k}$ yields

$$\begin{aligned}
L_2(w^k)L_{x^k} \to & \frac{1}{2}\begin{bmatrix} \xi_1 & \xi_2^{\mathsf{T}} \\ \xi_1\bar{w}_2 + 4(I - \bar{w}_2\bar{w}_2^{\mathsf{T}})s_2 & \bar{w}_2\xi_2^{\mathsf{T}} + 4(I - \bar{w}_2\bar{w}_2^{\mathsf{T}})s_1 \end{bmatrix} \\
= & \frac{1}{2}\begin{bmatrix} 1 \\ \bar{w}_2 \end{bmatrix}\xi^{\mathsf{T}} + 2\begin{bmatrix} 0 & 0 \\ (I - \bar{w}_2\bar{w}_2^{\mathsf{T}})s_2 & (I - \bar{w}_2\bar{w}_2^{\mathsf{T}})s_1 \end{bmatrix}, \\
L_2(w^k)L_{y^k} \to & \frac{1}{2}\begin{bmatrix} \eta_1 & \eta_2^{T} \\ \eta_1\bar{w}_2 + 4(I - \bar{w}_2\bar{w}_2^{\mathsf{T}})\omega_2 & \bar{w}_2\eta_2^{T} + 4(I - \bar{w}_2\bar{w}_2^{\mathsf{T}})\omega_1 \end{bmatrix} \\
= & \frac{1}{2}\begin{bmatrix} 1 \\ \bar{w}_2 \end{bmatrix}\eta^{\mathsf{T}} + 2\begin{bmatrix} 0 & 0 \\ (I - \bar{w}_2\bar{w}_2^{\mathsf{T}})\omega_2 & (I - \bar{w}_2\bar{w}_2^{\mathsf{T}})\omega_1 \end{bmatrix}
\end{aligned} \tag{3.47}$$

for some vectors $\xi = (\xi_1, \xi_2), \eta = (\eta_1, \eta_2) \in \mathbb{R} \times \mathbb{R}^{n-1}$ satisfying $|\xi_1| \le \|\xi_2\| \le 1$ and $|\eta_1| \le \|\eta_2\| \le 1$, $\bar{w}_2 \in \mathbb{R}^{n-1}$ satisfying $\|\bar{w}_2\| = 1$, and $s = (s_1, s_2), \omega = (\omega_1, \omega_2) \in \mathbb{R} \times \mathbb{R}^{n-1}$ satisfying $\|s\|^2 + \|\omega\|^2 \le 1/2$. Among others, ξ and η are some accumulation point of the sequences $\{\xi^k\}$ and $\{\eta^k\}$, respectively; and s and ω are some accumulation point of the sequences $\{s^k\}$ and $\{\omega^k\}$, respectively. From (3.43), (3.46) and (3.47), we obtain

$$(\phi_{\text{FB}})'_x(x^k, y^k) \to \frac{1}{2}\begin{bmatrix} 1 \\ \bar{w}_2 \end{bmatrix}\xi^{\mathsf{T}} + \frac{1}{2}\begin{bmatrix} 1 \\ -\bar{w}_2 \end{bmatrix}u^{\mathsf{T}} + 2\begin{bmatrix} 0 & 0 \\ (I - \bar{w}_2\bar{w}_2^{\mathsf{T}})s_2 & (I - \bar{w}_2\bar{w}_2^{\mathsf{T}})s_1 \end{bmatrix} - I,$$
$$(\phi_{\text{FB}})'_y(x^k, y^k) \to \frac{1}{2}\begin{bmatrix} 1 \\ \bar{w}_2 \end{bmatrix}\eta^{\mathsf{T}} + \frac{1}{2}\begin{bmatrix} 1 \\ -\bar{w}_2 \end{bmatrix}v^{\mathsf{T}} + 2\begin{bmatrix} 0 & 0 \\ (I - \bar{w}_2\bar{w}_2^{\mathsf{T}})\omega_2 & (I - \bar{w}_2\bar{w}_2^{\mathsf{T}})\omega_1 \end{bmatrix} - I.$$

This implies that as $k \to \infty$, $(\phi_{\text{FB}})'(x^k, y^k) \to [V_x - I \;\; V_y - I]$ with V_x and V_y satisfying (3.41). Combining with Case (1) then yields the desired result. □

By introducing the vector-valued function $\Phi_{\text{FB}} : \mathbb{R}^n \to \mathbb{R}^m$ defined below, it becomes evident that the SOCCP (3.4) can be reformulated as the following nonsmooth system of equations:

$$\Phi_{\rm FB}(\zeta) := \begin{pmatrix} \phi_{\rm FB}(F_1(\zeta), G_1(\zeta)) \\ \vdots \\ \phi_{\rm FB}(F_i(\zeta), G_i(\zeta)) \\ \vdots \\ \phi_{\rm FB}(F_q(\zeta), G_q(\zeta)) \end{pmatrix} = 0, \tag{3.48}$$

where $\phi_{\rm FB}$ is defined as in (3.10) with a suitable dimension. Consequently, its squared norm gives rise to a smooth merit function, defined as follows:

$$\Psi_{\rm FB}(\zeta) := \frac{1}{2}\|\Phi_{\rm FB}(\zeta)\|^2 = \sum_{i=1}^{q} \psi_{\rm FB}(F_i(\zeta), G_i(\zeta)). \tag{3.49}$$

Remark 3.1 When $x^2 + y^2 \in {\rm bd}(\mathcal{K}^l)$ with $(x, y) \neq (0, 0)$, using Lemma 3.2, we can also characterize V_x and V_y in Proposition 3.9(b) by

$$V_x \in \left\{ \frac{1}{\sqrt{2w_1}} \begin{bmatrix} x_1 & x_2^{\rm T} \\ x_2 & 2x_1 I - \frac{w_2 x_2^{\rm T}}{w_1} \end{bmatrix} + \frac{1}{2} \begin{bmatrix} 1 \\ \frac{-w_2}{\|w_2\|} \end{bmatrix} u^{\rm T} \right\}$$

$$V_y \in \left\{ \frac{1}{\sqrt{2w_1}} \begin{bmatrix} y_1 & y_2^{\rm T} \\ y_2 & 2y_1 I - \frac{w_2 y_2^{\rm T}}{w_1} \end{bmatrix} + \frac{1}{2} \begin{bmatrix} 1 \\ \frac{-w_2}{\|w_2\|} \end{bmatrix} v^{\rm T} \right\}$$

for some $u = (u_1, u_2), v = (v_1, v_2)$ satisfying $|u_1| \leq \|u_2\| \leq 1$ and $|v_1| \leq \|v_2\| \leq 1$.

To analyze the generalized Newton method, we begin by providing an estimate for the B-subdifferential of $\Phi_{\rm FB}$, along with a sufficient condition ensuring that all elements of the B-subdifferential of $\Phi_{\rm FB}$ at a solution are nonsingular. For convenience, throughout this section, let us denote for any $i \in \{1, 2, \ldots, q\}$ and $\zeta \in \mathbb{R}^n$:

$$\begin{aligned} F_i(\zeta) = (F_{i1}(\zeta), F_{i2}(\zeta)), \quad G_i(\zeta) = (G_{i1}(\zeta), G_{i2}(\zeta)) \in \mathbb{R} \times \mathbb{R}^{n_i - 1}, \\ w_i(\zeta) = (w_{i1}(\zeta), w_{i2}(\zeta)) = w(F_i(\zeta), G_i(\zeta)), \quad z_i(\zeta) = (z_{i1}(\zeta), z_{i2}(\zeta)) = z(F_i(\zeta), G_i(\zeta)), \end{aligned}$$

where $w(x, y)$ and $z(x, y)$ are the functions defined in (3.14). Note that $\Phi_{\rm FB}$ is (strongly) semismooth if and only if all its component functions are (strongly) semismooth. Moreover, since the composition of (strongly) semismooth functions remains (strongly) semismooth by [28, Theorem 19], the following proposition follows immediately from Proposition 3.3.

Proposition 3.10 *The operator $\Phi_{\rm FB} : \mathbb{R}^n \to \mathbb{R}^m$ defined by (3.48) is semismooth. Furthermore, it is strongly semismooth if F' and G' are locally Lipschitz continuous.*

Let $(\Phi_{\rm FB})_i$ denote the i-th component of the function $\Phi_{\rm FB}$. Notice that, for any $\zeta \in \mathbb{R}^n$,

$$\partial_B \Phi_{\rm FB}(\zeta)^{\rm T} \subseteq \partial_B (\Phi_{\rm FB})_1(\zeta)^{\rm T} \times \partial_B (\Phi_{\rm FB})_2(\zeta)^{\rm T} \times \cdots \times \partial_B (\Phi_{\rm FB})_q(\zeta)^{\rm T}, \tag{3.50}$$

where the latter denotes the set of all matrices whose $(n_{i-1}+1)$ to n_i-th columns belong to $\partial_B(\Phi_{\rm FB})_i(\zeta)^{\mathsf T}$ for $i=1,2,\ldots,q$ and $n_0=0$. From Proposition 3.9 and Remark 3.1, we immediately obtain the following estimate for $\partial_B\Phi_{\rm FB}(\zeta)^{\mathsf T}$.

Proposition 3.11 *Let $\Phi_{\rm FB}:\mathbb{R}^n\to\mathbb{R}^m$ be defined by (3.48). Then, for any $\zeta\in\mathbb{R}^n$,*

$$\partial_B\Phi_{\rm FB}(\zeta)^{\mathsf T}\subseteq\nabla F(\zeta)(A(\zeta)-I)+\nabla G(\zeta)(B(\zeta)-I), \tag{3.51}$$

where $A(\zeta)$ and $B(\zeta)$ are possibly multivalued $m\times m$ block diagonal matrices whose ith blocks $A_i(\zeta)$ and $B_i(\zeta)$ for $i=1,2,\ldots,q$ have the following representation:

(a) If $F_i(\zeta)^2+G_i(\zeta)^2\in{\rm int}\mathcal{K}^{n_i}$, then $A_i(\zeta)=L_{F_i(\zeta)}L^{-1}_{z_i(\zeta)}$ and $B_i(\zeta)=L_{G_i(\zeta)}L^{-1}_{z_i(\zeta)}$.

(b) If $F_i(\zeta)^2+G_i(\zeta)^2\in{\rm bd}\mathcal{K}^{n_i}$ and $(F_i(\zeta),G_i(\zeta))\neq(0,0)$, then

$$A_i(\zeta)\in\left\{\frac{1}{\sqrt{2w_{i1}(\zeta)}}\begin{bmatrix}F_{i1}(\zeta) & F_{i2}(\zeta)^{\mathsf T}\\ F_{i2}(\zeta) & 2F_{i1}(\zeta)I-\frac{F_{i2}(\zeta)w_{i2}(\zeta)^{\mathsf T}}{w_{i1}(\zeta)}\end{bmatrix}+\frac{1}{2}u_i(1,-\bar w_{i2}(\zeta)^{\mathsf T})\right\}$$

$$B_i(\zeta)\in\left\{\frac{1}{\sqrt{2w_{i1}(\zeta)}}\begin{bmatrix}G_{i1}(\zeta) & G_{i2}(\zeta)^{\mathsf T}\\ G_{i2}(\zeta) & 2G_{i1}(\zeta)I-\frac{G_{i2}(\zeta)w_{i2}(\zeta)^{\mathsf T}}{w_{i1}(\zeta)}\end{bmatrix}+\frac{1}{2}v_i(1,-\bar w_{i2}(\zeta)^{\mathsf T})\right\}$$

for some $u_i=(u_{i1},u_{i2})$, $v_i=(v_{i1},v_{i2})\in\mathbb{R}\times\mathbb{R}^{n_i-1}$ satisfying $|u_{i1}|\le\|u_{i2}\|\le 1$ and $|v_{i1}|\le\|v_{i2}\|\le 1$, where $\bar w_{i2}(\zeta)=w_{i2}(\zeta)/\|w_{i2}(\zeta)\|$.

(c) If $F_i(\zeta)=G_i(\zeta)=0$, then

$$A_i(\zeta)\in\left\{L_{\hat u_i}\right\}\cup\left\{\frac{1}{2}\xi_i\left(1,\bar w_{i2}^{\mathsf T}\right)+\frac{1}{2}u_i\left(1,-\bar w_{i2}^{\mathsf T}\right)+\begin{bmatrix}0 & 2s_{i2}^{\mathsf T}(I-\bar w_{i2}\bar w_{i2}^{\mathsf T})\\ 0 & 2s_{i1}(I-\bar w_{i2}\bar w_{i2}^{\mathsf T})\end{bmatrix}\right\}$$

$$B_i(\zeta)\in\left\{L_{\hat v_i}\right\}\cup\left\{\frac{1}{2}\eta_i\left(1,\bar w_{i2}^{\mathsf T}\right)+\frac{1}{2}v_i\left(1,-\bar w_{i2}^{\mathsf T}\right)+\begin{bmatrix}0 & 2\omega_{i2}^{\mathsf T}(I-\bar w_{i2}\bar w_{i2}^{\mathsf T})\\ 0 & 2\omega_{i1}(I-\bar w_{i2}\bar w_{i2}^{\mathsf T})\end{bmatrix}\right\}$$

for some $\hat u_i,\hat v_i\in\mathbb{R}^{n_i}$ satisfying $\|\hat u_i\|^2+\|\hat v_i\|^2=1$, $u_i=(u_{i1},u_{i2})$, $v_i=(v_{i1},v_{i2})$, $\xi_i=(\xi_{i1},\xi_{i2})$, $\eta_i=(\eta_{i1},\eta_{i2})\in\mathbb{R}\times\mathbb{R}^{n_i-1}$ satisfying $|u_{i1}|\le\|u_{i2}\|\le 1$, $|v_{i1}|\le\|v_{i2}\|\le 1$, $|\xi_{i1}|\le\|\xi_{i2}\|\le 1$ and $|\eta_{i1}|\le\|\eta_{i2}\|\le 1$, $\bar w_{i2}\in\mathbb{R}^{n_i-1}$ satisfying $\|\bar w_{i2}\|=1$, and $s_i=(s_{i1},s_{i2})$, $\omega_i=(\omega_{i1},\omega_{i2})\in\mathbb{R}\times\mathbb{R}^{n_i-1}$ such that $\|s_i\|^2+\|\omega_i\|^2\le 1/2$.

Lemma 3.12 *For any $\zeta\in\mathbb{R}^n$, let $A(\zeta)$ and $B(\zeta)$ be given as in Proposition 3.11. Then,*

(a) for all $i\in\{1,2,\ldots,q\}$ such that $F_i(\zeta)^2+G_i(\zeta)^2\in{\rm int}\mathcal{K}^{n_i}$, there holds that

$$\langle(A_i(\zeta)-I)v_i,(B_i(\zeta)-I)v_i\rangle\ge 0\quad\text{for any } v_i\in\mathbb{R}^{n_i};$$

(b) for all $i\in\{1,2,\ldots,q\}$, we have $\big\langle(A_i(\zeta)-I)(\Phi_{\rm FB})_i(\zeta),(B_i(\zeta)-I)(\Phi_{\rm FB})_i(\zeta)\big\rangle\ge 0$, and the inequality holds with equality if and only if $(\Phi_{\rm FB})_i(\zeta)=0$.

Proof (a) The proof is similar to that of [14, Lemma 6]. For completeness, we here include it. From Proposition 3.11(a), it follows that for any $\upsilon_i \in \mathbb{R}^{n_i}$,

$$\begin{aligned}\langle (A_i - I)\upsilon_i,\ (B_i - I)\upsilon_i\rangle &= \langle (L_{F_i}L_{z_i}^{-1} - I)\upsilon_i,\ (L_{G_i}L_{z_i}^{-1} - I)\upsilon_i\rangle \\ &= \langle (L_{F_i} - L_{z_i})L_{z_i}^{-1}\upsilon_i,\ (L_{G_i} - L_{z_i})L_{z_i}^{-1}\upsilon_i\rangle \\ &= \langle (L_{G_i} - L_{z_i})(L_{F_i} - L_{z_i})L_{z_i}^{-1}\upsilon_i,\ L_{z_i}^{-1}\upsilon_i\rangle, \quad (3.52)\end{aligned}$$

where, for convenience, we will omit the notation ζ in functions. Let S_i be the symmetric part of $(L_{G_i} - L_{z_i})(L_{F_i} - L_{z_i})$. Then, by computing, we have

$$\begin{aligned}S_i &= \frac{1}{2}\left[(L_{G_i} - L_{z_i})(L_{F_i} - L_{z_i}) + (L_{F_i} - L_{z_i})(L_{G_i} - L_{z_i})\right] \\ &= \frac{1}{2}(L_{z_i} - L_{F_i} - L_{G_i})^2 + \frac{1}{2}(L_{z_i}^2 - L_{F_i}^2 - L_{G_i}^2).\end{aligned}$$

Notice that $z_i = (F_i^2 + G_i^2)^{1/2} \in \text{int}\mathcal{K}^{n_i}$ and $z_i^2 - F_i^2 - G_i^2 = 0 \in \mathcal{K}^{n_i}$, and hence we have $L_{z_i}^2 - L_{F_i}^2 - L_{G_i}^2 \succeq O$ by [12, Proposition 3.4]. From (3.52), it then follows that

$$\begin{aligned}\langle (A_i - I)\upsilon_i,\ (B_i - I)\upsilon_i\rangle &= \langle S_i L_{z_i}^{-1}\upsilon_i, L_{z_i}^{-1}\upsilon_i\rangle \\ &\geq \frac{1}{2}\langle (L_{z_i} - L_{F_i} - L_{G_i})^2 L_{z_i}^{-1}\upsilon_i,\ L_{z_i}^{-1}\upsilon_i\rangle \\ &= \frac{1}{2}\|(L_{z_i} - L_{F_i} - L_{G_i})L_{z_i}^{-1}\upsilon_i\|^2 \geq 0\end{aligned}$$

for any $\upsilon_i \in \mathbb{R}^{n_i}$, where the first inequality is due to the fact that $L_{z_i}^2 - L_{F_i}^2 - L_{G_i}^2 \succeq O$.

(b) From Theorem 2.6.6 of [29] and the smoothness of $\psi_{\text{FB}}(x, y)$ (see [14]), we have

$$\nabla \psi_{\text{FB}}(x, y) \;=\; \partial_B \phi_{\text{FB}}(x, y)^{\mathsf{T}} \phi_{\text{FB}}(x, y) \quad \forall\, x, y \in \mathbb{R}^n,$$

which together with Propositions 3.9 and 3.11 implies that for $i = 1, 2, \ldots, q$,

$$\begin{aligned}\nabla_x \psi_{\text{FB}}(F_i(\zeta), G_i(\zeta)) &= (A_i(\zeta) - I)(\Phi_{\text{FB}})_i(\zeta), \\ \nabla_y \psi_{\text{FB}}(F_i(\zeta), G_i(\zeta)) &= (B_i(\zeta) - I)(\Phi_{\text{FB}})_i(\zeta).\end{aligned} \qquad (3.53)$$

Using Lemma 6(b) of [14], we immediately obtain the desired result. □

In what follows, we study under what conditions all elements of the B-subdifferential $\partial_B \Phi_{\text{FB}}(\zeta)$ at a solution are nonsingular. Given a solution ζ^* of the SOCCP, we call it *non-degeneracy* if $F_i(\zeta^*) + G_i(\zeta^*) \in \text{int}\mathcal{K}^{n_i}$ for all $i \in \{1, 2, \ldots, q\}$.

Remark 3.2 Let ζ^* be a solution of the SOCCP. From [30], we know that precisely one of the following six cases holds for each block pair $(F_i(\zeta), G_i(\zeta))$:

$F_i(\zeta^*)$	$G_i(\zeta^*)$	SC
$F_i(\zeta^*) \in \text{int}\mathcal{K}^{n_i}$	$G_i(\zeta^*) = 0$	yes
$F_i(\zeta^*) = 0$	$G_i(\zeta^*) \in \text{int}\mathcal{K}^{n_i}$	yes
$F_i(\zeta^*) \in \text{bd}^+\mathcal{K}^{n_i}$	$G_i(\zeta^*) \in \text{bd}^+\mathcal{K}^{n_i}$	yes
$F_i(\zeta^*) \in \text{bd}^+\mathcal{K}^{n_i}$	$G_i(\zeta^*) = 0$	no
$F_i(\zeta^*) = 0$	$G_i(\zeta^*) \in \text{bd}^+\mathcal{K}^{n_i}$	no
$F_i(\zeta^*) = 0$	$G_i(\zeta^*) = 0$	no,

where $\text{bd}^+\mathcal{K}^{n_i} = \text{bd}\mathcal{K}^{n_i} \setminus \{0\}$, and the last column indicates whether the strict complementarity, i.e., $F_i(\zeta^*) + G_i(\zeta^*) \in \text{int}\mathcal{K}^{n_i}$, holds or not. Particularly, when the i-th block pair satisfies the strict complementarity, $A_i(\zeta^*)$ and $B_i(\zeta^*)$ have an explicit expression as shown by Lemma 3.13 below.

Lemma 3.13 *Let ζ^* be a solution to the SOCCP* (3.4). *For any $i \in \{1, 2, \ldots, q\}$, we have*

(a) $A_i(\zeta^*) = 0$ and $B_i(\zeta^*) = I$ if $F_i(\zeta^*) = 0$ and $G_i(\zeta^*) \in \text{int}\mathcal{K}^{n_i}$;
(b) $A_i(\zeta^*) = I$ and $B_i(\zeta^*) = 0$ if $F_i(\zeta^*) \in \text{int}\mathcal{K}^{n_i}$ and $G_i(\zeta^*) = 0$;
(c) $A_i(\zeta^*) = L_{F_i(\zeta^*)}L^{-1}_{z_i(\zeta^*)}$ and $B_i(\zeta^*) = L_{G_i(\zeta^*)}L^{-1}_{z_i(\zeta^*)}$ if $F_i(\zeta^*), G_i(\zeta^*) \in \text{bd}^+\mathcal{K}^{n_i}$.

Proof (a) Since $F_i(\zeta^*)^2 + G_i(\zeta^*)^2 = G_i(\zeta^*)^2 \in \text{int}\mathcal{K}^{n_i}$, by Proposition 3.11(a),

$$A_i(\zeta^*) = L_{F_i(\zeta^*)}L^{-1}_{z_i(\zeta^*)} = 0 \quad \text{and} \quad B_i(\zeta^*) = L_{G_i(\zeta^*)}L^{-1}_{z_i(\zeta^*)} = L_{G_i(\zeta^*)}L^{-1}_{G_i(\zeta^*)} = I.$$

Similarly, we can prove that part(b) holds. Next we consider part(c). We claim that $F_i(\zeta^*)^2 + G_i(\zeta^*)^2 \in \text{int}\mathcal{K}^{n_i}$. Suppose not, then $F_i(\zeta^*)^2 + G_i(\zeta^*)^2 \in \text{bd}^+\mathcal{K}^{n_i}$, which by Lemma 3.3 implies that $F_{i1}(\zeta^*)G_{i1}(\zeta^*) = F_{i2}(\zeta^*)^T G_{i2}(\zeta^*)$. On the other hand, since $F_i(\zeta^*) \in \text{bd}^+\mathcal{K}^{n_i}$ and $G_i(\zeta^*) \in \text{bd}^+\mathcal{K}^{n_i}$, we have that

$$F_{i1}(\zeta^*) = \|F_{i2}(\zeta^*)\|, \quad G_{i1}(\zeta^*) = \|G_{i2}(\zeta^*)\|. \tag{3.54}$$

Combining the two sides then yields that $\|F_{i2}(\zeta^*)\| \cdot \|G_{i2}(\zeta^*)\| = F_{i2}(\zeta^*)^T G_{i2}(\zeta^*)$. This implies that $F_{i2}(\zeta^*) = \alpha G_{i2}(\zeta^*)$ for some $\alpha > 0$. Combining with (3.54) then yields $F_{i1}(\zeta^*) = \alpha G_{i1}(\zeta^*)$. Therefore, $F_i(\zeta^*) = \alpha G_i(\zeta^*)$. Noting that $F_i(\zeta^*)^T G_i(\zeta^*) = 0$ since ζ^* is a solution of the SOCCP, we have $F_i(\zeta^*) = G_i(\zeta^*) = 0$. This clearly contradicts the given assumption. Using Proposition 3.11(a), we then obtain the desired result. □

By Remark 3.2, if ζ^* is a nondegenerate solution of the SOCCP (3.4), then the index sets

$$\mathcal{I} := \left\{i \in \{1, 2, \ldots, q\} \mid F_i(\zeta^*) = 0,\ G_i(\zeta^*) \in \text{int}\mathcal{K}^{n_i}\right\},$$
$$\mathcal{B} := \left\{i \in \{1, 2, \ldots, q\} \mid F_i(\zeta^*) \in \text{bd}^+\mathcal{K}^{n_i},\ G_i(\zeta^*) \in \text{bd}^+\mathcal{K}^{n_i}\right\},$$

$$\mathcal{J} := \left\{ i \in \{1, 2, \ldots, q\} \mid F_i(\zeta^*) \in \text{int}\mathcal{K}^{n_i},\ G_i(\zeta^*) = 0 \right\} \tag{3.55}$$

form a partition of $\{1, 2, \ldots, q\}$. Thus, if $n = m$, by supposing that $\nabla G(\zeta^*)$ is invertible and rearranging the matrices appropriately, $P(\zeta^*) = \nabla G(\zeta^*)^{-1} \nabla F(\zeta^*)$ can be rewritten as

$$P(\zeta^*) = \begin{bmatrix} P(\zeta^*)_{\mathcal{I}\mathcal{I}} & P(\zeta^*)_{\mathcal{I}\mathcal{B}} & P(\zeta^*)_{\mathcal{I}\mathcal{J}} \\ P(\zeta^*)_{\mathcal{B}\mathcal{I}} & P(\zeta^*)_{\mathcal{B}\mathcal{B}} & P(\zeta^*)_{\mathcal{B}\mathcal{J}} \\ P(\zeta^*)_{\mathcal{J}\mathcal{I}} & P(\zeta^*)_{\mathcal{J}\mathcal{B}} & P(\zeta^*)_{\mathcal{J}\mathcal{J}} \end{bmatrix}.$$

We are now in a position to establish the following nonsingularity result, assuming that the given solution is nondegenerate.

Proposition 3.12 *Let ζ^* be a nondegenerate solution to the SOCCP* (3.4) *and $\mathcal{I}$, $\mathcal{B}$, $\mathcal{J}$ be index sets described by* (3.55). *Suppose that $n = m$ and $\nabla G(\zeta^*)$ is invertible. Let $P(\zeta^*) = \nabla G(\zeta^*)^{-1} \nabla F(\zeta^*)$. If $P(\zeta^*)_{\mathcal{I}\mathcal{I}}$ is nonsingular and its Schur-complement, denoted by $\widehat{P}(\zeta^*)_{\mathcal{I}\mathcal{I}}$, in the matrix*

$$\begin{bmatrix} P(\zeta^*)_{\mathcal{I}\mathcal{I}} & P(\zeta^*)_{\mathcal{I}\mathcal{B}} \\ P(\zeta^*)_{\mathcal{B}\mathcal{I}} & P(\zeta^*)_{\mathcal{B}\mathcal{B}} \end{bmatrix}$$

has the Cartesian P-property, then all $W \in \partial_B \Phi_{\rm FB}(\zeta^*)$ are nonsingular.

Proof Using (3.51) and noting that $\nabla G(\zeta^*)$ is invertible, it suffices to show that any matrix C belonging to $\nabla G(\zeta^*)^{-1} \nabla F(\zeta^*)(A(\zeta^*) - I) + (B(\zeta^*) - I)$ is invertible. By Lemma 3.13 and Proposition 3.11(a), C can be written in the following partitioned form

$$C = \begin{bmatrix} -P_{\mathcal{I}\mathcal{I}} & P_{\mathcal{I}\mathcal{B}}(A_{\mathcal{B}} - I_{\mathcal{B}}) & 0_{\mathcal{I}\mathcal{J}} \\ -P_{\mathcal{B}\mathcal{I}} & P_{\mathcal{B}\mathcal{B}}(A_{\mathcal{B}} - I_{\mathcal{B}}) + (B_{\mathcal{B}} - I_{\mathcal{B}}) & 0_{\mathcal{B}\mathcal{J}} \\ -P_{\mathcal{J}\mathcal{I}} & P_{\mathcal{J}\mathcal{B}}(A_{\mathcal{B}} - I_{\mathcal{B}}) & -I_{\mathcal{J}} \end{bmatrix},$$

where $I_{\mathcal{B}} = \text{diag}(I_i,\ i \in \mathcal{B})$ with I_i being an $n_i \times n_i$ identity matrix, $A_{\mathcal{B}} = \text{diag}(A_i,\ i \in \mathcal{B})$ and $B_{\mathcal{B}} = \text{diag}(B_i,\ i \in \mathcal{B})$. For simplicity, we here omit the notation ζ^* in the functions. It is not hard to see that these C are nonsingular if and only if

$$C_r = \begin{bmatrix} -P_{\mathcal{I}\mathcal{I}} & P_{\mathcal{I}\mathcal{B}}(A_{\mathcal{B}} - I_{\mathcal{B}}) \\ -P_{\mathcal{B}\mathcal{I}} & P_{\mathcal{B}\mathcal{B}}(A_{\mathcal{B}} - I_{\mathcal{B}}) + (B_{\mathcal{B}} - I_{\mathcal{B}}) \end{bmatrix}$$

is nonsingular. Showing that the matrix C_r is nonsingular is equivalent to proving that the only solution of the following system

$$-C_r y = -C_r \begin{bmatrix} y_{\mathcal{I}} \\ y_{\mathcal{B}} \end{bmatrix} = 0$$

is the zero vector. This system can be rewritten as

$$\begin{cases} P_{\mathcal{I}\mathcal{I}} y_{\mathcal{I}} + P_{\mathcal{I}\mathcal{B}}(I_{\mathcal{B}} - A_{\mathcal{B}}) y_{\mathcal{B}} = 0, \\ P_{\mathcal{B}\mathcal{I}} y_{\mathcal{I}} + P_{\mathcal{B}\mathcal{B}}(I_{\mathcal{B}} - A_{\mathcal{B}}) y_{\mathcal{B}} = -(I_{\mathcal{B}} - B_{\mathcal{B}}) y_{\mathcal{B}}. \end{cases}$$

Recalling that $P_{\mathcal{I}\mathcal{I}}$ is nonsingular, we obtain from the last system that

$$\begin{cases} y_{\mathcal{I}} = -P_{\mathcal{I}\mathcal{I}}^{-1} P_{\mathcal{I}\mathcal{B}}(I_{\mathcal{B}} - A_{\mathcal{B}}) y_{\mathcal{B}}, \\ (P_{\mathcal{B}\mathcal{B}} - P_{\mathcal{B}\mathcal{I}} P_{\mathcal{I}\mathcal{I}}^{-1} P_{\mathcal{I}\mathcal{B}})(I_{\mathcal{B}} - A_{\mathcal{B}}) y_{\mathcal{B}} = -(I_{\mathcal{B}} - B_{\mathcal{B}}) y_{\mathcal{B}}. \end{cases} \tag{3.56}$$

Therefore, establishing the nonsingularity of C_r reduces to demonstrating that the second equation admits only the zero vector as its solution. We proceed by contradiction: assume that $y_{\mathcal{B}} \neq 0$, and consider the following two cases.

Case (1): $(I_{\mathcal{B}} - A_{\mathcal{B}}) y_{\mathcal{B}} = 0$. Define $J_{\mathcal{B}} := \{i \in \mathcal{B} : (y_{\mathcal{B}})_i \neq 0\}$. Then $J_{\mathcal{B}} \neq \emptyset$. Moreover,

$$(I - A_i(\zeta^*))(y_{\mathcal{B}})_i = 0 \text{ and } (I - B_i(\zeta^*))(y_{\mathcal{B}})_i = 0 \text{ for all } i \in J_{\mathcal{B}},$$

where the second equality is from the second equation of (3.56). This means that

$$\left[(I - A_i(\zeta^*)) + (I - B_i(\zeta^*))\right] (y_{\mathcal{B}})_i = 0, \quad \forall i \in J_{\mathcal{B}}.$$

Note that $(y_{\mathcal{B}})_i \neq 0$ for all $i \in J_{\mathcal{B}}$, and hence the last equation implies that the matrix

$$\left[2I - A_i(\zeta^*) - B_i(\zeta^*)\right] \quad \forall i \in J_{\mathcal{B}}$$

is singular. On the other hand, from Lemma 3.13(c), it follows that

$$\begin{aligned} 2I - A_i(\zeta^*) - B_i(\zeta^*) &= 2I - L_{F_i(\zeta^*)} L_{z_i(\zeta^*)}^{-1} - L_{G_i(\zeta^*)} L_{z_i(\zeta^*)}^{-1} \\ &= \left[2L_{z_i(\zeta^*)} - L_{F_i(\zeta^*)} - L_{G_i(\zeta^*)}\right] L_{z_i(\zeta^*)}^{-1} \\ &= \left[L_{2z_i(\zeta^*)} - L_{F_i(\zeta^*)+G_i(\zeta^*)}\right] L_{z_i(\zeta^*)}^{-1}, \quad \forall i \in \mathcal{B}. \end{aligned} \tag{3.57}$$

Notice that $w_i(\zeta^*), z_i(\zeta^*) \in \text{int}\mathcal{K}^{n_i}$ for each $i \in \mathcal{B}$, and furthermore,

$$4z_i(\zeta^*)^2 - [F_i(\zeta^*) + G_i(\zeta^*)]^2 = 2w_i(\zeta^*) + [F_i(\zeta^*) - G_i(\zeta^*)]^2 \in \text{int}\mathcal{K}^{n_i}.$$

Using Proposition 3.4 of [12] then yields that $[2z_i(\zeta^*) - (F_i(\zeta^*) + G_i(\zeta^*))] \in \text{int}\mathcal{K}^{n_i}$, which implies that $L_{2z_i(\zeta^*)} - L_{F_i(\zeta^*)+G_i(\zeta^*)} \succ O$. Combining with (3.57), we obtain that $2I - A_i(\zeta^*) - B_i(\zeta^*)$ for each $i \in \mathcal{J}_{\mathcal{B}}$ is nonsingular. This leads to a contradiction.

Case (2): $(I_{\mathcal{B}} - A_{\mathcal{B}}) y_{\mathcal{B}} \neq 0$. Notice that $F_i(\zeta^*)^2 + G_i(\zeta^*)^2 \in \text{int}\mathcal{K}^{n_i}$ for each $i \in \mathcal{B}$ by Lemma 3.13(c), and hence applying Lemma 3.12(a) yields that

$$\langle [(I_{\mathcal{B}} - A_{\mathcal{B}}) y_{\mathcal{B}}]_i, [(B_{\mathcal{B}} - I_{\mathcal{B}}) y_{\mathcal{B}}]_i \rangle \leq 0 \quad \text{for } \forall i \in \mathcal{B}.$$

This together with the second equation in (3.56) means that

$$\left\langle [(I_{\mathcal{B}} - A_{\mathcal{B}}) y_{\mathcal{B}}]_i, [(P_{\mathcal{B}\mathcal{B}} - P_{\mathcal{B}\mathcal{I}} P_{\mathcal{I}\mathcal{I}}^{-1} P_{\mathcal{I}\mathcal{B}})(I_{\mathcal{B}} - A_{\mathcal{B}}) y_{\mathcal{B}}]_i \right\rangle \leq 0, \quad \forall i \in \mathcal{B}.$$

Since $P_{\mathcal{BB}} - P_{\mathcal{BI}} P_{\mathcal{II}}^{-1} P_{\mathcal{IB}}$ is exactly $\widehat{P}_{\mathcal{II}}$, using the Cartesian P-property of $\widehat{P}_{\mathcal{II}}$, this is only possible if $(I_{\mathcal{B}} - A_{\mathcal{B}})y_{\mathcal{B}} = 0$, and again we obtain a contradiction. $\square$

We now turn our attention to the stationary point property and the coerciveness of the function Ψ. In particular, we aim to present a condition, less restrictive than that in [14, Proposition 3], that ensures every stationary point of Ψ is a solution of the SOCCP. Furthermore, we establish that the function Ψ_{FB} associated with the SOCCP (3.4) is coercive under the assumption that F satisfies the uniform Cartesian P-property. To prove the first result, we begin with the following technical lemma.

Lemma 3.14 *Let $\psi_{\text{FB}} : \mathbb{R}^n \times \mathbb{R}^n \to \mathbb{R}_+$ be given by (3.11). Then, for any $x, y \in \mathbb{R}^n$,*

$$\phi_{\text{FB}}(x, y) \neq 0 \iff \nabla_x \psi_{\text{FB}}(x, y) \neq 0, \ \nabla_y \psi_{\text{FB}}(x, y) \neq 0.$$

Proof The equivalence is direct by Proposition 3.2. $\square$

Proposition 3.13 *Let $\Psi_{\text{FB}} : \mathbb{R}^n \to \mathbb{R}_+$ be given by (3.49). Suppose that $n = m$ and ∇G is invertible. If $\nabla G(\zeta)^{-1} \nabla F(\zeta)$ at any $\zeta \in \mathbb{R}^n$ has the Cartesian P_0-property, then every stationary point of Ψ_{FB} is a solution to the SOCCP.*

Proof Since Ψ_{FB} is continuously differentiable by Proposition 3.5 and Φ_{FB} is locally Lipschitz continuous, we have by Clarke [29] that for any $\zeta \in \mathbb{R}^n$ and any $V \in \partial \Phi(\zeta)^{\mathsf{T}}$

$$\nabla \Psi(\zeta) = V \Phi(\zeta).$$

Let ζ be an arbitrary stationary point of Ψ_{FB} and V be an element of $\partial_B \Phi_{\text{FB}}(\zeta)^{\mathsf{T}} (\subseteq \partial \Phi(\zeta)^{\mathsf{T}})$. From Eq. (3.50), it follows that there exist matrices $V_i \in \partial_B (\Phi_{\text{FB}})_i(\zeta)^{\mathsf{T}}$ such that

$$V = V_1 \times V_2 \times \cdots \times V_q.$$

In addition, for each $V_i \in \mathbb{R}^{n \times n_i}$, by Proposition 3.9 there exist matrices $A_i(\zeta) \in \mathbb{R}^{n_i \times n_i}$ and $B_i(\zeta) \in \mathbb{R}^{n_i \times n_i}$, as characterized by Proposition 3.11, such that

$$V_i = \nabla F_i(\zeta)(A_i(\zeta) - I) + \nabla G_i(\zeta)(B_i(\zeta) - I), \quad i = 1, 2, \ldots, q.$$

Let $A(\zeta) = \text{diag}(A_1(\zeta), \ldots, A_q(\zeta))$ and $B(\zeta) = \text{diag}(B_1(\zeta), \ldots, B_q(\zeta))$. Combining the last three equations, it then follows that

$$[\nabla F(\zeta)(A(\zeta) - I) + \nabla G(\zeta)(B(\zeta) - I)] \, \Phi_{\text{FB}}(\zeta) = 0,$$

which, by the invertibility of ∇G, is equivalent to

$$\left[\nabla G(\zeta)^{-1} \nabla F(\zeta)(A(\zeta) - I) + (B(\zeta) - I)\right] \Phi_{\text{FB}}(\zeta) = 0. \tag{3.58}$$

We next prove that $\Phi_{\text{FB}}(\zeta) = 0$. Suppose not, then there is an index $\nu \in \{1, 2, \ldots, q\}$ such that $(\Phi_{\text{FB}})_\nu(\zeta) = \phi_{\text{FB}}(F_\nu(\zeta), G_\nu(\zeta)) \neq 0$. From Propositions 3.9 and 3.11, we notice that

$$\begin{bmatrix} \nabla_x \psi(F_\nu(\zeta), G_\nu(\zeta)) \\ \nabla_y \psi(F_\nu(\zeta), G_\nu(\zeta)) \end{bmatrix} = \begin{bmatrix} (A_\nu(\zeta) - I)\Phi_\nu(\zeta) \\ (B_\nu(\zeta) - I)\Phi_\nu(\zeta) \end{bmatrix}.$$

Therefore, applying Lemma 3.14 yields

$$(A_\nu(\zeta) - I)(\Phi_{\rm FB})_\nu(\zeta) \neq 0 \quad \text{and} \quad (B_\nu(\zeta) - I)(\Phi_{\rm FB})_\nu(\zeta) \neq 0. \tag{3.59}$$

In addition, from (3.58), it follows that

$$\left[\nabla G(\zeta)^{-1} \nabla F(\zeta)(A(\zeta) - I)\Phi_{\rm FB}(\zeta)\right]_\nu + (B_\nu(\zeta) - I)(\Phi_{\rm FB})_\nu(\zeta) = 0.$$

Making the inner product with $(A_\nu(\zeta) - I)(\Phi_{\rm FB})_\nu(\zeta)$ on both sides, we obtain

$$\begin{aligned} &\left\langle (A_\nu(\zeta) - I)(\Phi_{\rm FB})_\nu(\zeta),\ \left[\nabla G(\zeta)^{-1} \nabla F(\zeta)(A(\zeta) - I)\Phi_{\rm FB}(\zeta)\right]_\nu \right\rangle \\ &+ \left\langle (A_\nu(\zeta) - I)(\Phi_{\rm FB})_\nu(\zeta),\ (B_\nu(\zeta) - I)(\Phi_{\rm FB})_\nu(\zeta) \right\rangle = 0. \end{aligned}$$

Notice that the first term on the left-hand side is nonnegative by (3.59) and the Cartesian P_0-property of $\nabla G(\zeta)^{-1} \nabla F(\zeta)$, and the second term is positive by Lemma 3.12(b) since $(\Phi_{\rm FB})_\nu(\zeta) \neq 0$. This leads to a contradiction. □

When ∇G is invertible, it follows from [14] that the column monotonicity of $\nabla F(\zeta)$ and $-\nabla G(\zeta)$ is equivalent to the condition $\nabla G(\zeta)^{-1} \nabla F(\zeta) \succeq O$, which clearly implies that $\nabla G(\zeta)^{-1} \nabla F(\zeta)$ possesses the Cartesian P_0-property. Consequently, the stationary point condition in Proposition 3.13 is weaker than the condition employed in [14, Proposition 3]. Moreover, for the SOCCP (3.4), this condition is equivalent to requiring that F satisfies the Cartesian P_0-property, which reduces to the familiar stationary point condition that F is a P_0-function in the case of the NCP.

Lemma 3.15 *Let $\psi_{\rm FB}$ be defined as in (3.11). For any sequence $\{(x^k, y^k)\} \subseteq \mathbb{R}^n \times \mathbb{R}^n$, let $\lambda_1^k \leq \lambda_2^k$ and $\mu_1^k \leq \mu_2^k$ denote the spectral values of x^k and y^k, respectively.*

(a) If $\lambda_1^k \to -\infty$ or $\mu_1^k \to -\infty$ as $k \to \infty$, then $\psi_{\rm FB}(x^k, y^k) \to +\infty$.
(b) If $\{\lambda_1^k\}$ and $\{\mu_1^k\}$ are bounded below, but $\lambda_2^k \to +\infty$, $\mu_2^k \to +\infty$, and $\frac{x^k}{\|x^k\|} \circ \frac{y^k}{\|y^k\|} \nrightarrow 0$ as $k \to \infty$, then $\psi_{\rm FB}(x^k, y^k) \to +\infty$.

Proof Part (a) is indeed Lemma 3.8. We next prove part (b). Suppose that $\{\phi_{\rm FB}(x^k, y^k)\}$ is bounded. Let $z^k = [(x^k)^2 + (y^k)^2]^{1/2}$ for each k. From the definition of $\phi_{\rm FB}$,

$$x^k + y^k = z^k - \phi(x^k, y^k), \quad \forall\, k.$$

Squaring two sides of the last equality then yields that

$$2x^k \circ y^k = -2z^k \circ \phi(x^k, y^k) + \phi(x^k, y^k)^2, \quad \forall\, k. \tag{3.60}$$

Since $\|x^k\| \to +\infty$ and $\|y^k\| \to +\infty$ by the given conditions, we have that

$$\lim_{k\to\infty} \frac{z^k}{\|x^k\|\|y^k\|} = \lim_{k\to\infty} \left[\frac{(x^k)^2}{\|x^k\|^2\|y^k\|^2} + \frac{(y^k)^2}{\|x^k\|^2\|y^k\|^2} \right]^{1/2} = 0,$$

which together with the boundedness of $\{\phi_{\rm FB}(x^k, y^k)\}$ implies that

$$\lim_{k\to\infty} \frac{-2z^k \circ \phi_{\rm FB}(x^k, y^k) + \phi_{\rm FB}(x^k, y^k)^2}{\|x^k\|\|y^k\|} = 0.$$

Using the equality (3.60), we obtain $\lim_{k\to\infty} \frac{x^k}{\|x^k\|} \circ \frac{y^k}{\|y^k\|} = 0$, which clearly contradicts the given assumption. Consequently, the conclusion follows. □

Lemma 3.15 extends the result of [14, Lemma 9(a)], which plays a pivotal role in establishing the coerciveness of the merit function Ψ. It is worth emphasizing that in Lemma 3.15(b), the condition $\frac{x^k}{\|x^k\|} \circ \frac{y^k}{\|y^k\|} \nrightarrow 0$ as $k \to \infty$ is indeed necessary, as demonstrated by the following counterexample.

Example 3.1 Consider the sequences $\{x^k\}$ and $\{y^k\}$ given as follows:

$$x^k = \begin{bmatrix} k \\ -(k+1) \\ 0 \end{bmatrix} \quad \text{and} \quad y^k = \begin{bmatrix} k \\ k-1 \\ 0 \end{bmatrix} \quad \text{for each } k.$$

It is easy to verify that $\lambda_1^k = -1,\ \mu_1^k = 1$ for each k, and $\lambda_2^k \to +\infty,\ \mu_2^k \to +\infty$, but

$$\frac{x^k}{\|x^k\|} \to \begin{bmatrix} 1/\sqrt{2} \\ -1/\sqrt{2} \\ 0 \end{bmatrix}, \quad \frac{y^k}{\|y^k\|} \to \begin{bmatrix} 1/\sqrt{2} \\ 1/\sqrt{2} \\ 0 \end{bmatrix}, \quad \text{and} \quad \frac{x^k}{\|x^k\|} \circ \frac{y^k}{\|y^k\|} \to 0.$$

That is, the sequences $\{x^k\}$ and $\{y^k\}$ do not satisfy the assumption $\frac{x^k}{\|x^k\|} \circ \frac{y^k}{\|y^k\|} \nrightarrow 0$. For such sequences, by a simple computation, we have

$$\phi_{\rm FB}(x^k, y^k) = \frac{1}{2} \begin{bmatrix} \sqrt{4k^2+2+4k} + \sqrt{4k^2+2-4k} - 4k \\ 4 - (\sqrt{4k^2+2+4k} - \sqrt{4k^2+2-4k}) \\ 0 \end{bmatrix}.$$

Since

$$\lim_{k\to\infty} \sqrt{4k^2+2+4k} + \sqrt{4k^2+2-4k} - 4k = 0,$$
$$\lim_{k\to\infty} 4 - (\sqrt{4k^2+2+4k} - \sqrt{4k^2+2-4k}) = 2,$$

we have $\lim_{k\to\infty} \|\phi_{\rm FB}(x^k, y^k)\| = 1$, i.e., the conclusion of Lemma 3.15(b) does not hold.

We are now ready to establish the coerciveness of $\Psi_{\rm FB}$ for the SOCCP (3.4), under the assumption that F satisfies the uniform Cartesian P-property, along with the following additional condition.

Condition A. For any sequence $\{\zeta^k\} \subseteq \mathbb{R}^n$ satisfying $\|\zeta^k\| \to +\infty$, if there exists an index $i \in \{1, 2, \ldots, q\}$ such that $\{\lambda_1(\zeta_i^k)\}$ and $\{\lambda_1(F_i(\zeta^k))\}$ are bounded below, and $\lambda_2(\zeta_i^k), \lambda_2(F_i(\zeta^k)) \to +\infty$, then

$$\limsup_{k\to\infty} \left\langle \frac{\zeta_i^k}{\|\zeta_i^k\|}, \frac{F_i(\zeta^k)}{\|F_i(\zeta^k)\|} \right\rangle > 0.$$

Proposition 3.14 *For the SOCCP (3.4), suppose that the mapping F has the uniform Cartesian P-property and satisfies Condition A. Then, the merit function $\Psi_{\rm FB}$ is coercive.*

Proof We shall prove that $\lim_{\|\zeta^k\|\to+\infty} \Psi(\zeta^k) = +\infty$. Let $\{\zeta^k\} \subseteq \mathbb{R}^n$ be a sequence such that $\|\zeta^k\| \to +\infty$, where $\zeta^k = (\zeta_1^k, \ldots, \zeta_q^k)$ with $\zeta_i^k \in \mathbb{R}^{n_i}$. Define the index set

$$J := \left\{ i \in \{1, 2, \ldots, q\} \mid \{\zeta_i^k\} \text{ is unbounded} \right\}.$$

Since $\{\zeta^k\}$ is unbounded, $J \neq \emptyset$. Let $\{\xi^k\}$ be a bounded sequence with $\xi^k = (\xi_1^k, \ldots, \xi_q^k)$ and $\xi_i^k \in \mathbb{R}^{n_i}$ for each k, where ξ_i^k is defined as follows:

$$\xi_i^k = \begin{cases} 0 & \text{if } i \in J, \\ \zeta_i^k & \text{otherwise,} \end{cases} \qquad i = 1, 2, \ldots, q.$$

By the uniform Cartesian P-property of F, there is a constant $\rho > 0$ such that

$$\begin{aligned} \rho\|\zeta^k - \xi^k\|^2 &\leq \max\nolimits_{i=1,\ldots,m} \left\langle \zeta_i^k - \xi_i^k, F_i(\zeta^k) - F_i(\xi^k) \right\rangle \\ &= \left\langle \zeta_\nu^k, F_\nu(\zeta^k) - F_\nu(\xi^k) \right\rangle \\ &\leq \|\zeta_\nu^k\| \|F_\nu(\zeta^k) - F_\nu(\xi^k)\| \quad \text{for each } k, \end{aligned} \tag{3.61}$$

where ν is an index from $\{1, 2, \ldots, q\}$ for which the maximum is attained which we have, without loss of generality, assumed to be independent of k. Clearly, $\nu \in J$, which means that $\{\zeta_\nu^k\}$ is unbounded. Consequently, there exists a subsequence, assumed to be $\{\zeta_\nu^k\}$ without loss of generality, such that $\|\zeta_\nu^k\| \to +\infty$. Notice that

$$\|\zeta^k - \xi^k\|^2 \geq \|\zeta_\nu^k - \xi_\nu^k\|^2 = \|\zeta_\nu^k\|^2, \quad \text{for each } k.$$

Dividing the both sides of (3.61) by $\|\zeta_\nu^k\|$ then yields that

$$\rho\|\zeta_\nu^k\| \leq \|F_\nu(\zeta^k) - F_\nu(\xi^k)\| \leq \|F_\nu(\zeta^k)\| + \|F_\nu(\xi^k)\|,$$

which implies $\|F_\nu(\zeta^k)\| \to +\infty$ since $\|\zeta_\nu^k\| \to +\infty$ and $\{F_\nu(\xi^k)\}$ is bounded. Thus,

$$\|\zeta_\nu^k\| \to +\infty \quad \text{and} \quad \|F_\nu(\zeta^k)\| \to +\infty. \tag{3.62}$$

If either $\lambda_1(\zeta_\nu^k) \to -\infty$ or $\lambda_1(F_\nu(\zeta^k)) \to -\infty$, then using Lemma 3.15(a) readily yields that $\psi_{\rm FB}(\zeta_\nu^k, F_\nu(\zeta^k)) \to +\infty$, and consequently, $\Psi_{\rm FB}(\zeta^k) \to +\infty$. Otherwise, (3.62) implies that $\{\lambda_1(\zeta_\nu^k)\}$ and $\{\lambda_1(F_\nu(\zeta^k))\}$ are bounded below, but $\lambda_2(\zeta_\nu^k) \to +\infty$ and $\lambda_2(F_\nu(\zeta^k)) \to +\infty$. Using Condition A, it then follows that

$$\limsup_{k\to\infty} \left\langle \frac{\zeta_\nu^k}{\|\zeta_\nu^k\|}, \frac{F_\nu(\zeta^k)}{\|F_\nu(\zeta^k)\|} \right\rangle > 0,$$

which in turn implies that

$$\limsup_{k\to\infty} \lambda_2 \left[\frac{\zeta_\nu^k}{\|\zeta_\nu^k\|} \circ \frac{F_\nu(\zeta^k)}{\|F_\nu(\zeta^k)\|} \right] > 0.$$

From this, we have $\frac{\zeta_\nu^k}{\|\zeta_\nu^k\|} \circ \frac{F_\nu(\zeta^k)}{\|F_\nu(\zeta^k)\|} \nrightarrow 0$. This shows that the sequences $\{\zeta_\nu^k\}$ and $\{F_\nu(\zeta^k)\}$ satisfy the conditions of Lemma 3.15(b), and therefore $\Psi_{\rm FB}(\zeta^k) \to +\infty$. □

When $n_1 = \cdots = n_q = 1$, Condition A is automatically satisfied, and the uniform Cartesian P-property of F reduces to the requirement that F is a uniform P-function. Hence, Proposition 3.14 recovers the corresponding result for the Fischer–Burmeister merit function in the context of the NCP; see [31, Theorem 4.2].

B. ϕ_τ and ψ_τ functions - variants of $\phi_{\rm FB}$ and $\psi_{\rm FB}$

We now proceed to examine the following one-parameter family of functions:

$$\psi_\tau(x, y) \;:=\; \frac{1}{2}\|\phi_\tau(x, y)\|^2, \tag{3.63}$$

where τ is a fixed parameter from $(0, 4)$ and $\phi_\tau : \mathbb{R}^n \times \mathbb{R}^n \to \mathbb{R}^n$ is defined by

$$\phi_\tau(x, y) := \left[(x - y)^2 + \tau(x \circ y)\right]^{1/2} - (x + y). \tag{3.64}$$

Using this class of SOC complementarity functions (3.64), the SOCCP (3.4) can be reformulated as the following nonsmooth system of equations:

$$\Phi_\tau(\zeta) := \begin{pmatrix} \phi_\tau(F_1(\zeta), G_1(\zeta)) \\ \vdots \\ \phi_\tau(F_i(\zeta), G_i(\zeta)) \\ \vdots \\ \phi_\tau(F_m(\zeta), G_m(\zeta)) \end{pmatrix} = 0. \tag{3.65}$$

The function Φ_τ naturally induces a merit function $\Psi_\tau : \mathbb{R}^n \to \mathbb{R}_+$, defined by

$$\Psi_\tau(\zeta) = \frac{1}{2}\|\Phi_\tau(\zeta)\|^2 = \sum_{i=1}^{m} \psi_\tau(F_i(\zeta), G_i(\zeta), \tag{3.66}$$

As will be shown, the function ψ_τ serves as a merit function associated with $\mathcal{K}$, and is continuously differentiable everywhere with explicitly computable gradient formulas (see Propositions 3.15–3.17). Consequently, the SOCCP can be reformulated as the following unconstrained smooth minimization problem:

$$\min_{\zeta \in \mathbb{R}^n} f_\tau(\zeta) := \psi_\tau(F(\zeta), G(\zeta)). \tag{3.67}$$

Moreover, we establish that every stationary point of f_τ is a solution to the SOCCP, provided that ∇F and $-\nabla G$ are column monotone (see Proposition 3.18). It is worth noting that ϕ_τ reduces to ϕ_{FB} when $\tau = 2$, and its limit as $\tau \to 0$ becomes a multiple of ϕ_{NR}. This reveals a close connection between this class of merit functions and two of the most prominent ones in the literature, thereby justifying a more detailed examination. Additionally, our investigation is motivated by the work of [32], in which ϕ_τ was employed to develop a nonsmooth Newton method for the NCP.

To proceed, we first establish that ψ_τ, as defined in (3.63), is a smooth merit function. By Lemma 3.1, both ϕ_τ and ψ_τ are well defined for all $x, y \in \mathbb{R}^n$, since the following identity holds:

$$\begin{aligned} (x-y)^2 + \tau(x \circ y) &= \left(x + \frac{\tau-2}{2}y\right)^2 + \frac{\tau(4-\tau)}{4}y^2 \\ &= \left(y + \frac{\tau-2}{2}x\right)^2 + \frac{\tau(4-\tau)}{4}x^2 \in \mathcal{K}^n. \end{aligned} \tag{3.68}$$

The following proposition confirms that ψ_τ is indeed a merit function associated with $\mathcal{K}^n$.

Proposition 3.15 *Let ψ_τ and ϕ_τ be given as in (3.63) and (3.64), respectively. Then, ψ_τ and ϕ_τ are C-functions associated with the SOC, that is,*

$$\psi_\tau(x, y) = 0 \iff \phi_\tau(x, y) = 0 \iff x \in \mathcal{K}^n, \quad y \in \mathcal{K}^n, \quad \langle x, y\rangle = 0.$$

Proof The first equivalence is clear by the definition of ψ_τ. We consider the second one.

"$\Leftarrow$". Since $x \in \mathcal{K}^n$, $y \in \mathcal{K}^n$ and $\langle x, y\rangle = 0$, we have $x \circ y = 0$. Substituting it into the expression of $\phi_\tau(x, y)$ then yields that $\phi_\tau(x, y) = (x^2 + y^2)^{1/2} - (x + y) = \phi_{\mathrm{FB}}(x, y)$. From Proposition 3.2, we immediately obtain $\phi_\tau(x, y) = 0$.

"$\Rightarrow$". Suppose that $\phi_\tau(x, y) = 0$. Then, $x + y = \left[(x - y)^2 + \tau(x \circ y)\right]^{1/2}$. Squaring both sides yields $x \circ y = 0$. This implies that $x + y = (x^2 + y^2)^{1/2}$, i.e., $\phi_{\mathrm{FB}}(x, y) = 0$. From Proposition 3.2, it then follows that $x \in \mathcal{K}^n$, $y \in \mathcal{K}^n$ and $\langle x, y\rangle = 0$. $\square$

In the following, we focus on establishing the smoothness of ψ_τ. To that end, we begin by introducing some notation that will be used throughout the analysis. For any $x = (x_1, x_2) \in \mathbb{R} \times \mathbb{R}^{n-1}$ and $y = (y_1, y_2) \in \mathbb{R} \times \mathbb{R}^{n-1}$, let

$$\begin{aligned} w &= (w_1, w_2) = w(x, y) := (x - y)^2 + \tau(x \circ y), \\ z &= (z_1, z_2) = z(x, y) := \left[(x - y)^2 + \tau(x \circ y)\right]^{1/2}. \end{aligned} \tag{3.69}$$

Then, $w \in \mathcal{K}^n$ and $z \in \mathcal{K}^n$. Moreover, by the definition of Jordan product,

$$\begin{aligned} w_1 &= w_1(x, y) = \|x\|^2 + \|y\|^2 + (\tau - 2)x^\top y, \\ w_2 &= w_2(x, y) = 2(x_1x_2 + y_1y_2) + (\tau - 2)(x_1y_2 + y_1x_2). \end{aligned} \tag{3.70}$$

Let $\lambda_1(w)$ and $\lambda_2(w)$ be the spectral values of w. By Lemma 3.1, we have

$$z_1 = z_1(x, y) = \frac{\sqrt{\lambda_2(w)} + \sqrt{\lambda_1(w)}}{2}, \quad z_2 = z_2(x, y) = \frac{\sqrt{\lambda_2(w)} - \sqrt{\lambda_1(w)}}{2}\bar{w}_2, \tag{3.71}$$

where $\bar{w}_2 := \frac{w_2}{\|w_2\|}$ if $w_2 \neq 0$ and otherwise $\bar{w}_2$ is any vector in $\mathbb{R}^{n-1}$ satisfying $\|\bar{w}_2\| = 1$.

Lemma 3.16 *For any $x = (x_1, x_2)$, $y = (y_1, y_2) \in \mathbb{R} \times \mathbb{R}^{n-1}$, if $w \notin \mathrm{int}(\mathcal{K}^n)$, then*

$$x_1^2 = \|x_2\|^2,\ y_1^2 = \|y_2\|^2,\ x_1y_1 = x_2^\top y_2,\ x_1y_2 = y_1x_2; \tag{3.72}$$

$$\begin{aligned} x_1^2 + y_1^2 + (\tau - 2)x_1y_1 &= \|x_1x_2 + y_1y_2 + (\tau - 2)x_1y_2\| \\ &= \|x_2\|^2 + \|y_2\|^2 + (\tau - 2)x_2^\top y_2. \end{aligned} \tag{3.73}$$

If, in addition, $(x, y) \neq (0, 0)$, then $w_2 \neq 0$, and furthermore,

$$x_2^\top \frac{w_2}{\|w_2\|} = x_1, \quad x_1\frac{w_2}{\|w_2\|} = x_2, \quad y_2^\top \frac{w_2}{\|w_2\|} = y_1, \quad y_1\frac{w_2}{\|w_2\|} = y_2. \tag{3.74}$$

Proof Since $w = (x - y)^2 + \tau(x \circ y) \notin \mathrm{int}(\mathcal{K}^n)$, using (3.68) and [14, Lemma 3.2] yields

$$\left(x_1 + \frac{\tau-2}{2}y_1\right)^2 = \left\|x_2 + \frac{\tau-2}{2}y_2\right\|^2, \quad y_1^2 - \|y_2\|^2,$$

$$\left(x_1 + \frac{\tau-2}{2}y_1\right)y_2 = \left(x_2 + \frac{\tau-2}{2}y_2\right)y_1, \quad \left(x_1 + \frac{\tau-2}{2}y_1\right)y_1 = \left(x_2 + \frac{\tau-2}{2}y_2\right)^{\mathsf{T}} y_2;$$

$$\left(y_1 + \frac{\tau-2}{2}x_1\right)^2 = \left\|y_2 + \frac{\tau-2}{2}x_2\right\|^2, \quad x_1^2 = \|x_2\|^2,$$

$$\left(y_1 + \frac{\tau-2}{2}x_1\right)x_2 = \left(y_2 + \frac{\tau-2}{2}x_2\right)x_1, \quad \left(y_1 + \frac{\tau-2}{2}x_1\right)x_1 = \left(y_2 + \frac{\tau-2}{2}x_2\right)^{\mathsf{T}} x_2.$$

From these equalities, we readily get the results in (3.72). Since $w \in \mathcal{K}^n$ but $w \notin \text{int}(\mathcal{K}^n)$, we have $\|x\|^2 + \|y\|^2 + (\tau-2)x^{\mathsf{T}}y = \|2x_1x_2 + 2y_1y_2 + (\tau-2)(x_1y_2 + y_1x_2)\|$ by $\lambda_1(w) = 0$. Applying the relations in (3.72) then gives the equalities in (3.73). If, in addition, $(x, y) \neq (0, 0)$, then it is clear that $\|x_1x_2 + y_1y_2 + (\tau-2)x_1y_2\| = x_1^2 + y_1^2 + (\tau-2)x_1y_1 \neq 0$. To prove the equalities in (3.74), it suffices to verify that $x_2^{\mathsf{T}}\frac{w_2}{\|w_2\|} = x_1$ and $x_1\frac{w_2}{\|w_2\|} = x_2$ by the symmetry of x and y in w. The verifications are straightforward by (3.73) and $x_1y_2 = y_1x_2$. □

Lemma 3.16 characterizes the behavior of x, y when $w = (x-y)^2 + \tau(x \circ y)$ lies on the boundary of $\mathcal{K}^n$. In fact, it can be regarded as an extension of [14, Lemma 3.2]. According to Lemma 3.16, when $w \notin \text{int}(\mathcal{K}^n)$, the spectral values of w are determined as follows:

$$\lambda_1(w) = 0, \quad \lambda_2(w) = 4\left(x_1^2 + y_1^2 + (\tau-2)x_1y_1\right). \tag{3.75}$$

If $(x, y) \neq (0, 0)$ also holds, then using Eqs. (3.71), (3.73) and (3.75) yields that

$$z_1(x, y) = \sqrt{x_1^2 + y_1^2 + (\tau-2)x_1y_1}, \quad z_2(x, y) = \frac{x_1x_2 + y_1y_2 + (\tau-2)x_1y_2}{\sqrt{x_1^2 + y_1^2 + (\tau-2)x_1y_1}}.$$

Thus, if $(x, y) \neq (0, 0)$ and $(x-y)^2 + \tau(x \circ y) \notin \text{int}(\mathcal{K}^n)$, $\phi_\tau(x, y)$ can be rewritten as

$$\phi_\tau(x, y) = z(x, y) - (x + y) = \begin{pmatrix} \sqrt{x_1^2 + y_1^2 + (\tau-2)x_1y_1} - (x_1 + y_1) \\ \dfrac{x_1x_2 + y_1y_2 + (\tau-2)x_1y_2}{\sqrt{x_1^2 + y_1^2 + (\tau-2)x_1y_1}} - (x_2 + y_2) \end{pmatrix}. \tag{3.76}$$

This specific expression (3.76) will be utilized in the proof of the following key result.

Lemma 3.17 *The function $z(x, y)$ defined by (3.69) (or (3.71) equivalently) is (continuously) differentiable at a point (x, y) if and only if $(x-y)^2 + \tau(x \circ y) \in \text{int}(\mathcal{K}^n)$, and furthermore,*

$$\nabla_x z(x, y) = L_{x + \frac{\tau-2}{2}y} L_z^{-1}, \quad \nabla_y z(x, y) = L_{y + \frac{\tau-2}{2}x} L_z^{-1},$$

where

$$L_z^{-1} = \begin{cases} \begin{pmatrix} b & c\bar{w}_2^T \\ c\bar{w}_2 & aI + (b-a)\bar{w}_2\bar{w}_2^T \end{pmatrix} & \text{if } w_2 \neq 0; \\ \left(1/\sqrt{w_1}\right) I & \text{if } w_2 = 0, \end{cases}$$

with

$$\begin{aligned} a &= \frac{2}{\sqrt{\lambda_2(w)} + \sqrt{\lambda_1(w)}}, \\ b &= \frac{1}{2}\left(\frac{1}{\sqrt{\lambda_2(w)}} + \frac{1}{\sqrt{\lambda_1(w)}}\right), \\ c &= \frac{1}{2}\left(\frac{1}{\sqrt{\lambda_2(w)}} - \frac{1}{\sqrt{\lambda_1(w)}}\right). \end{aligned} \tag{3.77}$$

Proof The proof follows similarly to that of Lemma 3.10 and is therefore omitted. □

Proposition 3.16 *The function ψ_τ given by (3.63) is differentiable at every $(x, y) \in \mathbb{R}^n \times \mathbb{R}^n$. Moreover, $\nabla_x\psi_\tau(0,0) = \nabla_y\psi_\tau(0,0) = 0$; if $(x-y)^2 + \tau(x \circ y) \in \text{int}(\mathcal{K}^n)$, then*

$$\begin{aligned} \nabla_x\psi_\tau(x, y) &= \left[L_{x+\frac{\tau-2}{2}y} L_z^{-1} - I\right]\phi_\tau(x, y), \\ \nabla_y\psi_\tau(x, y) &= \left[L_{y+\frac{\tau-2}{2}x} L_z^{-1} - I\right]\phi_\tau(x, y); \end{aligned} \tag{3.78}$$

if $(x, y) \neq (0, 0)$ and $(x-y)^2 + \tau(x \circ y) \notin \text{int}(\mathcal{K}^n)$, then $x_1^2 + y_1^2 + (\tau - 2)x_1y_1 \neq 0$ and

$$\begin{aligned} \nabla_x\psi_\tau(x, y) &= \left[\frac{x_1 + \frac{\tau-2}{2}y_1}{\sqrt{x_1^2 + y_1^2 + (\tau-2)x_1y_1}} - 1\right]\phi_\tau(x, y), \\ \nabla_y\psi_\tau(x, y) &= \left[\frac{y_1 + \frac{\tau-2}{2}x_1}{\sqrt{x_1^2 + y_1^2 + (\tau-2)x_1y_1}} - 1\right]\phi_\tau(x, y). \end{aligned} \tag{3.79}$$

Proof Case (1): $(x, y) = (0, 0)$. For any $u = (u_1, u_2)$, $v = (v_1, v_2) \in \mathbb{R} \times \mathbb{R}^{n-1}$, let μ_1, μ_2 be the spectral values of $(u-v)^2 + \tau(u \circ v)$ and $\xi^{(1)}, \xi^{(2)}$ be the spectral vectors. Then,

$$\begin{aligned} 2\left[\psi_\tau(u, v) - \psi_\tau(0, 0)\right] &= \left\| [u^2 + v^2 + (\tau-2)(u \circ v)]^{1/2} - u - v \right\|^2 \\ &= \left\| \sqrt{\mu_1}\,\xi^{(1)} + \sqrt{\mu_2}\,\xi^{(2)} - u - v \right\|^2 \\ &\leq \left(\sqrt{2\mu_2} + \|u\| + \|v\|\right)^2. \end{aligned}$$

In addition, from the definition of spectral value, it follows that

$$\begin{aligned}\mu_2 = & \ \|u\|^2 + \|v\|^2 + (\tau - 2)u^T v + 2\|(u_1 u_2 + v_1 v_2) + (\tau - 2)(u_1 v_2 + v_1 u_2)\| \\ \le & \ 2\|u\|^2 + 2\|v\|^2 + 3|\tau - 2|\|u\|\|v\| \le 5(\|u\|^2 + \|v\|^2).\end{aligned}$$

Now combining the last two equations, we have $\psi_\tau(u, v) - \psi_\tau(0, 0) = O(\|u\|^2 + \|v\|^2)$. This shows that ψ_τ is differentiable at $(0, 0)$ with $\nabla_x \psi_\tau(0, 0) = \nabla_y \psi_\tau(0, 0) = 0$.

Case (2): $(x - y)^2 + \tau(x \circ y) \in \mathrm{int}(\mathcal{K}^n)$. By Lemma 3.17, $z(x, y)$ defined by (3.71) is continuously differentiable at such (x, y), and consequently $\phi_\tau(x, y)$ is also continuously differentiable at such (x, y) since $\phi_\tau(x, y) = z(x, y) - (x + y)$. Notice that

$$z^2(x, y) = \left(x + \frac{\tau - 2}{2} y\right)^2 + \frac{\tau(4 - \tau)}{4} y^2,$$

which leads to $\nabla_x z(x, y) L_z = L_{x + \frac{\tau - 2}{2} y}$ by taking differentiation on both sides of x. Since $L_z \succ O$ by Lemma 3.17, it follows that $\nabla_x z(x, y) = L_{x + \frac{\tau - 2}{2} y} L_z^{-1}$. Consequently,

$$\nabla_x \phi_\tau(x, y) = \nabla_x z(x, y) - I = L_{x + \frac{\tau - 2}{2} y} L_z^{-1} - I.$$

This together with $\nabla_x \psi_\tau(x, y) = \nabla_x \phi_\tau(x, y) \phi_\tau(x, y)$ proves the first formula of (3.78). For the symmetry of x and y in ψ_τ, the second formula also holds.

Case (3): $(x, y) \neq (0, 0)$ and $(x - y)^2 + \tau(x \circ y) \notin \mathrm{int}(\mathcal{K}^n)$. For any $x' = (x_1', x_2')$, $y' = (y_1', y_2') \in \mathbb{R} \times \mathbb{R}^{n-1}$, it is easy to verify that

$$\begin{aligned} 2\psi_\tau(x', y') &= \left\| \left[x'^2 + y'^2 + (\tau - 2)(x' \circ y') \right]^{1/2} \right\|^2 + \|x' + y'\|^2 \\ &\quad -2 \left\langle \left[x'^2 + y'^2 + (\tau - 2)(x' \circ y') \right]^{1/2}, x' + y' \right\rangle \\ &= \|x'\|^2 + \|y'\|^2 + (\tau - 2)\langle x', y' \rangle + \|x' + y'\|^2 \\ &\quad -2 \left\langle \left[x'^2 + y'^2 + (\tau - 2)(x' \circ y') \right]^{1/2}, x' + y' \right\rangle, \end{aligned}$$

where the second equality uses the fact that $\|z\|^2 = \langle z^2, e \rangle$ for any $z \in \mathbb{R}^n$. Since $\|x'\|^2 + \|y'\|^2 + (\tau - 2)\langle x', y' \rangle + \|x' + y'\|^2$ is clearly differentiable in (x', y'), it suffices to show that $\langle [x'^2 + y'^2 + (\tau - 2)(x' \circ y')]^{1/2}, x' + y' \rangle$ is differentiable at $(x', y') = (x, y)$. By Lemma 3.16, $w_2 = w_2(x, y) \neq 0$, which implies $w_2' = w_2(x', y') = 2x_1' x_2' + 2y_1' y_2' + (\tau - 2)(x_1' y_2' + y_1' x_2') \neq 0$ for all $(x', y') \in \mathbb{R}^n \times \mathbb{R}^n$ sufficiently near to (x, y). Let μ_1, μ_2 be the spectral values of $x'^2 + y'^2 + (\tau - 2)(x' \circ y')$. Then we can compute that

$$
\begin{aligned}
&2\left\langle\left[x'^2 + y'^2 + (\tau - 2)(x' \circ y')\right]^{1/2}, x' + y'\right\rangle \\
&= \sqrt{\mu_2}\left[x_1' + y_1' + \frac{\left[2(x_1'x_2' + y_1'y_2') + (\tau - 2)(x_1'y_2' + y_1'x_2')\right]^T (x_2' + y_2')}{\|2(x_1'x_2' + y_1'y_2') + (\tau - 2)(x_1'y_2' + y_1'x_2')\|}\right] \\
&+ \sqrt{\mu_1}\left[x_1' + y_1' - \frac{\left[2(x_1'x_2' + y_1'y_2') + (\tau - 2)(x_1'y_2' + y_1'x_2')\right]^T (x_2' + y_2')}{\|2(x_1'x_2' + y_1'y_2') + (\tau - 2)(x_1'y_2' + y_1'x_2')\|}\right].
\end{aligned} \tag{3.80}
$$

Since $\lambda_2(w) > 0$ and $w_2(x, y) \neq 0$, the first term on the right-hand side of (3.80) is differentiable at $(x', y') = (x, y)$. Now, we claim that the second term is $o(\|x' - x\| + \|y' - y\|)$, i.e., it is differentiable at (x, y) with zero gradient. To see this, notice that $w_2(x, y) \neq 0$, and hence $\mu_1 = \|x'\|^2 + \|y'\|^2 + (\tau - 2)\langle x', y'\rangle - \|2(x_1'x_2' + y_1'y_2') + (\tau - 2)(x_1'y_2' + y_1'x_2')\|$, viewed as a function of (x', y'), is differentiable at $(x', y') = (x, y)$. Moreover, $\mu_1 = \lambda_1(w) = 0$ when $(x', y') = (x, y)$. Thus, the first-order Taylor's expansion of μ_1 at (x, y) yields

$$\mu_1 = O(\|x' - x\| + \|y' - y\|).$$

Also, since $w_2(x, y) \neq 0$, by the product and quotient rules for differentiation, the function

$$x_1' + y_1' - \frac{\left[2(x_1'x_2' + y_1'y_2') + (\tau - 2)(x_1'y_2' + y_1'x_2')\right]^{\mathrm{T}} (x_2' + y_2')}{\|2(x_1'x_2' + y_1'y_2') + (\tau - 2)(x_1'y_2' + y_1'x_2')\|} \tag{3.81}$$

is differentiable at $(x', y') = (x, y)$, and it has value 0 at $(x', y') = (x, y)$ due to

$$x_1 + y_1 - \frac{[x_1x_2 + y_1y_2 + (\tau - 2)x_1y_2]^{\mathrm{T}} (x_2 + y_2)}{\|x_1x_2 + y_1y_2 + (\tau - 2)x_1y_2\|} = x_1 - x_2^{\mathrm{T}}\frac{w_2}{\|w_2\|} + y_1 - y_2^{\mathrm{T}}\frac{w_2}{\|w_2\|} = 0.$$

Hence, the function in (3.81) is $O(\|x' - x\| + \|y' - y\|)$ in magnitude, which together with $\mu_1 = O(\|x' - x\| + \|y' - y\|)$ shows that the second term on the right-hand side of (3.80) is

$$O((\|x' - x\| + \|y' - y\|)^{3/2}) = o(\|x' - x\| + \|y' - y\|).$$

Thus, we have shown that ψ_τ is differentiable at (x, y). Moreover, we see that $2\nabla\psi_\tau(x, y)$ is the sum of the gradient of $\|x'\|^2 + \|y'\|^2 + (\tau - 2)\langle x', y'\rangle + \|x' + y'\|^2$ and the gradient of the first term on the right-hand side of (3.80), evaluated at $(x', y') = (x, y)$.

The gradient of $\|x'\|^2 + \|y'\|^2 + (\tau - 2)\langle x', y'\rangle + \|x' + y'\|^2$ with respect to x', evaluated at $(x', y') = (x, y)$, is $2x + (\tau - 2)y + 2(x + y)$. The derivative of the first term on the right-hand side of (3.80) with respect to x_1', evaluated at $(x', y') = (x, y)$, works out to be

$$\frac{1}{\sqrt{\lambda_2(w)}}\left[\left(x_1+\frac{\tau-2}{2}y_1\right)+\left(x_2+\frac{\tau-2}{2}y_2\right)^{\mathsf T}\frac{w_2}{\|w_2\|}\right]\left(x_1+y_1+(x_2+y_2)^{\mathsf T}\frac{w_2}{\|w_2\|}\right)$$
$$+\sqrt{\lambda_2(w)}\left[1+\frac{(x_2+\frac{\tau-2}{2}y_2)^{\mathsf T}(x_2+y_2)}{\|x_1x_2+y_1y_2+(\tau-2)x_1y_2\|}-\frac{w_2^T(x_2+y_2)\cdot w_2^{\mathsf T}(x_2+\frac{\tau-2}{2}y_2)}{\|x_1x_2+y_1y_2+(\tau-2)x_1y_2\|\cdot\|w_2\|^2}\right]$$
$$=\frac{2(x_1+\frac{\tau-2}{2}y_1)(x_1+y_1)}{\sqrt{x_1^2+y_1^2+(\tau-2)x_1y_1}}+2\sqrt{x_1^2+y_1^2+(\tau-2)x_1y_1},$$

where the equality follows from Lemma 3.16. Similarly, the gradient of the first term on the right of (3.80) with respect to x_2', evaluated at $(x', y') = (x, y)$, works out to be

$$\frac{1}{\sqrt{\lambda_2(w)}}\left[\left(x_2+\frac{\tau-2}{2}y_2\right)+\left(x_1+\frac{\tau-2}{2}y_1\right)\frac{w_2}{\|w_2\|}\right]\left(x_1+y_1+(x_2+y_2)^{\mathsf T}\frac{w_2}{\|w_2\|}\right)$$
$$+\sqrt{\lambda_2(w)}\left[\frac{(2x_1+(\tau-2)y_1)x_2+\frac{\tau}{2}(x_1+y_1)y_2}{\|x_1x_2+y_1y_2+(\tau-2)x_1y_2\|}-\frac{w_2^{\mathsf T}(x_2+y_2)\cdot(x_1+\frac{\tau-2}{2}y_1)w_2}{\|x_1x_2+y_1y_2+(\tau-2)x_1y_2\|\cdot\|w_2\|^2}\right]$$
$$=2\frac{(2x_1+(\tau-2)y_1)x_2+\frac{\tau}{2}(x_1+y_1)y_2}{\sqrt{x_1^2+y_1^2+(\tau-2)x_1y_1}}.$$

Then, combining the last two gradient expressions yields that

$$\begin{aligned}&2\nabla_x\psi_\tau(x,y)\\&=2x+(\tau-2)y+2(x+y)-\begin{bmatrix}2\sqrt{x_1^2+y_1^2+(\tau-2)x_1y_1}\\0\end{bmatrix}\\&\quad-\frac{2}{\sqrt{x_1^2+y_1^2+(\tau-2)x_1y_1}}\begin{bmatrix}(x_1+\frac{\tau-2}{2}y_1)(x_1+y_1)\\(2x_1+(\tau-2)y_1)x_2+\frac{\tau}{2}(x_1+y_1)y_2\end{bmatrix}.\end{aligned}$$

Using the fact that $x_1y_2 = y_1x_2$ and noting that ϕ_τ can be simplified as the one in (3.76) under this case, we readily rewrite the above expression for $\nabla_x\psi_\tau(x, y)$ in the form of (3.79). By symmetry, $\nabla_y\psi_\tau(x, y)$ also holds as the form of (3.79). □

Proposition 3.16 establishes that ψ_τ is differentiable and admits a computable gradient. To further demonstrate the continuity of this gradient, and thereby the smoothness of ψ_τ, we require the two essential technical lemmas.

Lemma 3.18 *For any $x = (x_1, x_2),\ y = (y_1, y_2) \in \mathbb{R} \times \mathbb{R}^{n-1}$, if $w_2 \neq 0$, then*

$$\left[\left(x_1+\frac{\tau-2}{2}y_1\right)+(-1)^i\left(x_2+\frac{\tau-2}{2}y_2\right)^{\mathsf T}\frac{w_2}{\|w_2\|}\right]^2$$

$$\leq \left\| \left(x_2 + \frac{\tau - 2}{2} y_2\right) + (-1)^i \left(x_1 + \frac{\tau - 2}{2} y_1\right) \frac{w_2}{\|w_2\|} \right\|^2$$
$$\leq \lambda_i(w)$$

for $i = 1, 2$. Furthermore, these relations also hold when interchanging x and y.

Proof The first inequality can be established by expanding the squares on both sides and applying the Cauchy–Schwarz inequality. It remains to verify the second inequality. To this end, observe that the left-hand side of the second inequality simplifies to

$$\|x\|^2 + \frac{(\tau - 2)^2}{4}\|y\|^2 + (\tau - 2)x^\top y + 2(-1)^i \left(x_1 + \frac{\tau - 2}{2} y_1\right)\left(x_2 + \frac{\tau - 2}{2} y_2\right)^\top \frac{w_2}{\|w_2\|},$$

whereas the right-hand side equals to

$$\|x\|^2 + \|y\|^2 + (\tau - 2)x^\top y + (-1)^i \|w_2\|,$$

we only need to prove the following inequality

$$(-1)^i \left[2\left(x_1 + \frac{\tau - 2}{2} y_1\right)\left(x_2 + \frac{\tau - 2}{2} y_2\right)^\top \frac{w_2}{\|w_2\|} - \|w_2\|\right] \leq \frac{\tau(4 - \tau)}{4}\|y\|^2.$$

Considering that $\frac{\tau(4-\tau)}{4} > 0$ and $\|w_2\| > 0$, the last inequality is actually equivalent to

$$\left|2\left(x_1 + \frac{\tau - 2}{2} y_1\right)\left(x_2 + \frac{\tau - 2}{2} y_2\right)^\top w_2 - \|w_2\|^2\right| \leq \frac{\tau(4 - \tau)}{4}\|y\|^2\|w_2\|. \tag{3.82}$$

By using $w_2 = 2(x_1x_2 + y_1y_2) + (\tau - 2)(x_1y_2 + y_1x_2)$, we can compute

$$\begin{aligned}
&2\Big(x_1 + ((\tau - 2)/2)y_1\Big)\Big(x_2 + ((\tau - 2)/2)y_2\Big)^\top w_2 \\
&= \left[4x_1y_1 + 4(\tau - 2)x_1^2 + 2(\tau - 2)y_1^2 + 3(\tau - 2)^2x_1y_1 + \frac{(\tau - 2)^3}{2}y_1^2\right] x_2^\top y_2 \\
&\quad + \left[2(\tau - 2)x_1y_1 + (\tau - 2)^2x_1^2 + (\tau - 2)^2y_1^2 + \frac{(\tau - 2)^3}{2}x_1y_1\right]\|y_2\|^2 \\
&\quad + \left[4x_1^2 + 4(\tau - 2)x_1y_1 + (\tau - 2)^2y_1^2\right]\|x_2\|^2
\end{aligned}$$

and

$$\|w_2\|^2 = \left[8x_1y_1 + 2(\tau-2)^2x_1y_1 + 4(\tau-2)x_1^2 + 4(\tau-2)y_1^2\right]x_2^{\mathrm{T}}y_2$$
$$+\left[4y_1^2 + (\tau-2)^2x_1^2 + 4(\tau-2)x_1y_1\right]\|y_2\|^2$$
$$+\left[4x_1^2 + 4(\tau-2)x_1y_1 + (\tau-2)^2y_1^2\right]\|x_2\|^2.$$

Applying these two equalities, it then follows that

$$2\Big(x_1 + ((\tau-2)/2)y_1\Big)\Big(x_2 + ((\tau-2)/2)y_2\Big)^{\mathrm{T}} w_2 - \|w_2\|^2$$
$$= \left[\Big((\tau-2)^2 - 4\Big)x_1y_1 + \Big((\tau-2)^3/2 - 2(\tau-2)\Big)y_1^2\right]x_2^{\mathrm{T}}y_2$$
$$+\left[\Big((\tau-2)^2 - 4\Big)y_1^2 + \Big((\tau-2)^3/2 - 2(\tau-2)\Big)x_1y_1\right]\|y_2\|^2$$
$$= (\tau^2 - 4\tau)\left[x_1y_1x_2^{\mathrm{T}}y_2 + \frac{\tau-2}{2}y_1^2x_2^{\mathrm{T}}y_2 + y_1^2\|y_2\|^2 + \frac{\tau-2}{2}x_1y_1\|y_2\|^2\right].$$

From this, to show the inequality in (3.82), it suffices to prove that

$$\left|4x_1y_1x_2^{\mathrm{T}}y_2 + 2(\tau-2)x_1y_1\|y_2\|^2 + 4y_1^2\|y_2\|^2 + 2(\tau-2)y_1^2x_2^{\mathrm{T}}y_2\right| \le \|y\|^2\|w_2\|. \tag{3.83}$$

Let L and R denote, respectively, the square of the left-hand side and the right-hand side of (3.83). We argue the assertion (3.83) by verifying that $R - L \ge 0$. Since

$$L = \Big(2x_1 + (\tau-2)y_1\Big)^2 4y_1^2(x_2^{\mathrm{T}}y_2)^2 + \Big(2y_1 + (\tau-2)x_1\Big)^2 4y_1^2\|y_2\|^4$$
$$+8y_1^2\|y_2\|^2x_2^{\mathrm{T}}y_2\Big(4x_1y_1 + 2(\tau-2)x_1^2 + 2(\tau-2)y_1^2 + (\tau-2)^2x_1y_1\Big),$$

and

$$R = \|y\|^4\left[(2x_1 + (\tau-2)y_1)^2\|x_2\|^2 + (2y_1 + (\tau-2)x_1)^2\|y_2\|^2\right]$$
$$+\|y\|^4\left[8x_1y_1 + 2x_1y_1(\tau-2)^2 + 4(\tau-2)(x_1^2 + y_1^2)\right]x_2^{\mathrm{T}}y_2.$$

Taking the difference between R and L leads to

$$R - L = \Big(2x_1 + (\tau-2)y_1\Big)^2\Big(\|y\|^4\|x_2\|^2 - 4y_1^2(x_2^{\mathrm{T}}y_2)^2\Big)$$
$$+ \Big(2y_1 + (\tau-2)x_1\Big)^2\Big(\|y\|^4\|y_2\|^2 - 4y_1^2\|y_2\|^4\Big) + 8x_1y_1x_2^{\mathrm{T}}y_2\Big(\|y\|^4 - 4y_1^2\|y_2\|^2\Big)$$
$$+ 4(\tau-2)x_2^{\mathrm{T}}y_2x_1^2\Big(\|y\|^4 - 4y_1^2\|y_2\|^2\Big) + 4(\tau-2)x_2^{\mathrm{T}}y_2y_1^2\Big(\|y\|^4 - 4y_1^2\|y_2\|^2\Big)$$

$$
\begin{aligned}
&+2(\tau-2)^2 x_1 y_1 x_2^\top y_2\left(\|y\|^4-4y_1^2\|y_2\|^2\right)\\
\geq&\left(\|y\|^4-4y_1^2\|y_2\|^2\right)\Big[\left(2x_1+(\tau-2)y_1\right)^2\|x_2\|^2+\left(2y_1+(\tau-2)x_1\right)^2\|y_2\|^2\\
&+8x_1y_1x_2^\top y_2+4(\tau-2)x_2^\top y_2x_1^2+4(\tau-2)x_2^\top y_2y_1^2+2(\tau-2)^2x_1y_1x_2^\top y_2\Big]\\
=&\left(y_1^2-\|y_2\|^2\right)^2\left\|2x_1x_2+(\tau-2)y_1x_2+2y_1y_2+(\tau-2)x_1y_2\right\|^2\\
\geq&0.
\end{aligned}
$$

By the symmetry between x and y, the above results remain valid upon interchanging x and y. □

Lemma 3.19 *For all (x, y) satisfying $(x-y)^2+\tau(x\circ y)\in \operatorname{int}(\mathcal{K}^n)$, there holds*

$$
\left\|L_{x+\frac{\tau-2}{2}y}L_z^{-1}\right\|_F\leq C \quad\text{and}\quad \left\|L_{y+\frac{\tau-2}{2}x}L_z^{-1}\right\|_F\leq C, \tag{3.84}
$$

where $C>0$ is a constant independent of x, y and τ, and $\|\cdot\|_F$ denotes the Frobenius norm.

Proof Due to the symmetry between x and y, it suffices to establish the first inequality in (3.84). Let $x=(x_1,x_2)\in\mathbb{R}\times\mathbb{R}^{n-1}$ and $y=(y_1,y_2)\in\mathbb{R}\times\mathbb{R}^{n-1}$, where the expression $(x-y)^2+\tau(x\circ y)$ lies within $\operatorname{int}(\mathcal{K}^n)$. We divide the proof into two cases: (1) $w_2=0$, and (2) $w_2\neq 0$.

Case (1): $w_2=0$. In this case, $z_2=0$ and $z_1=\sqrt{\|x\|^2+\|y\|^2+(\tau-2)x^\top y}>0$. Hence,

$$
L_{x+\frac{\tau-2}{2}y}L_z^{-1}=\frac{1}{\sqrt{\|x\|^2+\|y\|^2+(\tau-2)x^\top y}}\begin{bmatrix} x_1+\frac{\tau-2}{2}y_1 & x_2^\top+\frac{\tau-2}{2}y_2^\top\\ x_2+\frac{\tau-2}{2}y_2 & (x_1+\frac{\tau-2}{2}y_1)I\end{bmatrix}.
$$

Notice that $\|x\|^2+\|y\|^2+(\tau-2)x^\top y=\|x+\frac{\tau-2}{2}y\|^2+\frac{\tau(4-\tau)}{4}\|y\|^2$. Therefore,

$$
\frac{|x_1+\frac{\tau-2}{2}y_1|}{\sqrt{\|x\|^2+\|y\|^2+(\tau-2)x^\top y}}\leq 1 \quad\text{and}\quad \frac{\|x_2+\frac{\tau-2}{2}y_2\|}{\sqrt{\|x\|^2+\|y\|^2+(\tau-2)x^\top y}}\leq 1.
$$

This demonstrates that every entry of $L_{x+\frac{\tau-2}{2}y}L_z^{-1}$ is uniformly bounded, with the bound independent of x, y, and τ. Consequently, the first inequality in (3.84) holds in this case.

Case (2): $w_2\neq 0$. Now let λ_1 and λ_2 be the spectral values of w. By (3.71) and Lemma 3.17,

$$
L_{x+\frac{\tau-2}{2}y}L_z^{-1}=\begin{bmatrix} bs_1+cs_2^\top\bar{w}_2 & cs_1\bar{w}_2^\top+as_2^\top+(b-a)s_2^\top\bar{w}_2\bar{w}_2^\top\\ bs_2+cs_1\bar{w}_2 & cs_2\bar{w}_2^\top+as_1I+(b-a)s_1\bar{w}_2\bar{w}_2^\top\end{bmatrix},
$$

where $\bar{w}_2=\frac{w_2}{\|w_2\|}$, $s=(s_1,s_2)=x+\frac{\tau-2}{2}y$, and a, b and c are given by

$$
\begin{aligned}
a &= \frac{2}{\sqrt{\lambda_2}+\sqrt{\lambda_1}}, \\
b &= \frac{\sqrt{\lambda_2}+\sqrt{\lambda_1}}{2\sqrt{\lambda_2\lambda_1}}, \\
c &= \frac{\sqrt{\lambda_1}-\sqrt{\lambda_2}}{2\sqrt{\lambda_2\lambda_1}}.
\end{aligned}
$$

Using Lemma 3.18 and noting that $s_1 = x_1 + \frac{\tau-2}{2}y_1$ and $s_2 = x_2 + \frac{\tau-2}{2}y_2$, we have

$$
\begin{aligned}
\left|bs_1 + cs_2^{\mathrm{T}}\bar{w}_2\right| &\le \frac{1}{2\sqrt{\lambda_2}}|s_1 + s_2^{\mathrm{T}}\bar{w}_2| + \frac{1}{2\sqrt{\lambda_1}}|s_1 - s_2^{\mathrm{T}}\bar{w}_2| \le 1, \\
\|bs_2 + cs_1\bar{w}_2\| &\le \frac{1}{2\sqrt{\lambda_2}}\|s_2 + s_1\bar{w}_2\| + \frac{1}{2\sqrt{\lambda_1}}\|s_2 - s_1\bar{w}_2\| \le 1, \\
\left\|cs_1\bar{w}_2^{\mathrm{T}} + bs_2^{\mathrm{T}}\bar{w}_2\bar{w}_2^{\mathrm{T}}\right\| &= \left\|\frac{1}{2\sqrt{\lambda_2}}\left(s_1 + s_2^T\bar{w}_2\right)\bar{w}_2^{\mathrm{T}} - \frac{1}{2\sqrt{\lambda_1}}\left(s_1 - s_2^{\mathrm{T}}\bar{w}_2\right)\bar{w}_2^{\mathrm{T}}\right\| \\
&\le \frac{1}{2\sqrt{\lambda_2}}\left|s_1 + s_2^{\mathrm{T}}\bar{w}_2\right| + \frac{1}{2\sqrt{\lambda_1}}\left|s_1 - s_2^{\mathrm{T}}\bar{w}_2\right| \le 1, \\
\left\|as_2^{\mathrm{T}} - as_2^{\mathrm{T}}\bar{w}_2\bar{w}_2^{\mathrm{T}}\right\| &\le \left\|\frac{2s_2^{\mathrm{T}}}{\sqrt{\lambda_2}+\sqrt{\lambda_1}}\right\| \cdot \left\|I - \bar{w}_2\bar{w}_2^{\mathrm{T}}\right\|_F \le 2(n+1), \\
\left\|cs_2\bar{w}_2^{\mathrm{T}} + bs_1\bar{w}_2\bar{w}_2^T\right\|_F &= \left\|\frac{1}{2\sqrt{\lambda_2}}\left(s_2 + s_1\bar{w}_2\right)\bar{w}_2^{\mathrm{T}} - \frac{1}{2\sqrt{\lambda_1}}\left(s_2 - s_1\bar{w}_2\right)\bar{w}_2^{\mathrm{T}}\right\|_F \\
&\le \frac{1}{2\sqrt{\lambda_2}}\|s_2 + s_1\bar{w}_2\| + \frac{1}{2\sqrt{\lambda_1}}\|s_2 - s_1\bar{w}_2\| \le 1, \\
\left\|as_1 I - as_1\bar{w}_2\bar{w}_2^{\mathrm{T}}\right\|_F &\le \left\|\frac{2s_1}{\sqrt{\lambda_2}+\sqrt{\lambda_1}}\right\| \cdot \left\|I - \bar{w}_2\bar{w}_2^{\mathrm{T}}\right\|_F \le 2(n+1).
\end{aligned}
$$

The inequalities above imply that each entry of $L_{x+\frac{\tau-2}{2}y}L_z^{-1}$ is uniformly bounded, with the bound independent of x, y and τ. Therefore, the first inequality in (3.84) also holds in this case. □

Proposition 3.17 *The function ψ_τ defined by (3.63) is smooth everywhere on $\mathbb{R}^n \times \mathbb{R}^n$.*

Proof By Proposition 3.16 and the symmetry of x and y in $\nabla\psi_\tau$, it suffices to show that $\nabla_x\psi_\tau$ is continuous at every $(a, b) \in \mathbb{R}^n \times \mathbb{R}^n$. If $(a-b)^2 + \tau(a \circ b) \in \mathrm{int}(\mathcal{K}^n)$, the conclusion has been shown in Proposition 3.16. We next consider the other two cases.

Case (1): $(a, b) = (0, 0)$. By Proposition 3.16, we need to show that $\nabla_x\psi_\tau(x, y) \to 0$ as $(x, y) \to (0, 0)$. If $(x-y)^2 + \tau(x \circ y) \in \mathrm{int}(\mathcal{K}^n)$, then $\nabla_x\psi_\tau(x, y)$ is given by (2.20), whereas if $(x, y) \neq (0, 0)$ and $(x-y)^2 + \tau(x \circ y) \notin \mathrm{int}(\mathcal{K}^n)$, then $\nabla_x\psi_\tau(x, y)$ is given by (3.79). Notice that $L_{x+\frac{\tau-2}{2}y}L_z^{-1}$ and $\frac{x_1+\frac{\tau-2}{2}y_1}{\sqrt{x_1^2+y_1^2+(\tau-2)x_1y_1}}$ are bounded with bound independent of x, y and τ, using the continuity of $\phi_\tau(x, y)$ immediately yields the desired result.

Case (2): $(a, b) \neq (0, 0)$ and $(a-b)^2 + \tau(a \circ b) \notin \text{int}(\mathcal{K}^n)$. We will demonstrate that $\nabla_x \psi_\tau(x, y) \to \nabla_x \psi_\tau(a, b)$ by considering the following two subcases: (2a) $(x, y) \neq (0, 0)$ and $(x-y)^2 + \tau(x \circ y) \notin \text{int}(\mathcal{K}^n)$ and (2b) $(x-y)^2 + \tau(x \circ y) \in \text{int}(\mathcal{K}^n)$. In subcase (2a), $\nabla_x \psi_\tau(x, y)$ is given by (3.79). Noting that the right-hand side of (3.79) is continuous at (a, b), the desired result follows.

Next, we prove that $\nabla_x \psi_\tau(x, y) \to \nabla_x \psi_\tau(a, b)$ in subcase (2b). From (3.78), we have

$$\nabla_x \psi_\tau(x, y) = \left(x + \frac{\tau - 2}{2} y\right) - L_{x+\frac{\tau-2}{2}y} L_z^{-1}(x+y) - \phi_\tau(x, y). \tag{3.85}$$

On the other hand, since $(a, b) \neq (0, 0)$ and $(a-b)^2 + \tau(a \circ b) \notin \text{int}(\mathcal{K}^n)$,

$$\|a\|^2 + \|b\|^2 + (\tau - 2)a^\mathsf{T} b = \|2(a_1 a_2 + b_1 b_2) + (\tau - 2)(a_1 b_2 + b_1 a_2)\| \neq 0, \tag{3.86}$$

and moreover from (3.73) it follows that

$$\begin{aligned}\|a\|^2 + \|b\|^2 + (\tau - 2)a^\mathsf{T} b &= 2(a_1^2 + b_1^2 + (\tau - 2)a_1 b_1) \\ &= 2(\|a_2\|^2 + \|b_2\|^2 + (\tau - 2)a_2^\mathsf{T} b_2) \\ &= 2\|(a_1 a_2 + b_1 b_2) + (\tau - 2)a_1 b_2\|.\end{aligned} \tag{3.87}$$

Using the equalities in (3.87), it is not hard to verify that

$$\frac{a_1 + \frac{\tau-2}{2} b_1}{\sqrt{a_1^2 + b_1^2 + (\tau - 2)a_1 b_1}} \left((a-b)^2 + \tau(a \circ b)\right)^{1/2} = a + \frac{\tau - 2}{2} b.$$

This together with the expression of $\nabla_x \psi_\tau(a, b)$ given by (3.79) yields

$$\nabla_x \psi_\tau(a, b) = \left(a + \frac{\tau - 2}{2} b\right) - \frac{a_1 + \frac{\tau-2}{2} b_1}{\sqrt{a_1^2 + b_1^2 + (\tau - 2)a_1 b_1}} (a+b) - \phi_\tau(a, b). \tag{3.88}$$

Comparing (3.85) with (3.88), we see that if we wish to prove $\nabla_x \psi_\tau(x, y) \to \nabla_x \psi_\tau(a, b)$ as $(x, y) \to (a, b)$, it suffices to show that

$$L_{x+\frac{\tau-2}{2}y} L_z^{-1}(x+y) \to \frac{a_1 + \frac{\tau-2}{2} b_1}{\sqrt{a_1^2 + b_1^2 + (\tau - 2)a_1 b_1}} (a+b), \tag{3.89}$$

which is also equivalent to proving the following three relations

$$L_{x+\frac{\tau-2}{2}y}L_z^{-1}\left(x+\frac{\tau-2}{2}y\right) \to \frac{a_1+\frac{\tau-2}{2}b_1}{\sqrt{a_1^2+b_1^2+(\tau-2)a_1b_1}}\left(a+\frac{\tau-2}{2}b\right), \quad (3.90)$$

$$L_{y+\frac{\tau-2}{2}x}L_z^{-1}\left(y+\frac{\tau-2}{2}x\right) \to \frac{b_1+\frac{\tau-2}{2}a_1}{\sqrt{a_1^2+b_1^2+(\tau-2)a_1b_1}}\left(b+\frac{\tau-2}{2}a\right), \quad (3.91)$$

$$\frac{4-\tau}{2}L_{x-y}L_z^{-1}\left(y+\frac{\tau-2}{2}x\right) \to \frac{\frac{4-\tau}{2}(a_1-b_1)}{\sqrt{a_1^2+b_1^2+(\tau-2)a_1b_1}}\left(b+\frac{\tau-2}{2}a\right). \quad (3.92)$$

By the symmetry of x and y in (3.90) and (3.91), we only prove (3.90) and (3.92). Let

$$(\zeta_1,\zeta_2) := L_{x+\frac{\tau-2}{2}y}L_z^{-1}\left(x+\frac{\tau-2}{2}y\right), \quad (\xi_1,\xi_2) := L_{x-y}L_z^{-1}\left(y+\frac{\tau-2}{2}x\right). \quad (3.93)$$

Then showing (3.90) and (3.92) reduces to proving the following relations hold as $(x,y) \to (a,b)$:

$$\zeta_1 \to \frac{\left(a_1+\frac{\tau-2}{2}b_1\right)^2}{\sqrt{a_1^2+b_1^2+(\tau-2)a_1b_1}}, \quad \zeta_2 \to \frac{a_1+\frac{\tau-2}{2}b_1}{\sqrt{a_1^2+b_1^2+(\tau-2)a_1b_1}}\left(a_2+\frac{\tau-2}{2}b_2\right), \quad (3.94)$$

$$\xi_1 \to \frac{(a_1-b_1)\left(b_1+\frac{\tau-2}{2}a_1\right)}{\sqrt{a_1^2+b_1^2+(\tau-2)a_1b_1}}, \quad \xi_2 \to \frac{(a_1-b_1)}{\sqrt{a_1^2+b_1^2+(\tau-2)a_1b_1}}\left(b_2+\frac{\tau-2}{2}a_2\right). \quad (3.95)$$

To verify (3.94), we take (x,y) sufficiently near to (a,b). By (3.86), we may assume that $w_2 = w_2(x,y) \neq 0$. Let $s = (s_1,s_2) = x + \frac{\tau-2}{2}y$. Using Lemma 3.1(c) and (3.93), we can calculate that

$$\begin{aligned}
\zeta_1 &= \frac{1}{\det(z)}\left(s_1^2z_1 - 2s_1s_2^{\mathsf T}z_2 + \frac{\det(z)}{z_1}\|s_2\|^2 + \frac{(z_2^{\mathsf T}s_2)^2}{z_1}\right),\\
&= \frac{\|s_2\|^2}{z_1} + \frac{(s_1z_1 - s_2^{\mathsf T}z_2)^2}{z_1\det(z)}, \quad (3.96)\\
\zeta_2 &= \frac{1}{\det(z)}\left(s_1z_1s_2 - z_2^{\mathsf T}s_2s_2 - s_1^2z_2 + \frac{s_1\det(z)}{z_1}s_2 + \frac{s_1}{z_1}z_2^{\mathsf T}s_2z_2\right)\\
&= \frac{s_1}{z_1}s_2 + \frac{(s_1z_1 - s_2^{\mathsf T}z_2)}{\det(z)}\left(s_2 - \frac{s_1}{z_1}z_2\right). \quad (3.97)
\end{aligned}$$

Notice that, as $(x, y) \to (a, b)$, from Eqs. (3.70) and (3.86)–(3.87) it follows that

$$\lambda_1(w) \to 0 \quad \text{and} \quad \lambda_2(w) \to 4(a_1^2 + b_1^2 + (\tau - 2)a_1 b_1). \tag{3.98}$$

In addition, by the proof of Lemma 3.16, we also have

$$\left(a_1 + \frac{\tau - 2}{2} b_1\right)^2 = \left\| a_2 + \frac{\tau - 2}{2} b_2 \right\|^2 \quad \text{and} \quad b_1^2 = \|b_2\|^2.$$

Thus, from the last two equations and the expression of z given by (3.71), we have

$$\frac{\|s_2\|^2}{z_1} = \frac{2\|s_2\|^2}{\sqrt{\lambda_2(w)} + \sqrt{\lambda_1(w)}} \to \frac{\left(a_1 + \frac{\tau-2}{2} b_1\right)^2}{\sqrt{a_1^2 + b_1^2 + (\tau - 2)a_1 b_1}}. \tag{3.99}$$

On the other hand, for the second term in the right-hand side of (3.96), we can compute that

$$\begin{aligned} \frac{(s_1 z_1 - s_2^T z_2)^2}{z_1 \det(z)} = \; & \frac{1}{z_1 \sqrt{\lambda_2(w)}} \left[s_1^2 \sqrt{\lambda_1(w)} + s_1 \left(\sqrt{\lambda_2(w)} - \sqrt{\lambda_1(w)}\right) \left(s_1 - \frac{s_2^T w_2}{\|w_2\|}\right) \right. \\ & \left. + \frac{1}{4} \left(\sqrt{\lambda_2(w)} - \sqrt{\lambda_1(w)}\right)^2 \cdot \frac{1}{\sqrt{\lambda_1(w)}} \left(s_1 - \frac{s_2^T w_2}{\|w_2\|}\right)^2 \right]. \end{aligned} \tag{3.100}$$

Since $s_1^2 \sqrt{\lambda_1(w)}$, $s_1 - \frac{s_2^T w_2}{\|w_2\|} \to 0$ as $(x, y) \to (a, b)$ and $|s_1 - \frac{s_2^T w_2}{\|w_2\|}| \le \sqrt{\lambda_1(w)}$ by Lemma 3.18, the right-hand side of (3.100) tends to 0 as $(x, y) \to (a, b)$. Combining with (3.99), we prove the first relation in (3.94). We next prove the second relation in (3.94). Note that ζ_2 is given by (3.97). From (3.98) and (3.74), it follows that, as $(x, y) \to (a, b)$,

$$\frac{s_1}{z_1} s_2 = \frac{2 s_1 s_2}{\sqrt{\lambda_2(w)} + \sqrt{\lambda_1(w)}} \to \frac{a_1 + \frac{\tau-2}{2} b_1}{\sqrt{a_1^2 + b_1^2 + (\tau - 2)a_1 b_1}} \left(a_2 + \frac{\tau - 2}{2} b_2\right), \tag{3.101}$$

$$s_2 - \frac{s_1}{z_1} z_2 \to \left(a_2 + \frac{\tau - 2}{2} b_2\right) - \frac{\left(a_1^2 + b_1^2 + (\tau - 2)a_1 b_1\right)\left(a_2 + \frac{\tau-2}{2} b_2\right)}{a_1^2 + b_1^2 + (\tau - 2)a_1 b_1} = 0. \tag{3.102}$$

In addition, by the expression of z, we can compute that

$$\frac{(s_1 z_1 - s_2^T z_2)}{\det(z)} = \frac{s_1}{\sqrt{\lambda_2(w)}} + \frac{1 - \sqrt{\lambda_1(w)}/\sqrt{\lambda_2(w)}}{2\sqrt{\lambda_1(w)}} \left(s_1 - \frac{s_2^T w_2}{\|w_2\|}\right). \tag{3.103}$$

By (3.98), the first term on the right-hand side of (3.103) tends to $\frac{a_1+\frac{\tau-2}{2}b_1}{2\sqrt{a_1^2+b_1^2+(\tau-2)a_1b_2}}$ as $(x, y) \to (a, b)$, while the second term is bounded since $|s_1 - \frac{s_2^T w_2}{\|w_2\|}| \le \sqrt{\lambda_1(w)}$ by Lemma 3.18. Combining (3.101), (3.102), and (3.97) yields the second relation in (3.94). Consequently, (3.90) is established.

Now, we focus on the proof of (3.95). Let $u = x - y$ and $v = y + \frac{\tau-2}{2}x$. From Lemma 3.1(c) and (3.93), we know

$$
\begin{aligned}
\xi_1 &= \frac{1}{\det(z)}\left(u_1 z_1 v_1 - u_1 z_2^T v_2 - v_1 u_2^T z_2 + \frac{\det(z)}{z_1} u_2^T v_2 + \frac{u_2^T z_2 (z_2^T v_2)}{z_1}\right) \\
&= \frac{u_2^T v_2}{z_1} + \frac{(u_1 z_1 - u_2^T z_2)(v_1 z_1 - v_2^T z_2)}{z_1 \det(z)}, \qquad (3.104) \\
\xi_2 &= \frac{1}{\det(z)}\left(z_1 v_1 u_2 - z_2^T v_2 u_2 - u_1 v_1 z_2 + \frac{u_1 \det(z)}{z_1} v_2 + \frac{u_1}{z_1} z_2^T v_2 z_2\right) \\
&= \frac{u_1}{z_1} v_2 + \frac{(z_1 v_1 - z_2^T v_2)}{\det(z)}\left(u_2 - \frac{u_1}{z_1} z_2\right). \qquad (3.105)
\end{aligned}
$$

Since $(a - b)^2 + \tau(a \circ b) \notin \mathrm{int}(\mathcal{K}^n)$, we have $a_2^T b_2 = a_1 b_1$, $a_1^2 = \|a_2\|^2$ and $b_1^2 = \|b_2\|^2$ due to Lemma 3.16. This together with (3.98) implies that

$$
\frac{u_2^T v_2}{z_1} = \frac{2(x_2 - y_2)^T v_2}{\sqrt{\lambda_2(w)} + \sqrt{\lambda_1(w)}} \to \frac{(a_1 - b_1)\left(b_1 + \frac{\tau-2}{2}a_1\right)}{\sqrt{a_1^2 + b_1^2 + (\tau - 2)a_1 b_1}} \quad \text{as } (x, y) \to (a, b). \qquad (3.106)
$$

We next prove that the second term in the right-hand side of (3.104) tends to 0. By computing,

$$
\begin{aligned}
&\frac{(u_1 z_1 - u_2^T z_2)(v_1 z_1 - v_2^T z_2)}{z_1 \det(z)} \\
&= \frac{\sqrt{\lambda_1(w)\lambda_2(w)}}{z_1}\left[\frac{u_1}{\sqrt{\lambda_2(w)}} + \frac{\sqrt{\lambda_2(w)} - \sqrt{\lambda_1(w)}}{2\sqrt{\lambda_1(w)\lambda_2(w)}}\left(u_1 - \frac{u_2^T w_2}{\|w_2\|}\right)\right] \\
&\quad \left[\frac{v_1}{\sqrt{\lambda_2(w)}} + \frac{\sqrt{\lambda_2(w)} - \sqrt{\lambda_1(w)}}{2\sqrt{\lambda_1(w)\lambda_2(w)}}\left(v_1 - \frac{v_2^T w_2}{\|w_2\|}\right)\right].
\end{aligned}
$$

When $(x, y) \to (a, b)$, we have $\frac{1}{z_1}\sqrt{\lambda_1(w)\lambda_2(w)} \to 0$. In addition, by Lemma 3.18,

$$
\begin{aligned}
\left|u_1 - \frac{u_2^T w_2}{\|w_2\|}\right| = \frac{2}{4-\tau}\Bigg|&\left[\left(x_1 + \frac{\tau-2}{2}y_1\right) - \left(x_2 + \frac{\tau-2}{2}y_2\right)^T \frac{w_2}{\|w_2\|}\right] \\
&- \left[\left(y_1 + \frac{\tau-2}{2}x_1\right) - \left(y_2 + \frac{\tau-2}{2}x_2\right)^T \frac{w_2}{\|w_2\|}\right]\Bigg|
\end{aligned}
$$

$$\leq \frac{4\sqrt{\lambda_1(w)}}{4-\tau},$$

$$\left|v_1 - \frac{v_2^{\mathrm{T}} w_2}{\|w_2\|}\right| = \left|\left(y_1 + \frac{\tau-2}{2}x_1\right) - \left(y_2 + \frac{\tau-2}{2}x_2\right)^{\mathrm{T}} \frac{w_2}{\|w_2\|}\right| \leq \sqrt{\lambda_1(w)}.$$

This means that $\frac{1}{\sqrt{\lambda_1(w)}}\left(u_1 - \frac{u_2^{\mathrm{T}} w_2}{\|w_2\|}\right)$ and $\frac{1}{\sqrt{\lambda_1(w)}}\left(v_1 - \frac{v_2^{\mathrm{T}} w_2}{\|w_2\|}\right)$ are uniformly bounded. Notice that $\frac{u_1}{\sqrt{\lambda_2(w)}}$, $\frac{v_1}{\sqrt{\lambda_2(w)}}$ and $\frac{\sqrt{\lambda_2(w)}-\sqrt{\lambda_1(w)}}{\sqrt{\lambda_2(w)}}$ are also uniformly bounded. Therefore,

$$\frac{(u_1 z_1 - u_2^{\mathrm{T}} z_2)(v_1 z_1 - v_2^{\mathrm{T}} z_2)}{z_1 \det(z)} \to 0 \quad \text{as } (x, y) \to (a, b).$$

Together with (3.106), this establishes the first relation in (3.95). It remains to verify the second relation in (3.95). Note that ξ_2 is given by (3.105). When $(x, y) \to (a, b)$, from (3.98) and (3.74), there have

$$\frac{u_1}{z_1} v_2 \to \frac{(a_1 - b_1)}{\sqrt{a_1^2 + b_1^2 + (\tau-2)a_1 b_1}} \left(b_2 + \frac{\tau-2}{2}a_2\right), \tag{3.107}$$

$$u_2 - \frac{u_1}{z_1} z_2 \to (a_2 - b_2) - \frac{\left(a_1^2 + b_1^2 + (\tau-2)a_1 b_1\right)(a_2 - b_2)}{a_1^2 + b_1^2 + (\tau-2)a_1 b_1} = 0. \tag{3.108}$$

In addition, by the expression of z, we can compute that

$$\frac{(z_1 v_1 - z_2^{\mathrm{T}} v_2)}{\det(z)} = \frac{v_1}{\sqrt{\lambda_2(w)}} + \frac{1 - \sqrt{\lambda_1(w)}/\sqrt{\lambda_2(w)}}{2\sqrt{\lambda_1(w)}} \left(v_1 - \frac{v_2^{\mathrm{T}} w_2}{\|w_2\|}\right). \tag{3.109}$$

From (3.98), the first term on the right-hand side of (3.109) converges to $\frac{b_1 + \frac{\tau-2}{2}a_1}{2\sqrt{a_1^2+b_1^2+(\tau-2)a_1 b_2}}$ when $(x, y) \to (a, b)$, while the second term is bounded since $|v_1 - \frac{v_2^{\mathrm{T}} w_2}{\|w_2\|}| \leq \sqrt{\lambda_1(w)}$ by Lemma 3.18. Combining with (3.107), (3.108) and (3.105), we obtain the second relation in (3.95) which implies (3.92) holds. Thus, the proof is complete. □

We now turn our attention to the monotone SOCCP and show that every stationary point of the unconstrained minimization problem (3.67) is indeed a solution of the SOCCP. To begin, we establish the following key properties of $\nabla\psi_\tau$, which extend Proposition 3.6 to the general case where $\tau \in (0, 4)$.

Lemma 3.20 *For any* $x = (x_1, x_2),\ y = (y_1, y_2) \in \mathbb{R} \times \mathbb{R}^{n-1}$, *we have*

$$\langle x, \nabla_x \psi_\tau(x, y)\rangle + \langle y, \nabla_y \psi_\tau(x, y)\rangle = \|\phi_\tau(x, y)\|^2, \tag{3.110}$$

$$\langle \nabla_x \psi_\tau(x, y), \nabla_y \psi_\tau(x, y)\rangle \geq 0. \tag{3.111}$$

Furthermore, the equality in (3.111) holds if and only if $\phi_\tau(x, y) = 0$.

Proof When $(x, y) = (0, 0)$, it follows from Proposition 3.16 that $\nabla_x\psi_\tau(x, y) = \nabla_y\psi_\tau(x, y) = 0$, and the conclusion is immediate. We now proceed to examine the remaining two cases.

Case (1): $(x - y)^2 + \tau(x \circ y) \in \mathrm{int}(\mathcal{K}^n)$. By Proposition 3.16, we can compute that

$$\begin{aligned}
&\langle x, \nabla_x\psi_\tau(x, y)\rangle + \langle y, \nabla_y\psi_\tau(x, y)\rangle \\
&= \left\langle x, \left(L_{x+\frac{\tau-2}{2}y}L_z^{-1} - I\right)\phi_\tau\right\rangle + \left\langle y, \left(L_{y+\frac{\tau-2}{2}x}L_z^{-1} - I\right)\phi_\tau\right\rangle \\
&= \left\langle \left(L_z^{-1}L_{x+\frac{\tau-2}{2}y} - I\right)x, \phi_\tau\right\rangle + \left\langle \left(L_z^{-1}L_{y+\frac{\tau-2}{2}x} - I\right)y, \phi_\tau\right\rangle \\
&= \left\langle L_z^{-1}\left[(x^2 + y^2) + (\tau - 2)(x \circ y)\right] - (x + y), \phi_\tau\right\rangle \\
&= \langle L_z^{-1}z^2 - (x + y), \phi_\tau\rangle = \|\phi_\tau\|^2,
\end{aligned}$$

where, for simplicity, $\phi_\tau(x, y)$ is written as ϕ_τ. This proves (3.110). Notice that

$$\langle\nabla_x\psi_\tau(x, y), \nabla_y\psi_\tau(x, y)\rangle = \left\langle \left(L_{y+\frac{\tau-2}{2}x} - L_z\right)\left(L_{x+\frac{\tau-2}{2}y} - L_z\right)L_z^{-1}\phi_\tau, L_z^{-1}\phi_\tau\right\rangle.$$

Let S be the symmetric part of $(L_{y+\frac{\tau-2}{2}x} - L_z)(L_{x+\frac{\tau-2}{2}y} - L_z)$. Then,

$$\begin{aligned}
S &= \frac{1}{2}\left[\left(L_{y+\frac{\tau-2}{2}x} - L_z\right)\left(L_{x+\frac{\tau-2}{2}y} - L_z\right) + \left(L_{x+\frac{\tau-2}{2}y} - L_z\right)\left(L_{y+\frac{\tau-2}{2}x} - L_z\right)\right] \\
&= \frac{1}{2}\left[L_yL_x + \frac{\tau-2}{2}L_x^2 - L_zL_x + \frac{\tau-2}{2}L_y^2 + \frac{(\tau-2)^2}{4}L_xL_y - \frac{\tau-2}{2}L_zL_y\right. \\
&\quad -L_yL_z - \frac{\tau-2}{2}L_xL_z + L_z^2 + L_xL_y + \frac{\tau-2}{2}L_y^2 - L_zL_y \\
&\quad \left.+\frac{\tau-2}{2}L_x^2 + \frac{(\tau-2)^2}{4}L_yL_x - \frac{\tau-2}{2}L_zL_x - L_xL_z - \frac{\tau-2}{2}L_yL_z + L_z^2\right] \\
&= \frac{\tau}{4}(L_z - L_x - L_y)^2 + \frac{4-\tau}{4}(L_z^2 - L_{\widetilde{x}}^2 - L_{\widetilde{y}}^2),
\end{aligned}$$

where $\widetilde{x} := x + \frac{\tau-2}{2}y$ and $\widetilde{y} := \frac{1}{2}\sqrt{\tau(4-\tau)}\,y$. Noting that $z \in \mathcal{K}^n$ and $z^2 = \widetilde{x}^2 + \widetilde{y}^2$, we have $L_z^2 - L_{\widetilde{x}}^2 - L_{\widetilde{y}}^2 \succeq O$ by Proposition 3.4 of [12]. Consequently,

$$\begin{aligned}
\langle\nabla_x\psi_\tau(x, y), \nabla_y\psi_\tau(x, y)\rangle &= \langle SL_z^{-1}\phi_\tau, L_z^{-1}\phi_\tau\rangle \\
&\geq \frac{\tau}{4}\left\langle(L_z - L_x - L_y)^2L_z^{-1}\phi_\tau, L_z^{-1}\phi_\tau\right\rangle \qquad (3.112) \\
&= \frac{\tau}{4}\left\|L_{\phi_\tau}L_z^{-1}\phi_\tau\right\|^2,
\end{aligned}$$

where the equality is due to $L_z - L_x - L_y = L_{\phi_\tau}$. This implies (3.111). If the inequality in (3.111) holds with equality, then the above relation yields

$\|L_{\phi_\tau} L_z^{-1}\phi_\tau\|^2 = 0$, which says

$$L_{\phi_\tau} L_z^{-1}\phi_\tau = \phi_\tau \circ (L_z^{-1}\phi_\tau) = 0.$$

By the definition of Jordan product, $\langle \phi_\tau, L_z^{-1}\phi_\tau \rangle = 0$. This implies $\phi_\tau = 0$ since $L_z^{-1} \succ O$. Conversely, if $\phi_\tau = 0$, then it follows from (3.78) that $\langle \nabla_x \psi_\tau(x, y), \nabla_y \psi_\tau(x, y) \rangle = 0$.

Case (2): $(x, y) \neq (0, 0)$ and $(x - y)^2 + \tau(x \circ y) \notin \text{int}(\mathcal{K}^n)$. By (3.79), we can compute

$$\begin{aligned}
&\langle x, \nabla_x \psi_\tau(x, y) \rangle + \langle y, \nabla_y \psi_\tau(x, y) \rangle \\
&= \left\langle \frac{x_1 x + y_1 y + \frac{\tau-2}{2}(y_1 x + x_1 y)}{x_1^2 + y_1^2 + (\tau - 2)x_1 y_1} - (x + y),\ \phi_\tau(x, y) \right\rangle \\
&= \|\phi_\tau(x, y)\|^2,
\end{aligned}$$

where the last equality uses (3.76). This proves (3.110). Equation (3.111) holds since

$$\begin{aligned}
&\langle \nabla_x \psi_\tau(x, y), \nabla_y \psi_\tau(x, y) \rangle \\
&= \left[\frac{x_1 + \frac{\tau-2}{2} y_1}{\sqrt{x_1^2 + y_1^2 + (\tau - 2)x_1 y_1}} - 1 \right] \left[\frac{y_1 + \frac{\tau-2}{2} x_1}{\sqrt{x_1^2 + y_1^2 + (\tau - 2)x_1 y_1}} - 1 \right] \|\phi_\tau(x, y)\|^2 \\
&\geq 0,
\end{aligned}$$

where the inequality is due to $\frac{x_1 + \frac{\tau-2}{2} y_1}{\sqrt{x_1^2 + y_1^2 + (\tau-2)x_1 y_1}} \leq 1$ and $\frac{y_1 + \frac{\tau-2}{2} x_1}{\sqrt{x_1^2 + y_1^2 + (\tau-2)x_1 y_1}} \leq 1$. If (3.111) holds with equality, then either $\phi_\tau(x, y) = 0$ or $\frac{x_1 + \frac{\tau-2}{2} y_1}{\sqrt{x_1^2 + y_1^2 + (\tau-2)x_1 y_1}} = 1$ or $\frac{y_1 + \frac{\tau-2}{2} x_1}{\sqrt{x_1^2 + y_1^2 + (\tau-2)x_1 y_1}} = 1$. In the second case, we have $y_1 = 0$ and $x_1 \geq 0$, so that Lemma 3.16 yields $y_2 = 0$ and $x_1 = \|x_2\|$. In the third case, we have $x_1 = 0$ and $y_1 \geq 0$, so that Lemma 3.16 yields $x_2 = 0$ and $y_1 = \|y_2\|$. Thus, in the two cases, we have $\langle x, y \rangle = 0$, $x \in \mathcal{K}^n$, $y \in \mathcal{K}^n$. Consequently, $\phi_\tau(x, y) = 0$ by Proposition 3.15. Conversely, if $\phi_\tau = 0$, then from (3.78) it follows that $\langle \nabla_x \psi_\tau(x, y), \nabla_y \psi_\tau(x, y) \rangle = 0$. The proof is thus complete. □

We are now prepared to establish another key result stated as in Proposition 3.18: every stationary point of f_τ solves the SOCCP under the condition

$$\nabla F(\zeta) \text{ and } -\nabla G(\zeta) \text{ are column monotone for any } \zeta \in \mathbb{R}^n. \tag{3.113}$$

From [2, page 1014] or [33, page 222], $A, B \in \mathbb{R}^{n \times n}$ are column monotone if

$$Au + Bv = 0 \implies u^\top v \geq 0 \quad \text{for any } u, v \in \mathbb{R}^n.$$

In light of this, it is not hard to verify that, if $\nabla G(\zeta)$ is invertible, the condition (3.113) is equivalent to requiring $\nabla G(\zeta)^{-1}\nabla F(\zeta) \succeq O$ for any $\zeta \in \mathbb{R}^n$. This implies that, for the SOCCP (3.4), the condition (3.113) is actually equivalent to F being monotone.

Proposition 3.18 *Let f_τ be given by (3.67). If F and G satisfies the condition (3.113), then for every $\zeta \in \mathbb{R}^n$, either $f_\tau(\zeta) = 0$ or $\nabla f_\tau(\zeta) \neq 0$. If $\nabla f_\tau(\zeta) \neq 0$ and $\nabla G(\zeta)$ is invertible, then $\langle d(\zeta), \nabla f_\tau(\zeta)\rangle < 0$, where $d(\zeta) := -\left(\nabla G(\zeta)^{-1}\right)^T \nabla_x \psi_\tau(F(\zeta), G(\zeta))$.*

Proof By Lemma 3.20, applying the same arguments as in [14, Proposition 3], with ψ_{FB} and f_{FB} replaced by ψ_τ and f_τ, respectively, yields the desired result. We therefore omit the details. □

Lemma 3.21 *Let $\hat{z}(x, y, \varepsilon)$ be defined by*

$$\hat{z}(x, y, \varepsilon) := \left[(x-y)^2 + \tau(x \circ y) + \varepsilon e\right]^{1/2}.$$

Then, for any $\varepsilon > 0$, the function $\hat{z}(x, y, \varepsilon)$ is continuously differentiable everywhere, and there exists a scalar $C > 0$ such that

$$\|\nabla_x \hat{z}(x, y, \varepsilon)\|_F \le C, \quad \|\nabla_y \hat{z}(x, y, \varepsilon)\|_F \le C \tag{3.114}$$

for all $(x, y) \in \mathbb{R}^n \times \mathbb{R}^n$, where $\|A\|_F$ denotes the Frobenius norm of the matrix A.

Proof Since $(x-y)^2 + \tau(x \circ y) + \varepsilon e \in \text{int}(\mathcal{K}^n)$ for any $(x, y) \in \mathbb{R}^n \times \mathbb{R}^n$ and $\varepsilon > 0$, by Lemma 3.17 the function $\hat{z}(x, y, \varepsilon)$ is continuously differentiable everywhere and

$$\nabla_x \hat{z}(x, y, \epsilon) = \left(L_x + \frac{\tau-2}{2}L_y\right)L_{\hat{z}}^{-1}, \quad \nabla_y \hat{z}(x, y, \epsilon) = \left(L_y + \frac{\tau-2}{2}L_x\right)L_{\hat{z}}^{-1}. \tag{3.115}$$

We next prove the bound in (3.114) by the two cases: $w_2 \neq 0$ and $w_2 = 0$. Let

$$\hat{w} = (\hat{w}_1, \hat{w}_2) = \hat{w}(x, y, \varepsilon) := (x-y)^2 + \tau(x \circ y) + \varepsilon e.$$

Case (1). $w_2 \neq 0$. Then, $\hat{w}_2 \neq 0$ since $\hat{w}_2 = w_2$. Let $g = (g_1, g_2) := x + \frac{\tau-2}{2}y$. By (3.115) and the formula of $L_{\hat{z}}^{-1}$ given by Lemma 3.17, we can compute that

$$\nabla_x \hat{z}(x, y, \varepsilon) = \begin{bmatrix} \hat{b}g_1 + \hat{c}g_2^T\bar{w}_2 & \hat{c}g_1\bar{w}_2 + \hat{a}g_2^T + (\hat{b}-\hat{a})g_2^T\bar{w}_2\bar{w}_2^T \\ \hat{b}g_2 + \hat{c}g_1\bar{w}_2 & \hat{c}g_2\bar{w}_2^T + \hat{a}g_1 I + (\hat{b}-\hat{a})g_1\bar{w}_2\bar{w}_2 \end{bmatrix},$$

where $\hat{a}$, $\hat{b}$ and $\hat{c}$ are defined as in (3.77) with $w = \hat{w}$. Notice that

$$g_1 = x_1 + \frac{\tau-2}{2}y_1, \; g_2 = x_2 + \frac{\tau-2}{2}y_2; \; \lambda_1(\hat{w}) = \lambda_1(w) + \varepsilon, \; \lambda_2(\hat{w}) = \lambda_2(w) + \varepsilon.$$

Using the expression of $\hat{a}$, $\hat{b}$ and $\hat{c}$ and the result of Lemma 3.18 then yields that

$$\begin{aligned}
\left|\hat{b}g_1 + \hat{c}g_2^T\bar{w}_2\right| &\leq \frac{1}{2\sqrt{\lambda_2(w)}}\left|g_1 + g_2^T\bar{w}_2\right| + \frac{1}{2\sqrt{\lambda_1(w)}}\left|g_1 - g_2^T\bar{w}_2\right| \leq 1, \\
\left\|\hat{c}g_1\bar{w}_2^T + \hat{b}g_2^T\bar{w}_2\bar{w}_2^T\right\| &\leq \frac{1}{2\sqrt{\lambda_2(w)}}\left|g_1 + g_2^T\bar{w}_2\right| + \frac{1}{2\sqrt{\lambda_1(w)}}\left|g_1 - g_2^T\bar{w}_2\right| \leq 1, \\
\left\|\hat{a}g_2^T - \hat{a}g_2^T\bar{w}_2\bar{w}_2^T\right\| &\leq \frac{\|2g_2\|}{\sqrt{\|x\|^2 + \|y\|^2 + (\tau-2)x^T y}}(1 + \|\bar{w}_2\|) \leq 4, \\
\left\|\hat{b}g_2 + \hat{c}g_1\bar{w}_2\right\| &\leq \frac{1}{2\sqrt{\lambda_2(w)}}\|g_2 + g_1\bar{w}_2\| + \frac{1}{2\sqrt{\lambda_1(w)}}\|g_2 - g_1\bar{w}_2\| \leq 1, \\
\left\|\hat{c}g_2\bar{w}_2^T + \hat{b}g_1\bar{w}_2\bar{w}_2^T\right\|_F &\leq \frac{1}{2\sqrt{\lambda_2(w)}}\|g_2 + g_1\bar{w}_2\| + \frac{1}{2\sqrt{\lambda_1(w)}}\|g_2 - g_1\bar{w}_2\| \leq 1, \\
\left\|\hat{a}g_1 I - \hat{a}g_1\bar{w}_2\bar{w}_2^T\right\|_F &\leq \frac{2|g_1|}{\sqrt{\|x\|^2 + \|y\|^2 + (\tau-2)x^T y}} \cdot \left\|I - \bar{w}_2\bar{w}_2^T\right\|_F \leq 2(n-1).
\end{aligned}$$

The above inequalities imply that the first inequality in (3.114) holds under this case.

Case (2). $w_2 = 0$. In this case, from Lemma 3.17, it follows that

$$\nabla_x \hat{z}(x, y, \varepsilon) = \frac{1}{\sqrt{\hat{w}_1}}\left(L_x + \frac{\tau-2}{2}L_y\right) = \frac{1}{\sqrt{\hat{w}_1}}L_g.$$

Since $\hat{w}_1 = \|x + \frac{\tau-2}{2}y\|^2 + \frac{\tau(4-\tau)}{4}\|y\|^2 + \epsilon$, we have $|g_1|/\sqrt{\hat{w}_1} \leq 1$ and $\|g_2\|/\sqrt{\hat{w}_1} \leq 1$, which implies the first inequality in (3.114). Thus, we complete the proof for the first inequality. By the symmetry of x and y in $\hat{z}(x, y, \varepsilon)$, the second inequality clearly holds. □

Proposition 3.19 *The function ϕ_τ defined as in (3.64) has the following properties.*

(a) ϕ_τ is (continuously) differentiable at (x, y) if and only if $w(x, y) \in \text{int}(\mathcal{K}^n)$. Also,

$$\nabla_x \phi_\tau(x, y) = L_{x+\frac{\tau-2}{2}y}L_z^{-1} - I, \ \nabla_y \phi_\tau(x, y) = L_{y+\frac{\tau-2}{2}x}L_z^{-1} - I.$$

(b) ϕ_τ is globally Lipschitz continuous with the Lipschitz constant independent of τ.
(c) ϕ_τ is strongly semismooth at any $(x, y) \in \mathbb{R}^n \times \mathbb{R}^n$.

Proof (a) The proof directly follows from (3.78) and the following fact that

$$\phi_\tau(x, y) = z(x, y) - (x + y).$$

(b) It suffices to prove that $z(x, y)$ is globally Lipschitz continuous since $\phi_\tau(x, y) = z(x, y) - (x + y)$. To proceed, we denote

$$\hat{z} = \hat{z}(x, y, \varepsilon) := \left[(x - y)^2 + \tau(x \circ y) + \varepsilon e\right]^{1/2}$$

for any $\varepsilon > 0$ and $x = (x_1, x_2),\ y = (y_1, y_2) \in \mathbb{R} \times \mathbb{R}^{n-1}$. Then, applying Lemma 3.21 and the Mean-Value Theorem, we have

$$\begin{aligned}
\left\| z(x, y) - z(a, b) \right\| &= \left\| \lim_{\epsilon \to 0^+} \hat{z}(x, y, \epsilon) - \lim_{\epsilon \to 0^+} \hat{z}(a, b, \epsilon) \right\| \\
&\le \lim_{\epsilon \to 0^+} \left\| \hat{z}(x, y, \epsilon) - \hat{z}(a, y, \epsilon) + \hat{z}(a, y, \epsilon) - \hat{z}(a, b, \epsilon) \right\| \\
&\le \lim_{\epsilon \to 0^+} \left\| \int_0^1 \nabla_x \hat{z}(a + t(x - a), y, \epsilon)(x - a) dt \right\| \\
&\quad + \lim_{\epsilon \to 0^+} \left\| \int_0^1 \nabla_y \hat{z}(a, b + t(y - b), \epsilon)(y - b) dt \right\| \\
&\le \sqrt{2} C \|(x, y) - (a, b)\|
\end{aligned}$$

for any $(x, y), (a, b) \in \mathbb{R}^n \times \mathbb{R}^n$, where $C > 0$ is a constant independent of τ.

(c) From the definition of ϕ_τ and ϕ_{FB}, it is not hard to check that

$$\phi_\tau(x, y) = \phi_{\text{FB}} \left(x + \frac{\tau - 2}{2} y, \frac{\sqrt{\tau(4 - \tau)}}{2} y \right) + \frac{1}{2} \left(\tau - 4 + \sqrt{\tau(4 - \tau)} \right) y.$$

Note that ϕ_{FB} is strongly semismooth by Proposition 3.3, and the functions $x + \frac{\tau-2}{2} y$, $\frac{1}{2}\sqrt{\tau(4 - \tau)}y$ and $\frac{1}{2}(\tau - 4 + \sqrt{\tau(4 - \tau)})y$ are also strongly semismooth. Therefore, ϕ_τ is a strongly semismooth function since by [28, Theorem 19] the composition of strongly semismooth functions is strongly semismooth. □

Proposition 3.19(c) suggests that when a smoothing or nonsmooth Newton method is applied to solve the system (3.65), a fast convergence rate, at least superlinear, can be expected. To develop a semismooth Newton method for the SOCCP, it is essential to characterize the B-subdifferential $\partial_B \phi_\tau(x, y)$ at a general point (x, y). While the B-subdifferential of ϕ_{FB} has been discussed in [34], we extend the analysis here to ϕ_τ for any $\tau \in (0, 4)$.

Proposition 3.20 *Given a general point $(x, y) \in \mathbb{R} \times \mathbb{R}^{n-1}$, each element in $\partial_B \phi_\tau(x, y)$ is of the form $V = [V_x - I \;\; V_y - I]$ with V_x and V_y having the following representation:*

(a) If $(x - y)^2 + \tau(x \circ y) \in \text{int}(\mathcal{K}^n)$, then $V_x = L_z^{-1} L_{x + \frac{\tau-2}{2} y}$ and $V_y = L_z^{-1} L_{y + \frac{\tau-2}{2} x}$.

(b) If $(x - y)^2 + \tau(x \circ y) \in \text{bd}\mathcal{K}^n$ and $(x, y) \neq (0, 0)$, then

$$V_x \in \left\{ \frac{1}{2\sqrt{2w_1}} \begin{bmatrix} 1 & \bar{w}_2^{\mathrm{T}} \\ \bar{w}_2 & 4I - 3\bar{w}_2 \bar{w}_2^{\mathrm{T}} \end{bmatrix} \left(L_x + \frac{\tau - 2}{2} L_y \right) + \frac{1}{2} \begin{bmatrix} 1 \\ -\bar{w}_2 \end{bmatrix} u^{\mathrm{T}} \right\}$$

$$V_y \in \left\{ \frac{1}{2\sqrt{2w_1}} \begin{bmatrix} 1 & \bar{w}_2^{\mathsf{T}} \\ \bar{w}_2 & 4I - 3\bar{w}_2\bar{w}_2^{\mathsf{T}} \end{bmatrix} \left(L_y + \frac{\tau - 2}{2} L_x \right) + \frac{1}{2} \begin{bmatrix} 1 \\ -\bar{w}_2 \end{bmatrix} v^{\mathsf{T}} \right\} \tag{3.116}$$

for some $u = (u_1, u_2)$, $v = (v_1, v_2) \in \mathbb{R} \times \mathbb{R}^{n-1}$ *satisfying* $|u_1| \le \|u_2\| \le 1$ *and* $|v_1| \le \|v_2\| \le 1$, *where* $\bar{w}_2 = \frac{w_2}{\|w_2\|}$.

(c) *If* $(x, y) = (0, 0)$, *then* $V_x \in \{L_{\hat{u}}\}$, $V_y \in \{L_{\hat{v}}\}$ *for some* $\hat{u} = (\hat{u}_1, \hat{u}_2), \hat{v} = (\hat{v}_1, \hat{v}_2) \in \mathbb{R} \times \mathbb{R}^{l-1}$ *satisfying* $\|\hat{u}\|, \|\hat{v}\| \le 1$ *and* $\hat{u}_1\hat{v}_2 + \hat{v}_1\hat{u}_2 = 0$, *or*

$$\begin{aligned} V_x &\in \left\{ \frac{1}{2} \begin{bmatrix} 1 \\ \bar{w}_2 \end{bmatrix} \xi^{\mathsf{T}} + \frac{1}{2} \begin{bmatrix} 1 \\ -\bar{w}_2 \end{bmatrix} u^{\mathsf{T}} + 2 \begin{bmatrix} 0 & 0 \\ (I - \bar{w}_2\bar{w}_2^{\mathsf{T}})s_2 & (I - \bar{w}_2\bar{w}_2^{\mathsf{T}})s_1 \end{bmatrix} \right\} \\ V_y &\in \left\{ \frac{1}{2} \begin{bmatrix} 1 \\ \bar{w}_2 \end{bmatrix} \eta^{\mathsf{T}} + \frac{1}{2} \begin{bmatrix} 1 \\ -\bar{w}_2 \end{bmatrix} v^{\mathsf{T}} + 2 \begin{bmatrix} 0 & 0 \\ (I - \bar{w}_2\bar{w}_2^{\mathsf{T}})\omega_2 & (I - \bar{w}_2\bar{w}_2^{\mathsf{T}})\omega_1 \end{bmatrix} \right\} \end{aligned} \tag{3.117}$$

for some $u = (u_1, u_2)$, $v = (v_1, v_2)$, $\xi = (\xi_1, \xi_2)$, $\eta = (\eta_1, \eta_2) \in \mathbb{R} \times \mathbb{R}^{-1}$ *satisfying* $|u_1| \le \|u_2\| \le 1$, $|v_1| \le \|v_2\| \le 1$, $|\xi_1| \le \|\xi_2\| \le 1$ *and* $|\eta_1| \le \|\eta_2\| \le 1$, $\bar{w}_2 \in \mathbb{R}^{n-1}$ *satisfying* $\|\bar{w}_2\| = 1$, *and* $s = (s_1, s_2)$, $\omega = (\omega_1, \omega_2) \in \mathbb{R} \times \mathbb{R}^{n-1}$ *such that* $\|s\|^2 + \|\omega\|^2 \le 1$.

Proof Throughout the proof, let D_{ϕ_τ} denote the set of points where ϕ_τ is differentiable. Recall that this set is characterized by Proposition 3.19(a). For convenience, we write

$$\phi'_{\tau,x}(x, y) = \nabla_x \phi_\tau(x, y)^{\mathsf{T}} \quad \text{and} \quad \phi'_{\tau,y}(x, y) = \nabla_y \phi_\tau(x, y)^{\mathsf{T}}.$$

From Proposition 3.19(a), it then follows that for any $(x, y) \in D_{\phi_\tau}$,

$$\phi'_{\tau,x}(x, y) = L_z^{-1} L_{x + \frac{\tau-2}{2}y} - I, \quad \phi'_{\tau,x}(x, y) = L_z^{-1} L_{y + \frac{\tau-2}{2}x} - I. \tag{3.118}$$

Moreover, we observe from Lemma 3.1(c) that, when $w_2 \ne 0$, L_z^{-1} can be expressed as the sum of

$$L_1(w) = \frac{1}{2\sqrt{\lambda_1(w)}} \begin{bmatrix} 1 & -\bar{w}_2^{\mathsf{T}} \\ -\bar{w}_2 & \bar{w}_2\bar{w}_2^{\mathsf{T}} \end{bmatrix}$$

and

$$L_2(w) = \frac{1}{2\sqrt{\lambda_2(w)}} \begin{bmatrix} 1 & \bar{w}_2^{\mathsf{T}} \\ \bar{w}_2 & \dfrac{4\sqrt{\lambda_2(w)}(I - \bar{w}_2\bar{w}_2^{\mathsf{T}})}{\sqrt{\lambda_2(w)} + \sqrt{\lambda_1(w)}} + \bar{w}_2\bar{w}_2^{\mathsf{T}} \end{bmatrix},$$

and consequently $\phi'_{\tau,x}$ and $\phi'_{\tau,y}$ in (3.118) can be rewritten as

$$\begin{aligned} \phi'_{\tau,x}(x, y) &= (L_1(w) + L_2(w))L_{x + \frac{\tau-2}{2}y} - I, \\ \phi'_{\tau,x}(x, y) &= (L_1(w) + L_2(w))L_{y + \frac{\tau-2}{2}x} - I. \end{aligned} \tag{3.119}$$

(a) Under the given assumption, ϕ_τ is continuously differentiable at (x, y) by Proposition 3.19 (a). Consequently, the B-subdifferential $\partial_B\phi_\tau(x, y)$ consists of only one element,

$$\phi_\tau'(x, y) = \left[\phi_{\tau,x}'(x, y)\ \ \phi_{\tau,x}'(x, y)\right].$$

Substituting the formulas in (3.118) into it, we immediately obtain the conclusion.

(b) Assume that $(x, y) \neq (0, 0)$ satisfies $(x-y)^2 + \tau(x \circ y) \in \mathrm{bd}\mathcal{K}^n$. Let $\{(x^k, y^k)\} \subseteq D_{\phi_\tau}$ be an arbitrary sequence converging to (x, y). Let $w^k = (w_1^k, w_2^k) = w(x^k, y^k)$ and $z^k = z(x^k, y^k)$, where $w(x, y)$ and $z(x, y)$ are defined as in (3.69). From the given assumption on (x, y), we have $w \in \mathrm{bd}\mathcal{K}^n$ and $w_1 > 0$, which means that $\lambda_2(w) > \lambda_1(w) = 0$ and $\|w_2\| = w_1 > 0$. Hence, we assume without loss of generality that $w_2^k \neq 0$ for each k. Using the formulas in (3.119), it then follows that

$$\begin{aligned}
\phi_{\tau,x}'(x^k, y^k) &= \left(L_1(w^k) + L_2(w^k)\right) L_{x^k+\frac{\tau-2}{2}y^k} - I,\\
\phi_{\tau,y}'(x^k, y^k) &= \left(L_1(w^k) + L_2(w^k)\right) L_{y^k+\frac{\tau-2}{2}x^k} - I.
\end{aligned} \tag{3.120}$$

Notice that $\lim_{k\to+\infty} \lambda_2(w^k) = 2w_1 > 0$ and $\lim_{k\to+\infty} \lambda_1(w^k) = \lambda_1(w) = 0$, which, together with $\lim_{k\to+\infty} L_{x^k} = L_x$, $\lim_{k\to+\infty} L_{y^k} = L_y$ and $\lim_{k\to+\infty} w_2^k = w_2$, yields

$$\begin{aligned}
\lim_{k\to+\infty} L_2(w^k) L_{x^k+\frac{\tau-2}{2}y^k} &= C(w)\left(L_x + \frac{\tau-2}{2}L_y\right),\\
\lim_{k\to+\infty} L_2(w^k) L_{y^k+\frac{\tau-2}{2}x^k} &= C(w)\left(L_y + \frac{\tau-2}{2}L_x\right),
\end{aligned}$$

where $C(w)$ is defined as follows:

$$C(w) = \frac{1}{2\sqrt{2w_1}}\begin{bmatrix} 1 & \bar{w}_2^{\mathsf T} \\ \bar{w}_2 & 4I - 3\bar{w}_2\bar{w}_2^{\mathsf T} \end{bmatrix} \quad \text{with} \quad \bar{w}_2 = \frac{w_2}{\|w_2\|}.$$

In addition, by a simple computation, we have

$$\begin{aligned}
L_1(w^k) L_{x^k+\frac{\tau-2}{2}y^k} &= \frac{1}{2}\begin{bmatrix} u_1^k & (u_2^k)^{\mathsf T} \\ -u_1^k\bar{w}_2^k & -\bar{w}_2^k(u_2^k)^{\mathsf T} \end{bmatrix},\\
L_1(w^k) L_{y^k+\frac{\tau-2}{2}x^k} &= \frac{1}{2}\begin{bmatrix} v_1^k & (v_2^k)^{\mathsf T} \\ -v_1^k\bar{w}_2^k & -\bar{w}_2^k(v_2^k)^{\mathsf T} \end{bmatrix},
\end{aligned}$$

where $\bar{w}_2^k = w_2^k/\|w_2^k\|$ for each k, and

$$u_1^k = \frac{1}{\sqrt{\lambda_1(w^k)}}\left[\left(x_1^k + \frac{\tau-2}{2}y_1^k\right) - \left(x_2^k + \frac{\tau-2}{2}y_2^k\right)^{\mathsf T}\bar{w}_2^k\right],$$

$$u_2^k = \frac{1}{\sqrt{\lambda_1(w^k)}}\left[\left(x_2^k + \frac{\tau-2}{2}y_2^k\right) - \left(x_1^k + \frac{\tau-2}{2}y_1^k\right)\bar{w}_2^k\right],$$
$$v_1^k = \frac{1}{\sqrt{\lambda_1(w^k)}}\left[\left(y_1^k + \frac{\tau-2}{2}x_1^k\right) - \left(y_2^k + \frac{\tau-2}{2}x_2^k\right)^{\mathsf{T}}\bar{w}_2^k\right],$$
$$v_2^k = \frac{1}{\sqrt{\lambda_1(w^k)}}\left[\left(y_2^k + \frac{\tau-2}{2}x_2^k\right) - \left(y_1^k + \frac{\tau-2}{2}x_1^k\right)\bar{w}_2^k\right].$$

By Lemma 3.18, $|u_1^k| \le \|u_2^k\| \le 1$ and $|v_1^k| \le \|v_2^k\| \le 1$. Then, taking the limit (possibly on a subsequence) on $L_1(w^k)L_{x^k+\frac{\tau-2}{2}y^k}$ and $L_1(w^k)L_{y^k+\frac{\tau-2}{2}x^k}$, we have

$$\begin{aligned}
L_1(w^k)L_{x^k+\frac{\tau-2}{2}y^k} &\to \frac{1}{2}\begin{bmatrix} u_1 & u_2^{\mathsf{T}} \\ -u_1\bar{w}_2 & -\bar{w}_2u_2^{\mathsf{T}}\end{bmatrix} = \frac{1}{2}\begin{bmatrix} 1 \\ -\bar{w}_2\end{bmatrix}u^{\mathsf{T}} \\
L_1(w^k)L_{y^k+\frac{\tau-2}{2}x^k} &\to \frac{1}{2}\begin{bmatrix} v_1 & v_2^{\mathsf{T}} \\ -v_1\bar{w}_2 & -\bar{w}_2v_2^{\mathsf{T}}\end{bmatrix} = \frac{1}{2}\begin{bmatrix} 1 \\ -\bar{w}_2\end{bmatrix}v^{\mathsf{T}}
\end{aligned} \tag{3.121}$$

for some $u = (u_1, u_2), v = (v_1, v_2) \in \mathbb{R} \times \mathbb{R}^{n-1}$ with $|u_1| \le \|u_2\| \le 1$ and $|v_1| \le \|v_2\| \le 1$, where $\bar{w}_2 = w_2/\|w_2\|$. In fact, u and v are some accumulation point of the sequences $\{u^k\}$ and $\{v^k\}$, respectively. From (3.120)–(3.121), we obtain

$$\phi'_{\tau,x}(x^k, y^k) \to C(w)\left(L_x + \frac{\tau-2}{2}L_y\right) + \frac{1}{2}\begin{bmatrix} 1 \\ -\bar{w}_2\end{bmatrix}u^{\mathsf{T}} - I,$$
$$\phi'_{\tau,y}(x^k, y^k) \to C(w)\left(L_y + \frac{\tau-2}{2}L_x\right) + \frac{1}{2}\begin{bmatrix} 1 \\ -\bar{w}_2\end{bmatrix}v^{\mathsf{T}} - I.$$

This shows that as $k \to +\infty$, $\phi'_\tau(x^k, y^k) \to [V_x - I \;\; V_y - I]$ with V_x, V_y satisfying (3.116).

(c) Assume $(x, y) = (0, 0)$. Let $\{(x^k, y^k)\} \subseteq D_{\phi_\tau}$ be an arbitrary sequence converging to (x, y). Let $w^k = (w_1^k, w_2^k)$ and z^k be defined as in Case (b). From the given assumptions, we have $w = 0$. Therefore, we may assume without any loss of generality that $w_2^k = 0$ for all k or $w_2^k \neq 0$ for all k. We proceed the arguments by the two cases.

Case (1): $w_2^k = 0$ for all k. From Eq. (3.118) and Lemma 3.17, it follows that

$$\phi'_{\tau,x}(x^k, y^k) = \frac{1}{\sqrt{w_1^k}}\begin{bmatrix} x_1^k + \frac{\tau-2}{2}y_1^k & \left(x_2^k + \frac{\tau-2}{2}y_2^k\right)^{\mathsf{T}} \\ x_2^k + \frac{\tau-2}{2}y_2^k & \left(x_1^k + \frac{\tau-2}{2}y_1^k\right)I\end{bmatrix} - I,$$
$$\phi'_{\tau,y}(x^k, y^k) = \frac{1}{\sqrt{w_1^k}}\begin{bmatrix} y_1^k + \frac{\tau-2}{2}x_1^k & \left(y_2^k + \frac{\tau-2}{2}x_2^k\right)^{\mathsf{T}} \\ y_2^k + \frac{\tau-2}{2}x_2^k & \left(y_1^k + \frac{\tau-2}{2}x_1^k\right)I\end{bmatrix} - I.$$

Since

$$w_1^k = \left\| x^k + \frac{\tau - 2}{2} y^k \right\|^2 + \frac{\tau(4-\tau)}{4} \left\| y^k \right\|^2 = \left\| y^k + \frac{\tau - 2}{2} x^k \right\|^2 + \frac{\tau(4-\tau)}{4} \left\| x^k \right\|^2,$$

every element in the above $\phi'_{\tau,x}(x^k, y^k)$ and $\phi'_{\tau,y}(x^k, y^k)$ are bounded. Thus, taking limit (possibly on a subsequence) on $\phi'_{\tau,x}(x^k, y^k)$ and $\phi'_{\tau,y}(x^k, y^k)$, respectively, gives

$$\nabla_x \phi_\tau(x^k, y^k) \to \begin{bmatrix} \hat{u}_1 & \hat{u}_2^{\mathsf{T}} \\ \hat{u}_2 & \hat{u}_1 I \end{bmatrix} - I, \quad \nabla_y \phi_\tau(x^k, y^k) \to \begin{bmatrix} \hat{v}_1 & \hat{v}_2^{\mathsf{T}} \\ \hat{v}_2 & \hat{v}_1 I \end{bmatrix} - I$$

for some $\hat{u} = (\hat{u}_1, \hat{u}_2)$, $\hat{v} = (\hat{v}_1, \hat{v}_2) \in \mathbb{R} \times \mathbb{R}^{l-1}$ satisfying $\|\hat{u}\| \le 1$, $\|\hat{v}\| \le 1$ and $\hat{u}_1\hat{v}_2 + \hat{v}_1\hat{u}_2 = 0$. This shows that $\phi'_\tau(x^k, y^k) \to [V_x - I \quad V_y - I]$ with $V_x \in \{L_{\hat{u}}\}$, $V_y \in \{L_{\hat{v}}\}$.

Case (2): $w_2^k \neq 0$ for all k. Now $\phi'_{\tau,x}(x^k, y^k)$ and $\phi'_{\tau,y}(x^k, y^k)$ are given as in (3.120). Using the same arguments as part (b) and noting that $\{\bar{w}_2^k\}$ is bounded, we have

$$L_1(w^k) L_{x^k + \frac{\tau-2}{2} y^k} \to \frac{1}{2} \begin{bmatrix} 1 \\ -\bar{w}_2 \end{bmatrix} u^{\mathsf{T}}, \quad L_1(w^k) L_{y^k + \frac{\tau-2}{2} x^k} \to \frac{1}{2} \begin{bmatrix} 1 \\ -\bar{w}_2 \end{bmatrix} v^{\mathsf{T}} \tag{3.122}$$

for some vectors $u = (u_1, u_2)$, $v = (v_1, v_2) \in \mathbb{R} \times \mathbb{R}^{n-1}$ satisfying $|u_1| \le \|u_2\| \le 1$ and $|v_1| \le \|v_2\| \le 1$, and $\bar{w}_2 \in \mathbb{R}^{n-1}$ satisfying $\|\bar{w}_2\| = 1$. We next compute the limit of $L_2(w^k) L_{x^k + \frac{\tau-2}{2} y^k}$ and $L_2(w^k) L_{y^k + \frac{\tau-2}{2} x^k}$. By the definition of $L_2(w)$,

$$L_2(w^k) L_{x^k + \frac{\tau-2}{2} y^k} = \frac{1}{2} \begin{bmatrix} \xi_1^k & (\xi_2^k)^{\mathsf{T}} \\ \xi_1^k \bar{w}_2^k + 4(I - \bar{w}_2^k (\bar{w}_2^k)^{\mathsf{T}}) s_2^k & \bar{w}_2^k (\xi_2^k)^{\mathsf{T}} + 4(I - \bar{w}_2^k (\bar{w}_2^k)^{\mathsf{T}}) s_1^k \end{bmatrix},$$

$$L_2(w^k) L_{y^k + \frac{\tau-2}{2} x^k} = \frac{1}{2} \begin{bmatrix} \eta_1^k & (\eta_2^k)^{\mathsf{T}} \\ \eta_1^k \bar{w}_2^k + 4(I - \bar{w}_2^k (\bar{w}_2^k)^{\mathsf{T}}) \omega_2^k & \bar{w}_2^k (\eta_2^k)^{\mathsf{T}} + 4(I - \bar{w}_2^k (\bar{w}_2^k)^{\mathsf{T}}) \omega_1^k \end{bmatrix}$$

where

$$\xi_1^k = \frac{1}{\sqrt{\lambda_2(w^k)}} \left[\left(x_1^k + \frac{\tau-2}{2} y_1^k \right) + \left(x_2^k + \frac{\tau-2}{2} y_2^k \right)^{\mathsf{T}} \bar{w}_2^k \right],$$

$$\xi_2^k = \frac{1}{\sqrt{\lambda_2(w^k)}} \left[\left(x_2^k + \frac{\tau-2}{2} y_2^k \right) + \left(x_1^k + \frac{\tau-2}{2} y_1^k \right) \bar{w}_2^k \right],$$

$$\eta_1^k = \frac{1}{\sqrt{\lambda_2(w^k)}} \left[\left(y_1^k + \frac{\tau-2}{2} x_1^k \right) + \left(y_2^k + \frac{\tau-2}{2} x_2^k \right)^{\mathsf{T}} \bar{w}_2^k \right],$$

$$\eta_2^k = \frac{1}{\sqrt{\lambda_2(w^k)}} \left[\left(y_2^k + \frac{\tau-2}{2} x_2^k \right) + \left(y_1^k + \frac{\tau-2}{2} x_1^k \right) \bar{w}_2^k \right],$$

and

$$s_1^k = \frac{\left(x_1^k + \frac{\tau-2}{2} y_1^k\right)}{\sqrt{\lambda_2(w^k)} + \sqrt{\lambda_1(w^k)}}, \quad s_2^k = \frac{\left(x_2^k + \frac{\tau-2}{2} y_2^k\right)}{\sqrt{\lambda_2(w^k)} + \sqrt{\lambda_1(w^k)}};$$
$$\omega_1^k = \frac{\left(y_1^k + \frac{\tau-2}{2} x_1^k\right)}{\sqrt{\lambda_2(w^k)} + \sqrt{\lambda_1(w^k)}}, \quad \omega_2^k = \frac{\left(y_2^k + \frac{\tau-2}{2} x_2^k\right)}{\sqrt{\lambda_2(w^k)} + \sqrt{\lambda_1(w^k)}}.$$

By Lemma 3.18, $|\xi_1^k| \le \|\xi_2^k\| \le 1$ and $|\eta_1^k| \le \|\eta_2^k\| \le 1$. In addition,

$$\|s^k\|^2 + \|\omega^k\|^2 = \frac{\|x^k + \frac{\tau-2}{2} y^k\|^2 + \|y^k + \frac{\tau-2}{2} x^k\|^2}{2[\|x^k\|^2 + \|y^k\|^2 + (\tau-2)(x^k)^T y^k] + 2\sqrt{\lambda_2(w^k)}\sqrt{\lambda_1(w^k)}} \le 1.$$

Taking the limit on $L_2(w^k) L_{x^k + \frac{\tau-2}{2} y^k}$ and $L_2(w^k) L_{y^k + \frac{\tau-2}{2} x^k}$, we have

$$\begin{aligned} L_2(w^k) L_{x^k + \frac{\tau-2}{2} y^k} &\to \frac{1}{2} \begin{bmatrix} \xi_1 & \xi_2 \\ \xi_1 \bar{w}_2 + 4(I - \bar{w}_2 \bar{w}_2^{\mathsf T}) s_2 & \bar{w}_2 \xi_2^{\mathsf T} + 4(I - \bar{w}_2 \bar{w}_2^{\mathsf T}) s_1 \end{bmatrix} \\ &= \frac{1}{2} \begin{bmatrix} 1 \\ \bar{w}_2 \end{bmatrix} \xi^{\mathsf T} + 2 \begin{bmatrix} 0 & 0 \\ (I - \bar{w}_2 \bar{w}_2^{\mathsf T}) s_2 & (I - \bar{w}_2 \bar{w}_2^{\mathsf T}) s_1 \end{bmatrix} \quad (3.123) \\ L_2(w^k) L_{y^k + \frac{\tau-2}{2} x^k} &\to \frac{1}{2} \begin{bmatrix} \eta_1 & \eta_2 \\ \eta_1 \bar{w}_2^{\mathsf T} + 4(I - \bar{w}_2 \bar{w}_2^{\mathsf T}) \omega_2 & \bar{w}_2 \eta_2^{\mathsf T} + 4(I - \bar{w}_2 \bar{w}_2^{\mathsf T}) \omega_1 \end{bmatrix} \\ &= \frac{1}{2} \begin{bmatrix} 1 \\ \bar{w}_2 \end{bmatrix} \eta^{\mathsf T} + 2 \begin{bmatrix} 0 & 0 \\ (I - \bar{w}_2 \bar{w}_2^{\mathsf T}) \omega_2 & (I - \bar{w}_2 \bar{w}_2^{\mathsf T}) \omega_1 \end{bmatrix} \quad (3.124) \end{aligned}$$

for some vectors $\xi = (\xi_1, \xi_2), \eta = (\eta_1, \eta_2) \in \mathbb{R} \times \mathbb{R}^{n-1}$ satisfying $|\xi_1| \le \|\xi_2\| \le 1$ and $|\eta_1| \le \|\eta_2\| \le 1$, and $s = (s_1, s_2), \omega = (\omega_1, \omega_2) \in \mathbb{R} \times \mathbb{R}^{n-1}$ satisfying $\|s\|^2 + \|\omega\|^2 \le 1$. From Eqs. (3.122), (3.123) and (3.124), it follows that as $k \to +\infty$,

$$\phi'_{\tau,x}(x^k, y^k) \to \frac{1}{2} \begin{bmatrix} 1 \\ \bar{w}_2 \end{bmatrix} \xi^{\mathsf T} + \frac{1}{2} \begin{bmatrix} 1 \\ -\bar{w}_2 \end{bmatrix} u^{\mathsf T} + 2 \begin{bmatrix} 0 & 0 \\ (I - \bar{w}_2 \bar{w}_2^{\mathsf T}) s_2 & (I - \bar{w}_2 \bar{w}_2^{\mathsf T}) s_1 \end{bmatrix} - I,$$
$$\phi'_{\tau,x}(x^k, y^k) \to \frac{1}{2} \begin{bmatrix} 1 \\ \bar{w}_2 \end{bmatrix} \eta^{\mathsf T} + \frac{1}{2} \begin{bmatrix} 1 \\ -\bar{w}_2 \end{bmatrix} v^{\mathsf T} + 2 \begin{bmatrix} 0 & 0 \\ (I - \bar{w}_2 \bar{w}_2^{\mathsf T}) \omega_2 & (I - \bar{w}_2 \bar{w}_2^{\mathsf T}) \omega_1 \end{bmatrix} - I.$$

This shows that as $k \to +\infty$, $\phi'_\tau(x^k, y^k) \to [V_x - I \quad V_y - I]$ with V_x and V_y satisfying (3.117). Combining with Case (1), the desired result then follows. □

Proposition 3.21 *The operator $\Phi_\tau : \mathbb{R}^n \to \mathbb{R}^n$ given by (3.65) is semismooth. Moreover, it is strongly semismooth if F' and G' are locally Lipschitz continuous.*

Proof Note that Φ_τ is (strongly) semismooth if and only if each of its component functions is (strongly) semismooth. Since the composition of (strongly) semismooth functions remains (strongly) semismooth, as established in [28, Theorem 19], the desired conclusion follows directly from Proposition 3.19(c). □

To characterize the B-subdifferential of Φ_τ, we write $F_i(\zeta) = (F_{i1}(\zeta), F_{i2}(\zeta))$ and $G_i(\zeta) = (G_{i1}(\zeta), G_{i2}(\zeta))$, and denote w_i and z_i for $i = 1, 2, \ldots, m$ by

$$\begin{aligned} w_i &= (w_{i1}(\zeta), w_{i2}(\zeta)) = w(F_i(\zeta), G_i(\zeta)), \\ z_i &= (z_{i1}(\zeta), z_{i2}(\zeta)) = z(F_i(\zeta), G_i(\zeta)). \end{aligned}$$

For simplicity, we sometimes suppress in $F_i(\zeta)$ and $G_i(\zeta)$ the dependence on ζ.

Proposition 3.22 *Let $\Phi_\tau : \mathbb{R}^n \to \mathbb{R}^n$ be defined as in (3.65). Then, for any $\zeta \in \mathbb{R}^n$,*

$$\partial_B \Phi_\tau(\zeta)^{\mathsf{T}} \subseteq \nabla F(\zeta)\,(A(\zeta) - I) + \nabla G(\zeta)\,(B(\zeta) - I),$$

where $A(\zeta)$ and $B(\zeta)$ are possibly multivalued $n \times n$ block diagonal matrices whose ith blocks $A_i(\zeta)$ and $B_i(\zeta)$ for $i = 1, 2, \ldots, m$ have the following representation.

(a) If $(F_i(\zeta) - G_i(\zeta))^2 + \tau\,(F_i(\zeta) \circ G_i(\zeta)) \in \mathrm{int}\mathcal{K}^{n_i}$, then

$$A_i(\zeta) = L_{F_i + \frac{\tau-2}{2} G_i} L_{z_i}^{-1} \quad \text{and} \quad B_i(\zeta) = L_{G_i + \frac{\tau-2}{2} F_i} L_{z_i}^{-1}.$$

(b) If $(F_i(\zeta), G_i(\zeta)) \neq (0, 0)$ and $(F_i(\zeta) - G_i(\zeta))^2 + \tau\,(F_i(\zeta) \circ G_i(\zeta)) \in \mathrm{bd}\mathcal{K}^{n_i}$, then

$$A_i(\zeta) \in \left\{ \frac{1}{2\sqrt{2w_{i1}}} \left(L_{F_i} + \frac{\tau-2}{2} L_{G_i} \right) \begin{bmatrix} 1 & \bar{w}_{i2}^{\mathsf{T}} \\ \bar{w}_{i2} & 4I - 3\bar{w}_{i2}\bar{w}_{i2}^T \end{bmatrix} + \frac{1}{2} u_i (1, -\bar{w}_{i2}^{\mathsf{T}}) \right\}$$

$$B_i(\zeta) \in \left\{ \frac{1}{2\sqrt{2w_{i1}}} \left(L_{G_i} + \frac{\tau-2}{2} L_{F_i} \right) \begin{bmatrix} 1 & \bar{w}_{i2}^{\mathsf{T}} \\ \bar{w}_{i2} & 4I - 3\bar{w}_{i2}\bar{w}_{i2}^T \end{bmatrix} + \frac{1}{2} v_i (1, -\bar{w}_{i2}^{\mathsf{T}}) \right\}$$

for some $u_i = (u_{i1}, u_{i2}), v_i = (v_{i1}, v_{i2}) \in \mathbb{R} \times \mathbb{R}^{n_i - 1}$ satisfying $|u_{i1}| \le \|u_{i2}\| \le 1$ and $|v_{i1}| \le \|v_{i2}\| \le 1$, where $\bar{w}_{i2} = \frac{w_{i2}}{\|w_{i2}\|}$.

(c) If $(F_i(\zeta), G_i(\zeta)) = (0, 0)$, then

$$A_i(\zeta) \in \left\{ L_{\hat{u}_i} \right\} \cup \left\{ \frac{1}{2} \xi_i \left(1, \bar{w}_{i2}^{\mathsf{T}}\right) + \frac{1}{2} u_i \left(1, -\bar{w}_{i2}^{\mathsf{T}}\right) + \begin{bmatrix} 0 & 2s_{i2}^{\mathsf{T}}(I - \bar{w}_{i2}\bar{w}_{i2}^{\mathsf{T}}) \\ 0 & 2s_{i1}(I - \bar{w}_{i2}\bar{w}_{i2}^{\mathsf{T}}) \end{bmatrix} \right\}$$

$$B_i(\zeta) \in \left\{ L_{\hat{v}_i} \right\} \cup \left\{ \frac{1}{2} \eta_i \left(1, \bar{w}_{i2}^{\mathsf{T}}\right) + \frac{1}{2} v_i \left(1, -\bar{w}_{i2}^{\mathsf{T}}\right) + \begin{bmatrix} 0 & 2\omega_{i2}^{\mathsf{T}}(I - \bar{w}_{i2}\bar{w}_{i2}^{\mathsf{T}}) \\ 0 & 2\omega_{i1}(I - \bar{w}_{i2}\bar{w}_{i2}^{\mathsf{T}}) \end{bmatrix} \right\}$$

for some $\hat{u}_i = (\hat{u}_{i1}, \hat{u}_{i2}), \hat{v}_i = (\hat{v}_{i1}, \hat{v}_{i2}) \in \mathbb{R} \times \mathbb{R}^{n_i - 1}$ satisfying $\|\hat{u}_i\|, \|\hat{v}_i\| \le 1$ and $\hat{u}_{i1}\hat{v}_{i2} + \hat{v}_{i1}\hat{u}_{i2} = 0$, some $u_i = (u_{i1}, u_{i2}), v_i = (v_{i1}, v_{i2}), \xi_i = (\xi_{i1}, \xi_{i2}), \eta_i = (\eta_{i1}, \eta_{i2}) \in \mathbb{R} \times \mathbb{R}^{n_i - 1}$ with $|u_{i1}| \le \|u_{i2}\| \le 1$, $|v_{i1}| \le \|v_{i2}\| \le 1$, $|\xi_{i1}| \le \|\xi_{i2}\| \le 1$ and $|\eta_{i1}| \le \|\eta_{i2}\| \le 1$, $\bar{w}_{i2} \in \mathbb{R}^{n_i - 1}$ satisfying $\|\bar{w}_{i2}\| = 1$, and $s_i = (s_{i1}, s_{i2}), \omega_i = (\omega_{i1}, \omega_{i2}) \in \mathbb{R} \times \mathbb{R}^{n_i - 1}$ such that $\|s_i\|^2 + \|\omega_i\|^2 \le 1$.

Proof Let $\Phi_{\tau,i}(\zeta)$ denote the ith subvector of Φ_τ, i.e., $\Phi_{\tau,i}(\zeta) = \phi_\tau(F_i(\zeta), G_i(\zeta))$ for all $i = 1, 2, \ldots, m$. From [29, Proposition 2.6.2], it follows that

$$\partial_B \Phi_\tau(\zeta)^\top \subseteq \partial_B \Phi_{\tau,1}(\zeta)^\top \times \partial_B \Phi_{\tau,2}(\zeta)^\top \times \cdots \times \partial_B \Phi_{\tau,m}(\zeta)^\top, \tag{3.125}$$

where the latter denotes the set of all matrices whose $(n_{i-1}+1)$ to n_ith columns with $n_0 = 0$ belong to $\partial_B \Phi_{\tau,i}(\zeta)^\top$. Using the definition of B-subdifferential and the continuous differentiability of F and G, it is not difficult to verify that

$$\partial_B \Phi_{\tau,i}(\zeta)^\top = [\nabla F_i(\zeta)\ \ \nabla G_i(\zeta)]\partial_B \phi_\tau(F_i(\zeta), G_i(\zeta))^\top, \quad i = 1, \ldots, m \tag{3.126}$$

By applying Proposition 3.20 in conjunction with the preceding two equations, the desired result follows immediately. □

Proposition 3.23 *For any $\zeta \in \mathbb{R}^n$, let $A(\zeta)$ and $B(\zeta)$ be the multivalued block diagonal matrices given as in Proposition 3.22. Then, for any $i \in \{1, 2, \ldots, m\}$,*

$$\Big\langle (A_i(\zeta) - I)\Phi_{\tau,i}(\zeta),\ (B_i(\zeta) - I)\Phi_{\tau,i}(\zeta)\Big\rangle \geq 0,$$

and the equality holds if and only if $\Phi_{\tau,i}(\zeta) = 0$. Particularly, for the index i such that $(F_i(\zeta) - G_i(\zeta))^2 + \tau(F_i(\zeta) \circ G_i(\zeta) \in \mathrm{int}\mathcal{K}^{n_i}$, we have

$$\langle (A_i(\zeta) - I)\upsilon_i,\ (B_i(\zeta) - I)\upsilon_i\rangle \geq 0, \quad \text{for any } \upsilon_i \in \mathbb{R}^{n_i}.$$

Proof From [29, Theorem 2.6.6] and Proposition 3.19 (d), we have

$$\nabla \psi_\tau(x, y) = \partial_B \phi_\tau(x, y)^\top \phi_\tau(x, y).$$

Hence, for any $i = 1, 2, \ldots, m$, it follows that

$$\nabla \psi_\tau(F_i(\zeta), G_i(\zeta)) = \partial_B \phi_\tau(F_i(\zeta), G_i(\zeta))^\top \phi_\tau(F_i(\zeta), G_i(\zeta)).$$

In addition, from Propositions 3.20 and 3.22, it is not hard to see that

$$\big[A_i(\zeta)^\top - I\ \ B_i(\zeta)^\top - I\big] \in \partial_B \phi_\tau(F_i(\zeta), G_i(\zeta)).$$

Combining with the last two equations yields that for any $i = 1, 2, \ldots, m$,

$$\begin{aligned} \nabla_x \psi_\tau(F_i(\zeta), G_i(\zeta)) &= (A_i(\zeta) - I)\Phi_{\tau,i}(\zeta) \\ \nabla_y \psi_\tau(F_i(\zeta), G_i(\zeta)) &= (B_i(\zeta) - I)\Phi_{\tau,i}(\zeta). \end{aligned} \tag{3.127}$$

Consequently, the first part of the conclusions is direct by Proposition 4.1 of [35]. Notice that for any i such that $(F_i(\zeta) - G_i(\zeta))^2 + \tau(F_i(\zeta) \circ G_i(\zeta) \in \mathrm{int}\mathcal{K}^{n_i}$ and any $\upsilon_i \in \mathbb{R}^{n_i}$,

$$\begin{aligned} &\langle (A_i(\zeta) - I)\upsilon_i,\ (B_i(\zeta) - I)\upsilon_i\rangle \\ &= \Big\langle \Big(L_{F_i + \frac{\tau-2}{2}G_i} - L_{z_i}\Big)L_{z_i}^{-1}\upsilon_i,\ \Big(L_{G_i + \frac{\tau-2}{2}F_i} - L_{z_i}\Big)L_{z_i}^{-1}\upsilon_i\Big\rangle \end{aligned}$$

$$= \left\langle \left(L_{G_i + \frac{\tau-2}{2} F_i} - L_{z_i}\right)\left(L_{F_i + \frac{\tau-2}{2} G_i} - L_{z_i}\right) L_{z_i}^{-1} \upsilon_i,\ L_{z_i}^{-1} \upsilon_i \right\rangle.$$

Therefore, by employing the same reasoning as in Case (2) of [35, Proposition 4.1], we arrive at the second part of the conclusions. □

Lemma 3.22 *Let $\psi_\tau : \mathbb{R}^n \times \mathbb{R}^n \to \mathbb{R}_+$ be given by (3.63). Then, for any $x, y \in \mathbb{R}^n$,*

$$\phi_\tau(x, y) \neq 0 \iff \nabla_x \psi_\tau(x, y) \neq 0,\ \nabla_y \psi_\tau(x, y) \neq 0.$$

Proof The sufficiency follows directly from Proposition 3.16. Now suppose $\phi_\tau(x, y) \neq 0$. If either $\nabla_x \psi_\tau(x, y) = 0$ or $\nabla_y \psi_\tau(x, y) = 0$, then it follows that $\langle \nabla_x \psi_\tau(x, y), \nabla_y \psi_\tau(x, y) \rangle = 0$. However, by Proposition 3.15, this would imply $\phi_\tau(x, y) = 0$, leading to a contradiction. □

Proposition 3.24 *Let $\Psi_\tau : \mathbb{R}^n \to \mathbb{R}_+$ be given as (3.66). Suppose ∇G is invertible and $\nabla G(\zeta)^{-1} \nabla F(\zeta)$ at any $\zeta \in \mathbb{R}^n$ has the Cartesian P_0-property. Then, every stationary point of Ψ_τ is a solution of the SOCCP* (3.4).

Proof Let ζ be an arbitrary stationary point of $f_\tau(\zeta)$. Since Ψ_τ is continuously differentiable as established in Proposition 3.19(d), and Φ_τ is locally Lipschitz continuous, it follows from [29, Theorem 2.6.6] that for any $V \in \partial \Phi_\tau(\zeta)^\mathsf{T}$, we have

$$0 = \nabla \Psi_\tau(\zeta) = V \Phi_\tau(\zeta).$$

Let V be an element of $\partial_B \Phi_\tau(\zeta)^\mathsf{T} (\subseteq \partial \Phi_\tau(\zeta)^\mathsf{T})$. Then, from (3.125) it follows that there exist matrices $V_i \in \partial_B \Phi_{\tau,i}(\zeta)^\mathsf{T}$ such that

$$V = V_1 \times V_2 \times \cdots \times V_m.$$

In addition, for each $V_i \in \mathbb{R}^{n \times n_i}$, by Proposition 3.20 there exist matrices $A_i(\zeta) \in \mathbb{R}^{n_i \times n_i}$ and $B_i(\zeta) \in \mathbb{R}^{n_i \times n_i}$, as characterized by Proposition 3.22, such that

$$V_i = \nabla F_i(\zeta)(A_i(\zeta) - I) + \nabla G_i(\zeta)(B_i(\zeta) - I), \quad i = 1, 2, \ldots, m.$$

Let $A(\zeta) = \operatorname{diag}(A_1(\zeta), \ldots, A_m(\zeta))$ and $B(\zeta) = \operatorname{diag}(B_1(\zeta), \ldots, B_m(\zeta))$. Combining the last three equations, it then follows that

$$[\nabla F(\zeta)(A(\zeta) - I) + \nabla G(\zeta)(B(\zeta) - I)]\, \Phi_\tau(\zeta) = 0,$$

which, by the invertibility of $\nabla G(\zeta)$, is equivalent to

$$\left[\nabla G(\zeta)^{-1} \nabla F(\zeta)(A(\zeta) - I) + (B(\zeta) - I)\right] \Phi_\tau(\zeta) = 0. \tag{3.128}$$

Suppose that $\Phi_\tau(\zeta) \neq 0$. Then, there necessarily exists an index $\nu \in \{1, 2, \ldots, m\}$ such that $\Phi_{\tau,\nu}(\zeta) = \phi_\tau(F_\nu(\zeta), G_\nu(\zeta)) \neq 0$. Using Lemma 3.22 and Eq. (3.127) then yields

$$(A_\nu(\zeta) - I)\Phi_{\tau,\nu}(\zeta) \neq 0 \quad \text{and} \quad (B_\nu(\zeta) - I)\Phi_{\tau,\nu}(\zeta) \neq 0. \tag{3.129}$$

In addition, from (3.128) it follows that

$$\left[\nabla G(\zeta)^{-1}\nabla F(\zeta)(A(\zeta) - I)\Phi_\tau(\zeta)\right]_\nu + (B_\nu(\zeta) - I)\Phi_{\tau,\nu}(\zeta) = 0.$$

Making the inner product with $(A_\nu(\zeta) - I)\Phi_{\tau,\nu}(\zeta)$ on both sides, we obtain

$$\begin{aligned}&\left\langle (A_\nu(\zeta) - I)\Phi_{\tau,\nu}(\zeta),\ \left[\nabla G(\zeta)^{-1}\nabla F(\zeta)(A(\zeta) - I)\Phi_\tau(\zeta)\right]_\nu \right\rangle \\ &+ \left\langle (A_\nu(\zeta) - I)\Phi_{\tau,\nu}(\zeta),\ (B_\nu(\zeta) - I)\Phi_{\tau,\nu}(\zeta) \right\rangle = 0.\end{aligned}$$

Observe that the first term on the left-hand side is nonnegative by (3.129) and the assumption that $\nabla G(\zeta)^{-1}\nabla F(\zeta)$ possesses the Cartesian P_0-property at any $\zeta \in \mathbb{R}^n$. The second term is strictly positive by Lemma 3.23, given that $\Phi_{\tau,\nu}(\zeta) \neq 0$. This yields a contradiction. □

Remark 3.3 (i) It is easy to verify that $\nabla G(\zeta)^{-1}\nabla F(\zeta) \succeq O$ implies the Cartesian P_0-property of $\nabla G(\zeta)^{-1}\nabla F(\zeta)$. While, by [35], the column monotonicity of $\nabla F(\zeta)$ and $-\nabla G(\zeta)$ is now equivalent to $\nabla G(\zeta)^{-1}\nabla F(\zeta) \succeq O$. This means that the condition in Proposition 3.24 is weaker than the one (3.113) used in Proposition 3.18.

(ii) For the SOCCP (3.1), the condition stated in Proposition 3.24 is equivalent to requiring that F satisfies the Cartesian P_0-property. If $n_1 = n_2 = \cdots = n_m = 1$, this condition reduces to the classical requirement in the NCPs that F is a P_0-function.

Lemma 3.23 *Let ψ_τ be given by (3.63). Then, for any $(x, y) \in \mathbb{R}^n \times \mathbb{R}^n$, we have*

$$4\psi_\tau(x, y) \geq 2\,\|[\phi_\tau(x, y)]_+\|^2 \geq \frac{(4-\tau)^2}{4}\left[\|(-x)_+\|^2 + \|(-y)_+\|^2\right].$$

Proof Note that $z(x, y) - (x + \frac{\tau-2}{2}y) \in \mathcal{K}^n$ and $z(x, y) - (y + \frac{\tau-2}{2}x) \in \mathcal{K}^n$. Following the same proof line as Lemma 3.7 immediately yields the desired result. □

Lemma 3.24 *Let ψ_τ be defined as in (3.63). For any sequence $\{(x^k, y^k)\} \subseteq \mathbb{R}^n \times \mathbb{R}^n$, let $\lambda_1^k \leq \lambda_2^k$ and $\mu_1^k \leq \mu_2^k$ denote the spectral values of x^k and y^k, respectively.*

(a) If $\lambda_1^k \to -\infty$ or $\mu_1^k \to -\infty$, then $\psi_\tau(x^k, y^k) \to +\infty$.

(b) If $\{\lambda_1^k\}$ and $\{\mu_1^k\}$ are bounded below, but $\lambda_2^k \to +\infty$, $\mu_2^k \to +\infty$, and $\frac{x^k}{\|x^k\|} \circ \frac{y^k}{\|y^k\|} \nrightarrow 0$, then $\psi_\tau(x^k, y^k) \to +\infty$.

Proof Part (a) is direct by Lemma 3.23 and the following fact that

$$\left\|(-x^k)_+\right\|^2 = \frac{1}{2}\sum_{i=1}^{2}\left(\min\{0, \lambda_i^k\}\right)^2, \quad \left\|(-y^k)_+\right\|^2 = \frac{1}{2}\sum_{i=1}^{2}\left(\min\{0, \mu_i^k\}\right)^2.$$

We next prove part (b) by contradiction. Suppose that $\{\psi_\tau(x^k, y^k)\}$ is bounded. Since

$$x^k + y^k = z^k - \phi_\tau(x^k, y^k) \quad \forall k,$$

where $z^k = z(x^k, y^k)$ with $z(x, y)$ defined as in (3.69). Squaring the two sides of the last equality then yields that

$$(4-\tau)x^k \circ y^k = -2z^k \circ \phi_\tau(x^k, y^k) + (\phi_\tau(x^k, y^k))^2. \tag{3.130}$$

Noting that, for each k,

$$0 \le \frac{z_1^k}{\|x^k\|\|y^k\|} \le \frac{\sqrt{2w_1^k}}{\|x^k\|\|y^k\|} = \sqrt{\frac{\|x^k\|^2 + \|y^k\|^2 + (\tau-2)(x^k)^T y^k}{\|x^k\|^2\|y^k\|^2}},$$

we can verify that $\lim_{k\to+\infty} \frac{z_1^k}{\|x^k\|\|y^k\|} = 0$. Combining with $\frac{z^k}{\|x^k\|\|y^k\|} \in \mathcal{K}^l$ yields

$$\lim_{k\to+\infty} \frac{z^k}{\|x^k\|\|y^k\|} = 0.$$

Using Eq. (3.130) and the boundedness of $\{\phi_\tau(x^k, y^k)\}$, it then follows that

$$\lim_{k\to+\infty} \frac{x^k}{\|x^k\|} \circ \frac{y^k}{\|y^k\|} = 0,$$

which clearly contradicts the given assumption. The proof is complete. □

Now, invoking Lemma 3.24 and employing the same reasoning as in [34, Proposition 5.2], we can establish the boundedness of the level sets of $\Psi_\tau(\zeta)$ for the SOCCP (3.1), under the assumption that F possesses the uniform Cartesian P-property and satisfies the following condition:

Condition A. For any sequence $\{\zeta^k\} \subseteq \mathbb{R}^n$ such that $\|\zeta^k\| \to +\infty$, if there exists $i \in \{1, \ldots, m\}$ such that $\lambda_1(\zeta_i^k), \lambda_1(F_i(\zeta^k)) > -\infty$ and $\lambda_2(\zeta_i^k), \lambda_2(F_i(\zeta^k)) \to +\infty$, then

$$\limsup_{k\to+\infty} \left\langle \frac{\zeta_i^k}{\|\zeta_i^k\|}, \frac{F_i(\zeta^k)}{\|F_i(\zeta^k)\|} \right\rangle > 0.$$

Proposition 3.25 *For the SOCCP (3.1), if $F : \mathbb{R}^n \to \mathbb{R}^n$ has the uniform Cartesian P-property and satisfies Condition A, then the merit function Ψ_τ has bounded level sets.*

3.1.2 The Functions ϕ^p_{FB} and ψ^p_{FB} in SOC Setting

In this section, we study the generalized Fischer–Burmeister (FB) merit function associated with the SOC. Within this framework, it is natural to define

$$\psi^p_{\text{FB}}(x, y) := \frac{1}{2}\|\phi^p_{\text{FB}}(x, y)\|^2, \tag{3.131}$$

where p is a fixed real number from $(1, +\infty)$, and $\phi^p_{\text{FB}} : \mathbb{R}^n \times \mathbb{R}^n \to \mathbb{R}^n$ is defined by

$$\phi^p_{\text{FB}}(x, y) := \sqrt[p]{|x|^p + |y|^p} - (x + y) \tag{3.132}$$

with $|x|^p$ being the vector-valued SOC function (or Löwner function) associated with $|t|^p$ $(t \in \mathbb{R})$. In other words, given a real-valued function $g : \mathbb{R} \to \mathbb{R}$, recall that the vector-valued function $g^{\text{soc}} : \mathbb{R}^n \to \mathbb{R}^n$ by

$$g^{\text{soc}}(x) := g(\lambda_1(x))u_x^{(1)} + g(\lambda_2(x))u_x^{(2)}.$$

If g is defined on a subset of $\mathbb{R}$, then g^{soc} is defined on the corresponding subset of $\mathbb{R}^n$. The definition of g^{soc} is unambiguous whether $x_2 \neq 0$ or $x_2 = 0$. As mentioned, if we use the vector-valued functions associated with $|t|^p$ $(t \in \mathbb{R})$ and $\sqrt[p]{t}$ $(t \geq 0)$, then we obtain

$$\begin{aligned} |x|^p &:= |\lambda_1(x)|^p\, u_x^{(1)} + |\lambda_2(x)|^p\, u_x^{(2)} \quad \forall x \in \mathbb{R}^n, \\ \sqrt[p]{x} &:= \sqrt[p]{\lambda_1(x)}\, u_x^{(1)} + \sqrt[p]{\lambda_2(x)}\, u_x^{(2)} \quad \forall x \in \mathcal{K}^n, \end{aligned}$$

respectively. The two functions enhance that ϕ^p_{FB} in (3.132) is well defined for any $x, y \in \mathbb{R}^n$. Clearly, when $p = 2$, ψ^p_{FB} reduces to the FB merit function

$$\psi_{\text{FB}}(x, y) := \frac{1}{2}\|\phi_{\text{FB}}(x, y)\|^2,$$

where $\phi_{\text{FB}} : \mathbb{R}^n \times \mathbb{R}^n \to \mathbb{R}^n$ is the C-function associated with SOC defined by

$$\phi_{\text{FB}}(x, y) := \sqrt{x^2 + y^2} - (x + y).$$

Likewise, we denote $\Phi^p_{\text{FB}} : \mathbb{R}^n \to \mathbb{R}^m$ as

$$\Phi^p_{\text{FB}}(\zeta) := \begin{bmatrix} \phi^p_{\text{FB}}(F_1(\zeta), G_1(\zeta)) \\ \vdots \\ \phi^p_{\text{FB}}(F_i(\zeta), G_i(\zeta)) \\ \vdots \\ \phi^p_{\text{FB}}(F_q(\zeta), G_q(\zeta)) \end{bmatrix},$$

where ϕ^p_{FB} is defined as in (3.132) with a suitable dimension. Accordingly, its squared norm induces a merit function, given by

$$\Psi^p_{\text{FB}}(\zeta) := \frac{1}{2}\|\Phi^p_{\text{FB}}(\zeta)\|^2 = \sum_{i=1}^{q} \psi^p_{\text{FB}}(F_i(\zeta), G_i(\zeta)). \tag{3.133}$$

Lemma 3.25 *For any given $0 \le r \le 1$, $\xi^r \succeq_{\mathcal{K}^n} \eta^r$ whenever $\xi \succeq_{\mathcal{K}^n} \eta \succeq_{\mathcal{K}^n} 0$.*

Proof This is an immediate consequence of [36, Proposition 2.7] since $f(t) = t^r$ for $0 \le r \le 1$ is SOC-monotone on $[0, \infty)$. □

Lemma 3.26 *For any nonnegative real numbers a and b, the following results hold.*

(a) $(a+b)^\rho \ge a^\rho + b^\rho$ if $\rho > 1$, and the equality holds if and only if $ab = 0$;
(b) $(a+b)^\rho \le a^\rho + b^\rho$ if $0 < \rho < 1$, and the equality holds if and only if $ab = 0$.

Proof Without loss of generality, we assume that $a \le b$ and $b > 0$. Consider the function $f(t) = (t+1)^\rho - (t^\rho + 1)$ $(t \ge 0)$. It is easy to verify that f is increasing on $[0, +\infty)$ when $\rho > 1$. Hence, $f(a/b) \ge f(0) = 0$, i.e., $(a+b)^\rho \ge a^\rho + b^\rho$. Also, $f(a/b) = f(0)$ if and only if $a/b = 0$. That is, $(a+b)^\rho = a^\rho + b^\rho$ if and only if $ab = 0$. This proves part (a). Note that f is decreasing on $[0, +\infty)$ when $0 < \rho < 1$, and a similar argument leads to part(b). □

Lemma 3.27 *For any $\xi, \eta \in \mathcal{K}^n$, if $\xi + \eta \in \text{bd}\mathcal{K}^n$, then one of the following cases must hold: (i) $\xi = 0, \eta \in \text{bd}\mathcal{K}^n$; (ii) $\xi \in \text{bd}\mathcal{K}^n, \eta = 0$; (iii) $\xi = \gamma\eta$ for some $\gamma > 0$ with $\eta \in \text{bd}^+\mathcal{K}^n$.*

Proof Since $\xi, \eta \in \mathcal{K}^n$ and $\xi + \eta \in \text{bd}\mathcal{K}^n$, it follows that $\|\xi_2\| + \|\eta_2\| \ge \|\xi_2 + \eta_2\| = \xi_1 + \eta_1 \ge \|\xi_2\| + \|\eta_2\|$. This shows that $\xi_2 = 0$; or $\eta_2 = 0$; or $\xi_2 = \gamma\eta_2 \ne 0$ for some $\gamma > 0$. Substituting $\xi_2 = 0$, or $\eta_2 = 0$; or $\xi_2 = \gamma\eta_2$ into $\|\xi_2 + \eta_2\| = \xi_1 + \eta_1$ yields the result. □

Proposition 3.26 *Let ϕ^p_{FB} be defined by (3.132). Then, the function ϕ^p_{FB} is a C-function associated with the SOC. In other words, for any $x, y \in \mathbb{R}^n$, there holds*

$$\phi^p_{\text{FB}}(x, y) = 0 \iff x \in \mathcal{K}^n, \ y \in \mathcal{K}^n, \ \langle x, y \rangle = 0.$$

Proof "$\Leftarrow$". From [37, Proposition 6], there exists a Jordan frame $\{u^{(1)}, u^{(2)}\}$ such that $x = \lambda_1 u^{(1)} + \lambda_2 u^{(2)}$ and $y = \mu_1 u^{(1)} + \mu_2 u^{(2)}$ with $\lambda_i, \mu_i \ge 0$ for $i = 1, 2$. Then,

$$\begin{aligned}(x+y)^p &= (\lambda_1 + \mu_1)^p u^{(1)} + (\lambda_2 + \mu_2)^p u^{(2)},\\ x^p + y^p &= (\lambda_1^p + \mu_1^p) u^{(1)} + (\lambda_2^p + \mu_2^p) u^{(2)}.\end{aligned}$$

Since $0 = 2\langle x, y\rangle = \lambda_1\mu_1 + \lambda_2\mu_2$ implies $\lambda_1\mu_1 = \lambda_2\mu_2 = 0$, from the last two equalities and Lemma 3.26(a) we obtain $(x+y)^p = x^p + y^p$, and consequently $\phi^p_{\text{FB}}(x, y) = 0$.

"$\Rightarrow$". Since $\phi_p(x, y) = 0$, we have $x = \sqrt[p]{|x|^p + |y|^p} - y \succeq_{\mathcal{K}^n} |y| - y \in \mathcal{K}^n$, where the inequality is due to Lemma 3.25. Similarly, we have $y = \sqrt[p]{|x|^p + |y|^p} - x \succeq_{\mathcal{K}^n} |x| - x \in \mathcal{K}^n$. Now from $\phi_p(x, y) = 0$, we have $(x + y)^p = x^p + y^p$, and then

$$(\lambda_1(x + y))^p + (\lambda_2(x + y))^p = (\lambda_1(x))^p + (\lambda_2(x))^p + (\lambda_1(y))^p + (\lambda_2(y))^p.$$

Noting that $f(t) = (t_0 + t)^p + (t_0 - t)^p$ for a fixed $t_0 \geq 0$ is increasing on $[0, t_0]$, we also have

$$\begin{aligned}
\left[\lambda_1(x + y)\right]^p + \left[\lambda_2(x + y)\right]^p &\geq (x_1 + y_1 - \|x_2\| + \|y_2\|)^p + (x_1 + y_1 + \|x_2\| - \|y_2\|)^p \\
&= (\lambda_1(x) + \lambda_2(y))^p + (\lambda_2(x) + \lambda_1(y))^p \\
&\geq (\lambda_1(x))^p + (\lambda_2(y))^p + (\lambda_2(x))^p + (\lambda_1(y))^p, \quad (3.134)
\end{aligned}$$

where the last inequality is due to Lemma 3.26(a) and $x, y \in \mathcal{K}^n$. The last two equations imply that all the inequalities on the right-hand side of (3.134) become equalities. Therefore,

$$\|x_2 + y_2\| = \|x_2\| - \|y_2\|, \quad \lambda_1(x)\lambda_2(y) = 0, \quad \lambda_2(x)\lambda_1(y) = 0. \qquad (3.135)$$

Assume that $x_2 \neq 0$ and $y_2 \neq 0$. Since $x, y \in \mathcal{K}^n$, from the equalities in (3.135), we obtain $x_1 = \|x_2\|$, $y_1 = \|y_2\|$, and $x_2 = \widehat{\gamma} y_2$ for some $\widehat{\gamma} < 0$, which in turn implies $\langle x, y \rangle = 0$. When $x_2 = 0$ or $y_2 = 0$, using the continuity of the inner product yields $\langle x, y \rangle = 0$. □

Unless otherwise specified, throughout the remainder of this section, we assume that $\mathcal{K} = \mathcal{K}^n$ (all analysis is easily carried over to the general $\mathcal{K}$ as in (3.2)), $p > 1$ with $q = (1 - p^{-1})^{-1}$, and g^{soc} is the vector-valued function associated with $|t|^p$ ($t \in \mathbb{R}$), i.e., $g^{\text{soc}}(x) = |x|^p$. For brevity, we consistently write

$$w = w(x, y) := |x|^p + |y|^p \quad \text{and} \quad z = z(x, y) := \sqrt[p]{|x|^p + |y|^p} \quad \forall x, y \in \mathbb{R}^n.$$

By definitions of $|x|^p$ and $|y|^p$, clearly,

$$\begin{aligned}
w_1 := w_1(x, y) &= \frac{|\lambda_2(x)|^p + |\lambda_1(x)|^p}{2} + \frac{|\lambda_2(y)|^p + |\lambda_1(y)|^p}{2}, \\
w_2 := w_2(x, y) &= \frac{|\lambda_2(x)|^p - |\lambda_1(x)|^p}{2}\overline{x}_2 + \frac{|\lambda_2(y)|^p - |\lambda_1(y)|^p}{2}\overline{y}_2, \quad (3.136)
\end{aligned}$$

where $\overline{x}_2 = \frac{x_2}{\|x_2\|}$ if $x_2 \neq 0$, and otherwise $\overline{x}_2$ is an arbitrary vector in $\mathbb{R}^{n-1}$ with $\|\overline{x}_2\| = 1$, and $\overline{y}_2$ has the similar definition. Noting that $z(x, y) = \sqrt[p]{w(x, y)}$, we have

$$z_1 = z_1(x, y) = \frac{\sqrt[p]{\lambda_2(w)} + \sqrt[p]{\lambda_1(w)}}{2}, \quad z_2 = z_2(x, y) = \frac{\sqrt[p]{\lambda_2(w)} - \sqrt[p]{\lambda_1(w)}}{2}\overline{w}_2, \tag{3.137}$$

where $\overline{w}_2 = \frac{w_2}{\|w_2\|}$ if $w_2 \neq 0$, and otherwise $\overline{w}_2$ is an arbitrary vector in $\mathbb{R}^{n-1}$ with $\|\overline{w}_2\| = 1$.

To analyze the differentiability of ψ_{FB}^p, we present two essential lemmas. The first establishes key properties of points (x, y) for which $w(x, y) \in \text{bd}\mathcal{K}^n$, while the second offers a sufficient condition characterizing the points at which $z(x, y)$ is continuously differentiable.

Lemma 3.28 *For any (x, y) with $w(x, y) \in \text{bd}\mathcal{K}^n$, we have the following equalities:*

$$\begin{aligned} w_1(x, y) &= \|w_2(x, y)\| = 2^{p-1}(|x_1|^p + |y_1|^p), \\ x_1^2 = \|x_2\|^2, \ & y_1^2 = \|y_2\|^2, \ x_1 y_1 = x_2^T y_2, \ x_1 y_2 = y_1 x_2. \end{aligned} \tag{3.138}$$

If, in addition, $w_2(x, y) \neq 0$, the following equalities hold with $\overline{w}_2(x, y) = \frac{w_2(x,y)}{\|w_2(x,y)\|}$:

$$x_2^T \overline{w}_2(x, y) = x_1, \quad x_1 \overline{w}_2(x, y) = x_2, \quad y_2^T \overline{w}_2(x, y) = y_1, \quad y_1 \overline{w}_2(x, y) = y_2. \tag{3.139}$$

Proof Fix any (x, y) with $w(x, y) \in \text{bd}\mathcal{K}^n$. Since $|x|^p, |y|^p \in \mathcal{K}^n$, applying Lemma 3.27 with $\xi = |x|^p$ and $\eta = |y|^p$, we have $|x|^p \in \text{bd}\mathcal{K}^n$ and $|y|^p \in \text{bd}\mathcal{K}^n$. This means that $|\lambda_2(x)|^p \cdot |\lambda_1(x)|^p = 0$ and $|\lambda_2(y)|^p \cdot |\lambda_1(y)|^p = 0$. So, $x_1^2 = \|x_2\|^2$ and $y_1^2 = \|y_2\|^2$. Substituting this into $w_1(x, y)$, we readily obtain $w_1(x, y) = 2^{p-1}(|x_1|^p + |y_1|^p)$.

To prove other equalities in (3.138) and (3.139), we first consider the case where $x_1 + \|x_2\| = 0$ and $y_1 - \|y_2\| = 0$ with $x_2 \neq 0$ and $y_2 \neq 0$. Under this case,

$$w_1 = \frac{|\lambda_1(x)|^p + |\lambda_2(y)|^p}{2} = \left\| \frac{|\lambda_1(x)|^p}{2} \frac{x_2}{\|x_2\|} - \frac{|\lambda_2(y)|^p}{2} \frac{y_2}{\|y_2\|} \right\| = \|w_2\|,$$

which implies that $x_2^T y_2 = -\|x_2\| \|y_2\| = x_1 y_1$. Together with $x_1^2 = \|x_2\|^2$ and $y_1^2 = \|y_2\|^2$, we have that $x_1 y_2 = y_1 x_2$. From the definition of w_2, it then follows that

$$\begin{aligned} x_2^T w_2 &= -\frac{|\lambda_1(x)|^p}{2} \|x_2\| + \frac{|\lambda_2(y)|^p}{2} \frac{x_1 y_1}{\|y_2\|} = 2^{p-1} \left(|x_1|^p + |y_1|^p \right) x_1 = \|w_2\| x_1, \\ x_1 w_2 &= -\frac{|\lambda_1(x)|^p}{2} \frac{x_1 x_2}{\|x_2\|} + \frac{|\lambda_2(y)|^p}{2} \frac{y_1 x_2}{\|y_2\|} = 2^{p-1} \left(|x_1|^p + |y_1|^p \right) x_2 = \|w_2\| x_2. \end{aligned}$$

Similarly, we also have $y_2^T w_2 = \|w_2\| y_1$ and $y_1^T w_2 = \|w_2\| y_2$. The above arguments show that Eqs. (3.138) and (3.139) hold under the case where $x_1 = -\|x_2\|$, $y_1 = \|y_2\|$. Using the same arguments, we can prove that (3.138) and (3.139) hold under any one of the following cases: (1) $x_1 = \|x_2\|$, $y_1 = \|y_2\|$; (2) $x_1 = -\|x_2\|$, $y_1 = \|y_2\|$; (3) $x_1 = -\|x_2\|$, $y_1 = -\|y_2\|$. □

Lemma 3.29 *Let $z(x, y)$ be defined as in* (3.137). *Then, $z(x, y)$ is continuously differentiable at (x, y) with $w(x, y) \in \text{int}(\mathcal{K}^n)$, and*

$$\nabla_x z(x, y) = \nabla g^{\text{soc}}(x) \nabla g^{\text{soc}}(z)^{-1} \quad \text{and} \quad \nabla_y z(x, y) = \nabla g^{\text{soc}}(y) \nabla g^{\text{soc}}(z)^{-1},$$

where $\nabla g^{\text{soc}}(z)^{-1} = (p\sqrt[q]{w_1})^{-1} I$ *if* $w_2 = 0$, *and otherwise*

$$\nabla g^{\text{soc}}(z)^{-1} = \frac{1}{2p} \begin{bmatrix} \frac{1}{\sqrt[q]{\lambda_2(w)}} + \frac{1}{\sqrt[q]{\lambda_1(w)}} & \frac{\overline{w}_2^T}{\sqrt[q]{\lambda_2(w)}} - \frac{\overline{w}_2^T}{\sqrt[q]{\lambda_1(w)}} \\ \frac{\overline{w}_2}{\sqrt[q]{\lambda_2(w)}} - \frac{\overline{w}_2}{\sqrt[q]{\lambda_1(w)}} & \frac{2p(I - \overline{w}_2 \overline{w}_2^T)}{a(z)} + \frac{\overline{w}_2 \overline{w}_2^T}{\sqrt[q]{\lambda_2(w)}} + \frac{\overline{w}_2 \overline{w}_2^T}{\sqrt[q]{\lambda_1(w)}} \end{bmatrix}.$$

Proof Since $|t|^p$ $(t \in \mathbb{R})$ and $\sqrt[p]{t}$ $(t > 0)$ are continuously differentiable, by [12, Proposition 5.2] or [26, Proposition 5], the functions $g^{\text{soc}}(x)$ and $\sqrt{x}$ are continuously differentiable in $\mathbb{R}^n$ and $\text{int}(\mathcal{K}^n)$, respectively. This implies the first part of this lemma. A simple calculation gives the expression of $\nabla z(x, y)$. By the formula in [12, Proposition 5.2],

$$\nabla g^{\text{soc}}(x) = \begin{cases} p \,\text{sgn}(x_1)|x_1|^{p-1} I & \text{if } x_2 = 0; \\ \begin{bmatrix} b(x) & c(x)\overline{x}_2^T \\ c(x)\overline{x}_2 & a(x)I + (b(x) - a(x))\overline{x}_2 \overline{x}_2^T \end{bmatrix} & \text{if } x_2 \neq 0, \end{cases} \tag{3.140}$$

where

$$\begin{aligned} \overline{x}_2 &= \frac{x_2}{\|x_2\|}, \quad a(x) = \frac{|\lambda_2(x)|^p - |\lambda_1(x)|^p}{\lambda_2(x) - \lambda_1(x)}, \\ b(x) &= \frac{p}{2} \left[\text{sgn}(\lambda_2(x))|\lambda_2(x)|^{p-1} + \text{sgn}(\lambda_1(x))|\lambda_1(x)|^{p-1} \right], \\ c(x) &= \frac{p}{2} \left[\text{sgn}(\lambda_2(x))|\lambda_2(x)|^{p-1} - \text{sgn}(\lambda_1(x))|\lambda_1(x)|^{p-1} \right]. \end{aligned} \tag{3.141}$$

We next derive the formula of $\nabla g^{\text{soc}}(z)^{-1}$. When $w_2 = 0$, we have $\lambda_1(w) = \lambda_2(w) = w_1 > 0$, which by (3.137) implies $z_1 = \sqrt[p]{w_1}$ and $z_2 = 0$. From formula (3.140), it then follows that $\nabla g^{\text{soc}}(z) = p|z_1|^{p-1} I = p\sqrt[q]{w_1} I$. Consequently, $\nabla g^{\text{soc}}(z)^{-1} = \frac{1}{p\sqrt[q]{w_1}} I$. When $w_2 \neq 0$, since $\sqrt[p]{\lambda_2(w)} > \sqrt[p]{\lambda_1(w)}$, we have $z_2 \neq 0$ and $\overline{z}_2 = \frac{z_2}{\|z_2\|} = \overline{w}_2$ by (3.137). Using the expression of $\nabla g^{\text{soc}}(z)$, it is easy to verify that $b(z) + c(z)$ and $b(z) - c(z)$ are the eigenvalues of $\nabla g^{\text{soc}}(z)$ with $(1, \overline{w}_2)$ and $(1, -\overline{w}_2)$ being the corresponding eigenvectors, and $a(z)$ is the eigenvalue of multiplicity $n - 2$ with corresponding eigenvectors of the form $(0, \overline{v}_i)$, where $\overline{v}_1, \ldots, \overline{v}_{n-2}$ are any unit vectors in $\mathbb{R}^{n-1}$ that span the subspace orthogonal to w_2. Hence,

$$\nabla g^{\text{soc}}(z) = U \text{diag}\left(b(z) - c(z), a(z), \ldots, a(z), b(z) + c(z)\right) U^T,$$

where $U = \begin{bmatrix} u_1 & v_1 & \cdots & v_{n-2} & u_2 \end{bmatrix} \in \mathbb{R}^{n \times n}$ is an orthogonal matrix with

$$u_1 = \begin{bmatrix} 1 \\ -\overline{w}_2 \end{bmatrix}, \quad u_2 = \begin{bmatrix} 1 \\ \overline{w}_2 \end{bmatrix}, \quad v_i = \begin{bmatrix} 0 \\ \overline{v}_i \end{bmatrix} \quad \text{for } i = 1, \ldots, n - 2.$$

By this, we know that $\nabla g^{\rm soc}(z)^{-1}$ has the expression given as in the lemma. $\square$

Lemma 3.30 *Suppose that $w(x,y)$ is given by* (3.136) *and assume that $p \geq 2$. Let $x, y \in \mathbb{R}^n$ satisfy $w(x,y) \in {\rm bd}^+\mathcal{K}^n$ where ${\rm bd}^+\mathcal{K}^n = {\rm bd}\mathcal{K}^n\backslash\{0\}$. Then, we have*

$$\nabla_{x'} w_1(x',y')\,|_{(x',y')=(x,y)} = \nabla_{x'}\|w_2(x',y')\|\,|_{(x',y')=(x,y)} = \begin{bmatrix} 2^{p-2} p\,{\rm sgn}(x_1)|x_1|^{p-1} \\ 2^{p-2} p\,{\rm sgn}(x_1)|x_1|^{p-1}\overline{w}_2 \end{bmatrix}. \quad (3.142)$$

Proof Assume that $x_2 \neq 0$. By the expressions of $w_1(x',y')$ and $w_2(x',y')$, we calculate

$$\nabla_{x'} w_1(x',y')\,|_{(x',y')=(x,y)} = \frac{p}{2}\begin{pmatrix} {\rm sgn}(\lambda_2(x))|\lambda_2(x)|^{p-1} + {\rm sgn}(\lambda_1(x))|\lambda_1(x)|^{p-1} \\ \left({\rm sgn}(\lambda_2(x))|\lambda_2(x)|^{p-1} - {\rm sgn}(\lambda_1(x))|\lambda_1(x)|^{p-1}\right)\frac{x_2}{\|x_2\|} \end{pmatrix},$$

$$\nabla_{x'}\|w_2(x',y')\|\,|_{(x',y')=(x,y)} = \frac{p}{2}\begin{pmatrix} {\rm sgn}(\lambda_2(x))|\lambda_2(x)|^{p-1} - {\rm sgn}(\lambda_1(x))|\lambda_1(x)|^{p-1} \\ \left({\rm sgn}(\lambda_2(x))|\lambda_2(x)|^{p-1} + {\rm sgn}(\lambda_1(x))|\lambda_1(x)|^{p-1}\right)\frac{x_2}{\|x_2\|} \end{pmatrix}$$

$$\frac{x_2^T w_2}{\|x_2\|\|w_2\|} + \frac{|\lambda_2(x)|^p - |\lambda_1(x)|^p}{2}\begin{pmatrix} 0 \\ \frac{w_2}{\|x_2\|\|w_2\|} - \frac{x_2 x_2^T w_2}{\|x_2\|^3\|w_2\|} \end{pmatrix}.$$

Using the equalities in (3.139), the last two equalities can be simplified as

$$\nabla_{x'} w_1(x',y')\,|_{(x',y')=(x,y)} = \nabla_{x'}\|w_2(x',y')\|\,|_{(x',y')=(x,y)} = \begin{bmatrix} 2^{p-2} p\,{\rm sgn}(x_1)|x_1|^{p-1} \\ 2^{p-2} p\,{\rm sgn}(x_1)|x_1|^{p-1}\overline{w}_2 \end{bmatrix}.$$

If $x_2 = 0$, using the result for $x_2 \neq 0$ and the continuity of $\nabla w_1(x',y')$ and $\nabla\|w_2(x',y')\|$ at (x,y), we easily obtain Eq. (3.142). $\square$

Proposition 3.27 *Let $\psi_{\rm FB}^p$ be defined by (3.131). Then, the function $\psi_{\rm FB}^p$ for $p \in (1,4)$ is differentiable everywhere. For any given $x, y \in \mathbb{R}^n$, if $w(x,y) = 0$, then $\nabla_x \psi_{\rm FB}^p(x,y) = \nabla_y \psi_{\rm FB}^p(x,y) = 0$; if $w(x,y) \in {\rm int}(\mathcal{K}^n)$, then*

$$\begin{aligned} \nabla_x \psi_{\rm FB}^p(x,y) &= \left(\nabla g^{\rm soc}(x)\nabla g^{\rm soc}(z)^{-1} - I\right)\phi_{\rm FB}^p(x,y), \\ \nabla_y \psi_{\rm FB}^p(x,y) &= \left(\nabla g^{\rm soc}(y)\nabla g^{\rm soc}(z)^{-1} - I\right)\phi_{\rm FB}^p(x,y); \end{aligned} \quad (3.143)$$

and if $w(x,y) \in {\rm bd}^+\mathcal{K}^n$, then

$$\begin{aligned} \nabla_x \psi_{\rm FB}^p(x,y) &= \left(\frac{{\rm sgn}(x_1)|x_1|^{p-1}}{\sqrt[q]{|x_1|^p + |y_1|^p}} - 1\right)\phi_{\rm FB}^p(x,y), \\ \nabla_y \psi_{\rm FB}^p(x,y) &= \left(\frac{{\rm sgn}(y_1)|y_1|^{p-1}}{\sqrt[q]{|x_1|^p + |y_1|^p}} - 1\right)\phi_{\rm FB}^p(x,y). \end{aligned} \quad (3.144)$$

Proof Fix any $(x,y) \in \mathbb{R}^n \times \mathbb{R}^n$. If $w(x,y) \in {\rm int}(\mathcal{K}^n)$, the result is direct by Lemma 3.29 since $\phi_{\rm FB}^p(x,y) = z(x,y) - (x+y)$. In fact, in this case, $\psi_{\rm FB}^p$ is continuously differentiable at (x,y). Hence, it suffices to consider the cases $w(x,y) = 0$ and $w(x,y) \in {\rm bd}^+\mathcal{K}^n$. In the following arguments, x' and y' are arbitrary vectors in $\mathbb{R}^n$, and $\mu_1(x',y')$, $\mu_2(x',y')$ are the spectral values of $w(x',y')$ with $\xi^{(1)}, \xi^{(2)} \in \mathbb{R}^n$ being the corresponding spectral vectors.

Case (1): $w(x, y) = 0$. Since $(x, y) = (0, 0)$ now, we only need to prove, for any $x', y' \in \mathbb{R}^n$,

$$\psi^p_{\rm FB}(x', y') - \psi^p_{\rm FB}(0, 0) = \frac{1}{2}\left\|z(x', y') - (x' + y')\right\|^2 = O(\|(x', y')\|), \tag{3.145}$$

which shows that ψ_p is differentiable at $(0, 0)$ with $\nabla_x \psi^p_{\rm FB}(0, 0) = \nabla_y \psi^p_{\rm FB}(0, 0) = 0$. Indeed,

$$\begin{aligned}\left\|z(x', y') - (x' + y')\right\| &= \left\|\sqrt[p]{\mu_1(x', y')}\,\xi^{(1)} + \sqrt[p]{\mu_2(x', y')}\,\xi^{(2)} - (x' + y')\right\| \\ &\le \sqrt{2}\sqrt[p]{\mu_2(x', y')} + \|x'\| + \|y'\|.\end{aligned} \tag{3.146}$$

From the definition of $w_1(x, y)$ and $w_2(x, y)$, it is easy to obtain that

$$\mu_2(x', y') = w_1(x', y') + w_2(x', y') \le |\lambda_2(x')|^p + |\lambda_1(x')|^p + |\lambda_2(y')|^p + |\lambda_1(y')|^p.$$

Using the nondecreasing of $\sqrt[p]{t}$ and Lemma 3.26(b), it then follows that

$$\begin{aligned}\sqrt[p]{\mu_2(x', y')} &\le \left(|\lambda_2(x')|^p + |\lambda_1(x')|^p + |\lambda_2(y')|^p + |\lambda_1(y')|^p\right)^{1/p} \\ &\le |\lambda_2(x')| + |\lambda_1(x')| + |\lambda_2(y')| + |\lambda_1(y')| \le 2(\|x'\| + \|y'\|).\end{aligned}$$

This, together with (3.146), implies that Eq. (3.145) holds.

Case (2): $w(x, y) \in \text{bd}^+\mathcal{K}^n$. Now $w_1(x, y) = \|w_2(x, y)\| \neq 0$, and one of x_2 and y_2 must be nonzero by Eq. (3.139). We proceed the arguments by three steps as shown below.

Step 1: to prove that $w_1(x', y')$ and $w_2(x', y')$ are $\lceil p \rceil$ times differentiable at $(x', y') = (x, y)$, where $\lceil p \rceil$ denotes the maximum integer not greater than p. Since one of x_2 and y_2 is nonzero, we prove this result by considering three possible cases: (i) $x_2 \neq 0$, $y_2 \neq 0$; (ii) $x_2 = 0$, $y_2 \neq 0$; and (iii) $x_2 \neq 0$, $y_2 = 0$. For case (i), since $\frac{x_2'}{\|x_2'\|}$, $\frac{y_2'}{\|y_2'\|}$, $\lambda_2(x')$, $\lambda_1(x')$, $\lambda_2(y')$ and $\lambda_1(y')$ are infinite times differentiable at (x, y), and $|t|^p$ is $\lceil p \rceil$ times continuously differentiable in $\mathbb{R}$, it follows that $w_1(x', y')$ and $w_2(x', y')$ are $\lceil p \rceil$ times differentiable at (x, y). Now assume that case (ii) is satisfied. From the arguments in case (i), we know that

$$\frac{|\lambda_2(y')|^p + |\lambda_1(y')|^p}{2} \quad \text{and} \quad \frac{|\lambda_2(y')|^p - |\lambda_1(y')|^p}{2}\frac{y_2'}{\|y_2'\|}$$

are $\lceil p \rceil$ times differentiable at (x, y). In addition, since $|\lambda_i(x')|^p \le 2^{\frac{p}{2}}\|x'\|^p$ for $i = 1, 2$, and $x = 0$ in this case, we have that $|\lambda_2(x')|^p + |\lambda_1(x')|^p$ and $\frac{1}{2}(|\lambda_2(x')|^p - |\lambda_1(x')|^p)\overline{x}_2'$ are $\lceil p \rceil$ times differentiable at x with the first $\lceil p \rceil - 1$ order derivatives being zero. Thus, $w_1(x', y')$ and $w_2(x', y')$ are $\lceil p \rceil$ times differentiable at (x, y). By the symmetry of x', y' in $w(x', y')$ and the arguments in case (ii), the result also holds for case (iii).

<u>**Step 2:**</u> to show that $\psi_{\rm FB}^p$ is differentiable at (x, y). By the definition of $\psi_{\rm FB}^p$, we have

$$2\psi_{\rm FB}^p(x', y') = \|x' + y'\|^2 + \|z(x', y')\|^2 - 2\left\langle z(x', y'), x' + y'\right\rangle.$$

Since $\|x' + y'\|^2$ is differentiable, it suffices to argue that the last two terms on the right-hand side are differentiable at (x, y). By formula (1.8), it is not hard to calculate that

$$\begin{aligned} 2\left\|z(x', y')\right\|^2 &= (\mu_2(x', y'))^{\frac{2}{p}} + (\mu_1(x', y'))^{\frac{2}{p}}, \\ 2\left\langle z(x', y'), x' + y'\right\rangle &= \sqrt[p]{\mu_2(x', y')}\left(x_1' + y_1' + \frac{(w_2(x', y'))^{\rm T}(x_2' + y_2')}{\|w_2(x', y')\|}\right) \\ &\quad + \sqrt[p]{\mu_1(x', y')}\left(x_1' + y_1' - \frac{(w_2(x', y'))^{\rm T}(x_2' + y_2')}{\|w_2(x', y')\|}\right). \end{aligned} \quad (3.147)$$

Since $w_2(x, y) \neq 0$, $\mu_2(x, y) = \lambda_2(w) > 0$, and $w_1(x', y')$ and $w_2(x', y')$ are differentiable at (x, y) by Step 1, we have that $(\mu_2(x', y'))^{\frac{2}{p}}$ and the first term on the right-hand side of (3.147) are differentiable at (x, y). Thus, it suffices to prove that $(\mu_1(x', y'))^{\frac{2}{p}}$ and the last term on the right-hand side of (3.147) are differentiable at (x, y).

We first argue that $(\mu_1(x', y'))^{\frac{2}{p}}$ is differentiable at (x, y). Since $w_2(x, y) \neq 0$, and $w_1(x', y')$ and $w_2(x', y')$ are $\lceil p \rceil$ times differentiable at (x, y) by Step 1, the function $\mu_1(x', y')$ is $\lceil p \rceil$ times differentiable at (x, y). When $p < 2$, by the mean-value theorem and $\mu_1(x, y) = \lambda_1(w) = 0$, it follows that $\mu_1(x', y') = O(\|x' - x\| + \|y' - y\|)$ for any (x', y') sufficiently close to (x, y), and so $(\mu_1(x', y'))^{\frac{2}{p}} = O[(\|x' - x\| + \|y' - y\|)^{\frac{2}{p}}]$. This shows that $(\mu_1(x', y'))^{\frac{2}{p}}$ is differentiable at (x, y) with zero derivative. When $p \geq 2$, $\mu_1(x', y')$ is infinite times differentiable at (x, y), and its first-derivative equals zero by the result in the Appendix. From the second-order Taylor expansion of $\mu_1(x', y')$ at (x, y), it follows that

$$(\mu_1(x', y'))^{\frac{2}{p}} = O\left[(\|x' - x\| + \|y' - y\|)^{\frac{4}{p}}\right].$$

This implies that $(\mu_1(x', y'))^{\frac{2}{p}}$ is differentiable at (x, y) with zero gradient when $2 \leq p < 4$. Thus, we prove that $(\mu_1(x', y'))^{\frac{2}{p}}$ is differentiable at (x, y) with zero gradient when $p \in (1, 4)$.

We next consider the last term on the right-hand side of (3.147). Observe that

$$x_1' + y_1' - \frac{(w_2(x', y'))^{\rm T}(x_2' + y_2')}{\|w_2(x', y')\|}$$

is differentiable at (x, y), and its function value at (x, y) equals zero by (3.139). Hence, this term is $O(\|x' - x\| + \|y' - y\|)$, which, along with $\mu_1(x', y') =$

$O(\|x'-x\|+\|y'-y\|)$, means that the last term of (3.147) is $O((\|x'-x\|+\|y'-y\|)^{1+\frac{1}{p}}) = o(\|x'-x\|+\|y'-y\|)$. This shows that the last term of (3.147) is differentiable at (x, y) with zero derivative.

Step 3: to derive the formula of $\nabla_x \psi_{\mathrm{FB}}^p(x, y)$. From Step 2, we see that $2\nabla\psi_{\mathrm{FB}}^p(x, y)$ equals the difference between the gradient of $\frac{1}{2}(\mu_2(x', y'))^{\frac{2}{p}} + \|x'+y'\|^2$ and that of the first term on the right side of (3.147), evaluated at (x, y). By Lemma 3.30, the gradients of $(\mu_2(x', y'))^{1/p}$ and $(\mu_2(x', y'))^{2/p}$ with respect to x', evaluated at $(x', y') = (x, y)$, are

$$\nabla_{x'}(\mu_2(x', y'))^{1/p}|_{(x',y')=(x,y)} = (\lambda_2(w))^{\frac{1}{p}-1} 2^{p-1}\,\mathrm{sgn}(x_1)|x_1|^{p-1}\begin{bmatrix} 1 \\ \overline{w}_2 \end{bmatrix}, \tag{3.148}$$

$$\nabla_{x'}(\mu_2(x', y'))^{2/p}|_{(x',y')=(x,y)} = (\lambda_2(w))^{\frac{2}{p}-1} 2^{p}\,\mathrm{sgn}(x_1)|x_1|^{p-1}\begin{bmatrix} 1 \\ \overline{w}_2 \end{bmatrix}. \tag{3.149}$$

By the product and quotient rules for differentiation, the gradient of $x_1' + y_1' + \frac{(w_2(x',y'))^{\mathrm{T}}(x_2'+y_2')}{\|w_2(x',y')\|}$ with respect to x', evaluated at $(x', y') = (x, y)$, works out to be

$$\begin{bmatrix} 1 \\ \overline{w}_2 \end{bmatrix} + \nabla_{x'}w_2(x', y')|_{(x',y')=(x,y)}\left(\frac{x_2+y_2}{\|w_2\|} - \frac{\overline{w}_2\overline{w}_2^{\mathrm{T}}(x_2+y_2)}{\|w_2\|}\right) = \begin{bmatrix} 1 \\ \overline{w}_2 \end{bmatrix},$$

where the equality is using (3.139). Along with (3.148), the gradient of the first term on the right side of (3.147) with respect to x', evaluated at $(x', y') = (x, y)$, is

$$(\lambda_2(w))^{\frac{1}{p}-1}(x_1+y_1)2^p\,\mathrm{sgn}(x_1)|x_1|^{p-1}\begin{bmatrix} 1 \\ \overline{w}_2 \end{bmatrix} + (\lambda_2(w))^{\frac{1}{p}}\begin{bmatrix} 1 \\ \overline{w}_2 \end{bmatrix}. \tag{3.150}$$

In addition, the gradient of $\|x'+y'\|^2$ with respect to x', evaluated at $(x', y') = (x, y)$, is $2(x+y)$. Together with Eqs. (3.149)–(3.150), we obtain that

$$\begin{aligned}
&2\nabla_x\psi_{\mathrm{FB}}^p(x, y) \\
&= 2(x+y) + (\lambda_2(w))^{\frac{2}{p}-1}2^{p-1}\,\mathrm{sgn}(x_1)|x_1|^{p-1}\begin{bmatrix} 1 \\ \overline{w}_2 \end{bmatrix} \\
&\quad -(\lambda_2(w))^{\frac{1}{p}-1}(x_1+y_1)2^p\,\mathrm{sgn}(x_1)|x_1|^{p-1}\begin{bmatrix} 1 \\ \overline{w}_2 \end{bmatrix} - (\lambda_2(w))^{\frac{1}{p}}\begin{bmatrix} 1 \\ \overline{w}_2 \end{bmatrix}.
\end{aligned}$$

Since $\lambda_1(w) = 0$, from (3.137) it follows that

$$\phi_{\mathrm{FB}}^p(x, y) = z(x, y) - (x+y) = \frac{1}{2}(\lambda_2(w))^{\frac{1}{p}}\begin{bmatrix} 1 \\ \overline{w}_2 \end{bmatrix} - (x+y).$$

Combining the last two equations and using $x_1\overline{w}_2 = x_2$ and $y_1\overline{w}_2 = y_2$, we obtain

$$
\begin{aligned}
2\nabla_x \psi_{_{\rm FB}}^p(x, y) &= (\lambda_2(w))^{\frac{1}{p}-1} 2^p \operatorname{sgn}(x_1)|x_1|^{p-1}(\phi_{_{\rm FB}}^p(x, y) + (x + y)) \\
&\quad -(\lambda_2(w))^{\frac{1}{p}-1} 2^p \operatorname{sgn}(x_1)|x_1|^{p-1}(x + y) - 2\phi_{_{\rm FB}}^p(x, y) \\
&= 2\left[\frac{\operatorname{sgn}(x_1)|x_1|^{p-1}}{(|x_1|^p + |y_1|^p)^{1/q}} - 1\right]\phi_{_{\rm FB}}^p(x, y),
\end{aligned}
$$

where the last equality is from $\lambda_2(w) = 2w_1 = 2^p(|x_1|^p + |y_1|^p)$. This proves the first equality in (3.144). By the symmetry of x and y in $\psi_{_{\rm FB}}^p$, the second equality in (3.144) also holds. □

Based on the analysis in Step 2 of Case (2), we observe that the function $\psi_{_{\rm FB}}^p$ is differentiable for all $p \geq 4$ provided that the first $\lceil \frac{p}{2} \rceil$ derivatives of $\mu_1(x', y') := w_1(x', y') - |w_2(x', y')|$, evaluated at $(x', y') = (x, y)$, vanish. At present, however, it remains unclear whether this condition indeed holds. Proposition 3.27 confirms that $\psi_{_{\rm FB}}^p$ is differentiable for all $p \in (1, 4)$. A natural question then arises: is the gradient $\nabla\psi_{_{\rm FB}}^p$ continuous? In what follows, we answer this question affirmatively by establishing three technical lemmas, each valid for all $p > 1$.

Lemma 3.31 *There exists a constant $c_1 > 0$ such that for all (x, y) with $w(x, y) \in \operatorname{int}(\mathcal{K}^n)$,*

$$
\left\| L_{|x|^{p-1}} L_{z^{p-1}}^{-1} \right\|_F \leq c_1 \quad \text{and} \quad \left\| L_{|y|^{p-1}} L_{z^{p-1}}^{-1} \right\|_F \leq c_1,
$$

where c is independent of x and y, and $\|A\|_F$ means the Frobenius norm of matrix A.

Proof Due to the symmetry of x and y in $z(x, y)$, it suffices to prove the first inequality. To this end, we first prove that for any (x, y) with $w(x, y) \in \operatorname{int}(\mathcal{K}^n)$, it holds that

$$
0 \leq \lambda\left(L_{|x|^{p-1}} L_{z^{p-1}}^{-1}\right) \leq \mathbf{1}, \tag{3.151}
$$

where, for a matrix $A \in \mathbb{R}^{n\times n}$, $\lambda(A) \in \mathbb{R}^n$ denotes the vector of eigenvalues of A, and $\mathbf{1}$ means a vector with all components being 1. Indeed, since $z \succ_{\mathcal{K}^n} 0$ and $|x|^{p-1} \succeq_{\mathcal{K}^n} 0$, we have $L_z \succ 0$ and $L_{|x|^{p-1}} \succeq 0$. Applying [38, Theorem 7.6.3] with $A = L_{z^{p-1}}^{-1}$ and $B = L_{|x|^{p-1}}$ yields that $\lambda(L_{z^{p-1}}^{-1} L_{|x|^{p-1}}) \geq 0$, and then $\lambda(L_{|x|^{p-1}} L_{z^{p-1}}^{-1}) \geq 0$. In addition, since $z^p \succeq_{\mathcal{K}^n} |x|^p$, from Lemma 3.25 it follows that $(z^p)^{\frac{p-1}{p}} \succeq_{\mathcal{K}^n} (|x|^p)^{\frac{p-1}{p}}$, i.e., $z^{p-1} \succeq_{\mathcal{K}^n} |x|^{p-1}$. Then $L_{z^{p-1}} - L_{|x|^{p-1}} \succeq 0$. Applying the result of Exercise 7 in [38, Page 468] with $A = L_{z^{p-1}}$ and $B = -L_{|x|^{p-1}}$, we have that $\lambda\left(-L_{z^{p-1}}^{-1} L_{|x|^{p-1}}\right) \geq -\mathbf{1}$. Consequently, $\lambda\left(L_{|x|^{p-1}} L_{z^{p-1}}^{-1}\right) \leq \mathbf{1}$. Together with $\lambda(L_{|x|^{p-1}} L_{z^{p-1}}^{-1}) \geq 0$, we prove that (3.151) holds.

Next we prove that there exists a constant $c_1 > 0$ such that for all (x, y) satisfying $w(x, y) \in \operatorname{int}(\mathcal{K}^n)$, $\|L_{|x|^{p-1}} L_{z^{p-1}}^{-1}\|_F \leq c_1$ where c_1 is independent of x and y. Suppose on the contrary that such c_1 does not exist. Then, there exists a sequence $\{(x^k, y^k)\} \subset \mathbb{R}^n \times \mathbb{R}^n$ with $w(x^k, y^k) \in \operatorname{int}(\mathcal{K}^n)$ such that $\|L_{|x^k|^{p-1}} L_{(z^k)^{p-1}}^{-1}\|_F$ is unbounded. We assume (taking a subsequence if necessary) that $\lim_{k\to\infty} \|L_{|x^k|^{p-1}} L^{-1}_{(z^k)^{p-1}}\|_F = +\infty$. For each k, let $A^k = L_{|x^k|^{p-1}}$ and $B^k = L^{-1}_{(z^k)^{p-1}}$. Subsequencing if necessary, we may assume that

$$\lim_{k\to\infty} \frac{A^k}{\|A^k\|_F} = A^* \quad \text{and} \quad \lim_{k\to\infty} \frac{B^k}{\|B^k\|_F} = B^*.$$

In the following arguments, for any $A, B \in \mathbb{R}^{n\times n}$ with all eigenvalues in $\mathbb{R}$, we let $\lambda^{\downarrow}(A)$ and $\lambda^{\uparrow}(A)$ be the vectors obtained by rearranging the coordinates of $\lambda(A)$ in the decreasing and increasing orders, respectively. That is, if $\lambda^{\downarrow}(A) = (\lambda_1^{\downarrow}(A), \cdots, \lambda_n^{\downarrow}(A))$, then $\lambda_1^{\downarrow}(A) \geq \cdots \geq \lambda_n^{\downarrow}(A)$. Similarly, if $\lambda^{\uparrow}(A) = (\lambda_1^{\uparrow}(A), \cdots, \lambda_n^{\uparrow}(A))$, then $\lambda_1^{\uparrow}(A) \leq \cdots \leq \lambda_n^{\uparrow}(A)$. We write $\lambda(A) \prec \lambda(B)$ if $\sum_{j=1}^{l} \lambda_j^{\downarrow}(A) \leq \sum_{j=1}^{l} \lambda_j^{\downarrow}(B)$ for any $1 \leq l \leq n$ and $\sum_{j=1}^{n} \lambda_j^{\downarrow}(A) = \sum_{j=1}^{n} \lambda_j^{\downarrow}(B)$. Since $A^k \succ 0$ and $B^k \succeq 0$ for each k, applying the result of [39, Problem III.6.14] gives that

$$\lambda^{\downarrow}\left(\frac{A^k}{\|A^k\|_F}\right) \cdot \lambda^{\uparrow}\left(\frac{B^k}{\|B^k\|_F}\right) \prec \lambda\left(\frac{A^k B^k}{\|A^k\|_F \|B^k\|_F}\right) \prec \lambda^{\downarrow}\left(\frac{A^k}{\|A^k\|_F}\right) \cdot \lambda^{\downarrow}\left(\frac{B^k}{\|B^k\|_F}\right),$$

where "$\cdot$" denotes the componentwise product of vectors. Since $\lim_{k\to\infty} \|A^k\|_F \|B^k\|_F = +\infty$, taking the limit $k \to +\infty$ and using Eq. (3.151) and the continuity of $\lambda(\cdot)$, we obtain

$$\lambda^{\downarrow}(A^*) \cdot \lambda^{\uparrow}(B^*) \prec 0 \prec \lambda^{\downarrow}(A^*) \cdot \lambda^{\downarrow}(B^*). \tag{3.152}$$

Since $A^* \succeq 0$ and $B^* \succeq 0$, each component of $\lambda^{\downarrow}(A^*)$ and $\lambda^{\uparrow}(B^*)$ is nonnegative, and the first relation of (3.152) then implies $\lambda^{\downarrow}(A^*) \cdot \lambda^{\uparrow}(B^*) = 0$. Note that for each k, all eigenvalues of A^k and B^k are respectively given as follows:

$$|\lambda_1(x^k)|^{p-1},\ \underbrace{\left[|\lambda_1(x^k)|^{p-1} + |\lambda_2(x^k)|^{p-1}\right], \cdots, \left[|\lambda_1(x^k)|^{p-1} + |\lambda_2(x^k)|^{p-1}\right]}_{n-2},\ |\lambda_2(x^k)|^{p-1},$$

$$\frac{1}{[\lambda_1(z^k)]^{p-1}},\ \underbrace{\frac{1}{[\lambda_1(z^k)]^{p-1} + [\lambda_2(z^k)]^{p-1}}, \cdots, \frac{1}{[\lambda_1(z^k)]^{p-1} + [\lambda_2(z^k)]^{p-1}}}_{n-2},\ \frac{1}{[\lambda_2(z^k)]^{p-1}},$$

which, by the positive homogeneousness of the eigenvalue function, means that

$$\lambda_1^{\downarrow}(A^*) \geq \lambda_2^{\downarrow}(A^*) = \cdots = \lambda_{n-1}^{\downarrow}(A^*) \geq \lambda_n^{\downarrow}(A^*) \geq 0,$$
$$0 \leq \lambda_1^{\uparrow}(B^*) \leq \lambda_2^{\uparrow}(B^*) = \cdots = \lambda_{n-1}^{\uparrow}(B^*) \leq \lambda_n^{\uparrow}(B^*).$$

Then, from $\lambda^{\downarrow}(A^*) \cdot \lambda^{\uparrow}(B^*) = 0$, we deduce that $\lambda_1^{\uparrow}(B^*) = 0$ and $\lambda_1^{\downarrow}(A^*) > 0$ (*If not, we will have* $\lambda_1^{\downarrow}(A^*) = 0$, *which implies* $\lambda(A^*) = 0$, *and then* $A^* = 0$ *follows* *by the positive semidefiniteness of* A^*. *This contradicts the fact that* $\|A^*\|_F = 1$). Similarly, we can deduce that $\lambda_n^\uparrow(B^*) > 0$ and $\lambda_n^\downarrow(A^*) = 0$. Also, either of $\lambda_2^\downarrow(A^*)$ and $\lambda_2^\uparrow(B^*)$ is zero. Without loss of generality, we assume that $\lambda_2^\downarrow(A^*) = 0$. Thus, the above arguments show that

$$\lambda_1^\downarrow(A^*) > \lambda_2^\downarrow(A^*) = 0 = \cdots = 0 = \lambda_n^\downarrow(A^*),$$
$$\lambda_n^\uparrow(B^*) \geq \lambda_{n-1}^\uparrow(B^*) = \cdots = \lambda_2^\uparrow(B^*) \geq \lambda_1^\uparrow(B^*) = 0 \text{ and } \lambda_n^\uparrow(B^*) > 0.$$

However, from the second relation of (3.152) and the last two equations, we have

$$0 = \textstyle\sum_{j=1}^n \lambda_j^\downarrow(A^*)\lambda_j^\downarrow(B^*) = \lambda_1^\downarrow(A^*)\lambda_n^\uparrow(B^*) + (n-1)\lambda_2^\downarrow(A^*)\lambda_2^\uparrow(B^*) + \lambda_n^\downarrow(A^*)\lambda_1^\uparrow(B^*)$$
$$= \lambda_1^\downarrow(A^*)\lambda_n^\uparrow(B^*) > 0,$$

which is clearly impossible. Thus, the constant c_1 satisfying the requirement exists. □

Lemma 3.31 extends the result of Lemma 3.4 for the case $p = 2$, where the conclusion was previously obtained via direct computation. In contrast, our current approach employs a different proof technique. By combining Lemma 3.31 with the explicit forms of $L_{|x|^{p-1}}L_{z^{p-1}}^{-1}$ and $L_{|y|^{p-1}}L_{z^{p-1}}^{-1}$, we arrive at the following result.

Lemma 3.32 *For any* x, y *with* $w(x, y) \in \text{int}(\mathcal{K}^n)$, *let* $\widetilde{x} = |x|^{p-1}$ *and* $\widetilde{y} = |y|^{p-1}$. *Then,*

$$\frac{\widetilde{x}_1 + (-1)^i \widetilde{x}_2^T \overline{w}_2}{\sqrt[q]{\lambda_i(w)}} = O(1), \quad \frac{\widetilde{x}_2 + (-1)^i \widetilde{x}_1 \overline{w}_2}{\sqrt[q]{\lambda_i(w)}} = O(1),$$
$$\frac{\widetilde{y}_1 + (-1)^i \widetilde{y}_2^T \overline{w}_2}{\sqrt[q]{\lambda_i(w)}} = O(1), \quad \frac{\widetilde{y}_2 + (-1)^i \widetilde{y}_1 \overline{w}_2}{\sqrt[q]{\lambda_i(w)}} = O(1) \tag{3.153}$$

for $i = 1, 2$, *where* $\overline{w}_2 = \frac{w_2(x,y)}{\|w_2(x,y)\|}$, *and* $O(1)$ *denotes a term that is uniformly bounded.*

Proof Fix any (x, y) satisfying $w \in \text{int}(\mathcal{K}^n)$. We write $\lambda_1 = \lambda_1(w)$ and $\lambda_2 = \lambda_2(w)$ for simplicity. From (3.136), we have $\sqrt[q]{\lambda_2} \geq \sqrt[q]{w_1} \geq \sqrt[q]{\frac{|\lambda_2(x)|^p + |\lambda_1(x)|^p}{2}}$. Note that

$$\sqrt[q]{\frac{|\lambda_2(x)|^p + |\lambda_1(x)|^p}{2}} \geq \left(\frac{\max(|\lambda_2(x)|, |\lambda_1(x)|)}{\sqrt[p]{2}}\right)^{\frac{p}{q}} \geq \left(\frac{\sqrt{|\lambda_2(x)|^2 + |\lambda_1(x)|^2}}{2^{\frac{1}{p}+1}}\right)^{\frac{p}{q}} = \frac{\|x\|^{\frac{p}{q}}}{2^{\frac{p+2}{2q}}}$$

for $p > 2$, and for $1 < p \leq 2$,

$$\sqrt[q]{\frac{|\lambda_2(x)|^p + |\lambda_1(x)|^p}{2}} \geq \sqrt[q]{\frac{(|\lambda_2(x)|^2 + |\lambda_1(x)|^2)^{\frac{p}{2}}}{2}} = 2^{\frac{p-2}{2q}}\|x\|^{\frac{p}{q}}.$$

Therefore,

$$\sqrt[q]{\lambda_2} \geq \begin{cases} 2^{-\frac{p+2}{2q}} \|x\|^{\frac{p}{q}} & \text{if } p > 2; \\ 2^{\frac{p-2}{2q}} \|x\|^{\frac{p}{q}} & \text{if } p \in (1, 2]. \end{cases} \tag{3.154}$$

Since $\widetilde{x}_1 = \frac{1}{2}(|\lambda_2(x)|^{p-1} + |\lambda_1(x)|^{p-1})$ and $\widetilde{x}_2 = \frac{1}{2}(|\lambda_2(x)|^{p-1} - |\lambda_1(x)|^{p-1})$, we have

$$\widetilde{x}_1 = \frac{1}{2}(|\lambda_2(x)|^{\frac{p}{q}} + |\lambda_1(x)|^{\frac{p}{q}}) \leq \|x\|^{\frac{p}{q}} \text{ and } \|\widetilde{x}_2\| \leq \frac{1}{2}(|\lambda_2(x)|^{\frac{p}{q}} + |\lambda_1(x)|^{\frac{p}{q}}) \leq \|x\|^{\frac{p}{q}}. \tag{3.155}$$

Together with (3.154), we obtain the first two relations in (3.153) for $i = 2$. Notice that

$$z^{p-1} = \sqrt[q]{w} = \left(\frac{\sqrt[q]{\lambda_2} + \sqrt[q]{\lambda_1}}{2}, \frac{\sqrt[q]{\lambda_2} - \sqrt[q]{\lambda_1}}{2}\overline{w}_2\right).$$

By Lemma 3.1(c) and $\widetilde{x} = |x|^{p-1}$, we calculate that $L_{|x|^{p-1}}L^{-1}_{z^{p-1}}$ equals

$$\begin{bmatrix} \frac{\widetilde{x}_1 + \widetilde{x}_2^T\overline{w}_2}{2\sqrt[q]{\lambda_2}} + \frac{\widetilde{x}_1 - \widetilde{x}_2^T\overline{w}_2}{2\sqrt[q]{\lambda_1}} & \left(\frac{\widetilde{x}_1\overline{w}_2^T}{2\sqrt[q]{\lambda_2}} - \frac{\widetilde{x}_1\overline{w}_2^T}{2\sqrt[q]{\lambda_1}}\right) + \frac{2\widetilde{x}_2^T}{\sqrt[q]{\lambda_2} + \sqrt[q]{\lambda_1}} + \frac{\frac{\sqrt[q]{\lambda_2}}{\sqrt[q]{\lambda_1}} - 2 + \frac{\sqrt[q]{\lambda_1}}{\sqrt[q]{\lambda_2}}}{2(\sqrt[q]{\lambda_2} + \sqrt[q]{\lambda_1})}\widetilde{x}_2^T\overline{w}_2\overline{w}_2^T \\ \frac{\widetilde{x}_2 + \widetilde{x}_1\overline{w}_2}{2\sqrt[q]{\lambda_2}} + \frac{\widetilde{x}_2 - \widetilde{x}_1\overline{w}_2}{2\sqrt[q]{\lambda_1}} & \left(\frac{\widetilde{x}_2\overline{w}_2^T}{2\sqrt[q]{\lambda_2}} - \frac{\widetilde{x}_2\overline{w}_2^T}{2\sqrt[q]{\lambda_1}}\right) + \frac{2\widetilde{x}_1 I}{\sqrt[q]{\lambda_2} + \sqrt[q]{\lambda_1}} + \frac{\frac{\sqrt[q]{\lambda_2}}{\sqrt[q]{\lambda_1}} - 2 + \frac{\sqrt[q]{\lambda_1}}{\sqrt[q]{\lambda_2}}}{2(\sqrt[q]{\lambda_2} + \sqrt[q]{\lambda_1})}\widetilde{x}_1\overline{w}_2\overline{w}_2^T \end{bmatrix}.$$

Substituting the first two relations in (3.153) for $i = 2$ into the last equation and noting that

$$\frac{\widetilde{x}_1\overline{w}_2^T}{2\sqrt[q]{\lambda_2}}, \quad \frac{\widetilde{x}_2\overline{w}_2^T}{2\sqrt[q]{\lambda_2}}, \quad \frac{\widetilde{x}_2^T}{\sqrt[q]{\lambda_2} + \sqrt[q]{\lambda_1}}, \quad \text{and} \quad \frac{\widetilde{x}_1}{\sqrt[q]{\lambda_2} + \sqrt[q]{\lambda_1}}$$

are all uniformly bounded by Eqs. (3.154)–(3.155), we obtain that

$$\begin{aligned} L_{|x|^{p-1}}L^{-1}_{z^{p-1}} &= \begin{bmatrix} O(1) + \frac{\widetilde{x}_1 - \widetilde{x}_2^T\overline{w}_2}{2\sqrt[q]{\lambda_1}} & O(1) - \frac{\widetilde{x}_1\overline{w}_2^T}{2\sqrt[q]{\lambda_1}} + \frac{\sqrt[q]{\lambda_2}}{2(\sqrt[q]{\lambda_2} + \sqrt[q]{\lambda_1})\sqrt[q]{\lambda_1}}\widetilde{x}_2^T\overline{w}_2\overline{w}_2^T \\ O(1) + \frac{\widetilde{x}_2 - \widetilde{x}_1\overline{w}_2}{2\sqrt[q]{\lambda_1}} & O(1) - \frac{\widetilde{x}_2\overline{w}_2^T}{2\sqrt[q]{\lambda_1}} + \frac{\sqrt[q]{\lambda_2}}{2(\sqrt[q]{\lambda_2} + \sqrt[q]{\lambda_1})\sqrt[q]{\lambda_1}}\widetilde{x}_1\overline{w}_2\overline{w}_2^T \end{bmatrix} \\ &= \begin{bmatrix} O(1) + \frac{\widetilde{x}_1 - \widetilde{x}_2^T\overline{w}_2}{2\sqrt[q]{\lambda_1}} & O(1) - \frac{\widetilde{x}_1\overline{w}_2^T}{2(\sqrt[q]{\lambda_2} + \sqrt[q]{\lambda_1})} - \frac{\sqrt[q]{\lambda_2}(\widetilde{x}_1 - \widetilde{x}_2^T\overline{w}_2)}{2(\sqrt[q]{\lambda_2} + \sqrt[q]{\lambda_1})\sqrt[q]{\lambda_1}}\overline{w}_2^T \\ O(1) + \frac{\widetilde{x}_2 - \widetilde{x}_1\overline{w}_2}{2\sqrt[q]{\lambda_1}} & O(1) - \frac{\widetilde{x}_2\overline{w}_2^T}{2(\sqrt[q]{\lambda_2} + \sqrt[q]{\lambda_1})} + \frac{\sqrt[q]{\lambda_2}(\widetilde{x}_2 - \widetilde{x}_1\overline{w}_2)}{2(\sqrt[q]{\lambda_2} + \sqrt[q]{\lambda_1})\sqrt[q]{\lambda_1}}\overline{w}_2^T \end{bmatrix}. \end{aligned}$$

This, along with Lemma 3.31, implies that the first two relations in (3.153) hold for $i = 1$. By the symmetry of x and y in $w(x, y)$, the last two relations in (3.153) also hold. □

Remark 3.4 The first relation of (3.153) for $i = 1$ is equivalent to saying that

$$\frac{|\lambda_2(x)|^{p-1}(1-\overline{x}_2^T\overline{w}_2)+|\lambda_1(x)|^{p-1}(1+\overline{x}_2^T\overline{w}_2)}{\sqrt[q]{\lambda_1(w)}} = O(1), \tag{3.156}$$

whereas the second relation for $i = 1$ is equivalent to saying that

$$\begin{aligned}&\left\|\frac{(|\lambda_2(x)|^{p-1}-|\lambda_1(x)|^{p-1})\overline{x}_2-(|\lambda_2(x)|^{p-1}+|\lambda_1(x)|^{p-1})\overline{w}_2}{\sqrt[q]{\lambda_1(w)}}\right\|^2,\\ =&\frac{|\lambda_2(x)|^{2p-2}(1-\overline{x}_2^T\overline{w}_2)+|\lambda_1(x)|^{2p-2}(1+\overline{x}_2^T\overline{w}_2)}{(\sqrt[q]{\lambda_1(w)})^2} = O(1).\end{aligned} \tag{3.157}$$

Equations (3.156) and (3.157) play an important role in the proof of the following lemma.

Lemma 3.33 *There exists a constant $c_2 > 0$ such that for all (x, y) with $w(x, y) \in \operatorname{int}(\mathcal{K}^n)$,*

$$\left\|\nabla g^{\text{soc}}(x)\nabla g^{\text{soc}}(z)^{-1}\right\|_F \le c_2 \quad\text{and}\quad \left\|\nabla g^{\text{soc}}(y)\nabla g^{\text{soc}}(z)^{-1}\right\|_F \le c_2,$$

where c_2 is independent of x and y.

Proof By the symmetry of x and y in $\nabla g^{\text{soc}}(z)$, it suffices to prove the first inequality. Fix any (x, y) with $w = w(x, y) \in \operatorname{int}(\mathcal{K}^n)$. Suppose $w_2 \neq 0$ and $x_2 \neq 0$. By the expressions of $\nabla g^{\text{soc}}(x)$ and $\nabla g^{\text{soc}}(z)^{-1}$ given by Lemma 3.29, it is not hard to calculate that

$$2p\nabla g^{\text{soc}}(x)\nabla g^{\text{soc}}(z)^{-1} = \begin{bmatrix} a_1(x,z) & a_2^T(x,z) \\ b_2(x,z) & A_1(x,z), \end{bmatrix}$$

where

$$\begin{aligned}
a_1(x,z) &= \frac{1}{\sqrt[q]{\lambda_2(w)}}\left(b(x)+c(x)\overline{x}_2^T\overline{w}_2\right)+\frac{1}{\sqrt[q]{\lambda_1(w)}}\left(b(x)-c(x)\overline{x}_2^T\overline{w}_2\right),\\
a_2(x,z) &= \frac{\left(b(x)+c(x)\overline{x}_2^T\overline{w}_2\right)\overline{w}_2}{\sqrt[q]{\lambda_2(w)}}-\frac{\left(b(x)-c(x)\overline{x}_2^T\overline{w}_2\right)\overline{w}_2}{\sqrt[q]{\lambda_1(w)}}+\frac{2pc(x)\left(\overline{x}_2-\overline{x}_2^T\overline{w}_2\overline{w}_2\right)}{a(z)},\\
b_2(x,z) &= \frac{1}{\sqrt[q]{\lambda_2(w)}}\left[c(x)\overline{x}_2+a(x)\overline{w}_2+(b(x)-a(x))\overline{x}_2^T\overline{w}_2\overline{x}_2\right]\\
&\quad+\frac{1}{\sqrt[q]{\lambda_1(w)}}\left[c(x)\overline{x}_2-a(x)\overline{w}_2-(b(x)-a(x))\overline{x}_2^T\overline{w}_2\overline{x}_2\right],\\
A_1(x,z) &= \frac{1}{\sqrt[q]{\lambda_2(w)}}\left[c(x)\overline{x}_2\overline{w}_2^T+a(x)\overline{w}_2\overline{w}_2^T+(b(x)-a(x))\overline{x}_2^T\overline{w}_2\overline{x}_2\overline{w}_2^T\right]\\
&\quad-\frac{1}{\sqrt[q]{\lambda_1(w)}}\left[c(x)\overline{x}_2\overline{w}_2^T-a(x)\overline{w}_2\overline{w}_2^T-(b(x)-a(x))\overline{x}_2^T\overline{w}_2\overline{x}_2\overline{w}_2^T\right]\\
&\quad+\frac{2p}{a(z)}\left[a(x)(I-\overline{w}_2\overline{w}_2^T)+(b(x)-a(x))\left(\overline{x}_2\overline{x}_2^T-\overline{x}_2^T\overline{w}_2\overline{x}_2\overline{w}_2^T\right)\right].
\end{aligned}$$

From the definitions of $a(x)$, $b(x)$ and $c(x)$ in (3.141), it follows that

$$\max(|b(x)|, |c(x)|) \le \frac{p}{2}\left(|\lambda_2(x)|^{p-1} + |\lambda_1(x)|^{p-1}\right),$$
$$|a(x)| = p\left|t_1\lambda_2(x) + (1-t_1)\lambda_1(x)\right|^{p-1} \le p\max(|\lambda_2(x)|^{p-1}, |\lambda_1(x)|^{p-1}) \tag{3.158}$$

for some $t_1 \in (0, 1)$, where the equality is using the mean-value theorem. Therefore,

$$|a(x)| \le p\|x\|^{\frac{p}{q}}, \quad |b(x)| \le p\|x\|^{\frac{p}{q}} \text{ and } |c(x)| \le p\|x\|^{\frac{p}{q}}. \tag{3.159}$$

Noting that $0 \le \lambda_1(w)/\lambda_2(w) < 1$ and $\sqrt[p]{\lambda_1(w)/\lambda_2(w)} \ge \lambda_1(w)/\lambda_2(w)$, we have

$$a(z) = \frac{\lambda_2(w) - \lambda_1(w)}{\sqrt[p]{\lambda_2(w)} - \sqrt[p]{\lambda_1(w)}} = \sqrt[q]{\lambda_2(w)}\frac{1 - \lambda_1(w)/\lambda_2(w)}{1 - \sqrt[p]{\lambda_1(w)/\lambda_2(w)}} \ge \sqrt[q]{\lambda_2(w)}. \tag{3.160}$$

By Eqs. (3.159), (3.160) and (3.154), we can simplify $a_1(x, z)$, $a_2(x, z)$, $b_1(x, z)$ and $A_1(x, z)$ as

$$\begin{aligned}
a_1(x,z) &= O(1) + \frac{1}{\sqrt[q]{\lambda_1(w)}}\left(b(x) - c(x)\overline{x}_2^T\overline{w}_2\right),\\
a_2(x,z) &= O(1) - \frac{1}{\sqrt[q]{\lambda_1(w)}}\left(b(x) - c(x)\overline{x}_2^T\overline{w}_2\right)\overline{w}_2,\\
b_2(x,z) &= O(1) + \frac{1}{\sqrt[q]{\lambda_1(w)}}\left[\left(c(x) - b(x)\overline{x}_2^T\overline{w}_2\right)\overline{x}_2 + a(x)(\overline{x}_2^T\overline{w}_2\overline{x}_2 - \overline{w}_2)\right],\\
A_1(x,z) &= O(1) - \frac{1}{\sqrt[q]{\lambda_1(w)}}\left[\left(c(x) - b(x)\overline{x}_2^T\overline{w}_2\right)\overline{x}_2\overline{w}_2^T + a(x)(\overline{x}_2^T\overline{w}_2\overline{x}_2\overline{w}_2^T - \overline{w}_2\overline{w}_2^T)\right].
\end{aligned} \tag{3.161}$$

By the definitions of $b(x)$ and $c(x)$, it is easy to verify that

$$\left|b(x) - c(x)\overline{x}_2^T\overline{w}_2\right| \le \frac{p}{2}\left[|\lambda_2(x)|^{p-1}(1 - \overline{x}_2^T\overline{w}_2) + |\lambda_1(x)|^{p-1}(1 + \overline{x}_2^T\overline{w}_2)\right],$$
$$\left|c(x) - b(x)\overline{x}_2^T\overline{w}_2\right| \le \frac{p}{2}\left[|\lambda_2(x)|^{p-1}(1 - \overline{x}_2^T\overline{w}_2) + |\lambda_1(x)|^{p-1}(1 + \overline{x}_2^T\overline{w}_2)\right],$$

which, together with (3.156), implies

$$\frac{b(x) - c(x)\overline{x}_2^T\overline{w}_2}{\sqrt[q]{\lambda_1(w)}} = O(1), \quad \frac{c(x) - b(x)\overline{x}_2^T\overline{w}_2}{\sqrt[q]{\lambda_1(w)}} = O(1).$$

In addition, it is easy to compute that

$$\|a(x)(\overline{x}_2^T\overline{w}_2\overline{x}_2 - \overline{w}_2)\|^2 = a^2(x)(1 - \overline{x}_2^T\overline{w}_2)(1 + \overline{x}_2^T\overline{w}_2),$$
$$\left\|a(x)(\overline{x}_2^T\overline{w}_2\overline{x}_2\overline{w}_2^T - \overline{w}_2\overline{w}_2^T)\right\|_F^2 \le a^2(x)(1 - \overline{x}_2^T\overline{w}_2)(1 + \overline{x}_2^T\overline{w}_2).$$

By Eq. (3.158), we have $a^2(x) \le p^2 \max(|\lambda_2(x)|^{2p-2}, |\lambda_1(x)|^{2p-2})$. Using (3.157) and noting that $0 \le 1 - \overline{x}_2^T \overline{w}_2 \le 2$ and $0 \le 1 + \overline{x}_2^T \overline{w}_2 \le 2$, we obtain

$$\frac{\|a(x)(\overline{x}_2^T \overline{w}_2 \overline{x}_2 - \overline{w}_2)\|}{\sqrt[q]{\lambda_1(w)}} = O(1) \text{ and } \frac{\|a(x)(\overline{x}_2^T \overline{w}_2 \overline{x}_2 \overline{w}_2^T - \overline{w}_2 \overline{w}_2^T)\|_F}{\sqrt[q]{\lambda_1(w)}} = O(1). \tag{3.162}$$

From (3.161)–(3.162), $a_1(x,z)$, $a_2(x,z)$, $b_2(x,z)$ and $A_1(x,z)$ are all uniformly bounded, and hence there exists a constant $C_2 > 0$ such that $\left\|\nabla g^{\text{soc}}(x)[\nabla g^{\text{soc}}(z)]^{-1}\right\|_F \le C_2$.

Suppose $x_2 = 0$ or $w_2 = 0$. Then there exists a sequence $\{(x^k, y^k)\} \subset \mathbb{R}^n \times \mathbb{R}^n$ with $x_2^k \ne 0$, $w_2(x^k, y^k) \ne 0$ and $w(x^k, y^k) \in \text{int}(\mathcal{K}^n)$ for all k such that $x^k \to x$ and $w^k \to w$ as $k \to \infty$. From the above result, $\left\|\nabla g^{\text{soc}}(x^k)\nabla g^{\text{soc}}(z^k)^{-1}\right\|_F \le c_2$ for all k. Noting that $\nabla g^{\text{soc}}(x)$ is continuous since $|t|^p$ is continuously differentiable, and $\nabla g^{\text{soc}}(z)^{-1}$ is continuous at any $z(x,y) \in \text{int}(\mathcal{K}^n)$, we have $\left\|\nabla g^{\text{soc}}(x)\nabla g^{\text{soc}}(z)^{-1}\right\|_F \le c_2$. The proof is complete. □

Proposition 3.28 *Let ψ_{FB}^p be defined by (3.131). Then, the function ψ_{FB}^p with $p \in (1,4)$ is smooth everywhere on $\mathbb{R}^n \times \mathbb{R}^n$.*

Proof By Proposition 3.27 and the symmetry of x and y in $\nabla\psi_{\text{FB}}^p$, it suffices to prove that $\nabla_x \psi_{\text{FB}}^p$ is continuous at every $(x,y) \in \mathbb{R}^n \times \mathbb{R}^n$. Choose a point $(x,y) \in \mathbb{R}^n \times \mathbb{R}^n$ arbitrarily. When $w(x,y) \in \text{int}(\mathcal{K}^n)$, the conclusion has been shown in Proposition 3.27. We next consider the other two cases where $w(x,y) = 0$ and $w(x,y) \in \text{bd}^+\mathcal{K}^n$, respectively.

Case (1): $w(x,y) = 0$. Now we have $(x,y) = (0,0)$, and $\nabla_x \psi_{\text{FB}}^p(0,0) = 0$ by Proposition 3.27. Thus, it suffices to show that $\nabla_{x'} \psi_{\text{FB}}^p(x',y') \to 0$ as $(x',y') \to (0,0)$. If $w(x',y') \in \text{int}(\mathcal{K}^n)$, then $\nabla_{x'} \psi_{\text{FB}}^p(x',y')$ is given by (3.143); and if $w(x',y') \in \text{bd}^+\mathcal{K}^n$, then $\nabla_{x'} \psi_{\text{FB}}^p(x',y')$ is given by (3.144). Since $\nabla g^{\text{soc}}(x')\nabla g^{\text{soc}}(z')^{-1} - I$ and $\frac{\text{sgn}(x_1')|x_1'|^{p-1}}{\sqrt[q]{|x_1'|^p + |y_1'|^p}}$ are uniformly bounded, where the uniform boundedness of the former is due to Lemma 3.33, using the continuity of ϕ_{FB}^p and noting that $\phi_{\text{FB}}^p(0,0) = 0$ immediately yields that $\nabla_{x'} \psi_{\text{FB}}^p(x',y') \to 0$ as $(x',y') \to (0,0)$.

Case (2): $w(x,y) \in \text{bd}^+\mathcal{K}^n$. For any (x',y') sufficiently close to (x,y), in order to prove that $\nabla_x \psi_{\text{FB}}^p(x',y') \to \nabla_x \psi_{\text{FB}}^p(x,y)$, we only need to consider the cases where $w(x',y') \in \text{int}(\mathcal{K}^n)$ and $w(x',y') \in \text{bd}^+\mathcal{K}^n$. When $w(x',y') \in \text{bd}^+\mathcal{K}^n$, $\nabla_x \psi_{\text{FB}}^p(x',y')$ has an expression of (3.144) which is continuous at (x,y) since $|x_1|^p + |y_1|^p > 0$ by Lemma 3.28, and then $\nabla_x \psi_{\text{FB}}^p(x',y') \to \nabla_x \psi_{\text{FB}}^p(x,y)$. We next concentrate on the case $w(x',y') \in \text{int}(\mathcal{K}^n)$, for which case

$$\begin{aligned} &\nabla_x \psi_{\text{FB}}^p(x',y') \\ &= \nabla g^{\text{soc}}(x')\nabla g^{\text{soc}}(z')^{-1}z' - \nabla g^{\text{soc}}(x')\nabla g^{\text{soc}}(z')^{-1}(x'+y') - \phi_{\text{FB}}^p(x',y'). \end{aligned} \tag{3.163}$$

We next proceed to the arguments by the following two subcases: $x_2 \neq 0$ and $x_2 = 0$.

Subcase (2.1): $x_2 \neq 0$. Under this case, by the expression of $\nabla_x \psi^p_{\rm FB}(x,y)$ in (3.144), we have

$$\begin{aligned}\nabla_x \psi^p_{\rm FB}(x,y) &= \frac{\operatorname{sgn}(x_1)|x_1|^{p-1}}{\sqrt[q]{|x_1|^p+|y_1|^p}}\phi^p_{\rm FB}(x,y) - \phi^p_{\rm FB}(x,y) \\ &= \frac{\operatorname{sgn}(x_1)|x_1|^{p-1}\sqrt[p]{|x_1|^p+|y_1|^p}}{\sqrt[q]{|x_1|^p+|y_1|^p}}\begin{bmatrix}1\\ \overline{w}_2\end{bmatrix} - \frac{\operatorname{sgn}(x_1)|x_1|^{p-1}}{\sqrt[q]{|x_1|^p+|y_1|^p}}(x+y) - \phi^p_{\rm FB}(x,y)\end{aligned}$$

where the second equality is by $z(x,y) = \sqrt[p]{|x_1|^p+|y_1|^p}\begin{bmatrix}1\\ \overline{w}_2\end{bmatrix}$. Comparing it with (3.163), we see, to prove $\nabla_x \psi^p_{\rm FB}(x',y') \to \nabla_x \psi^p_{\rm FB}(x,y)$ as $(x',y') \to (x,y)$, it suffices to argue that

$$\nabla g^{\rm soc}(x')\nabla g^{\rm soc}(z')^{-1}z' \to \frac{\operatorname{sgn}(x_1)|x_1|^{p-1}\sqrt[p]{|x_1|^p+|y_1|^p}}{\sqrt[q]{|x_1|^p+|y_1|^p}}\begin{bmatrix}1\\ \overline{w}_2\end{bmatrix} \quad \text{as } (x',y') \to (x,y) \tag{3.164}$$

and

$$\begin{aligned}\nabla g^{\rm soc}(x')\nabla g^{\rm soc}(z')^{-1}x' &\to \frac{\operatorname{sgn}(x_1)|x_1|^{p-1}}{\sqrt[q]{|x_1|^p+|y_1|^p}}x \quad \text{as } (x',y') \to (x,y), \\ \nabla g^{\rm soc}(x')\nabla g^{\rm soc}(z')^{-1}y' &\to \frac{\operatorname{sgn}(x_1)|x_1|^{p-1}}{\sqrt[q]{|x_1|^p+|y_1|^p}}y \quad \text{as } (x',y') \to (x,y). \end{aligned}\tag{3.165}$$

First of all, let us prove (3.164). Since $w(x,y) \in {\rm bd}^+\mathcal{K}^n$ implies $|x|^p \in {\rm bd}\mathcal{K}^n$, we have

$$a(x) = 2^{p-1}\operatorname{sgn}(x_1)|x_1|^{p-1}, \quad b(x) = 2^{p-2}p\operatorname{sgn}(x_1)|x_1|^{p-1}, \quad c(x) = 2^{p-2}p|x_1|^{p-1}. \tag{3.166}$$

Since $w_2(x,y) \neq 0$ (if not, $w(x,y)=0$), we have $w_2' = w_2(x',y') \neq 0$. By the expressions of $\nabla g^{\rm soc}(x')$ and $\nabla g^{\rm soc}(z')^{-1}$, it is not hard to calculate that $\nabla g^{\rm soc}(x')\nabla g^{\rm soc}(z')^{-1}z'$ equals

$$\frac{\lambda_2(w')^{\frac{1}{p}-\frac{1}{q}}}{2p}\nabla g^{\rm soc}(x')\begin{bmatrix}1\\ \overline{w}_2'\end{bmatrix} + \frac{\sqrt[p]{\lambda_1(w')}}{2}\nabla g^{\rm soc}(x')\nabla g^{\rm soc}(z')^{-1}\begin{bmatrix}1\\ -\overline{w}_2'\end{bmatrix}$$

where $z' = z(x',y')$, $w' = w(x',y')$ and $\overline{w}_2' = \frac{w_2'}{\|w_2'\|}$. Note that $\lambda_1(w') \to 0$ and $\overline{w}_2' \to \overline{w}_2$ with $\overline{w}_2 = \frac{w_2}{\|w_2\|}$ as $(x',y') \to (x,y)$. By Lemma 3.33, the last term on the right-hand side tends to 0, whereas by the continuity of $\nabla g^{\rm soc}$ the first term approaches $\frac{1}{2p}\lambda_2(w)^{\frac{1}{p}-\frac{1}{q}}\nabla g^{\rm soc}(x)\begin{bmatrix}1\\ \overline{w}_2\end{bmatrix}$. Thus, together with $\lambda_2(w) = 2w_1 = 2^p(|x_1|^p+|y_1|^p)$, it holds that as $(x',y') \to (x,y)$,

$$\nabla g^{\rm soc}(x')\nabla g^{\rm soc}(z')^{-1}z' \to p^{-1}2^{-\frac{p}{q}}(|x_1|^p+|y_1|^p)^{\frac{1}{p}-\frac{1}{q}}\nabla g^{\rm soc}(x)\begin{bmatrix}1\\ \overline{w}_2\end{bmatrix}. \tag{3.167}$$

In addition, using Eqs. (3.166) and (3.140), we readily obtain that

$$\nabla g^{\rm soc}(x) = 2^{p-2}p\,\mathrm{sgn}(x_1)|x_1|^{p-1}\begin{bmatrix}1 & \dfrac{x_2^{\rm T}}{x_1}\\ \dfrac{x_2}{x_1} & \dfrac{2}{p}I+\left(1-\dfrac{2}{p}\right)\dfrac{x_2x_2^{\rm T}}{x_1^2}\end{bmatrix}.$$

This along with (3.167) means that as $(x',y')\to(x,y)$, $\nabla g^{\rm soc}(x')\nabla g^{\rm soc}(z')^{-1}z'$ approaches

$$\begin{aligned}&\frac{1}{2}\mathrm{sgn}(x_1)|x_1|^{p-1}(|x_1|^p+|y_1|^p)^{\frac{1}{p}-\frac{1}{q}}\begin{bmatrix}1 & \dfrac{x_2^{\rm T}}{x_1}\\ \dfrac{x_2}{x_1} & \dfrac{2}{p}I+\left(1-\dfrac{2}{p}\right)\dfrac{x_2x_2^{\rm T}}{x_1^2}\end{bmatrix}\begin{bmatrix}1\\ \overline{w}_2\end{bmatrix},\\ &= \mathrm{sgn}(x_1)|x_1|^{p-1}(|x_1|^p+|y_1|^p)^{\frac{1}{p}-\frac{1}{q}}\begin{bmatrix}1\\ \overline{w}_2\end{bmatrix},\end{aligned}$$

where the equality is using Lemma 3.28. This shows that Eq. (3.164) holds.

Next, we prove the second relation of (3.165), and an analogous argument can be used to prove the first relation. Let $(\zeta_1,\zeta_2) := \nabla g^{\rm soc}(x')\nabla g^{\rm soc}(z')^{-1}y'$. We only need to establish

$$\zeta_1 \to \frac{\mathrm{sgn}(x_1)|x_1|^{p-1}}{\sqrt[q]{|x_1|^p+|y_1|^p}}y_1,\ \ \zeta_2 \to \frac{\mathrm{sgn}(x_1)|x_1|^{p-1}}{\sqrt[q]{|x_1|^p+|y_1|^p}}y_2 \ \text{ as } (x',y')\to(x,y). \tag{3.168}$$

Note that $x_2'\neq 0$ for (x',y') sufficiently close to (x,y). By Eq. (3.140) and the expression of $\nabla g^{\rm soc}(z')^{-1}$ in Lemma 3.29, a direct calculation yields that

$$\begin{aligned}2p\,\zeta_1 =\ & \frac{1}{\sqrt[q]{\lambda_2(w')}}\left[b(x')+c(x')(\overline{x}_2')^{\rm T}\overline{w}_2'\right][y_1'+(\overline{w}_2')^{\rm T}y_2']\\ &+\frac{1}{\sqrt[q]{\lambda_1(w')}}\left[b(x')-c(x')(\overline{x}_2')^{\rm T}\overline{w}_2'\right][y_1'-(\overline{w}_2')^{\rm T}y_2']\\ &+\frac{2p}{a(z')}c(x')\left[(\overline{x}_2')^{\rm T}y_2'-(\overline{x}_2')^{\rm T}\overline{w}_2'(\overline{w}_2')^{\rm T}y_2'\right],\end{aligned} \tag{3.169}$$

$$\begin{aligned}2p\,\zeta_2 =\ & \frac{1}{\sqrt[q]{\lambda_2(w')}}\left[c(x')\overline{x}_2'+a(x')\overline{w}_2'+(b(x')-a(x'))(\overline{x}_2')^{\rm T}\overline{w}_2'\overline{x}_2'\right]y_1'\\ &+\frac{1}{\sqrt[q]{\lambda_2(w')}}\left[c(x')\overline{x}_2'(\overline{w}_2')^{\rm T}+a(x')\overline{w}_2'(\overline{w}_2')^{\rm T}+(b(x')-a(x'))(\overline{x}_2')^{\rm T}\overline{w}_2'\overline{x}_2'(\overline{w}_2')^{\rm T}\right]y_2'\\ &+\frac{2p}{a(z')}\left[a(x')(I-\overline{w}_2'(\overline{w}_2')^{\rm T})+(b(x')-a(x'))\left(\overline{x}_2'(\overline{x}_2')^{\rm T}-(\overline{x}_2')^{\rm T}\overline{w}_2'\overline{x}_2'(\overline{w}_2')^{\rm T}\right)\right]y_2'\\ &+\frac{1}{\sqrt[q]{\lambda_1(w')}}\left[c(x')\overline{x}_2'-a(x')\overline{w}_2'-(b(x')-a(x'))(\overline{x}_2')^{\rm T}\overline{w}_2'\overline{x}_2'\right]y_1',\end{aligned} \tag{3.170}$$

$$-\frac{1}{\sqrt[q]{\lambda_1(w')}}\left[c(x')\overline{x}_2'(\overline{w}_2')^{\mathsf{T}} - a(x')\overline{w}_2'(\overline{w}_2')^{\mathsf{T}} - (b(x') - a(x'))(\overline{x}_2')^{\mathsf{T}}\overline{w}_2'\overline{x}_2'(\overline{w}_2')^{\mathsf{T}}\right] y_2',$$

where $a(x')$, $b(x')$ and $c(x')$ are defined as in (3.141) with x replaced by x'. Since $\sqrt[q]{\lambda_2(w')}$, $b(x')$, $c(x')$, $\overline{x}_2'$ and $\overline{w}_2'$ are continuous at (x, y), it follows that

$$\sqrt[q]{\lambda_2(w')} \;\to\; \sqrt[q]{2w_1} = 2^{\frac{p}{q}}(|x_1|^p + |y_1|^p)^{\frac{1}{q}},$$
$$\left[b(x') + c(x')(\overline{x}_2')^{\mathsf{T}}\overline{w}_2'\right][y_1' + (\overline{w}_2')^{\mathsf{T}} y_2'] \to \left(b(x) + c(x)\overline{x}_2^{\mathsf{T}}\overline{w}_2\right)(y_1 + y_2^{\mathsf{T}}\overline{w}_2)$$

as $(x', y') \to (x, y)$. This, along with Lemma 3.28 and Eq. (3.166), implies that the first term on the right-hand side of (3.169) tends to $\frac{2p\ \mathrm{sgn}(x_1)|x_1|^{p-1}}{\sqrt[q]{|x_1|^p+|y_1|^p}} y_1$. Since

$$y_1' - (\overline{w}_2')^{\mathsf{T}} y_2' \to y_1 - y_2^{\mathsf{T}}\overline{w}_2 = 0,$$
$$(\overline{x}_2')^{\mathsf{T}} y_2' - (\overline{x}_2')^{\mathsf{T}}\overline{w}_2'(\overline{w}_2')^{\mathsf{T}} y_2' \to \overline{x}_2^{\mathsf{T}} y_2 - \overline{x}_2^{\mathsf{T}}\overline{w}_2\overline{w}_2^{\mathsf{T}} y_2 = 0,$$

whereas $(b(x') - c(x')(\overline{x}_2')^{\mathsf{T}}\overline{w}_2')/\sqrt[q]{\lambda_1(w')}$ and $2pc(x')/a(z')$ are uniformly bounded by the proof of Lemma 3.32, the last two terms of (3.169) tend to 0 as $(x', y') \to (x, y)$, and we prove the first relation in (3.168). We next prove the second relation of (3.168). From the above discussions, the first three terms on the right-hand side of (3.170) respectively tend to

$$\frac{1}{\sqrt[q]{2w_1}}\left[c(x)\overline{x}_2 + a(x)\overline{w}_2 + (b(x) - a(x))\overline{x}_2^{\mathsf{T}}\overline{w}_2\overline{x}_2\right] y_1,$$
$$\frac{1}{\sqrt[q]{2w_1}}\left[c(x)\overline{x}_2 + a(x)\overline{w}_2 + (b(x) - a(x))\overline{x}_2^{T}\overline{w}_2\overline{x}_2\right]\overline{w}_2^{T} y_2,$$
$$\frac{2p}{a(z)}\left[a(x)(I - \overline{w}_2\overline{w}_2^{\mathsf{T}}) + (b(x) - a(x))\left(\overline{x}_2\overline{x}_2^{\mathsf{T}} - \overline{x}_2^{\mathsf{T}}\overline{w}_2\overline{x}_2\overline{w}_2^{\mathsf{T}}\right)\right] y_2,$$

as $(x', y') \to (x, y)$, whose sum, by Lemma 3.28 and formula (3.166), can be simplified as

$$\frac{2}{\sqrt[q]{2w_1}}\left[\mathrm{sgn}(x_1)c(x) + b(x)\right] y_2 = \frac{2p\,\mathrm{sgn}(x_1)|x_1|^{p-1}}{\sqrt[q]{|x_1|^p + |y_1|^p}} y_2.$$

Observe that the sum of the last two terms on the right side of (3.170) can be rewritten as

$$\frac{1}{\sqrt[q]{\lambda_1(w')}}\left[c(x')\overline{x}_2' - a(x')\overline{w}_2' - (b(x') - a(x'))(\overline{x}_2')^{\mathsf{T}}\overline{w}_2'\overline{x}_2'\right](y_1' - (\overline{w}_2')^{\mathsf{T}} y_2'),$$

which clearly tends to zero as $(x', y') \to (x, y)$, since the first term is uniformly bounded by the proof of Lemma 3.32, whereas the term $y_1' - (\overline{w}_2')^T y_2' \to y_1 - \overline{w}_2^T y_2 = 0$. Thus, we complete the proof of the second relation in (3.168). Consequently, the second relation in (3.164) follows. This shows that $\nabla_{x'} \psi_p(x', y') \to \nabla_x \psi_p(x, y)$ as $(x', y') \to (x, y)$.

Subcase (2.2): $x_2 = 0$. Now we have $x = 0$ from $|x|^p \in \mathrm{bd}\mathcal{K}^n$, and $\nabla g^{\mathrm{soc}}(x) = 0$. Hence,

$$\nabla_x \psi_{\mathrm{FB}}^p(x, y) = \frac{\mathrm{sgn}(x_1)|x_1|^{p-1}}{\sqrt[q]{|x_1|^p + |y_1|^p}} \phi_{\mathrm{FB}}^p(x, y) - \phi_{\mathrm{FB}}^p(x, y) = -\phi_{\mathrm{FB}}^p(0, y). \quad (3.171)$$

On the other hand, since $\nabla g^{\mathrm{soc}}(x) = 0$, it follows from (3.167) that

$$\nabla g^{\mathrm{soc}}(x') \nabla g^{\mathrm{soc}}(z')^{-1} z' \to 0 \quad \text{as} \quad (x', y') \to (x, y);$$

while using Lemma 3.33 and $x = 0$, we have

$$\nabla g^{\mathrm{soc}}(x') \nabla g^{\mathrm{soc}}(z')^{-1} x' \to 0 \quad \text{as} \quad (x', y') \to (x, y);$$

and from the continuity of ϕ_{FB}^p and $x = 0$, it follows that

$$\phi_{\mathrm{FB}}^p(x', y)' \to \phi_{\mathrm{FB}}^p(0, y) \quad \text{as} \quad (x', y') \to (x, y).$$

Using the last three equations and comparing (3.163) with (3.171), we see, in order to prove that $\nabla_x \psi_{\mathrm{FB}}^p(x', y') \to \nabla_x \psi_{\mathrm{FB}}^p(x, y)$ as $(x', y') \to (x, y)$, it suffices to show that

$$\nabla g^{\mathrm{soc}}(x') \nabla g^{\mathrm{soc}}(z')^{-1} y' \to 0 \quad \text{as} \quad (x', y') \to (x, y). \quad (3.172)$$

Next, for any (x', y') sufficiently close to (x, y), we write $(\zeta_1, \zeta_2) := \nabla g^{\mathrm{soc}}(x') \nabla g^{\mathrm{soc}}(z')^{-1} y'$.

If $x_2' \neq 0$, then $2p\zeta_1$ and $2p\zeta_2$ are given by (3.169) and (3.170), respectively. Using the same arguments as Subcase (2.1), we have that the second term of (3.169) and the sum of the last two terms of (3.170) tend to 0 as $(x', y') \to (x, y)$. Since $\sqrt[q]{\lambda_2(w')}$, $a(z')$, $b(x')$, $c(x')$ and $\overline{w}_2'$ are continuous at (x, y), $\|\overline{x}_2'\| = 1$, and $b(x) = c(x) = 0$ from (3.166) and $x = 0$, the first term and the third term of (3.169) also tend to 0. This proves that $2p\zeta_1 \to 0$ as $(x', y') \to (x, y)$. We next prove that the first three terms of (3.170) also tend to 0. From the mean-value theorem, $|a(x')| = p\left|t_1\lambda_2(x') + (1 - t_1)\lambda_1(x')\right|^{p-1}$ for some $t_1 \in (0, 1)$. Note that the function $|t|^{p-1}$ $(p > 1)$ is continuous on $\mathbb{R}$, whereas $\lambda_2(x') \to 0$ and $\lambda_1(x') \to 0$ as $(x', y') \to (x, y)$. So, $|a(x')| \to 0$ when $(x', y') \to (x, y)$. In addition, as $(x', y') \to (x, y)$,

$$b(x') \to 0, \quad c(x') \to 0, \quad \sqrt[q]{\lambda_2(w')} \to \sqrt[q]{2w_1} > 0, \quad \text{and} \quad a(z') \to a(z) > 0.$$

This implies that the first three terms of (3.170) also tend to 0. Consequently, $2p\zeta_2 \to 0$ as $(x', y') \to (x, y)$. Thus, (3.172) holds for this case.

If $x_2' = 0$, then using (3.140) and the expression of $\nabla g^{\mathrm{soc}}(z')^{-1}$ in Lemma 3.29, we have

$$\begin{aligned}
\zeta_1 &= \frac{p\,\mathrm{sgn}(x_1')|x_1'|^{p-1}}{\sqrt[q]{\lambda_2(w')}}\left[y_1' + (\overline{w}_2')^{\mathsf{T}} y_2'\right] + \frac{p\,\mathrm{sgn}(x_1')|x_1'|^{p-1}}{\sqrt[q]{\lambda_1(w')}}\left[y_1' - (\overline{w}_2')^{\mathsf{T}} y_2'\right],\\
\zeta_2 &= \frac{p\,\mathrm{sgn}(x_1')|x_1'|^{p-1}}{\sqrt[q]{\lambda_2(w')}}\left[y_1' + (\overline{w}_2')^{\mathsf{T}} y_2'\right]\overline{w}_2' - \frac{p\,\mathrm{sgn}(x_1')|x_1'|^{p-1}}{\sqrt[q]{\lambda_1(w')}}\left[y_1' - (\overline{w}_2')^{\mathsf{T}} y_2'\right]\overline{w}_2'\\
&\quad + \frac{2p^2\,\mathrm{sgn}(x_1')|x_1'|^{p-1}}{a(z')}\left[y_2' - \overline{w}_2'(\overline{w}_2')^{\mathsf{T}} y_2'\right].
\end{aligned}$$

Since $\mathrm{sgn}(x_1')|x_1'|^{p-1}$ is continuous and $\sqrt[q]{\lambda_2(w)} > 0$, we have, as $(x', y') \to (x, y)$,

$$\frac{p\,\mathrm{sgn}(x_1')|x_1'|^{p-1}}{\sqrt[q]{\lambda_2(w')}}[y_1' + (\overline{w}_2')^{\mathsf{T}} y_2'] \to \frac{p\mathrm{sgn}(x_1)|x_1|^{p-1}}{\sqrt[q]{\lambda_2(w)}}(y_1 + \overline{w}_2^{\mathsf{T}} y_2) = 0.$$

In addition, $|x_1'|^{p-1}/\sqrt[q]{\lambda_1(w')}$ is bounded with the bound independent of x' and y' because $\lambda_1(w') = w_1' - \|w_2'\| \geq |x_1'|^p$ by (3.136) when $x_2' = 0$. Besides, $y_1' - (\overline{w}_2')^T y_2' \to y_1 - \overline{w}_2^T y_2 = 0$ as $(x', y') \to (x, y)$, where the equality is due to Lemma 3.28. Hence we have

$$\frac{p\,\mathrm{sgn}(x_1')|x_1'|^{p-1}}{\sqrt[q]{\lambda_1(w')}}(y_1' - (\overline{w}_2')^{\mathsf{T}} y_2') \to 0 \quad \text{as} \quad (x', y') \to (x, y).$$

Thus, we prove that $\zeta_1 \to 0$ and the first two terms of ζ_2 tend to 0 as $(x', y') \to (x, y)$. Since $a(z') = \frac{\lambda_2(w') - \lambda_1(w')}{\sqrt[p]{\lambda_2(w')} - \sqrt[p]{\lambda_1(w')}}$ and $(y_2' - \overline{w}_2'(\overline{w}_2')^{\mathsf{T}} y_2')$ are continuous, we have

$$a(z') \to \frac{\lambda_2(w) - \lambda_1(w)}{\sqrt[p]{\lambda_2(w)} - \sqrt[p]{\lambda_1(w)}} = \sqrt[q]{\lambda_2(w)} \text{ and } \left[y_2' - \overline{w}_2'(\overline{w}_2')^{\mathsf{T}} y_2'\right] \to y_2 - \overline{w}_2\overline{w}_2^{\mathsf{T}} y_2 = 0$$

as $(x', y') \to (x, y)$, where the last equality is due to Lemma 3.28. This, together with $\mathrm{sgn}(x_1')|x_1'|^{p-1} \to \mathrm{sgn}(x_1)|x_1|^{p-1} = 0$, means that the last term of ζ_2 also tends to 0. Thus, we show that ζ_2 tends to 0 as $(x', y') \to (x, y)$. Consequently, (3.172) holds in this case. □

Remark 3.5 It is worth noting that the proof of Proposition 3.28 remains valid for all $p \geq 4$. Consequently, if the differentiability of the function ψ_p for $p \geq 4$ can be established, a question that remains open, then it would follow that ψ_p is continuously differentiable.

We now examine the conditions under which Ψ^p_{FB} is coercive; that is,

$$\limsup_{\|x\|\to\infty} \Psi^p_{\text{FB}}(x) = \infty.$$

Establishing this property is essential for analyzing the global convergence of both merit function methods and equation-based approaches built upon ϕ^p_{FB}. To this end, we first present two technical lemmas that form the foundation of our analysis.

Lemma 3.34 *Let ϕ^p_{FB} and ψ^p_{FB} be given by (3.132) and (3.131), respectively. Then, for any $x, y \in \mathbb{R}^n$, there hold*

(a)

$$\langle x, \nabla_x \psi_p(x, y)\rangle + \langle y, \nabla_y \psi_p(x, y)\rangle = \|\phi_p(x, y)\|^2.$$

(b)

$$4\psi^p_{\text{FB}}(x, y) \geq 2\left\|[\phi^p_{\text{FB}}(x, y)]_+\right\|^2 \geq \max\left(\|(-x)_+\|^2, \|(-y)_+\|^2\right),$$

where $(\cdot)_+$ means the minimum Euclidean distance projection onto $\mathcal{K}^n$.

Proof Noting the fact that $\sqrt[p]{|x|^p + |y|^p} - x \in \mathcal{K}^n$ and $\sqrt[p]{|x|^p + |y|^p} - y \in \mathcal{K}^n$, the proof of part(a) is similar to Proposition 3.6 whereas the proof of part(b) is similar to Lemma 3.7. We omit them here. □

Lemma 3.35 *Assume that $\{(x^k, y^k)\} \subseteq \mathbb{R}^n \times \mathbb{R}^n$ satisfies either of the conditions*

(a) $\lambda_1(x^k) \to -\infty$ or $\lambda_1(y^k) \to -\infty$;
(b) $\{\lambda_1(x^k)\}$ and $\{\lambda_1(y^k)\}$ are bounded below, $\lambda_2(x^k), \lambda_2(y^k) \to +\infty$, and $\langle \frac{x^k}{\|x^k\|}, \frac{y^k}{\|y^k\|}\rangle \nrightarrow 0$.

Then, when p is a rational number, it holds that $\limsup_{k\to\infty} \psi^p_{\text{FB}}(x^k, y^k) = +\infty$.

Proof If $\{(x^k, y^k)\}$ satisfies condition (a), the result follows from Lemma 3.34 and the fact

$$2\|(-x^k)_+\|^2 = \min\left(0, \lambda_1(x^k)\right)^2 + \min\left(0, \lambda_2(x^k)\right)^2.$$

It remains to consider the case where $\{(x^k, y^k)\}$ satisfies condition(b). Now from the given assumptions we have (taking a subsequence if necessary) $x_1^k \to +\infty$ and $y_1^k \to +\infty$. Without loss of generality, we assume (subsequencing if necessary) that

$$\lim_{k\to\infty} x^k/\|x^k\| = x^* \quad \text{and} \quad \lim_{k\to\infty} y^k/\|y^k\| = y^*. \tag{3.173}$$

Since $\{\lambda_1(x^k)\}$ and $\{\lambda_1(y^k)\}$ are bounded below, there exists a fixed element $d \in \mathbb{R}^n$ such that $x^k - d \in \text{int}(\mathcal{K}^n)$ and $y^k - d \in \text{int}(\mathcal{K}^n)$ for each k (*Indeed, letting γ be the lower bound of $\{\lambda_1(x^k)\}$ and $\{\lambda_1(y^k)\}$, we have $x^k - (\gamma - 1)e \in \text{int}(\mathcal{K}^n)$ and $y^k - (\gamma - 1)e \in \text{int}(\mathcal{K}^n)$ since $\lambda_1(z^k - (\gamma - 1)e) \geq \lambda_1(z^k) + \lambda_1((1 - \gamma)e) \geq \gamma + 1 - \gamma = 1$ for $z^k = x^k$ or y^k*). Thus, $\frac{x^k - d}{\|x^k\|} \in \text{int}(\mathcal{K}^n)$ and $\frac{y^k - d}{\|y^k\|} \in \text{int}(\mathcal{K}^n)$ for each k.

This implies that $\frac{x^k}{\|x^k\|} \in \mathcal{K}^n$ and $\frac{y^k}{\|y^k\|} \in \mathcal{K}^n$, and consequently $x^k \in \mathcal{K}^n$ and $y^k \in \mathcal{K}^n$, for all sufficiently large k. We will proceed the arguments by three cases as shown below, where all k are assumed to be sufficiently large.

Case (1): the sequence $\{\|x^k\|/\|y^k\|\}$ is unbounded. Since p is a rational number, we may write $p = n/m$ with n, m being natural numbers and $n > m$. Suppose that the conclusion does not hold, i.e., $\{\phi^p_{\text{FB}}(x^k, y^k)\}$ is bounded. From the definition of ϕ^p_{FB} and $x^k, y^k \in \mathcal{K}^n$, we have $(x^k)^{\frac{n}{m}} + (y^k)^{\frac{n}{m}} = \left[x^k + y^k + \phi^p_{\text{FB}}(x^k, y^k)\right]^{\frac{n}{m}}$, which is equivalent to

$$\left[(x^k)^{\frac{n}{m}} + (y^k)^{\frac{n}{m}}\right]^m = \left[x^k + y^k + \phi^p_{\text{FB}}(x^k, y^k)\right]^n. \tag{3.174}$$

Since $\{\|x^k\|/\|y^k\|\}$ is unbounded, $\|x^k\| \to +\infty$, $\|y^k\| \to +\infty$ and $n > m$, by expanding $\left[(x^k)^{\frac{n}{m}} + (y^k)^{\frac{n}{m}}\right]^m = \underbrace{\left[(x^k)^{\frac{n}{m}} + (y^k)^{\frac{n}{m}}\right] \circ \cdots \circ \left[(x^k)^{\frac{n}{m}} + (y^k)^{\frac{n}{m}}\right]}_{m}$, we obtain that the left-hand side of (3.174) is $(x^k)^n + (y^k)^n + o(\|x^k\|^{n-1}\|y^k\|)$, whereas by expanding $\left[x^k + y^k + \phi^p_{\text{FB}}(x^k, y^k)\right]^n$ and noting that $\{\phi^p_{\text{FB}}(x^k, y^k)\}$ is bounded and $\{\|x^k\|/\|y^k\|\}$ is unbounded, the right-hand side of (3.174) is $(x^k + y^k)^n + o(\|x^k\|^{n-1}\|y^k\|)$, which can be further written as

$$\begin{aligned}(x^k)^n + (y^k)^n + (x^k)^{n-1} \circ y^k + (x^k) \circ ((x^k)^{n-2} \circ y^k) + \cdots + x^k \circ (x^k \circ (\cdots ((x^k)^2 \circ y^k \underbrace{)\cdots)}_{n-3} \\ +2x^k \circ (x^k \circ (x^k \circ (\cdots (x^k \circ y^k \underbrace{)\cdots)}_{n-2} + o(\|x^k\|^{n-1}\|y^k\|).\end{aligned}$$

Here, $o(\|x^k\|^{n-1}\|y^k\|)$ denotes the term e_k satisfying $\lim_{k\to\infty} \frac{\|e_k\|}{\|x^k\|^{n-1}\|y^k\|} = 0$. Therefore,

$$\begin{aligned}(x^k)^{n-1} \circ y^k + (x^k) \circ ((x^k)^{n-2} \circ y^k) + \cdots + x^k \circ (x^k \circ (\cdots ((x^k)^2 \circ y^k \underbrace{)\cdots)}_{n-3} \\ +2x^k \circ (x^k \circ (x^k \circ (\cdots (x^k \circ y^k \underbrace{)\cdots)}_{n-2} = o(\|x^k\|^{n-1}\|y^k\|).\end{aligned}$$

Making the inner product with the unit element e for the both sides then gives

$$n\langle (x^k)^{n-1}, y^k\rangle = o(\|x^k\|^{n-1}\|y^k\|).$$

Dividing the both sides by $\|x^k\|^{n-1}\|y^k\|$ and taking the limit $k \to \infty$, we obtain $\langle (x^*)^{n-1}, y^*\rangle = 0$. Noting that $x^*, y^* \in \mathcal{K}^n$ and $\|x^*\| = \|y^*\| = 1$, from $\langle (x^*)^{n-1}, y^*\rangle = 0$ we deduce that

$$y_1^* = \|y_2^*\| \text{ and } (x^*)^{n-1} = \alpha(y_1^*, -y_2^*) \text{ for some } \alpha > 0.$$

From this, it is easy to get $\langle x^*, y^* \rangle = 0$, which by (3.173) contradicts the given condition that $\langle \frac{x^k}{\|x^k\|}, \frac{y^k}{\|y^k\|} \rangle \nrightarrow 0$. Thus, we prove that the conclusion $\limsup_{k\to\infty} \psi_p(x^k, y^k) = +\infty$ holds.

<u>**Case (2)**</u>: the sequence $\{\|y^k\|/\|x^k\|\}$ is unbounded. By the symmetry of x and y in $\phi_p(x, y)$, using the same arguments as in Case (1) leads to the desired result.

<u>**Case (3)**</u>: the sequences $\{\|y^k\|/\|x^k\|\}$ and $\{\|x^k\|/\|y^k\|\}$ are bounded. In this case, taking subsequences of $\{x^k\}$ and $\{y^k\}$ if necessary, we may assume that $\lim_{k\to\infty} \frac{\|y^k\|}{\|x^k\|} = c$ for some $0 < c < +\infty$. By the definition of ϕ^p_{FB} and $x^k, y^k \in \mathcal{K}^n$, we have

$$(x^k)^p + (y^k)^p = \left[x^k + y^k + \phi^p_{\text{FB}}(x^k, y^k)\right]^p .$$

Suppose that the conclusion does not hold. Then, dividing the both sides of last equality by $\|x^k\|$ and taking the limit $k \to \infty$, it is not hard to obtain

$$(x^*)^p + (cy^*)^p = (x^* + cy^*)^p,$$

which is equivalent to saying that $\phi^p_{\text{FB}}(x^*, cy^*) = 0$ since $x^*, y^* \in \mathcal{K}^n$. Therefore, $x^* \circ cy^* = 0$. This clearly contradicts the given condition $\langle \frac{x^k}{\|x^k\|}, \frac{y^k}{\|y^k\|} \rangle \nrightarrow 0$, and the result follows. □

Remark 3.6 At present, we are unable to prove Lemma 3.35 for irrational values of p, despite attempts to exploit the density of rational numbers in $\mathbb{R}$. Nevertheless, numerical evidence strongly supports the validity of the result in such cases.

Proposition 3.29 *Let Ψ^p_{FB} be defined as in* (3.133). *Suppose that G is an identity mapping, and F has the uniform Jordan P-property and satisfies the linear growth (see Definition 1.11(d) and (e)). Then, $\Psi^p_{\text{FB}}(\zeta)$ is coercive for a rational p.*

Proof Suppose on the contrary that there is a constant $\gamma > 0$ and a sequence $\{\zeta^k\} \subset \mathbb{R}^n$ with $\|\zeta^k\| \to \infty$ such that $\Psi^p_{\text{FB}}(\zeta^k) \le \gamma$ for all k. Let $\zeta^k = (\zeta^k_1, \dots, \zeta^k_m)$ with $\zeta^k_i \in \mathbb{R}^{n_i}$ for $i = 1, 2, \dots, m$. Let $\mathcal{I} := \left\{i \in \{1, 2, \dots, m\} \mid \{\zeta^k_i\} \text{ is unbounded}\right\}$. Clearly, $\mathcal{I} \neq \emptyset$. Define

$$\xi^k_i = \begin{cases} 0 & \text{if } i \in \mathcal{I}; \\ \zeta^k_i & \text{otherwise,} \end{cases} \qquad i = 1, 2, \dots, m.$$

Then, the sequence $\{\xi^k\} \subseteq \mathbb{R}^n$ is bounded. Since F has the uniform Jordan P-property (see Definition 1.11(d)), there holds

$$\lambda_2\left[(\zeta^k - \xi^k) \circ (F(\zeta^k) - F(\xi^k))\right] \ge \varrho \|\zeta^k - \xi^k\|^2$$

for some $\varrho > 0$. Let $z^k = (\zeta^k - \xi^k) \circ (F(\zeta^k) - F(\xi^k))$ for each k. Suppose that each z^k has the spectral decomposition $\lambda_1(z^k)u^k_1 + \lambda_2(z^k)u^k_2$. Then, from the last inequality,

$$\varrho\|\zeta^k - \xi^k\|^2 \le 2\langle z^k, u_2^k\rangle = 2\left\langle(\zeta^k - \xi^k)\circ(F(\zeta^k) - F(\xi^k)), u_2^k\right\rangle \\ \le 2\|\zeta^k - \xi^k\|\|F(\zeta^k) - F(\xi^k)\|. \tag{3.175}$$

This implies $\|F(\zeta^k)\| \to \infty$. Since $\Psi^p_{\rm FB}(\zeta^k) \le \gamma$ for each k, from Lemma 3.35(a) it follows that $\{\lambda_1(\zeta^k)\}$ and $\{\lambda_1(F(\zeta^k))\}$ are bounded below. Together with $\|\zeta^k\| \to \infty$ and $\|F(\zeta^k)\| \to \infty$, we obtain $\lambda_2(\zeta^k), \lambda_2(F(\zeta^k)) \to +\infty$. In addition, from (3.175) and the linear growth of F, we necessarily have $\lim_{k\to\infty} \frac{\zeta^k}{\|\zeta^k\|} \circ \frac{F(\zeta^k)}{\|F(\zeta^k)\|} \neq 0$. If not, on one hand, from the boundedness of $\{\xi^k\}$, we have $\lim_{k\to\infty} \frac{(\zeta^k-\xi^k)\circ(F(\zeta^k)-F(\xi^k))}{\|\zeta^k\|\|F(\zeta^k)\|} = 0$; and on the other hand

$$\lim_{k\to\infty} \frac{\varrho\|\zeta^k - \xi^k\|^2}{\|\zeta^k\|\|F(\zeta^k)\|} \ge \lim_{k\to\infty} \frac{\varrho\|\zeta^k - \xi^k\|^2}{\|\zeta^k\|(\|F(0)\| + c\|\zeta^k\|)} = \frac{\varrho}{c} > 0,$$

which is impossible by (3.175). By Lemma 3.35(b), $\limsup_{k\to\infty} \|\psi^p_{\rm FB}(\zeta^k, F(\zeta^k))\| = \infty$. This gives a contradiction to $\Psi^p_{\rm FB}(\zeta^k) \le \gamma$ for all k. Thus, we prove that $\Psi^p_{\rm FB}$ is coercive. □

3.1.3 The Functions $\phi^p_{\rm NR}$ and $\phi^p_{\rm D-FB}$ in SOC Setting

Another widely adopted C-function in the SOC setting is the vector-valued natural residual function , defined as

$$\phi_{\rm NR}(x, y) := x - (x - y)_+$$

where $(\cdot)_+$ denotes the Euclidean projection onto $\mathcal{K}^n$. This function induces the natural residual merit function $\psi_{\rm NR}$, given by

$$\psi_{\rm NR}(x, y) := \frac{1}{2}\|\phi_{\rm NR}(x, y)\|^2,$$

where $\phi_{\rm NR} : \mathbb{R}^n \times \mathbb{R}^n \to \mathbb{R}^n$. According to Proposition 1.3, $\phi_{\rm NR}$ qualifies as a C-function associated with SOC and has been further analyzed in [12, 13] in the context of smoothing methods for SOCCPs. In comparison with the Fischer–Burmeister (FB) merit function $\psi_{\rm FB}$, a notable limitation of $\psi_{\rm NR}$ is its lack of differentiability. A natural generalization of $\phi_{\rm NR}$ is the function $\phi^p_{\rm NR} : \mathbb{R}^n \times \mathbb{R}^n \to \mathbb{R}^n$, defined by

$$\phi^p_{\rm NR}(x, y) = x^p - [(x - y)_+]^p. \tag{3.176}$$

Again, it is based on the idea of "discrete generalization"; and $p > 1$ needs to be a positive integer. Applying the same idea back to the Fischer–Burmeister function, we can define $\phi^p_{\rm D-FB} : \mathbb{R}^n \times \to \mathbb{R}^n$ as

$$\phi^p_{\text{D-FB}}(x, y) = \left(\sqrt{x^2 + y^2}\right)^p - (x + y)^p, \tag{3.177}$$

where $p > 1$ is a positive odd integer, $(x, y) \in \mathbb{R}^n \times \mathbb{R}^n$, $x^2 = x \circ x$ is the Jordan product of x with itself and $\sqrt{x}$ with $x \in \mathcal{K}^n$ being the unique vector such that $\sqrt{x} \circ \sqrt{x} = x$. Notice that when $p = 1$, $\phi^p_{\text{D-FB}}$ reduces to the Fischer–Burmeister function. In other words, we extend the new function ϕ^p_{NR} and $\phi^p_{\text{D-FB}}$, constructed by discrete generalization in Sect. 2.2, to the SOC setting. In particular, we show that the function $\phi^p_{\text{D-FB}}$ and ϕ^p_{NR} are C-functions associated with $\mathcal{K}^n$. In addition, we present the computing formulas for their Jacobian matrices.

Lemma 3.36 *For $p = 2m + 1$ with $m = 1, 2, 3, \cdots$ and $x = (x_1, x_2)$, $y = (y_1, y_2) \in \mathbb{R} \times \mathbb{R}^{n-1}$, suppose that x^p and y^p represent $(x \circ x \circ \cdots \circ x)$ and $(y \circ y \circ \cdots \circ y)$ for p-times, respectively. Then, $x^p = y^p$ if and only if $x = y$.*

Proof "$\Leftarrow$" This direction is trivial.

"$\Rightarrow$" Suppose that $x^p = y^p$. By the spectral decomposition (1.8), we write

$$x = \lambda_1(x)u_x^{(1)} + \lambda_2(x)u_x^{(2)},$$
$$y = \lambda_1(y)u_y^{(1)} + \lambda_2(y)u_y^{(2)}.$$

Then, $x^p = (\lambda_1(x))^p u_x^{(1)} + (\lambda_2(x))^p u_x^{(2)}$ and $y^p = (\lambda_1(y))^p u_y^{(1)} + (\lambda_2(y))^p u_y^{(2)}$. Since $x^p = y^p$ and eigenvalues are unique, we obtain $(\lambda_1(x))^p = (\lambda_1(y))^p$ and $(\lambda_2(x))^p = (\lambda_2(y))^p$. By Lemma 2.16, this implies $\lambda_1(x) = \lambda_1(y)$ and $\lambda_2(x) = \lambda_2(y)$. Moreover, $\{u_x^{(1)}, u_x^{(2)}\}$ and $\{u_y^{(1)}, u_y^{(2)}\}$ are Jordan frames, we have $u_x^{(1)} + u_x^{(2)} = u_y^{(1)} + u_y^{(2)} = e$, where e is the identity element. From $x^p = y^p$ and $u_x^{(1)} + u_x^{(2)} = u_y^{(1)} + u_y^{(2)}$, we obtain

$$\left[(\lambda_1(x))^p - (\lambda_2(x))^p\right](u_x^{(1)} - u_y^{(1)}) = 0.$$

If $(\lambda_1(x))^p = (\lambda_2(x))^p$, we have $\lambda_1(x) = \lambda_2(x)$ and $\lambda_1(y) = \lambda_2(y)$, that is, $x = \lambda_1(x)e = y$. Otherwise, if $(\lambda_1(x))^p \neq (\lambda_2(x))^p$, we must have $u_x^{(1)} = u_y^{(1)}$, which implies $u_x^{(2)} = u_y^{(2)}$. □

Proposition 3.30 *Let $\phi^p_{\text{D-FB}}$ be defined by* (3.177)*. Then, the function $\phi^p_{\text{D-FB}}$ is a C-function associated with $\mathcal{K}^n$, i.e., it satisfies*

$$\phi^p_{\text{D-FB}}(x, y) = 0 \iff x \in \mathcal{K}^n,\ y \in \mathcal{K}^n,\ \langle x, y\rangle = 0,$$

where $p > 1$ is a positive odd integer.

Proof Since $\phi^p_{\text{D-FB}}(x, y) = 0$, we have $\left(\sqrt{x^2 + y^2}\right)^p = (x + y)^p$. Using p being a positive odd integer and applying Lemma 3.36 yield

$$\left(\sqrt{x^2 + y^2}\right)^p = (x + y)^p \iff \sqrt{x^2 + y^2} = x + y.$$

By Proposition 3.2, it is known that $\phi_{\rm FB}(x, y) := \sqrt{x^2 + y^2} - (x + y)$ is a complementarity function associated with $\mathcal{K}^n$. This indicates that $\phi^p_{\rm D-FB}$ is a complementarity function associated with $\mathcal{K}^n$. □

With similar technique, we can prove that $\phi^p_{\rm NR}$ can be extended as a C-function in SOC setting.

Proposition 3.31 *Let $\phi^p_{\rm NR}$ be defined as in* (3.176). *Then, the function $\phi^p_{\rm NR}$ is a C-function associated with $\mathcal{K}^n$, i.e., it satisfies*

$$\phi^p_{\rm NR}(x, y) = 0 \iff x \in \mathcal{K}^n,\ y \in \mathcal{K}^n,\ \langle x, y\rangle = 0,$$

where $p > 1$ is a positive odd integer.

Proof From Lemma 3.36, we see that $\phi^p_{\rm NR}(x, y) = 0$ if and only if $x = (x - y)_+$. On the other hand, by Proposition 1.3, it is known that $\phi_{\rm NR}(x, y) = x - (x - y)_+$ is a complementarity function associated with $\mathcal{K}^n$, which implies $x - (x - y)_+ = 0$ if and only if $x \in \mathcal{K}^n$, $y \in \mathcal{K}^n$, and $\langle x, y\rangle = 0$. Hence, $\phi^p_{\rm NR}$ is a C-function associated with $\mathcal{K}^n$. □

In order to compute the Jacobian of $\phi^p_{\rm D-FB}$, we need to introduce some notations for convenience. For any $x = (x_1, x_2) \in \mathbb{R} \times \mathbb{R}^{n-1}$ and $y = (y_1, y_2) \in \mathbb{R} \times \mathbb{R}^{n-1}$, we define

$$w(x, y) := x^2 + y^2 = (w_1(x, y), w_2(x, y)) \in \mathbb{R} \times \mathbb{R}^{n-1} \text{ and } v(x, y) := x + y.$$

Then, it is clear that $w(x, y) \in \mathcal{K}^n$ and $\lambda_i(w) \geq 0$, $i = 1, 2$.

Proposition 3.32 *Let $\phi^p_{\rm D-FB}$ be defined as in* (3.177) *and $g^{\rm soc}(x) = (\sqrt{|x|})^p$, $h^{\rm soc}(x) = x^p$ are the vector-valued functions corresponding to $g(t) = |t|^{\frac{p}{2}}$ and $h(t) = t^p$ for $t \in \mathbb{R}$, respectively. Then, $\phi^p_{\rm D-FB}$ is continuously differentiable at any $(x, y) \in \mathbb{R}^n \times \mathbb{R}^n$. Moreover, we have*

$$\begin{aligned}
\nabla_x \phi^p_{\rm D-FB}(x, y) &= 2L_x \nabla g^{\rm soc}(w) - \nabla h^{\rm soc}(v),\\
\nabla_y \phi^p_{\rm D-FB}(x, y) &= 2L_y \nabla g^{\rm soc}(w) - \nabla h^{\rm soc}(v),
\end{aligned}$$

where $w := w(x, y) = x^2 + y^2$, $v := v(x, y) = x + y$, $t \mapsto \operatorname{sign}(t)$ is the sign function, and

$$\nabla g^{\rm soc}(w) = \begin{cases} \dfrac{p}{2}|w_1|^{\frac{p}{2}-1} \cdot \operatorname{sign}(w_1) I & \text{if } w_2 = 0;\\ \begin{bmatrix} b_1(w) & c_1(w)\bar{w}_2^{\rm T} \\ c_1(w)\bar{w}_2 & a_1(w)I + (b_1(w) - a_1(w))\,\bar{w}_2\bar{w}_2^{\rm T} \end{bmatrix} & \text{if } w_2 \neq 0; \end{cases}$$

$$\bar{w}_2 = \frac{w_2}{\|w_2\|},$$

$$a_1(w) = \frac{|\lambda_2(w)|^{\frac{p}{2}} - |\lambda_1(w)|^{\frac{p}{2}}}{\lambda_2(w) - \lambda_1(w)},$$
$$b_1(w) = \frac{p}{4}\left[|\lambda_2(w)|^{\frac{p}{2}-1} + |\lambda_1(w)|^{\frac{p}{2}-1}\right],$$
$$c_1(w) = \frac{p}{4}\left[|\lambda_2(w)|^{\frac{p}{2}-1} - |\lambda_1(w)|^{\frac{p}{2}-1}\right],$$

and

$$\nabla h^{\rm soc}(v) = \begin{cases} p v_1^{p-1} I & \text{if } v_2 = 0; \\ \begin{bmatrix} b_2(v) & c_2(v)\bar{v}_2^T \\ c_2(v)\bar{v}_2 & a_2(v)I + (b_2(v) - a_2(v))\,\bar{v}_2\bar{v}_2^T \end{bmatrix} & \text{if } v_2 \neq 0; \end{cases} \tag{3.178}$$

$$\begin{aligned} \bar{v}_2 &= \frac{v_2}{\|v_2\|}, \\ a_2(v) &= \frac{(\lambda_2(v))^p - (\lambda_1(v))^p}{\lambda_2(v) - \lambda_1(v)}, \\ b_2(v) &= \frac{p}{2}\left[(\lambda_2(v))^{p-1} + (\lambda_1(v))^{p-1}\right], \\ c_2(v) &= \frac{p}{2}\left[(\lambda_2(v))^{p-1} - (\lambda_1(v))^{p-1}\right], \end{aligned} \tag{3.179}$$

Proof From the definition of $\phi^p_{\text{D-FB}}$, it is clear to see that for any $(x, y) \in \mathbb{R}^n \times \mathbb{R}^n$,

$$\begin{aligned} \phi^p_{\text{D-FB}}(x, y) &= \left(\sqrt{x^2 + y^2}\right)^p - (x + y)^p \\ &= \left(\sqrt{|x^2 + y^2|}\right)^p - (x + y)^p \\ &= \left[|\lambda_1(w)|^{\frac{p}{2}} u^{(1)}(w) + |\lambda_2(w)|^{\frac{p}{2}} u^{(2)}(w)\right] \\ &\quad - \left[(\lambda_1(v))^p u^{(1)}(v) + (\lambda_2(v))^p u^{(2)}(v)\right] \\ &= g^{\rm soc}(w) - h^{\rm soc}(v). \end{aligned} \tag{3.180}$$

For $p \geq 3$, since both $|t|^{\frac{p}{2}}$ and t^p are continuously differentiable on $\mathbb{R}$, by [26, Proposition 5] and [12, Proposition 5.2], we know that the function $g^{\rm soc}$ and $h^{\rm soc}$ are continuously differentiable on $\mathbb{R}^n$. Moreover, it is clear that $w(x, y) = x^2 + y^2$ is continuously differentiable on $\mathbb{R}^n \times \mathbb{R}^n$, then we conclude that $\phi^p_{\text{D-FB}}$ is continuously differentiable. Moreover, from the formula in [26, Proposition 4] and [12, Proposition 5.2], we have

$$\nabla g^{\rm soc}(w) = \begin{cases} \frac{p}{2}|w_1|^{\frac{p}{2}-1} \cdot \text{sign}(w_1) I & \text{if } w_2 = 0; \\ \begin{bmatrix} b_1(w) & c_1(w)\bar{w}_2^T \\ c_1(w)\bar{w}_2 & a_1(w)I + (b_1(w) - a_1(w))\,\bar{w}_2\bar{w}_2^T \end{bmatrix} & \text{if } w_2 \neq 0; \end{cases}$$

$$\nabla h^{\rm soc}(v) = \begin{cases} p v_1^{p-1} I & \text{if } \ v_2 = 0; \\ \begin{bmatrix} b_2(v) & c_2(v)\bar{v}_2^{\rm T} \\ c_2(v)\bar{v}_2 & a_2(v) I + (b_2(v) - a_2(v))\, \bar{v}_2 \bar{v}_2^{\rm T} \end{bmatrix} & \text{if } \ v_2 \neq 0; \end{cases}$$

where

$$\begin{aligned} \bar{w}_2 &= \frac{w_2}{\|w_2\|}, & \bar{v}_2 &= \frac{v_2}{\|v_2\|} \\ a_1(w) &= \frac{|\lambda_2(w)|^{\frac{p}{2}} - |\lambda_1(w)|^{\frac{p}{2}}}{\lambda_2(w) - \lambda_1(w)}, & a_2(v) &= \frac{(\lambda_2(v))^p - (\lambda_1(v))^p}{\lambda_2(v) - \lambda_1(v)}, \\ b_1(w) &= \frac{p}{4}\left[|\lambda_2(w)|^{\frac{p}{2}-1} + |\lambda_1(w)|^{\frac{p}{2}-1}\right], & b_2(v) &= \frac{p}{2}\left[(\lambda_2(v))^{p-1} + (\lambda_1(v))^{p-1}\right], \\ c_1(w) &= \frac{p}{4}\left[|\lambda_2(w)|^{\frac{p}{2}-1} - |\lambda_1(w)|^{\frac{p}{2}-1}\right], & c_2(v) &= \frac{p}{2}\left[(\lambda_2(v))^{p-1} - (\lambda_1(v))^{p-1}\right]. \end{aligned}$$

By taking differentiation on both sides about x and y for (3.180), respectively, and applying the chain rule for differentiation, it follows that

$$\begin{aligned} \nabla_x \phi^p_{\rm D-FB}(x, y) &= 2L_x \nabla g^{\rm soc}(w) - \nabla h^{\rm soc}(v), \\ \nabla_y \phi^p_{\rm D-FB}(x, y) &= 2L_y \nabla g^{\rm soc}(w) - \nabla h^{\rm soc}(v). \end{aligned}$$

Hence, we complete the proof. □

With Lemma 3.36 and Proposition 3.30, we can construct more C-functions associated with SOC, which are variants of $\phi^p_{\rm D-FB}(x, y)$. More specifically, consider that k and m are positive integers and $f^{\rm soc}(x, y) : \mathbb{R}^n \times \mathbb{R}^n \to \mathbb{R}^n$ is the vector-valued function corresponding to a given real-valued function f, the following functions are new variants of $\phi^p_{\rm D-FB}(x, y)$.

$$\begin{aligned} \widetilde{\phi}_1(x, y) &= \left[\sqrt{x^2 + y^2} + f^{\rm soc}(x, y)\right]^{\frac{2k+1}{2m+1}} - \left[x + y + f^{\rm soc}(x, y)\right]^{\frac{2k+1}{2m+1}}. \\ \widetilde{\phi}_2(x, y) &= \left[\sqrt{x^2 + y^2} - x - y\right]^{\frac{k}{m}}. \\ \widetilde{\phi}_3(x, y) &= \left[\sqrt{x^2 + y^2} - x + f^{\rm soc}(x, y)\right]^{\frac{2k+1}{2m+1}} - \left[y + f^{\rm soc}(x, y)\right]^{\frac{2k+1}{2m+1}}. \\ \widetilde{\phi}_4(x, y) &= \left[\sqrt{x^2 + y^2} - y + f^{\rm soc}(x, y)\right]^{\frac{2k+1}{2m+1}} - \left[x + f^{\rm soc}(x, y)\right]^{\frac{2k+1}{2m+1}}. \end{aligned}$$

Proposition 3.33 *All the above functions $\widetilde{\phi}_i$ for $i \in \{1, 2, 3, 4\}$ are C-functions associated with $\mathcal{K}^n$.*

Proof The results follow from applying Lemma 3.36 and Proposition 3.30. □

In general, for complementarity functions associated with $\mathcal{K}^n$, we have the following parallel result to Proposition 2.70 in the NCP setting.

Proposition 3.34 *Suppose that $\phi(x, y) = \varphi_1(x, y) - \varphi_2(x, y)$ is a C-function associated with $\mathcal{K}^n$ on $\mathbb{R}^n \times \mathbb{R}^n$, and k, m are positive integers. Then, $\left[\phi(x, y)\right]^{\frac{k}{m}}$ and $\left[\varphi_1(x, y)\right]^{\frac{2k+1}{2m+1}} - \left[\varphi_2(x, y)\right]^{\frac{2k+1}{2m+1}}$ are C-functions associated with $\mathcal{K}^n$.*

Proof According to k and m are positive integers and by using Lemma 3.36, we have

$$\begin{aligned} & \left[\phi(x,y)\right]^{\frac{k}{m}} = 0 \\ \Longleftrightarrow & \left\{\left[\phi(x,y)\right]^{\frac{k}{m}}\right\}^{m} = 0 \\ \Longleftrightarrow & \left[\phi(x,y)\right]^{k} = 0 \\ \Longleftrightarrow & \phi(x,y) = 0. \end{aligned}$$

Similarly, we have

$$\begin{aligned} & \left[\varphi_1(x,y)\right]^{\frac{2k+1}{2m+1}} - \left[\varphi_2(x,y)\right]^{\frac{2k+1}{2m+1}} = 0 \\ \Longleftrightarrow & \left[\varphi_1(x,y)\right]^{\frac{2k+1}{2m+1}} = \left[\varphi_2(x,y)\right]^{\frac{2k+1}{2m+1}} \\ \Longleftrightarrow & \left\{\left[\varphi_1(x,y)\right]^{\frac{2k+1}{2m+1}}\right\}^{2m+1} = \left\{\left[\varphi_2(x,y)\right]^{\frac{2k+1}{2m+1}}\right\}^{2m+1} \\ \Longleftrightarrow & \left[\varphi_1(x,y)\right]^{2k+1} = \left[\varphi_2(x,y)\right]^{2k+1} \\ \Longleftrightarrow & \varphi_1(x,y) = \varphi_2(x,y) \\ \Longleftrightarrow & \phi(x,y) = 0. \end{aligned}$$

From the above arguments and the assumption, the proof is complete. □

Remark 3.7 We elaborate more about Proposition 3.34.

(a) Based on existing complementarity functions, we can construct new C-functions associated with $\mathcal{K}^n$ in light of Proposition 3.34.
(b) When k is a positive odd integer, $\phi(x,y)^k$ is a C-function associated with $\mathcal{K}^n$. This means that perturbing the odd integer parameter k, we obtain the new complementarity functions associated with $\mathcal{K}^n$. In addition, if $\phi(x,y)$ is a C-function, then for any positive integer m, $\left[\phi(x,y)\right]^{\frac{k}{m}}$ is also a C-function. We can determine nice complementarity functions associated with $\mathcal{K}^n$ among these functions by their numerical performance.

Finally, we establish formula for Jacobian of ϕ^p_{NR} and the smoothness of ϕ^p_{NR}. To this aim, we need the following technical lemma.

Lemma 3.37 *Let $p > 1$. Then, the real-valued function $f(t) = (t_+)^p$ is continuously differentiable with $f'(t) = p(t_+)^{p-1}$ where $t_+ = \max\{0, t\}$.*

Proof By the definition of t_+, we have

$$f(t) = (t_+)^p = \begin{cases} t^p & \text{if } t \geq 0, \\ 0 & \text{if } t < 0, \end{cases}$$

which implies

$$f'(t) = \begin{cases} p\,t^{p-1} & \text{if } t \geq 0, \\ 0 & \text{if } t < 0. \end{cases}$$

Then, it is easy to see that $f'(t) = p(t_+)^{p-1}$ is continuous for $p > 1$. □

Proposition 3.35 *Let ϕ^p_{NR} be defined as in (3.176) and $h^{\text{soc}}(x) = x^p$, $l^{\text{soc}}(x) = (x_+)^p$ be the vector-valued functions corresponding to the real-valued functions $h(t) = t^p$ and $l(t) = (t_+)^p$, respectively. Then, ϕ^p_{NR} is continuously differentiable at any $(x, y) \in \mathbb{R}^n \times \mathbb{R}^n$, and its Jacobian is given by*

$$\begin{aligned} \nabla_x \phi^p_{\text{NR}}(x, y) &= \nabla h^{\text{soc}}(x) - \nabla l^{\text{soc}}(x - y), \\ \nabla_y \phi^p_{\text{NR}}(x, y) &= \nabla l^{\text{soc}}(x - y), \end{aligned}$$

where ∇h^{soc} satisfies (3.178)–(3.179) *and*

$$\nabla l^{\text{soc}}(u) = \begin{cases} p((u_1)_+)^{p-1} I & \text{if } u_2 = 0; \\ \begin{bmatrix} b_3(u) & c_3(u)\bar{u}_2^{\mathrm{T}} \\ c_3(u)\bar{u}_2 & a_3(u)I + (b_3(u) - a_3(u))\,\bar{u}_2\bar{u}_2^{\mathrm{T}} \end{bmatrix} & \text{if } u_2 \neq 0; \end{cases}$$

$$\begin{aligned} \bar{u}_2 &= \frac{u_2}{\|u_2\|}, \\ a_3(u) &= \frac{(\lambda_2(u)_+)^p - (\lambda_1(u)_+)^p}{\lambda_2(u) - \lambda_1(u)}, \\ b_3(u) &= \frac{p}{2}\left[(\lambda_2(u)_+)^{p-1} + (\lambda_1(u)_+)^{p-1}\right], \\ c_3(u) &= \frac{p}{2}\left[(\lambda_2(u)_+)^{p-1} - (\lambda_1(u)_+)^{p-1}\right]. \end{aligned}$$

Proof In light of [26, Proposition 5] and [12, Proposition 5.2], the results follow from applying Lemma 3.37 and using the chain rule for differentiation. □

3.1.4 Other C-Functions in SOC Setting

A. YF type of merit functions

It was also shown in the paper [14] that, like the NCP case, ψ_{FB} is smooth and, when ∇F is positive semidefinite, every stationary point of (3.7) solves SOCCP. For SDCP, which is a natural extension of NCP, where $\mathbb{R}^n_+$ is replaced by the cone of positive semidefinite matrices $\mathcal{S}^n_+$ and the partial order $\leq$ is also changed by $\preceq_{\mathcal{S}^n_+}$ (a partial order associated with $\mathcal{S}^n_+$ where $A \preceq_{\mathcal{S}^n_+} B$ means $B - A \in \mathcal{S}^n_+$) accordingly, the above features hold for the following analog of the SDCP merit function studied by Yamashita and Fukushima [40]:

$$\psi_{\text{YF}}(x, y) := \psi_1(\langle x, y\rangle) + \psi_{\text{FB}}(x, y), \tag{3.181}$$

where $\psi_1 : \mathbb{R} \to \mathbb{R}_+$ is any smooth function satisfying

$$\psi_1(t) = 0 \ \forall t \leq 0 \quad \text{and} \quad \psi_1'(t) > 0 \ \forall t > 0. \tag{3.182}$$

In [40], $\psi_1(t) = \frac{1}{4}(\max\{0, t\})^4$ was considered. In fact, the function ψ_{YF}, which was recently studied in [14], is also a SOCCP version merit function that enjoys favorable properties as what ψ_{YF} has and possesses additional properties including bounded level sets and error bound.

Proposition 3.36 *Let ψ_{YF} be defined as in* (3.181)–(3.182). *Then, the function ψ_{YF} is a smooth C-function in the SOC setting.*

Proof From the definition, it is clear that ψ_{YF} is a C-function. By Proposition 3.5, ψ_{FB} is smooth and ψ_1 is smooth due to (3.182), hence ψ_{YF} is smooth. □

In order to show properties of error bound and bounded level set, we need its merit function as below:

$$f_{\text{YF}}(\zeta) := \psi_{\text{YF}}(F(\zeta), G(\zeta)). \tag{3.183}$$

Proposition 3.37 *Let ψ_{YF} be defined as in* (3.181)–(3.182) *and f_{YF} be given by* (3.183). *Then, for every $\zeta \in \mathbb{R}^n$ where $\nabla F(\zeta), -\nabla G(\zeta)$ are column monotone, either (i) $f_{\text{YF}}(\zeta) = 0$ or (ii) $\nabla f_{\text{YF}}(\zeta) \neq 0$. In case (ii), if $\nabla G(\zeta)$ is invertible, then $\langle d_{\text{YF}}(\zeta), \nabla f_{\text{YF}}(\zeta)\rangle < 0$, where*

$$d_{\text{YF}}(\zeta) := -(\nabla G(\zeta)^{-1})^{\mathsf{T}}\Big(\psi_1'(\langle F(\zeta), G(\zeta)\rangle)G(\zeta) + \nabla_x \psi_{\text{FB}}(F(\zeta), G(\zeta))\Big).$$

Proof Fix any $\zeta \in \mathbb{R}^n$ where $\nabla F(\zeta), -\nabla G(\zeta)$ are column monotone. By Proposition 3.36, we know that ψ_{YF} is smooth. Then, the chain rule for differentiation yields

$$\begin{aligned}\nabla f_{\text{YF}}(\zeta) = \alpha\Big(\nabla F(\zeta)G(\zeta) + \nabla G(\zeta)F(\zeta)\Big) \\ +\nabla F(\zeta)\nabla_x \psi_{\text{FB}}(F(\zeta), G(\zeta)) + \nabla G(\zeta)\nabla_y \psi_{\text{FB}}(F(\zeta), G(\zeta)),\end{aligned}$$

where we let $\alpha := \psi_1'\Big(\langle F(\zeta), G(\zeta)\rangle\Big)$. In what follows we consider the case of $N = 1$, i.e., $\psi_{\text{FB}}(x, y) = \frac{1}{2}\left\|\phi_{\text{FB}}(x, y)\right\|^2$. Extending the proof to the case of $N \geq 2$ is straightforward.

Suppose $\nabla f_{\text{YF}}(\zeta) = 0$. Then, dropping the argument "(ζ)" for simplicity, we have

$$\alpha\Big(\nabla F G + \nabla G F\Big) + \nabla F\, \nabla_x \psi_{\text{FB}}(F, G) + \nabla G\, \nabla_y \psi_{\text{FB}}(F, G) = 0.$$

The column monotone property of $\nabla F, -\nabla G$ gives

$$\big\langle \alpha G + \nabla_x \psi_{\text{FB}}(F, G, \alpha F + \nabla_y \psi_{\text{FB}}(F, G)\big\rangle \le 0.$$

Upon collecting terms on the left-hand side, we have

$$\alpha^2 \langle F, G\rangle + \alpha \left(\big\langle F, \nabla_x \psi_{\text{FB}}(F, G)\big\rangle + \big\langle G, \nabla_y \psi_{\text{FB}}(F, G)\big\rangle\right) + \big\langle \nabla_x \psi_{\text{FB}}(F, G), \nabla_y \psi_{\text{FB}}(F, G)\big\rangle \le 0.$$

Our assumption (3.182) on ψ_1 implies the first term is nonnegative. By Proposition 3.6, the second and the third terms are also nonnegative. Thus, the third term must be zero, so Proposition 3.6(b) implies $\phi_{\text{FB}}(F, G) = 0$. Thus, $f_{\text{FB}}(\zeta) = \frac{1}{2}\left\|\phi_{\text{FB}}(F(\zeta), G(\zeta))\right\|^2 = 0$.

Suppose $\nabla f_{\text{YF}}(\zeta) \neq 0$ and $\nabla G(\zeta)$ is invertible. Again, we drop the argument "(ζ)" for simplicity. Then,

$$\begin{aligned}
&\langle d_{\text{YF}}, \nabla f_{\text{YF}}\rangle \\
&= \Big\langle -(\nabla G^{-1})^{\mathsf{T}}(\alpha G + \nabla_x \psi_{\text{FB}}(F, G)),\ \nabla F(\alpha G + \nabla_x \psi_{\text{FB}}(F, G)) + \nabla G(\alpha F + \nabla_y \psi_{\text{FB}}(F, G))\Big\rangle \\
&= -\Big\langle \alpha G + \nabla_x \psi_{\text{FB}}(F, G),\ \nabla G^{-1} \nabla F(\alpha G + \nabla_x \psi_{\text{FB}}(F, G))\Big\rangle \\
&\quad -\Big\langle \alpha G + \nabla_x \psi_{\text{FB}}(F, G),\ \alpha F + \nabla_y \psi_{\text{FB}}(F, G)\Big\rangle \\
&\le -\Big\langle \alpha G + \nabla_x \psi_{\text{FB}}(F, G),\ \alpha F + \nabla_y \psi_{\text{FB}}(F, G)\Big\rangle \\
&= -\alpha^2 \langle F, G\rangle - \alpha\Big(\langle F,\ \nabla_x \psi_{\text{FB}}(F, G)\rangle + \langle G,\ \nabla_y \psi_{\text{FB}}(F, G)\rangle\Big) - \langle \nabla_x \psi_{\text{FB}}(F, G),\ \nabla_y \psi_{\text{FB}}(F, G)\rangle,
\end{aligned}$$

where the first inequality follows from $\nabla G^{-1} \nabla F \succeq 0$. We argued earlier that all three terms on the right-hand side are non-positive. Moreover, by Proposition 3.6(b), the third term is zero if and only if $\phi_{\text{FB}}(F, G) = 0$, i.e., ζ is a global minimum of f_{YF} and hence a stationary point of f_{YF}. Since $\nabla f_{\text{YF}}(\zeta) \neq 0$, the right-hand side cannot equal zero, so it must be negative. □

Proposition 3.38 *Suppose that F and G are jointly strongly monotone mappings from $\mathbb{R}^n$ to $\mathbb{R}^n$. Also, suppose that the general SOCCP* (3.4) *has a solution ζ^*. Then, there exists a scalar $\tau > 0$ such that*

$$\tau \|\zeta - \zeta^*\|^2 \le \max\{0, \langle F(\zeta), G(\zeta)\rangle\} + \|(-F(\zeta))_+\| + \|(-G(\zeta))_+\| \quad \forall \zeta \in \mathbb{R}^n. \tag{3.184}$$

Moreover,

$$\tau \|\zeta - \zeta^*\|^2 \le \psi_1^{-1}\left(f_{\text{YF}}(\zeta)\right) + 2\sqrt{2} f_{\text{YF}}(\zeta)^{1/2} \quad \forall \zeta \in \mathbb{R}^n, \tag{3.185}$$

where f_{YF} is given by (3.183) with $N = 1$, $\psi_1 : \mathbb{R} \to [0, \infty)$ is a smooth function satisfying (3.182), and ψ_1^{-1} denotes the inverse function of ψ_1 on $[0, \infty)$.

Proof First, we observe that ψ_1^{-1} is well defined since, by (3.182), ψ_1 is strictly increasing on $[0, \infty)$. Because F and G are jointly strongly monotone, there exists a scalar $\rho > 0$ such that, for any $\zeta \in \mathbb{R}^n$,

$$\begin{aligned}
&\rho\|\zeta - \zeta^*\|^2 \\
&\leq \langle F(\zeta) - F(\zeta^*),\ G(\zeta) - G(\zeta^*)\rangle \\
&= \langle F(\zeta), G(\zeta)\rangle + \langle -F(\zeta), G(\zeta^*)\rangle + \langle F(\zeta^*), -G(\zeta)\rangle \\
&\leq \max\{0, \langle F(\zeta), G(\zeta)\rangle\} + \langle (-F(\zeta))_+, G(\zeta^*)\rangle + \langle F(\zeta^*), (-G(\zeta))_+\rangle \\
&\leq \max\{0, \langle F(\zeta), G(\zeta)\rangle\} + \|(-F(\zeta))_+\|\ \|G(\zeta^*)\| + \|F(\zeta^*)\|\ \|(-G(\zeta))_+\| \\
&\leq \max\{1, \|F(\zeta^*)\|, \|G(\zeta^*)\|\}\Big(\max\{0, \langle F(\zeta), G(\zeta)\rangle\} + \|(-F(\zeta))_+\| + \|(-G(\zeta))_+\| \Big),
\end{aligned}$$

where the second inequality uses Lemma 1.1(b). Setting $\tau := \dfrac{\rho}{\max\{1, \|F(\zeta^*)\|, \|G(\zeta^*)\|\}}$ yields (3.184). Moreover, using (3.181), (3.182) and (3.183), we have

$$\max\{0, \langle F(\zeta), G(\zeta)\rangle\} \leq \psi_1^{-1}\left(f_{\text{YF}}(\zeta)\right) \quad \text{and} \quad \psi_{\text{FB}}(F(\zeta), G(\zeta)) \leq f_{\text{YF}}(\zeta).$$

Using Lemma 3.7 and the second inequality, we have

$$\begin{aligned}
\|(-F(\zeta))_+\| + \|(-G(\zeta))_+\| &\leq \sqrt{2}\left(\|(-F(\zeta))_+\|^2 + \|(-G(\zeta))_+\|^2\right)^{1/2} \\
&\leq 2\sqrt{2}\, \psi_{\text{FB}}(F(\zeta), G(\zeta))^{1/2} \\
&\leq 2\sqrt{2}\, f_{\text{YF}}(\zeta)^{1/2}.
\end{aligned}$$

Thus, there holds

$$\max\{0, \langle F(\zeta), G(\zeta)\rangle\} + \|(-F(\zeta))_+\| + \|(-G(\zeta))_+\| \leq \psi_1^{-1}\left(f_{\text{YF}}(\zeta)\right) + 2\sqrt{2} f_{\text{YF}}(\zeta)^{1/2}.$$

This together with (3.184) yields (3.185). □

If in addition F is continuous and $G(\zeta) = \zeta$ for all $\zeta \in \mathbb{R}^n$, then the assumption that the SOCCP has a solution can be dropped from Proposition 3.38, see, e.g., [2, Proposition 2.2.7]. Moreover, the exponent 2 in the definition of joint strong monotonicity can be replaced by any $q > 1$, and Proposition 3.38 would generalize accordingly.

By using Lemma 3.8 and Proposition 3.37, we have the following analog of [40, Theorem 4.1] on solution existence and boundedness of the level sets of f_{YF}.

Proposition 3.39 *Suppose that F and G are differentiable, jointly monotone mappings from $\mathbb{R}^n$ to $\mathbb{R}^n$ satisfying*

$$\lim_{\|\zeta\|\to\infty} \|F(\zeta)\| + \|G(\zeta)\| = \infty. \tag{3.186}$$

Suppose also that SOCCP is strictly feasible, i.e., there exists $\bar{\zeta} \in \mathbb{R}^n$ such that $F(\bar{\zeta}), G(\bar{\zeta}) \in \text{int}(\mathcal{K}^n)$. Then, the level set

$$\mathcal{L}(\gamma) := \{\zeta \in \mathbb{R}^n \mid f_{\text{YF}}(\zeta) \le \gamma\}$$

is nonempty and bounded for all $\gamma \ge 0$, where f_{YF} is given by (3.181)–(3.182) *with $N = 1$, and $\psi_1 : \mathbb{R} \to [0, \infty)$ is a smooth function satisfying* (3.182).

Proof For any $\gamma \ge 0$, if $\{\zeta^k\}_{k=1}^{\infty} \subseteq \mathcal{L}(\gamma)$, then $f_{\text{YF}}(\zeta^k)\}$ is bounded and the joint monotonicity of F and G yields

$$\langle F(\zeta^k), G(\bar{\zeta})\rangle + \langle F(\bar{\zeta}), G(\zeta^k)\rangle \le \langle F(\zeta^k), G(\zeta^k)\rangle + \langle F(\bar{\zeta}), G(\bar{\zeta})\rangle, \quad k = 1, 2, \ldots$$

Using this together with Lemma 3.8 and an argument analogous to the proof of [40, Theorem 4.1], we obtain that $\{\|F(\zeta^k)\| + \|G(\zeta^k)\|\}$ is bounded. Then, (3.186) implies $\{\zeta^k\}$ is bounded. This shows that $\mathcal{L}(\gamma)$ is bounded.

The proof of $\mathcal{L}(\gamma) \neq \emptyset$ uses Proposition 3.37 and is nearly identical to the proof of [40, Theorem 4.1]. □

We point out that there is a result presented in Sect. 4.1 (see Lemma 4.1), which is analogous to Proposition 3.39 and uses R_{01}-function. The condition of R_{01}-function is weaker than strong monotonicity, and it is also weaker than monotonicity plus strict feasibility in certain sense, see [41, 42].

As below, we make a slight modification of ψ_{YF}, for which ψ_1 is replaced by the mapping $\psi_0 : \mathbb{R}^n \times \mathbb{R}^n \to \mathbb{R}_+$ that is given by

$$\psi_0(x, y) := \frac{1}{2}\|(x \circ y)_+\|^2, \tag{3.187}$$

where $(\cdot)_+$ denotes the orthogonal projection onto $\mathcal{K}^n$. If we observe closely, we may see there is some relation between ψ_0 and ψ_1 : both are smooth functions. Moreover, if we let $\hat{\psi}_1 : \mathbb{R}^n \times \mathbb{R}^n \to \mathbb{R}$ be $\hat{\psi}_1(x, y) := \psi_1(x, y)$, then the graphs of ψ_0 and $\hat{\psi}_1$ share similar features. In other words, our new merit function $\psi_{\text{new}} : \mathbb{R}^n \times \mathbb{R}^n \to \mathbb{R}_+$ is defined as

$$\psi_{\text{new}}(x, y) := \alpha\, \psi_0(x, y) + \psi_{\text{FB}}(x, y), \tag{3.188}$$

where $\alpha > 0$. When $\alpha = 0$, ψ_{new} reduces to ψ_{FB} which is the squared norm of Fischer–Burmeister function (3.11). Thus, this new merit function can be viewed as the extension of the squared norm of Fischer–Burmeister function. We shall show that the SOCCP (3.1) is equivalent to the following global minimization via the new merit function ψ_{new}:

$$\min_{\zeta \in \mathbb{R}^n} f_{\text{new}}(\zeta) \quad \text{where} \quad f_{\text{new}}(\zeta) := \psi_{\text{new}}(F(\zeta), \zeta). \tag{3.189}$$

Indeed, this new merit function ψ_{new} was studied by Yamada, Yamashita, and Fukushima in [43] for the NCP setting. We are motivated by their work and wish to explore its extension to the SOCCP (3.1). Analogous to the additional properties that ψ_{YF}, given as (3.181)–(3.182), possesses and as will be seen later, if F is strongly monotone [2] then f_{new} provides a global error bound which plays an important role in analyzing the convergence rate of some iterative methods for solving the SOCCP (3.1); and if F is monotone and a strictly feasible solution exists then f_{new} has bounded level sets, which will ensure that the sequence generated by a descent algorithm has at least one accumulation point. All these properties will make it possible to construct a descent algorithm for solving the equivalent unconstrained reformulation of the SOCCP (3.1). In contrast, the merit function induced by ψ_{FB} lacks these properties. In addition, we will show that ψ_{new} is continuously differentiable and its gradient has a computable formula. All the aforementioned features are significant reasons for choosing and studying this new merit function ψ_{new}.

Lemma 3.38 *Let $\psi_0 : \mathbb{R}^n \times \mathbb{R}^n \to \mathbb{R}_+$ be given by (3.187). Then, ψ_0 is continuously differentiable and*

$$\begin{aligned} \nabla_x \psi_0(x, y) &= L_y \cdot (x \circ y)_+, \\ \nabla_y \psi_0(x, y) &= L_x \cdot (x \circ y)_+. \end{aligned} \tag{3.190}$$

Proof For any $z \in \mathbb{R}^n$, we can factor z as $z = \lambda_1 u^{(1)} + \lambda_2 u^{(2)}$. Then let $g : \mathbb{R}^n \to \mathbb{R}^n$ be defined as

$$g(z) := \frac{1}{2}((z)_+)^2 = \hat{g}(\lambda_1)u^{(1)} + \hat{g}(\lambda_2)u^{(2)},$$

where $\hat{g} : \mathbb{R} \to \mathbb{R}$ is given by $\hat{g}(\lambda) := \frac{1}{2}(\max(0, \lambda))^2$. From the continuous differentiability of $\hat{g}$ and [26, Proposition 5.2], the vector-valued function g is also continuously differentiable. Hence, the first component $g_1(z) := \frac{1}{2}\|(z)_+\|^2$ of $g(z)$ is continuously differentiable as well. By an easy computation, we have $\nabla g_1(z) = (z)_+$. Now, let

$$z(x, y) := x \circ y = (\langle x, y \rangle \ , \ x_1 y_2 + y_1 x_2),$$

then we have $\psi_0(x, y) = g_1(z(x, y))$. Applying the chain rule, we obtain

$$\begin{aligned} \nabla_x \psi_0 &= \nabla_x z \cdot \nabla g_1(z) = L_y \cdot (x \circ y)_+, \\ \nabla_y \psi_0 &= \nabla_y z \cdot \nabla g_1(z) = L_x \cdot (x \circ y)_+, \end{aligned}$$

where

$$\nabla_x z(x, y) = \begin{bmatrix} y_1 & y_2^T \\ y_2 & y_1 I \end{bmatrix} = L_y \quad \text{and} \quad \nabla_y z(x, y) = \begin{bmatrix} x_1 & x_2^T \\ x_2 & x_1 I \end{bmatrix} = L_x.$$

Thus, the desired result (3.190) is achieved. □

Proposition 3.40 *Let $\psi_{\text{new}} : \mathbb{R}^n \times \mathbb{R}^n \to \mathbb{R}_+$ be defined as in (3.187)–(3.188). Then, the following results hold.*

(a) $\psi_{\text{new}}(x, y) \geq 0$ *for all* $(x, y) \in \mathbb{R}^n \times \mathbb{R}^n$.

(b) $\psi_{\text{new}}(x, y) = 0 \iff x \in \mathcal{K}^n,\ y \in \mathcal{K}^n,\ x \circ y = 0 \iff x \in \mathcal{K}^n,\ y \in \mathcal{K}^n,\ \langle x, y\rangle = 0$.

(c) ψ_{new} *is continuously differentiable at every* $(x, y) \in \mathbb{R}^n \times \mathbb{R}^n$. *Moreover,* $\nabla_x \psi_{\text{new}}(0, 0) = \nabla_y \psi_{\text{new}}(0, 0) = 0$. *If* $(x, y) \neq (0, 0)$ *and* $x^2 + y^2 \in \text{int}(\mathcal{K}^n)$, *then*

$$\begin{cases} \nabla_x \psi_{\text{new}}(x, y) = \alpha\, L_y \cdot (x \circ y)_+ + \left(L_x L^{-1}_{(x^2+y^2)^{1/2}} - I \right) \phi_{\text{FB}}(x, y), \\ \nabla_y \psi_{\text{new}}(x, y) = \alpha\, L_x \cdot (x \circ y)_+ + \left(L_y L^{-1}_{(x^2+y^2)^{1/2}} - I \right) \phi_{\text{FB}}(x, y). \end{cases}$$

If $(x, y) \neq (0, 0)$ *and* $x^2 + y^2 \notin \text{int}(\mathcal{K}^n)$, *then* $x_1^2 + y_1^2 \neq 0$ *and*

$$\begin{cases} \nabla_x \psi_{\text{new}}(x, y) = 2\alpha\, |x_1| \cdot (y)_+^2 + \left(\dfrac{x_1}{\sqrt{x_1^2 + y_1^2}} - 1 \right) \phi_{\text{FB}}(x, y), \\ \nabla_y \psi_{\text{new}}(x, y) = 2\alpha\, |y_1| \cdot (x)_+^2 + \left(\dfrac{y_1}{\sqrt{x_1^2 + y_1^2}} - 1 \right) \phi_{\text{FB}}(x, y). \end{cases}$$

Proof (a) It is clear by definition.

(b) We only need to prove the first equivalence since the second one is a known result in [12]. Suppose $\psi_{\text{new}}(x, y) = 0$, it yields $\psi_{\text{FB}}(x, y) = 0$. Thus, the desirable result follows by Proposition 3.4(a). On the other hand, $x \in \mathcal{K}^n$, $y \in \mathcal{K}^n$, $x \circ y = 0$ imply $\psi_{\text{FB}}(x, y) = 0$; and $\psi_0(x, y) = 0$ from $x \circ y = 0$. Therefore, $\psi_{\text{new}}(x, y) = 0$.

(c) If $(x, y) = (0, 0)$, it is easy to know $\nabla_x \psi_0(0, 0) = \nabla_y \psi_0(0, 0) = 0$ by Lemma 3.38. Hence $\nabla_x \psi_{\text{new}}(0, 0) = \nabla_y \psi_{\text{new}}(0, 0) = 0$. If $(x, y) \neq (0, 0)$ and $x^2 + y^2 \in \text{int}(\mathcal{K}^n)$, then the results follow by Proposition 3.4(b) and Lemma 3.38. If $(x, y) \neq (0, 0)$ and $x^2 + y^2 \notin \text{int}(\mathcal{K}^n)$, then by applying Lemma 3.2, we have

$$\begin{aligned} x \circ y &= (\langle x, y\rangle\ ,\ x_1 y_2 + y_1 x_2) \\ &= (x_1 y_1 + x_2^T y_2\ ,\ x_1 y_2 + y_1 x_2) \\ &= (2x_1 y_1\ ,\ 2x_1 y_2) \\ &= 2x_1 y. \end{aligned}$$

Therefore,

$$L_y \cdot (x \circ y)_+ = L_y \cdot (2x_1 y)_+ = 2|x_1|\, L_y \cdot (y)_+ = 2|x_1| \cdot (y \circ (y)_+) = 2|x_1| \cdot (y)_+^2,$$

where the last equality is due to

$$y \circ y_+ = [(y)_+ + (y)_-] \circ (y)_+ = (y)_+^2 + (y)_- \circ (y)_+ = (y)_+^2.$$

Similarly, we have $L_x \cdot (x \circ y)_+ = 2|y_1| \cdot (x)_+^2$. This together with Proposition 3.4 leads to the desired results. □

Proposition 3.41 *Let f_{new} be defined as (3.187)–(3.189). Then, f_{new} is smooth with $f_{\text{new}}(\zeta) \geq 0$ for all $\zeta \in \mathbb{R}^n$ and $f_{\text{new}}(\zeta) = 0$ if and only if ζ solves the SOCCP. Moreover, suppose that the SOCCP* (3.1) *has at least one solution. Then, ζ is a global minimization of f_{new} if and only if ζ solves the SOCCP* (3.1).

Proof The results follow Proposition 3.40 and definition of f_{new}. □

The error bound is an important concept that indicates how close an arbitrary point is to the solution set of the SOCCP (3.1). Thus, an error bound may be used to provide stopping criterion for an iterative method. As below, we establish a proposition about the error bound of f_{new} given as (3.187)–(3.189). We need the next technical lemma to prove the error bound property.

Proposition 3.42 *Suppose that F is strongly monotone mapping from $\mathbb{R}^n$ to $\mathbb{R}^n$. Also, suppose that the SOCCP* (3.1) *has a solution ζ^*. Then, there exists a scalar $\tau > 0$ such that*

$$\tau \|\zeta - \zeta^*\|^2 \leq \|(F(\zeta) \circ \zeta)_+\| + \|(-F(\zeta))_+\| + \|(-\zeta)_+\| \quad \forall \zeta \in \mathbb{R}^n. \tag{3.191}$$

Moreover,

$$\tau \|\zeta - \zeta^*\|^2 \leq \sqrt{2}\left(\frac{1}{\alpha} + 2\right) f_{\text{new}}(\zeta)^{1/2} \quad \forall \zeta \in \mathbb{R}^n, \tag{3.192}$$

where $\alpha > 0$, and f_{new} is given by (3.187)–(3.189).

Proof Since F is strongly monotone, there exists a scalar $\rho > 0$ such that, for any $\zeta \in \mathbb{R}^n$,

$$\begin{aligned}
&\rho \|\zeta - \zeta^*\|^2 \\
\leq\, &\langle F(\zeta) - F(\zeta^*),\ \zeta - \zeta^* \rangle \\
=\, &\langle F(\zeta), \zeta \rangle + \langle -F(\zeta), \zeta^* \rangle + \langle F(\zeta^*), -\zeta \rangle \\
\leq\, &\langle F(\zeta), \zeta \rangle + \langle (-F(\zeta))_+, \zeta^* \rangle + \langle F(\zeta^*), (-\zeta)_+ \rangle \\
\leq\, &\langle F(\zeta), \zeta \rangle + \|(-F(\zeta))_+\|\, \|\zeta^*\| + \|F(\zeta^*)\|\, \|(-\zeta)_+\| \\
\leq\, &\sqrt{2}\|(F(\zeta) \circ \zeta)_+\| + \|(-F(\zeta))_+\|\, \|\zeta^*\| + \|F(\zeta^*)\|\, \|(-\zeta)_+\| \\
\leq\, &\max\{\sqrt{2}, \|F(\zeta^*)\|, \|\zeta^*\|\}\Big(\|(F(\zeta) \circ \zeta)_+\| + \|(-F(\zeta))_+\| + \|(-\zeta)_+\| \Big),
\end{aligned}$$

where the second inequality uses Lemma 1.1(b) while the fourth inequality is from (3.18). Then, setting $\tau := \dfrac{\rho}{\max\{\sqrt{2}, \|F(\zeta^*)\|, \|\zeta^*\|\}}$ yields (3.191).

Moreover, we have

$$\|(F(\zeta) \circ \zeta)_+\| = \sqrt{2}\, \psi_0(F(\zeta), \zeta)^{1/2} \le \frac{\sqrt{2}}{\alpha} f_{\text{new}}(\zeta)^{1/2},$$

and

$$\begin{aligned} \|(-F(\zeta))_+\| + \|(-\zeta)_+\| &\le \sqrt{2}\left(\|(-F(\zeta))_+\|^2 + \|(-\zeta)_+\|^2\right)^{1/2} \\ &\le 2\sqrt{2}\, \psi_{\text{FB}}(F(\zeta), \zeta)^{1/2} \\ &\le 2\sqrt{2}\, f_{\text{new}}(\zeta)^{1/2}, \end{aligned}$$

where the second inequality is true by Lemma 3.7. Thus,

$$\|(F(\zeta) \circ \zeta)_+\| + \|(-F(\zeta))_+\| + \|(-\zeta)_+\| \le \sqrt{2}\left(\frac{1}{\alpha} + 2\right) f_{\text{new}}(\zeta)^{1/2}.$$

This together with (3.191) yield (3.192). □

The boundedness of level sets of a merit function is also important since it ensures that the sequence generated by a descent method has at least one accumulation. The following proposition gives conditions under which f_{new} has bounded level sets. Similar properties based on other slightly modified merit functions of ψ_{YF} and ψ_{new} can be found in [44].

Proposition 3.43 *Suppose that F is a monotone mapping from $\mathbb{R}^n$ to $\mathbb{R}^n$ and that SOCCP is strictly feasible, i.e., there exists $\hat{\zeta} \in \mathbb{R}^n$ such that $F(\hat{\zeta}), \hat{\zeta} \in \text{int}(\mathcal{K}^n)$. Then, the level set*

$$\mathcal{L}(\gamma) := \{\zeta \in \mathbb{R}^n \mid f_{\text{new}}(\zeta) \le \gamma\}$$

is bounded for all $\gamma \ge 0$, where f_{new} is given by (3.187)–(3.189) with $\alpha > 0$.

Proof We will prove this result by contradiction. Suppose there exists an unbounded sequence $\{\zeta^k\} \subset \mathcal{L}(\gamma)$ for some $\gamma \ge 0$. It can be seen that the sequence of the smaller spectral values of $\{\zeta^k\}$ and $\{F(\zeta^k)\}$ is bounded below. In fact, if not, it follows form Lemma 3.8(a) that $f_{\text{new}}(\zeta^k) \to \infty$, which contradicts $\{\zeta^k\} \subset \mathcal{L}(\gamma)$. Therefore, the unboundedness of $\{\zeta^k\}$ leads to the sequence of the bigger spectral values of $\{\zeta^k\}$ tends to infinity. Now, let $\hat{\zeta}$ be a strictly feasible solution of the SOCCP. Since F is monotone, we have

$$\langle F(\zeta^k) - F(\hat{\zeta})\, ,\ \zeta^k - \hat{\zeta}\rangle \ge 0,$$

which yields

$$\left\langle F(\zeta^k), \hat{\zeta} \right\rangle + \left\langle F(\hat{\zeta}), \zeta^k \right\rangle \le \left\langle F(\zeta^k), \zeta^k \right\rangle + \left\langle F(\hat{\zeta}), \hat{\zeta} \right\rangle. \tag{3.193}$$

Then, by Lemma 3.8(b) and $F(\hat{\zeta}), \hat{\zeta} \in \text{int}(\mathcal{K}^n)$, we obtain $\langle F(\zeta^k), \hat{\zeta} \rangle + \langle F(\hat{\zeta}), \zeta^k \rangle \to \infty$, which together with (3.193) lead to $\left\langle F(\zeta^k), \zeta^k \right\rangle \to \infty$. Thus, by Lemma 3.6(c) and (3.188)–(3.189), we have

$$\|(F(\zeta^k) \circ \zeta^k)_+\| \to \infty \quad \Longrightarrow \quad \psi_{\text{new}}(F(\zeta^k), \zeta^k) \to \infty \quad \Longrightarrow \quad f_{\text{new}}(\zeta^k) \to \infty.$$

But, this contradicts $\{\zeta^k\} \subset \mathcal{L}(\gamma)$. Therefore, the proof is complete. □

B. LT type of merit functions.

Next, we study another two classes of merit functions for the SOCCP (3.4). The first class is

$$f_{\text{LT}}(\zeta) := \psi_0(\langle F(\zeta), G(\zeta) \rangle) + \psi(F(\zeta), G(\zeta)), \tag{3.194}$$

where $\psi_0 : \mathbb{R} \to \mathbb{R}_+$ satisfies

$$\psi_0(t) = 0 \quad \forall t \le 0 \quad \text{and} \quad \psi_0'(t) > 0 \quad \forall t > 0, \tag{3.195}$$

and $\psi : \mathbb{R}^n \times \mathbb{R}^n \to \mathbb{R}_+$ satisfies

$$\psi(x, y) = 0, \ \langle x, y \rangle \le 0 \quad \Longleftrightarrow \quad (x, y) \in \mathcal{K}^n \times \mathcal{K}^n, \ \langle x, y \rangle = 0. \tag{3.196}$$

The function f_{LT} was proposed by Luo and Tseng for the NCP case in [33] and was extended to the SDCP case by Tseng in [17]. In addition, we make a slight modification of f_{LT} which forms another class of merit function as below.

$$\widehat{f_{\text{LT}}}(\zeta) := \psi_0^*(F(\zeta) \circ G(\zeta)) + \psi(F(\zeta), G(\zeta)), \tag{3.197}$$

where $\psi_0^* : \mathbb{R}^n \to \mathbb{R}_+$ is given as

$$\psi_0^*(w) = \frac{1}{2}\|(w)_+\|^2. \tag{3.198}$$

and $\psi : \mathbb{R}^n \times \mathbb{R}^n \to \mathbb{R}_+$ satisfies (3.196). We notice that ψ_0^* possesses the following property:

$$\psi_0^*(w) = 0 \quad \Longleftrightarrow \quad w \preceq_{\mathcal{K}^n} 0,$$

which is a similar feature to (3.195) in some sense. Examples of ψ_0 and ψ will be given later. The second class of merit functions for the SDCP case was recently studied in [45] and a variant of $\widehat{f_{\text{LT}}}$ was also studied by the author in [46].

We will show that both f_{LT} and $\widehat{f_{\text{LT}}}$ provide global error bound (Propositions 3.48 and 3.49), which plays an important role in analyzing the convergence rate of some iterative methods for solving the SOCCP, if F and G are jointly strongly monotone. We will also prove that if F and G are jointly monotone and a strictly feasible solution exists, then both f_{LT} and $\widehat{f_{\text{LT}}}$ have bounded level sets (Propositions 3.50 and 3.51), which will ensure that the sequence generated by a descent algorithm has at least an accumulation point. All these properties will make it possible to construct a descent algorithm for solving the equivalent unconstrained reformulation of the SOCCP. In contrast, the merit function induced by ψ_{FB} lacks these properties. In addition, we will show that both f_{LT} and $\widehat{f_{\text{LT}}}$ are differentiable and their gradients have computable formulas. All the aforementioned features are significant reasons for choosing and studying these new merit functions.

First, we notice that ψ_0 is differentiable and strictly increasing on $[0, \infty)$. An example of ψ_0 is $\psi_0(t) = \frac{1}{4}(\max\{0, t\})^4$. Let Ψ_+ (we adopt the notation used as in [17]) denote the collection of $\psi : \mathbb{R}^n \times \mathbb{R}^n \to \mathbb{R}_+$ satisfying (3.196) that are differentiable and satisfy the following conditions:

$$\begin{cases} \langle \nabla_x \psi(x, y), \nabla_y \psi(x, y) \rangle \geq 0, & \forall (x, y) \in \mathbb{R}^n \times \mathbb{R}^n. \\ \langle x, \nabla_x \psi(x, y) \rangle + \langle y, \nabla_y \psi(x, y) \rangle \geq 0, & \forall (x, y) \in \mathbb{R}^n \times \mathbb{R}^n. \end{cases} \tag{3.199}$$

We will give an example of ψ belonging to Ψ_+ in Proposition 3.44. Before that, we need couple technical lemmas, which will be used for proving Propositions 3.44 and 3.45.

Proposition 3.44 *Let $\psi_1 : \mathbb{R}^n \times \mathbb{R}^n \to \mathbb{R}_+$ be given by*

$$\psi_1(x, y) := \frac{1}{2}\Big(\|(-x)_+\|^2 + \|(-y)_+\|^2 \Big). \tag{3.200}$$

Then, the following results hold.

(a) ψ_1 satisfies (3.196).
(b) ψ_1 is convex and differentiable at every $(x, y) \in \mathbb{R}^n \times \mathbb{R}^n$ with $\nabla_x \psi_1(x, y) = (x)_-$ and $\nabla_y \psi_1(x, y) = (y)_-$.
(c) For every $(x, y) \in \mathbb{R}^n \times \mathbb{R}^n$, we have

$$\langle \nabla_x \psi_1(x, y), \nabla_y \psi_1(x, y) \rangle \geq 0.$$

(d) For every $(x, y) \in \mathbb{R}^n \times \mathbb{R}^n$, we have

$$\langle x, \nabla_x \psi_1(x, y) \rangle + \langle y, \nabla_y \psi_1(x, y) \rangle = \|(x)_-\|^2 + \|(y)_-\|^2.$$

(e) ψ_1 belongs to Ψ_+.

Proof (a) Suppose $\psi_1(x, y) = 0$ and $\langle x, y \rangle \leq 0$. Then by definition of ψ_1 as (3.200), we have $(-x)_+ = 0,\ (-y)_+ = 0$ which implies $x \in \mathcal{K}^n$, $y \in \mathcal{K}^n$. Since $\mathcal{K}^n$ is self-dual, $x, y \in \mathcal{K}^n$ leads to $\langle x, y \rangle \geq 0$ by (3.17). This together with $\langle x, y \rangle \leq 0$ yields

$\langle x, y\rangle = 0$. The other direction is clear from the above arguments. Hence, we proved that ψ_1 satisfies (3.196).

(b) For any $x \in \mathbb{R}^n$, we have the decomposition $x = (x)_+ + (x)_- = (x)_+ - (-x)_+$. Hence,

$$\frac{1}{2}\|(-x)_+\|^2 = \frac{1}{2}\|(x)_+ - x\|^2 = \min_{w\in\mathcal{K}^n} \frac{1}{2}\|w - x\|^2,$$

which is convex and differentiable in x (see [47, page 255]). Moreover, the chain rule gives

$$\nabla_x\left[\frac{1}{2}\|(-x)_+\|^2\right] = -(-x)_+ = (x)_-.$$

Similar formula holds for y. Thus, ψ_1 is convex and differentiable at every $(x, y) \in \mathbb{R}^n \times \mathbb{R}^n$ with $\nabla_x\psi_1(x, y) = -(-x)_+ = (x)_-$ and $\nabla_y\psi_1(x, y) = -(-y)_+ = (y)_-$.

(c) From part(b), we have

$$\langle\nabla_x\psi_1(x, y), \nabla_y\psi_1(x, y)\rangle = \langle(x)_-, (y)_-\rangle = \langle(-x)_+, (-y)_+\rangle \geq 0,$$

where the inequality is true by (3.17).

(d) By applying Lemma 3.6(a), we obtain

$$\langle x, \nabla_x\psi_1(x, y)\rangle = \langle x, (x)_-\rangle = \|(x)_-\|^2.$$

Similarly, $\langle y, \nabla_x\psi_1(x, y)\rangle = \|(y)_-\|^2$ and hence the desired result holds.

(e) This is an immediate consequence of (a) through (d). □

Next, we consider a further restriction on ψ. Let Ψ_{++} denote the collection of $\psi \in \Psi_+$ satisfying the following conditions:

$$\psi(x, y) = 0 \quad \forall (x, y) \in \mathbb{R}^n \times \mathbb{R}^n \quad \text{whenever} \quad \langle\nabla_x\psi(x, y), \nabla_y\psi(x, y)\rangle = 0. \tag{3.201}$$

We notice that the ψ_1 defined as (3.200) in Proposition 3.44 does not belong to Ψ_{++}. An example of such ψ belonging to Ψ_{++} is given in Proposition 3.45.

Proposition 3.45 *Let $\psi_2 : \mathbb{R}^n \times \mathbb{R}^n \to \mathbb{R}_+$ be given by*

$$\psi_2(x, y) := \frac{1}{2}\|\phi_{\text{FB}}(x, y)_+\|^2, \tag{3.202}$$

where ϕ_{FB} is defined as (3.10). Then, the following results hold.

(a) ψ_2 satisfies (3.196).
(b) ψ_2 is differentiable at every $(x, y) \in \mathbb{R}^n \times \mathbb{R}^n$ Moreover, $\nabla_x\psi_2(0, 0) = \nabla_y\psi_2(0, 0) = 0$. If $(x, y) \neq (0, 0)$ and $x^2 + y^2 \in \text{int}(\mathcal{K}^n)$, then

$$\nabla_x \psi_2(x, y) = \left(L_x L^{-1}_{(x^2+y^2)^{1/2}} - I \right) \phi_{\text{FB}}(x, y)_+,$$
$$\nabla_y \psi_2(x, y) = \left(L_y L^{-1}_{(x^2+y^2)^{1/2}} - I \right) \phi_{\text{FB}}(x, y)_+.$$

If $(x, y) \neq (0, 0)$ *and* $x^2 + y^2 \notin \text{int}(\mathcal{K}^n)$*, then* $x_1^2 + y_1^2 \neq 0$ *and*

$$\nabla_x \psi_2(x, y) = \left(\frac{x_1}{\sqrt{x_1^2 + y_1^2}} - 1 \right) \phi_{\text{FB}}(x, y)_+,$$
$$\nabla_y \psi_2(x, y) = \left(\frac{y_1}{\sqrt{x_1^2 + y_1^2}} - 1 \right) \phi_{\text{FB}}(x, y)_+. \tag{3.203}$$

(c) *For every* $(x, y) \in \mathbb{R}^n \times \mathbb{R}^n$*, we have*

$$\langle \nabla_x \psi_2(x, y), \nabla_y \psi_2(x, y) \rangle \geq 0,$$

and the equality holds whenever $\psi_2(x, y) = 0$.

(d) *For every* $(x, y) \in \mathbb{R}^n \times \mathbb{R}^n$*, we have*

$$\langle x, \nabla_x \psi_2(x, y) \rangle + \langle y, \nabla_y \psi_2(x, y) \rangle = \|\phi_{\text{FB}}(x, y)_+\|^2.$$

(e) ψ_2 *belongs to* Ψ_{++}.

Proof (a) Suppose $\psi_2(x, y) = 0$ and $\langle x, y \rangle \leq 0$. Let $z := -\phi_{\text{FB}}(x, y)$. Then $(-z)_+ = \phi_{\text{FB}}(x, y)_+ = 0$ which says $z \in \mathcal{K}^n$. Since $x + y = (x^2 + y^2)^{1/2} + z$, squaring both sides and simplifying yield

$$2(x \circ y) = 2\left((x^2 + y^2)^{1/2} \circ z \right) + z^2.$$

Now, taking trace of both sides and using the fact $tr(x \circ y) = 2\langle x, y \rangle$, we obtain

$$4\langle x, y \rangle = 4\langle (x^2 + y^2)^{1/2}, z \rangle + 2\|z\|^2. \tag{3.204}$$

Since $(x^2 + y^2)^{1/2} \in \mathcal{K}^n$ and $z \in \mathcal{K}^n$, then we know $\langle (x^2 + y^2)^{1/2}, z \rangle \geq 0$ by Lemma 3.6(b). Thus, the right-hand side of (3.204) is nonnegative, which togethers with $\langle x, y \rangle \leq 0$ implies $\langle x, y \rangle = 0$. Therefore, with this, the Eq. (3.204) says $z = 0$ which is equivalent to $\phi_{\text{FB}}(x, y) = 0$. Then by Proposition 3.2, we have $x, y \in \mathcal{K}^n$. Conversely, if $x, y \in \mathcal{K}^n$ and $\langle x, y \rangle = 0$, then again Proposition 3.2 yields $\phi_{\text{FB}}(x, y) = 0$. Thus, $\psi_2(x, y) = 0$ and $\langle x, y \rangle \leq 0$.

(b) For the proof of part(b), we need to discuss three cases.

Case (1): If $(x, y) = (0, 0)$, then for any $h, k \in \mathbb{R}^n$, let $\mu_1 \leq \mu_2$ be the spectral values and let $v^{(1)}, v^{(2)}$ be the corresponding spectral vectors of $h^2 + k^2$. Hence, by Lemma 3.1(b),

$$\begin{aligned} \|(h^2 + k^2)^{1/2} - h - k\| &= \|\sqrt{\mu_1}v^{(1)} + \sqrt{\mu_2}v^{(2)} - h - k\| \\ &\leq \sqrt{\mu_1}\|v^{(1)}\| + \sqrt{\mu_2}\|v^{(2)}\| + \|h\| + \|k\| \\ &= (\sqrt{\mu_1} + \sqrt{\mu_2})/\sqrt{2} + \|h\| + \|k\|. \end{aligned}$$

Also

$$\begin{aligned} \mu_1 \leq \mu_2 &= \|h\|^2 + \|k\|^2 + 2\|h_1 h_2 + k_1 k_2\| \\ &\leq \|h\|^2 + \|k\|^2 + 2|h_1|\|h_2\| + 2|k_1|\|k_2\| \\ &\leq 2\|h\|^2 + 2\|k\|^2. \end{aligned}$$

Combining the above two inequalities yields

$$\begin{aligned} \psi_2(h, k) - \psi_2(0, 0) &= \frac{1}{2}\|\phi_{\text{FB}}(h, k)_+\|^2 \\ &\leq \|\phi_{\text{FB}}(h, k)\|^2 \\ &= \|(h^2 + k^2)^{1/2} - h - k\|^2 \\ &\leq \left((\sqrt{\mu_1} + \sqrt{\mu_2})/\sqrt{2} + \|h\| + \|k\|\right)^2 \\ &\leq \left(2\sqrt{2\|h\|^2 + 2\|k\|^2}/\sqrt{2} + \|h\| + \|k\|\right)^2 \\ &= O(\|h\|^2 + \|k\|^2), \end{aligned}$$

where the first inequality is from Lemma 3.7. This shows that ψ_2 is differentiable at $(0, 0)$ with

$$\nabla_x \psi_2(0, 0) = \nabla_y \psi_2(0, 0) = 0.$$

Case (2): If $(x, y) \neq (0, 0)$ and $x^2 + y^2 \in \text{int}(\mathcal{K}^n)$, let z be factored as $z = \lambda_1 u^{(1)} + \lambda_2 u^{(2)}$ for any $z \in \mathbb{R}^n$. Now, let $g : \mathbb{R}^n \to \mathbb{R}^n$ be defined as

$$g(z) := \frac{1}{2}((z)_+)^2 = \hat{g}(\lambda_1)u^{(1)} + \hat{g}(\lambda_2)u^{(2)},$$

where $\hat{g} : \mathbb{R} \to \mathbb{R}$ is given by $\hat{g}(\lambda) := \frac{1}{2}(\max(0, \lambda))^2$. From the continuous differentiability of $\hat{g}$ and Prop. 5.2 of [26], the vector-valued function g is also continuously differentiable. Hence, the first component $g_1(z) = \frac{1}{2}\|(z)_+\|^2$ of $g(z)$ is continuously differentiable as well. By an easy computation, we have $\nabla g_1(z) = (z)_+$.

Since $\psi_2(x, y) = g_1(\phi_{\rm FB}(x, y))$ and $\phi_{\rm FB}$ is differentiable at $(x, y) \neq (0, 0)$ with $x^2 + y^2 \in \mathrm{int}(\mathcal{K}^n)$ (see [12, Corollary 5.2]). Hence, the chain rule yields

$$\nabla_x \psi_2(x, y) = \nabla_x \phi_{\rm FB}(x, y) \nabla g_1(\phi_{\rm FB}(x, y)) = \left(L_x L^{-1}_{(x^2+y^2)^{1/2}} - I \right) \phi_{\rm FB}(x, y)_+,$$

$$\nabla_y \psi_2(x, y) = \nabla_y \phi_{\rm FB}(x, y) \nabla g_1(\phi_{\rm FB}(x, y)) = \left(L_y L^{-1}_{(x^2+y^2)^{1/2}} - I \right) \phi_{\rm FB}(x, y)_+.$$

Case (3): If $(x, y) \neq (0, 0)$ and $x^2 + y^2 \notin \mathrm{int}(\mathcal{K}^n)$, by direct computation, we know $\|x\|^2 + \|y\|^2 = 2\|x_1 x_2 + y_1 y_2\|$ under this case. Since $(x, y) \neq (0, 0)$, this also implies $x_1 x_2 + y_1 y_2 \neq 0$. We notice that we can not apply the chain rule as in case(2) since $\phi_{\rm FB}$ is no longer differentiable at such (x, y) of case(3). By the spectral factorization, we observe that

$$\begin{aligned} \phi_{\rm FB}(x, y)_+ = \phi_{\rm FB}(x, y) &\Longleftrightarrow \phi_{\rm FB}(x, y) \in \mathcal{K}^n \\ \phi_{\rm FB}(x, y)_+ = 0 &\Longleftrightarrow \phi_{\rm FB}(x, y) \in -\mathcal{K}^n \\ \phi_{\rm FB}(x, y)_+ = \lambda_2 u^{(2)} &\Longleftrightarrow \phi_{\rm FB}(x, y) \notin \mathcal{K}^n \cup -\mathcal{K}^n, \end{aligned} \tag{3.205}$$

where λ_2 is the bigger spectral value of $\phi_{\rm FB}(x, y)$ and $u^{(2)}$ is the corresponding spectral vector. Indeed, by applying Lemma 3.2, under this case, we have (as in [14, Eq. (26)])

$$\phi_{\rm FB}(x, y) = \left(\sqrt{x_1^2 + y_1^2} - (x_1 + y_1), \frac{x_1 x_2 + y_1 y_2}{\sqrt{x_1^2 + y_1^2}} - (x_2 + y_2) \right). \tag{3.206}$$

Therefore, λ_2 and $u^{(2)}$ are given as below:

$$\begin{aligned} \lambda_2 &= \sqrt{x_1^2 + y_1^2} - (x_1 + y_1) + \|w_2\|, \\ u^{(2)} &= \frac{1}{2}\left(1, \frac{w_2}{\|w_2\|}\right), \end{aligned} \tag{3.207}$$

where $w_2 = \frac{x_1 x_2 + y_1 y_2}{\sqrt{x_1^2 + y_1^2}} - (x_2 + y_2)$. To prove the differentiability of ψ_2 under this case, we shall discuss the following three subcases according to the above observation (3.205).

(i) If $\phi_{\rm FB}(x, y) \notin \mathcal{K}^n \cup -\mathcal{K}^n$ then $\phi_{\rm FB}(x, y)_+ = \lambda_2 u^{(2)}$ where λ_2 and $u^{(2)}$ are given as in (3.207). From the fact that $\|u^{(2)}\| = \frac{1}{\sqrt{2}}$, we obtain

$$\begin{aligned} \psi_2(x, y) &= \frac{1}{2}\|\phi_{\rm FB}(x, y)_+\|^2 = \frac{1}{4}\lambda_2^2 \\ &= \frac{1}{4}\left[\left(\sqrt{x_1^2 + y_1^2} - (x_1 + y_1) \right)^2 + 2\left(\sqrt{x_1^2 + y_1^2} - (x_1 + y_1) \right) \cdot \|w_2\| + \|w_2\|^2 \right]. \end{aligned}$$

Since $(x, y) \neq (0, 0)$ in this case, ψ_2 is differentiable clearly. Moreover, using the product rule and chain rule for differentiation, the derivative of ψ_2 with respect to x_1 works out to be

$$\begin{aligned}
\frac{\partial}{\partial x_1}\psi_2(x, y) &= \frac{1}{4}\Bigg[2\Big(\sqrt{x_1^2 + y_1^2} - (x_1 + y_1)\Big)\Big(\frac{x_1}{\sqrt{x_1^2 + y_1^2}} - 1\Big) + 2\Big(\frac{x_1}{\sqrt{x_1^2 + y_1^2}} - 1\Big)\|w_2\| \\
&\quad + 2\Big(\sqrt{x_1^2 + y_1^2} - (x_1 + y_1)\Big) \cdot \frac{w_2^T \nabla_{x_1} w_2}{\|w_2\|} + 2w_2^T \nabla_{x_1} w_2\Bigg] \\
&= \frac{1}{2}\left[\Big(\frac{x_1}{\sqrt{x_1^2 + y_1^2}} - 1\Big)\Big(\sqrt{x_1^2 + y_1^2} - (x_1 + y_1) + \|w_2\|\Big)\right].
\end{aligned}$$

The last equality of the above expression is true because of

$$\begin{aligned}
\nabla_{x_1} w_2 &= \frac{x_2 \cdot \sqrt{x_1^2 + y_1^2} - (x_1 x_2 + y_1 y_2) \cdot \frac{x_1}{\sqrt{x_1^2+y_1^2}}}{(x_1^2 + y_1^2)} \\
&= \frac{\frac{1}{\sqrt{x_1^2+y_1^2}}\left[x_2(x_1^2 + y_1^2) - (x_1^2 x_2 + x_1 y_1 y_2)\right]}{(x_1^2 + y_1^2)} \\
&= \frac{x_1^2 x_2 + y_1^2 x_2 - x_1^2 x_2 - x_1 y_1 y_2}{(\sqrt{x_1^2 + y_1^2})^3} \\
&= 0,
\end{aligned}$$

where the last equality holds by Lemma 3.2. Similarly, the gradient of ψ_2 with respect to x_2 works out to be

$$\begin{aligned}
\nabla_{x_2}\psi_2(x, y) &= \frac{1}{4}\left[2\Big(\sqrt{x_1^2 + y_1^2} - (x_1 + y_1)\Big)\frac{\nabla_{x_2} w_2 \cdot w_2}{\|w_2\|} + 2\nabla_{x_2} w_2 \cdot w_2\right] \\
&= \frac{1}{2}\left[\Big(\sqrt{x_1^2 + y_1^2} - (x_1 + y_1)\Big)\Big(\frac{x_1}{\sqrt{x_1^2 + y_1^2}} - 1\Big)\frac{w_2}{\|w_2\|} + \Big(\frac{x_1}{\sqrt{x_1^2 + y_1^2}} - 1\Big)w_2\right] \\
&= \frac{1}{2}\left[\Big(\frac{x_1}{\sqrt{x_1^2 + y_1^2}} - 1\Big)\Big(\sqrt{x_1^2 + y_1^2} - (x_1 + y_1) + \|w_2\|\Big)\frac{w_2}{\|w_2\|}\right].
\end{aligned}$$

Then, we can rewrite $\nabla_x \psi_2(x, y)$ as

$$\begin{aligned}
\nabla_x \psi_2(x, y) &= \begin{bmatrix} \frac{\partial}{\partial x_1}\psi_2(x, y) \\ \nabla_{x_2}\psi_2(x, y) \end{bmatrix} \\
&:= \begin{bmatrix} \Xi_1 \\ \Xi_2 \end{bmatrix}
\end{aligned}$$

$$= \left(\frac{x_1}{\sqrt{x_1^2 + y_1^2}} - 1 \right) \lambda_2 u^{(2)}$$

$$= \left(\frac{x_1}{\sqrt{x_1^2 + y_1^2}} - 1 \right) \phi_{\rm FB}(x, y)_+, \tag{3.208}$$

where

$$\Xi_1 := \frac{1}{2} \left(\frac{x_1}{\sqrt{x_1^2 + y_1^2}} - 1 \right) \left(\sqrt{x_1^2 + y_1^2} - (x_1 + y_1) + \|w_2\| \right) \in \mathbb{R}$$

$$\Xi_2 := \frac{1}{2} \left(\frac{x_1}{\sqrt{x_1^2 + y_1^2}} - 1 \right) \left(\sqrt{x_1^2 + y_1^2} - (x_1 + y_1) + \|w_2\| \right) \frac{w_2}{\|w_2\|} \in \mathbb{R}^{n-1}.$$

(ii) If $\phi_{\rm FB}(x, y) \in \mathcal{K}^n$ then $\phi_{\rm FB}(x, y)_+ = \phi_{\rm FB}(x, y)$ and hence $\psi_2(x, y) = \frac{1}{2}\|\phi_{\rm FB}(x, y)_+\|^2 = \frac{1}{2}\|\phi_{\rm FB}(x, y)\|^2$. Thus, by [14, Prop. 3.1(b)], we know that the gradient of ψ_2 under this subcase is as below:

$$\nabla_x \psi_2(x, y) = \left(\frac{x_1}{\sqrt{x_1^2 + y_1^2}} - 1 \right) \phi_{\rm FB}(x, y) = \left(\frac{x_1}{\sqrt{x_1^2 + y_1^2}} - 1 \right) \phi_{\rm FB}(x, y)_+$$
$$\tag{3.209}$$
$$\nabla_y \psi_2(x, y) = \left(\frac{y_1}{\sqrt{x_1^2 + y_1^2}} - 1 \right) \phi_{\rm FB}(x, y) = \left(\frac{y_1}{\sqrt{x_1^2 + y_1^2}} - 1 \right) \phi_{\rm FB}(x, y)_+.$$

If there is (x', y') such that $\phi_{\rm FB}(x', y') \notin \mathcal{K}^n \cup -\mathcal{K}^n$ and $\phi_{\rm FB}(x', y') \to \phi_{\rm FB}(x, y) \in \mathcal{K}^n$ (the neighborhood of point belonging to this subcase). From (3.208) and (3.209), it can be seen that

$$\nabla_x \psi_2(x', y') \to \nabla_x \psi_2(x, y), \quad \nabla_y \psi_2(x', y') \to \nabla_y \psi_2(x, y).$$

Thus, ψ_2 is differentiable under this subcase.
(iii) If $\phi_{\rm FB}(x, y) \in -\mathcal{K}^n$ then $\phi_{\rm FB}(x, y)_+ = 0$. Thus, $\psi_2(x, y) = \frac{1}{2}\|\phi_{\rm FB}(x, y)_+\|^2 = 0$ and it is clear that its gradient under this subcase is

$$\nabla_x \psi_2(x, y) = 0 = \left(\frac{x_1}{\sqrt{x_1^2 + y_1^2}} - 1 \right) \phi_{\rm FB}(x, y)_+, \tag{3.210}$$

$$\nabla_y \psi_2(x, y) = 0 = \left(\frac{y_1}{\sqrt{x_1^2 + y_1^2}} - 1 \right) \phi_{\rm FB}(x, y)_+.$$

Again, if there is (x', y') such that $\phi_{\text{FB}}(x', y') \notin \mathcal{K}^n \cup -\mathcal{K}^n$ and $\phi_{\text{FB}}(x', y') \to \phi_{\text{FB}}(x, y) \in -\mathcal{K}^n$ (the neighborhood of point belonging to this subcase). From (3.208) and (3.210), it can be seen that

$$\nabla_x \psi_2(x', y') \to 0 = \nabla_x \psi_2(x, y), \quad \nabla_y \psi_2(x', y') \to 0 = \nabla_y \psi_2(x, y).$$

Thus, ψ_2 is differentiable under this subcase.
From the above, we complete the proof of this case and therefore the proof for part(b) is done.

(c) We wish to show that $\langle \nabla_x \psi_2(x, y), \nabla_y \psi_2(x, y) \rangle \geq 0$ and the equality holds if and only if $\psi_2(x, y) = 0$. We follow the three cases as above.

Case (1): If $(x, y) = (0, 0)$, by part (b), we know $\nabla_x \psi_2(x, y) = \nabla_y \psi_2(x, y) = 0$. Therefore, the desired equality holds.

Case (2): If $(x, y) \neq (0, 0)$ and $x^2 + y^2 \in \text{int}(\mathcal{K}^n)$, by part (b), we have

$$\begin{aligned}
\langle \nabla_x \psi_2(x, y), \nabla_y \psi_2(x, y) \rangle = & \quad \langle (L_x L_z^{-1} - I)(\phi_{\text{FB}})_+, (L_y L_z^{-1} - I)(\phi_{\text{FB}})_+ \rangle \\
= & \quad \langle (L_x - L_z) L_z^{-1}(\phi_{\text{FB}})_+, (L_y - L_z) L_z^{-1}(\phi_{\text{FB}})_+ \rangle \\
= & \quad \langle (L_y - L_z)(L_x - L_z) L_z^{-1}(\phi_{\text{FB}})_+, L_z^{-1}(\phi_{\text{FB}})_+ \rangle.
\end{aligned} \tag{3.211}$$

Let S be the symmetric part of $(L_y - L_z)(L_x - L_z)$. Then

$$\begin{aligned}
S &= \frac{1}{2}\Big((L_y - L_z)(L_x - L_z) + (L_x - L_z)(L_y - L_z)\Big) \\
&= \frac{1}{2}\Big(L_x L_y + L_y L_x - L_z(L_x + L_y) - (L_x + L_y)L_z + 2L_z^2\Big) \\
&= \frac{1}{2}(L_z - L_x - L_y)^2 + \frac{1}{2}(L_z^2 - L_x^2 - L_y^2).
\end{aligned}$$

Since $z \in \mathcal{K}^n$ and $z^2 = x^2 + y^2$, Lemma 3.5 implies $L_z^2 - L_x^2 - L_y^2 \succeq O$. Then (3.211) yields

$$\begin{aligned}
&\langle \nabla_x \psi_2(x, y), \nabla_y \psi_2(x, y) \rangle \\
=& \langle S L_z^{-1}(\phi_{\text{FB}})_+, L_z^{-1}(\phi_{\text{FB}})_+ \rangle \\
=& \frac{1}{2}\langle (L_z - L_x - L_y)^2 L_z^{-1}(\phi_{\text{FB}})_+, L_z^{-1}(\phi_{\text{FB}})_+ \rangle + \frac{1}{2}\langle (L_z^2 - L_x^2 - L_y^2) L_z^{-1}(\phi_{\text{FB}})_+, L_z^{-1}(\phi_{\text{FB}})_+ \rangle \\
\geq& \frac{1}{2}\langle (L_z - L_x - L_y)^2 L_z^{-1}(\phi_{\text{FB}})_+, L_z^{-1}(\phi_{\text{FB}})_+ \rangle \\
=& \frac{1}{2}\| L_{\phi_{\text{FB}}} L_z^{-1}(\phi_{\text{FB}})_+ \|^2,
\end{aligned}$$

where the last equality uses $L_z - L_x - L_y = L_{z-x-y} = L_{\phi_{\rm FB}}$. If the equality holds, then the above relation yields $\|L_{\phi_{\rm FB}} L_z^{-1}(\phi_{\rm FB})_+\|^2 = 0$ and, by Lemma 3.1(d),

$$L_{\phi_{\rm FB}} L_z^{-1}(\phi_{\rm FB})_+ = \phi_{\rm FB} \circ (L_z^{-1}(\phi_{\rm FB})_+) = L_z^{-1}(\phi_{\rm FB})_+ \circ \phi_{\rm FB} = 0.$$

Since $z = (x^2 + y^2)^{1/2} \in \mathrm{int}(\mathcal{K}^n)$ so that $L_z^{-1} \succ O$ (see Lemma 3.1(d)), multiplying L_z^{-1} both sides gives $\phi_{\rm FB} \circ (\phi_{\rm FB})_+ = 0$. From definition of Jordan product (1.2) and Lemma 3.6(a), it implies $(\phi_{\rm FB})_+ = 0$; and hence $\psi_2 = 0$. Conversely, if $(\phi_{\rm FB})_+ = 0$, then it is clear that $\langle \nabla_x \psi_2(x, y), \nabla_y \psi_2(x, y)\rangle = 0$.

Case (3): If $(x, y) \neq (0, 0)$ and $x^2 + y^2 \notin \mathrm{int}(\mathcal{K}^n)$, by part (b), we have

$$\langle \nabla_x \psi_2(x, y), \nabla_y \psi_2(x, y)\rangle = \left(\frac{x_1}{\sqrt{x_1^2 + y_1^2}} - 1\right)\left(\frac{y_1}{\sqrt{x_1^2 + y_1^2}} - 1\right) \|\phi_{\rm FB}(x, y)_+\|^2 \geq 0.$$

If the equality holds, then either $\phi_{\rm FB}(x, y)_+ = 0$ or $\frac{x_1}{\sqrt{x_1^2+y_1^2}} = 1$ or $\frac{y_1}{\sqrt{x_1^2+y_1^2}} = 1$. In the second case, we have $y_1 = 0$ and $x_1 \geq 0$, so that Lemma 3.2 yields $y_2 = 0$ and $x_1 = \|x_2\|$. In the third case, we have $x_1 = 0$ and $y_1 \geq 0$, so that Lemma 3.2 yields $x_2 = 0$ and $y_1 = \|y_2\|$. Thus, in these two cases, we have $x \circ y = 0,\ x \in \mathcal{K}^n,\ y \in \mathcal{K}^n$. Then, by (3.196), $\psi_2(x, y) = 0$.

(d) Again, we need to discuss the three cases as below.

Case (1): If $(x, y) = (0, 0)$, by part (b), we know $\nabla_x \psi_2(x, y) = \nabla_y \psi_2(x, y) = 0$. Therefore, the desired equality holds.

Case (2): If $(x, y) \neq (0, 0)$ and $x^2 + y^2 \in \mathrm{int}(\mathcal{K}^n)$, by part (b), we have

$$\nabla_x \psi_2(x, y) = \left(L_x L_z^{-1} - I\right)\phi_{\rm FB}(x, y)_+,$$
$$\nabla_y \psi_2(x, y) = \left(L_y L_z^{-1} - I\right)\phi_{\rm FB}(x, y)_+,$$

where we let $z := (x^2 + y^2)^{1/2}$. For simplicity, we will write $\phi(x, y)_+$ as ϕ_+. Thus,

$$\begin{aligned}
\langle x, \nabla_x \psi_2(x, y)\rangle + \langle y, \nabla_y \psi_2(x, y)\rangle &= \langle x, (L_x L_z^{-1} - I)(\phi_{\rm FB})_+\rangle + \langle y, (L_y L_z^{-1} - I)(\phi_{\rm FB})_+\rangle \\
&= \langle (L_z^{-1} L_x - I)x, (\phi_{\rm FB})_+\rangle + \langle (L_z^{-1} L_y - I)y, (\phi_{\rm FB})_+\rangle \\
&= \langle L_z^{-1} L_x x + L_z^{-1} L_y y - x - y, (\phi_{\rm FB})_+\rangle \\
&= \langle L_z^{-1}(x^2 + y^2) - x - y, (\phi_{\rm FB})_+\rangle \\
&= \langle L_z^{-1} z^2 - x - y, (\phi_{\rm FB})_+\rangle \\
&= \langle z - x - y, (\phi_{\rm FB})_+\rangle \\
&= \|(\phi_{\rm FB})_+\|^2,
\end{aligned}$$

where the next-to-last equality follows from $L_z z = z^2$, so that $L_z^{-1} z^2 = z$ and the last equality is from Lemma 3.6(a).

Case (3): If $(x, y) \neq (0, 0)$ and $x^2 + y^2 \notin \text{int}(\mathcal{K}^n)$, by part(b), we have

$$\nabla_x \psi_2(x, y) = \left(\frac{x_1}{\sqrt{x_1^2 + y_1^2}} - 1 \right) \phi_{\text{FB}}(x, y)_+,$$

$$\nabla_y \psi_2(x, y) = \left(\frac{y_1}{\sqrt{x_1^2 + y_1^2}} - 1 \right) \phi_{\text{FB}}(x, y)_+.$$

Thus,

$$\begin{aligned}
&\langle x, \nabla_x \psi_2(x, y)\rangle + \langle y, \nabla_y \psi_2(x, y)\rangle \\
&= \left(\frac{x_1}{\sqrt{x_1^2 + y_1^2}} - 1 \right) \langle x, (\phi_{\text{FB}})_+\rangle + \left(\frac{y_1}{\sqrt{x_1^2 + y_1^2}} - 1 \right) \langle y, (\phi_{\text{FB}})_+\rangle \\
&= \left\langle \left(\frac{x_1}{\sqrt{x_1^2 + y_1^2}} - 1 \right) x + \left(\frac{y_1}{\sqrt{x_1^2 + y_1^2}} - 1 \right) y, (\phi_{\text{FB}})_+ \right\rangle \\
&= \left\langle \frac{x_1 x + y_1 y}{\sqrt{x_1^2 + y_1^2}} - x - y, (\phi_{\text{FB}})_+ \right\rangle \\
&= \langle \phi_{\text{FB}}, (\phi_{\text{FB}})_+\rangle \\
&= \|(\phi_{\text{FB}})_+\|^2,
\end{aligned}$$

where the next-to-last equality uses (3.206) and the last equality is from Lemma 3.6(a) again.

(e) This is an immediate consequence of (a) through (d). □

We notice that (3.203) can be rewritten as

$$\nabla_x \psi_2(x, y) = L_z^{-1} \Big[[z - x - y]_+ \Big] \circ (x - z),$$

$$\nabla_y \psi_2(x, y) = L_z^{-1} \Big[[z - x - y]_+ \Big] \circ (y - z),$$

where $z = (x^2 + y^2)^{1/2}$. This is a similar form as in [17, Lemma 7.2]. Nonetheless, (3.203) cannot be rewritten as the above form since L_z^{-1} does not exist whenever $x^2 + y^2$ is on the boundary of $\mathcal{K}^n$. The next proposition is a result which is an extension of [17, Proposition 7.1] for SDCP to the case of SOCCP. Though the ideas for arguments are similar, we present the proof for completion.

Proposition 3.46 *Let $f_{\text{LT}} : \mathbb{R}^n \to \mathbb{R}_+$ be given as (3.194) with ψ_0 satisfying (3.195) and ψ satisfying (3.196). Then, the following results hold.*

(a) For all $\zeta \in \mathbb{R}^n$, we have $f_{\text{LT}}(\zeta) \geq 0$ and $f_{\text{LT}}(\zeta) = 0$ if and only if ζ solves the SOCCP.

(b) If ψ_0, ψ and F, G are differentiable, then so is f_{LT} and

$$\begin{aligned}\nabla f_{\text{LT}}(\zeta) = {} & \psi_0'(\langle F(\zeta), G(\zeta)\rangle)\Big[\nabla F(\zeta)G(\zeta) + \nabla G(\zeta)F(\zeta)\Big] \\ & + \nabla F(\zeta)\nabla_x \psi(F(\zeta), G(\zeta)) \\ & + \nabla G(\zeta)\nabla_y \psi(F(\zeta), G(\zeta)).\end{aligned}$$

(c) Assume F, G are differentiable on $\mathbb{R}^n$ and ψ belongs to Ψ_+ (respectively, Ψ_{++}). Then, for every $\zeta \in \mathbb{R}^n$ where $\nabla G(\zeta)^{-1}\nabla F(\zeta)$ is positive definite (respectively, positive semidefinite), either (i) $f_{\text{LT}}(\zeta) = 0$ or (ii) $\nabla f_{\text{LT}}(\zeta) \neq 0$ with $\langle d(\zeta), \nabla f_{\text{LT}}(\zeta)\rangle < 0$, where

$$d(\zeta) := -(\nabla G(\zeta)^{-1})^{\mathsf{T}}\Big[\psi_0'(\langle F(\zeta), G(\zeta)\rangle)G(\zeta) + \nabla_x \psi(F(\zeta), G(\zeta))\Big].$$

Proof (a) This consequence follows from (3.194) and (3.195)–(3.196).

(b) By direct computation and chain rule, the result follows.

(c) First, we consider the case of $\psi \in \Psi_{++}$ and fix $\zeta \in \mathbb{R}^n$ where $\nabla G(\zeta)^{-1}\nabla F(\zeta)$ is positive semidefinite. Let $\alpha := \psi_0'(\langle F(\zeta), G(\zeta)\rangle)$ and drop the argument "(ζ)" for simplicity. Then

$$\begin{aligned}& \langle d, \nabla f_{\text{LT}}\rangle \\ = & \langle -(\nabla G^{-1})^T(\alpha G + \nabla_x \psi(F, G)), \nabla F(\alpha G + \nabla_x \psi(F, G)) + \nabla G(\alpha F + \nabla_y \psi(F, G))\rangle \\ = & - \langle \alpha G + \nabla_x \psi(F, G), \nabla G^{-1}\nabla F(\alpha G + \nabla_x \psi(F, G))\rangle \\ & - \langle \alpha G + \nabla_x \psi(F, G), \alpha F + \nabla_y \psi(F, G)\rangle \\ \leq & - \langle \alpha G + \nabla_x \psi(F, G), \alpha F + \nabla_y \psi(F, G)\rangle \\ = & - \alpha^2 \langle F, G\rangle - \alpha\Big(\langle F, \nabla_x \psi(F, G)\rangle + \langle G, \nabla_y \psi(F, G)\rangle\Big) - \langle \nabla_x \psi(F, G), \nabla_y \psi(F, G)\rangle \\ = & - \alpha^2 \langle F, G\rangle - \langle \nabla_x \psi(F, G), \nabla_y \psi(F, G)\rangle,\end{aligned}$$

where the first inequality holds since $\nabla G^{-1}\nabla F$ is positive semidefinite and the inequality follows from $\alpha \geq 0$ and Eq. (3.199). Now, we observe that $t\psi_0'(t) > 0$ if and only if $t > 0$ since ψ_0 is strictly increasing on $[0, \infty)$. Therefore, the first term on the right-hand side is non-positive and equals zero if $\langle F, G\rangle \leq 0$. In addition, by Eqs. (3.199) and (3.201), the second term on the right-hand side is non-positive and equals zero only if $\psi(F, G) = 0$. Thus, we have $\langle d(\zeta), \nabla f_{\text{LT}}(\zeta)\rangle \leq 0$ and the equality holds only when $\langle F(\zeta), G(\zeta)\rangle \le 0$ and $\psi(F(\zeta), G(\zeta)) = 0$, in which Eq. (3.196) implies ζ satisfies (3.4), i.e., $f_{\text{LT}}(\zeta) = 0$.

Similar arguments can be applied for the case of $\psi \in \Psi_+$ and $\nabla G(\zeta)^{-1}\nabla F(\zeta)$ being positive definite. □

Next, we further consider another class of merit functions by modifying f_{LT} a bit where ψ_0 is replaced by $\psi_0^* : \mathbb{R}^n \to \mathbb{R}_+$ given as (3.198), i.e., $\psi_0^*(w) = \frac{1}{2}\|(w)_+\|^2$. It is known that the function ψ_0^* given in (3.198) is continuously differentiable (see [47, p. 255]) with $\nabla\psi_0^*(w) = [w]_+$ (by the chain rule). In other words, we will study $\widehat{f_{\text{LT}}} : \mathbb{R}^n \to \mathbb{R}_+$ defined as (3.197)–(3.198):

$$\widehat{f_{\text{LT}}}(\zeta) := \psi_0^*(F(\zeta) \circ G(\zeta)) + \psi(F(\zeta), G(\zeta)),$$

where ψ_0^* is given as (3.198) and ψ satisfies (3.196). By imitating the steps for proving Proposition 3.46 and using Lemma 3.38 as below, we obtain Proposition 3.47, which is a result analogous to Proposition 3.46. We omit its proof.

Proposition 3.47 *Let $\widehat{f_{\text{LT}}} : \mathbb{R}^n \to \mathbb{R}_+$ be given as (3.197)–(3.198). Then, the following results hold.*

(a) For all $x \in \mathbb{R}^n$, we have $\widehat{f_{\text{LT}}}(\zeta) \ge 0$ and $\widehat{f_{\text{LT}}}(\zeta) = 0$ if and only if ζ solves the SOCCP.
(b) If ψ_0^*, ψ and F, G are differentiable, then so is $\widehat{f_{\text{LT}}}$ and

$$\begin{aligned}\nabla\widehat{f_{\text{LT}}}(\zeta) = &\Big[\nabla F(\zeta)L_{G(\zeta)} + \nabla G(\zeta)L_{F(\zeta)}\Big](F(\zeta) \circ G(\zeta))_+ \\ &+\nabla F(\zeta)\nabla_x\psi(F(\zeta), G(\zeta)) \\ &+\nabla G(\zeta)\nabla_y\psi(F(\zeta), G(\zeta)).\end{aligned}$$

We originally thought there should have parallel results to Proposition 3.46(c) for $\widehat{f_{\text{LT}}}$ and whose proofs are also similar. In other words, we wish to have the following:

Assume F, G are differentiable on $\mathbb{R}^n$ and ψ belongs to Ψ_+ (respectively, Ψ_{++}). Then, for every $\zeta \in \mathbb{R}^n$ where $\nabla G(\zeta)^{-1}\nabla F(\zeta)$ is positive definite (respectively, positive semidefinite), either (i) $\widehat{f_{\text{LT}}}(\zeta) = 0$ or (ii) $\nabla\widehat{f_{\text{LT}}}(\zeta) \ne 0$ with $\langle d(\zeta), \nabla\widehat{f_{\text{LT}}}(\zeta)\rangle < 0$, where

$$d(\zeta) := -(\nabla G(\zeta)^{-1})^{\mathrm{T}}\Big[L_{G(\zeta)} \cdot (F(\zeta) \circ G(\zeta))_+ + \nabla_x\psi(F(\zeta), G(\zeta))\Big].$$

However, we are not able to complete the arguments even though ψ_0^* is in relation to ψ_0 in certain sense. We thank a referee for pointing this out. We suspect that there needs more subtle properties of ψ_0^* to finish it.

The error bound is an important concept that indicates how close an arbitrary point is to the solution set of SOCCP. Thus, an error bound may be used to provide stopping criterion for an iterative method. As below, we establish propositions about the error bound properties of f_{LT}, $\widehat{f_{\text{LT}}}$ given as (3.194) and (3.197). We need some technical lemmas as below to prove the error bound properties.

Lemma 3.39 *Let ψ_1, ψ_2 be given as (3.200) and (3.202), respectively. Then, ψ_1 and ψ_2 satisfy the following inequality.*

$$\psi_i(x, y) \geq \alpha\left(\|(-x)_+\|^2 + \|(-y)_+\|^2 \right) \quad \forall (x, y) \in \mathbb{R}^n \times \mathbb{R}^n, \tag{3.212}$$

for some positive constant α and $i = 1, 2$.

Proof For ψ_1, it is clear by definition (3.200) where $\alpha = \frac{1}{2}$. For ψ_2, the inequality is still true, where $\alpha = \frac{1}{4}$, due to Lemma 3.7. □

Lemma 3.40 *Let ψ_0^* be given as (3.198). Then, ψ_0^* satisfies*

$$\psi_0^*(w) \geq \beta\|(w)_+\|^2 \quad \forall w \in \mathbb{R}^n,$$

for some positive constant β.

Proof It is clear by definition of ψ_0^* given as (3.198) where $\beta = \frac{1}{2}$. □

Proposition 3.48 *Let f_{LT} be given by (3.194)–(3.196) with ψ satisfying (3.212). Suppose that F and G are jointly strongly monotone mapping from $\mathbb{R}^n$ to $\mathbb{R}^n$ and SOCCP has a solution ζ^*. Then, there exists a scalar $\tau > 0$ such that*

$$\tau\|\zeta - \zeta^*\|^2 \leq \max\{0, \langle F(\zeta), G(\zeta)\rangle\} + \|(-F(\zeta))_+\| + \|(-G(\zeta))_+\| \quad \forall \zeta \in \mathbb{R}^n. \tag{3.213}$$

Moreover,

$$\tau\|\zeta - \zeta^*\|^2 \leq \psi_0^{-1}(f_{\text{LT}}(\zeta)) + \frac{\sqrt{2}}{\sqrt{\alpha}} f_{\text{LT}}(\zeta)^{1/2} \quad \forall \zeta \in \mathbb{R}^n, \tag{3.214}$$

where α is a positive constant.

Proof Since F and G are jointly strongly monotone, there exists a scalar $\rho > 0$ such that, for any $\zeta \in \mathbb{R}^n$,

$$\begin{aligned}
&\rho\|\zeta - \zeta^*\|^2 \\
\leq &\langle F(\zeta) - F(\zeta^*),\ G(\zeta) - G(\zeta^*)\rangle \\
= &\langle F(\zeta), G(\zeta)\rangle + \langle -F(\zeta), G(\zeta^*)\rangle + \langle F(\zeta^*), -G(\zeta)\rangle \\
\leq &\max\{0, \langle F(\zeta), G(\zeta)\rangle\} + \langle (-F(\zeta))_+, G(\zeta^*)\rangle + \langle F(\zeta^*), (-G(\zeta))_+\rangle \\
\leq &\max\{0, \langle F(\zeta), G(\zeta)\rangle\} + \|(-F(\zeta))_+\|\ \|G(\zeta^*)\| + \|F(\zeta^*)\|\ \|(-G(\zeta))_+\|
\end{aligned}$$

$$\leq \max\{1, \|F(\zeta^*)\|, \|G(\zeta^*)\|\}\Big(\max\{0, \langle F(\zeta), G(\zeta)\rangle\} + \|(-F(\zeta))_+\| + \|(-G(\zeta))_+\| \Big),$$

where the second inequality uses Lemma 1.1(b). Setting $\tau := \dfrac{\rho}{\max\{1, \|F(\zeta^*)\|, \|G(\zeta^*)\|\}}$ yields (3.213).

Notice that ψ_0^{-1} is well defined by (3.195), and by using that ψ_0 is strictly increasing on $[0, \infty)$, we thus have

$$\max\{0, \langle F(\zeta), G(\zeta)\rangle\} \leq \psi_0^{-1}\left(f_{\text{LT}}(\zeta)\right).$$

In addition, it is clear that

$$\psi(F(\zeta), G(\zeta)) \leq f_{\text{LT}}(\zeta).$$

Now using Lemma 3.39 and the above inequality, we obtain

$$\begin{aligned}
\|(-F(\zeta))_+\| + \|(-G(\zeta))_+\| &\leq \sqrt{2}\left(\|(-F(\zeta))_+\|^2 + \|(-G(\zeta))_+\|^2\right)^{1/2} \\
&\leq \frac{\sqrt{2}}{\sqrt{\alpha}}\, \psi(F(\zeta), G(\zeta))^{1/2} \\
&\leq \frac{\sqrt{2}}{\sqrt{\alpha}}\, f_{\text{LT}}(\zeta)^{1/2}.
\end{aligned}$$

Thus,

$$\max\{0, \langle F(\zeta), G(\zeta)\rangle\} + \|(-F(\zeta))_+\| + \|(-G(\zeta))_+\| \leq \psi_0^{-1}\left(f_{\text{LT}}(\zeta)\right) + \frac{\sqrt{2}}{\sqrt{\alpha}}\, f_{\text{LT}}(\zeta)^{1/2}.$$

This together with (3.213) yields (3.214). □

Proposition 3.49 *Let $\widehat{f_{\text{LT}}}$ be given by (3.197)–(3.198) with ψ satisfying (3.212). Suppose that F and G are jointly strongly monotone mapping from $\mathbb{R}^n$ to $\mathbb{R}^n$ and the SOCCP has a solution ζ^*. Then, there exists a scalar $\tau > 0$ such that*

$$\tau\|\zeta - \zeta^*\|^2 \leq \|(F(\zeta) \circ G(\zeta))_+\| + \|(-F(\zeta))_+\| + \|(-G(\zeta))_+\| \quad \forall \zeta \in \mathbb{R}^n. \tag{3.215}$$

Moreover,

$$\tau\|\zeta - \zeta^*\|^2 \leq \left(\frac{1}{\sqrt{\beta}} + \frac{\sqrt{2}}{\sqrt{\alpha}}\right)\widehat{f_{\text{LT}}}(\zeta)^{1/2} \quad \forall \zeta \in \mathbb{R}^n, \tag{3.216}$$

where α and β are positive constants.

Proof Since F and G are jointly strongly monotone, there exists a scalar $\rho > 0$ such that, for any $\zeta \in \mathbb{R}^n$,

$$\begin{aligned}
&\rho\|\zeta - \zeta^*\|^2 \\
\leq&\langle F(\zeta) - F(\zeta^*),\ G(\zeta) - G(\zeta^*)\rangle \\
=&\langle F(\zeta), G(\zeta)\rangle + \langle -F(\zeta), G(\zeta^*)\rangle + \langle F(\zeta^*), -G(\zeta)\rangle \\
\leq&\langle F(\zeta), G(\zeta)\rangle + \langle (-F(\zeta))_+, G(\zeta^*)\rangle + \langle F(\zeta^*), (-G(\zeta))_+\rangle \\
\leq&\langle F(\zeta), G(\zeta)\rangle + \|(-F(\zeta))_+\|\ \|G(\zeta^*)\| + \|F(\zeta^*)\|\ \|(-G(\zeta))_+\| \\
\leq&\sqrt{2}\|(F(\zeta)\circ G(\zeta))_+\| + \|(-F(\zeta))_+\|\ \|G(\zeta^*)\| + \|F(\zeta^*)\|\ \|(-G(\zeta))_+\| \\
\leq&\max\{\sqrt{2}, \|F(\zeta^*)\|, \|G(\zeta^*)\|\}\Big(\|(F(\zeta)\circ G(\zeta))_+\| + \|(-F(\zeta))_+\| + \|(-G(\zeta))_+\|\Big),
\end{aligned}$$

where the second inequality uses Lemma 1.1(b) while the fourth inequality is from (3.18). Then, setting $\tau := \dfrac{\rho}{\max\{\sqrt{2}, \|F(\zeta^*)\|, \|G(\zeta^*)\|\}}$ yields (3.215).

Moreover, by Lemma 3.40, we have

$$\|(F(\zeta)\circ G(\zeta))_+\| \leq \frac{1}{\sqrt{\beta}}\,\psi_0^*(F(\zeta)\circ G(\zeta))^{1/2} \leq \frac{1}{\sqrt{\beta}}\,\widehat{f_{\rm LT}}(\zeta)^{1/2},$$

and (as in Proposition 3.48)

$$\begin{aligned}
\|(-F(\zeta))_+\| + \|(-G(\zeta))_+\| \leq&\sqrt{2}\left(\|(-F(\zeta))_+\|^2 + \|(-G(\zeta))_+\|^2\right)^{1/2} \\
\leq&\frac{\sqrt{2}}{\sqrt{\alpha}}\,\psi(F(\zeta), G(\zeta))^{1/2} \\
\leq&\frac{\sqrt{2}}{\sqrt{\alpha}}\,\widehat{f_{\rm LT}}(\zeta)^{1/2},
\end{aligned}$$

where the second inequality is true by Lemma 3.39. Thus,

$$\|(F(\zeta)\circ G(\zeta))_+\| + \|(-F(\zeta))_+\| + \|(-G(\zeta))_+\| \leq \left(\frac{1}{\sqrt{\beta}} + \frac{\sqrt{2}}{\sqrt{\alpha}}\right)\widehat{f_{\rm LT}}(\zeta)^{1/2}.$$

This together with (3.215) yields (3.216). □

Now, we give conditions under which $f_{\rm LT}$, $\widehat{f_{\rm LT}}$ has bounded level sets in Propositions 3.50 and 3.51, respectively. We need the next lemma which is key to proving the properties of bounded level sets.

Lemma 3.41 *Let ψ_1, ψ_2 be given by (3.200) and (3.202), respectively. For any $\{(x^k, y^k)\}_{k=1}^{\infty} \subseteq \mathbb{R}^n \times \mathbb{R}^n$, let $\lambda_1^k \leq \lambda_2^k$ and $\mu_1^k \leq \mu_2^k$ denote the spectral values of x^k and y^k, respectively. Then, the following results hold.*

(a) If $\lambda_1^k \to -\infty$ or $\mu_1^k \to -\infty$, then $\psi_i(x^k, y^k) \to \infty$, for $i = 1, 2$.
(b) Suppose that $\{\lambda_1^k\}$ and $\{\mu_1^k\}$ are bounded below. If $\lambda_2^k \to \infty$ or $\mu_2^k \to \infty$, then $\langle x, x^k\rangle + \langle y, y^k\rangle \to \infty$ for any $x, y \in \mathrm{int}(\mathcal{K}^n)$.

Proof (a) For ψ_1, the proof follows by the fact that

$$2\|(-x^k)_+\|^2 = \sum_{i=1}^{2} \left(\max\{0, -\lambda_i^k\}\right)^2$$

and similarly for $\|(-y^k)_+\|^2$; see [12, Property 2.2 and Proposition 3.3].

For ψ_2, using the same fact plus Lemma 3.7 leads to the desired result.

(b) Fix any $x = (x_1, x_2), y = (y_1, y_2) \in \mathbb{R} \times \mathbb{R}^{n-1}$ with $\|x_2\| < x_1$, $\|y_2\| < y_1$. Using the spectral decomposition

$$x^k = \left(\frac{\lambda_1^k + \lambda_2^k}{2}, \frac{\lambda_2^k - \lambda_1^k}{2} w_2^k\right) \quad \text{with } \|w_2^k\| = 1,$$

we have

$$\langle x, x^k\rangle = \left(\frac{\lambda_1^k + \lambda_2^k}{2}\right) x_1 + \left(\frac{\lambda_2^k - \lambda_1^k}{2}\right) x_2^T w_2^k = \frac{\lambda_1^k}{2}(x_1 - x_2^T w_2^k) + \frac{\lambda_2^k}{2}(x_1 + x_2^T w_2^k). \tag{3.217}$$

Since $\|w_2^k\| = 1$, we have $x_1 - x_2^T w_2^k \geq x_1 - \|x_2\| > 0$ and $x_1 + x_2^T w_2^k \geq x_1 - \|x_2\| > 0$. Since $\{\lambda_1^k\}$ is bounded below, the first term on the right-hand side of (3.217) is bounded below. If $\{\lambda_2^k\} \to \infty$, then the second term on the right-hand side of (3.217) tends to infinity. Hence, $\langle x, x^k\rangle \to \infty$. A similar argument shows that $\langle y, y^k\rangle$ is bounded below. Thus, $\langle x, x^k\rangle + \langle y, y^k\rangle \to \infty$. If $\{\mu_2^k\} \to \infty$, the argument is symmetric to the one above. □

Proposition 3.50 *Let $f_{\rm LT}$ be given as (3.194)–(3.196) with ψ satisfying the condition of Lemma 3.41(a). Suppose that F, G are differentiable, jointly monotone mappings from $\mathbb{R}^n$ to $\mathbb{R}^n$ satisfying*

$$\lim_{\|\zeta\| \to \infty} \left(\|F(\zeta)\| + \|G(\zeta)\|\right) = \infty. \tag{3.218}$$

Suppose also that SOCCP is strictly feasible, i.e., there exists $\bar{\zeta} \in \mathbb{R}^n$ such that $F(\bar{\zeta}), G(\bar{\zeta}) \in \mathrm{int}(\mathcal{K}^n)$. Then, the level set

$$\mathcal{L}(\gamma) := \{\zeta \in \mathbb{R}^n \mid f_{\rm LT}(\zeta) \leq \gamma\}$$

is bounded for all $\gamma \geq 0$.

Proof Suppose there exists an unbounded sequence $\{\zeta^k\} \subseteq \mathcal{L}(\gamma)$ for some $\gamma \geq 0$. It can be seen that the sequence of the smaller spectral values of $\{F(\zeta^k)\}$ and $\{G(\zeta^k)\}$ is bounded below. In fact, if not, it follows from Lemma 3.41(a) that $\psi(F(\zeta^k), G(\zeta^k)) \to \infty$. Thus, we have $f_{\rm LT}(\zeta^k) \to \infty$, which contradicts $\{\zeta^k\} \subseteq \mathcal{L}(\gamma)$. Therefore, the unboundedness of $\{\zeta^k\}$ and (3.218) yield that the sequence of the bigger spectral values of $\{F(\zeta^k)\}$ or $\{G(\zeta^k)\}$ tends to infinity. Since F, G are jointly monotone, we have

$$\langle F(\zeta^k) - F(\bar{\zeta})\ ,\ G(\zeta^k) - G(\bar{\zeta})\rangle \geq 0,$$

which is equivalent to

$$\langle F(\zeta^k), G(\bar{\zeta})\rangle + \langle F(\bar{\zeta}), G(\zeta^k)\rangle \leq \langle F(\zeta^k), G(\zeta^k)\rangle + \langle F(\bar{\zeta}), G(\bar{\zeta})\rangle. \tag{3.219}$$

Then, by Lemma 3.41(b) and $F(\bar{\zeta}), G(\bar{\zeta}) \in \text{int}(\mathcal{K}^n)$, we obtain $\langle F(\zeta^k), G(\bar{\zeta})\rangle + \langle F(\bar{\zeta}), G(\zeta^k)\rangle \to \infty$, which together with (3.219) lead to $\langle F(\zeta^k), G(\zeta^k)\rangle \to \infty$. Thus, $f_{\rm LT}(\zeta^k) \to \infty$. But, this contradicts $\{\zeta^k\} \subseteq \mathcal{L}(\gamma)$. Hence, we proved that $\mathcal{L}(\gamma)$ is bounded. □

Proposition 3.51 *Let $\widehat{f_{\rm LT}}$ be given as (3.197)–(3.198) with ψ satisfying the condition of Lemma 3.41(a). Suppose that F, G are differentiable, jointly monotone mappings from $\mathbb{R}^n$ to $\mathbb{R}^n$ satisfying*

$$\lim_{\|\zeta\|\to\infty} \left(\|F(\zeta)\| + \|G(\zeta)\| \right) = \infty.$$

Suppose also that the SOCCP is strictly feasible, i.e., there exists $\bar{\zeta} \in \mathbb{R}^n$ such that $F(\bar{\zeta}), G(\bar{\zeta}) \in \text{int}(\mathcal{K}^n)$. Then, the level set

$$\mathcal{L}(\gamma) := \{\zeta \in \mathbb{R}^n \mid \widehat{f_{\rm LT}}(\zeta) \leq \gamma\}$$

is bounded for all $\gamma \geq 0$.

Proof The arguments are similar to those in Proposition 3.50, so we omit the proof. □

3.2 Complementarity Functions Associated with Positive Semidefinite Cone

The study of C-functions associated with the positive semidefinite cone has received considerable attention in the literature. A foundational contribution is due to Tseng [17], who investigated several important instances, including the gap function, the regularized gap function, the implicit Lagrangian function, the squared Fischer–Burmeister (FB) function, and the LT-type functions. Further developments include the work of Yamashita and Fukushima [40], who explored a variant of the LT-type C-function, and Kanzow and Nagel [48], who examined FB-type variants tailored to the positive semidefinite cone. The differential properties of these functions were subsequently analyzed by Zhang, Zhang, and Pang [49]. Notably, Sun and Sun [24] established the strong semismoothness of the FB function, while Bonnans, Pang, and Cominetti [50] discussed nonsingularity conditions for semidefinite programming based on FB-type C-functions.

In recent years, however, there has been limited advancement in the construction or extension of C-functions specifically for the positive semidefinite cone. This is largely because many developments within the broader framework of symmetric cones in the past decade naturally encompass the positive semidefinite case as a special instance. In what follows, we present a selection of representative C-functions within this setting.

Let $\mathbb{S}^n$ be the space of $n \times n$ real symmetric matrices endowed with the inner product

$$\langle X, Y\rangle := \operatorname{tr}(XY), \qquad \text{for any } X, Y \in \mathbb{S}^n,$$

where "tr" denotes the trace, that is, the sum of the diagonal entries. Then, the SDCP (standing for positive semidefinite complementarity problem) is to find a matrix $X \in \mathbb{S}^n$ such that

$$G(X) \in \mathbb{S}^n_+, \quad F(X) \in \mathbb{S}^n_+, \quad \langle G(X), F(X)\rangle = 0, \tag{3.220}$$

where $G, F : \mathbb{S}^n \to \mathbb{S}^n$ and $\mathbb{S}^n_+$ represents the positive semidefinite cone. If G is the identity mapping, then (3.220) reduces to

$$X \in \mathbb{S}^n_+, \quad F(X) \in \mathbb{S}^n_+, \quad \langle X, F(X)\rangle = 0. \tag{3.221}$$

When $F : \mathbb{S}^n \to \mathbb{S}^n$ is affine, the SDCP (3.221) is called the positive semidefinite linear complementarity problem (SDLCP). Moreover, when $\mathbb{S}^n$ is restricted to the space of diagonal matrices, the SDCP (3.221) reduces to an NCP again.

Proposition 3.52 *Let the gap function $\psi_{\text{gap}} : \mathbb{S}^n \to \mathbb{R}$ be defined as*

$$\psi_{\text{gap}}(X) = \max_{Z \in \mathbb{S}^n_+} \{\langle F(X), X - Z\rangle\}. \tag{3.222}$$

Then, the function ψ_{gap} defined as in (3.222) is a C-function associated with positive semidefinite cone. In other words,

$$\psi_{\text{gap}}(X) = 0 \iff X \in \mathbb{S}^n_+, \quad F(X) \in \mathbb{S}^n_+, \quad \langle X, F(X)\rangle = 0.$$

Proof Please see Propositions 3.1 and 3.2 in [17]. □

Proposition 3.53 *For any $\alpha \in (0, \infty)$, define the regularized gap function $\psi_{\text{r-gap}} : \mathbb{S}^n \times \mathbb{S}^n \to \mathbb{R}$ as*

$$\psi_{\text{r-gap}}(X, Y) = \max_{Z \in \mathbb{S}^n_+} \left\{ \langle X, Y - Z \rangle - \frac{1}{2\alpha} \|Y - Z\|^2 \right\}. \tag{3.223}$$

Then, the function $\psi_{\text{r-gap}}$ defined as in (3.223) is a differentiable C-function associated with positive semidefinite cone. In other words,

$$\psi_{\text{r-gap}}(X, Y) = 0 \iff X \in \mathbb{S}^n_+, \quad Y \in \mathbb{S}^n_+, \quad \langle X, Y \rangle = 0.$$

Proof Please see Lemma 4.1 and Proposition 4.1 in [17]. □

Proposition 3.54 *For any $\alpha \in (0, \infty)$, define the Implicit Lagrangian function $\psi_{\text{MS}} : \mathbb{S}^n \times \mathbb{S}^n \to \mathbb{R}$ as*

$$\begin{aligned} &\psi_{\text{MS}}(X, Y) \\ = &\max_{Z_1, Z_2 \in \mathbb{S}^n_+} \left\{ \langle X, Y - Z_1 \rangle - \langle Z_2, Y \rangle - \frac{1}{2\alpha} \left(\|X - Z_2\|^2 + \|Y - Z_1\|^2 \right) \right\} \\ = &\langle X, Y \rangle + \frac{1}{2\alpha} \left(\|(X - \alpha Y)_+\|^2 - \|X\|^2 + \|(Y - \alpha X)_+\|^2 - \|Y\|^2 \right). \end{aligned} \tag{3.224}$$

Then, the function ψ_{MS} defined as in (3.224) is a differentiable C-function associated with positive semidefinite cone. In other words,

$$\psi_{\text{MS}}(X, Y) = 0 \iff X \in \mathbb{S}^n_+, \quad Y \in \mathbb{S}^n_+, \quad \langle X, Y \rangle = 0.$$

Proof Please see Lemma 5.1 and Proposition 5.1 in [17]. □

Proposition 3.55 *Let the Fischer–Burmeister function $\phi_{\text{FB}} : \mathbb{S}^n \times \mathbb{S}^n \to \mathbb{S}^n$ be defined as*

$$\phi_{\text{FB}}(X, Y) = \sqrt{X^2 + Y^2} - (X + Y), \quad \forall X, Y \in \mathbb{S}^n. \tag{3.225}$$

Its induced merit function $\psi_{\text{FB}} : \mathbb{S}^n \times \mathbb{S}^n \to \mathbb{R}$ is given by

$$\psi_{\text{FB}}(X, Y) = \frac{1}{2} \left\| \phi_{\text{FB}}(X, Y) \right\|^2. \tag{3.226}$$

Then, both function ϕ_{FB} and ψ_{FB} are C-function associated with positive semidefinite cone. In other words,

$$\psi_{\text{FB}}(X, Y) = 0 \iff \phi_{\text{FB}}(X, Y) = 0 \iff X \in \mathbb{S}^n_+, \quad Y \in \mathbb{S}^n_+, \quad \langle X, Y \rangle = 0.$$

Proof The proof is first proposed by Tseng, please see Lemma 6.1 and Proposition 6.1 in [17]. There is an alternative proof by using the simple result that for $X, Y \in \mathbb{S}^n_+$, there holds

$$XY = 0 \iff X \bullet Y = 0,$$

where $X \bullet Y = \text{tr}(XY) = \frac{1}{2}(XY + YX)$. Please refer to [48, Proposition 2.2] for more details. □

The directional derivatives, the B-subdifferential, and the generalized Jacobian of the Fischer–Burmeister function ϕ_{FB}, as defined in (3.225), were thoroughly characterized in [49]. These results provide a foundational basis for analyzing the convergence behavior of nonsmooth function approaches to solving the semidefinite complementarity problem (SDCP) (3.220). Furthermore, the equivalent conditions for the nonsingularity of the generalized Jacobian of ϕ_{FB} were established in [50]. In addition, the associated merit function ψ_{FB}, defined as in (3.226), was shown in [51] to be an LC^1 function, that is, it is continuously differentiable and possesses a Lipschitz continuous gradient mapping.

Proposition 3.56 *Let the LT-type function* $\psi_{\text{LT}} : \mathbb{S}^n \times \mathbb{S}^n \to \mathbb{R}$ *be defined by*

$$\psi_{\text{LT}}(X, Y) = \psi_0 \left(\langle X, Y\rangle\right) + \psi(X, Y), \tag{3.227}$$

where $\psi_0 : \mathbb{R} \to \mathbb{R}_+$ *satisfies*

$$\psi_0(t) = 0 \iff t \le 0, \tag{3.228}$$

and $\psi : \mathbb{S}^n \times \mathbb{S}^n \to \mathbb{R}_+$ *satisfies*

$$\psi(X, Y) = 0,\ \langle X, Y\rangle \le 0 \iff X \in \mathbb{S}^n_+,\ Y \in \mathbb{S}^n_+,\ \langle X, Y\rangle = 0. \tag{3.229}$$

Then, the function ψ_{LT} *defined as in* (3.227)–(3.228) *is a differentiable* C*-function associated with positive semidefinite cone. In other words,*

$$\psi_{\text{LT}}(X, Y) = 0 \iff X \in \mathbb{S}^n_+, \quad Y \in \mathbb{S}^n_+, \quad \langle X, Y\rangle = 0.$$

Proof Two examples of ψ satisfying (3.229) were considered in [17]:

$$\psi(X, Y) = \frac{1}{2}\left(\|(X)_-\|^2 + \|(Y)_-\|^2\right), \quad \psi(X, Y) = \frac{1}{2}\left\|\left(\phi_{\text{FB}}(X, Y)\right)_+\right\|^2.$$

Although both functions satisfy (3.229), they still have different features. For instance, while both choices are differentiable, only the first one is convex. For detailed arguments, please see Lemmas 7.1, 7.2, and Proposition 5.1 in [17]. □

Proposition 3.57 *Let the function* $\psi_{\text{YF}} : \mathbb{S}^n \times \mathbb{S}^n \to \mathbb{R}$ *be defined by*

$$\psi_{\text{YF}}(X, Y) = \frac{1}{4}\max\left\{0, \langle X, Y\rangle\right\}^4 + \frac{1}{2}\left\|\phi_{\text{FB}}(X, Y)\right\|^2. \tag{3.230}$$

Then, the function ψ_{YF} *defined as in* (3.230) *is a differentiable* C*-function associated with positive semidefinite cone. In other words,*

$$\psi_{\text{YF}}(X, Y) = 0 \iff X \in \mathbb{S}^n_+, \quad Y \in \mathbb{S}^n_+, \quad \langle X, Y\rangle = 0.$$

Proof Please see Lemmas 2.2, 2.3, Theorems 3.1, and 3.2 in [40]. □

It was noted in [40] that if the second term in (3.230) is replaced by $\frac{1}{2}\left|\phi_{\mathrm{FB}}(X, Y)+\right|^2$, the resulting function falls within the class of LT-type functions as defined in (3.227). In the context of NCP, the function ψ_{LT} is known to exhibit convexity under certain conditions [33]. However, this favorable property does not extend to the setting of the positive semidefinite cone. From the authors' perspective, the function ψ_{YF} is structurally simpler than ψ_{LT}, as it avoids the projection onto the cone of positive semidefinite matrices, $\mathbb{S}^n+$. Moreover, [40] also investigated the boundedness of level sets and established error bound results based on the merit function ψ_{FB}.

Proposition 3.58 *Let the function* $\phi_{\mathrm{NR}} : \mathbb{S}^n \times \mathbb{S}^n \to \mathbb{S}^n$ *be defined as*

$$\phi_{\mathrm{NR}}(X, Y) = X - (X - Y)_+. \tag{3.231}$$

Then, the function ϕ_{NR} *defined as in* (3.231) *satisfies*

$$\phi_{\mathrm{NR}}(X, Y) = 0 \iff X \in \mathbb{S}^n_+, \quad Y \in \mathbb{S}^n_+, \quad XY = 0.$$

Moreover, if we define $\phi^{\tau}_{\mathrm{NR}} : \mathbb{S}^n \times \mathbb{S}^n \to \mathbb{S}^n$ *as*

$$\phi^{\tau}_{\mathrm{NR}}(X, Y) = \sqrt{(X - Y)^2 + 4\tau^2 I} - (X + Y), \quad \tau > 0. \tag{3.232}$$

Then, the function $\phi^{\tau}_{\mathrm{NR}}$ *defined as in* (3.232) *satisfies*

$$\phi^{\tau}_{\mathrm{NR}}(X, Y) = 0 \iff X \succ O, \quad Y \succ O, \quad XY = \tau^2 I.$$

Proof The function $\phi^{\tau}_{\mathrm{NR}}$ is a smoothed NR function, which is also called Chen–Harker–Kanzow–Smale smoothing function in the literature. It is defined by observing

$$\varphi(a, b) = 2\min\{a, b\} = (a + b) - |a - b| = (a + b) - \sqrt{(a - b)^2}.$$

Please see Propositions 2.3 and 2.4 in [48] for the proof. □

As noted in Sect. 3.1, the second-order cone complementarity problem (SOCCP) can be reformulated as a semidefinite complementarity problem (SDCP) by observing that, for any $x = (x_1, x_2) \in \mathbb{R} \times \mathbb{R}^{n-1}$, the condition $x \in \mathcal{K}^n$ holds if and only if

$$L_x := \begin{bmatrix} x_1 & x_2^{\mathrm{T}} \\ x_2 & x_1 I \end{bmatrix}$$

is positive semidefinite (see also [12, p. 437] and [23]). However, this reformulation comes at the cost of a dimensional increase from n to $n(n+1)/2$, and it remains unclear whether this increase can be efficiently alleviated by exploiting the special "arrow" structure of L_x. For this reason, it remains meaningful, particularly from the standpoint of applications, to study C-functions tailored separately to the second-order cone and the cone of positive semidefinite matrices $\mathbb{S}^n_+$.

3.3 Complementarity Functions Associated with Symmetric Cone

It is natural to pursue the extension of the C-functions discussed in Chap. 2 and Sect. 3.1 to the broader setting of general symmetric cones. At first glance, such extensions may appear straightforward, given the unifying framework provided by Euclidean Jordan algebras, which encompass second-order cones, positive semidefinite cones, and symmetric cones. However, the analytical challenges are often greater than expected, primarily due to the lack of an explicit spectral decomposition formula in the general symmetric cone setting. When an extension avoids reliance on spectral decomposition, the analysis is relatively tractable. In what follows, we begin with a few merit functions and C-functions that are more amenable to such treatment.

Several researchers have contributed to the development of merit and C-functions for symmetric cones. Notably, Liu, Zhang, and Wang [41] extended a class of merit functions originally proposed in [52] to the symmetric cone complementarity problem (SCCP) (3.234). Kong, Tuncel, and Xiu [18] studied the extension of the implicit Lagrangian function introduced by Mangasarian and Solodov [53] to the symmetric cone setting. In addition, Kong, Sun, and Xiu [54] proposed a regularized smoothing method for solving the SCCP (3.234), based on the natural residual complementarity function associated with symmetric cones.

3.3.1 *Existing C-Functions Associated with Symmetric Cone*

Given a Euclidean Jordan algebra $(\mathbb{V}, \circ, \langle\cdot,\cdot\rangle)$ where $\mathbb{V}$ is a finite-dimensional vector space over the real field $\mathbb{R}$ endowed with the inner product $\langle\cdot,\cdot\rangle$ and "$\circ$" denotes the Jordan product. Let $\mathcal{K}$ be a symmetric cone in $\mathbb{V}$ and $G, F : \mathbb{V} \to \mathbb{V}$ be nonlinear transformations assumed to be continuously differentiable throughout this section. Consider the symmetric cone complementarity problem (SCCP) of finding $\zeta \in \mathbb{V}$ such that

$$G(\zeta) \in \mathcal{K}, \quad F(\zeta) \in \mathcal{K}, \quad \langle G(\zeta), F(\zeta)\rangle = 0. \tag{3.233}$$

If G is the identity mapping, then the SCCP (3.233) reduces to

$$\zeta \in \mathcal{K}, \ F(\zeta) \in \mathcal{K}, \ \langle \zeta, F(\zeta)\rangle = 0. \tag{3.234}$$

The model provides a simple, natural, and unified framework for various existing complementarity problems such as the nonnegative orthant nonlinear complementarity problem (NCP), the second-order cone complementarity problem (SOCCP), and the semidefinite complementarity problem (SDCP). In addition, the model itself is closely related to the KKT optimality conditions for the convex symmetric cone program (CSCP):

$$\begin{aligned} &\min\ g(x) \\ &\text{s.t. } \langle a_i, x\rangle = b_i, \quad i = 1, 2, \ldots, m, \\ &\qquad x \in \mathcal{K}, \end{aligned}$$

where $a_i \in \mathbb{V},\ b_i \in \mathbb{R}$ for $i = 1, 2, \ldots, m$, and $g : \mathbb{V} \to \mathbb{R}$ is a convex twice continuously differentiable function. Therefore, the SCCP has wide applications in engineering, economics, management science, and other fields; see [2, 7, 30, 55] and references therein.

Recall the differentiable NCP function ϕ_{MS} introduced in (2.11), originally proposed by Mangasarian and Solodov. This function also serves as a merit function and is defined as

$$\phi_{\text{MS}}(a, b) = ab + \frac{1}{2\alpha}\left([a - \alpha b]_+^2 - a^2 + [b - \alpha a]_+^2 - b^2\right), \quad \alpha > 1.$$

As in the SOC setting, see (3.8) and (3.9), there are two natural approaches to extend this function to the framework of symmetric cones. In this context, for any $\alpha > 0$ with $\alpha \neq 1$, we define the associated real-valued implicit Lagrangian function $\psi_{\text{MS}} : \mathbb{R}^n \times \mathbb{R}^n \to \mathbb{R}_+$ as follows:

$$\psi_{\text{MS}}(x, y) = \langle x, y\rangle + \frac{1}{2\alpha}\left(\|(x - \alpha y)_+\|^2 - \|x\|^2 + \|(y - \alpha x)_+\|^2 - \|y\|^2\right). \tag{3.235}$$

and the vector-valued implicit Lagrangian function, $\phi_{\text{MS}} : \mathbb{R}^n \times \mathbb{R}^n \to \mathbb{R}^n$, by

$$\phi_{\text{MS}}(x, y) := x \circ y + \frac{1}{2\alpha}\left[(x - \alpha y)_+^2 - x^2 + (y - \alpha x)_+^2 - y^2\right]. \tag{3.236}$$

Here, $[\cdot]_+$ denotes the metric projection onto $\mathcal{K}$ and $x^2 = x \circ x$. The requirement of the parameter $\alpha > 0$ with $\alpha \neq 1$ is due to the below observation in [18]. When $\alpha = 1$, it is noted that

$$\begin{aligned} \phi_{\text{MS}}(x, y) &= x \circ y + \frac{1}{2}\left[(x - y)_+^2 - x^2 + (y - x)_+^2 - y^2\right] \\ &= x \circ y + \frac{1}{2}\left[(x - y)_+^2 - x^2 + (x - y)_-^2 - y^2\right] \\ &= x \circ y + \frac{1}{2}\left[(x - y)^2 - x^2 - y^2\right] \\ &= 0. \end{aligned}$$

Proposition 3.59 establishes that both the real-valued and vector-valued implicit Lagrangian functions qualify as C-functions associated with the symmetric cone. It also outlines several of their key properties.

Proposition 3.59 *Let ψ_{MS} and ϕ_{MS} be defined as in* (3.235) *and* (3.236)*, respectively. Then, the following hold.*

(a) Both ψ_{MS} and ϕ_{MS} are C-functions associated with symmetric cone. In other words,

$$\psi_{\text{MS}}(x, y) = 0 \iff \phi_{\text{MS}}(x, y) = 0 \iff x \in \mathcal{K},\ y \in \mathcal{K},\ \langle x, y\rangle = 0.$$

(b) For $\alpha > 1$, $\psi_{\text{MS}}(x, y) \geq 0$, whereas $\psi_{\text{MS}}(x, y) \leq 0$, for $0 < \alpha < 1$.
(c) $\psi_{\text{MS}}(x, y) = \langle e, \phi_{\text{MS}}(x, y)\rangle$, where e is the identity element of $\mathbb{V}$.
(d) ϕ_{MS} is strongly semismooth.
(e) ψ_{MS} is continuously differentiable with $\nabla\psi_{\text{MS}}(x, y) = \nabla\phi_{\text{MS}}(x, y)^{\mathsf{T}} e$.

Proof Part(a) uses the fact that $(x - \alpha y)_+$ is the unique solution to the problem:

$$\min_{z \in \mathcal{K}} \langle \alpha y, z - x\rangle + \frac{1}{2}\|z - x\|^2.$$

Then, in terms of the so-called regularized gap function, the arguments proceed. Please see [18, Theorem 3.2] and [18, Theorem 4.1] for details.

Parts (b)–(c) are derived from [18, Theorem 4.1], part (d) from [18, Theorem 3.4], and part (e) from [18, Lemma 4.2]. □

Likewise, the vector-valued Fischer–Burmeister function in the setting of symmetric cone is defined by

$$\phi_{\text{FB}}(x, y) := (x^2 + y^2)^{1/2} - (x + y), \tag{3.237}$$

and the vector-valued natural residual function is

$$\phi_{\text{NR}}(x, y) := x - (x - y)_+, \tag{3.238}$$

where $(\cdot)_+$ denotes the metric projection on $\mathcal{K}$. Here, $x^2 = x \circ x$, $x^{1/2}$ is a vector such that $(x^{1/2})^2 = x$, and $x + y$ means the usual componentwise addition of vectors. Note that ϕ_{NR} is already a C-function associated with symmetric cone due to Proposition 1.3. Nonetheless, both ϕ_{FB} and ϕ_{NR} functions were shown to be C-functions in [37, Proposition 6].

Proposition 3.60 *[37, Proposition 6] Let ϕ_{FB} and ϕ_{NR} be defined as in* (3.237) *and* (3.238)*, respectively. Then, both ϕ_{FB} and ϕ_{NR} are C-functions associated with symmetric cone. In other words,*

$$\phi_{\text{FB}}(x, y) = 0 \iff \phi_{\text{NR}}(x, y) = 0 \iff x \in \mathcal{K},\ y \in \mathcal{K},\ \langle x, y\rangle = 0.$$

Proof Using the following equivalent properties [37, Proposition 6]:

$$\begin{aligned} & x \in \mathcal{K},\ y \in \mathcal{K},\ \langle x, y\rangle = 0 \\ \Longleftrightarrow & x \in \mathcal{K},\ y \in \mathcal{K},\ x \circ y = 0 \\ \Longleftrightarrow & x + y = (x^2 + y^2)^{1/2} \\ \Longleftrightarrow & x = [x - y]_+. \end{aligned}$$

Then, the desired results follow. □

Proposition 3.61 *[56, Theorem 3.1] Let ϕ_{FB} and ϕ_{NR} be defined as in* (3.237) *and* (3.238)*, respectively. Then, there holds*

$$(2 - \sqrt{2}) \left\|\phi_{\text{NR}}(x, y)\right\| \le \left\|\phi_{\text{FB}}(x, y)\right\| \le (2 + \sqrt{2}) \left\|\phi_{\text{NR}}(x, y)\right\|.$$

Proof The proof proceeds by considering various values of the rank r associated with the Euclidean Jordan algebra. For detailed arguments, please refer to [56, Theorem 3.1]. □

The strong semismoothness of ϕ_{NR} has already been established in [57], whereas the corresponding property for ϕ_{FB} remains an open question. In particular, Chang, Chen, and Pan investigated this issue in [58] and demonstrated that ϕ_{FB} is strongly semismooth in the Euclidean Jordan algebras $\mathbb{L}^n$, $\mathbb{S}^n$, $\mathbb{H}^n$, and $\mathbb{Q}^n$. This constitutes an almost complete resolution, with the sole exception being the algebra $\mathbb{O}^3$, the space of 3×3 Hermitian matrices over octonions, also known as the Albert algebra. The inability to draw a definitive conclusion in this case arises from the non-associativity of the octonion algebra $\mathbb{O}$, which prevents representing its elements as real matrices. For further details, the reader is referred to [58].

According to ϕ_{NR}, ϕ_{FB}, and ψ_{MS} given as in (3.238), (3.237), and (3.235), respectively, there induce merit functions $\psi_{\text{NR}}(\zeta)$, $\psi_{\text{FB}}(\zeta)$, and $\psi_{\text{MS}}(\zeta)$ as below:

$$\psi_{\text{NR}}(\zeta) := \|\phi_{\text{NR}}(\zeta, F(\zeta))\|^2, \tag{3.239}$$

$$\begin{aligned} \psi_{\text{FB}}(\zeta) &:= \frac{1}{2} \left\|\phi_{\text{FB}}(\zeta, F(\zeta))\right\|^2, \\ \psi_{\text{MS}}(\zeta) &:= 2\alpha\langle\zeta, F(\zeta)\rangle + \Big\{\|(-\alpha F(\zeta) + \zeta)_+\|^2 - \|\zeta\|^2 \\ &\quad + \|(-\alpha\zeta + F(\zeta))_+\|^2 - \|F(\zeta)\|^2\Big\}. \end{aligned} \tag{3.240}$$

Proposition 3.62 *Let ψ_{NR} and ψ_{MS} be defined as in* (3.239) *and* (3.240)*, respectively. For each $\alpha > 1$, the following holds:*

$$2(\alpha - 1)\psi_{\text{NR}}(\zeta) \le \psi_{\text{MS}}(\zeta) \le 2\alpha(\alpha - 1)\psi_{\text{NR}}(\zeta), \quad \forall \zeta \in \mathbb{V}. \tag{3.241}$$

Proof For convenience, we denote

$$f(\zeta, \alpha) := -\langle\alpha F(\zeta), (\zeta - \alpha F(\zeta))_+ - \zeta\rangle - \frac{1}{2}\|(\zeta - \alpha F(\zeta))_+ - \zeta\|^2.$$

We point out that there is another expression for $f(x,\alpha)$ as given below, see [59, Theorem 3.1].

$$
\begin{aligned}
f(x,\alpha) &= \max_{y\in\mathcal{K}} -\left\langle \alpha F(\zeta) + \frac{1}{2}(y-\zeta),\, y-\zeta \right\rangle \\
&= -\left\langle \alpha F(\zeta) + \frac{1}{2}((\zeta-\alpha F(\zeta))_+ - \zeta),\, (\zeta-\alpha F(\zeta))_+ - \zeta \right\rangle \qquad (3.242)\\
&\geq -\left\langle \alpha F(\zeta) + \frac{1}{2}((\zeta- F(\zeta))_+ - \zeta),\, (\zeta- F(\zeta))_+ - \zeta \right\rangle .
\end{aligned}
$$

Now, we compute

$$
\begin{aligned}
\frac{1}{\alpha} f(\zeta,\alpha) &= -\langle F(\zeta), (\zeta-\alpha F(\zeta))_+ - \zeta\rangle - \frac{1}{2\alpha}\|(\zeta-\alpha F(\zeta))_+ - \zeta\|^2 \\
&= \langle \zeta, F(\zeta)\rangle + \frac{1}{\alpha}\langle \zeta-\alpha F(\zeta), (\zeta-\alpha F(\zeta))_+\rangle - \frac{1}{2\alpha}\|(\zeta-\alpha F(\zeta))_+\|^2 - \frac{1}{2\alpha}\|\zeta\|^2 \\
&= \langle \zeta, F(\zeta)\rangle + \frac{1}{2\alpha}\left(\|(\zeta-\alpha F(\zeta))_+\|^2 - \|\zeta\|^2\right).
\end{aligned}
$$

Likewise,

$$
f(\zeta,1) = -\left\langle F(\zeta) + \frac{1}{2}((\zeta- F(\zeta))_+ - \zeta),\, (\zeta- F(\zeta))_+ - \zeta \right\rangle
$$

and

$$
\alpha f(\zeta,\frac{1}{\alpha}) = -\frac{1}{2\alpha}\left(\|(-\alpha\zeta + F(\zeta))_+\|^2 - \|F(\zeta)\|^2\right).
$$

Combining the above two equations, we obtain an identity for $\psi_{\text{MS}}(x)$

$$
\psi_{\text{MS}}(\zeta) = 2\alpha\left(\frac{1}{\alpha} f(\zeta,\alpha) - \alpha f(\zeta,\frac{1}{\alpha})\right). \qquad (3.243)
$$

To show the desired two inequalities, we proceed by two steps. The first step is to verify the left-hand side of (3.241). To see this,

$$
\begin{aligned}
\psi_{\text{MS}}(\zeta) &= 2\alpha\left(\frac{1}{\alpha} f(\zeta,\alpha) - \alpha f(\zeta,\frac{1}{\alpha})\right) \\
&= 2\alpha\left(\frac{1}{\alpha} f(\zeta,\alpha) - f(\zeta,1)\right) + 2\alpha\left(f(\zeta,1) - \alpha f(\zeta,\frac{1}{\alpha})\right) \\
&\geq 2\alpha\Big[-\langle F(\zeta), (\zeta-F(\zeta))_+ - \zeta\rangle - \frac{1}{2\alpha}\|(\zeta-F(\zeta))_+ - \zeta\|^2 \\
&\quad + \langle F(\zeta), (\zeta-F(\zeta))_+ - \zeta\rangle + \frac{1}{2}\|(\zeta-F(\zeta))_+ - \zeta\|^2\Big] + 2\alpha\left(f(\zeta,1) - \alpha f(\zeta,\frac{1}{\alpha})\right) \\
&= 2\alpha\frac{\alpha-1}{2\alpha}\psi_{\text{NR}}(\zeta) + 2\alpha\left(f(\zeta,1) - \alpha f(\zeta,\frac{1}{\alpha})\right)
\end{aligned}
$$

$$
\begin{aligned}
&= (\alpha-1)\psi_{\rm NR}(\zeta) + 2\alpha\Big[-\langle F(\zeta), (\zeta - F(\zeta))_+ - \zeta\rangle - \frac{1}{2}\|(\zeta - F(\zeta))_+ - \zeta\|^2 \\
&\quad + \langle F(\zeta), (\zeta - \frac{1}{\alpha}F(\zeta))_+ - \zeta\rangle + \frac{\alpha}{2}\|(\zeta - \frac{1}{\alpha}F(\zeta))_+ - \zeta\|^2\Big] \\
&\geq (\alpha-1)\psi_{\rm NR}(\zeta) + 2\alpha\frac{\alpha-1}{2\alpha}\psi_{\rm NR}(\zeta) \\
&= 2(\alpha-1)\psi_{\rm NR}(\zeta),
\end{aligned}
$$

where the first inequality follows from (3.242). Next, we verify the right-hand side of (3.241). To this end, wc observe two things:

$$
\begin{aligned}
&\frac{1}{\alpha}f(\zeta,\alpha) - f(x,1) \\
&= -\langle F(\zeta), (\zeta - \alpha F(\zeta))_+ - \zeta\rangle - \frac{1}{2\alpha}\|(\zeta - \alpha F(\zeta))_+ - \zeta\|^2 \\
&\quad + \langle F(\zeta), (\zeta - F(\zeta))_+ - \zeta\rangle + \frac{1}{2}\|(\zeta - F(\zeta))_+ - \zeta\|^2 \\
&= \frac{\alpha-1}{2\alpha}\psi_{\rm NR}(\zeta) + \frac{1}{2\alpha}\psi_{\rm NR}(\zeta) - \frac{1}{2\alpha}\|(\zeta - \alpha F(\zeta))_+ - \zeta\|^2 \\
&\quad + \langle F(\zeta), (\zeta - F(\zeta))_+ - (\zeta - \alpha F(\zeta))_+\rangle \\
&= \frac{\alpha-1}{2}\psi_{\rm NR}(\zeta) - \frac{(\alpha-1)^2}{2\alpha}\psi_{\rm NR}(\zeta) \\
&\quad - \frac{1}{2\alpha}\|(\zeta - \alpha F(\zeta))_+ - (\zeta - F(\zeta))_+\|^2 + \frac{1}{\alpha}\|(\zeta - F(\zeta))_+ - \zeta\|^2 \\
&\quad - \frac{1}{\alpha}\langle(\zeta - \alpha F(\zeta))_+ - \zeta, (\zeta - F(\zeta))_+ - \zeta\rangle \\
&\quad + \langle F(\zeta), (\zeta - F(\zeta))_+ - (\zeta - \alpha F(\zeta))_+\rangle \\
&= \frac{\alpha-1}{2}\psi_{\rm NR}(\zeta) \\
&\quad - \frac{1}{2\alpha}\|(\alpha-1)(\zeta - (\zeta - F(\zeta))_+) + (\zeta - \alpha F(\zeta))_+ - (\zeta - F(\zeta))_+\|^2 \\
&\quad - \langle(\zeta - F(\zeta))_+ - \zeta + F(\zeta), (\zeta - \alpha F(\zeta))_+ - (\zeta - F(\zeta))_+\rangle \\
&\leq \frac{\alpha-1}{2}\psi_{\rm NR}(\zeta)
\end{aligned}
$$

and

$$
\begin{aligned}
&f(\zeta,1) - \alpha f(\zeta,\frac{1}{\alpha}) \\
&= -\langle F(\zeta), (x - F(\zeta))_+ - \zeta\rangle - \frac{1}{2}\|(\zeta - F(\zeta))_+ - \zeta\|^2 \\
&\quad + \langle F(\zeta), (x - \frac{1}{\alpha}F(\zeta))_+ - \zeta\rangle + \frac{\alpha}{2}\left\|(\zeta - \frac{1}{\alpha}F(\zeta))_+ - \zeta\right\|^2
\end{aligned}
$$

$$
\begin{aligned}
&= \max_{y\in\mathcal{K}} -\langle F(\zeta) + \frac{1}{2}(y-\zeta), y-\zeta\rangle + \alpha \min_{y\in\mathcal{K}} \langle \frac{1}{\alpha}F(\zeta) + \frac{1}{2}(y-\zeta), y-\zeta\rangle \\
&\le -\left\langle F(\zeta) + \frac{1}{2}((\zeta - F(\zeta))_+ - \zeta), (\zeta - F(\zeta))_+ - \zeta \right\rangle \\
&\quad + \left\langle F(\zeta) + \frac{\alpha}{2}((\zeta - F(\zeta))_+ - \zeta), (\zeta - F(\zeta))_+ - \zeta \right\rangle \\
&= \frac{\alpha - 1}{2}\|(\zeta - F(\zeta))_+ - \zeta\|^2 \\
&= \frac{\alpha - 1}{2}\psi_{\text{NR}}(\zeta).
\end{aligned}
$$

The above two expressions together with the identity (3.243) yield

$$
\begin{aligned}
\psi_{\text{MS}}(\zeta) &\le 2\alpha\left(\frac{\alpha-1}{2}\psi_{\text{NR}}(\zeta) + \frac{\alpha-1}{2}\psi_{\text{NR}}(\zeta)\right) \\
&= 2\alpha(\alpha-1)\psi_{\text{NR}}(\zeta).
\end{aligned}
$$

Thus, the proof is complete. □

Propositions 3.61 and 3.62 show that ψ_{FB}, ψ_{NR}, and ψ_{MS} exhibit similar growth behavior in the general symmetric cone setting. These functions can also be effectively used to characterize the boundedness of level sets and to establish local error bounds for the SCCPs.

Definition 3.1 For the residual function $r(\zeta) = \|\phi_{\text{NR}}(\zeta, F(\zeta))\|$, the function $r(\zeta)$ is said to be a local error bound if there exist constants $c > 0$ and $\delta > 0$ such that for each $\zeta \in \{\zeta \in \mathbb{V} \,|\, d(\zeta, S) \le \delta\}$, there holds

$$
d(\zeta, S) \le c\, r(\zeta),
$$

where S denote the solution set of the SCCP (3.234) and $d(\zeta, S) = \inf_{y\in S}\|\zeta - y\|$.

Lemma 3.42 *Let ϕ_{FB} be defined as in (3.237). Then, for any $x, y \in \mathbb{V}$,*

$$
\left\|(\phi_{\text{FB}}(x,y))_+\right\|^2 \ge \frac{1}{2}\left(\|(-x)_+\|^2 + \|(-y)_+\|^2\right).
$$

Proof This is a special case of Lemma 3.51 when $\tau = 2$. □

Lemma 3.43 *Let ϕ_{NR} be defined as in (3.238). Then, for any $x, y \in \mathbb{V}$, there is a constant $\beta > 0$ such that*

$$
\left\|\phi_{\text{NR}}(x,y)\right\|^2 \ge \frac{\beta}{2}\left(\|(-x)_+\|^2 + \|(-y)_+\|^2\right).
$$

Proof By applying Proposition 3.61 and Lemma 3.42, the desired result is obtained immediately. □

Proposition 3.63 *Consider the residual function* $r(\zeta) = \|\phi_{\text{NR}}(\zeta, F(\zeta))\|$. *If* F *is an* R_0^w*-function, then the level set* $\mathcal{L}(\gamma) := \{\zeta \in \mathbb{V} \mid r(\zeta) \leq \gamma\}$ *is bounded for all* $\gamma \geq 0$.

Proof Suppose there is an unbounded sequence $\{\zeta_k\} \subseteq \mathcal{L}(\gamma)$ for some $\gamma > 0$. If $\limsup \omega((-\zeta_k)_+) = \infty$, then (through a subsequence) $\|(-\zeta_k)_+\| \to \infty$, by Lemma 3.43, which implies that $r(\zeta_k) \to \infty$. This contradicts the boundness of $\mathcal{L}(\gamma)$. A similar contradiction ensues if $\limsup \omega((F(-\zeta_k))_+) = \infty$. Thus, for the specified unbounded sequence $\{\zeta_k\}$ satisfying the condition in Definition 1.14, by Definition 1.14, we also obtain that $\omega(\phi_{\text{NR}}(\zeta_k, F(\zeta_k))) \to \infty$. With $r(\zeta_k) = \|\phi_{\text{NR}}(\zeta_k, F(\zeta_k))\|$, it is easy to see that $r(\zeta_k) \to \infty$. This leads to a contradiction. Consequently, the level set $\mathcal{L}(\gamma) := \{\zeta \in \mathbb{V} \mid r(\zeta) \leq \gamma\}$ is bounded for all $\gamma \geq 0$. □

Proposition 3.63 establishes that the residual function $r(\zeta)$ possesses the bounded level set property under the R_0-type condition. However, this condition alone is insufficient for $r(\zeta)$ to serve as a local error bound, even in the simpler case of the NCPs, which are special cases of the SCCPs. An illustrative counterexample is provided in [60], showing that $r(x)$ fails to act as a local error bound for an R_0-type NCP. Specifically, consider the mapping $F : \mathbb{R} \to \mathbb{R}$ defined by $F(\zeta) = \zeta^3$. It is straightforward to verify that F is an R_0^s-function and that the corresponding NCP has a bounded solution set $S = 0$. Nonetheless, $r(\zeta)$ does not qualify as a local error bound. This raises a natural question: under what additional condition can $r(x)$ serve as a local error bound for the SCCPs? The next proposition addresses this by providing a sufficient condition.

Proposition 3.64 *Consider the residual function* $r(\zeta) = \|\phi_{\text{NR}}(\zeta, F(\zeta))\|$. *Suppose that the solution set* S *of the SCCPs is nonempty and that* ϕ_{NR} *is* BD*-regular at all solutions of the SCCPs. Then,* $r(\zeta)$ *is a local error bound if it has a local bounded level set.*

Proof Since $r(\zeta)$ has a local bounded level set, there exists $\varepsilon > 0$ such that the level set $\mathcal{L}(\varepsilon) = \{\zeta \mid r(\zeta) \leq \varepsilon\}$ is bounded. Thus, the set $\mathcal{L}(\varepsilon) = \{\zeta \mid r(\zeta) \leq \varepsilon\}$ is compact. Suppose that the conclusion is wrong. Then, there exists a sequence $\{\zeta_k\} \subset \mathcal{L}(\varepsilon)$ such that

$$\frac{r(\zeta_k)}{\text{dist}(\zeta_k, S)} \to 0 \quad \text{as} \quad k \to \infty.$$

Here $\text{dist}(\zeta_k, S)$ denotes the distance between ζ_k and S. Therefore, $r(\zeta_k) \to 0$ and it follows from compactness of $\mathcal{L}(\varepsilon)$ that there is a convergent subsequence. Without loss of generality, let $\{\zeta_k\}$ be a convergent sequence, and $\bar{\zeta}$ be its limit, that is, $\zeta_k \to \bar{\zeta} \in \mathcal{L}(\varepsilon)$. Then, $r(\bar{\zeta}) = 0$, which implies $\bar{\zeta} \in S$. It turns out that

$$\frac{r(\zeta_k)}{\|\zeta_k - \bar{\zeta}\|} \to 0 \quad \text{as} \quad k \to \infty. \tag{3.244}$$

From [57], we know that $\phi_{\text{NR}}(\zeta, F(\zeta))$ is semismooth. By applying [61, Proposition 3] and BD-regular property of $\phi_{\text{NR}}(\zeta, F(\zeta))$, there exist constants $c > 0$ and $\delta > 0$ such that $r(\zeta) \geq c\|\zeta - \bar{\zeta}\|$ for any ζ with $\|\zeta - \bar{\zeta}\| < \delta$. This contradicts (3.244). Consequently, the residual function $r(\zeta)$ is a local error bound for the SCCPs. □

Results analogous to Proposition 3.64 can also be established for the other two merit functions. In light of Propositions 3.61 and 3.62, we may conclude that both ψ_{FB} and ψ_{MS} also serve as local error bounds for the SCCPs.

We now turn to the derivation of a global error bound for SCCPs by leveraging an R_0-type condition together with a BD-regularity condition. To this end, we introduce the following definition and a technical lemma.

Definition 3.2 For the residual function $r(\zeta) = \|\phi_{\text{NR}}(\zeta, F(\zeta))\|$, the function $r(\zeta)$ is said to be a global error bound if there exist constant $c > 0$ such that for each $\zeta \in \mathbb{V}$,

$$d(\zeta, S) \leq c\, r(\zeta),$$

where S denote the solution set of the SCCP (3.234) and $d(\zeta, S) = \inf_{y \in S} \|\zeta - y\|$.

Lemma 3.44 *Let $\{\zeta_k\}$ be any sequence such that $\|\zeta_k\| \to \infty$. If F is an R_0^s-function, then*

$$\liminf_{k\to\infty} \frac{r(\zeta_k)}{\|\zeta_k\|} > 0.$$

Proof Suppose that the result is false. There exists a subsequence ζ_{n_k} with $\|\zeta_{n_k}\| \to \infty$ such that

$$\frac{r(\zeta_{n_k})}{\|\zeta_{n_k}\|} \to 0. \tag{3.245}$$

From Lemma 3.43, it follows that

$$\frac{(-\zeta_{n_k})_+}{\|\zeta_{n_k}\|} \to 0 \quad \text{and} \quad \frac{(-F(\zeta_{n_k}))_+}{\|\zeta_{n_k}\|} \to 0.$$

This together with the definition of R_0^s-function implies

$$\liminf_{k\to\infty} \frac{\omega(\phi_{\text{NR}}(\zeta_{n_k}, F(\zeta_{n_k})))}{\|\zeta_{n_k}\|} > 0,$$

which contradicts formula (3.245). Consequently, we have the desired result. □

Proposition 3.65 *Suppose that F is an R_0^s-function and that ϕ_{NR} is BD-regular at all solutions of SCCPs. Then, there exists a $\kappa > 0$ such that for any $\zeta \in \mathbb{V}$*

$$\text{dist}(\zeta, S) \leq \kappa\, r(\zeta),$$

where S is the solution set of SCCPs, $\text{dist}(\zeta, S)$ *denotes the distance between ζ and S.*

Proof By the definition of R_0^s-function, Propositions 3.63 and 3.64, we claim that $r(x)$ is a local error bound so there exist $c > 0$ and $\delta > 0$ such that

$$r(\zeta) < \delta \quad \Longrightarrow \quad d(\zeta, S) \leq c\, r(\zeta).$$

Suppose $r(\zeta)$ does not have the global error bound property. Then, there exists ζ_k such that for any fixed $\bar{\zeta} \in S$,

$$\|\zeta_k - \bar{\zeta}\| \geq \mathrm{dist}(\zeta_k, S) > k\, r(\zeta_k)$$

for all k. Clearly, the inequality $r(\zeta_k) < \delta$ cannot hold for infinitely many $k's$, else $kr(\zeta_k) < d(\zeta_k, S) \leq c\, r(\zeta_k)$ implies that $k \leq c$ for infinitely many $k's$. Therefore, $r(\zeta_k) \geq \delta$ for all large k. Now,

$$\|\zeta_k - \bar{\zeta}\| \geq d(\zeta_k, S) \geq k\, r(\zeta_k) \geq k\delta$$

for infinitely many $k's$. This implies that $\|\zeta_k\| \to \infty$. Now divide the inequality and take the limit $k \to \infty$, we have

$$1 = \lim_{k\to\infty} \frac{\|\zeta_k - \bar{\zeta}\|}{\|\zeta_k\|} \geq \lim_{k\to\infty} k \frac{r(\zeta_k)}{\|\zeta_k\|} \to \infty,$$

where the last implication holds because F is an R_0^s-function and Lemma 3.44. This clearly is a contradiction. □

Proposition 3.66 *Under the same conditions as in Proposition 3.65, both the merit function* $\psi_{\rm FB}(\zeta)$ *and the implicit Lagrangian function* $\psi_{\rm MS}(\zeta)$ *are global error bounds for SCCPs.*

When $F : \mathbb{V} \to \mathbb{V}$ is a linear mapping of the form $F(\zeta) = L(\zeta) + q$ with $q \in \mathbb{V}$, and the linear operator L possesses the R_0-property, then the residual function $r(\zeta)$ not only serves as a local error bound but can, in fact, be strengthened to a global error bound for the SCLCPs, as shown in the result below.

Proposition 3.67 *Suppose that* $r(\zeta)$ *is a local error bound for SCLCPs and the linear transformation* L *has* R_0*-property. Then, there exists* $k > 0$ *such that* $\mathrm{dist}(\zeta, S) \leq k\, r(\zeta)$ *for every* $\zeta \in \mathbb{V}$.

Proof Suppose that the conclusion is false. Then, for any integer $k > 0$, there exists an $\zeta_k \in \mathbb{R}^n$ such that $\mathrm{dist}(\zeta_k, S) > k\, r(\zeta_k)$. Let $z(\zeta_k)$ denote the closest solution of SCLCPs to ζ_k. Choosing a fixed solution $\zeta_0 \in S$, we have

$$\|\zeta_k - \zeta_0\| \geq \|\zeta_k - z(\zeta_k)\| \geq \mathrm{dist}(\zeta_k, S) > k\, r(\zeta_k). \tag{3.246}$$

Since $r(\zeta)$ is a local error bound, it implies that there exist some integer $K > 0$ and $\delta > 0$ such that for all $k > K$, $r(\zeta_k) > \delta$. If not, then for every integer $K > 0$ and any $\delta > 0$, there exist some $k > K$ such that $r(\zeta_k) \le \delta$. By property of local error bound of $r(\zeta)$, we have

$$\frac{\delta}{k}\|\zeta_k - z(\zeta_k)\| > \delta\, r(\zeta_k) \ge \|\zeta_k - z(\zeta_k)\|.$$

Thus, we obtain $\frac{\delta}{k} > 1$. As k goes to infinity, this leads to a contradiction. Consequently, $r(x_k) > \delta$. This together with (3.246) implies that $\|\zeta_k - \zeta_0\| \ge \|\zeta_k - z(\zeta_k)\| > k\delta$ which says that $\|\zeta_k\| \to \infty$ as $k \to \infty$. Now, we consider the sequence $\{\frac{\zeta_k}{\|\zeta_k\|}\}$. There exist a subsequence $\{\zeta_{k_i}\}$ such that

$$\lim_{i\to\infty} \frac{\zeta_{k_i}}{\|\zeta_{k_i}\|} = \zeta.$$

Hence, it follows from (3.246) that

$$\begin{aligned}
1 &= \lim_{i\to\infty} \frac{\|\zeta_{k_i} - \zeta_0\|}{\|\zeta_{k_i}\|} \\
&\ge \lim_{i\to\infty} k_i \frac{r(\zeta_{k_i})}{\|\zeta_{k_i}\|} \\
&= \lim_{i\to\infty} k_i \left\| \frac{\zeta_{k_i}}{\|\zeta_{k_i}\|} - \left(\frac{\zeta_{k_i}}{\|\zeta_{k_i}\|} - L\frac{\zeta_{k_i}}{\|\zeta_{k_i}\|} - \frac{q}{\|\zeta_{k_i}\|} \right)_+ \right\| \\
&= \lim_{i\to\infty} k_i \|\zeta - (\zeta - L(\zeta))_+\|.
\end{aligned}$$

This implies that $\|\zeta - (\zeta - L(\zeta))_+\| = 0$, which shows that ζ is a nonzero solution of SCLCPs with $q \in \mathbb{V}$. It contradicts the R_0-property of L. Then, the proof is complete. □

It is worth noting an important point: if the R_0-property of the linear transformation L is replaced with the weaker condition of monotonicity, the conclusion of Proposition 3.67 may no longer hold. This limitation can be illustrated through the following example, which employs the implicit Lagrangian function ψ_{MS}.

Example 3.2 Let $L : \mathbb{R}^2 \to \mathbb{R}^2$ be defined as

$$L := \begin{bmatrix} \frac{1}{\sqrt{2}} & -\frac{1}{\sqrt{2}} \\ -\frac{1}{\sqrt{2}} & \frac{1}{\sqrt{2}} \end{bmatrix} \quad \text{and} \quad q := \begin{bmatrix} 2 \\ 0 \end{bmatrix}.$$

It is easy to prove that the symmetric cone is $\mathbb{R}^2_+$ and the corresponding SCLCP has a unique solution $x^* = (0, 0)^{\mathsf{T}}$. Choosing $\zeta_k = \left(\frac{k}{\sqrt{2}}, \frac{k}{\sqrt{2}}\right)^{\mathsf{T}}$, $k \ge 0$ gives $F(\zeta_k) = L(\zeta_k) + q = (2, 0)^{\mathsf{T}}$. Then, for any $k > 2\sqrt{2}\alpha$ with $\alpha > 1$, we have

$$\psi_{\rm MS}(\zeta_k) = 4\alpha\left(\frac{k}{\sqrt{2}}\right) + \left(-2\alpha + \frac{k}{\sqrt{2}}\right)^2 + \left(\frac{k}{\sqrt{2}}\right)^2 - 2\left(\frac{k}{\sqrt{2}}\right)^2 - 4$$
$$= 4\left(\alpha^2 - 1\right).$$

However, $\text{dist}(\zeta_k, S) = \|\zeta_k\| = k$. This implies $\text{dist}(\zeta_k, S) > \psi_{\rm MS}(\zeta_k)$ as $k \to \infty$, which explains that $\psi_{\rm MS}(\zeta)$ cannot serve as global error bound for the SCLCPs.

Building on Proposition 3.60, Kum and Lim further demonstrated that their penalized functions remain C-functions associated with the symmetric cone; see [62, Theorems 3.4 and 3.6].

Proposition 3.68 *For* $\lambda \in (0, 1)$, *we define*

$$\phi_{\rm FB}^{\lambda}(x, y) = \lambda\phi_{\rm FB}(x, y) + (1 - \lambda)(x_+ \circ y_+), \tag{3.247}$$

and

$$\phi_{\rm NR}^{\lambda}(x, y) = \lambda\phi_{\rm NR}(x, y) + (1 - \lambda)(x_+ \circ y_+). \tag{3.248}$$

Then, $\phi_{\rm FB}^{\lambda}$ *and* $\phi_{\rm NR}^{\lambda}$ *are* C*-functions associated with symmetric cone.*

Proof The results follow directly from the application of the Lowner–Heinz inequality and Proposition 3.60. For detailed arguments, see Theorems 3.4 and 3.6 in [62]. □

Following this direction, we consider the extension of the one-parameter class of functions originally proposed by Kanzow and Kleinmichel [32], along with a corresponding class of regularized functions. Specifically, we define the one-parameter family of vector-valued functions $\phi_\tau : \mathbb{V} \times \mathbb{V} \to \mathbb{V}$ as follows:

$$\phi_\tau(x, y) := \left(x^2 + y^2 + (\tau - 2)(x \circ y)\right)^{1/2} - (x + y), \tag{3.249}$$

where $\tau \in (0, 4)$ is an arbitrary but fixed parameter. Consequently, its squared norm yields a merit function associated with $\mathcal{K}$

$$\psi_\tau(x, y) \ := \ \frac{1}{2}\|\phi_\tau(x, y)\|^2, \tag{3.250}$$

where $\|\cdot\|$ is the norm induced by $\langle\cdot, \cdot\rangle$, and the SCCP can be reformulated as

$$\min_{\zeta\in\mathbb{V}} \ f_\tau(\zeta) := \psi_\tau(G(\zeta), F(\zeta)). \tag{3.251}$$

When $\tau = 2$, the function ϕ_τ reduces to the vector-valued FB function given in (3.237), while in the limit as $\tau \to 0$, it becomes a scalar multiple of the vector-valued natural residual function defined in (3.238). In this sense, the one-parameter family of vector-valued functions ϕ_τ unifies two widely used C-functions associated with the symmetric cone $\mathcal{K}$. We shall now prove that for any $\tau \in (0, 4)$, both ϕ_τ and its corresponding merit function ψ_τ are C-functions associated with $\mathcal{K}$. To see this, we use the definition of the Jordan product to derive the identity

$$\begin{aligned} x^2 + y^2 + (\tau - 2)(x \circ y) &= \left(x + \frac{\tau - 2}{2} y\right)^2 + \frac{\tau(4 - \tau)}{4} y^2 \\ &= \left(y + \frac{\tau - 2}{2} x\right)^2 + \frac{\tau(4 - \tau)}{4} x^2 \in \mathcal{K} \quad (3.252) \end{aligned}$$

for any $x, y \in \mathbb{V}$. This confirms that the function ϕ_τ is well defined. These functions ϕ_τ and ψ_τ were previously extended to the SOC setting in Sect. 3.1; see (3.64) and (3.63). Although the results below resemble those in the SOC context, the analysis here differs significantly: rather than relying on direct computations, it is grounded in the structure of Euclidean Jordan algebras and their associated properties.

Proposition 3.69 *For any $x, y \in \mathbb{V}$ and $\tau \in (0, 4)$, let ϕ_τ and ψ_τ be given by (3.249) and (3.250), respectively. Then, ϕ_τ and ψ_τ are C-functions associated with the symmetric cone. In other words,*

$$\psi_\tau(x, y) = 0 \iff \phi_\tau(x, y) = 0 \iff x \in \mathcal{K},\ y \in \mathcal{K},\ \langle x, y\rangle = 0.$$

Proof The first equivalence is clear by the definition of ψ_τ, and we only need to prove the second equivalence. Suppose that $\phi_\tau(x, y) = 0$. Then,

$$\left[x^2 + y^2 + (\tau - 2)(x \circ y)\right]^{1/2} = (x + y). \tag{3.253}$$

Squaring the two sides of (3.253) yields

$$x^2 + y^2 + (\tau - 2)(x \circ y) = x^2 + y^2 + 2(x \circ y),$$

which implies $x \circ y = 0$ since $\tau \in (0, 4)$. Substituting $x \circ y = 0$ into (3.253), we have

$$x = \left(x^2 + y^2\right)^{1/2} - y \quad \text{and} \quad y = \left(x^2 + y^2\right)^{1/2} - x.$$

Since $x^2 + y^2 \in \mathcal{K}$, $x^2 \in \mathcal{K}$ and $y^2 \in \mathcal{K}$, from [37, Proposition 8] or [63, Corollary 9] it follows that $x, y \in \mathcal{K}$. Consequently, the necessity holds. For the other direction, suppose $x, y \in \mathcal{K}$ and $x \circ y = 0$. Then, $(x + y)^2 = x^2 + y^2$. This, together with $x \circ y = 0$, implies

$$\left[x^2 + y^2 + (\tau - 2)(x \circ y)\right]^{1/2} - (x + y) = 0.$$

Consequently, the sufficiency follows. The proof is thus complete. □

Lemma 3.45 *For any $x, y \in \mathbb{V}$, let $u(x, y) := (x^2 + y^2)^{1/2}$. Then, the function $u(x, y)$ is continuously differentiable at any point (x, y) such that $x^2 + y^2 \in \operatorname{int}(\mathcal{K})$. Furthermore,*

$$\nabla_x u(x, y) = \mathcal{L}(x)\mathcal{L}^{-1}(u(x, y)) \quad \text{and} \quad \nabla_y u(x, y) = \mathcal{L}(y)\mathcal{L}^{-1}(u(x, y)) \tag{3.254}$$

Proof The first part is due to Theorem 1.4, and hence it remains to derive the formulas in (3.254). From the definition of $u(x, y)$, it follows that

$$u^2(x, y) = x^2 + y^2, \quad \forall\, x, y \in \mathbb{V}. \tag{3.255}$$

By formula (1.20), it is easy to verify that $\nabla_x(x^2) = 2\mathcal{L}(x)$. Differentiating on both sides of (3.255) with respect to x then yields that

$$2\nabla_x u(x, y)\mathcal{L}(u(x, y)) = 2\mathcal{L}(x).$$

This implies that $\nabla_x u(x, y) = \mathcal{L}(x)\mathcal{L}^{-1}(u(x, y))$ since, by $u(x, y) \in \text{int}(\mathcal{K})$, $\mathcal{L}(u(x, y))$ is positive definite on $\mathbb{V}$. Similarly, we have that $\nabla_y u(x, y) = \mathcal{L}(y)\mathcal{L}^{-1}(u(x, y))$. □

To present another lemma, we first introduce some related notations. For any $0 \neq z \in \mathcal{K}$ and $z \notin \text{int}(\mathcal{K})$, suppose that z has the spectral decomposition $z = \sum_{j=1}^{r} \lambda_j(z)c_j$, where $\{c_1, c_2, \cdots, c_r\}$ is a Jordan frame and $\lambda_1(z), \cdots, \lambda_r(z)$ are the eigenvalues arranged in the decreasing order $\lambda_1(z) \geq \lambda_2(z) \geq \cdots \geq \lambda_r(z) = 0$. Define the index

$$j^* := \min\left\{ j \,\middle|\, \lambda_j(z) = 0, \ j = 1, 2, \cdots, r \right\}$$

and let

$$c_J := \sum_{l=1}^{j^*-1} c_l.$$

Clearly, j^* and c_J are well defined since $0 \neq z \in \mathcal{K}$ and $z \notin \text{int}(\mathcal{K})$. Since c_J is an idempotent and $c_J \neq 0$ (otherwise $z = 0$), $\mathbb{V}$ can be decomposed as the orthogonal direct sum of the subspaces $\mathbb{V}(c_J, 1)$, $\mathbb{V}(c_J, \frac{1}{2})$ and $\mathbb{V}(c_J, 0)$. In the sequel, we write $P_1(c_J)$, $P_{\frac{1}{2}}(c_J)$ and $P_0(c_J)$ as the orthogonal projection onto $\mathbb{V}(c_J, 1)$, $\mathbb{V}(c_J, \frac{1}{2})$ and $\mathbb{V}(c_J, 0)$, respectively. From [41], we know that $\mathcal{L}(z)$ is positive definite on $\mathbb{V}(c_J, 1)$ and is a one-to-one mapping from $\mathbb{V}(c_J, 1)$ to $\mathbb{V}(c_J, 1)$. This means that $\mathcal{L}(z)$ an inverse $\mathcal{L}^{-1}(z)$ on $\mathbb{V}(c_J, 1)$, i.e., for any $u \in \mathbb{V}(c_J, 1)$, $\mathcal{L}^{-1}(z)u$ is the unique $v \in \mathbb{V}(c_J, 1)$ such that $z \circ v = u$.

Lemma 3.46 *For any $x, y \in \mathbb{V}$, let $z : \mathbb{V} \times \mathbb{V} \to \mathbb{V}$ be the mapping defined as*

$$z = z(x, y) := \left[x^2 + y^2 + (\tau - 2)(x \circ y)\right]^{1/2}. \tag{3.256}$$

If $(x, y) \neq (0, 0)$ such that $z(x, y) \notin \text{int}(\mathcal{K})$, then the following results hold:

(a) The elements x, y, $x + y$, $x + \frac{\tau-2}{2}y$ and $y + \frac{\tau-2}{2}x$ belong to the subspace $\mathbb{V}(c_J, 1)$.

(b) *For any* $h \in \mathbb{V}$ *such that* $z^2(x, y) + h \in \mathcal{K}$, *let* $w = w(x, y) := [z^2(x, y) + h]^{1/2} - z(x, y)$. *Then,* $P_1(c_J)w = \frac{1}{2}\mathcal{L}^{-1}(z(x, y))[P_1(c_J)h] + o(\|h\|)$.

Proof From identity (3.252) and the definition of z, it is evident that $z(x, y) \in \mathcal{K}$ for all $x, y \in \mathbb{V}$. Therefore, by applying arguments similar to those in [41, Lemma 11], the desired result follows. □

Proposition 3.70 *The function* ψ_τ *defined by (3.250) is differentiable everywhere on* $\mathbb{V} \times \mathbb{V}$. *Furthermore,* $\nabla_x \psi_\tau(0, 0) = \nabla_y \psi_\tau(0, 0) = 0$, *and if* $(x, y) \neq (0, 0)$, *then*

$$\begin{aligned} \nabla_x \psi_\tau(x, y) &= \left[\mathcal{L}\Big(x + \frac{\tau - 2}{2}y\Big)\mathcal{L}^{-1}(z(x, y)) - \mathcal{I}\right]\phi_\tau(x, y), \\ \nabla_y \psi_\tau(x, y) &= \left[\mathcal{L}\Big(y + \frac{\tau - 2}{2}x\Big)\mathcal{L}^{-1}(z(x, y)) - \mathcal{I}\right]\phi_\tau(x, y), \end{aligned} \tag{3.257}$$

where $z(x, y)$ *is given by (3.256).*

Proof We prove the conclusion by the following three cases.

Case (1): $(x, y) = (0, 0)$. For any $u, v \in \mathbb{V}$, suppose that $u^2 + v^2 + (\tau - 2)(u \circ v)$ has the spectrum decomposition $u^2 + v^2 + (\tau - 2)(u \circ v) = \sum_{j=1}^r \mu_j d_j$, where $\{d_1, d_2, \ldots, d_r\}$ is the corresponding Jordan frame. Then, for $j = 1, 2, \ldots, r$, we have

$$\begin{aligned} \mu_j = \frac{1}{\|d_j\|^2}\Big\langle \textstyle\sum_{j=1}^r \mu_j d_j, d_j\Big\rangle &= \langle u^2 + v^2 + (\tau - 2)(u \circ v), d_j\rangle \\ &= \left\langle \Big(u + \frac{\tau - 2}{2}v\Big)^2 + \frac{\tau(4 - \tau)}{4}v^2, d_j\right\rangle \\ &\le \left\langle \Big(u + \frac{\tau - 2}{2}v\Big)^2 + \frac{\tau(4 - \tau)}{4}v^2, e\right\rangle \\ &= \|u\|^2 + (\tau - 2)\langle u, v\rangle + \|v\|^2 \\ &\le (\tau/2)(\|u\|^2 + \|v\|^2), \end{aligned} \tag{3.258}$$

where the second equality is by $\|d_j\| = 1$, the first inequality is due to $e = \sum_{j=1}^r d_j$ and $d_j \in \mathcal{K}$ for $j = 1, 2, \ldots, r$, and the last inequality is due to

$$\langle x, y\rangle \le \frac{1}{2}(\|x\|^2 + \|y\|^2) \quad \text{and} \quad \|x + y\|^2 \le 2(\|x\|^2 + \|y\|^2). \tag{3.259}$$

Therefore, we have

$$\begin{aligned} \psi_\tau(u, v) - \psi_\tau(0, 0) &= \frac{1}{2}\left\| [u^2 + v^2 + (\tau - 2)(u \circ v)]^{1/2} - (u + v)\right\|^2 \\ &= \frac{1}{2}\left\| \textstyle\sum_{j=1}^r \sqrt{\mu_j}\, d_j - (u + v)\right\|^2 \end{aligned}$$

$$
\begin{aligned}
&\leq \left\| \sum_{j=1}^{r} \sqrt{\mu_j}\, d_j \right\|^2 + \|u+v\|^2 \\
&\leq \sum_{j=1}^{r} \mu_j \|d_j\|^2 + 2(\|u\|^2 + \|v\|^2) \\
&\leq \left(\frac{1}{2}\tau r + 2 \right) (\|u\|^2 + \|v\|^2),
\end{aligned}
$$

where the first two inequalities are due to (3.259), and the last one is from (3.258). This shows that ψ_τ is differentiable at $(0,0)$ with $\nabla_x \psi_\tau(0,0) = \nabla_y \psi_\tau(0,0) = 0$.

Case (2): $z(x,y) \in \mathrm{int}(\mathcal{K})$. Since $\phi_\tau(x,y) = z(x,y) - (x+y)$, we have from Theorem 1.4 that ϕ_τ is continuously differentiable under this case. Notice that

$$
\psi_\tau(x,y) = \frac{1}{2} \left\langle e, \phi_\tau^2(x,y) \right\rangle,
$$

and hence the function ψ_τ is continuously differentiable. Applying the chain rule yields

$$
\nabla_x \psi_\tau(x,y) = \nabla_x \phi_\tau(x,y) \mathcal{L}(\phi_\tau(x,y)) e = \nabla_x \phi_\tau(x,y) \phi_\tau(x,y). \tag{3.260}
$$

On the other hand, from (3.252) it follows that

$$
\phi_\tau(x,y) = \left[\left(x + \frac{\tau-2}{2} y \right)^2 + \frac{\tau(4-\tau)}{4} y^2 \right]^{1/2} - (x+y),
$$

and therefore using the formulas in (3.254) gives

$$
\nabla_x \phi_\tau(x,y) = \mathcal{L}\left(x + \frac{\tau-2}{2} y \right) \mathcal{L}^{-1}(z(x,y)) - \mathcal{I}.
$$

This, together with (3.260), immediately implies

$$
\nabla_x \psi_\tau(x,y) = \left[\mathcal{L}\left(x + \frac{\tau-2}{2} y \right) \mathcal{L}^{-1}(z(x,y)) - \mathcal{I} \right] \phi_\tau(x,y).
$$

For symmetry of x and y in $\psi_\tau(x,y)$, we also have

$$
\nabla_y \psi_\tau(x,y) = \left[\mathcal{L}\left(y + \frac{\tau-2}{2} x \right) \mathcal{L}^{-1}(z(x,y)) - \mathcal{I} \right] \phi_\tau(x,y).
$$

Case (3): $(x,y) \neq (0,0)$ and $z(x,y) \notin \mathrm{int}(\mathcal{K})$. For any $u, v \in \mathbb{V}$, define

$$
\hat{z} := 2\hat{x} \circ u + 2\hat{y} \circ v + u^2 + v^2 + (\tau - 2) u \circ v
$$

with $\hat{x} = x + \frac{\tau-2}{2}y$ and $\hat{y} = y + \frac{\tau-2}{2}x$. It is not difficult to verify that

$$\begin{aligned} z^2(x,y) + \hat{z} &= \left((x+u) + \frac{\tau-2}{2}(y+v)\right)^2 + \frac{\tau(4-\tau)}{4}(y+v)^2 \\ &= z^2(x+u, y+v) \in \mathcal{K}. \end{aligned}$$

Let

$$w(x,y) := \left(z^2(x,y) + \hat{z}\right)^{1/2} - z(x,y).$$

From the definitions of ψ_τ and $z(x,y)$, it then follows that

$$\begin{aligned} &\psi_\tau(x+u, y+v) - \psi_\tau(x,y) \\ &= \frac{1}{2}\left[\left\|[z^2(x,y)+\hat{z}]^{1/2} - (x+u+y+v)\right\|^2 - \|z(x,y)-(x+y)\|^2\right] \quad (3.261) \\ &= \frac{1}{2}\left[\langle \hat{z}, e\rangle + \|u+v\|^2\right] - \langle w(x,y), x+u+y+v\rangle + \langle x+y-z(x,y), u+v\rangle \\ &= -\langle w(x,y), x+y\rangle + \langle x+y-z(x,y), u+v\rangle + \langle \hat{x}, u\rangle + \langle \hat{y}, v\rangle + o(\|(u,v)\|). \end{aligned}$$

By Lemma 3.46(a), $x + y \in \mathbb{V}(c_J, 1)$. Thus, using Lemma 3.46(b), we have

$$\begin{aligned} \langle w(x,y), x+y\rangle &= \langle P_1(c_J)w(x,y),\ x+y\rangle \\ &= \left\langle \frac{1}{2}\mathcal{L}^{-1}(z(x,y))[P_1(c_J)\hat{z}] + o(\|\hat{z}\|),\ x+y\right\rangle \\ &= \frac{1}{2}\left\langle P_1(c_J)\hat{z},\ \mathcal{L}^{-1}(z(x,y))[x+y]\right\rangle + o(\|\hat{z}\|) \\ &= \left\langle P_1(c_J)\left[\hat{x}\circ u + \hat{y}\circ v\right], \mathcal{L}^{-1}(z(x,y))[x+y]\right\rangle + o(\|(u,v)\|) \\ &= \left\langle \hat{x}\circ u + \hat{y}\circ v,\ P_1(c_J)[\mathcal{L}^{-1}(z(x,y))(x+y)]\right\rangle + o(\|(u,v)\|) \\ &= \left\langle \hat{x}\circ u + \hat{y}\circ v,\ \mathcal{L}^{-1}(z(x,y))(x+y)\right\rangle + o(\|(u,v)\|) \\ &= \left\langle \left[\mathcal{L}^{-1}(z(x,y))(x+y)\right]\circ\hat{x}, u\right\rangle \\ &\quad + \left\langle \left[\mathcal{L}^{-1}(z(x,y))(x+y)\right]\circ\hat{y}, v\right\rangle + o(\|(u,v)\|), \quad (3.262) \end{aligned}$$

where the first equality is since $\mathbb{V} = \mathbb{V}(c_J, 1) \oplus \mathbb{V}(c_J, \frac{1}{2}) \oplus \mathbb{V}(c_J, 0)$, the fifth one is due to $P_1(c_J) = P_1^*(c_J)$, and the sixth is from the fact that $\mathcal{L}^{-1}(z(x,y))(x+y) \in \mathbb{V}(c_J, 1)$. Combining (3.261) with (3.262), we obtain that

$$\begin{aligned} &\psi_\tau(x+u, y+v) - \psi_\tau(x,y) \\ &= \left\langle \hat{x} + x + y - z(x,y) - \left[\mathcal{L}^{-1}(z(x,y))(x+y)\right]\circ\hat{x}, u\right\rangle \\ &\quad + \left\langle \hat{y} + x + y - z(x,y) - \left[\mathcal{L}^{-1}(z(x,y))(x+y)\right]\circ\hat{y}, v\right\rangle + o(\|(u,v)\|). \end{aligned}$$

This implies that the function ψ_τ is differentiable at (x, y), and furthermore,

$$\nabla_x \psi_\tau(x, y) = \hat{x} + x + y - z(x, y) - \left[\mathcal{L}^{-1}(z(x, y))(x + y)\right] \circ \hat{x},$$
$$\nabla_y \psi_\tau(x, y) = \hat{y} + x + y - z(x, y) - \left[\mathcal{L}^{-1}(z(x, y))(x + y)\right] \circ \hat{y}.$$

Notice that

$$\begin{aligned}
&\hat{x} + x + y - z(x, y) - \left[\mathcal{L}^{-1}(z(x, y))(x + y)\right] \circ \hat{x} \\
&= \hat{x} - \phi_\tau(x, y) - \left[\mathcal{L}^{-1}(z(x, y))(x + y)\right] \circ \left(x + \frac{\tau - 2}{2} y\right) \\
&= x + \frac{\tau - 2}{2} y - \phi_\tau(x, y) - \mathcal{L}\left(x + \frac{\tau - 2}{2} y\right)\left[\mathcal{L}^{-1}(z(x, y))(x + y)\right] \\
&= \mathcal{L}\left(x + \frac{\tau - 2}{2} y\right)\mathcal{L}^{-1}(z(x, y))[z(x, y) - x - y] - \phi_\tau(x, y) \\
&= \left[\mathcal{L}\Big(x + \frac{\tau - 2}{2} y\Big)\mathcal{L}^{-1}(z(x, y)) - \mathcal{I}\right]\phi_\tau(x, y),
\end{aligned}$$

where the third equality is due to $\mathcal{L}^{-1}(z(x, y))z(x, y) = e$ and the fact that

$$x + \frac{\tau - 2}{2} y = \mathcal{L}\left(x + \frac{\tau - 2}{2} y\right) e = \mathcal{L}\left(x + \frac{\tau - 2}{2} y\right)\mathcal{L}^{-1}(z(x, y))z(x, y).$$

Therefore,

$$\nabla_x \psi_\tau(x, y) = \left[\mathcal{L}\Big(x + \frac{\tau - 2}{2} y\Big)\mathcal{L}^{-1}(z(x, y)) - \mathcal{I}\right]\phi_\tau(x, y).$$

Similarly, we also have

$$\nabla_y \psi_\tau(x, y) = \left[\mathcal{L}\left(y + \frac{\tau - 2}{2} x\right)\mathcal{L}^{-1}(z(x, y)) - \mathcal{I}\right]\phi_\tau(x, y).$$

This shows that the conclusion holds under this case. The proof is thus complete. □

It should be pointed out that formula (3.257) is well defined even if $z(x, y) \notin \text{int}(\mathcal{K})$ since in this case $\phi_\tau(x, y) \in \mathbb{V}(c_J, 1)$ by Lemma 3.46(a). When $\mathbb{V}$ is specified as the Lorentz algebra $(\mathbb{R}^n, \circ, \langle\cdot, \cdot\rangle_{\mathbb{R}^n})$, the formula reduces to the one of [35, Proposition 3.2]; whereas when $\mathbb{V}$ is specified as $(\mathbb{S}^n, \circ, \langle\cdot, \cdot\rangle_{\mathbb{S}^n})$ and $\tau = 2$, the formula is same as the one in [17, Lemma 6.3(b)] by noting that $z(x, y) = (x^2 + y^2)^{1/2}$ and

$$\begin{aligned}
\nabla_x \psi_\tau(x, y) &= \mathcal{L}(x)\mathcal{L}^{-1}(z(x, y))\phi_{\text{FB}}(x, y) - \phi_{\text{FB}}(x, y) \\
&= x \circ [\mathcal{L}^{-1}(z(x, y))\phi_{\text{FB}}(x, y)] - \mathcal{L}(z(x, y))\mathcal{L}^{-1}(z(x, y))\phi_{\text{FB}}(x, y)
\end{aligned}$$

$$= x \circ [\mathcal{L}^{-1}(z(x,y))\phi_{\text{FB}}(x,y)] - z(x,y) \circ [\mathcal{L}^{-1}(z(x,y))\phi_{\text{FB}}(x,y)]$$
$$= [\mathcal{L}^{-1}(z(x,y))\phi_{\text{FB}}(x,y)] \circ (x - z(x,y)).$$

Thus, formula (3.257) provides a unified framework for the SOCCP and the SDCP settings.

From Proposition 3.70, we immediately derive several properties of the gradient $\nabla\psi_\tau$, which were previously established in the contexts of the NCPs [32] and the SOCCPs [35], respectively.

Proposition 3.71 *Let ψ_τ be given as in (3.250). Then, for any $(x, y) \in \mathbb{V} \times \mathbb{V}$, we have*

(a) $\langle x, \nabla_x\psi_\tau(x,y)\rangle + \langle y, \nabla_y\psi_\tau(x,y)\rangle = \|\phi_\tau(x,y)\|^2$.
(b) $\nabla\psi_\tau(x,y) = 0$ *if and only if* $x \in \mathcal{K},\ y \in \mathcal{K},\ \langle x, y\rangle = 0$.

Proof (a) If $(x, y) = (0, 0)$, the result is clear. Otherwise, from (3.257) it follows that

$$\begin{aligned}
&\langle x, \nabla_x\psi_\tau(x,y)\rangle + \langle y, \nabla_y\psi_\tau(x,y)\rangle \\
&= \left\langle x, \left(x + \frac{\tau-2}{2}y\right) \circ [\mathcal{L}^{-1}(z(x,y))\phi_\tau(x,y)]\right\rangle - \langle x, \phi_\tau(x,y)\rangle \\
&\quad + \left\langle y, \left(y + \frac{\tau-2}{2}x\right) \circ [\mathcal{L}^{-1}(z(x,y))\phi_\tau(x,y)]\right\rangle - \langle y, \phi_\tau(x,y)\rangle \\
&= \left\langle x \circ \left(x + \frac{\tau-2}{2}y\right), \mathcal{L}^{-1}(z(x,y))\phi_\tau(x,y)\right\rangle - \langle x, \phi_\tau(x,y)\rangle \\
&\quad + \left\langle y \circ \left(y + \frac{\tau-2}{2}x\right), \mathcal{L}^{-1}(z(x,y))\phi_\tau(x,y)\right\rangle - \langle y, \phi_\tau(x,y)\rangle \\
&= \langle z^2(x,y), \mathcal{L}^{-1}(z(x,y))\phi_\tau(x,y)\rangle - \langle x + y, \phi_\tau(x,y)\rangle \\
&= \langle z(x,y), \phi_\tau(x,y)\rangle - \langle x + y, \phi_\tau(x,y)\rangle \\
&= \|\phi_\tau(x,y)\|^2,
\end{aligned}$$

where the next-to-last equality is by $z^2 = \mathcal{L}(z)z$ and the symmetry of $\mathcal{L}(z)$.

(b) The proof is direct by part(a), Propositions 3.69 and 3.70. □

Next, we investigate the continuity of the gradients $\nabla_x\psi_\tau(x,y)$ and $\nabla_y\psi_\tau(x,y)$. To this end, for any $\varepsilon > 0$, we define the mapping $z_\varepsilon : \mathbb{V} \times \mathbb{V} \to \mathbb{V}$ by

$$z_\varepsilon = z_\varepsilon(x,y) := \left(x^2 + y^2 + (\tau-2)(x \circ y) + \varepsilon e\right)^{1/2}. \tag{3.263}$$

From (3.252), clearly, $z_\varepsilon(x,y) \in \text{int}(\mathcal{K})$ for any $x, y \in \mathbb{V}$, and hence the operator $\mathcal{L}(z_\varepsilon(x,y))$ is positive definite on $\mathbb{V}$. Since the spectral function induced by $\varphi(t) = \sqrt{t}$ $(t \geq 0)$ is continuous, therefore from Theorem 1.4, it follows that $z_\varepsilon(x,y) \to z(x,y)$ as $\varepsilon \to 0^+$ for any $(x,y) \in \mathbb{V} \times \mathbb{V}$, where $z(x,y)$ is given by (3.256). This means that $\mathcal{L}(z_\varepsilon(x, y)) \to \mathcal{L}(z(x, y))$ as $\varepsilon \to 0^+$. In what follows, we prove that the gradients $\nabla_x \psi_\tau(x, y)$ and $\nabla_y \psi_\tau(x, y)$ are Lipschitz continuous by arguing the Lipschitz continuity of $z_\varepsilon(x, y)$ and the mapping

$$H_\varepsilon(x, y) := \mathcal{L}\Big(x + \frac{\tau - 2}{2} y\Big)\mathcal{L}^{-1}(z_\varepsilon(x, y))(x + y). \tag{3.264}$$

To establish the Lipschitz continuity of $z_\varepsilon(x, y)$, we need the following crucial lemma.

Lemma 3.47 *For any $(x, y) \in \mathbb{V} \times \mathbb{V}$ and $\varepsilon > 0$, let $z_\varepsilon(x, y)$ be defined as in (3.263). Then the function $z_\varepsilon(x, y)$ is continuously differentiable everywhere with*

$$\begin{aligned} \nabla_x z_\varepsilon(x, y) &= \mathcal{L}\Big(x + \frac{\tau - 2}{2} y\Big)\mathcal{L}^{-1}(z_\varepsilon(x, y)), \\ \nabla_y z_\varepsilon(x, y) &= \mathcal{L}\Big(y + \frac{\tau - 2}{2} x\Big)\mathcal{L}^{-1}(z_\varepsilon(x, y)). \end{aligned} \tag{3.265}$$

Furthermore, there exists a constant $C > 0$, independent of x, y and ϵ, τ, such that

$$\|\nabla_x z_\varepsilon(x, y)\| \le C \quad \text{and} \quad \|\nabla_y z_\varepsilon(x, y)\| \le C.$$

Proof The first part follows from Lemma 3.45 and the following fact that

$$\begin{aligned} z_\varepsilon(x, y) &= \left[\Big(x + \frac{\tau - 2}{2} y\Big)^2 + \frac{\tau(4 - \tau)}{4} y^2 + \varepsilon e\right]^{1/2} \\ &= \left[\Big(y + \frac{\tau - 2}{2} x\Big)^2 + \frac{\tau(4 - \tau)}{4} x^2 + \varepsilon e\right]^{1/2}. \end{aligned}$$

We next prove that the operator $\nabla_x z_\varepsilon(x, y)$ is bounded for any $x, y \in \mathbb{V}$ and $\varepsilon > 0$. Let $\{u_1, u_2, \ldots, u_n\}$ be an orthonormal basis of $\mathbb{V}$. For any $x, y \in \mathbb{V}$, let $L(z^2)$, $L(x + \frac{\tau-2}{2} y)$, $L(z_\varepsilon)$ and $L((x + \frac{\tau-2}{2} y)^2)$ be the corresponding matrix representation of the operators $\mathcal{L}(z^2)$, $\mathcal{L}(x + \frac{\tau-2}{2} y)$, $\mathcal{L}(z_\varepsilon)$ and $\mathcal{L}((x + \frac{\tau-2}{2} y)^2)$ with respect to the basis $\{u_1, u_2, \ldots, u_n\}$. Then, by formula (3.265), it suffices to prove that the matrix $L(x + \frac{\tau-2}{2} y)L^{-1}(z_\varepsilon)$ is bounded for any $x, y \in \mathbb{V}$ and $\varepsilon > 0$. The verifications are given as below.

Suppose that $z = z(x, y)$ has the spectral decomposition $z = \sum_{j=1}^r \lambda_j(z)c_j$, where $\lambda_1(z) \ge \lambda_2(z) \ge \cdots \ge \lambda_r(z) \ge 0$ are the eigenvalue of z and $\{c_1, c_2, \ldots, c_r\}$ is the corresponding Jordan frame. From Theorem 1.5, $\mathcal{L}(z)$ has the spectral decomposition

$$\mathcal{L}(z) = \sum_{j=1}^r \lambda_j(z)\mathcal{C}_{jj}(z) + \sum_{1 \le j < l \le r} \frac{1}{2}\big(\lambda_j(z) + \lambda_l(z)\big)\,\mathcal{C}_{jl}(z)$$

with the spectrum $\sigma(\mathcal{L}(z))$ consisting of all distinct numbers in $\{\frac{1}{2}(\lambda_j(z) + \lambda_l(z)) \mid j, l = 1, 2, \ldots, r\}$, and $\mathcal{L}(z^2)$ has the spectral decomposition

$$\mathcal{L}(z^2) = \sum_{j=1}^{r} \lambda_j^2(z) \mathcal{C}_{jj}(z) + \sum_{1\le j<l\le r} \frac{1}{2}\left(\lambda_j^2(z) + \lambda_l^2(z)\right) \mathcal{C}_{jl}(z) \tag{3.266}$$

with $\sigma(\mathcal{L}(z^2))$ consisting of all distinct numbers in $\{\frac{1}{2}\left(\lambda_j^2(z) + \lambda_l^2(z)\right) \mid j, l = 1, 2, \ldots, r\}$. By the definition of $z_\varepsilon(x, y)$, it is easy to verify that $z_\varepsilon = \sum_{j=1}^{r} \sqrt{\lambda_j^2(z) + \varepsilon}\ c_j$, and consequently the symmetric operator $\mathcal{L}(z_\varepsilon)$ has the spectral decomposition

$$\mathcal{L}(z_\varepsilon) = \sum_{j=1}^{r} \sqrt{\lambda_j^2(z) + \varepsilon}\, \mathcal{C}_{jj}(z) + \sum_{1\le j<l\le r} \frac{1}{2}\left(\sqrt{\lambda_j^2(z) + \varepsilon} + \sqrt{\lambda_l^2(z) + \varepsilon}\,\right)\mathcal{C}_{jl}(z) \tag{3.267}$$

with the spectrum $\sigma(\mathcal{L}(z_\varepsilon))$ consisting of all distinct numbers in

$$\left\{\frac{1}{2}\left(\sqrt{\lambda_j^2(z) + \varepsilon} + \sqrt{\lambda_l^2(z) + \varepsilon}\,\right) \mid j, l = 1, 2, \ldots, r\right\}.$$

We first prove that the matrix $L(x + \frac{\tau-2}{2}y)\left(L(z^2) + \epsilon I\right)^{-1/2}$ is bounded for any $x, y \in \mathbb{V}$ and $\epsilon > 0$. For this purpose, let P be an $n \times n$ orthogonal matrix such that

$$PL(z^2)P^{\mathsf{T}} = \mathrm{diag}\left(\lambda_1(L(z^2)), \lambda_2(L(z^2)), \cdots, \lambda_n(L(z^2))\right), \tag{3.268}$$

where $\lambda_1(L(z^2)) \ge \lambda_2(L(z^2)) \cdots \ge \lambda_n(L(z^2)) \ge 0$ are the eigenvalues of $L(z^2)$. Then, it is not hard to verify that for any $\varepsilon > 0$,

$$P\left(L(z^2) + \varepsilon I\right)^{-1/2} P^{\mathsf{T}} = \mathrm{diag}\left(\frac{1}{\sqrt{\lambda_1(L(z^2)) + \varepsilon}}, \cdots, \frac{1}{\sqrt{\lambda_n(L(z^2)) + \varepsilon}}\right).$$

Denote $\widetilde{U} := PL\left(x + \frac{\tau-2}{2}y\right) P^{\mathsf{T}}$. We can compute that

$$\begin{aligned} & L\left(x + \frac{\tau-2}{2}y\right)\left(L(z^2) + \varepsilon I\right)^{-1/2} \\ &= P^{\mathsf{T}}\widetilde{U}\mathrm{diag}\left(\frac{1}{\sqrt{\lambda_1(L(z^2)) + \varepsilon}}, \cdots, \frac{1}{\sqrt{\lambda_n(L(z^2)) + \varepsilon}}\right) P \\ &= P^{\mathsf{T}}\left[\frac{\widetilde{U}_{ik}}{\sqrt{\lambda_k(L(z^2)) + \varepsilon}}\right]_{\substack{1\le i\le n\\ 1\le k\le n}} P. \end{aligned} \tag{3.269}$$

Since $\mathcal{L}(z^2) = \mathcal{L}\left((x + \frac{\tau-2}{2}y)^2\right) + \mathcal{L}\left(\frac{\tau(4-\tau)}{4}y^2\right)$ and $\mathcal{L}(y^2)$ is positive semidefinite, we obtain

$$L(z^2) - L\left((x + \frac{\tau-2}{2}y)^2\right) \succeq O.$$

In addition, by Proposition 1.2, $\mathcal{L}[(x+\frac{\tau-2}{2}y)^2]-\mathcal{L}(x+\frac{\tau-2}{2}y)\mathcal{L}(x+\frac{\tau-2}{2}y)$ is positive semidefinite, and hence we have

$$L\left(\left(x+\frac{\tau-2}{2}y\right)^2\right)-L\left(x+\frac{\tau-2}{2}y\right)L\left(x+\frac{\tau-2}{2}y\right)\succeq O.$$

The last two equations thus imply

$$L(z^2)-L\left(x+\frac{\tau-2}{2}y\right)L\left(x+\frac{\tau-2}{2}y\right)\succeq O. \tag{3.270}$$

Now, for any given $k\in\{1,2,\ldots,n\}$, from (3.268) and (3.270), it follows that

$$\begin{aligned}\lambda_k(L(z^2)) &= \left[PL(z^2)P^{\mathsf T}\right]_{kk}\\ &\geq \left[PL\Big(x+\frac{\tau-2}{2}y\Big)L\Big(x+\frac{\tau-2}{2}y\Big)P^{\mathsf T}\right]_{kk}\\ &= [\widetilde{U}\widetilde{U}]_{kk}=\sum_{i=1}^{n}\widetilde{U}_{ik}^2,\end{aligned}$$

where the inequality is by the fact that the diagonal entries of a positive semidefinite matrix are nonnegative. This immediately yields

$$\sqrt{\lambda_k(L(z^2))+\varepsilon}\geq\sqrt{\textstyle\sum_{i=1}^n\widetilde{U}_{ik}^2}\geq\widetilde{U}_{ik}\quad\forall\, i=1,2,\ldots,n.$$

Combining with Eq. (3.269), there exists a constant $C_1>0$ such that

$$\left\|L\left(x+\frac{\tau-2}{2}y\right)\left(L(z^2)+\varepsilon I\right)^{-1/2}\right\|\leq C_1\quad\forall\, x,y\in\mathbb{V}\text{ and }\varepsilon>0. \tag{3.271}$$

Next, we prove that the matrix $\left(L(z^2)+\varepsilon I\right)^{1/2}L^{-1}(z_\varepsilon)$ is bounded for any $x,y\in\mathbb{V}$ and $\varepsilon>0$. Let $C_{jl}(z)$ for $1\leq j,l\leq r$ be the matrix representation of $\mathcal{C}_{jl}(z)$ with respect to the basis $\{u_1,u_2,\ldots,u_n\}$. From equations (3.266)–(3.267), it then follows that

$$(L(z^2)+\varepsilon I)^{1/2}=\sum_{j=1}^{r}\sqrt{\lambda_j^2(z)+\varepsilon}\,C_{jj}(z)+\sum_{1\leq j<l\leq r}\frac{1}{2}\sqrt{2(\lambda_j^2(z)+\lambda_l^2(z)+2\varepsilon)}\,C_{jl}(z),$$

$$L^{-1}(z_\varepsilon)=\sum_{j=1}^{r}\frac{1}{\sqrt{\lambda_j^2(z)+\varepsilon}}C_{jj}(z)+\sum_{1\leq j<l\leq r}\frac{1}{\left(\sqrt{\lambda_j^2(z)+\varepsilon}+\sqrt{\lambda_l^2(z)+\varepsilon}\right)/2}C_{jl}(z).$$

Using the last two equalities and (1.21), it is easy to compute

$$\left(L(z^2)+\epsilon I\right)^{1/2}L^{-1}(z_\varepsilon)=\sum_{j=1}^{r}C_{jj}(z)+\sum_{1\le j<l\le r}\frac{\sqrt{2(\lambda_j^2(z)+\lambda_l^2(z)+2\varepsilon)}}{\sqrt{\lambda_j^2(z)+\varepsilon}+\sqrt{\lambda_l^2(z)+\varepsilon}}C_{jl}(z). \tag{3.272}$$

Notice that the projection matrix $C_{jl}(z)$ with $1\le j,l\le r$ is bounded for any $x,y\in\mathbb{V}$, and for any $x,y\in\mathbb{V}$ and $\varepsilon>0$

$$\sqrt{\lambda_j^2(z)+\varepsilon}+\sqrt{\lambda_l^2(z)+\varepsilon}\ge\sqrt{\lambda_j^2(z)+\lambda_l^2(z)+2\varepsilon}\quad\forall\,1\le j,l\le r.$$

Hence, from (3.272) we can deduce that $(L(z^2)+\varepsilon I)^{1/2}L^{-1}(z_\varepsilon)$ is bounded for any $x,y\in\mathbb{V}$ and $\varepsilon>0$, i.e., there exists a positive constant C_2 such that

$$\left\|\left(L(z^2)+\varepsilon I\right)^{1/2}L^{-1}(z_\varepsilon)\right\|\le C_2\quad\forall\,x,y\in\mathbb{V}\text{ and }\varepsilon>0. \tag{3.273}$$

Combining (3.273) and (3.271), we have that the matrix $L\left(x+\frac{\tau-2}{2}y\right)L^{-1}(z_\varepsilon)$ is bounded for any $x,y\in\mathbb{V}$ and $\varepsilon>0$, because

$$\begin{aligned}
&\left\|L\left(x+\frac{\tau-2}{2}y\right)L^{-1}(z_\varepsilon)\right\|\\
=&\left\|L\left(x+\frac{\tau-2}{2}y\right)\left[\left(L(z^2)+\varepsilon I\right)^{-1/2}\left(L(z^2)+\varepsilon I\right)^{1/2}\right]L^{-1}(z_\varepsilon)\right\|\\
=&\left\|\left[L\left(x+\frac{\tau-2}{2}y\right)\left(L(z^2)+\varepsilon I\right)^{-1/2}\right]\left[\left(L(z^2)+\varepsilon I\right)^{1/2}L^{-1}(z_\varepsilon)\right]\right\|\\
\le&\left\|L\left(x+\frac{\tau-2}{2}y\right)\left(L(z^2)+\varepsilon I\right)^{-1/2}\right\|\cdot\left\|\left(L(z^2)+\varepsilon I\right)^{1/2}L^{-1}(z_\varepsilon)\right\|\\
\le&C_1C_2,\quad\forall\,x,y\in\mathbb{V}\text{ and }\varepsilon>0.
\end{aligned}$$

Consequently, there exists a constant $C>0$ such that $\|\nabla_x z_\varepsilon(x,y)\|\le C$ for any $x,y\in\mathbb{V}$ and $\varepsilon>0$. For the symmetry, $\|\nabla_y z_\varepsilon(x,y)\|\le C$ also holds for any $x,y\in\mathbb{V}$ and $\varepsilon>0$. From the discussions above, we see that the constant C is also independent of τ. □

Invoking Lemma 3.47 and the Mean-Value Theorem, we establish global Lipschitz continuity of the mapping $z_\varepsilon(x,y)$, as stated in the following proposition.

Proposition 3.72 *For any $x,y\in\mathbb{V}$ and $\varepsilon>0$, let $z_\varepsilon(x,y)$ be defined as in (3.263). Then, the function $z_\varepsilon(x,y)$ is globally Lipschitz continuous.*

Proof For any $(x,y),(a,b)\in\mathbb{V}\times\mathbb{V}$, by applying Mean-Value Theorem yields

$$\begin{aligned}
\|z_\varepsilon(x,y)-z_\varepsilon(a,b)\|&=\|[z_\varepsilon(x,y)-z_\varepsilon(a,y)]+[z_\varepsilon(a,y)-z_\varepsilon(a,b)]\|\\
&=\left\|\int_0^1\nabla_x z_\varepsilon(a+t(x-a),y)(x-a)dt\right.
\end{aligned}$$

$$
\begin{aligned}
&+ \int_0^1 \nabla_y z_\varepsilon(a, b + t(y-b))(y-b)dt\Bigg\| \\
&\le \sqrt{2}\int_0^1 \|\nabla_x z_\varepsilon(a + t(x-a), y)\| \cdot \|x-a\|dt \\
&\quad + \sqrt{2}\int_0^1 \|\nabla_y z_\varepsilon(a, b + t(y-b))\| \cdot \|y-b\|dt \\
&\le \sqrt{2}C(\|x-a\| + \|y-b\|) \\
&\le 2C\|(x, y) - (a, b)\|,
\end{aligned}
$$

where the last two inequalities are respectively by Lemma 3.47 and (3.259). This shows that the function $z_\varepsilon(x, y)$ is globally Lipschitz continuous. □

We now turn our attention to the Lipschitz continuity of the mapping H_ε defined in (3.264). To this end, we show that the partial derivatives $\nabla_x H_\varepsilon(x, y)$ and $\nabla_y H_\varepsilon(x, y)$ remain bounded for all $x, y \in \mathbb{V}$ and $\varepsilon > 0$. The computation of these derivatives relies on the following lemma.

Lemma 3.48 *For any $x, y \in \mathbb{V}$ and $\varepsilon > 0$, let $h \in \mathbb{V}$ be such that $z_\varepsilon^2(x, y) + h \in \mathcal{K}$ and write $w := [z_\varepsilon^2(x, y) + h]^{1/2} - z_\varepsilon(x, y)$. Then, $w = \frac{1}{2}\mathcal{L}^{-1}\big(z_\varepsilon(x, y)\big)h + o(\|h\|)$.*

Proof From the definition of w, it immediately follows that

$$[w + z_\varepsilon(x, y)]^2 = z_\varepsilon^2(x, y) + h,$$

which is equivalent to saying

$$w^2 + 2w \circ z_\varepsilon(x, y) = h \tag{3.274}$$

or

$$h = 2\mathcal{L}\big(z_\varepsilon(x, y)\big)w + w^2. \tag{3.275}$$

We claim that, as $\|h\| \to 0$, there must have $\|w\| \to 0$. Indeed, let $\|h\| \to 0$, then we obtain from (3.274) that $w^2 + 2w \circ z_\varepsilon(x, y) = 0$. Adding $z_\varepsilon^2(x, y)$ to both sides gives

$$\big(w + z_\varepsilon(x, y)\big)^2 = z_\varepsilon^2(x, y).$$

This, by the fact that $w + z_\varepsilon(x, y) \in \mathcal{K}$ and $z_\varepsilon(x, y) \in \mathrm{int}(\mathcal{K})$, implies

$$w + z_\varepsilon(x, y) = z_\varepsilon(x, y),$$

and hence $w = 0$. Since $\mathcal{L}(z_\varepsilon(x, y))$ is invertible on $\mathbb{V}$ and $\|w\| \to 0$ as $\|h\| \to 0$, using the implicit function theorem and Eq. (3.275) yields

$$w_\varepsilon = \frac{1}{2}\mathcal{L}^{-1}\big(z_\varepsilon(x, y)\big)h + o\big(\|h\|\big).$$

Thus, the proof is complete. □

Lemma 3.49 *For any $x, y \in \mathbb{V}$ and $\epsilon > 0$, let $H_\varepsilon(x, y)$ be given as in (3.264). Then, $H_\varepsilon(x, y)$ is differentiable everywhere. Moreover, for any given $u, v \in \mathbb{V}$,*

$$\begin{aligned}
\nabla_x H_\varepsilon(x, y)u &= \left[\mathcal{L}^{-1}\big(z_\varepsilon(x, y)\big)(x+y)\right] \circ u + \mathcal{L}\Big(x + \frac{\tau-2}{2}y\Big)\mathcal{L}^{-1}(z_\varepsilon(x, y)) \\
&\quad \left[u - \mathcal{L}^{-1}\big(z_\varepsilon(x, y)\big)(x+y) \circ \mathcal{L}^{-1}(z_\varepsilon(x, y))\mathcal{L}\Big(x + \frac{\tau-2}{2}y\Big)u\right], \\
\nabla_y H_\varepsilon(x, y)v &= \frac{\tau-2}{2}\left[\mathcal{L}^{-1}\big(z_\varepsilon(x, y)\big)(x+y)\right] \circ v + \mathcal{L}\Big(x + \frac{\tau-2}{2}y\Big)\mathcal{L}^{-1}(z_\varepsilon(x, y)) \\
&\quad \left[v - \mathcal{L}^{-1}\big(z_\varepsilon(x, y)\big)(x+y) \circ \mathcal{L}^{-1}(z_\varepsilon(x, y))\mathcal{L}\Big(y + \frac{\tau-2}{2}x\Big)v\right].
\end{aligned} \tag{3.276}$$

Proof For any $x, y \in \mathbb{V}$ and any given $u, v \in \mathbb{V}$, let $x' = x + \frac{\tau-2}{2}y$, $y' = y + \frac{\tau-2}{2}x$ and

$$h := 2x' \circ u + 2y' \circ v + u^2 + v^2 + (\tau - 2)u \circ v.$$

It is easy to verify that

$$\begin{aligned}
z_\varepsilon^2(x, y) + h &= (x+u)^2 + (y+v)^2 + (\tau-2)[(x+u) \circ (y+v)] + \varepsilon e \\
&= z_\varepsilon^2(x+u, y+v) \in \operatorname{int}(\mathcal{K}).
\end{aligned}$$

Let

$$w := \left[z_\varepsilon^2(x, y) + h\right]^{1/2} - z_\varepsilon(x, y).$$

Then,

$$w + z_\varepsilon(x, y) = \left[z_\varepsilon^2(x, y) + h\right]^{1/2} = z_\varepsilon(x+u, y+v) \in \operatorname{int}(\mathcal{K}).$$

Applying Lemma 3.48 then leads to

$$w = \frac{1}{2}\mathcal{L}^{-1}(z_\varepsilon(x, y))h + o(\|(u, v)\|), \tag{3.277}$$

which implies that $w \to 0$ as $u \to 0$, $v \to 0$ and $w = O(\|(u, v)\|)$. Write

$$g := \mathcal{L}^{-1}(z_\varepsilon(x, y))(x+y) \quad \text{and} \quad g + s := \mathcal{L}^{-1}(z_\varepsilon(x, y) + w)(x+u+y+v). \tag{3.278}$$

We next express s in terms of g, w, u, v and $z_\varepsilon(x, y)$. By (3.278), it is clear that

$$\mathcal{L}(z_\varepsilon(x, y))g = x + y \quad \text{and} \quad \mathcal{L}(z_\varepsilon(x, y) + w)(g + s) = x + u + y + v,$$

which in turn implies

$$\mathcal{L}(z_\varepsilon(x, y))s = u + v - w \circ g - w \circ s,$$

and

$$s = \mathcal{L}^{-1}(z_\varepsilon(x, y))(u + v - w \circ g - w \circ s). \tag{3.279}$$

Using (3.277) and (3.279), we have that $\|s\| \to 0$ as $\|(u, v)\| \to 0$. This, together with (3.277), means that $w \circ s = o(\|w\|) = o(\|(u, v)\|)$. Therefore,

$$\mathcal{L}^{-1}(z_\varepsilon(x, y))(w \circ s) = o(\|(u, v)\|)$$

and

$$s = \mathcal{L}^{-1}(z_\varepsilon(x, y))(u + v - w \circ g) + o(\|(u, v)\|).$$

Now, from the above discussions and the definition of H_ε, it follows that

$$\begin{aligned}
&H_\varepsilon(x + u, y + v) - H_\varepsilon(x, y)\\
&= \mathcal{L}\left(x + u + \frac{\tau - 2}{2}(y + v)\right)\mathcal{L}^{-1}(z_\varepsilon(x, y) + w)(x + u + y + v)\\
&\quad -\mathcal{L}\left(x + \frac{\tau - 2}{2}y\right)\mathcal{L}^{-1}(z_\varepsilon(x, y))(x + y)\\
&= \mathcal{L}\left(x + u + \frac{\tau - 2}{2}(y + v)\right)(g + s) - \mathcal{L}\left(x + \frac{\tau - 2}{2}y\right)g\\
&= \mathcal{L}\left(x + \frac{\tau - 2}{2}y\right)s + \mathcal{L}\left(u + \frac{\tau - 2}{2}v\right)(g + s)\\
&= \mathcal{L}\left(x + \frac{\tau - 2}{2}y\right)\left[\mathcal{L}^{-1}(z_\varepsilon(x, y))(u + v - g \circ w)\right] + \mathcal{L}\left(u + \frac{\tau - 2}{2}v\right)g + o(\|(u, v)\|)\\
&= \mathcal{L}\left(x + \frac{\tau - 2}{2}y\right)\mathcal{L}^{-1}(z_\varepsilon(x, y))(u + v) - \mathcal{L}\left(x + \frac{\tau - 2}{2}y\right)\mathcal{L}^{-1}(z_\varepsilon(x, y))\\
&\quad\left[\mathcal{L}^{-1}(z_\varepsilon(x, y))(x + y) \circ \left(\mathcal{L}^{-1}(z_\varepsilon(x, y))\mathcal{L}(x')u + \mathcal{L}^{-1}(z_\varepsilon(x, y))\mathcal{L}(y')v\right)\right]\\
&\quad +\mathcal{L}\left(u + \frac{\tau - 2}{2}v\right)\mathcal{L}^{-1}(z_\varepsilon(x, y))(x + y) + o(\|(u, v)\|).
\end{aligned}$$

This means that H_ε is differentiable at the point (x, y). Also, the formulas of $\nabla_x H_\varepsilon(x, y)u$ and $\nabla_x H_\varepsilon(x, y)v$ are exactly given by (3.276). The proof is then complete. □

Lemma 3.50 *For any $x, y \in \mathbb{V}$ and $\varepsilon > 0$, let $H_\varepsilon(x, y)$ be defined as in (3.264). Then, for any given $u, v \in \mathbb{V}$, there exists a constant $C > 0$ independent of x, y and ε, τ such that*

$$\|\nabla_x H_\varepsilon(x, y)u\| \le C\tau^{-1}\|u\| \quad \text{and} \quad \|\nabla_y H_\varepsilon(x, y)v\| \le C\tau^{-1}\|v\|.$$

Proof By Lemma 3.47, there exists a constant $\bar{C} > 0$ independent of x, y, ε, τ such that

$$\left\| \mathcal{L}\left(x + \frac{\tau - 2}{2} y\right) \mathcal{L}^{-1}(z_\varepsilon(x, y)) \right\| \le \bar{C} \text{ and } \left\| \mathcal{L}\left(y + \frac{\tau - 2}{2} x\right) \mathcal{L}^{-1}(z_\varepsilon(x, y)) \right\| \le \bar{C}.$$

Hence, their adjoint operators $\mathcal{L}^{-1}(z_\varepsilon(x, y))\mathcal{L}(x + \frac{\tau-2}{2} y)$ and $\mathcal{L}^{-1}(z_\varepsilon(x, y))\mathcal{L}(y + \frac{\tau-2}{2} x)$ are also bounded for any $x, y \in \mathbb{V}$ and $\varepsilon > 0$, i.e.,

$$\left\| \mathcal{L}^{-1}\big(z_\varepsilon(x, y)\big) \mathcal{L}\left(x + \frac{\tau - 2}{2} y\right) \right\| \le \bar{C} \text{ and } \left\| \mathcal{L}^{-1}\big(z_\varepsilon(x, y)\big) \mathcal{L}\left(y + \frac{\tau - 2}{2} x\right) \right\| \le \bar{C}.$$

Noting that

$$\mathcal{L}^{-1}\big(z_\varepsilon(x, y)\big)(x + y) = \frac{2}{\tau} \mathcal{L}^{-1}(z_\varepsilon(x, y)) \left[\mathcal{L}\left(x + \frac{\tau - 2}{2} y\right) e + \mathcal{L}\left(y + \frac{\tau - 2}{2} x\right) e \right],$$

we also have

$$\left\| \mathcal{L}^{-1}\big(z_\varepsilon(x, y)\big)(x + y) \right\| \le 4\bar{C}\tau^{-1}. \tag{3.280}$$

Thus, by the formulas of $\nabla_x H_\varepsilon(x, y)u$ and $\nabla_y H_\varepsilon(x, y)v$, we get the desired result. □

By applying Lemmas 3.49 and 3.50, and following the same reasoning as in the proof of Proposition 3.72, we establish the global Lipschitz continuity of $H_\varepsilon(x, y)$, as summarized in Proposition 3.73. Moreover, the Lipschitz continuities of $\nabla_x \psi_\tau$ and $\nabla_y \psi_\tau$ are also shown in Proposition 3.74.

Proposition 3.73 *For any $x, y \in \mathbb{V}$ and $\varepsilon > 0$, let $H_\varepsilon(x, y)$ be defined as in (3.264). Then the function $H_\varepsilon(x, y)$ is globally Lipschitz continuous with the Lipschitz constant being $C\tau^{-1}$, where $C > 0$ is independent of x, y and ε, τ.*

Proposition 3.74 *The function ψ_τ has a Lipschitz continuous gradient with the Lipschitz constant being positive multiple of $1 + \tau^{-1}$, i.e., there exists a constant $C > 0$ such that*

$$\begin{aligned} \|\nabla_x \psi_\tau(x, y) - \nabla_x \psi_\tau(a, b)\| &\le C(1 + \tau^{-1})\|(x, y) - (a, b)\|, \\ \|\nabla_y \psi_\tau(x, y) - \nabla_y \psi_\tau(a, b)\| &\le C(1 + \tau^{-1})\|(x, y) - (a, b)\|, \end{aligned}$$

for any $(x, y), (a, b) \in \mathbb{V} \times \mathbb{V}$, where C is independent of $(x, y), (a, b)$ and ε, τ. In other words, ψ_τ is an LC^1 function.

Proof For the symmetry, we only need to prove the first inequality. By (3.257), we know

$$\nabla_x \psi_\tau(x, y) = 2x + \frac{\tau}{2} y - z(x, y) - \mathcal{L}\left(x + \frac{\tau - 2}{2} y\right) \mathcal{L}^{-1}(z(x, y))(x + y).$$

For any $\varepsilon > 0$, let $G_\varepsilon : \mathbb{V} \times \mathbb{V} \to \mathbb{V}$ be the mapping defined by

$$G_\varepsilon(x, y) := 2x + \frac{\tau}{2} y - z_\varepsilon(x, y) - H_\varepsilon(x, y).$$

Then, from Propositions 3.72 and 3.73, it follows that $G_\varepsilon(x, y)$ is globally Lipschitz continuous with the Lipschitz constant being $C(1 + \tau^{-1})$, i.e., for all $(x, y), (a, b) \in \mathbb{V} \times \mathbb{V}$,

$$\|G_\varepsilon(x, y) - G_\varepsilon(a, b)\| \leq C(1 + \tau^{-1})\|(x, y) - (a, b)\|. \tag{3.281}$$

We next show that for any $(x, y) \in \mathbb{V} \times \mathbb{V}$, $G_\varepsilon(x, y) \to \nabla_x \psi_\tau(x, y)$ as $\varepsilon \to 0^+$. Indeed, if $(x, y) = (0, 0)$, then $H_\varepsilon(0, 0) \to 0$ by (3.264) and (3.280), and so $G_\varepsilon(0, 0) \to 0 = \nabla_x \psi_\tau(0, 0)$. If $(x, y) \neq (0, 0)$, noting that $z_\varepsilon(x, y) \to z(x, y)$ as $\varepsilon \to 0^+$, it suffices to prove that

$$\mathcal{L}^{-1}\big(z_\varepsilon(x, y)\big)(x + y) \;\to\; \mathcal{L}^{-1}(z(x, y))(x + y). \tag{3.282}$$

If $(x, y) \neq (0, 0)$ such that $z \in \text{int}(\mathcal{K})$, then $\mathcal{L}(z)$ is positive definite on $\mathbb{V}$. If $(x, y) \neq (0, 0)$ such that $z \notin \text{int}(\mathcal{K})$, then from Sect. 3 it follows that $\mathcal{L}(z)$ is positive definite on the subspace $\mathbb{V}(c_J, 1)$. By the proof of [64, Lemma 4.1(ii)], $\mathcal{L}^{-1}(z)$ is then continuous on $\mathbb{V}$ or $\mathbb{V}(c_J, 1)$, which implies the result of (3.282). Thus, we show that $G_\varepsilon(x, y) \to \nabla_x \psi_\tau(x, y)$ as $\varepsilon \to 0^+$ for any $(x, y) \in \mathbb{V} \times \mathbb{V}$. Now taking $\varepsilon \to 0^+$ in (3.281) and applying the relation between $G_\varepsilon(x, y)$ and $\nabla_x \psi_\tau(x, y)$ shown as above, we get the desired result. □

Besides (3.251), we also consider a class of regularized functions for f_τ defined as

$$\widehat{f_\tau}(\zeta) := \psi_0(G(\zeta) \circ F(\zeta)) + \psi_\tau(G(\zeta), F(\zeta)), \tag{3.283}$$

where $\psi_0 : \mathbb{V} \to \mathbb{R}_+$ is continuously differentiable and satisfies

$$\psi_0(u) = 0 \quad \forall u \in -\mathcal{K} \quad \text{and} \quad \psi_0(u) \geq \beta\|(u)_+\| \quad \forall u \in \mathbb{V} \tag{3.284}$$

for some constant $\beta > 0$. Using the properties of ψ_0 in (3.284), it is not hard to verify that $\widehat{f_\tau}$ is a merit function for the SCCP (3.233). The class of functions will reduce to the one studied in [41] if $\tau = 2$ and G degenerates into an identity transformation. As below, we show that the class of merit functions provide a global error bound for the solution of the SCCP under the condition that G and F have the joint uniform Cartesian P-property.

Lemma 3.51 *For any $x, y \in \mathbb{V}$, let ψ_τ be defined as in (3.250). Then, there holds*

$$4\psi_\tau(x, y) \geq 2\,\|[\phi_\tau(x, y)]_+\|^2 \geq \frac{4 - \tau}{2}\Big[\|(-x)_+\|^2 + \|(-y)_+\|^2\Big].$$

Proof The first inequality is due to Lemma 1.1(a) and the definition of ψ_τ. We next prove the second inequality. From (3.252) and Lemma 1.1(b), it follows that

$$\left[x^2+y^2+(\tau-2)(x\circ y)\right]^{1/2}-\left(x+\frac{\tau-2}{2}y\right)\in\mathcal{K},$$
$$\left[x^2+y^2+(\tau-2)(x\circ y)\right]^{1/2}-\left(y+\frac{\tau-2}{2}x\right)\in\mathcal{K}.$$

Combining with Lemma 1.1(c), we then obtain

$$\begin{aligned}
&2\left\|\left[(x^2+y^2+(\tau-2)x\circ y)^{1/2}-x-y\right]_+\right\|^2\\
&=\left\|\left[\left(x^2+y^2+(\tau-2)(x\circ y)\right)^{1/2}-\left(x+\frac{\tau-2}{2}y\right)-\frac{4-\tau}{2}y\right]_+\right\|^2\\
&\quad+\left\|\left[\left(x^2+y^2+(\tau-2)(x\circ y)\right)^{1/2}-\left(y+\frac{\tau-2}{2}x\right)-\frac{4-\tau}{2}x\right]_+\right\|^2\\
&\geq\frac{4-\tau}{2}\|(-y)_+\|^2+\frac{4-\tau}{2}\|(-x)_+\|^2.
\end{aligned}$$

Thus, the proof is complete. □

We now demonstrate that the regularized merit function $\widehat{f}_\tau$ furnishes a global error bound for the solution of the SCCP, provided that the pair (G, F) satisfies the joint uniform Cartesian P-property.

Proposition 3.75 *Let $\widehat{f}_\tau$ be defined as in (3.283)–(3.284). Suppose that G and F have the joint uniform Cartesian P-property and the SCCP* (3.233) *has a solution, denoted by ζ^*. Then, there exists a constant $\kappa > 0$ such that for any $\zeta \in \mathbb{V}$,*

$$\kappa\|\zeta-\zeta^*\|^2\leq\beta^{-1}\widehat{f}_\tau(\zeta)+\frac{4}{\sqrt{4-\tau}}\left(\widehat{f}_\tau(\zeta)\right)^{1/2}. \tag{3.285}$$

Proof Since G and F have the joint uniform Cartesian P-property, there exists a constant $\rho > 0$ such that, for any $\zeta \in \mathbb{V}$, there is an index $i \in \{1, 2, \ldots, m\}$ such that

$$\begin{aligned}
\rho\|\zeta-\zeta^*\|^2&\leq\langle G_i(\zeta)-G_i(\zeta^*), F_i(\zeta)-F_i(\zeta^*)\rangle\\
&=\langle G_i(\zeta), F_i(\zeta)\rangle+\langle F_i(\zeta^*), -G_i(\zeta)\rangle+\langle -F_i(\zeta), G_i(\zeta^*)\rangle\\
&\leq\langle G_i(\zeta), F_i(\zeta)\rangle+\langle [-G_i(\zeta)]_+, F_i(\zeta^*)\rangle+\langle [-F_i(\zeta)]_+, G_i(\zeta^*)\rangle\\
&\leq\lambda_{\max}[G_i(\zeta)\circ F_i(\zeta)]+\|F_i(\zeta^*)\|\|[-G_i(\zeta)]_+\|+\|G_i(\zeta^*)\|\|[-F_i(\zeta)]_+\|\\
&\leq\max\left\{1, \|G_i(\zeta^*)\|, \|F_i(\zeta^*)\|\right\}\\
&\qquad\times\left[\lambda_{\max}[(G_i(\zeta)\circ F_i(\zeta))_+]+\|[-G_i(\zeta)]_+\|+\|[-F_i(\zeta)]_+\|\right]\\
&\leq\max\left\{1, \|G(\zeta^*)\|, \|F(\zeta^*)\|\right\}\\
&\qquad\times\left[\|[G_i(\zeta)\circ F_i(\zeta)]_+\|+\|[-G_i(\zeta)]_+\|+\|[-F_i(\zeta)]_+\|\right],
\end{aligned}$$

where the equality is since $\langle G_i(\zeta^*), F_i(\zeta^*)\rangle = 0$, the second inequality is due to Lemma 1.1(b), and the third one follows from Proposition 2.1 of [42] and the Cauchy–Schwarz inequality. Setting $\kappa := \dfrac{\rho}{\max\{1, \|G(\zeta^*)\|, \|F(\zeta^*)\|\}}$, we immediately obtain

$$\kappa\|\zeta - \zeta^*\|^2 \le \|[G_i(\zeta) \circ F_i(\zeta)]_+\| + \|[-G_i(\zeta)]_+\| + \|[-F_i(\zeta)]_+\|.$$

From the conditions given by (3.284), clearly, for any $i \in \{1, 2, \dots, m\}$, we have

$$\|[G_i(\zeta) \circ F_i(\zeta)]_+\| \le \beta^{-1}\psi_0 (G_i(\zeta) \circ F_i(\zeta)) \le \beta^{-1}\widehat{f}_\tau(\zeta).$$

In addition, applying Lemma 3.51, we achieve

$$\begin{aligned}
\|[-G_i(\zeta)]_+\| + \|[-F_i(\zeta)]_+\| &\le \sqrt{2}\left(\|[-G(\zeta)]_+\|^2 + \|[-F(\zeta)]_+\|^2\right)^{1/2} \\
&\le \frac{4}{\sqrt{4-\tau}}\psi_\tau(G(\zeta), F(\zeta))^{1/2} \\
&\le \frac{4}{\sqrt{4-\tau}}\left(\widehat{f}_\tau(\zeta)\right)^{1/2}.
\end{aligned}$$

Combining the last three inequalities immediately yields the desired result (3.285). □

For the NCP, Kanzow and Kleinmichel [32] established that the merit function f_τ provides a global error bound when the mapping F is a Lipschitz continuous uniform Pfunction. In contrast, Proposition 3.75 reveals that the regularized merit function $\widehat{f}_\tau$ does not require the Lipschitz continuity of F. This distinction arises because the former relies on the global error bound property of $\psi_{\rm NR}$, along with the similar growth behavior shared by ψ_τ and $\psi_{\rm NR}$, whereas the regularized merit function $\widehat{f}_\tau$ achieves the same goal through the inclusion of the regularization term ψ_0. When the SCCP (3.233) reduces to the special case (3.234), that is,

$$\zeta \in \mathcal{K}, \;\; F(\zeta) \in \mathcal{K}, \;\; \langle \zeta, F(\zeta)\rangle = 0.$$

Proposition 3.2 and Corollary 3.1 of [42] indicate that the assumption of the existence of a solution x^* can be omitted from Proposition 3.75, due to the fact that the uniform Cartesian P-property implies the uniform Jordan P-property. Moreover, we note that for the SCCP (3.234), a similar global error bound result was previously established in [41] under the stronger assumption that F possesses the uniform P^*-property. This condition can be more restrictive than the uniform Cartesian P-property in certain settings. For instance, when F is an affine mapping of the form $F(\zeta) = M\zeta + q$, and $\mathbb{V}$ is the Lorentz algebra with $\dim(\mathbb{V}) \ge 5$, Sun showed that F satisfies the uniform P^*-property if and only if M is positive definite. It is evident that the positive definiteness of M ensures the Cartesian P-property of F, but the converse does not necessarily hold. For example, for $\mathbb{V} = \mathbb{V}_1 \times \mathbb{V}_2 \times \mathbb{V}_3$ with $\dim(\mathbb{V}_1) = \dim(\mathbb{V}_2) =$

2 and $\dim(\mathbb{V}_3) = 1$, let M be a block diagonal matrix composed of $\begin{bmatrix} 1 & 0 \\ 0 & 1 \end{bmatrix}$, $\begin{bmatrix} 1 & 1 \\ -1 & 1 \end{bmatrix}$ and 1. It is easy to verify that $F(\zeta) = M\zeta + q$ for any $q \in \mathbb{V}$ has the Cartesian P-property, but M is not positive definite.

Now, we provide a condition to guarantee the boundedness of the level sets

$$\mathcal{L}_{\widehat{f}_\tau}(\gamma) := \left\{\zeta \in \mathbb{V} \,\middle|\, \widehat{f}_\tau(\zeta) \le \gamma\right\},$$

for any $\gamma \ge 0$. Specifically, we will prove that the following condition is sufficient.

Assumption 3.1 For any sequence $\{\zeta^k\} \subseteq \mathbb{V}$ such that

$$\|\zeta^k\| \to +\infty, \quad \left\|[-G(\zeta^k)]_+\right\| < +\infty, \quad \left\|[-F(\zeta^k)]_+\right\| < +\infty, \tag{3.286}$$

there holds that

$$\max_{1\le i\le m} \lambda_{\max}\left[G_i(\zeta^k) \circ F_i(\zeta^k)\right] \to +\infty. \tag{3.287}$$

Proposition 3.76 *If the mappings G and F satisfy Assumption 3.1, then the level sets $\mathcal{L}_{\widehat{f}_\tau}(\gamma)$ of $\widehat{f}_\tau$ for all $\gamma \ge 0$ are bounded.*

Proof Assume on the contrary that there is an unbounded sequence $\{\zeta^k\} \subseteq \mathcal{L}_{\widehat{f}_\tau}(\gamma)$ for some $\gamma \ge 0$. Then, $f_\tau(\zeta^k) \le \widehat{f}_\tau(\zeta^k) \le \gamma$ for all k. By Lemma 3.51, G and F satisfy (3.286). Hence, there is $\nu \in \{1, \ldots, m\}$ such that $\lambda_{\max}\left[G_\nu(\zeta^k) \circ F_\nu(\zeta^k)\right] \to +\infty$. Noting that

$$\lambda_{\max}\left[G_\nu(\zeta^k) \circ F_\nu(\zeta^k)\right] \le \lambda_{\max}\left[(G_\nu(\zeta^k) \circ F_\nu(\zeta^k))_+\right] \le \left\|[G_\nu(\zeta^k) \circ F_\nu(\zeta^k)]_+\right\| \le \beta^{-1}\widehat{f}_\tau(\zeta^k),$$

we have $\widehat{f}_\tau(\zeta^k) \to +\infty$. This contradicts the fact that $\{\zeta^k\} \subseteq \mathcal{L}_{\widehat{f}_\tau}(\gamma)$. □

Assumption 3.1 is a relatively mild condition that nonetheless ensures the boundedness of the level sets of $\widehat{f}_\tau$. In fact, this assumption is satisfied by SCCPs involving jointly monotone mappings with a strictly feasible point, as well as those possessing the joint Cartesian R_{02}-property. To substantiate this claim, we present the following technical lemma, which may be viewed as an extension of Lemma 3.8(b) to the context of symmetric cones.

Lemma 3.52 *Let $\{x^k\} \subseteq \mathbb{V}$ be any sequence satisfying $\|x^k\| \to +\infty$. If the sequence $\{\lambda_{\min}(x^k)\}$ is bounded below, then $\langle (x^k)_+, \hat{x}\rangle \to +\infty$ for any $\hat{x} \in \operatorname{int}(\mathcal{K})$.*

Proof For every k, let x^k have the spectral decomposition $x^k = \sum_{j=1}^r \lambda_j(x^k)q_j^k$ with $\{q_1^k, \ldots, q_r^k\}$ being the corresponding Jordan frame. Let $\hat{x}$ have the spectral decomposition $x = \sum_{j=1}^r \lambda_j(\hat{x})c_j$ with $\{c_1, \ldots, c_r\}$ being the corresponding Jordan frame. Without loss of generality, suppose that $\lambda_{l_k}(x^k) = \lambda_{\max}(x^k)$, where $1 \le l_k \le r$. Then, for every k,

$$\begin{aligned}
\langle (x^k)_+, \hat{x} \rangle &= \left\langle \sum_{j=1}^{r} (\lambda_j(x^k))_+ q_j^k,\ \sum_{j=1}^{r} \lambda_j(\hat{x}) c_j \right\rangle \\
&\geq \lambda_{\max}((x^k)_+) \lambda_{\min}(\hat{x}) \left\langle q_{l_k}^k,\ \textstyle\sum_{j=1}^{r} c_j \right\rangle \\
&= \lambda_{\max}((x^k)_+) \lambda_{\min}(\hat{x}) \langle q_{l_k}^k, e \rangle, \qquad (3.288)
\end{aligned}$$

where the inequality holds since $q_j^k, c_j \in \mathcal{K}$ and $\lambda_j((x^k)_+), \lambda_j(\hat{x}) \geq 0$ for all $j = 1, 2, \ldots, r$. Notice that $\|(x^k)_-\| < +\infty$ as $k \to \infty$ since $\{\lambda_{\min}(x^k)\}$ is bounded below. Using the fact that $\|x^k\|^2 = \|(x^k)_+\|^2 + \|(x^k)_-\|^2$ and $\|x^k\| \to +\infty$, we then have that $\|(x^k)_+\| \to +\infty$. This, together with $(x^k)_+ \in \mathcal{K}$, immediately implies

$$\lambda_{\max}((x^k)_+) \to +\infty. \qquad (3.289)$$

Since $\{q_{l_k}^k\}$ is bounded, we assume (subsequencing if necessary) that $\lim_{k\to+\infty} q_{l_k}^k = q^*$. By the closedness of $\mathcal{K}$ and $\|q_{l_k}^k\| = 1$ for each k, we have $q^* \in \mathcal{K} \setminus \{0\}$. From [1, Proposition I.1.4], it then follows that $\langle q^*, e \rangle > 0$ since $e \in \text{int}(\mathcal{K})$. Thus, taking the limit on the both sides of (3.288) and using the Eq. (3.289), we readily obtain that $\langle (x^k)_+, \hat{x} \rangle \to +\infty$. □

Proposition 3.77 *Assumption 3.1 is satisfied if one of the following statements holds:*

(a) G and F are jointly monotone mappings with $\|G(\zeta)\| + \|F(\zeta)\| \to +\infty$ as $\|\zeta\| \to +\infty$ and there exists a point $\hat{\zeta} \in \mathcal{V}$ such that $G(\hat{\zeta}), F(\hat{\zeta}) \in \text{int}(\mathcal{K})$;
(b) G and F have the joint Cartesian R_{02}-property.

Proof (a) Let $\{\zeta^k\}$ be a sequence satisfying (3.286). Since G and F are jointly monotone,

$$\left\langle G(\zeta^k) - G(\hat{\zeta}), F(\zeta^k) - F(\hat{\zeta}) \right\rangle \geq 0,$$

which by Lemma 1.1(a) is equivalent to

$$\begin{aligned}
\left\langle G(\zeta^k), F(\zeta^k) \right\rangle + \left\langle G(\hat{\zeta}), F(\hat{\zeta}) \right\rangle \geq & \left\langle [G(\zeta^k)]_+, F(\hat{\zeta}) \right\rangle + \left\langle [G(\zeta^k)]_-, F(\hat{\zeta}) \right\rangle \\
& + \left\langle [F(\zeta^k)]_+, G(\hat{\zeta}) \right\rangle + \left\langle [F(\zeta^k)]_-, G(\hat{\zeta}) \right\rangle.
\end{aligned}$$

Notice that the sequences $\{\lambda_{\min}(G(\zeta^k))\}$ and $\{\lambda_{\min}(F(\zeta^k))\}$ are bounded below by (3.286), $\|G(\zeta^k)\| + \|F(\zeta^k)\| \to +\infty$ and $G(\hat{\zeta}), F(\hat{\zeta}) \in \text{int}(\mathcal{K})$. Using Lemma 3.52 then yields

$$\left\langle [G(\zeta^k)]_+, F(\hat{\zeta}) \right\rangle + \left\langle [F(\zeta^k)]_+, G(\hat{\zeta}) \right\rangle \to +\infty.$$

In addition, by (3.286) it is easy to verify

$$\left\langle [G(\zeta^k)]_-, F(\hat{\zeta}) \right\rangle > -\infty \quad \text{and} \quad \left\langle [F(\zeta^k)]_-, G(\hat{\zeta}) \right\rangle > -\infty.$$

Therefore, from the last three equations it follows that

$$\sum_{i=1}^{m} \left\langle G_i(\zeta^k), F_i(\zeta^k) \right\rangle = \left\langle G(\zeta^k), F(\zeta^k) \right\rangle \to +\infty,$$

which in turn implies that there exists an index ν such that $\langle G_\nu(\zeta^k), F_\nu(\zeta^k) \rangle \to +\infty$. By [42, Proposition 2.1(ii)], we have $\lambda_{\max}\left[G_\nu(\zeta^k) \circ F_\nu(\zeta^k)\right] \to +\infty$, which implies (3.287).

(b) The proof is direct by Definition 1.15. □

When $G(\zeta) = \zeta$ for any $\zeta \in \mathbb{V}$, Liu, Zhang and Wang [41] established the boundness of the level sets of $L_{\widehat{f_\tau}}(\gamma)$ for $\tau = 2$ under the condition that F is a R_{02}-function. The condition, in view of Proposition 1.10, is stronger than the one of Proposition 3.77(b). Thus, Proposition 3.77(b) generalizes the result of [41, Theorem 7].

Regarding the LT-type complementarity functions associated with SOC, f_{LT} and $\widehat{f_{\text{LT}}}$ studied in (3.194) and (3.197), there was an extension to symmetric cone setting in [41]. More specifically, Liu, Zhang, and Wang considered the following merit function:

$$\phi_{\text{LT}}^{\text{sc}}(\zeta) := \varphi(\zeta \circ F(\zeta)) + \psi(\zeta, F(\zeta)), \tag{3.290}$$

where $\varphi : \mathbb{V} \to \mathbb{R}_+$ satisfies

$$\varphi(t) = 0 \iff -t \in \mathcal{K}, \tag{3.291}$$

and $\psi : \mathbb{V} \times \mathbb{V} \to \mathbb{R}_+$ satisfies

$$\psi(u, v) = 0,\ u \circ v \in -\mathcal{K} \iff u \in \mathcal{K},\ v \in \mathcal{K},\ \langle u, v \rangle = 0. \tag{3.292}$$

Proposition 3.78 *[41, Lemma 3.1] Let $\phi_{\text{LT}}^{\text{sc}} : \mathbb{V} \times \mathbb{V} \to \mathbb{R}_+$ be defined as*

$$\phi_{\text{LT}}^{\text{sc}}(x, y) := \varphi(x \circ y) + \psi(x, y), \tag{3.293}$$

where φ and ψ satisfy (3.291) *and* (3.292)*, respectively. Then, $\phi_{\text{LT}}^{\text{sc}}$ is a C-function associated with symmetric cone.*

Proof "⇒" Suppose that $\phi_{\text{LT}}^{\text{sc}}(x, y) = 0$ holds. From (3.293), it is clear that $\varphi(x, y) = 0$ and $\psi(x, y) = 0$. Then, applying (3.291) and (3.292) leads to $x \in \mathcal{K}$, $y \in \mathcal{K}$ and $\langle x, y \rangle = 0$.

"⇐" Suppose that $x \in \mathcal{K}$, $y \in \mathcal{K}$ and $\langle x, y \rangle = 0$. Then, from (3.292), it yields $\psi(x, y) = 0$ and $x \circ y \in -\mathcal{K}$. This together with (3.291) says that $\varphi(x, y) = 0$. Thus, we conclude that $\phi_{\text{LT}}^{\text{sc}}(x, y) = 0$. □

An example of φ is given in [41], which is

$$\varphi(t) = \frac{1}{2} \|[t]_+\|^2,$$

and three examples of ψ are provided:

$$\begin{aligned}
\psi_1(x, y) &= \frac{1}{2} \left(\|[-x]_+\|^2 + \|[-y]_+\|^2\right) \\
\psi_2(x, y) &= \frac{1}{2} \left(\left\|\sqrt{x^2 + y^2} - (x + y)\right\|^2\right) \\
\psi_3(x, y) &= \frac{1}{2} \left(\left\|\left[\sqrt{x^2 + y^2} - (x + y)\right]_+\right\|^2\right).
\end{aligned}$$

The function $\phi^{\rm sc}_{\rm LT}(\zeta)$ by plugging the above φ and ψ_1 into (3.290) was proved continuously differentiable in [41, Theorem 6.1] provided F is continuously differentiable, whereas the functions $\phi^{\rm sc}_{\rm LT}(\zeta)$ by plugging the above φ with ψ_1 and ψ_2 were shown differentiable in [41, Theorem 6.2] provided F is differentiable. Please refer to [41] for their detailed gradient expressions.

There is another type of merit functions, similar to the aforementioned LT-type C-function associated with symmetric cone. It is the so-called EP-type functions, which were originally proposed by Evtushenko and Purtov [65]. Kong and Xiu [66] extended them to symmetric cone setting. In particular, they define

$$\phi^{\rm sc}_{\alpha-{\rm EP}}(x, y) := -(x \circ y) + \frac{1}{2\alpha}\left([x + y]_-\right)^2, \quad 0 < \alpha \le 1, \tag{3.294}$$

$$\phi^{\rm sc}_{\beta-{\rm EP}}(x, y) := -(x \circ y) + \frac{1}{2\beta}\left([x]_-^2 + [y]_-^2\right), \quad 0 < \beta < 1. \tag{3.295}$$

Note that the parameter α could be 1 in defining $\phi^{\rm sc}_{\alpha-{\rm EP}}$, but $\beta \neq 1$ is needed for defining $\phi^{\rm sc}_{\beta-{\rm EP}}$. This is because, by choosing $0 \neq x = y \in -\mathcal{K}$, there occurs

$$\phi^{\rm sc}_{1-{\rm EP}}(x, y) = -(x \circ x) + \frac{1}{2\beta}\left([x]_-^2 + [x]_-^2\right) = -x^2 + [x]_-^2 = 0.$$

Proposition 3.79 *[66, Theorem 3.2] Let $\phi^{\rm sc}_{\alpha-{\rm EP}}$ and $\phi^{\rm sc}_{\beta-{\rm EP}}$ be defined as in* (3.294) *and* (3.295)*, respectively. Then, $\phi^{\rm sc}_{\alpha-{\rm EP}}$ and $\phi^{\rm sc}_{\beta-{\rm EP}}$ are smooth C-functions associated with symmetric cone.*

Proof The projection formula is unknown for general symmetric cone, so the Peirce Decomposition Theorem (Theorem 1.2) is employed. Please refer to [66, Theorem 3.2] for the arguments for showing they are C-functions, and see [66, Theorem 3.3] for their smoothness. □

In light of $\phi_{\rm MS}$ given as in (3.236), $\phi^{\rm sc}_{\alpha-{\rm EP}}$ given as in (3.294), and $\phi^{\rm sc}_{\beta-{\rm EP}}$ given as in (3.295), there induce the following functions:

$$f_{\text{MS}}(\zeta) := \frac{1}{2}\left\|\phi_{\text{MS}}(\zeta, F(\zeta))\right\|^2, \tag{3.296}$$

$$f_\alpha(\zeta) := \frac{1}{2}\left\|\phi^{\text{sc}}_{\alpha\text{-EP}}(\zeta, F(\zeta))\right\|^2, \tag{3.297}$$

$$f_\beta(\zeta) := \frac{1}{2}\left\|\phi^{\text{sc}}_{\beta\text{-EP}}(\zeta, F(\zeta))\right\|^2. \tag{3.298}$$

We will study the growth behavior of these three functions, for which we need a few technical lemmas.

Lemma 3.53 *For a given Jordan frame $\{c_1, c_2, \ldots, c_q\}$, if $z \in \mathbb{V}$ can be written as*

$$z = \sum_{i=1}^{q} z_i c_i + \sum_{1\le i<j\le q} z_{ij}$$

with $z_i \in \mathbb{R}$ for $i = 1, 2, \ldots, q$ and $z_{ij} \in \mathbb{V}_{ij}$ for $1 \le i < j \le q$, then

$$z_+ = \sum_{i=1}^{q} s_i c_i + \sum_{1\le i<j\le q} s_{ij}, \quad z_- = \sum_{i=1}^{q} w_i c_i + \sum_{1\le i<j\le q} w_{ij},$$

where $s_i \ge (z_i)_+ \ge 0, 0 \ge (z_i)_- \ge w_i$ with $s_i + w_i = z_i$ for $i = 1, \ldots, q$, and $s_{ij}, w_{ij} \in \mathbb{V}_{ij}$ with $s_{ij} + w_{ij} = z_{ij}$ for $1 \le i < j \le q$.

Proof Please see [66, Lemma 3.1]. □

Lemma 3.54 *For any $x, y \in \mathbb{V}$, the following inequalities always hold:*

(a) $\lambda_{\min}(x)\|c\|^2 \le \langle x, c\rangle \le \lambda_{\max}(x)\|c\|^2$ for any nonzero idempotent c;
(b) $|\lambda_{\max}(x+y) - \lambda_{\max}(x)| \le \|y\|$ and $|\lambda_{\min}(x+y) - \lambda_{\min}(x)| \le \|y\|$;
(c) $\lambda_{\max}(x+y) \le \lambda_{\max}(x) + \lambda_{\max}(y)$ and $\lambda_{\min}(x+y) \ge \lambda_{\min}(x) + \lambda_{\min}(y)$.

Proof Please see [67, Lemma 14] and [42, Proposition 2.1]. □

Proposition 3.80 *Let $\phi^{\text{sc}}_{\alpha\text{-EP}}$ and $\phi^{\text{sc}}_{\beta\text{-EP}}$ be defined as in (3.294) and (3.295), respectively. Then, for any $x, y \in \mathbb{V}$, there hold*

$$\left\|\phi^{\text{sc}}_{\alpha\text{-EP}}(x, y)\right\| \ge \left(\frac{2\alpha - \alpha^2}{2\alpha}\right) \max\left\{[(\lambda_{\min}(x))_-]^2,\ [(\lambda_{\min}(y))_-]^2\right\}, \tag{3.299}$$

$$\left\|\phi^{\text{sc}}_{\beta\text{-EP}}(x, y)\right\| \ge \left(\frac{1 - \beta^2}{2\beta}\right) \max\left\{[(\lambda_{\min}(x))_-]^2,\ [(\lambda_{\min}(y))_-]^2\right\}. \tag{3.300}$$

Proof Suppose that x has the spectral decomposition $x = \sum_{i=1}^{q} x_i c_i$ with $x_i \in \mathbb{R}$ and $\{c_1, c_2, \ldots, c_q\}$ being a Jordan frame. From Theorem 1.2, $y \in \mathbb{V}$ can be expressed by

$$y = \sum_{i=1}^{q} y_i c_i + \sum_{1\le i<j\le q} y_{ij}, \tag{3.301}$$

where $y_i \in \mathbb{R}$ for $i = 1, 2, \ldots, q$ and $y_{ij} \in \mathbb{V}_{ij}$. Therefore, for any $l \in \{1, 2, \ldots, q\}$,

$$\begin{aligned}\langle c_l, x \circ y \rangle = \langle c_l \circ x, y \rangle &= \left\langle x_l c_l, \sum_{i=1}^{q} y_i c_i + \sum_{1\le i<j\le q} y_{ij} \right\rangle \\ &= x_l \left\langle c_l, \sum_{i=1}^{q} y_i c_i \right\rangle + x_l \left\langle c_l, \sum_{1\le i<j\le q} y_{ij} \right\rangle \\ &= x_l y_l, \end{aligned} \tag{3.302}$$

where the last equality is since $\langle c_l, \sum_{1\le i<j\le q} y_{ij} \rangle = 0$ by the orthogonality of $\mathbb{V}_{ij}$ $(i \le j)$.

We next prove the inequality (3.299). From (3.301) and the spectral decomposition of x,

$$x + y = \sum_{i=1}^{q} (x_i + y_i) c_i + \sum_{1\le i<j\le q} y_{ij},$$

which together with Lemma 3.53 implies

$$(x + y)_- = \sum_{i=1}^{q} u_i c_i + \sum_{1\le i<j\le q} u_{ij},$$

where $u_i \le (x_i + y_i)_- \le 0$ for $i = 1, 2, \ldots, q$ and $u_{ij} \in \mathbb{V}_{ij}$. By this, we can compute

$$\begin{aligned}\left\langle c_l, [(x + y)_-]^2 \right\rangle &= \left\langle c_l \circ \left(\sum_{i=1}^{q} u_i c_i + \sum_{1\le i<j\le q} u_{ij} \right), (x + y)_- \right\rangle \\ &= \left\langle u_l c_l + \left(c_l \circ \sum_{1\le i<j\le q} u_{ij} \right), \sum_{i=1}^{q} u_i c_i + \sum_{1\le i<j\le q} u_{ij} \right\rangle \\ &= u_l^2 + u_l \left\langle c_l, \sum_{1\le i<j\le q} u_{ij} \right\rangle + \left\langle \sum_{1\le i<j\le q} u_{ij}, c_l \circ \sum_{i=1}^{q} u_i c_i \right\rangle \\ &\quad + \left\langle c_l \circ \sum_{1\le i<j\le q} u_{ij}, \sum_{1\le i<j\le q} u_{ij} \right\rangle \\ &= u_l^2 + \left\langle c_l, \left(\sum_{1\le i<j\le q} u_{ij} \right)^2 \right\rangle, \quad \forall\, l = 1, 2, \ldots, q, \end{aligned} \tag{3.303}$$

where the last equality is since $\langle c_l, \sum_{1\le i<j\le q} u_{ij} \rangle = 0$ by the orthogonality of $\mathbb{V}_{ij}$ $(i \le j)$. Now, using Eqs. (3.302)–(3.303), we achieve

$$\begin{aligned}\langle c_l, -\phi^{\rm sc}_{\alpha-{\rm EP}}(x,y)\rangle &= \left\langle c_l, x\circ y - (1/2\alpha)\left[(x+y)_-\right]^2\right\rangle \\ &= x_l y_l - (1/2\alpha)\left[u_l^2 + \left\langle c_l, \left(\sum_{1\le i<j\le q} u_{ij}\right)^2\right\rangle\right] \\ &\le x_l y_l - (1/2\alpha)\left[(x_l+y_l)_-\right]^2, \quad \forall\, l = 1,2,\ldots,q, \end{aligned} \tag{3.304}$$

where the inequality is due to the following facts

$$u_l \le (x_l + y_l)_- \le 0 \quad \text{and} \quad \left\langle c_l, \left(\sum_{1\le i<j\le q} u_{ij}\right)^2\right\rangle \ge 0.$$

On the other hand, from Lemma 3.54(a) we have

$$\langle c_l, -\phi^{\rm sc}_{\alpha-{\rm EP}}(x,y)\rangle \ge \lambda_{\min}\big(-\phi^{\rm sc}_{\alpha-{\rm EP}}(x,y)\big)\|c_l\|^2 = \lambda_{\min}\big(-\phi^{\rm sc}_{\alpha-{\rm EP}}(x,y)\big), \quad \forall\, l = 1,2,\ldots,q. \tag{3.305}$$

Thus, combining (3.304) with (3.305), it follows that

$$2\alpha\lambda_{\min}(-\phi^{\rm sc}_{\alpha-{\rm EP}}(x,y)) \le 2\alpha x_l y_l - [(x_l+y_l)_-]^2, \quad \forall\, l = 1,2,\ldots,q.$$

Let $\lambda_{\min}(x) = x_\nu$ with $\nu \in \{1,2,\ldots,q\}$. Then, we particularly have that

$$2\alpha\lambda_{\min}(-\phi^{\rm sc}_{\alpha-{\rm EP}}(x,y)) \le 2\alpha\lambda_{\min}(x)y_\nu - [(\lambda_{\min}(x)+y_\nu)_-]^2. \tag{3.306}$$

We next proceed the proof by the two cases: $\lambda_{\min}(x) \le 0$ and $\lambda_{\min}(x) > 0$.

Case (i): $\lambda_{\min}(x) \le 0$. Under this case, we will prove the below inequality:

$$2\alpha\lambda_{\min}(x)y_\nu - [(\lambda_{\min}(x)+y_\nu)_-]^2 \le -(2\alpha-\alpha^2)[(\lambda_{\min}(x))_-]^2, \tag{3.307}$$

which, together with (3.306), immediately implies

$$\left\|\phi^{\rm sc}_{\alpha-{\rm EP}}(x,y)\right\| \ge \left|\lambda_{\min}(-\phi^{\rm sc}_{\alpha-{\rm EP}}(x,y))\right| \ge \left[(2\alpha-\alpha^2)/(2\alpha)\right][(\lambda_{\min}(x))_-]^2. \tag{3.308}$$

In fact, if $\lambda_{\min}(x) + y_\nu \ge 0$, then we can deduce that

$$2\alpha\lambda_{\min}(x)y_\nu - [(\lambda_{\min}(x)+y_\nu)_-]^2 = 2\alpha(\lambda_{\min}(x))_-(y_\nu)_+ \le -(2\alpha-\alpha^2)[(\lambda_{\min}(x))_-]^2;$$

and otherwise we will have

$$\begin{aligned} &2\alpha\lambda_{\min}(x)y_\nu - [(\lambda_{\min}(x)+y_\nu)_-]^2 \\ &= 2\alpha\lambda_{\min}(x)y_\nu - [(\lambda_{\min}(x)+y_\nu)]^2 \\ &\le -(2\alpha-\alpha^2)[\lambda_{\min}(x)]^2 \end{aligned}$$

$$= -(2\alpha - \alpha^2)[(\lambda_{\min}(x))_-]^2.$$

Case (ii): $\lambda_{\min}(x) > 0$. Under this case, the inequality (3.308) clearly holds.

Summing up the above discussions, the inequality (3.308) holds for any $x, y \in \mathbb{V}$. In view of the symmetry of x and y in $\phi^{\rm sc}_{\alpha-{\rm EP}}(x, y)$, we also have

$$\left\|\phi^{\rm sc}_{\alpha-{\rm EP}}(x, y)\right\| \geq \left[(2\alpha - \alpha^2)/(2\alpha)\right][(\lambda_{\min}(y))_-]^2$$

for any $x, y \in \mathbb{V}$. Thus, the proof of the inequality (3.299) is complete.

We next prove the inequality (3.300). By the spectral decomposition of x, we have that $(x_-)^2 = \sum_{i=1}^q [(x_i)_-]^2 c_i$, which in turn implies

$$\langle c_l, (x_-)^2\rangle = [(x_l)_-]^2, \quad \forall\, l = 1, 2, \ldots, q. \tag{3.309}$$

In addition, from Lemma 3.53 and the expression of y given by (3.301), it follows that

$$y_- = \sum_{i=1}^q v_i c_i + \sum_{1\leq i<j\leq q} v_{ij},$$

where $v_i \leq (y_i)_- \leq 0$ for $i = 1, 2, \ldots, q$ and $v_{ij} \in \mathbb{V}_{ij}$. By the same arguments as (3.303),

$$\langle c_l, (y_-)^2\rangle = v_l^2 + \left\langle c_l, \left(\sum_{1\leq i<j\leq q} v_{ij}\right)^2\right\rangle, \quad \forall\, l = 1, 2, \ldots, q. \tag{3.310}$$

Now, from Eqs. (3.302), (3.309) and (3.310), it follows that

$$\begin{aligned}
\left\langle c_l, -\phi^{\rm sc}_{\beta-{\rm EP}}(x, y)\right\rangle &= \left\langle c_l, x \circ y - (1/2\beta)\left[(x_-)^2 + (y_-)^2\right]\right\rangle \\
&= x_l y_l - (1/2\beta)\left[((x_l)_-)^2 + v_l^2 + \left\langle c_l, \left(\textstyle\sum_{1\leq i<j\leq q} v_{ij}\right)^2\right\rangle\right] \\
&\leq x_l y_l - (1/2\beta)\left[((x_l)_-)^2 + (v_l)^2\right] \\
&\leq x_l y_l - (1/2\beta)\left[((x_l)_-)^2 + ((y_l)_-)^2\right], \quad \forall\, l = 1, 2, \ldots, q,
\end{aligned}$$

where the first inequality is due to the nonnegativity of $\langle c_l, (\sum_{1\leq i<j\leq q} v_{ij})^2\rangle$, and the second one is due to $v_l \leq (y_l)_- \leq 0$. On the other hand, by Lemma 3.54(a),

$$\left\langle c_l, -\phi^{\rm sc}_{\beta-{\rm EP}}(x, y)\right\rangle \geq \lambda_{\min}(-\phi^{\rm sc}_{\beta-{\rm EP}}(x, y))\|c_l\|^2 = \lambda_{\min}(-\phi^{\rm sc}_{\beta-{\rm EP}}(x, y)), \quad \forall l = 1, 2, \ldots, q.$$

Combining the last two inequalities immediately leads to

$$\lambda_{\min}(-\phi^{\rm sc}_{\beta-\rm EP}(x,y)) \le x_l y_l - (1/2\beta)\left[((x_l)_-)^2 + ((y_l)_-)^2\right], \quad \forall\, l = 1, 2, \ldots, q.$$

Let $\lambda_{\min}(x) = x_\nu$ with $\nu \in \{1, 2, \ldots, q\}$ and suppose that $\lambda_{\min}(x) \le 0$. Then,

$$\begin{aligned}
\lambda_{\min}(-\phi^{\rm sc}_{\beta-\rm EP}(x,y)) \le& \lambda_{\min}(x) y_\nu - (1/2\beta)\left[((\lambda_{\min}(x))_-)^2 + ((y_\nu)_-)^2\right] \\
\le& [(\lambda_{\min}(x))_-][(y_\nu)_-] - (1/2\beta)\left[((\lambda_{\min}(x))_-)^2 + ((y_\nu)_-)^2\right] \\
=& -(1/2\beta)\left\{\left[\beta(\lambda_{\min}(x))_- - (y_\nu)_-\right]^2 + (1-\beta^2)\left[(\lambda_{\min}(x))_-\right]^2\right\} \\
\le& -\left(\frac{1-\beta^2}{2\beta}\right)\left[(\lambda_{\min}(x))_-\right]^2,
\end{aligned}$$

which in turn implies

$$\left\|\phi^{\rm sc}_{\beta-\rm EP}(x,y)\right\| \ge \left|\lambda_{\min}(-\phi^{\rm sc}_{\beta-\rm EP}(x,y))\right| \ge \left(\frac{1-\beta^2}{2\beta}\right)\left[(\lambda_{\min}(x))_-\right]^2. \tag{3.311}$$

If $\lambda_{\min}(x) = x_\nu > 0$, then the inequality (3.311) is obvious. Thus, (3.311) holds for any $x, y \in \mathcal{V}$. In view of the symmetry of x and y in $\phi^{\rm sc}_{\beta-\rm EP}(x,y)$, we also have

$$\left\|\phi^{\rm sc}_{\beta-\rm EP}(x,y)\right\| \ge \left|\lambda_{\min}(-\phi^{\rm sc}_{\beta-\rm EP}(x,y))\right| \ge \left(\frac{1-\beta^2}{2\beta}\right)\left[(\lambda_{\min}(y))_-\right]^2$$

for any $x, y \in \mathbb{V}$. Consequently, the desired result follows. □

The following proposition characterizes an important property for the smooth EP-type C-functions $\phi^{\rm sc}_{\alpha-\rm EP}$ and $\phi^{\rm sc}_{\beta-\rm EP}$ under a unified framework.

Proposition 3.81 *Let $\phi^{\rm sc}_{\alpha-\rm EP}$ and $\phi^{\rm sc}_{\beta-\rm EP}$ be defined as in* (3.294) *and* (3.295), *respectively. Suppose that $\{x^k\} \subset \mathbb{V}$ and $\{y^k\} \subset \mathbb{V}$ are the sequences satisfying one of the following conditions:*

(i) either $\lambda_{\min}(x^k) \to -\infty$ or $\lambda_{\min}(y^k) \to -\infty$;
(ii) $\lambda_{\min}(x^k), \lambda_{\min}(y^k) > -\infty$, $\lambda_{\max}(x^k), \lambda_{\max}(y^k) \to +\infty$ and $\|x^k \circ y^k\| \to +\infty$.

Then, $\|\phi^{\rm sc}_{\alpha-\rm EP}(x^k, y^k)\| \to +\infty$ and $\left\|\phi^{\rm sc}_{\beta-\rm EP}(x^k, y^k)\right\| \to +\infty$.

Proof If Case (i) is satisfied, then the assertion is direct by Proposition 3.80. In what follows, we will prove the assertion under Case (ii). Notice that in this case the sequences $\{x^k\}$, $\{y^k\}$ and $\{x^k + y^k\}$ are all bounded below since $\lambda_{\min}(x^k), \lambda_{\min}(y^k) > -\infty$ and $\lambda_{\min}(x^k + y^k) \ge \lambda_{\min}(x^k) + \lambda_{\min}(y^k) > -\infty$. Therefore, the sequences $\{[(x^k + y^k)_-]^2\}$, $\left\{\left([x^k]_-\right)^2\right\}$ and $\left\{\left([y^k]_-\right)^2\right\}$ are bounded. In addition, we have $\lambda_{\min}(x^k \circ y^k) \to -\infty$ or $\lambda_{\max}(x^k \circ y^k) \to +\infty$ since $\|x^k \circ y^k\| \to +\infty$.

If $\lambda_{\min}(x^k \circ y^k) \to -\infty$ as $k \to \infty$, then by Lemma 3.54(c) there hold

$$\begin{aligned}
\lambda_{\min}(-\phi^{\mathrm{sc}}_{\alpha-\mathrm{EP}}(x,y)) &= \lambda_{\min}\left[(x^k \circ y^k) - (1/2\alpha)((x^k+y^k)_-)^2\right] \\
&\le \lambda_{\min}(x^k \circ y^k) + (1/2\alpha)\left\|((x^k+y^k)_-)^2\right\|, \\
\lambda_{\min}(-\phi^{\mathrm{sc}}_{\beta-\mathrm{EP}}(x,y)) &= \lambda_{\min}\left[(x^k \circ y^k) - (1/2\beta)\left(((x^k)_-)^2 + ((y^k)_-)^2\right)\right] \\
&\le \lambda_{\min}(x^k \circ y^k) + (1/2\beta)\left\|((x^k)_-)^2 + ((y^k)_-)^2\right\|,
\end{aligned}$$

which, together with the boundedness of $\|((x^k+y^k)_-)^2\|$ and $\|((x^k)_-)^2 + ((y^k)_-)^2\|$, implies $\lambda_{\min}(-\phi^{\mathrm{sc}}_{\alpha-\mathrm{EP}}(x^k,y^k)) \to -\infty$ and $\lambda_{\min}(-\phi^{\mathrm{sc}}_{\beta-\mathrm{EP}}(x^k,y^k)) \to -\infty$. Since

$$\left\|\phi^{\mathrm{sc}}_{\alpha-\mathrm{EP}}(x^k,y^k)\right\| \ge \left|\lambda_{\min}(-\phi^{\mathrm{sc}}_{\alpha-\mathrm{EP}}(x,y))\right| \quad \text{and} \quad \left\|\phi^{\mathrm{sc}}_{\beta-\mathrm{EP}}(x^k,y^k)\right\| \ge \left|\lambda_{\min}(-\phi^{\mathrm{sc}}_{\beta-\mathrm{EP}}(x,y))\right|,$$

we immediately obtain $\left\|\phi^{\mathrm{sc}}_{\alpha-\mathrm{EP}}(x^k,y^k)\right\| \to +\infty$ and $\left\|\phi^{\mathrm{sc}}_{\beta-\mathrm{EP}}(x^k,y^k)\right\| \to +\infty$.

If $\lambda_{\max}(x^k \circ y^k) \to +\infty$ as $k \to \infty$, from Lemma 3.54(c) it then follows that

$$\begin{aligned}
\lambda_{\max}(-\phi^{\mathrm{sc}}_{\alpha-\mathrm{EP}}(x,y)) &= \lambda_{\max}\left[(x^k \circ y^k) - (1/2\alpha)((x^k+y^k)_-)^2\right] \\
&\ge \lambda_{\max}(x^k \circ y^k) - (1/2\alpha)\left\|((x^k+y^k)_-)^2\right\|, \\
\lambda_{\max}(-\phi^{\mathrm{sc}}_{\beta-\mathrm{EP}}(x,y)) &= \lambda_{\max}\left[(x^k \circ y^k) - (1/2\beta)\left(((x^k)_-)^2 + ((y^k)_-)^2\right)\right] \\
&\ge \lambda_{\max}(x^k \circ y^k) - (1/2\beta)\left\|((x^k)_-)^2 + ((y^k)_-)^2\right\|,
\end{aligned}$$

which, by the boundedness of $\|((x^k+y^k)_-)^2\|$ and $\|((x^k)_-)^2 + ((y^k)_-)^2\|$, implies that $\lambda_{\max}(-\phi^{\mathrm{sc}}_{\alpha-\mathrm{EP}}(x^k,y^k)) \to +\infty$ and $\lambda_{\max}(-\phi^{\mathrm{sc}}_{\beta-\mathrm{EP}}(x^k,y^k)) \to +\infty$. Noting that

$$\left\|\phi^{\mathrm{sc}}_{\alpha-\mathrm{EP}}(x^k,y^k)\right\| \ge \left|\lambda_{\max}(-\phi^{\mathrm{sc}}_{\alpha-\mathrm{EP}}(x^k,y^k))\right| \quad \text{and} \quad \left\|\phi^{\mathrm{sc}}_{\beta-\mathrm{EP}}(x^k,y^k)\right\| \ge \left|\lambda_{\max}(-\phi^{\mathrm{sc}}_{\beta-\mathrm{EP}}(x^k,y^k))\right|,$$

we readily obtain $\left\|\phi^{\mathrm{sc}}_{\alpha-\mathrm{EP}}(x^k,y^k)\right\| \to +\infty$ and $\left\|\phi^{\mathrm{sc}}_{\beta-\mathrm{EP}}(x^k,y^k)\right\| \to +\infty$. □

When $\mathbb{V} = \mathbb{R}^n$ with "$\circ$" being the componentwise product of the vectors, $\|x^k \circ y^k\| \to +\infty$ automatically holds if $\lambda_{\max}(x^k), \lambda_{\max}(y^k) \to +\infty$, and Proposition 3.81 reduces to the result of [68, Lemma 2.5] for the NCPs. However, for the general Euclidean Jordan algebra, this condition is necessary as illustrated by the following example.

Example 3.3 Consider the Lorentz algebra $\mathbb{L}^n = (\mathbb{R}^n, \circ, \langle\cdot,\cdot\rangle_{\mathbb{R}^n})$. Assume that $n = 3$ and take the sequences $\{x^k\}$ and $\{y^k\}$ as follows:

$$x^k = \begin{pmatrix} k \\ k \\ 0 \end{pmatrix} \quad \text{and} \quad y^k = \begin{pmatrix} k \\ -k \\ 0 \end{pmatrix} \quad \text{for each } k.$$

It is easy to verify that $\lambda_{\min}(x^k) = 0$, $\lambda_{\min}(y^k) = 0$, $\lambda_{\max}(x^k), \lambda_{\max}(y^k) \to +\infty$, but $\|x^k \circ y^k\| \nrightarrow +\infty$. For such $\{x^k\}$ and $\{y^k\}$, by computation we have

$\left\|\phi^{\rm sc}_{\alpha-\rm EP}(x^k, y^k)\right\| = 0$ and $\left\|\phi^{\rm sc}_{\beta-\rm EP}(x^k, y^k)\right\| = 0$, i.e., the conclusion of Proposition 3.81 does not hold.

Lemma 3.55 *[69, Lemma 4.1] Let $\{x^k\}$ and $\{y^k\}$ be the sequences such that $x^k \to \bar{x}$ and $y^k \to \bar{y}$ when $k \to \infty$. Then, we have that $x^k \circ y^k \to \bar{x} \circ \bar{y}$.*

Proof Please see [69, Lemma 4.1] for detailed arguments. □

Now we are in a position to establish the coerciveness of f_α and f_β. Assume that $(\mathbb{V}, \circ, \langle\cdot,\cdot\rangle)$ is a general Euclidean Jordan algebra. We first consider the SCLCP case.

Proposition 3.82 *Let f_α and f_β be defined as in* (3.297) *and* (3.298), *respectively. If $F(\zeta) = L(\zeta) + b$ with the linear transformation L having the P-property, then f_α and f_β are coercive.*

Proof Let $\{\zeta^k\}$ be a sequence such that $\|\zeta^k\| \to +\infty$. We only need to prove that

$$f_\alpha(\zeta^k) \to +\infty \quad \text{and} \quad f_\beta(\zeta^k) \to +\infty. \tag{3.312}$$

By passing to a subsequence if necessary, we assume that $\zeta^k/\|\zeta^k\| \to \bar{\zeta}$, and consequently $(L(\zeta^k) + b)/\|\zeta^k\| \to L(\bar{\zeta})$. If $\lambda_{\min}(\zeta^k) \to -\infty$, then from Proposition 3.81 it follows that $\left\|\phi^{\rm sc}_{\alpha-\rm EP}(\zeta^k, L(\zeta^k)+b)\right\| \to +\infty$, and $\left\|\phi^{\rm sc}_{\beta-\rm EP}(\zeta^k, L(\zeta^k)+b)\right\| \to +\infty$, which in turn implies (3.312).

Now assume that $\{\zeta^k\}$ is bounded below. We argue that the sequence $\{L(\zeta^k) + b\}$ is unbounded by contradiction. Suppose that $\{L(\zeta^k) + b\}$ is bounded. Then,

$$L(\bar{\zeta}) = \lim_{k\to\infty} \left[(L(\zeta^k) + b)/\|\zeta^k\|\right] = 0 \in \mathcal{K}.$$

Since $\{\zeta^k\}$ is bounded below and $\lambda_{\max}(\zeta^k) \to +\infty$ by $\|\zeta^k\| \to +\infty$, there is an element $\bar{d} \in \mathbb{V}$ such that $(\zeta^k - \bar{d})/\|\zeta^k - \bar{d}\| \in \mathcal{K}$ for each k. Noting that $\mathcal{K}$ is closed, and we have

$$\lim_{k\to\infty} (\zeta^k - \bar{d})/\|\zeta^k - \bar{d}\| = \bar{\zeta}/\|\bar{\zeta}\| = \bar{\zeta} \in \mathcal{K}.$$

Thus, $\bar{\zeta} \in \mathcal{K}$, $L(\bar{\zeta}) \in \mathcal{K}$ and $\bar{\zeta} \circ L(\bar{\zeta}) = 0$. From [37, Proposition 6], it follows that $\bar{\zeta}$ and $L(\bar{\zeta})$ operator commute. This, together with $\bar{\zeta} \circ L(\bar{\zeta}) = 0 \in -\mathcal{K}$ and the P-property of L, implies that $\bar{\zeta} = 0$, yielding a contradiction to $\|\bar{\zeta}\| = 1$. Hence, the sequence $\{L(\zeta^k) + b\}$ is unbounded. Without loss of generality, assume that $\|L(\zeta^k) + b\| \to +\infty$.

If $\lambda_{\min}(L(\zeta^k) + b) \to -\infty$, then using Proposition 3.81 yields the desired result of (3.312). We next assume that the sequence $\{L(\zeta^k) + b\}$ is bounded below. We prove that

$$(\zeta^k/\|\zeta^k\|) \circ \left[(L(\zeta^k) + b)/\|\zeta^k\|\right] \nrightarrow 0. \tag{3.313}$$

Suppose that (3.313) does not hold, then from Lemma 3.55, it follows that

$$\bar{\zeta} \circ L(\bar{\zeta}) = \lim_{k \to +\infty} \left[(\zeta^k - d)/\|\zeta^k\|\right] \circ \left[(L(\zeta^k) + b - d)/\|\zeta^k\|\right] = 0 \quad \forall d \in \mathbb{V}. \tag{3.314}$$

Since $\{\zeta^k\}$ and $\{L(\zeta^k) + b\}$ are bounded below and $\lambda_{\max}(\zeta^k), \lambda_{\max}(L(\zeta^k) + b) \to +\infty$, there is an element $\tilde{d}$ such that $\zeta^k - \tilde{d} \in \mathcal{K}$ and $L(\zeta^k) + b - \tilde{d} \in \mathcal{K}$ for each k. Therefore,

$$\left[(\zeta^k - \tilde{d})/\|\zeta^k\|\right] \in \mathcal{K} \quad \text{and} \quad \left[(L(\zeta^k) + b - \tilde{d})/\|\zeta^k\|\right] \in \mathcal{K}, \quad \forall\, k.$$

Noting that $\mathcal{K}$ is closed and $\bar{\zeta} = \lim_{k\to\infty}(\zeta^k - \tilde{d})/\|\zeta^k\|$ and $L(\bar{\zeta}) = \lim_{k\to\infty}$ $[(L(\zeta^k) + b - \tilde{d})/\|\zeta^k\|]$, we have

$$\bar{\zeta} \in \mathcal{K} \quad \text{and} \quad L(\bar{\zeta}) \in \mathcal{K}. \tag{3.315}$$

From (3.314) and (3.315) and [37, Proposition 6], it follows that $\bar{\zeta}$ and $L(\bar{\zeta})$ operator commute. Using the P-property of L and noting that $\bar{\zeta} \circ L(\bar{\zeta}) = 0 \in -\mathcal{K}$, we then obtain $\bar{\zeta} = 0$, which clearly contradicts $\|\bar{\zeta}\| = 1$. Therefore, (3.313) holds. Since $\|\zeta^k\| \to +\infty$, we have $\|\zeta^k \circ (L(\zeta^k) + b)\| \to +\infty$. Combining with $\lambda_{\min}(\zeta^k), \lambda_{\min}(L(\zeta^k) + b) > -\infty$ and $\|\zeta^k\|, \|L(\zeta^k) + b\| \to +\infty$, it follows that the sequences $\{\zeta^k\}$ and $\{L(\zeta^k) + b\}$ satisfy condition(ii) of Proposition 3.81. This means that the result (3.312) holds. □

Proposition 3.83 *Let f_α and f_β be defined as in* (3.297) *and* (3.298)*, respectively. If the mapping F has the uniform Jordan P-property, then f_α and f_β are coercive.*

Proof The proof technique is similar to that in [70, Theorem 4.1]. For completeness, we include it. Let $\{\zeta^k\}$ be a sequence such that $\|\zeta^k\| \to +\infty$. Corresponding to the Cartesian structure of $\mathbb{V}$, let $\zeta^k = (\zeta_1^k, \ldots, \zeta_m^k)$ with $\zeta_i^k \in \mathcal{V}_i$ for each k. Define

$$J := \left\{i \in \{1, 2, \ldots, m\} \mid \{\zeta_i^k\} \text{ is unbounded}\right\}.$$

Clearly, the set $J \neq \emptyset$ since $\{\zeta^k\}$ is unbounded. Let $\{\xi^k\}$ be a bounded sequence with $\xi^k = (\xi_1^k, \ldots, \xi_m^k)$ and $\xi_i^k \in \mathbb{V}_i$ for $i = 1, 2, \ldots, m$, where ξ_i^k for each k is defined as follows:

$$\xi_i^k = \begin{cases} 0 & \text{if } i \in J; \\ \zeta_i^k & \text{otherwise}, \end{cases} \qquad i = 1, 2, \ldots, m.$$

Since F has the uniform Jordan P-property, there is a constant $\rho > 0$ such that

$$\begin{aligned} \rho\|\zeta^k - \xi^k\|^2 &\le \max_{i=1,\ldots,m} \lambda_{\max}\left[(\zeta_i^k - \xi_i^k) \circ (F_i(\zeta^k) - F_i(\xi^k))\right] \\ &= \lambda_{\max}\left[\zeta_\nu^k \circ (F_\nu(\zeta^k) - F_\nu(\xi^k))\right] \\ &\le \|\zeta_\nu^k \circ (F_\nu(\zeta^k) - F_\nu(\xi^k))\| \\ &\le \|\zeta_\nu^k\| \|F_\nu(\zeta^k) - F_\nu(\xi^k)\|, \end{aligned} \tag{3.316}$$

where ν is an index from $\{1, 2, \ldots, m\}$ for which the maximum is attained and the last inequality is due to (1.5). Clearly, $\nu \in J$ by the definition of $\{\xi^k\}$, and consequently, $\{\zeta_\nu^k\}$ is unbounded. Without loss of generality, we assume that

$$\|\zeta_\nu^k\| \to +\infty. \tag{3.317}$$

Since

$$\|\zeta^k - \xi^k\|^2 \geq \|\zeta_\nu^k - \xi_\nu^k\|^2 = \|\zeta_\nu^k\|^2, \quad \text{for each } k, \tag{3.318}$$

dividing the both sides of (3.316) by $\|\zeta_\nu^k\|$ then yields that

$$\rho\|\zeta_\nu^k\| \leq \|F_\nu(\zeta^k) - F_\nu(\xi^k)\| \leq \|F_\nu(\zeta^k)\| + \|F_\nu(\xi^k)\|.$$

Notice that $\{F(\xi^k)\}$ is bounded since the mapping F is continuous and $\{\xi^k\}$ is bounded. Hence, the last inequality immediately implies

$$\|F_\nu(\zeta^k)\| \to +\infty. \tag{3.319}$$

In addition, we can verify by contradiction that

$$\|\zeta_\nu^k \circ F_\nu(\zeta^k)\| \to +\infty. \tag{3.320}$$

In fact, if $\{\|\zeta_\nu^k \circ F_\nu(\zeta^k)\|\}$ is bounded, then on the one hand, we have

$$\lim_{k\to\infty} \|\zeta_\nu^k \circ (F_\nu(\zeta^k) - F_\nu(\xi^k))\| / \|\zeta_\nu^k\|^2 = 0.$$

But, on the other hand, the inequality (3.318) yields

$$\lim_{k\to+\infty} \rho\|\zeta^k - \xi^k\|^2 / \|\zeta_\nu^k\|^2 \geq \rho > 0,$$

which clearly contradicts the third inequality in (3.316). Thus, from Eqs. (3.317), (3.319) and (3.320), the sequences $\{\zeta_\nu^k\}$ and $\{F_\nu(\zeta^k)\}$ satisfy the conditions of Proposition 3.1. Therefore, there necessarily holds that $\left\|\left(\phi^{\text{sc}}_{\alpha-\text{EP}}\right)^{(\nu)}(\zeta_\nu^k, F_\nu(\zeta^k))\right\| \to +\infty$ and $\left\|\left(\phi^{\text{sc}}_{\beta-\text{EP}}\right)^{(\nu)}(\zeta_\nu^k, F_\nu(\zeta^k))\right\| \to +\infty$, which in turn implies that $f_\alpha(\zeta^k) \to +\infty$ and $f_\beta(\zeta^k) \to +\infty$ as $k \to \infty$. □

From Definition 1.10 and Lemma 3.54(a), clearly, the uniform Cartesian P-property implies the uniform Jordan P-property. Hence, the functions f_α and f_β are also coercive if F has the uniform Cartesian P-property. In addition, when $\mathbb{V} = \mathbb{R}^n$ with "$\circ$" being the componentwise product of the vectors, the uniform Cartesian P-property and the uniform Jordan P-property of F are equivalent to saying that F is a uniform P-function; (see [2, Page 299] and discussions in Sect. 1.4), and now Proposition 3.83 recovers the known result [71, Theorem 2.3].

In order to establish the coerciveness of the implicit Lagrangian merit function f_{MS}, we need the help of the natural residual complementarity function over symmetric cones

$$r_{\alpha}(x, y) := x - (x - (1/\alpha)y)_{+}, \quad \forall\, x, y \in \mathbb{V} \text{ and } \alpha > 0. \tag{3.321}$$

To this end, we first characterize the growth behavior of the residual function r_{α}.

Lemma 3.56 *Let r_{α} be defined as in (3.321). Then, for any $x, y \in \mathbb{V}$, we have*

$$\lambda_{\min}(r_{\alpha}(x, y)) \leq \min\left\{\lambda_{\min}(x),\ (1/\alpha)\lambda_{\min}(y)\right\}.$$

Proof For any $x, y \in \mathbb{V}$, from the definition of r_{α} in (3.321) and Lemma 3.54(c), we have

$$\begin{aligned}\lambda_{\min}(x) &= \lambda_{\min}\left[r_{\alpha}(x, y) + (x - (1/\alpha)y)_{+}\right] \\ &\geq \lambda_{\min}(r_{\alpha}(x, y)) + \lambda_{\min}\left[(x - (1/\alpha)y)_{+}\right],\end{aligned}$$

which implies

$$\lambda_{\min}(r_{\alpha}(x, y)) \leq \lambda_{\min}(x) - \lambda_{\min}\left[(x - (1/\alpha)y)_{+}\right] \leq \lambda_{\min}(x). \tag{3.322}$$

On the other hand, we notice that the function r_{α} can be rewritten as

$$r_{\alpha}(x, y) = (x - (1/\alpha)y)_{-} + (1/\alpha)y.$$

Consequently, it leads to

$$\begin{aligned}(1/\alpha)\lambda_{\min}(y) &= \lambda_{\min}\left[r_{\alpha}(x, y) - (x - (1/\alpha)y)_{-}\right] \\ &\geq \lambda_{\min}(r_{\alpha}(x, y)) + \lambda_{\min}\left[-(x - (1/\alpha)y)_{-}\right] \\ &= \lambda_{\min}(r_{\alpha}(x, y)) + \lambda_{\min}\left[(-x + (1/\alpha)y)_{+}\right].\end{aligned}$$

This implies that

$$\lambda_{\min}(r_{\alpha}(x, y)) \leq (1/\alpha)\lambda_{\min}(y) - \lambda_{\min}\left[(-x + (1/\alpha)y)_{+}\right] \leq (1/\alpha)\lambda_{\min}(y). \tag{3.323}$$

From Eqs. (3.322) and (3.323), we prove the desired inequality. □

Proposition 3.84 *Let r_{α} be defined as in (3.321). Suppose $\{x^k\} \subset \mathbb{V}$ and $\{y^k\} \subset \mathbb{V}$ be the sequences satisfying one of the following conditions*

(i) either $\lambda_{\min}(x^k) \to -\infty$ or $\lambda_{\min}(y^k) \to -\infty$;
(ii) $\lambda_{\min}(x^k), \lambda_{\min}(y^k) > -\infty$, $\quad \lambda_{\max}(x^k), \lambda_{\max}(y^k) \to +\infty \quad$ and $\quad (x^k/\|x^k\|) \circ (y^k/\|y^k\|) \nrightarrow 0$.

Then, $\left\|r_{\alpha}(x^k, y^k)\right\| \to +\infty$.

Proof If Case (i) holds, the result is direct by Lemma 3.56 and the fact that

$$\|r_\alpha(x^k, y^k)\| \geq |\lambda_{\min}[r_\alpha(x^k, y^k)]|.$$

It remains to prove the desired result under Case (ii). Suppose that the sequence $\{r_\alpha(x^k, y^k)\}$ is bounded. From the definition of r_α, we have

$$\begin{aligned} r_\alpha(x^k, y^k) &= x^k - (1/2)\left(x^k - (1/\alpha)y^k\right) - (1/2)\left|x^k - (1/\alpha)y^k\right| \\ &= (1/2)\left(x^k + (1/\alpha)y^k\right) - (1/2)\left|x^k - (1/\alpha)y^k\right|. \end{aligned}$$

Therefore,

$$\left|x^k - (1/\alpha)y^k\right| = \left(x^k + (1/\alpha)y^k\right) - 2r_\alpha(x^k, y^k).$$

Squaring two sides of the last equation then yields

$$(1/\alpha)x^k \circ y^k = r_\alpha(x^k, y^k) \circ \left(x^k + (1/\alpha)y^k\right) - [r_\alpha(x^k, y^k)]^2.$$

Dividing the two sides by $\|x^k\|\,\|y^k\|$ and using the boundedness of $\{r_\alpha(x^k, y^k)\}$, we obtain

$$\lim_{k\to\infty} (x^k/\|x^k\|) \circ (y^k/\|y^k\|) = 0.$$

This contradicts the given assumption that $(x^k/\|x^k\|) \circ (y^k/\|y^k\|) \nrightarrow 0$. □

When $\mathbb{V} = \mathbb{R}^n$ with "$\circ$" being the componentwise product of the vectors, the condition $\lambda_{\max}(x^k), \lambda_{\max}(y^k) \to +\infty$ implies $(x^k/\|x^k\|) \circ (y^k/\|y^k\|) \nrightarrow 0$, and consequently Proposition 3.84 gives an important property of the natural residual NCP function or the minimum NCP function; see [68, Lemma 2.5]. But, for the general Euclidean Jordan algebra, the following example illustrates that $(x^k/\|x^k\|) \circ (y^k/\|y^k\|) \nrightarrow 0$ is necessary.

Example 3.4 Consider the Lorentz algebra $\mathbb{L}^n = (\mathbb{R}^n, \circ, \langle\cdot,\cdot\rangle_{\mathbb{R}^n})$ with $n = 3$. Take the sequences $\{x^k\}$ and $\{y^k\}$ as follows:

$$x^k = \begin{pmatrix} k \\ -(k+1) \\ (1/\alpha) \end{pmatrix} \quad \text{and} \quad y^k = \begin{pmatrix} k \\ k-1 \\ 1 \end{pmatrix} \quad \text{for each } k.$$

It is easy to verify that $\lambda_{\min}(x^k) \to -1$, $\lambda_{\min}(y^k) = 1$ and $\lambda_{\max}(x^k), \lambda_{\max}(y^k) \to +\infty$, but

$$x^k/\|x^k\| \to \begin{pmatrix} 1/\sqrt{2} \\ -1/\sqrt{2} \\ 0 \end{pmatrix}, \quad y^k/\|y^k\| \to \begin{pmatrix} 1/\sqrt{2} \\ 1/\sqrt{2} \\ 0 \end{pmatrix}, \quad \text{and} \quad (x^k/\|x^k\|) \circ (y^k/\|y^k\|) \to 0.$$

Therefore, the sequences $\{x^k\}$ and $\{y^k\}$ do not satisfy the assumption $(x^k/\|x^k\|) \circ (y^k/\|y^k\|) \nrightarrow 0$. For such sequences, by computation, we have that

$$r_\alpha(x^k, y^k) = \begin{pmatrix} k \\ -(k+1) \\ (1/\alpha) \end{pmatrix} - \begin{pmatrix} k + (1/2) - (1/2\alpha) \\ -k - (1/2) + (1/2\alpha) \\ 0 \end{pmatrix} = \begin{pmatrix} (1/2\alpha) - (1/2) \\ -(1/2\alpha) - (1/2) \\ (1/\alpha) \end{pmatrix}.$$

Clearly, $\|r_\alpha(x^k, y^k)\| \nrightarrow +\infty$, i.e., the conclusion of Proposition 3.84 does not hold.

Lemma 3.57 *Let $\phi_{\rm MS}$ and r_α be defined as in* (3.236) *and (3.321), respectively. Then, for any $x, y \in \mathbb{V}$, there holds*

$$\|\phi_{\rm MS}(x, y)\| \geq \max\left\{\left(\frac{\alpha^2 - 1}{2\alpha\|e\|}\right)\|r_\alpha(x, y)\|^2,\ \left(\frac{1-\alpha^2}{2\alpha\|e\|}\right)\|r_{1/\alpha}(x, y)\|^2\right\}.$$

Proof First, for any $x, y \in \mathbb{V}$, the following identity always holds:

$$\begin{aligned}\langle e, \phi_{\rm MS}(x, y)\rangle &= \langle x, y\rangle + (1/2\alpha)\left\{\|(x-\alpha y)_+\|^2 - \|x\|^2 + \|(y-\alpha x)_+\|^2 - \|y\|^2\right\} \\ &= \langle y, (x - (1/\alpha)y)_+\rangle + (\alpha/2)\|x - (x-(1/\alpha)y)_+\|^2 \\ &\quad - \langle y, (x-\alpha y)_+\rangle - (1/2\alpha)\,\|x - (x-\alpha y)_+\|^2. \end{aligned} \tag{3.324}$$

In fact, for any $x, y \in \mathbb{V}$, we can compute

$$\begin{aligned}&\langle y, (x - (1/\alpha)y)_+\rangle + (\alpha/2)\|x - (x-(1/\alpha)y)_+\|^2 \\ &= \langle y, (1/\alpha)(\alpha x - y)_+ - x\rangle + \langle y, x\rangle + (\alpha/2)\|(1/\alpha)(\alpha x - y)_+ - x\|^2 \\ &= (\alpha/2)\|(1/\alpha)(\alpha x - y)_+ - x + (1/\alpha)y\|^2 + \langle y, x\rangle - (1/2\alpha)\|y\|^2 \\ &= (1/2\alpha)\| - (y-\alpha x)_- + (y - \alpha x)\|^2 + \langle y, x\rangle - (1/2\alpha)\|y\|^2 \\ &= (1/2\alpha)\|(y-\alpha x)_+\|^2 + \langle y, x\rangle - (1/2\alpha)\|y\|^2\end{aligned}$$

and

$$\begin{aligned}&\langle y, (x-\alpha y)_+\rangle + (1/2\alpha)\,\|x - (x-\alpha y)_+\|^2 \\ &= (1/2\alpha)\|(x-\alpha y)_+\|^2 - (1/\alpha)\,\langle x - \alpha y, (x-\alpha y)_+\rangle + (1/2\alpha)\|x\|^2 \\ &= -(1/2\alpha)\,\|(x-\alpha y)_+\|^2 + (1/2\alpha)\|x\|^2.\end{aligned}$$

These two equalities immediately implies (3.324). Now consider the optimization problem

$$\min_{z\in\mathcal{K}}\ \langle y, z\rangle + (1/2\alpha)\langle z - x, z - x\rangle.$$

It is easy to verify that $z^* = (x - \alpha y)_+$ is the unique optimal solution, whereas $(x - (1/\alpha)y)_+$ is a feasible solution. Therefore, we have

$$\langle y, (x-\alpha y)_+\rangle + \frac{1}{2\alpha}\|x - (x-\alpha y)_+\|^2 \leq \langle y, (x - (1/\alpha)y)_+\rangle + \frac{1}{2\alpha}\|x - (x - (1/\alpha)y)_+\|^2.$$

Combining this inequality with (3.324) yields

$$\langle e, \phi_{\rm MS}(x, y)\rangle \geq \left(\frac{\alpha^2 - 1}{2\alpha}\right) \|x - (x - (1/\alpha)y)_+\|^2,$$

which implies

$$\left\|\phi_{\rm MS}(x, y)\right\| \geq \langle e/\|e\|, \phi_{\rm MS}(x, y)\rangle \geq \left(\frac{\alpha^2 - 1}{2\alpha\|e\|}\right) \|r_\alpha(x, y)\|^2. \tag{3.325}$$

In addition, consider the following strictly convex optimization problem

$$\min_{z \in \mathcal{K}} \ \langle y, z\rangle + (\alpha/2)\langle z - x, z - x\rangle.$$

We can verify that $z^* = (x - (1/\alpha)y)_+$ is the unique optimal solution, whereas $(x - \alpha y)_+$ is a feasible solution. Consequently, we have

$$\langle y, (x - (1/\alpha)y)_+\rangle + \frac{\alpha}{2}\|x - (x - (1/\alpha)y)_+\|^2 \leq \langle y, (x - \alpha y)_+\rangle + \frac{\alpha}{2}\|x - (x - \alpha y)_+\|^2.$$

Combining this inequality with (3.324) then yields

$$\langle e, \phi_{\rm MS}(x, y)\rangle \leq \left(\frac{\alpha^2 - 1}{2\alpha}\right) \|x - (x - \alpha y)_+\|^2,$$

which in turn implies

$$\|\phi_{\rm MS}(x, y)\| \geq -\left\langle e/\|e\|, \phi_{\rm MS}(x, y)\right\rangle \geq \left(\frac{1 - \alpha^2}{2\alpha\|e\|}\right) \left\|r_{\frac{1}{\alpha}}(x, y)\right\|^2. \tag{3.326}$$

From (3.325) and (3.326), we establish the desired result. The proof is thus complete. □

Note that in Lemma 3.57 there holds $\|e\| = \sqrt{q}$ since the rank of $\mathbb{V}$ is assume to be q. Now, in light of Proposition 3.84 and Lemma 3.57, we have the following property of $\phi_{\rm MS}$.

Proposition 3.85 *Let $\phi_{\rm MS}$ be defined as in* (3.236). *Suppose $\{x^k\} \subset \mathbb{V}$ and $\{y^k\} \subset \mathbb{V}$ be the sequences satisfying one of the following conditions*

(i) either $\lambda_{\min}(x^k) \to -\infty$ or $\lambda_{\min}(y^k) \to -\infty$;
(ii) $\lambda_{\min}(x^k), \lambda_{\min}(y^k) > -\infty$, $\lambda_{\max}(x^k), \lambda_{\max}(y^k) \to +\infty$ and $(x^k/\|x^k\|) \circ (y^k/\|y^k\|) \nrightarrow 0$.

Then, $\left\|\phi_{\rm MS}(x^k, y^k)\right\| \to +\infty$.

Similar to Proposition 3.84, when $\mathbb{V} = \mathbb{R}^n$ with $\circ$ being the componentwise product, the assumption $(x^k/\|x^k\|) \circ (y^k/\|y^k\|) \nrightarrow 0$ is automatically satisfied, and from

Proposition 3.85, we readily obtain the result [72, Lemma 6.2] for the NCPs. However, for the general Euclidean Jordan algebra, the following example shows that the assumption $(x^k/\|x^k\|) \circ (y^k/\|y^k\|) \nrightarrow 0$ is also necessary.

Example 3.5 Consider the Lorentz algebra $\mathbb{L}^n = (\mathbb{R}^n, \circ, \langle \cdot, \cdot \rangle_{\mathbb{R}^n})$ with $n = 3$ and take the sequences $\{x^k\}$ and $\{y^k\}$ as follows:

$$x^k = \begin{pmatrix} k \\ -k \\ 0 \end{pmatrix} \quad \text{and} \quad y^k = \begin{pmatrix} k^2 \\ k^2+1 \\ 0 \end{pmatrix} \quad \text{for each } k.$$

It is easy to verify that $\lambda_{\min}(x^k) = 0$, $\lambda_{\min}(y^k) = -1$ and $\lambda_{\max}(x^k), \lambda_{\max}(y^k) \to +\infty$, but

$$\left(x^k/\|x^k\|\right) \to \frac{1}{\sqrt{2}}\begin{pmatrix} 1 \\ -1 \\ 0 \end{pmatrix}, \quad \left(y^k/\|y^k\|\right) \to \frac{1}{\sqrt{2}}\begin{pmatrix} 1 \\ 1 \\ 0 \end{pmatrix}, \quad \left(x^k/\|x^k\|\right) \circ \left(y^k/\|y^k\|\right) \to 0.$$

This shows that the sequences $\{x^k\}$ and $\{y^k\}$ do not satisfy the assumption $(x^k/\|x^k\|) \circ (y^k/\|y^k\|) \nrightarrow 0$. For such $\{x^k\}$ and $\{y^k\}$, we can compute

$$((x^k - \alpha y^k)_+)^2 - (x^k)^2 = \begin{pmatrix} 2k\alpha + (\alpha^2/2) \\ -2k\alpha - (\alpha^2/2) \\ 0 \end{pmatrix}, \quad ((y^k - \alpha x^k)_+)^2 - (y^k)^2 = \begin{pmatrix} -(1/2) \\ (1/2) \\ 0 \end{pmatrix},$$

and

$$\begin{aligned} \phi_{\text{MS}}(x^k, y^k) &= \begin{pmatrix} -k \\ k \\ 0 \end{pmatrix} + (1/2\alpha)\left[\begin{pmatrix} 2k\alpha + (\alpha^2/2) \\ -2k\alpha - (\alpha^2/2) \\ 0 \end{pmatrix} + \begin{pmatrix} -(1/2) \\ (1/2) \\ 0 \end{pmatrix}\right] \\ &= \begin{pmatrix} (\alpha/4) - (1/4\alpha) \\ -(\alpha/4) + (1/4\alpha) \\ 0 \end{pmatrix}. \end{aligned}$$

Clearly, $\|\phi_{\text{MS}}(x^k, y^k)\| \nrightarrow \infty$, i.e., the result of Proposition 3.85 does not hold.

Now assume that $(\mathbb{V}, \circ, \langle \cdot, \cdot \rangle)$ is a general Euclidean Jordan algebra. We establish the coercive properties of the merit function f_{MS} for the SCLCP and the SCCP.

Proposition 3.86 *Let f_{MS} be given by (3.297). If $F(\zeta) = L(\zeta) + b$ with the linear transformation L having the P-property, then the function f_{MS} is coercive.*

Proof Let $\{\zeta^k\}$ be a sequence such that $\|\zeta^k\| \to +\infty$. By passing to a subsequence if necessary, we can assume that $\zeta^k/\|\zeta^k\| \to \bar{\zeta}$, and hence $(L(\zeta^k) + b)/\|\zeta^k\| \to L(\bar{\zeta})$. By the proof of Proposition 3.82, $L(\bar{\zeta}) \neq 0$ and $\{L(\zeta^k) + b\}$ is unbounded. Without loss of generality, assume that $\|L(\zeta^k) + b\| \to +\infty$.

If $\lambda_{\min}(\zeta^k) \to -\infty$ or $\lambda_{\min}(L(\zeta^k)+b) \to -\infty$, then using Proposition 3.85 yields

$$\|\phi_{\text{MS}}(\zeta^k, L(\zeta^k)+b)\| \to +\infty \quad \text{and} \quad f_{\text{MS}}(\zeta^k) \to +\infty.$$

We next assume that the sequences $\{\zeta^k\}$ and $\{L(\zeta^k)+b\}$ are bounded below. Since $\lambda_{\max}(\zeta^k), \lambda_{\max}(L(\zeta^k)+b) \to +\infty$ by $\|\zeta^k\|, \|L(\zeta^k)+b\| \to +\infty$, there is necessarily an element d such that $\zeta^k - d \in \mathcal{K}$ and $L(\zeta^k)+b-d \in \mathcal{K}$ for each k, which implies

$$(\zeta^k - d)/\|\zeta^k\| \in \mathcal{K} \quad \text{and} \quad (L(\zeta^k)+b-d)/\|L(\zeta^k)+b\| \in \mathcal{K} \quad \text{for each } k.$$

Using the fact that $\mathcal{K}$ is a closed convex cone and noting that

$$\bar{\zeta} = \lim_{k\to\infty} (\zeta^k - d)/\|\zeta^k\|, \quad L(\bar{\zeta})/\|L(\bar{\zeta})\| = \lim_{k\to\infty} (L(\zeta^k)+b-d)/\|L(\zeta^k)+b\|,$$

we have $\bar{\zeta} \in \mathcal{K}$ and $L(\bar{\zeta})/\|L(\bar{\zeta})\| \in \mathcal{K}$. Suppose $(\zeta^k/\|\zeta^k\|) \circ (L(\zeta^k)+b)/\|L(\zeta^k)+b\| \to 0$. Then, from Lemma 3.55, it follows that $\bar{\zeta} \circ (L(\bar{\zeta})/\|L(\bar{\zeta})\|) = 0$. Consequently,

$$\bar{\zeta} \in \mathcal{K}, \quad L(\bar{\zeta}) \in \mathcal{K} \quad \text{and} \quad \bar{\zeta} \circ L(\bar{\zeta}) = 0.$$

By [37, Proposition 6], $\bar{\zeta}$ and $L(\bar{\zeta})$ operator commute. This, together with $\bar{\zeta} \circ L(\bar{\zeta}) = 0 \in -\mathcal{K}$ and the P-property of L, means that $\bar{\zeta} = 0$, which is impossible since $\|\bar{\zeta}\| = 1$. Thus, $(\zeta^k/\|\zeta^k\|) \circ (L(\zeta^k)+b)/\|L(\zeta^k)+b\|) \nrightarrow 0$. Notice that $\lambda_{\min}(\zeta^k), \lambda_{\min}(L(\zeta^k)+b) > -\infty$ and $\|\zeta^k\|, \|L(\zeta^k)+b\| \to +\infty$, and hence the sequences $\{\zeta^k\}$ and $\{L(\zeta^k)+b\}$ satisfy the condition(ii) of Proposition 3.85, which implies that $f_{\text{MS}}(\zeta^k) \to +\infty$. □

Proposition 3.87 *Let f_{MS} be given by (3.297). The function f_{MS} is coercive under one of the following conditions:*

(C.1) the mapping F has the uniform Jordan P-property and the Lipschitz continuity;
(C.2) F has the uniform Jordan P-property and, for any $\{\zeta^k\}$, if there exists an index $i \in \{1, 2, \ldots, m\}$ such that $\lambda_{\max}(\zeta_i^k) \to +\infty$ and $\lambda_{\max}(F_i(\zeta^k)) \to +\infty$, then

$$\limsup_{k\to\infty} \left\langle \frac{\zeta_i^k}{\|\zeta_i^k\|}, \frac{F_i(\zeta^k)}{\|F_i(\zeta^k)\|} \right\rangle > 0. \tag{3.327}$$

Proof The proof is similar to that for [70, Theorem 4.1], and we here include it for completeness. Let $\{\zeta^k\} \subset \mathbb{V}$ be any sequence such that $\|\zeta^k\| \to +\infty$. Corresponding to the structure of $\mathbb{V}$, we write $\zeta^k = (\zeta_1^k, \ldots, \zeta_m^k)$ with $\zeta_i^k \in \mathbb{V}_i$ for each k. Define

$$J := \left\{ i \in \{1, 2, \ldots, m\} \mid \{\zeta_i^k\} \text{ is unbounded} \right\}.$$

Clearly, the set $J \neq \emptyset$ since $\{\zeta^k\}$ is unbounded. Let $\{\xi^k\}$ be a bounded sequence with $\xi^k = (\xi_1^k, \dots, \xi_m^k)$ and $\xi_i^k \in \mathbb{V}_i$ for $i = 1, 2, \dots, m$, where ξ_i^k for each k is defined as:

$$\xi_i^k = \begin{cases} 0 & \text{if } i \in J, \\ \zeta_i^k & \text{otherwise,} \end{cases} \quad i = 1, 2, \dots, m.$$

If Condition C.1 holds, then by the uniform Jordan P-property, there is $\rho > 0$ such that

$$\begin{aligned} \rho\|\zeta^k - \xi^k\|^2 &\le \max_{i=1,\dots,m} \lambda_{\max}\left[(\zeta_i^k - \xi_i^k) \circ (F_i(\zeta^k) - F_i(\xi^k))\right] \\ &= \lambda_{\max}\left[\zeta_\nu^k \circ (F_\nu(\zeta^k) - F_\nu(\xi^k))\right] \\ &\le \|\zeta_\nu^k \circ (F_\nu(\zeta^k) - F_\nu(\xi^k))\| \\ &\le \|\zeta_\nu^k\|\|F_\nu(\zeta^k) - F_\nu(\xi^k)\|, \end{aligned} \tag{3.328}$$

where ν is an index from $\{1, 2, \dots, m\}$ for which the maximum is attained, and by the definition of $\{\xi^k\}$, clearly, $\nu \in J$, and the last inequality is due to (1.5). Since $\nu \in J$, $\{\zeta_\nu^k\}$ is unbounded. Without loss of generality, assume that

$$\|\zeta_\nu^k\| \to +\infty. \tag{3.329}$$

Notice that

$$\|\zeta^k - \xi^k\|^2 \ge \|\zeta_\nu^k - \xi_\nu^k\|^2 = \|\zeta_\nu^k\|^2, \quad \forall k.$$

Dividing the both sides of (3.328) by $\|\zeta_\nu^k\|$ then yields

$$\rho\|\zeta_\nu^k\| \le \|F_\nu(\zeta^k) - F_\nu(\xi^k)\| \le \|F_\nu(\zeta^k)\| + \|F_\nu(\xi^k)\|,$$

which, together with the boundedness of $\{F_\nu(\xi^k)\}$, implies

$$\|F_\nu(\zeta^k)\| \to +\infty. \tag{3.330}$$

From Eqs. (3.329) and (3.330), we thus obtain

$$\|\zeta_\nu^k\| \to +\infty, \quad \|F_\nu(\zeta^k)\| \to +\infty. \tag{3.331}$$

We next show that $(\zeta_\nu^k/\|\zeta_\nu^k\|) \circ (F_\nu(\zeta^k)/\|F_\nu(\zeta^k)\|) \not\to 0$. If it does not hold, by the continuity of $\lambda_{\max}(\cdot)$, we will have that $\lambda_{\max}\left[(\zeta_\nu^k/\|\zeta_\nu^k\|) \circ (F_\nu(\zeta^k)/\|F_\nu(\zeta^k)\|\right] \to 0$. Consequently,

$$\begin{aligned} &\lim_{k\to\infty} \lambda_{\max}\left[\zeta_\nu^k \circ (F_\nu(\zeta^k) - F_\nu(\xi^k))\right] / [\|\zeta_\nu^k\|\|F_\nu(\zeta^k)\|] \\ \le &\lim_{k\to\infty} \lambda_{\max}\left[(\zeta_\nu^k/\|\zeta_\nu^k\|) \circ (F_\nu(\zeta^k)/\|F_\nu(\zeta^k)\|)\right] \\ &+ \lim_{k\to\infty} \lambda_{\max}\left[-\zeta_\nu^k \circ F_\nu(\xi^k)\right] / \left[\|\zeta_\nu^k\|\|F_\nu(\zeta^k)\|\right] \\ = &\,0, \end{aligned} \tag{3.332}$$

where the inequality is due to Lemma 3.54(c). On the other hand, from the Lipschitz continuity of the mapping F, there exists a scalar $\gamma > 0$ such that

$$\|F(\zeta^k) - F(0)\| \leq \gamma\|\zeta^k - 0\| = \gamma\|\zeta^k\| \quad \text{for each } k,$$

which in turn implies

$$\|F_\nu(\zeta^k)\| \leq \|F_\nu(\zeta^k) - F_\nu(0)\| + \|F_\nu(0)\| \leq \gamma\|\zeta^k\| + \|F_\nu(0)\|, \quad \forall k.$$

From the last inequality, we obtain

$$\begin{aligned}
&\lim_{k\to\infty} \rho\|\zeta^k - \xi^k\|^2/[\|\zeta_\nu^k\|\|F_\nu(\zeta^k)\|] \\
&\geq \lim_{k\to\infty} \rho\|\zeta^k - \xi^k\|^2/[\|\zeta^k\|(\gamma\|\zeta^k\| + \|F_\nu(0)\|)] = \frac{\rho}{\gamma} > 0.
\end{aligned}$$

This, together with (3.332), gives a contradiction to the first inequality of (3.328). Thus, the sequences $\{\zeta_\nu^k\}$ and $\{F_\nu(\zeta^k)\}$ satisfy the conditions of Proposition 3.85. Consequently,

$$\left\|\phi_{\text{MS}}^{(\nu)}(\zeta_\nu^k, F_\nu(\zeta^k))\right\| \to +\infty \quad \text{and} \quad f_{\text{MS}}(\zeta^k) \to +\infty.$$

If Condition C.2 is satisfied, then from the above discussions we see that Eqs. (3.328)–(3.331) still hold. If $\lambda_{\min}(\zeta_\nu^k) \to -\infty$ or $\lambda_{\min}(F_\nu(\zeta^k)) \to -\infty$, then using Lemmas 3.56 and 3.57 gives that $\phi_{\text{MS}}^{(\nu)}(\zeta_\nu^k, F_\nu(\zeta^k)) \to +\infty$, and hence $f_{\text{MS}}(\zeta^k) \to +\infty$. Otherwise, by Eq. (3.331), we will have $\lambda_{\max}(\zeta_\nu^k) \to +\infty$ and $\lambda_{\max}(F_\nu(\zeta^k)) \to +\infty$. From the given assumption, it then follows that

$$\limsup_{k\to\infty} \left\langle \zeta_\nu^k/\|\zeta_\nu^k\|, F_\nu(\zeta^k)/\|F_\nu(\zeta^k)\|\right\rangle > 0,$$

which, by Lemma 3.54(a), leads to

$$\limsup_{k\to\infty} \lambda_{\max}\left[(\zeta_\nu^k/\|\zeta_\nu^k\|) \circ (F_\nu(\zeta^k)/\|F_\nu(\zeta^k)\|)\right] > 0.$$

This shows that $(\zeta_\nu^k/\|\zeta_\nu^k\|) \circ (F_\nu(\zeta^k)/\|F_\nu(\zeta^k)\|) \nrightarrow 0$. Hence, the sequences $\{\zeta_\nu^k\}$ and $\{F_\nu(\zeta^k)\}$ satisfy the conditions of Proposition 3.85. Consequently, $\left\|\phi_{\text{MS}}^{(\nu)}(\zeta_\nu^k, F_\nu(\zeta^k))\right\| \to +\infty$ and $f_{\text{MS}}(\zeta^k) \to +\infty$. The proof is then complete. □

Notice that, when $\mathbb{V} = \mathbb{R}^n$ with $\circ$ being the componentwise product of the vectors, the assumption (3.327) is automatically satisfied and the uniform Jordan P-property of F is equivalent to saying that F is a uniform P-function. Thus, Proposition 3.87 reduces to the known result [70, Theorem 4.1] for the NCPs. However, for the general Euclidean Jordan algebra, besides the uniform Jordan P-property of F, it requires that F is Lipschitz continuous or satisfies the assumption (3.327) so that $(\zeta_\nu^k/\|\zeta_\nu^k\|) \circ (F_\nu(\zeta^k)/\|F_\nu(\zeta^k)\|) \nrightarrow 0$.

In addition, using Proposition 3.84 and the same arguments as in Proposition 3.86 and Proposition 3.87, we can achieve the coerciveness of the natural residual merit function for the SCCP:

$$R_\alpha(\zeta) := \frac{1}{2}\left\|r_\alpha(\zeta, F(\zeta))\right\|^2. \tag{3.333}$$

Proposition 3.88 *The function R_α defined by (3.333) is coercive under Condition C.1 or C.2 of Proposition 3.87. If $F(\zeta) = L(\zeta) + b$ with the linear transformation L having the P-property, then R_α is also coercive.*

Furthermore, from Lemma 3.57, we conclude that the growth rate of $f_{\rm MS}$ is higher than that of the natural residual merit function R_α.

Proposition 3.89 *Let $\{\zeta^k\}$ be a sequence such that $\|\zeta^k\| \to +\infty$. If F satisfies Condition C.1 or C.2 of Proposition 3.87, then $R_\alpha(\zeta^k) \to +\infty$, $f_{\rm MS}(\zeta^k) \to +\infty$ and*

$$\frac{f_{\rm MS}(\zeta^k)}{\left[R_\alpha(\zeta^k)\right]^{1+\sigma}} \to +\infty \ \text{ with } \ 0 \le \sigma < 1.$$

3.3.2 *Constructions of C-Functions Associated with Symmetric Cone*

Building upon the preceding discussions, two natural questions arise concerning the construction of C-functions for the symmetric cone complementarity problem:

(i) Is there a systematic framework for constructing complementarity functions associated with symmetric cones?
(ii) Can existing NCP functions be adapted to generate C-functions in the symmetric cone setting?

These questions have long stood as central challenges in the study of complementarity functions. In this section, we offer affirmative answers to both. Specifically, we propose two distinct methods for constructing C-functions tailored to symmetric cones. The first approach is inspired by a class of NCP functions originally examined by Mangasarian in [73], as detailed below.

Property 3.1 Assume that $\theta : \mathbb{R} \to \mathbb{R}$ is a strictly increasing function, that is, $a > b \Longleftrightarrow \theta(a) > \theta(b)$, and let $\theta(0) = 0$. Then, the function

$$\phi(a, b) := \theta(|a - b|) - \theta(a) - \theta(b)$$

is an NCP function.

In [73], Mangasarian provided two examples of θ, namely $\theta(z) = z|z|$ and $\theta(z) = z$. Accordingly, they induce the following NCP functions:

$$\phi_{\text{Man1}}(a, b) = (a - b)^2 - b|b| - a|a|,$$
$$\phi_{\text{Man2}}(a, b) = |a - b| - b - a.$$

Motivated by Property 3.1, as will be demonstrated later, we introduce a class of vector-valued functions designed to induce C-functions associated with symmetric cones. Furthermore, we develop several compositional forms of such C-functions, expanding the repertoire of available constructions.

The second method builds upon existing NCP functions. As noted earlier, numerous researchers have explored the extension of NCP functions to serve as C-functions for the symmetric cone complementarity problem (SCCP). Our novel approach lies in utilizing these existing NCP functions, originally defined as real-valued functions, to construct vector-valued C-functions within the symmetric cone framework. With nearly sixty NCP functions documented in the literature, this idea introduces a powerful and versatile mechanism for generating a rich variety of C-functions. We believe this contribution represents a significant breakthrough, laying a solid foundation for future analytical developments on the SCCP via NCP-based techniques. In particular, we present general formulations of C-functions derived from NCP functions and apply this framework to two prominent symmetric cones: the second-order cone and the positive semidefinite cone. These constructions are based on explicit expressions of the inner (Jordan) product, further highlighting the potential of this methodology. This innovative direction opens new avenues for addressing complementarity problems through minimization approaches grounded in NCP function theory.

Lemma 3.58 *For any $x, y \in \mathcal{K}$, if $x \succeq_{\mathcal{K}} 0$, $y \succeq_{\mathcal{K}} 0$ and $x \succeq_{\mathcal{K}} y$, then $x^{1/2} \succeq_{\mathcal{K}} y^{1/2}$.*

Proof Please see [37, Proposition 8]. □

Lemma 3.59 *Let X, Y be $n \times n$ matrices in $\mathbb{S}^{n\times n}$. Then, the following hold:*

(a) $X \succeq 0 \Rightarrow UXU^{\mathsf{T}} \succeq 0$ *for any orthogonal matrix U.*
(b) $X \succeq 0,\ Y \succeq 0 \Rightarrow \langle X, Y\rangle \geq 0.$
(c) $X \succeq 0,\ Y \succeq 0,\ \langle X, Y\rangle = 0 \Rightarrow XY = YX = 0.$
(d) *If $X \succeq 0$, $Y \succeq 0$, then $\langle X, Y\rangle = 0 \Longleftrightarrow XY = 0$.*
(e) *Given X and Y in $\mathbb{S}^{n\times n}$ with $XY = YX$, there exists an orthogonal matrix U, diagonal matrices D and E such that $X = UDU^{\mathsf{T}}$ and $Y = UEU^{\mathsf{T}}$.*

Proof Please see [74, 75]. □

Lemma 3.60 *Let $x = (x_1, \bar{x}_2) \in \mathbb{R} \times \mathbb{R}^{n-1}$ and $y = (y_1, \bar{y}_2) \in \mathbb{R} \times \mathbb{R}^{n-1}$. Then,*

$$x \succeq_{\mathbb{L}^n_+} 0,\ \ y \succeq_{\mathbb{L}^n_+} 0 \ \text{ and } \ x \circ y = 0$$

if and only if the following hold

(i) If $\bar{x}_2 \neq 0$ and $\bar{y}_2 \neq 0$, then x, y are both on the boundary of $\mathcal{L}^n_+$, share the same spectral vectors, and can be expressed as

$$x = \lambda_2(x) \cdot u_x^{(2)} = 2x_1 \cdot \frac{1}{2}\left(1, \frac{\bar{x}_2}{||\bar{x}_2||}\right),$$

$$y = \lambda_2(y) \cdot u_y^{(2)} = 2y_1 \cdot \frac{1}{2}\left(1, -\frac{\bar{x}_2}{||\bar{x}_2||}\right),$$

with $\langle u_x^{(2)}, u_y^{(2)} \rangle = 0$ *or* $u_x^{(2)} \circ u_y^{(2)} = 0$.

(ii) If $\bar{x}_2 = 0$ *or* $\bar{y}_2 = 0$*, then it goes to the trivial cases that* $x = 0$ *and* $y \in \mathbb{L}_+^n$ *or* $x \in \mathbb{L}_+^n$ *and* $y = 0$.

Proof The proof follows an approach similar to that of [12, Proposition 2.1]. For the sake of completeness, we include the full details below.

"$\Leftarrow$" The proof of this direction is trivial.

"$\Rightarrow$" From $x \succeq_{\mathbb{L}_+^n} 0$, $y \succeq_{\mathbb{L}_+^n} 0$ and $x \circ y = (\langle x, y \rangle, x_1\bar{y}_2 + y_1\bar{x}_2) = 0$, we have

$$\langle x, y \rangle = x_1 y_1 + \bar{x}_2^{\mathsf{T}}\bar{y}_2 = 0, \quad x_1 \geq ||\bar{x}_2||, \quad y_1 \geq ||\bar{y}_2||. \tag{3.334}$$

To proceed, we discuss two cases.

(i) If $\bar{x}_2 \neq 0$ and $\bar{y}_2 \neq 0$, then Eq. (3.334) implies $-\bar{x}_2^{\mathsf{T}}\bar{y}_2 = x_1 y_1 \geq ||\bar{x}_2||||\bar{y}_2||$. Since $-\bar{x}_2^{\mathsf{T}}\bar{y}_2 \leq ||\bar{x}_2||||\bar{y}_2||$, it leads to $x_1 y_1 = -\bar{x}_2^{\mathsf{T}}\bar{y}_2 = ||\bar{x}_2||||\bar{y}_2||$. Hence $x_1 = ||\bar{x}_2||$, $y_1 = ||\bar{y}_2||$; otherwise, if $x \in \text{int}(\mathbb{L}_+^n)$ or $y \in \text{int}(\mathbb{L}_+^n)$ then $x_1 y_1 > ||\bar{x}_2||||\bar{y}_2||$, which is impossible. This means x and y are both on the boundary of $\mathbb{L}_+^n$. Using the facts that the second component of $x \circ y$ is zero, i.e., $x_1\bar{y}_2 + y_1\bar{x}_2 = 0$, and the fact that $x_1 = ||\bar{x}_2||$, $y_1 = ||\bar{y}_2||$, these yield that

$$x = \lambda_2(x) \cdot u_x^{(2)} = (x_1 + ||\bar{x}_2||) \cdot \frac{1}{2}(1, \frac{\bar{x}_2}{||\bar{x}_2||}) = 2x_1 \cdot \frac{1}{2}(1, \frac{\bar{x}_2}{||\bar{x}_2||})$$

and

$$y = \lambda_2(y) \cdot u_y^{(2)} = (y_1 + ||\bar{y}_2||) \cdot \frac{1}{2}(1, \frac{\bar{y}_2}{||\bar{y}_2||}) = 2y_1 \cdot \frac{1}{2}(1, -\frac{\bar{x}_2}{||\bar{x}_2||}),$$

where x and y can be viewed as sharing the same spectral vectors $\{u_x^{(2)}, u_y^{(2)}\}$ with $u_x^{(2)} = \frac{1}{2}(1, \frac{\bar{x}_2}{||\bar{x}_2||})$, $u_y^{(2)} = \frac{1}{2}(1, -\frac{\bar{x}_2}{||\bar{x}_2||}) = u_x^{(1)}$ and $\langle u_x^{(2)}, u_y^{(2)} \rangle = u_x^{(2)} \circ u_y^{(2)} = 0$.

(ii) If $\bar{x}_2 = 0$, from Eq. (3.334), we obtain $x_1 y_1 = 0$. It leads to $x_1 = 0$ or $y_1 = 0$. For $x_1 = 0$, then we have $x = 0$ and y can be any element in $\mathbb{L}_+^n$. For $y_1 = 0$, then $\bar{y}_2$ must be 0 from the third inequality of (3.334), which means $y = 0$ and x can be any element in $\mathbb{L}_+^n$ in this case. Similar to the case $\bar{y}_2 = 0$. □

Lemma 3.61 *Let* $x = (x_1, \bar{x}_2) \in \mathbb{R} \times \mathbb{R}^{n-1}$ *and* $y = (y_1, \bar{y}_2) \in \mathbb{R} \times \mathbb{R}^{n-1}$ *with* $\bar{x}_2 \neq 0$, $\bar{y}_2 \neq 0$. *Then,*

$$x \succeq_{\mathbb{L}_+^n} 0, \; y \succeq_{\mathbb{L}_+^n} 0 \text{ and } x \circ y = 0$$

if and only if $x_1 = \|\bar{x}_2\|$, $y_1 = \|\bar{y}_2\|$*, and* $x_1\bar{y}_2 + y_1\bar{x}_2 = 0$.

Proof This is an immediate consequence of Lemma 3.60. □

Lemma 3.62 *Let* $x = (x_1, \bar{x}_2) \in \mathbb{R} \times \mathbb{R}^{n-1}$ *and* $y = (y_1, \bar{y}_2) \in \mathbb{R} \times \mathbb{R}^{n-1}$ *with* $\bar{x}_2 \neq 0$, $\bar{y}_2 \neq 0$. *If* $x \succeq_{\mathbb{L}^n_+} 0$, $y \succeq_{\mathbb{L}^n_+} 0$ *and* $x \circ y = 0$, *then* $\bar{y}_2 = -m\bar{x}_2$, *where* $m := \frac{\|\bar{y}_2\|}{\|\bar{x}_2\|}$. *Moreover,*

$$\bar{y}_2 = -m\bar{x}_2 \Longleftrightarrow \text{there exists } k \in \{2, \cdots, n\} \text{ such that } y_k = -mx_k \neq 0 \quad \text{and } y_l x_k = x_l y_k \text{ for all } l \in \{2, \cdots, n\}. \tag{3.335}$$

Proof From case (i) in the proof of Lemma 3.60, we see that $\bar{x}_2^T \bar{y}_2 = -||\bar{x}_2||||\bar{y}_2||$, which further implies

$$\frac{\bar{x}_2^T \bar{y}_2}{||\bar{y}_2||} = -||\bar{x}_2|| \Longleftrightarrow \frac{\bar{x}_2^T \bar{y}_2}{||\bar{y}_2||} = -\frac{\bar{x}_2^T \bar{x}_2}{||\bar{x}_2||} \Longleftrightarrow \frac{\bar{y}_2}{||\bar{y}_2||} = -\frac{\bar{x}_2}{||\bar{x}_2||} \Longleftrightarrow \bar{y}_2 = -\frac{||\bar{y}_2||}{||\bar{x}_2||}\bar{x}_2.$$

Letting $m := \frac{\|\bar{y}_2\|}{\|\bar{x}_2\|}$, it implies $\bar{y}_2 = -m\bar{x}_2$.

Next, we prove the relation (3.335).

"$\Rightarrow$" Since $\bar{y}_2 = -m\bar{x}_2$, and $\bar{x}_2 \neq 0$, $\bar{y}_2 \neq 0$, there exists $k \in \{2, \cdots, n\}$ such that $x_k \neq 0$, $y_k \neq 0$ and $y_k = -mx_k$. In addition, $y_l = -mx_l$ for all $l \in \{2, \cdots, n\}$. Multiplying by $-mx_k$ both sides of this equation, we have

$$y_l(-mx_k) = -mx_l(-mx_k) = -mx_l y_k.$$

Thus, we prove that $y_l x_k = x_l y_k$.

"$\Leftarrow$" Since $y_l x_k = x_l y_k$ and $y_k = -mx_k \neq 0$, it yields $y_l x_k = x_l(-mx_k)$. This implies that $y_l = -mx_l$ for all $l \in \{2, \cdots, n\}$. Hence, $\bar{y}_2 = -m\bar{x}_2$. □

A. First construction method of C-functions.

As discussed in Chap. 2, several systematic approaches exist for NCP functions, typically relying on the property that $a \geq 0, b \geq 0, ab = 0$ implies either $a = 0$ or $b = 0$. Unfortunately, this implication does not hold in the setting of symmetric cones. Recall from (1.28) (see also [37, Proposition 6]) that

$$x \succeq_{\mathcal{K}} 0,\ y \succeq_{\mathcal{K}} 0,\ x \circ y = 0 \quad \Longleftrightarrow \quad x \succeq_{\mathcal{K}} 0,\ y \succeq_{\mathcal{K}} 0,\ \langle x, y \rangle = 0.$$

The core difficulty with symmetric cones $\mathcal{K}$ lies in the fact that

$$x \succeq_{\mathcal{K}} 0,\ y \succeq_{\mathcal{K}} 0,\ x \circ y = 0 \text{ does not imply that } x = 0 \text{ or } y = 0.$$

Nonetheless, as we shall see, the following assumption may help to compensate for this limitation.

Assumption 3.2 A function $\theta : \mathbb{R}^n \to \mathbb{R}^n$ is said to satisfy Assumption 3.2 if

(i) $x \succeq_{\mathcal{K}} 0$ if and only if $\theta(x) \succeq_{\mathcal{K}} 0$.
(ii) for any $x, y \succeq_{\mathcal{K}} 0$, $x \circ y = 0$ if and only if $\theta(x) \circ \theta(y) = 0$.

Assumption 3.2(i) is a slightly weaker than the strictly increasing property mentioned in Property 3.1, whereas Assumption 3.2(ii) is used to adjust the expression in a general symmetric cone setting.

Proposition 3.90 *Suppose that* $\theta : \mathbb{R}^n \to \mathbb{R}^n$ *satisfies Assumption 3.2. Then, the function* $\varphi : \mathbb{R}^n \times \mathbb{R}^n \to \mathbb{R}^n$ *defined by*

$$\varphi(x, y) := |\theta(x) - \theta(y)| - \theta(x) - \theta(y)$$

is a C-function in the symmetric cone setting.

Proof It suffices to verify that $\varphi(x, y) = 0$ if and only if $x \succeq_{\mathcal{K}} 0,\ y \succeq_{\mathcal{K}} 0,\ x \circ y = 0$.

"$\Rightarrow$" Assume that $\varphi(x, y) = 0$, we observe

$$\begin{aligned} & \varphi(x, y) = |\theta(x) - \theta(y)| - \theta(x) - \theta(y) = 0 \\ \Longleftrightarrow\ & |\theta(x) - \theta(y)| = \theta(x) + \theta(y) \\ \Longleftrightarrow\ & |\theta(x) - \theta(y)|^2 = (\theta(x) + \theta(y))^2 \\ \Longleftrightarrow\ & \theta(x)^2 - 2\theta(x) \circ \theta(y) + \theta(y)^2 = \theta(x)^2 + 2\theta(x) \circ \theta(y) + \theta(y)^2 \\ \Longleftrightarrow\ & \theta(x) \circ \theta(y) = 0. \end{aligned} \tag{3.336}$$

Letting $\omega = |\theta(x) - \theta(y)|$ gives $\omega^2 = \theta(x)^2 - 2\theta(x) \circ \theta(y) + \theta(y)^2 = \theta(x)^2 + \theta(y)^2$. Thus, we have $\omega^2 \succeq_{\mathcal{K}} \theta(x)^2$ and $\omega^2 \succeq_{\mathcal{K}} \theta(y)^2$. This leads to $\omega \succeq_{\mathcal{K}} \theta(x)$ and $\omega \succeq_{\mathcal{K}} \theta(y)$ by applying Lemma 3.58. Since $\varphi(x, y) = 0$, $\omega = \theta(x) + \theta(y)$, it follows that $\theta(x) = \omega - \theta(y) \succeq_{\mathcal{K}} 0$ and $\theta(y) = \omega - \theta(x) \succeq_{\mathcal{K}} 0$. Using Assumption 3.2(i) of θ, we obtain $x, y \succeq_{\mathcal{K}} 0$. Then, we further have $x \circ y = 0$ from Assumption 3.2(ii).

"$\Leftarrow$" Suppose that $x \succeq_{\mathcal{K}} 0$, $y \succeq_{\mathcal{K}} 0$, $x \circ y = 0$ and θ satisfies Assumption 3.2. Then it is clear to see that $\theta(x) \succeq_{\mathcal{K}} 0$, $\theta(y) \succeq_{\mathcal{K}} 0$ and $\theta(x) \circ \theta(y) = 0$. This fact together with (3.336) shows $\varphi(x, y) = 0$. □

What are some examples of $\theta(\cdot)$ function that satisfy Assumption 3.2? Indeed, in light of Theorem 1.1 and note that $x \in \mathcal{K}$ if and only if $\lambda_i(x) \geq 0$ for all $i = 1, \cdots, r$, we can confirm that the following functions satisfy Assumption 3.2 in their domain:

$$\begin{aligned} \theta_1(z) &= z, \\ \theta_2(z) &= z^p, \text{ where } p \text{ is positive odd integer}, \\ \theta_3(z) &= z|z|, \\ \theta_4(z) &= z^{1/2}, \text{ where } \theta_4 : \mathcal{K} \to \mathcal{K}. \end{aligned}$$

Hence, by Proposition 3.90, these functions corresponds to C-functions φ_1, φ_2, φ_3, and φ_4 which are listed below.

$$\begin{aligned}
\varphi_1(x,y) &= |x-y|-(x+y) = -\frac{1}{2}\phi_{\rm NR}(x,y);\\
\varphi_2(x,y) &= |x^p-y^p|-x^p-y^p,\ \text{where } p \text{ is positive odd integer;}\\
\varphi_3(x,y) &= \big|x|x|-y|y|\big|-x|x|-y|y|;\\
\varphi_4(x,y) &= \big|x^{1/2}-y^{1/2}\big|-x^{1/2}-y^{1/2},\ \text{where } \varphi_4:\mathcal{K}\times\mathcal{K}\to\mathcal{K}.
\end{aligned}$$

Next, we explore composition forms of C-functions. More specifically, given a $\theta(\cdot)$ function satisfying Assumption 3.2 and any C-function φ, the composition function $\varphi(\theta(x),\theta(y))$ is a C-function as well.

Proposition 3.91 *Suppose that $\theta:\mathbb{R}^n\to\mathbb{R}^n$ satisfies Assumption 3.2. Then, for any C-function $\varphi:\mathbb{R}^n\times\mathbb{R}^n\to\mathbb{R}^n$, the composition function $\varphi(\theta(x),\theta(y))$ is also a C-function.*

Proof "$\Leftarrow$" If $x\succeq_{\mathcal{K}}0$, $y\succeq_{\mathcal{K}}0$, $x\circ y=0$ and θ satisfies Assumption 3.2, we have $\theta(x)\succeq_{\mathcal{K}}0$ and $\theta(y)\succeq_{\mathcal{K}}0$ by Assumption 3.2(i) and $\theta(x)\circ\theta(y)=0$ by Assumption 3.2(ii). Then, it follows that $\varphi(\theta(x),\theta(y))=0$ since φ is a C-function.

"$\Rightarrow$" If $\varphi(\theta(x),\theta(y))=0$, we have $\theta(x),\theta(y)\succeq_{\mathcal{K}}0$ and $\theta(x)\circ\theta(y)=0$ since φ is a C-function. Again, applying Assumption 3.2 yields $x,y\succeq_{\mathcal{K}}0$ and $x\circ y=0$. □

Since those functions $\theta_1,\theta_2,\theta_3,\theta_4$ satisfy Assumption 3.2, we can employ them and apply Theorem 3.91 to obtain more C-functions. For example, if we take the Fischer–Burmeister function

$$\varphi_{\rm FB}(x,y)=(x^2+y^2)^{1/2}-(x+y),$$

then we achieve the following C-functions accordingly:

$$\begin{aligned}
\tilde{\varphi}_1(x,y) &= \phi_{\rm FB}(x,y);\\
\tilde{\varphi}_2(x,y) &= (x^{2p}+y^{2p})^{1/2}-(x^p+y^p),\ \text{where } p \text{ is positive odd integer;}\\
\tilde{\varphi}_3(x,y) &= ((x|x|)^2+(y|y|)^2)^{1/2}-(x|x|+y|y|);\\
\tilde{\varphi}_4(x,y) &= (x+y)^{1/2}-(x^{1/2}+y^{1/2}),\ \text{where } \tilde{\varphi}_4:\mathcal{K}\times\mathcal{K}\to\mathcal{K}.
\end{aligned}$$

In fact, items (i) and (ii) in Assumption 3.2 can be combined together as a complementarity property, which is slightly weaker than Assumption 3.2.

Assumption 3.3 A function $\theta:\mathbb{R}^n\to\mathbb{R}^n$ is said to satisfy Assumption 3.3 if

$$x\succeq_{\mathcal{K}}0,\ y\succeq_{\mathcal{K}}0,\ x\circ y=0 \iff \theta(x)\succeq_{\mathcal{K}}0,\ \theta(y)\succeq_{\mathcal{K}}0,\ \theta(x)\circ\theta(y)=0.$$

It is clear that Assumption 3.2 implies Assumption 3.3, but the reverse direction is not true. It is noted that Assumption 3.3 is sufficient for Proposition 3.91. The following is a weaker version of the composition form.

Proposition 3.92 *Suppose that $\theta : \mathbb{R}^n \to \mathbb{R}^n$ satisfies Assumption 3.3. Then, for any C-function $\varphi : \mathbb{R}^n \times \mathbb{R}^n \to \mathbb{R}^n$, the composition function $\varphi(\theta(x), \theta(y))$ is also a C-function.*

Proof The proof is straightforward. Since φ is a C-function and θ satisfies Assumption 3.3, we have

$$\begin{aligned} &\varphi(\theta(x), \theta(y)) = 0 \\ \Longleftrightarrow\ &\theta(x) \succeq_{\mathcal{K}} 0,\ \theta(y) \succeq_{\mathcal{K}} 0,\ \theta(x) \circ \theta(y) = 0 \\ \Longleftrightarrow\ &x \succeq_{\mathcal{K}} 0,\ \ y \succeq_{\mathcal{K}} 0,\ \ x \circ y = 0. \end{aligned}$$

Hence, $\varphi(\theta(x), \theta(y))$ is also a C-function. □

If we choose $\theta(z) = z$, then the composition function $\varphi(\theta(x), \theta(y))$ in Proposition 3.92 goes back to the original C-function $\varphi(x, y)$. If we choose $\varphi_1(x, y) = x - (x - y)_+ = \phi_{\rm NR}(x, y)$, $\varphi_2(x, y) = (x^2 + y^2)^{1/2} - (x + y) = \phi_{\rm FB}(x, y)$, composing them with different $\theta(\cdot)$ leads to various C-functions.

1. Let $\theta(z) = z^p$ where p is positive odd integer. Then, applying Proposition 3.92 implies

$$\begin{aligned} \varphi_1(\theta(x), \theta(y)) &= x^p - (x^p - y^p)_+, \\ \varphi_2(\theta(x), \theta(y)) &= \left(x^{2p} + y^{2p}\right)^{1/2} - (x^p + y^p) \end{aligned}$$

   are also C-functions.
2. Let $\theta(z) = z|z|$. Then, applying Proposition 3.92 implies

$$\begin{aligned} \varphi_1(\theta(x), \theta(y)) &= x|x| - \left(x|x| - y|y|\right)_+, \\ \varphi_2(\theta(x), \theta(y)) &= \left((x|x|)^2 + (y|y|)^2\right)^{1/2} - (x|x| + y|y|) \end{aligned}$$

   are also C-functions.

We now introduce a special class of functions that also satisfy Assumption 3.3. This enables us to generate a broad family of functions $\theta(\cdot)$, which can be effectively employed in conjunction with Proposition 3.92.

Proposition 3.93 *For any real-valued function $f : \mathbb{R} \to \mathbb{R}$ with the following properties:*

(i) $t \geq 0$ if and only if $f(t) \geq 0$;
(ii) $t = 0$ if and only if $f(t) = 0$,

the vector-valued function $f^{\rm sc} : \mathbb{R}^n \to \mathbb{R}^n$ associated with $\mathcal{K}$, defined by

$$f^{\rm sc}(x) = f(\lambda_1(x))e_1 + \cdots + f(\lambda_r(x))e_r \quad \forall x \in \mathbb{V},$$

satisfies Assumption 3.3. Here, $\lambda_i(x)$ and $\{e_i\}$ for $i = 1, 2, \cdots, r$ are the spectral values and the spectral vectors of x, respectively.

Proof Let $x, y \in \mathbb{V}$, the spectral decompositions of x and y are given by

$$x = \sum_{i=1}^{r} \lambda_i(x) e_i \quad \text{and} \quad y = \sum_{i=1}^{r} \lambda_i(y) f_i.$$

Then, we have

$$f^{sc}(x) = \sum_{i=1}^{r} f(\lambda_i(x)) e_i \quad \text{and} \quad f^{sc}(y) = \sum_{i=1}^{r} f(\lambda_i(y)) f_i.$$

From the above properties (i)–(ii) of f, we obtain

$$\begin{aligned}
& x \succeq_{\mathcal{K}} 0,\ y \succeq_{\mathcal{K}} 0,\ x \circ y = 0 \\
\Longleftrightarrow\ & x \succeq_{\mathcal{K}} 0,\ y \succeq_{\mathcal{K}} 0,\ \langle x, y \rangle = 0 \\
\Longleftrightarrow\ & \lambda_i(x) \geq 0,\ \lambda_i(y) \geq 0,\ \sum_{i,j}^{r} \lambda_i(x)\lambda_j(y) \langle e_i, f_j \rangle = 0 \\
\Longleftrightarrow\ & \lambda_i(x) \geq 0,\ \lambda_i(y) \geq 0,\ \lambda_i(x)\lambda_j(y) = 0 \text{ or } \langle e_i, f_j \rangle = 0 \\
\Longleftrightarrow\ & f(\lambda_i(x)) \geq 0,\ f(\lambda_i(y)) \geq 0,\ f(\lambda_i(x)) f(\lambda_j(y)) = 0 \text{ or } \langle e_i, f_j \rangle = 0 \\
\Longleftrightarrow\ & f^{sc}(x) \succeq_{\mathcal{K}} 0,\ f^{sc}(y) \succeq_{\mathcal{K}} 0,\ \sum_{i,j}^{r} f(\lambda_i(x)) f(\lambda_j(y)) \langle e_i, f_j \rangle = 0 \\
\Longleftrightarrow\ & f^{sc}(x) \succeq_{\mathcal{K}} 0,\ f^{sc}(y) \succeq_{\mathcal{K}} 0,\ \langle f^{sc}(x), f^{sc}(y) \rangle = 0 \\
\Longleftrightarrow\ & f^{sc}(x) \succeq_{\mathcal{K}} 0,\ f^{sc}(y) \succeq_{\mathcal{K}} 0,\ f^{sc}(x) \circ f^{sc}(y) = 0,
\end{aligned}$$

where $i = 1, 2, \cdots, r$ and $j = 1, 2, \cdots, r$. Thus, it is clear to see that Assumption 3.3 is satisfied and the proof is complete. □

We list a couple of examples of f mentioned in Proposition 3.93. The first one is $f(t) = t^p$ with a positive odd number p. It is clear that the properties (i) and (ii) are held. Hence, its corresponding SC-function reduces to the regular function $f^{sc}(x) = x^p$. The second one is $f(t) = \dfrac{t}{t^2+1}$, which also possesses (i) $t \geq 0$ if and only if $f(t) \geq 0$; and (ii) $t = 0$ if and only if $f(t) = 0$. Then, in light of Proposition 3.93, its SC-function satisfies Assumption 3.3. This means we can employ this f^{sc} function as a choice of $\theta(\cdot)$ function in Proposition 3.92 to generate C-functions below:

$$\theta(x) = f^{sc}(x) = \frac{\lambda_1(x)}{\lambda_1(x)^2+1} e_1 + \cdots + \frac{\lambda_r(x)}{\lambda_r(x)^2+1} e_r,$$

where $x \in \mathbb{V}$, $\lambda_i(x)$ for $i = 1, 2, \cdots, r$ are spectral values of x, and $\{e_i\}_{i=1}^r$ is a Jordan frame. Note that the expression is not explicit.

Indeed, by applying Lemma 3.60, we can omit property(ii) in Proposition 3.93 in the special case of the second-order cone.

Proposition 3.94 *For any real-valued function* $f : \mathbb{R} \to \mathbb{R}$ *with* $t \geq 0$ *if and only if* $f(t) \geq 0$*, the following vector-valued function associated with* $\mathbb{L}^n_+$ *(also called SOC function for short), defined by*

$$f^{\mathrm{soc}}(x) = f(\lambda_1(x))u_x^{(1)} + f(\lambda_2(x))u_x^{(2)},$$

satisfies Assumption 3.3. Here, $\lambda_i(x)$ *and* $u_x^{(i)}$*, for* $i = 1, 2$ *are the spectral values and the spectral vectors of* $x = (x_1, \bar{x}_2) \in \mathbb{R} \times \mathbb{R}^{n-1}$.

Proof Let $x = (x_1, \bar{x}_2) \in \mathbb{R} \times \mathbb{R}^{n-1}$ and $y = (y_1, \bar{y}_2) \in \mathbb{R} \times \mathbb{R}^{n-1}$.

For $\bar{x}_2 = 0$ or $\bar{y}_2 = 0$, it can be seen that $x = 0$ or $y = 0$ by Lemma 3.60. It follows that $f^{\mathrm{soc}}(x) = 0$ or $f^{\mathrm{soc}}(y) = 0$ when $\bar{x}_2 = 0$ or $\bar{y}_2 = 0$. Hence, $f^{\mathrm{soc}}(y)$ automatically satisfies Assumption 3.3.

For $\bar{x}_2 \neq 0$ and $\bar{y}_2 \neq 0$, from Lemma 3.60, we have

$$\begin{aligned}
& x \succeq_{\mathbb{L}^n_+} 0,\ y \succeq_{\mathbb{L}^n_+} 0,\ x \circ y = 0 \\
\Longleftrightarrow\ & \lambda_i(x) \geq 0,\ \lambda_i(y) \geq 0,\ i = 1, 2,\ \text{and } x = \lambda_2(x)u_x^{(2)},\ y = \lambda_2(y)u_x^{(1)} \\
\Longleftrightarrow\ & f(\lambda_i(x)) \geq 0,\ f(\lambda_i(y)) \geq 0,\ i = 1, 2,\ \text{and } f^{\mathrm{soc}}(x) = f(\lambda_2(x))u_x^{(2)},\ f^{\mathrm{soc}}(y) = f(\lambda_2(y))u_x^{(1)} \\
\Longleftrightarrow\ & f^{\mathrm{soc}}(x) \succeq_{\mathbb{L}^n_+} 0,\ f^{\mathrm{soc}}(y) \succeq_{\mathbb{L}^n_+} 0,\ f^{\mathrm{soc}}(x) \circ f^{\mathrm{soc}}(y) = 0,
\end{aligned}$$

where the desired result follows. Thus, the proof is complete. □

A trivial example of f mentioned in Proposition 3.94 is $f(t) = t^3$, where it is easy to check $t \geq 0$ if and only if $f(t) \geq 0$. Therefore, from Proposition 3.94, its SOC function becomes

$$f^{\mathrm{soc}}(x) = (\lambda_1(x))^3 u_x^{(1)} + (\lambda_2(x))^3 u_x^{(2)}$$

and satisfies Assumption 3.3. Note that this expression is explicit due to (1.9)–(1.10). Again, we can plug in $\theta(x) := f^{\mathrm{soc}}(x)$ in Proposition 3.92 to construct C-functions in the SOC setting.

In fact, Assumption 3.3 can be extended to the following two-function version.

Assumption 3.4 The functions $\theta_1, \theta_2 : \mathbb{R}^n \to \mathbb{R}^n$ is said to satisfy Assumption 3.4 if

$$x \succeq_{\mathcal{K}} 0,\ y \succeq_{\mathcal{K}} 0,\ x \circ y = 0 \quad \Longleftrightarrow \quad \theta_1(x) \succeq_{\mathcal{K}} 0,\ \theta_2(y) \succeq_{\mathcal{K}} 0,\ \theta_1(x) \circ \theta_2(y) = 0.$$

By invoking Assumption 3.4, Proposition 3.92 can be naturally extended to a more general setting, as presented below.

Proposition 3.95 *Suppose that $\theta_1, \theta_2 : \mathbb{R}^n \to \mathbb{R}^n$ satisfy Assumption 3.4. Then, for any C-function $\varphi : \mathbb{R}^n \times \mathbb{R}^n \to \mathbb{R}^n$, the composition function $\varphi\big(\theta_1(x), \theta_2(y)\big)$ is also a C-function.*

Proof The proof is straightforward. Since φ is a C-function and θ_1, θ_2 satisfy Assumption 3.4, it is easy to verify that

$$\begin{aligned} & \varphi(\theta_1(x), \theta_2(y)) = 0 \\ \Longleftrightarrow\ & \theta_1(x) \succeq_{\mathcal{K}} 0,\ \theta_2(y) \succeq_{\mathcal{K}} 0,\ \theta_1(x) \circ \theta_2(y) = 0 \\ \Longleftrightarrow\ & x \succeq_{\mathcal{K}} 0,\ y \succeq_{\mathcal{K}} 0,\ x \circ y = 0. \end{aligned}$$

Hence, we show that $\varphi\big(\theta_1(x), \theta_2(y)\big)$ is also a C-function. □

Here are examples of $\theta_1(\cdot)$ and $\theta_2(\cdot)$ in Proposition 3.95:

$$\theta_1(x) = x^3 + x \quad \text{and} \quad \theta_2(y) = y|y|.$$

Composing these two functions with the natural residual function $\varphi_{\rm NR}(x, y) = x - (x - y)_+$ yields

$$\varphi_{\rm NR}(\theta_1(x), \theta_2(y)) = x^3 + x - (x^3 + x - y|y|)_+$$

which is a C-function due to Proposition 3.95. Note that it is true that if we exchange the position of θ_1 and θ_2 in the composition. There is another surprising result that if we switch the roles of φ and θ in Proposition 3.95, the goal is still achieved.

Proposition 3.96 *Suppose that $\theta : \mathbb{R}^n \to \mathbb{R}^n$ satisfies $z = 0$ if and only if $\theta(z) = 0$. Then, for any C-function $\varphi : \mathbb{R}^n \times \mathbb{R}^n \to \mathbb{R}^n$, the composition function $\theta\big(\varphi(\cdot, \cdot)\big)$ is also a C-function.*

Proof Since φ is a C-function and θ satisfies $z = 0$ if and only if $\theta(z) = 0$, we have

$$\theta\big(\varphi(x, y)\big) = 0 \quad \Longleftrightarrow \quad \varphi(x, y) = 0 \quad \Longleftrightarrow \quad x \succeq_{\mathcal{K}} 0,\ y \succeq_{\mathcal{K}} 0,\ x \circ y = 0.$$

This proves that $\theta(\varphi(x, y))$ is also a C-function. □

Below are several examples of functions $\theta(\cdot)$ as referenced in Proposition 3.96:

1. $\theta(z) = z^p$, where p is a positive integer;
2. $\theta(z) = |z|$;
3. $\theta(z) = f^{\rm sc}(z)$ where $f^{\rm sc}(z)$ is the SC function induced from a real-valued function f with $t = 0$ if and only if $f(t) = 0$.

B. Second construction method of C-functions.

The central idea of the second construction method for C-functions lies in utilizing existing NCP functions, originally real-valued, to generate C-functions, which are vector-valued. This represents a novel and promising direction, revealing that the extensive collection of known NCP functions (approximately sixty) can be systematically employed to produce a rich variety of C-functions.

It is important to emphasize that a C-function is vector-valued, whereas an NCP function is typically real-valued. The challenge of extending an NCP function to a C-function has remained an open problem for several decades. In what follows, we present a detailed exposition of how this extension can be achieved in the setting of symmetric cones.

Proposition 3.97 *Let $\phi : \mathbb{R}^2 \to \mathbb{R}$ be an NCP function. For any $x \in \mathbb{V}$ and $y \in \mathbb{V}$, the following $\Phi : \mathbb{V} \times \mathbb{V} \to \mathbb{V}$ defined by*

$$\Phi(x, y) := \sum_{i,j=1}^{r} \phi^2(\lambda_i(x), \lambda_j(y)) e_i \circ f_j$$

is a C-function, where $\{e_i\}_{i=1}^r$, $\{f_j\}_{j=1}^r$ are Jordan frames of x and y, respectively.

Proof Let $x \in \mathbb{V}$ and $y \in \mathbb{V}$, the spectral decompositions of x and y are given by

$$x = \sum_{i=1}^{r} \lambda_i(x) e_i \quad \text{and} \quad y = \sum_{j=1}^{r} \lambda_j(y) f_j.$$

By definition of C-function, it suffices to show that $\Phi(x, y) = 0 \iff x \in \mathcal{K},\ y \in \mathcal{K},\ \langle x, y \rangle = 0$.

"$\Rightarrow$" Since $\langle e_i, f_j \rangle \geq 0$, it yields

$$\begin{aligned}
\Phi(x, y) = 0 &\iff \sum_{i,j=1}^{r} \phi^2(\lambda_i(x), \lambda_j(y)) e_i \circ f_j = 0 \\
&\implies \left\langle \sum_{i,j=1}^{r} \phi^2(\lambda_i(x), \lambda_j(y)) e_i \circ f_j, e \right\rangle = 0 \\
&\iff \sum_{i,j=1}^{r} \phi^2(\lambda_i(x), \lambda_j(y)) \langle e_i, f_j \rangle = 0 \\
&\iff \phi^2(\lambda_i(x), \lambda_j(y)) = 0 \text{ or } \langle e_i, f_j \rangle = 0 \\
&\iff \phi(\lambda_i(x), \lambda_j(y)) = 0 \text{ or } \langle e_i, f_j \rangle = 0 \\
&\iff \lambda_i(x) \geq 0,\ \lambda_j(y) \geq 0, \lambda_i(x)\lambda_j(y) = 0 \text{ or } \langle e_i, f_j \rangle = 0 \\
&\iff x \in \mathcal{K},\ y \in \mathcal{K},\ \sum_{i,j=1}^{r} \lambda_i(x)\lambda_j(y) \langle e_i, f_j \rangle = 0 \\
&\iff x \in \mathcal{K},\ y \in \mathcal{K},\ \langle x, y \rangle = 0.
\end{aligned}$$

"$\Leftarrow$" By the above equivalences, we obtain

$$\begin{aligned}
& x \in \mathcal{K},\ y \in \mathcal{K},\ \langle x, y\rangle = 0 \\
\Longleftrightarrow\ & \phi^2(\lambda_i(x), \lambda_j(y)) = 0 \text{ or } \langle e_i, f_j\rangle = 0 \\
\Longleftrightarrow\ & \phi^2(\lambda_i(x), \lambda_j(y)) = 0 \text{ or } e_i \circ f_j = 0 \\
\Longrightarrow\ & \sum_{i,j=1}^{r} \phi^2(\lambda_i(x), \lambda_j(y)) e_i \circ f_j = 0 \\
\Longleftrightarrow\ & \Phi(x, y) = 0.
\end{aligned}$$

Thus, we achieve the desired result. □

In fact, if $\mathbb{V} \equiv \mathbb{R}$, then a C-function $\Phi(x, y)$ reduces to an NCP function $\phi^2(x, y)$. It is clear that we can write out the components of $\Phi(x, y)$ shown as in Proposition 3.97 in the second-order cone setting. Let $x = (x_1, \bar{x}_2) \in \mathbb{R} \times \mathbb{R}^{n-1}$ and $y = (y_1, \bar{y}_2) \in \mathbb{R} \times \mathbb{R}^{n-1}$.

$$\Phi(x, y) = \begin{pmatrix} a + b\bar{u}_2^{\mathrm{T}}\bar{v}_2 \\ c\bar{u}_2 + d\bar{v}_2 \end{pmatrix}, \tag{3.337}$$

where

$$\bar{u}_2 = \begin{cases} \frac{\bar{x}_2}{\|\bar{x}_2\|} & \text{if } \bar{x}_2 \neq 0 \\ \omega & \text{otherwise,} \end{cases} \qquad \bar{v}_2 = \begin{cases} \frac{\bar{y}_2}{\|\bar{y}_2\|} & \text{if } \bar{y}_2 \neq 0 \\ \vartheta & \text{otherwise,} \end{cases}$$

with any vector $\omega, \vartheta \in \mathbb{R}^{n-1}$ such that $\|\omega\| = 1$, $\|\vartheta\| = 1$, and

$$\begin{aligned}
a &= \frac{\phi^2(\lambda_1(x), \lambda_1(y)) + \phi^2(\lambda_1(x), \lambda_2(y)) + \phi^2(\lambda_2(x), \lambda_1(y)) + \phi^2(\lambda_2(x), \lambda_2(y))}{4}, \\
b &= \frac{\phi^2(\lambda_1(x), \lambda_1(y)) - \phi^2(\lambda_1(x), \lambda_2(y)) - \phi^2(\lambda_2(x), \lambda_1(y)) + \phi^2(\lambda_2(x), \lambda_2(y))}{4}, \\
c &= \frac{-\phi^2(\lambda_1(x), \lambda_1(y)) - \phi^2(\lambda_1(x), \lambda_2(y)) + \phi^2(\lambda_2(x), \lambda_1(y)) + \phi^2(\lambda_2(x), \lambda_2(y))}{4}, \\
d &= \frac{-\phi^2(\lambda_1(x), \lambda_1(y)) + \phi^2(\lambda_1(x), \lambda_2(y)) - \phi^2(\lambda_2(x), \lambda_1(y)) + \phi^2(\lambda_2(x), \lambda_2(y))}{4}.
\end{aligned}$$

Example 3.6 Consider the Fischer–Burmeister function $\phi_{\text{FB}}(a, b) = \sqrt{a^2 + b^2} - (a + b)$ for all $(a, b) \in \mathbb{R} \times \mathbb{R}$. Then the corresponding C-function is

$$\Phi_{\text{FB}}(x, y) = \sum_{i,j=1}^{r} \phi_{\text{FB}}^2(\lambda_i(x), \lambda_j(y)) e_i \circ f_j.$$

It is easy to see that

$$\Phi_{\text{FB}}(x, y) = 0 \iff \phi_{\text{FB}}(x, y) = (x^2 + y^2)^{1/2} - (x + y) = 0.$$

As noted in [76, Sect. 3], the component-wise expression of $\phi_{\text{FB}}(x, y)$ is quite intricate, which implies that its subgradient formula is also complex. In contrast, by employing the explicit formula for $\Phi(x, y)$ provided in (3.337), the computation of the subgradient of $\Phi_{\text{FB}}(x, y)$ becomes more tractable. Consequently, $\Phi_{\text{FB}}(x, y)$ may offer greater ease of implementation in numerical experiments compared to $\phi_{\text{FB}}(x, y)$ when solving the SOCCP.

Proposition 3.98 *Let $\phi : \mathbb{R}^2 \to \mathbb{R}$ be an NCP function. Suppose that $x \in \mathbb{V}$ and $y \in \mathbb{V}$ with $\{e_i\}_{i=1}^r, \{f_j\}_{j=1}^r$ as their corresponding Jordan frames, respectively. Then, the following $\Phi^1, \Phi^2 : \mathbb{V} \times \mathbb{V} \to \mathbb{V}$ defined by*

$$\Phi^1(x, y) := \sum_{(i,j)\notin I}^{r} \phi^2(\lambda_i(x), \lambda_j(y)) e_i \circ f_j$$

$$\Phi^2(x, y) := \sum_{(i,j)\notin I}^{r} \phi^2(\lambda_i(x), \lambda_j(y)) e_i$$

are C-functions, where $I = \{(i, j) \in \{1, \cdots, r\} \mid \langle e_i, f_j\rangle = 0\}$.

Proof Using the proof of Proposition 3.97, it is easy to show that $\Phi^1(x, y)$ is a C-function. We will now prove $\Phi^2(x, y)$ is a C-function. Note the fact that $e_i \in \mathcal{K}$ and $\langle e_i, e_i\rangle > 0$ for all $i = 1, \cdots, r$. Therefore, we have

$$\begin{aligned}
&\Phi^2(x, y) = 0\\
\iff &\sum_{(i,j)\notin I}^{r} \phi^2(\lambda_i(x), \lambda_j(y)) e_i = 0\\
\iff &\left\langle \sum_{(i,j)\notin I}^{r} \phi^2(\lambda_i(x), \lambda_j(y)) e_i, e \right\rangle = 0\\
\iff &\left\langle \sum_{(i,j)\notin I}^{r} \phi^2(\lambda_i(x), \lambda_j(y)) e_i, \sum_{i=1}^{r} e_i \right\rangle = 0\\
\iff &\sum_{(i,j)\notin I}^{r} \phi^2(\lambda_i(x), \lambda_j(y)) \langle e_i, e_i\rangle = 0\\
\iff &\phi^2(\lambda_i(x), \lambda_j(y)) = 0,\ (i, j) \notin I\\
\iff &\phi(\lambda_i(x), \lambda_j(y)) = 0,\ (i, j) \notin I\\
\iff &\lambda_i(x) \geq 0,\ \lambda_j(y) \geq 0,\ \lambda_i(x)\lambda_j(y) = 0,\ (i, j) \notin I
\end{aligned}$$

$$
\begin{aligned}
&\Longleftrightarrow x \in \mathcal{K},\ y \in \mathcal{K},\ \sum_{(i,j)\notin I}^{r} \lambda_i(x)\lambda_j(y)\big\langle e_i, f_j\big\rangle = 0 \\
&\Longleftrightarrow x \in \mathcal{K},\ y \in \mathcal{K},\ \sum_{(i,j)\notin I}^{r} \lambda_i(x)\lambda_j(y)\big\langle e_i, f_j\big\rangle + \sum_{(i,j)\in I}^{r} \lambda_i(x)\lambda_j(y)\big\langle e_i, f_j\big\rangle = 0 \\
&\Longleftrightarrow x \in \mathcal{K},\ y \in \mathcal{K},\ \sum_{i,j=1}^{r} \lambda_i(x)\lambda_j(y)\big\langle e_i, f_j\big\rangle = 0 \\
&\Longleftrightarrow x \in \mathcal{K},\ y \in \mathcal{K},\ \langle x, y\rangle = 0.
\end{aligned}
$$

Then, the desired result follows. □

Note that from $\Phi^2(x, y)$ in Proposition 3.98, we obtain that

$$
\Phi_1^2(x, y) := \sum_{(i,j)\notin I}^{r} \phi^2(\lambda_i(x), \lambda_j(y)) f_j \quad \text{and} \quad \Phi_2^2(x, y) := \sum_{(i,j)\notin I}^{r} \phi^2(\lambda_i(x), \lambda_j(y))(e_i + f_j)
$$

are also C-functions.

We now establish C-functions for the special case of two commutative operators x and y which recover the existing C-functions. In particular, x and y share the same Jordan frame, that is,

$$
x = \lambda_1(x)e_1 + \cdots + \lambda_r(x)e_r \quad \text{and} \quad y = \lambda_{\sigma(1)}(y)e_1 + \cdots + \lambda_{\sigma(r)}(y)e_r,
$$

where $\sigma : \{1, \cdots, r\} \to \{1, \cdots, r\}$. Moreover, $x^p = \lambda_1^p(x)e_1 + \cdots + \lambda_r^p(x)e_r$ for any positive number p.

Proposition 3.99 *Let $\phi : \mathbb{R}^2 \to \mathbb{R}$ be an NCP function. For any $x \in \mathbb{V}$ and $y \in \mathbb{V}$, the following $\widetilde{\Phi} : \mathbb{V} \times \mathbb{V} \to \mathbb{V}$ defined by*

$$
\widetilde{\Phi}(x, y) := \sum_{i=1}^{r} \phi(\lambda_i(x), \lambda_{\sigma(i)}(y))e_i
$$

is a C-function, where $\{e_1, e_2, \cdots, e_r\}$ is a Jordan frame of x and y.

Proof It suffices to show that $\widetilde{\Phi}(x, y) = 0 \Longleftrightarrow x \in \mathcal{K},\ y \in \mathcal{K},\ \langle x, y\rangle = 0$. Indeed, we have

$$
\begin{aligned}
&\widetilde{\Phi}(x, y) = 0 \\
\Longleftrightarrow\ & \sum_{i=1}^{r} \phi(\lambda_i(x), \lambda_{\sigma(i)}(y))e_i = 0 \\
\Longleftrightarrow\ & \left\langle \sum_{i=1}^{r} \phi(\lambda_i(x), \lambda_{\sigma(i)}(y))e_i, \sum_{i=1}^{r} \phi(\lambda_i(x), \lambda_{\sigma(i)}(y))e_i \right\rangle = 0
\end{aligned}
$$

$$
\begin{aligned}
&\Longleftrightarrow \sum_{i=1}^{r} \phi^2(\lambda_i(x), \lambda_{\sigma(i)}(y)) \langle e_i, e_i \rangle = 0 \\
&\Longleftrightarrow \phi^2(\lambda_i(x), \lambda_{\sigma(i)}(y)) = 0, \ i = 1, \cdots, r \\
&\Longleftrightarrow \phi(\lambda_i(x), \lambda_{\sigma(i)}(y)) = 0, \ i = 1, \cdots, r \\
&\Longleftrightarrow \lambda_i(x) \geq 0, \ \lambda_i(y) \geq 0, \lambda_i(x)\lambda_{\sigma(i)}(y) = 0, \ i = 1, \cdots, r \\
&\Longleftrightarrow x \in \mathcal{K}, \ y \in \mathcal{K}, \ \sum_{i=1}^{r} \lambda_i(x)\lambda_{\sigma(i)}(y) \langle e_i, e_i \rangle = 0 \\
&\Longleftrightarrow x \in \mathcal{K}, \ y \in \mathcal{K}, \ \langle x, y \rangle = 0,
\end{aligned}
$$

where $\langle e_i, e_i \rangle > 0$ and $\langle e_i, e_j \rangle = 0$ whenever $i \neq j$. Then, the proof is complete. □

Note that if $\sigma(i) = i$ for $i = 1, \cdots, r$ then

$$y = \lambda_1(y)e_1 + \cdots + \lambda_r(y)e_r.$$

Since $\langle e_i, e_i \rangle > 0$, we have

$$x \in \mathcal{K}, \ y \in \mathcal{K}, \ \langle x, y \rangle = 0 \quad \Longleftrightarrow \quad \lambda_i(x) \geq 0 \, \lambda_i(y) \geq 0, \ \lambda_i(x)\lambda_i(y) = 0 \quad \Longrightarrow \quad x = 0 \text{ or } y = 0.$$

Based on Proposition 3.99, we will show that $\widetilde{\Phi}(x, y)$ retrieves the existing C-functions in the special case of two commutative operators x and y. In particular, we focus on two popular NCP functions, which are the FB and NR functions

$$
\begin{aligned}
\phi_{\text{FB}}(a, b) &= \left(a^2 + b^2\right)^{1/2} - (a + b), \\
\phi_{\text{NR}}(a, b) &= a - (a - b)_+
\end{aligned}
$$

for any $(a, b) \in \mathbb{R} \times \mathbb{R}$. The corresponding C-functions are

$$
\begin{aligned}
\widetilde{\Phi}_{\text{FB}}(x, y) &= \sum_{i=1}^{r} \phi_{\text{FB}}(\lambda_i(x), \lambda_{\sigma(i)}(y))e_i, \\
\widetilde{\Phi}_{\text{NR}}(x, y) &= \sum_{i=1}^{r} \phi_{\text{NR}}(\lambda_i(x), \lambda_{\sigma(i)}(y))e_i.
\end{aligned}
$$

Then, there have

$$
\begin{aligned}
\widetilde{\Phi}_{\text{FB}}(x, y) &\equiv \varphi_{\text{FB}}(x, y), \\
\widetilde{\Phi}_{\text{NR}}(x, y) &\equiv \varphi_{\text{NR}}(x, y).
\end{aligned}
$$

Since x and y operator commute, it implies

$$x^2 + y^2 = \sum_{i=1}^{r} \lambda_i^2(x)e_i + \sum_{i=1}^{r} \lambda_{\sigma(i)}^2(y)e_i \quad \text{and} \quad \left(x^2 + y^2\right)^{1/2} = \sum_{i=1}^{r} \left(\lambda_i^2(x) + \lambda_{\sigma(i)}^2(y)\right)^{1/2} e_i,$$

and

$$(x - y)_+ = \sum_{i=1}^{r} \left(\lambda_i(x) - \lambda_{\sigma(i)}(y)\right)_+ e_i.$$

Hence, we obtain

$$\begin{aligned}
\widetilde{\Phi}_{\text{FB}}(x, y) &= \sum_{i=1}^{r} \left(\left(\lambda_i^2(x) + \lambda_{\sigma(i)}^2(y)\right)^{1/2} - (\lambda_i(x) + \lambda_{\sigma(i)}(y))\right) e_i \\
&= \sum_{i=1}^{r} \left(\lambda_i^2(x) + \lambda_{\sigma(i)}^2(y)\right)^{1/2} e_i - \left(\sum_{i=1}^{r} \lambda_i(x)e_i + \sum_{i=1}^{r} \lambda_{\sigma(i)}(y))e_i\right) \\
&= \left(x^2 + y^2\right)^{1/2} - (x + y) = \varphi_{\text{FB}}(x, y),
\end{aligned}$$

and

$$\begin{aligned}
\widetilde{\Phi}_{\text{NR}}(x, y) &= \sum_{i=1}^{r} \left(\lambda_i(x) - \left(\lambda_i(x) - \lambda_{\sigma(i)}(y)\right)_+\right) e_i \\
&= \sum_{i=1}^{r} \lambda_i(x)e_i - \sum_{i=1}^{r} \left(\lambda_i(x) - \lambda_{\sigma(i)}(y)\right)_+ e_i \\
&= x - (x - y)_+ = \varphi_{\text{NR}}(x, y).
\end{aligned}$$

Similar arguments apply for other existing C-functions in the literature.

Remark 3.8 We point out a few comments to understand more about the construction of C-functions by using NCP functions.

(i) From Proposition 3.99 and [37, Proposition 6], we obtain that for any $x \in \mathbb{V}$, $y \in \mathbb{V}$, $x \neq 0$, $y \neq 0$, there holds

$$x \in \mathcal{K},\ y \in \mathcal{K},\ \langle x, y\rangle = 0 \iff \lambda_1(x) = 0,\ \lambda_1(y) = 0,\ \langle x, y\rangle = 0, \tag{3.338}$$

where $\lambda_i(x)$ and $\lambda_i(y)$, $i = 1, \cdots, r$ are arranged in the increasing order $\lambda_1(x) \leq \cdots \leq \lambda_r(x)$ and $\lambda_1(y) \leq \cdots \leq \lambda_r(y)$, respectively. Indeed, it is enough to prove that $x \in \mathcal{K},\ y \in \mathcal{K},\ \langle x, y\rangle = 0 \Longrightarrow \lambda_1(x) = 0,\ \lambda_1(y) = 0,\ \langle x, y\rangle = 0$. According to [37, Proposition 6], we have that x and y operator commute which together with the proof of Proposition 3.99 indicate

$$\begin{aligned}
&\lambda_i(x) \geq 0,\ \lambda_i(y) \geq 0,\ \lambda_i(x)\lambda_{\sigma(i)}(y) = 0,\ i = 1, \cdots, r \\
\Longrightarrow &\lambda_1(x)\lambda_{\sigma(1)}(y) + \cdots + \lambda_r(x)\lambda_{\sigma(r)}(y) = 0.
\end{aligned}$$

Using the rearrangement inequality, we obtain

$$0 = \lambda_1(x)\lambda_{\sigma(1)}(y) + \cdots + \lambda_r(x)\lambda_{\sigma(r)}(y) \geq \lambda_1(x)\lambda_r(y) + \cdots + \lambda_r(x)\lambda_1(y) \geq 0,$$

which yields

$$\lambda_1(x)\lambda_r(y) = 0,\ \lambda_r(x)\lambda_1(y) = 0 \Longrightarrow \lambda_1(x) = 0,\ \lambda_1(y) = 0,$$

where $\lambda_r(x) > 0$ and $\lambda_r(y) > 0$ due to $x \neq 0$, $y \neq 0$.

(ii) Using the relation (3.338), for any $x \in \mathbb{V}$ and $y \in \mathbb{V}$ and assume that $\lambda_i(x)$ and $\lambda_i(y)$, $i = 1, \cdots, r$ are listed in the increasing order, the following holds

$$\begin{aligned} & x \in \mathcal{K},\ y \in \mathcal{K},\ \langle x, y\rangle = 0 \\ \Longleftrightarrow & \phi(\lambda_1(x), \lambda_1(y)) = 0,\ \langle x, y\rangle = 0 \ \text{ or} \\ & \phi(\lambda_1(x),\ \lambda_r(y)) = 0,\ \phi(\lambda_r(x), \lambda_1(y)) = 0,\ \langle x, y\rangle = 0, \end{aligned} \tag{3.339}$$

where ϕ is an NCP function.

(iii) We observe that constructing a general class of C-functions based on NCP functions opens a novel avenue for addressing the SCCP, leveraging spectral eigenvalues and spectral vectors (Jordan frames). In particular, we have identified a new direction for solving the SOCCP and SDCP by formulating them as minimization problems via the relation (3.339). Moreover, in the cases of two special symmetric cones, the second-order cone and the positive semidefinite cone, the Jordan product admits explicit expressions. This enables the construction of simplified C-functions for these cases through the same relation (3.339).

As noted in Remark 3.8(iii), a simplified form of C-functions can be constructed for both the second-order cone and the positive semidefinite cone by exploiting the relation (3.339). To illustrate this approach, we begin with the second-order cone setting, employing Lemmas 3.60, 3.61, and 3.62.

Proposition 3.100 *Let $\phi : \mathbb{R}^2 \to \mathbb{R}$ be an NCP function. For any $x = (x_1, \bar{x}_2) \in \mathbb{R} \times \mathbb{R}^{n-1}$ and $y = (y_1, \bar{y}_2) \in \mathbb{R} \times \mathbb{R}^{n-1}$, the following two vector-valued functions $\Phi^1, \Phi^2 : \mathbb{R}^n \times \mathbb{R}^n \to \mathbb{R}^n$ defined by*

$$\Phi^1(x, y) := \begin{pmatrix} \phi(\lambda_1(x), \lambda_1(y)) \\ x_1\bar{y}_2 + y_1\bar{x}_2 \end{pmatrix}$$

$$\Phi^2(x, y) := \begin{pmatrix} \phi(\lambda_1(x), \lambda_2(y)) \\ \phi(\lambda_2(x), \lambda_1(y)) \\ \bar{y}_3x_k - \bar{x}_3y_k \end{pmatrix}$$

are C-functions in the second-order cone setting. Here, $k \in \{2, \cdots, n\}$ and

$$\bar{x}_3 := (x_2, \cdots, x_{k-1}, x_{k+1}, \cdots, x_n)^\mathsf{T} \in \mathbb{R}^{n-2}, \quad \bar{y}_3 := (y_2, \cdots, y_{k-1}, y_{k+1}, \cdots, y_n)^\mathsf{T} \in \mathbb{R}^{n-2};$$

and $\lambda_i(x)$, $\lambda_i(y)$ for $i = 1, 2$ are the spectral values of x and y associated with second-order cone, respectively. In particular, there holds $\|\bar{x}_2\| y_k = -\|\bar{y}_2\| x_k \neq 0$ for some $k \in \{2, \cdots, n\}$ when $\bar{x}_2 \neq 0$ and $\bar{y}_2 \neq 0$.

Proof For $\bar{x}_2 = 0$ or $\bar{y}_2 = 0$, from Lemma 3.60, we know that $x = 0$ or $y = 0$. Then, it is easy to verify

$$x \succeq_{\mathbb{L}^n_+} 0, \ y \succeq_{\mathbb{L}^n_+} 0, \ x \circ y = 0 \iff \Phi^1(x, y) = 0 \text{ and } \Phi^2(x, y) = 0.$$

Therefore, we only focus on the case of $\bar{x}_2 \neq 0$ and $\bar{y}_2 \neq 0$.

(i) We first prove that $\Phi^1(x, y)$ is a C-function. To proceed, we note a fact that for any $x \in \mathbb{L}^n_+$, $y \in \mathbb{L}^n_+$, there holds

$$\lambda_1(x) = 0, \ x_1 \bar{y}_2 + y_1 \bar{x}_2 = 0 \iff \lambda_1(y) = 0, \ x_1 \bar{y}_2 + y_1 \bar{x}_2 = 0.$$

This fact together with Lemma 3.61 yields

$$\begin{aligned}
\Phi^1(x, y) = 0 &\iff \begin{cases} \phi(\lambda_1(x), \lambda_1(y)) = 0 \\ x_1 \bar{y}_2 + y_1 \bar{x}_2 = 0 \end{cases} \\
&\iff \begin{cases} \lambda_1(x)\lambda_1(y) = 0, \ \lambda_1(x) \geq 0, \ \lambda_1(y) \geq 0 \\ x_1 \bar{y}_2 + y_1 \bar{x}_2 = 0 \end{cases} \\
&\iff \begin{cases} \lambda_1(x) = 0, \ \lambda_1(y) = 0 \\ x_1 \bar{y}_2 + y_1 \bar{x}_2 = 0 \end{cases} \\
&\iff x \succeq_{\mathbb{L}^n_+} 0, \ y \succeq_{\mathbb{L}^n_+} 0, \ x \circ y = 0.
\end{aligned}$$

Thus, $\Phi^1(x, y)$ is a C-function.

(ii) We now show that $\Phi^2(x, y)$ is a C-function. Applying Lemmas 3.61 and 3.62, it follows that

$$\begin{aligned}
& x \succeq_{\mathbb{L}^n_+} 0, \ y \succeq_{\mathbb{L}^n_+} 0, \ x \circ y = 0 \\
\Longrightarrow & \begin{cases} \lambda_1(x) = 0 \\ \lambda_1(y) = 0 \\ \bar{y}_2 = -\frac{\|\bar{y}_2\|}{\|\bar{x}_2\|} \bar{x}_2 \end{cases} \\
\Longrightarrow & \begin{cases} \phi(\lambda_1(x), \lambda_2(y)) = 0 \\ \phi(\lambda_2(x), \lambda_1(y)) = 0 \\ \|\bar{x}_2\| y_k = -\|\bar{y}_2\| x_k \neq 0 \text{ for some } k \in \{2, \cdots, n\} \\ y_l x_k = x_l y_k \ \text{ for all } l \in \{2, \cdots, n\} \end{cases}
\end{aligned}$$

$$\Longrightarrow \begin{cases} \phi(\lambda_1(x), \lambda_2(y)) = 0 \\ \phi(\lambda_2(x), \lambda_1(y)) = 0 \\ \|\bar{x}_2\| y_k = -\|\bar{y}_2\| x_k \neq 0 \text{ for some } k \in \{2, \cdots, n\} \\ \bar{y}_3 x_k - \bar{x}_3 y_k = 0 \end{cases}$$
$$\Longrightarrow \Phi^2(x, y) = 0.$$

Conversely, suppose that $\Phi^2(x, y) = 0$. Due to ϕ being an NCP function, we obtain

$$\begin{cases} \phi(\lambda_1(x), \lambda_2(y)) = 0 \\ \phi(\lambda_2(x), \lambda_1(y)) = 0 \\ \bar{y}_3 x_k - \bar{x}_3 y_k = 0 \end{cases}$$
$$\Longrightarrow \begin{cases} \lambda_1(x) \geq 0, \ \lambda_2(x) \geq 0, \ \lambda_1(y) \geq 0, \ \lambda_2(y) \geq 0 \\ \lambda_1(x)\lambda_2(y) = 0 \\ \lambda_2(x)\lambda_1(y) = 0 \end{cases}$$
$$\Longrightarrow \begin{cases} x \in \mathbb{L}^n_+, \ y \in \mathbb{L}^n_+ \\ \lambda_1(x)\lambda_2(y) + \lambda_2(x)\lambda_1(y) = 0. \end{cases} \tag{3.340}$$

Note that $\lambda_i(x) = x_1 + (-1)^i \|\bar{x}_2\|$ and $\lambda_i(y) = y_1 + (-1)^i \|\bar{y}_2\|$ for $i = 1, 2$. Hence, we have

$$\lambda_1(x)\lambda_2(y) = x_1 y_1 - \|\bar{x}_2\| \|\bar{y}_2\| + x_1 \|\bar{y}_2\| - y_1 \|\bar{x}_2\|,$$
$$\lambda_2(x)\lambda_1(y) = x_1 y_1 - \|\bar{x}_2\| \|\bar{y}_2\| - x_1 \|\bar{y}_2\| + y_1 \|\bar{x}_2\|.$$

This fact together with (3.340) leads to

$$\lambda_1(x)\lambda_2(y) + \lambda_2(x)\lambda_1(y) = 2(x_1 y_1 - \|\bar{x}_2\| \|\bar{y}_2\|) = 2(x_1 y_1 + \bar{x}_2^T \bar{y}_2) = 0,$$

which says that $\langle x, y \rangle = 0$. Thus, $\Phi^2(x, y)$ is a C-function. □

Note that in Proposition 3.100, the component $x_1 \bar{y}_2 + y_1 \bar{x}_2$ of $\Phi^1(x, y)$ is a vector in $\mathbb{R}^{n-1}$ while the component $\bar{y}_3 x_k - \bar{x}_3 y_k$ of $\Phi^2(x, y)$ is a vector in $\mathbb{R}^{n-2}$. Therefore, both ranges of $\Phi^1(x, y)$ and $\Phi^2(x, y)$ are in $\mathbb{R}^n$. It is well known that there have plenty of NCP functions in the literature. According to Proposition 3.100, we can convert them into C-functions associated with second-order cone. We illustrate this using two NCP functions in the following example.

Example 3.7 We consider two popular NCP functions as follows:

$$\phi_{\rm FB}(a, b) = \sqrt{a^2 + b^2} - (a + b) \quad \text{and} \quad \phi_{\rm NR}(a, b) = a - (a - b)_+, \ \ \forall (a, b) \in \mathbb{R} \times \mathbb{R}.$$

In light of Proposition 3.100, it is not hard to see that

$$\Phi^1_{\rm FB}(x, y) = \begin{pmatrix} \phi_{\rm FB}(\lambda_1(x), \lambda_1(y)) \\ x_1 \bar{y}_2 + y_1 \bar{x}_2 \end{pmatrix}, \quad \Phi^1_{\rm NR}(x, y) = \begin{pmatrix} \phi_{\rm NR}(\lambda_1(x), \lambda_1(y)) \\ x_1 \bar{y}_2 + y_1 \bar{x}_2 \end{pmatrix}$$

and

$$\Phi^2_{\text{FB}}(x,y) = \begin{pmatrix} \phi_{\text{FB}}(\lambda_1(x), \lambda_2(y)) \\ \phi_{\text{FB}}(\lambda_2(x), \lambda_1(y)) \\ y_l x_k - x_l y_k \end{pmatrix}, \quad \Phi^2_{\text{NR}}(x,y) = \begin{pmatrix} \phi_{\text{NR}}(\lambda_1(x), \lambda_2(y)) \\ \phi_{\text{NR}}(\lambda_2(x), \lambda_1(y)) \\ \bar{y}_3 x_k - \bar{x}_3 y_k \end{pmatrix}$$

are C-functions, where $\|\bar{x}_2\| y_k = -\|\bar{y}_2\| x_k \neq 0$ for some $k \in \{2, \cdots, n\}$ when $\bar{x}_2 \neq 0$ and $\bar{y}_2 \neq 0$,

$$\bar{x}_3 = (x_2, \cdots, x_{k-1}, x_{k+1}, \cdots, x_n)^\top \in \mathbb{R}^{n-2}, \quad \bar{y}_3 = (y_2, \cdots, y_{k-1}, y_{k+1}, \cdots, y_n)^\top \in \mathbb{R}^{n-2},$$

and $\lambda_i(x), \lambda_i(y)$ for $i = 1, 2$ are spectral values of x and y, respectively.

Indeed, we can further conclude that

$$\Phi^i_{\text{FB}}(x,y) = 0, \ i = 1, 2 \iff \varphi_{\text{FB}}(x,y) = (x^2 + y^2)^{1/2} - (x+y) = 0$$

and

$$\Phi^i_{\text{NR}}(x,y) = 0, \ i = 1, 2 \iff \varphi_{\text{NR}}(x,y) = x - (x-y)_+ = 0.$$

To see this, by definition of C-function and Lemma 3.61, for $\bar{x}_2 \neq 0$ and $\bar{y}_2 \neq 0$, we have

$$\begin{aligned} \varphi_{\text{FB}}(x,y) = 0 &\iff x \in \mathbb{L}^n_+, \ y \in \mathbb{L}^n_+, \ x \circ y = 0 \\ &\iff \lambda_1(x) = 0, \ \lambda_1(y) = 0, \ x_1\bar{y}_2 + y_1\bar{x}_2 = 0 \\ &\iff \Phi^1_{\text{FB}}(x,y) = 0. \end{aligned}$$

For $\bar{x}_2 = 0$ or $\bar{y}_2 = 0$, it is easy to check $\varphi_{\text{FB}}(x,y) = 0 \iff \Phi^1_{\text{FB}}(x,y) = 0$ by definition of C-function and Lemma 3.60. Similar arguments apply for other cases. The above discussions indicate that $\Phi^i_{\text{FB}}(x,y)$ are C-functions and equivalent to the traditional complementarity function $\varphi_{\text{FB}}(x,y)$; $\Phi^i_{\text{NR}}(x,y)$ are C-functions and equivalent to the traditional complementarity function $\varphi_{\text{NR}}(x,y)$.

Remark 3.9 We elaborate more about Proposition 3.100 as below.

(i) In Proposition 3.100, if ϕ is a continuously differentiable NCP function, then $\Phi^1(x,y)$ and $\Phi^2(x,y)$ are continuously differentiable C-functions when $\bar{x}_2 \neq 0$ and $\bar{y}_2 \neq 0$. Let $y = F(x)$, where $F : \mathbb{R}^n \to \mathbb{R}^n$ is continuously differentiable. Then, the first row of the Jacobian $J\Phi^1(x, F(x))$ and the first and second row of the Jacobian $J\Phi^2(x, F(x))$ are described by

$$\begin{aligned} \left(J\Phi^1(x, F(x))\right)_1 &= \frac{\partial \phi}{\partial \lambda_1(x)} \nabla \lambda_1(x)^\top + \frac{\partial \phi}{\partial \lambda_1(F(x))} \left(DF(x) \nabla \lambda_1(F(x))\right)^\top, \\ \left(J\Phi^2(x, F(x))\right)_1 &= \frac{\partial \phi}{\partial \lambda_1(x)} \nabla \lambda_1(x)^\top + \frac{\partial \phi}{\partial \lambda_2(F(x))} \left(DF(x) \nabla \lambda_2(F(x))\right)^\top, \end{aligned}$$

$$\left(J\Phi^2(x, F(x))\right)_2 = \frac{\partial \phi}{\partial \lambda_2(x)} \nabla \lambda_2(x)^\top + \frac{\partial \phi}{\partial \lambda_1(F(x))} \left(DF(x)\nabla\lambda_1(F(x))\right)^\top,$$

when $x \in \text{bd}(\mathbb{L}^n_+)\backslash\{0\}$ and $F(x) \in \text{bd}(\mathbb{L}^n_+)\backslash\{0\}$. Since ϕ is continuously differentiable, it can be seen that

$$\frac{\partial \phi}{\partial \lambda_1(x)}(0,0) = 0, \quad \frac{\partial \phi}{\partial \lambda_1(F(x))}(0,0) = 0,$$
$$\frac{\partial \phi}{\partial \lambda_2(x)}(\lambda_2(x), 0) \neq 0, \quad \text{and} \quad \frac{\partial \phi}{\partial \lambda_2(F(x))}(0, \lambda_2(F(x))) \neq 0,$$

when $x \in \text{bd}(\mathbb{L}^n_+)\backslash\{0\}$ and $F(x) \in \text{bd}(\mathbb{L}^n_+)\backslash\{0\}$. Thus, for $x \in \text{bd}(\mathbb{L}^n_+)\backslash\{0\}$ and $F(x) \in \text{bd}(\mathbb{L}^n_+)\backslash\{0\}$, the first row of the Jacobian $J\Phi^1(x, F(x))$ is zero and the first and second row of the Jacobian $J\Phi^2(x, F(x))$ are nonzero. In summary, when we apply Newton method to solve the SOCCP, $\Phi^2(x, F(x))$ is a better choice than $\Phi^1(x, F(x))$.

(ii) It is generally difficult to derive explicit component-wise formulas for many of the existing C-functions in the literature. However, Proposition 3.100 provides explicit expressions for the components of $\Phi^1(x, y)$ and $\Phi^2(x, y)$, making the computation of their subgradients significantly more tractable. This contrasts with the complex formulation of the B-subgradient of the Fischer–Burmeister C-function, as shown in [34, Proposition 3.1]. Consequently, employing $\Phi^1(x, y)$ and $\Phi^2(x, y)$ for solving the SOCCP may facilitate easier implementation in numerical simulations.

(iii) Regarding Remark 3.8(ii)–(iii), we propose a new direction to tackle the SOCCP which can be solved by the following unconstrained minimization problem

$$\min_{x \in \mathbb{R}^n} \phi^2(\lambda_1(x), \lambda_1(F(x))) + \langle x, F(x)\rangle^2$$

or

$$\min_{x \in \mathbb{R}^n} \phi^2(\lambda_1(x), \lambda_2(F(x))) + \phi^2(\lambda_2(x), \lambda_1(F(x))) + \langle x, F(x)\rangle^2,$$

where $F : \mathbb{R}^n \to \mathbb{R}^n$ is a map.

Similar to Proposition 3.94, the SOC function defined in (1.11) plays a pivotal role in our second construction approach. The significance of certain special SOC functions within this framework is demonstrated in Proposition 3.101.

Proposition 3.101 *Let $\phi : \mathbb{R}^2 \to \mathbb{R}$ be an NCP function. Suppose that $F(x)$ is a SOC function induced from function $f : \mathbb{R} \to \mathbb{R}$, which means $F(x)$ can be written:*

$$F(x) = f(\lambda_1(x))u_x^{(1)} + f(\lambda_2(x))u_x^{(2)} \quad \text{or} \quad F(x) = f(\lambda_2(x))u_x^{(1)} + f(\lambda_1(x))u_x^{(2)}.$$

Then, there holds

$$x \in \mathbb{L}^n_+, \; F(x) \in \mathbb{L}^n_+, \; \langle x, F(x)\rangle = 0$$

$$\Longleftrightarrow \Phi^3(x, F(x)) := \begin{pmatrix} \phi(\lambda_1(x), f(\lambda_1(x))) \\ \phi(\lambda_2(x), f(\lambda_2(x))) \\ 0 \end{pmatrix} = 0 \text{ or}$$

$$\Phi^4(x, F(x)) := \begin{pmatrix} \phi(\lambda_1(x), f(\lambda_2(x))) \\ \phi(\lambda_2(x), f(\lambda_1(x))) \\ 0 \end{pmatrix} = 0,$$

where $x = (x_1, \bar{x}_2) \in \mathbb{R} \times \mathbb{R}^{n-1}$ *and* $\lambda_i(x)$, $u_x^{(i)}$ *for* $i = 1, 2$ *are the spectral values and the spectral vectors of* x, *respectively.*

Proof We will prove for the case $\Phi^3(x, F(x))$. Assume that $F(x)$ can be written as

$$F(x) = f(\lambda_1(x))u_x^{(1)} + f(\lambda_2(x))u_x^{(2)}.$$

Hence, we have

$$\begin{aligned}
& x \in \mathbb{L}^n_+, \ F(x) \in \mathbb{L}^n_+, \ \langle x, F(x)\rangle = 0 \\
\Longleftrightarrow\ & \lambda_i(x) \geq 0, \ f(\lambda_i(x)) \geq 0, \ \lambda_1(x) f(\lambda_1(x)) + \lambda_2(x) f(\lambda_2(x)) = 0 \\
\Longleftrightarrow\ & \lambda_i(x) \geq 0, \ f(\lambda_i(x)) \geq 0, \ \lambda_1(x) f(\lambda_1(x)) = 0, \ \lambda_2(x) f(\lambda_2(x)) = 0 \\
\Longleftrightarrow\ & \phi(\lambda_1(x), f(\lambda_1(x))) = 0, \ \phi(\lambda_2(x), f(\lambda_2(x))) = 0 \\
\Longleftrightarrow\ & \Phi^3(x, F(x)) = 0.
\end{aligned}$$

Similar arguments apply to the case when $\Phi^4(x, F(x)) = 0$. □

Next, by using Lemma 3.59 and noting that $\mathbb{S}^{n\times n} \cong \mathbb{R}^{\frac{n(n+1)}{2}}$, we show how to construct C-functions based on given NCP functions in the setting of positive semidefinite cone. We introduce the following notations for convenience. For any $X, Y \in \mathbb{S}^{n\times n}$, we denote

$$X := \left[\mathbf{x}_1 \mid \cdots \mid \mathbf{x}_n\right], \quad Y := \left[\mathbf{y}_1 \mid \cdots \mid \mathbf{y}_n\right],$$

where $\mathbf{x}_i$ and $\mathbf{y}_i$ for $i = 1, \cdots, n$ are column vectors of matrices X and Y, respectively.

Proposition 3.102 *Let* $\phi : \mathbb{R}^2 \to \mathbb{R}$ *be an NCP function. For any* $X, Y \in \mathbb{S}^{n\times n}$, *the following two functions* $\Phi^i : \mathbb{S}^{n\times n} \times \mathbb{S}^{n\times n} \to \mathbb{R}^{\frac{n(n+1)}{2}}$, $i = 1, 2$, *given by*

$$\Phi^1(X, Y) := \begin{pmatrix} \phi(\lambda_1(X), \lambda_1(Y)) \\ \mathbf{x}_1^{\mathsf{T}}\mathbf{y}_1 \\ \vdots \\ \mathbf{x}_n^{\mathsf{T}}\mathbf{y}_n \\ \mathbf{0} \end{pmatrix}$$

$$\Phi^2(X,Y) := \begin{pmatrix} \phi(\lambda_1(X), \lambda_n(Y)) \\ \phi(\lambda_n(X), \lambda_1(Y)) \\ \mathbf{x}_1^\mathsf{T}\mathbf{y}_1 \\ \vdots \\ \mathbf{x}_n^\mathsf{T}\mathbf{y}_n \\ \mathbf{0} \end{pmatrix}$$

are C-functions. Here, the zero vector in $\Phi^1(X,Y)$ *belongs to* $\mathbb{R}^{\frac{(n+1)(n-2)}{2}}$ *whereas the zero vector in* $\Phi^2(X,Y)$ *belongs to* $\mathbb{R}^{\frac{n^2-n-4}{2}}$. *In addition,* $\lambda_i(X)$, $\lambda_i(Y)$ *for* $i = 1, \cdots, n$ *are eigenvalues of matrices* X, Y, *which are arranged in the increasing order* $\lambda_1(X) \le \cdots \le \lambda_n(X)$ *and* $\lambda_1(Y) \le \cdots \le \lambda_n(Y)$, *respectively.*

Proof First, according to Lemma 3.59 and $\lambda_1(X) \le \cdots \le \lambda_n(X)$, $\lambda_1(Y) \le \cdots \le \lambda_n(Y)$, we have

$$\begin{aligned} & X \succeq 0,\ Y \succeq 0,\ \langle X, Y\rangle = 0 \\ \Longleftrightarrow\ & X \succeq 0,\ Y \succeq 0,\ XY = 0 \\ \Longleftrightarrow\ & \lambda_1(X) \ge 0,\ \lambda_1(Y) \ge 0,\ \text{and } XY = 0. \end{aligned} \tag{3.341}$$

Suppose that $X = 0$ or $Y = 0$, it is easy to see that

$$X \succeq 0,\ Y \succeq 0,\ \langle X, Y\rangle = 0 \quad \Longleftrightarrow \quad \Phi^1(X,Y) = 0 \text{ and } \Phi^2(X,Y) = 0.$$

Therefore, it suffices to consider the case of $X \ne 0$ and $Y \ne 0$. Suppose that $\Phi^1(X,Y) = 0$. Noting that $\langle X, Y\rangle = \mathrm{tr}(XY) = \sum_{i=1}^n \mathbf{x}_i^\mathsf{T}\mathbf{y}_i$, we have

$$\begin{aligned} \Phi^1(X,Y) = 0 \Longrightarrow & \begin{cases} \phi(\lambda_1(X), \lambda_1(Y)) = 0 \\ \mathbf{x}_i^\mathsf{T}\mathbf{y}_i = 0,\ i = 1, \cdots, n \end{cases} \\ \Longrightarrow & \begin{cases} \lambda_1(X) \ge 0, \lambda_1(Y) \ge 0 \\ \sum_{i=1}^n \mathbf{x}_i^\mathsf{T}\mathbf{y}_i = 0 \end{cases} \\ \Longrightarrow & X \succeq 0,\ Y \succeq 0,\ \langle X, Y\rangle = 0. \end{aligned}$$

Conversely, from (3.341), we know

$$X \succeq 0, Y \succeq 0, \langle X, Y\rangle = 0 \quad \Longrightarrow \quad \lambda_1(X) \ge 0,\ \lambda_1(Y) \ge 0,\ \text{and } XY = 0. \tag{3.342}$$

We now claim that

$$(3.342) \quad \Longrightarrow \quad \lambda_1(X) = 0,\ \lambda_1(Y) = 0,\ \text{and } XY = 0. \tag{3.343}$$

By contradiction, suppose that $\lambda_1(X) > 0$. Hence, $\lambda_i(X) > 0$ for all $i = 1, \cdots, n$. It follows that $\det(X) = \lambda_1(X) \cdots \lambda_n(X) > 0$, that is, X is nonsingular matrix. Multiplying both sides of $XY = 0$ by X^{-1} leads to $Y = 0$, which contradicts the fact that $Y \neq 0$. Thus, $\lambda_1(X) = 0$. Similarly, we can argue that $\lambda_1(Y) = 0$. This says that $\lambda_1(X)\lambda_1(Y) = 0$, and then $\phi(\lambda_1(X), \lambda_1(Y)) = 0$. On the other hand, since $XY = 0$, $\mathbf{x}_i^T \mathbf{y}_i = 0$ for all $i = 1, \cdots, n$. All the above concludes $\Phi^1(X, Y) = 0$.

For the case of $\Phi^2(X, Y)$, likewise, we also have

$$\begin{aligned}
\Phi^2(X, Y) = 0 &\Longrightarrow \begin{cases} \phi(\lambda_1(X), \lambda_n(Y)) = 0 \\ \phi(\lambda_n(X), \lambda_1(Y)) = 0 \\ \mathbf{x}_i^T \mathbf{y}_i = 0, \ i = 1, \cdots, n \end{cases} \\
&\Longrightarrow \begin{cases} \lambda_1(X) \geq 0, \lambda_1(Y) \geq 0 \\ \sum_{i=1}^n \mathbf{x}_i^T \mathbf{y}_i = 0 \end{cases} \\
&\Longrightarrow X \succeq 0, \ Y \succeq 0, \ \langle X, Y \rangle = 0.
\end{aligned}$$

Conversely, suppose that $X \succeq 0$, $Y \succeq 0$, and $\langle X, Y \rangle = 0$. Hence, $\lambda_n(X) > 0$ and $\lambda_n(Y) > 0$. From (3.342) and (3.343), we have

$$\lambda_1(X) = 0, \ \lambda_1(Y) = 0, \ \text{and } XY = 0.$$

This yields $\lambda_1(X)\lambda_n(Y) = 0$ and $\lambda_n(X)\lambda_1(Y) = 0$, which further imply that $\phi(\lambda_1(X), \lambda_n(Y)) = 0$ and $\phi(\lambda_n(X), \lambda_1(Y)) = 0$. Moreover, since $XY = 0$, $\mathbf{x}_i^T \mathbf{y}_i = 0$ for all $i = 1, \cdots, n$. Thus, we conclude that $\Phi^2(X, Y) = 0$. □

Note that both $\Phi^1(X, Y)$ and $\Phi^2(X, Y)$ yield vectors in $\mathbb{R}^{\frac{n(n+1)}{2}}$. Therefore, they could be viewed as matrix-valued function. In fact, there exist a lot of matrix expressions for $\Phi^1(X, Y)$ and $\Phi^2(X, Y)$. For instance,

$$\Phi^1(X, Y) \equiv \begin{pmatrix} \mathbf{x}_1^T \mathbf{y}_1 & \phi(\lambda_1(X), \lambda_1(Y)) & 0 & \cdots & 0 \\ \phi(\lambda_1(X), \lambda_1(Y)) & \mathbf{x}_2^T \mathbf{y}_2 & 0 & \cdots & 0 \\ 0 & 0 & \mathbf{x}_3^T \mathbf{y}_3 & \cdots & 0 \\ \vdots & \vdots & \vdots & \ddots & \vdots \\ 0 & 0 & 0 & \cdots & \mathbf{x}_n^T \mathbf{y}_n \end{pmatrix}$$

$$\Phi^2(X, Y) \equiv \begin{pmatrix} \mathbf{x}_1^T \mathbf{y}_1 & \phi(\lambda_1(X), \lambda_n(Y)) & \phi(\lambda_n(X), \lambda_1(Y)) & \cdots & 0 \\ \phi(\lambda_1(X), \lambda_n(Y)) & \mathbf{x}_2^T \mathbf{y}_2 & 0 & \cdots & 0 \\ \phi(\lambda_n(X), \lambda_1(Y)) & 0 & \mathbf{x}_3^T \mathbf{y}_3 & \cdots & 0 \\ \vdots & \vdots & \vdots & \ddots & \vdots \\ 0 & 0 & 0 & \cdots & \mathbf{x}_n^T \mathbf{y}_n \end{pmatrix}.$$

Example 3.8 We consider the FB function $\phi_{\text{FB}}(a, b) = \sqrt{a^2 + b^2} - (a + b)$ for all $(a, b) \in \mathbb{R} \times \mathbb{R}$. Their corresponding C-functions are

$$\Phi^1_{\text{FB}}(X, Y) = \begin{pmatrix} \phi_{\text{FB}}(\lambda_1(X), \lambda_1(Y)) \\ \mathbf{x}_1^T \mathbf{y}_1 \\ \vdots \\ \mathbf{x}_n^T \mathbf{y}_n \\ \mathbf{0} \end{pmatrix}$$

$$\Phi^2_{\text{FB}}(X, Y) = \begin{pmatrix} \phi_{\text{FB}}(\lambda_1(X), \lambda_n(Y)) \\ \phi_{\text{FB}}(\lambda_n(X), \lambda_1(Y)) \\ \mathbf{x}_1^T \mathbf{y}_1 \\ \vdots \\ \mathbf{x}_n^T \mathbf{y}_n \\ \mathbf{0} \end{pmatrix},$$

where $\lambda_i(X), \lambda_i(Y)$ for $i = 1, \cdots, n$ are eigenvalues of matrices X, Y, which are arranged in the increasing order $\lambda_1(X) \leq \cdots \leq \lambda_n(X)$ and $\lambda_1(Y) \leq \cdots \leq \lambda_n(Y)$, respectively.

Likewise, in the setting of positive semidefinite cone, it is easy to see that

$$\Phi^i_{\text{FB}}(X, Y) = 0, \; i = 1, 2 \iff \phi_{\text{FB}}(X, Y) = (X^2 + Y^2)^{1/2} - (X + Y) = 0.$$

This feature indicates that $\Phi^i_{\text{FB}}(X, Y)$ are C-functions and equivalent to the traditional complementarity functions $\phi_{\text{FB}}(X, Y)$.

Remark 3.10 There are some other possible forms equivalent to $\Phi^1(X, Y)$ and $\Phi^2(X, Y)$ in Proposition 3.102 without having a lot of zeros. For instance, we could define

$$\widetilde{\Phi}^1(X, Y) := \begin{pmatrix} \phi(\lambda_1(X), \lambda_1(Y)) \\ \mathbf{x}_1^T \mathbf{y}_1 \\ \vdots \\ \mathbf{x}_n^T \mathbf{y}_n \\ \mathbf{v}^1_{XY} \end{pmatrix}$$

$$\widetilde{\Phi}^2(X, Y) := \begin{pmatrix} \phi(\lambda_1(X), \lambda_n(Y)) \\ \phi(\lambda_n(X), \lambda_1(Y)) \\ \mathbf{x}_1^T \mathbf{y}_1 \\ \vdots \\ \mathbf{x}_n^T \mathbf{y}_n \\ \mathbf{v}^2_{XY} \end{pmatrix},$$

where $\lambda_i(X), \lambda_i(Y)$ for $i = 1, \cdots, n$ are eigenvalues of matrices X, Y, which are arranged in the increasing order. Here, $\mathbf{v}^1_{XY} \in \mathbb{R}^{\frac{(n+1)(n-2)}{2}}$ and $\mathbf{v}^2_{XY} \in \mathbb{R}^{\frac{n^2-n-4}{2}}$ may have many alternative forms, one pair of them is

$$\mathbf{v}^1_{XY} := \left(\mathbf{x}_1^\mathsf{T}\mathbf{y}_2, \cdots, \mathbf{x}_1^\mathsf{T}\mathbf{y}_n, \mathbf{x}_2^\mathsf{T}\mathbf{y}_3, \cdots, \mathbf{x}_{n-2}^\mathsf{T}\mathbf{y}_n\right)^\mathsf{T},$$

$$\mathbf{v}^2_{XY} := \left(\mathbf{x}_1^\mathsf{T}\mathbf{y}_2, \cdots, \mathbf{x}_1^\mathsf{T}\mathbf{y}_n, \mathbf{x}_2^\mathsf{T}\mathbf{y}_3, \cdots, \mathbf{x}_{n-2}^\mathsf{T}\mathbf{y}_{n-1}\right)^\mathsf{T}.$$

Again, there are many matrix forms for $\widetilde{\Phi}^1(X, Y)$ and $\widetilde{\Phi}^2(X, Y)$. We hereby provide one matrix form as follows:

$$\widetilde{\Phi}^1(X, Y) \equiv \begin{pmatrix} \mathbf{x}_1^\mathsf{T}\mathbf{y}_1 & \phi(\lambda_1(X), \lambda_1(Y)) & \mathbf{x}_1^\mathsf{T}\mathbf{y}_2 & \cdots & \mathbf{x}_1^\mathsf{T}\mathbf{y}_{n-1} \\ \phi(\lambda_1(X), \lambda_1(Y)) & \mathbf{x}_2^\mathsf{T}\mathbf{y}_2 & \mathbf{x}_1^\mathsf{T}\mathbf{y}_n & \cdots & \mathbf{x}_2^\mathsf{T}\mathbf{y}_{n-1} \\ \mathbf{x}_1^\mathsf{T}\mathbf{y}_2 & \mathbf{x}_1^\mathsf{T}\mathbf{y}_n & \mathbf{x}_3^\mathsf{T}\mathbf{y}_3 & \cdots & \mathbf{x}_3^\mathsf{T}\mathbf{y}_{n-1} \\ \vdots & \vdots & \vdots & \ddots & \vdots \\ \mathbf{x}_1^\mathsf{T}\mathbf{y}_{n-1} & \mathbf{x}_2^\mathsf{T}\mathbf{y}_{n-1} & \mathbf{x}_3^\mathsf{T}\mathbf{y}_{n-1} & \cdots & \mathbf{x}_n^\mathsf{T}\mathbf{y}_n \end{pmatrix}$$

$$\widetilde{\Phi}^2(X, Y) \equiv \begin{pmatrix} \mathbf{x}_1^\mathsf{T}\mathbf{y}_1 & \phi(\lambda_1(X), \lambda_n(Y)) & \phi(\lambda_n(X), \lambda_1(Y)) & \cdots & \mathbf{x}_1^\mathsf{T}\mathbf{y}_{n-2} \\ \phi(\lambda_1(X), \lambda_n(Y)) & \mathbf{x}_2^\mathsf{T}\mathbf{y}_2 & \mathbf{x}_1^\mathsf{T}\mathbf{y}_{n-1} & \cdots & \mathbf{x}_2^\mathsf{T}\mathbf{y}_{n-2} \\ \phi(\lambda_n(X), \lambda_1(Y)) & \mathbf{x}_1^\mathsf{T}\mathbf{y}_{n-1} & \mathbf{x}_3^\mathsf{T}\mathbf{y}_3 & \cdots & \mathbf{x}_3^\mathsf{T}\mathbf{y}_{n-2} \\ \vdots & \vdots & \vdots & \ddots & \vdots \\ \mathbf{x}_1^\mathsf{T}\mathbf{y}_{n-2} & \mathbf{x}_2^\mathsf{T}\mathbf{y}_{n-2} & \mathbf{x}_3^\mathsf{T}\mathbf{y}_{n-2} & \cdots & \mathbf{x}_n^\mathsf{T}\mathbf{y}_n \end{pmatrix}.$$

Note that it might be difficult in using $\Phi^1(X, Y)$ and $\Phi^2(X, Y)$ to define a merit function $\frac{1}{2}\|\Phi(X, Y)\|^2$ for solving the SDCP due to the implicitness of eigenvalues of a real symmetric matrix. Thus, we propose a new direction to deal with the SDCP through NCP functions. More precisely, we will present a form of optimization problem for the SDCP. Let $F : \mathbb{S}^{n\times n} \to \mathbb{S}^{n\times n}$ be a mapping. The SDCP is to find a matrix $X \in \mathbb{S}^{n\times n}$ such that

$$X \in \mathbb{S}^{n\times n}_+, \ F(X) \in \mathbb{S}^{n\times n}_+, \ \langle X, F(X)\rangle = 0. \tag{3.344}$$

According to the relation (3.338), the SDCP (3.344) is equivalent to find a matrix $X \in \mathcal{S}^{n\times n}$ such that

$$\lambda_1(X) = 0, \ \lambda_1(F(X)) = 0, \ \langle X, F(X)\rangle = 0$$

when $X \neq 0$ and $F(X) \neq 0$. Using the fact that

$$\lambda_1(X) = \min_{\|u\|=1} u^\mathsf{T} X u \quad \text{and} \quad \lambda_1(F(X)) = \min_{\|v\|=1} v^\mathsf{T} F(X) v.$$

Then, for the case $X \in \mathrm{bd}(\mathbb{S}^{n\times n}_+)$ and $F(X) \in \mathrm{bd}(\mathbb{S}^{n\times n}_+)$, the SDCP (3.344) becomes the following bilevel optimization problem:

$$\begin{cases} \min\ f(X, \lambda_1(X), \lambda_1(F(X))) := (\lambda_1(X))^2 + (\lambda_1(F(X)))^2 + \langle X, F(X)\rangle^2 \\ \text{s.t. } \lambda_1(X) = \min\limits_{\|u\|=1} u^\top X u \ \text{ and } \ \lambda_1(F(X)) = \min\limits_{\|v\|=1} v^\top F(X) v,\ X \in \mathbb{S}^{n\times n}. \end{cases}$$

If the minimal value is zero, then there exists a matrix $X \in \mathbb{S}^{n\times n}$ satisfying

$$\lambda_1(X) = 0,\ \lambda_1(F(X)) = 0,\ \langle X, F(X)\rangle = 0$$

which is a solution of the SDCP. We see that the above problem does not provide the solution for the cases $X = 0$ and $F(X) \in \text{int}(\mathbb{S}^{n\times n}_+)$ or $X \in \text{int}(\mathbb{S}^{n\times n}_+)$ and $F(X) = 0$. However, this will not happen if we use the same technique for $\Phi^1(X, F(X))$ and $\Phi^2(X, F(X))$. Note that

$$\Phi^1(X, F(X)) = 0 \iff \begin{array}{c} \langle X, F(X)\rangle = 0 \\ \text{and } \phi(\lambda_1(X), \lambda_1(F(X))) = 0, \end{array}$$

or

$$\Phi^2(X, F(X)) = 0 \iff \begin{array}{c} \langle X, F(X)\rangle = 0, \\ \phi(\lambda_1(X), \lambda_n(F(X))) = 0 \text{ and } \phi(\lambda_n(X), \lambda_1(F(X))) = 0, \end{array}$$

where ϕ is a given NCP function. Then, we have the corresponding bilevel optimization problems:

$$\begin{cases} \min\ f(X, \lambda_1(X), \lambda_1(F(X))) := (\phi(\lambda_1(X), \lambda_1(F(X))))^2 + \langle X, F(X)\rangle^2 \\ \text{s.t. } \lambda_1(X) = \min\limits_{\|u\|=1} u^\top X u \ \text{ and } \ \lambda_1(F(X)) = \min\limits_{\|v\|=1} v^\top F(X) v,\ X \in \mathbb{S}^{n\times n}. \end{cases}$$

or

$$\begin{cases} \min\ f(X, \lambda_1(X), \lambda_1(F(X))) := (\phi(\lambda_1(X), \lambda_n(F(X))))^2 + (\phi(\lambda_n(X), \lambda_1(F(X))))^2 + \langle X, F(X)\rangle^2 \\ \text{s.t. } \lambda_1(X) = \min\limits_{\|u\|=1} u^\top X u,\ \lambda_n(X) = \max\limits_{\|u\|=1} u^\top X u,\ \lambda_1(F(X)) = \min\limits_{\|v\|=1} v^\top F(X) v, \\ \quad \text{and } \lambda_n(F(X)) = \max\limits_{\|u\|=1} u^\top F(X) u,\ X \in \mathbb{S}^{n\times n}. \end{cases}$$

Therefore, if the minimal value is zero, then there exists a matrix $X \in \mathbb{S}^{n\times n}$ satisfying $\lambda_1(X) \geq 0,\ \lambda_1(F(X)) \geq 0,\ \langle X, F(X)\rangle = 0$ which means that we can obtain the solution on the boundary and interior of $\mathbb{S}^{n\times n}_+$.

At last, we introduce C-functions based on a special type of matrix-valued functions. For a real-valued function $f : \mathbb{R} \to \mathbb{R}$, recall from (1.14) that there is a corresponding matrix-valued function defined by

$$f^{\text{mat}}(X) = f(\lambda_1(X))U_1 + \cdots + f(\lambda_n(X))U_n,$$

where X has the spectral decomposition $X = \lambda_1(X)U_1 + \cdots + \lambda_n(X)U_n$. For more details regarding this special matrix-valued functions, please refers to [77].

Proposition 3.103 *Let $\phi : \mathbb{R}^2 \to \mathbb{R}$ be an NCP function. Suppose that $F(X)$ is the matrix-valued function induced by a function $f : \mathbb{R} \to \mathbb{R}$, that is, $F(X)$ is written* *as*

$$F(X) = f(\lambda_1(X))U_1 + \cdots + f(\lambda_n(X))U_n,$$

with $X = \lambda_1(X)U_1 + \cdots + \lambda_n(X)U_n$, *where* $\lambda_i(X)$, $i = 1, \cdots, n$ *are eigenvalues of* X *and* $\{U_i\}_{i=1}^n$ *is a Jordan frame. Then, for any* $X \in \mathbb{S}^{n\times n}$, *the following function* $\Phi^3 : \mathbb{S}^{n\times n} \times \mathbb{S}^{n\times n} \to \mathbb{R}^{\frac{n(n+1)}{2}}$ *given by*

$$\Phi^3(X, F(X)) := \begin{pmatrix} \phi(\lambda_1(X), f(\lambda_1(X))) \\ \vdots \\ \phi(\lambda_n(X), f(\lambda_n(X))) \\ \mathbf{0} \end{pmatrix}$$

is a C-function, where the zero vector belongs to $\mathbb{R}^{\frac{n(n-1)}{2}}$.

Proof Again, applying Lemma 3.59 yields

$$\begin{aligned} & X \succeq 0,\ F(X) \succeq 0,\ \langle X, F(X)\rangle = 0 \\ \Longleftrightarrow\ & X \succeq 0,\ F(X) \succeq 0,\ XF(X) = 0 \\ \Longleftrightarrow\ & \lambda_i(X) \geq 0,\ f(\lambda_i(X)) \geq 0,\ i = 1, \cdots, n, \text{ and } \sum_{i=1}^n \lambda_i(X) f(\lambda_i(X))U_i = 0 \\ \Longleftrightarrow\ & \lambda_i(X) \geq 0,\ f(\lambda_i(X)) \geq 0,\ \lambda_i(X) f(\lambda_i(X)) = 0,\ i = 1, \cdots, n \\ \Longleftrightarrow\ & \phi(\lambda_i(X),\ f(\lambda_i(X))) = 0,\ i = 1, \cdots, n \\ \Longleftrightarrow\ & \Phi^3(X, F(X)) = 0. \end{aligned}$$

This clearly proves that $\Phi^3(X, F(X))$ is a C-function. □

To close this section, we point out that there exists matrix forms for $\Phi^3(X, F(X))$ in Proposition 3.103, one of them is

$$\Phi^3(X, F(X)) \equiv \begin{pmatrix} \phi(\lambda_1(X), f(\lambda_1(X))) & 0 & \cdots & 0 \\ 0 & \phi(\lambda_2(X), f(\lambda_2(X))) & \cdots & 0 \\ \vdots & \vdots & \ddots & \vdots \\ 0 & 0 & \cdots & \phi(\lambda_n(X), f(\lambda_n(X))) \end{pmatrix}.$$

References

1. J. Faraut, A. Koranyi, *Analysis on Symmetric Cones*. Oxford Mathematical Monographs (Oxford University Press, New York, 1994)
2. F. Facchinei, J.-S. Pang, *Finite-Dimensional Variational Inequalities and Complementarity Problems*, Volumes I and II (Springer, New York, 2003)

3. M.C. Ferris, O.L. Mangasarian, J.-S. Pang (eds.), *Complementarity: Applications, Algorithms and Extensions* (Kluwer Academic Publishers, Dordrecht, 2001)
4. M.C. Ferris, J.-S. Pang (eds.), *Complementarity and Variational Problems: State of the Art* (SIAM Publications, Philadelphia, 1996)
5. M.C. Ferris, J.-S. Pang, Engineering and economic applications of complementarity problems. SIAM Rev. **39**, 669–713 (1997)
6. F. Alizadeh, S. Schmieta, *Symmetric Cones, Potential Reduction Methods*, in Handbook of Semidefinite Programming, ed. by H. Wolkowicz, R. Saigal, L. Vandenberghe (Kluwer, Boston, 2000), pp. 195-233
7. M.S. Lobo, L. Vandenberghe, S. Boyd, H. Lebret, Application of second-order cone programming. Linear Algebra Appl. **284**, 193–228 (1998)
8. R.D.C. Monteiro, T. Tsuchiya, Polynomial convergence of primal-dual algorithms for the second-order cone programs based on the MZ-family of directions. Math. Program. **88**, 61–83 (2000)
9. S. Schmieta, F. Alizadeh, Associative and Jordan algebras, and polynomial time interior-point algorithms for symmetric cones. Math. Oper. Res. **26**, 543–564 (2001)
10. T. Tsuchiya, A convergence analysis of the scaling-invariant primal-dual path-following algorithms for second-order cone programming. Optim. Methods Softw. **11**, 141–182 (1999)
11. X.D. Chen, D. Sun, J. Sun, Complementarity functions and numerical experiments for second-order cone complementarity problems. Comput. Optim. Appl. **25**, 39–56 (2003)
12. M. Fukushima, Z.-Q. Luo, P. Tseng, Smoothing functions for second-order cone complementarity problems. Comput. Optim. Appl. **17**, 159–182 (2000)
13. S. Hayashi, N. Yamashita, M. Fukushima, A combined smoothing and regularization method for monotone second-order cone complementarity problems. SIAM J. Optim. **15**, 593–615 (2005)
14. J.-S. Chen, P. Tseng, An unconstrained smooth minimization reformulation of the second-order cone complementarity problem. Math. Program. **104**, 293–327 (2005)
15. D.P. Bertsekas, *Nonlinear Programming*, 2nd edn. (Athena Scientific, Massachusetts, 1999)
16. R. Fletcher, *Practical Methods of Optimization*, 2nd edn. (Wiley-Interscience, Chichester, 1987)
17. P. Tseng, Merit function for semidefinite complementarity problems. Math. Program. **83**, 159–185 (1998)
18. L.C. Kong, L. Tuncel, N.H. Xiu, Vector-valued implicit Lagrangian for symmetric cone complementarity problems. Asia-Pacific J. Oper. Res. **26**, 199–233 (2009)
19. P. Tseng, N. Yamashita, M. Fukushima, Equivalence of complementarity problems to differentiable minimization: a unified approach. SIAM J. Optim. **6**, 446–460 (1996)
20. N. Yamashita, M. Fukushima, On the stationary points of the Implicit Lagrangian for nonlinear complementarity problems. J. Optim. Theory Appl. **84**, 653–663 (1995)
21. O.L. Mangasarian, M.V. Solodov, A linearly convergent derivative-free descent method for strongly monotone complementarity problems. Comput. Optim. Appl. **64**, 5–16 (1999)
22. J.M. Peng, Equivalence of varational inequality problems to unconstrained minimization. Math. Program. **78**, 347–355 (1997)
23. C.K. Sim, G. Zhao, A note on treating a second order cone program as a special case of a semidefinite program. Math. Program. **102**(3), 609–613 (2005)
24. D. Sun, J. Sun, Strong semismoothness of Fischer-Burmeister SDC and SOC function. Math. Program. **103**, 575–581 (2005)
25. A. Korányi, Monotone functions on formally real Jordan algebras. Math. Ann. **269**, 73–76 (1984)
26. J.-S. Chen, X. Chen, P. Tseng, Analysis of nonsmooth vector-valued functions associated with second-order cones. Math. Program. **101**, 95–117 (2004)
27. J.M. Ortega, W.C. Rheinboldt, *Iterative Solution of Nonlinear Equations in Several Variables* (SIAM, Philadelphia, 2000)
28. A. Fischer, Solution of monotone complementarity problems with locally Lipschitzian functions. Math. Program. **76**, 513–532 (1997)

29. F.H. Clarke, *Optimization and Nonsmooth Analysis* (Wiley, New York, 1983)
30. F. Alizadeh, D. Goldfarb, Seond-order cone programming. Math. Program. **95**, 3–51 (2003)
31. F. Facchinei, J. Soares, A new merit function for nonlinear complementarity problems and a related algorithm. SIAM J. Optim. **7**, 225–247 (1997)
32. C. Kanzow, H. Kleinmichel, A new class of semismooth Newton method for nonlinear complementarity problems. Comput. Optim. Appl. **11**, 227–251 (1998)
33. Z.Q. Luo, P. Tseng, A new class of merit functions for the nonlinear complementarity problem, in *Complementarity and Variational Problems: State of the Art*, ed. by M.C. Ferris, J.S. Pang (SIAM, Philadelphia, 1997), pp. 204–225
34. S.-H. Pan, J.-S. Chen, A damped Gauss-Newton method for the second-order cone complementarity problem. Appl. Math. Optim. **59**, 293–318 (2009)
35. J.-S. Chen, S.-H. Pan, A one-parametric class of merit functions for the second-order cone complementarity problem. Comput. Optim. Appl. **45**(3), 581–606 (2010)
36. J.-S. Chen, *SOC Functions and Their Applications*, Springer Optimization and Its Applications, vol. 143 (Springer, Singapore, 2019)
37. M.S. Gowda, R. Sznajder, J. Tao, Some P-properties for linear transformations on Euclidean Jordan algebras. Linear Algebra Appl. **393**, 203–232 (2004)
38. R.A. Horn, C.R. Johnson, *Matrix Analysis* (Cambridge University Press, Cambridge, 1985)
39. R. Bhatia, *Matrix Analysis* (Springer, New York, 1997)
40. N. Yamashita, M. Fukushima, A new merit function and a descent method for semidefinite complementarity problems, in *Reformulation - Nonsmooth, Piecewise Smooth, Semismooth and Smoothing Methods*, ed. by M. Fukushima, L. Qi (Kluwer Academic Publishers, Boston, 1999), pp. 405–420
41. Y. Liu, L. Zhang, Y. Wang, Some properties of a class of merit functions for symmetric cone complementarity problems. Asia-Pacific J. Oper. Res. **23**, 473–496 (2006)
42. J. Tao, M.S. Gowda, Some P-properties for nonlinear transformations on Euclidean Jordan algebras. Math. Oper. Res. **4**, 985–1004 (2005)
43. K. Yamada, N. Yamashita, M. Fukushima, A new derivative-free descent method for the nonlinear complementarity problems, in *Nonlinear Optimization and Related Topics*, ed. by G.D. Pillo, F. Giannessi (Netherlands, Kluwer Academic Publishers, 2000), pp. 463–489
44. J.-S. Chen, Conditions for error bounds and bounded level sets of some merit functions for the SOCCP. J. Optim. Theory Appl. **135**, 459–473 (2007)
45. R.M.B. Goes, P.R. Oliveira, A new class of merit functions for the semidefinite complementarity problem. *Annals of the Brazilian Workshop on Continuous Optimization* (2002), pp. 1–18
46. J.-S. Chen, A new merit function and its related properties for the second-order cone complementarity problem. Pac. J. Optim. **2**, 167–179 (2006)
47. R.T. Rockafellar, *Convex Analysis* (Princeton University Press, Princeton, New Jersey, 1970)
48. C. Kanzow, C. Nagel, Semidefinite programs: new search directions, smoothing-type methods, and numerical results. SIAM J. Optim. **13**(1), 1–23 (2002)
49. L. Zhang, N. Zhang, L. Pang, Differential properties of the symmetric matrix-valued Fischer-Burmeister function. J. Optim. Theory Appl. **153**, 436–460 (2012)
50. S.-J. Bi, S.-H. Pan, J.-S. Chen, Nonsingular conditions for the Fischer-Burmeister system of nonlinear SDPs. SIAM J. Optim. **21**(4), 1392–1417 (2011)
51. C.K. Sun, J. Sun, D. Ralph, A note on the Lipschitz continuity of the gradient of the squared norm of the matrix-valued Fischer-Burmeister function. Math. Program. **107**, 547–553 (2006)
52. C. Kanzow, Y. Yamashita, M. Fukushima, New NCP functions and their properties. J. Optim. Theory Appl. **97**, 115–135 (1997)
53. O.L. Mangasarian, M.V. Solodov, Nonlinear complementarity as unconstrained and constrained minimization. Math. Program. **62**, 2777–297 (1993)
54. L.C. Kong, J. Sun, N.H. Xiu, A regularized smoothing Newton method for symmetric cone complementarity problems. SIAM J. Optim. **19**, 1028–1047 (2008)
55. L. Vandenberghe, S. Boyd, A primal-dual potential reduction method for problems involving matrix inequalities. Math. Program. **69**, 205–236 (1995)

56. S.-J. Bi, S.-H. Pan, J.-S. Chen, The same growth of FB and NR symmetric cone complementarity functions. Optim. Lett. **6**(1), 153–162 (2012)
57. D. Sun, J. Sun, Löwner's operator and spectral functions in Euclidean Jordan algebras. Math. Oper. Res. **33**, 421–445 (2008)
58. Y.-L. Chang, J.-S. Chen, S.-H. Pan, Strong semismoothness of Fischer-Burmeister complementarity function associated with symmetric cones. J. Nonlinear Convex Anal. **13**(4), 799–806 (2012)
59. M. Fukushima, Equivalent differentiable optimization problems and descent methods for asymmetric variational inequality problems. Math. Program. **53**(1), 99–110 (1992)
60. B. Chen, Error bounds for R_0-type and monotone nonlinear complementarity problems. J. Optim. Theory Appl. **108**, 297–316 (2001)
61. J.-S. Pang, L. Qi, Nonsmooth equation: motivation and algorithms. SIAM J. Optim. **3**, 443–465 (1993)
62. S. Kum, Y. Lim, Penalized complementarity functions on symmetric cones. J. Glob. Optim. **46**, 475–485 (2010)
63. Y. Lim, Applications of geometric means on symmetric cones. Math. Ann. **319**, 457–468 (2001)
64. Z.-H. Huang, X.H. Liu, Extension of smoothing Newton algorithms to solve linear programming over symmetric cones. J. Syst. Sci. Complex. **24**(1), 195–206 (2011)
65. Y.G. Evtushenko, V.A. Purtov, Sufficient conditions for a minimum for nonlinear programming problems. Sov. Math.-Dokl. **30**, 313–316 (1984)
66. L.C. Kong, N.H. Xiu, New smooth C-function for symmetric cone complementarity problems. Optim. Lett. **1**, 391–400 (2007)
67. S.H. Schmieta, F. Alizadeh, Extension of primal-dual interior point algorithms for symmetric cones. Math. Program. **96**, 409–438 (2003)
68. C. Kanzow, Global convergence properties of some iterative methods for linear complementarity problems. SIAM J. Optim. **6**, 326–341 (1996)
69. Z.-H. Huang, T. Ni, Smoothing algorithms for complementarity problems over symmetric cones. Comput. Optim. Appl. **45**, 557–579 (2010)
70. C. Kanzow, M. Fukushima, Theoretical and numerical investigation of the D-gap function for box constrained variational inequalities. Math. Program. **83**, 55–87 (1998)
71. P. Tseng, Global behaviour of a class of merit functions for the nonlinear complementarity problem. J. Optim. Theory Appl. **89**, 17–37 (1996)
72. C. Kanzow, M. Fukushima, Equivalence of the generalized complementarity problem to differentiable unconstrained minimization. J. Optim. Theory Appl. **90**, 581–603 (1996)
73. O.L. Mangasarian, Equivalence of the complementarity problem to a system of nonlinear equations. SIAM J. Appl. Math. **31**, 89–92 (1976)
74. M.S. Gowda, Y. Song, On semidefinite linear complementarity problems. Math. Program. **88**, 575–587 (2000)
75. M.V. Ramana, An exact duality theory for semidefinite programming and its complexity implications. Math. Program. **77**, 129–162 (1997)
76. S.-H. Pan, S. Kum, Y. Lim, J.-S. Chen, On the generalized Fischer-Burmeister merit function for the second-order cone complementarity problem. Math. Comput. **83**(287), 1143–1171 (2014)
77. X. Chen, H. Qi, P. Tseng, Analysis of nonsmooth symmetric-matrix-valued functions with applications to semidefinite complementarity problems. SIAM J. Optim. **13**, 960–985 (2003)

Chapter 4
Optimization Algorithms Using Complementarity Functions

Abstract In this chapter, we present several optimization algorithms that utilize complementarity functions. As discussed in Chap. 2, there are four well-established approaches for solving the NCP, each of which can be extended to broader settings such as the SOCCP, the SDCP, and the SCCP. Accordingly, the chapter is organized into four sections, with each section illustrating one or two representative algorithms within a given approach, accompanied by various complementarity functions to highlight the algorithmic applications of C-functions. Numerous related algorithms employing C-functions can also be found in the literature; for further reference, see [1–9].

4.1 Merit Function Approach

In this section, we present the merit function approach for solving the SOCCP (3.1). As will be shown, the problem is reformulated as an unconstrained minimization of an appropriately defined merit function over $\mathbb{R}^n$. We then introduce a descent method to solve this unconstrained reformulation. Recall the SOCCP (3.1): the goal is to find $\zeta \in \mathbb{R}^n$ such that

$$\langle F(\zeta), \zeta \rangle = 0, \quad F(\zeta) \in \mathcal{K}, \quad \zeta \in \mathcal{K},$$

where $\langle \cdot, \cdot \rangle$ is the Euclidean inner product, $F : \mathbb{R}^n \to \mathbb{R}^n$ is a smooth (i.e., continuously differentiable) mapping, and $\mathcal{K}$ is the Cartesian product of second-order cones. In other words,

$$\mathcal{K} = \mathcal{K}^{n_1} \times \cdots \times \mathcal{K}^{n_m},$$

J. Chen, *Complementarity Functions in Optimization*, Springer Optimization and Its Applications 228, https://doi.org/10.1007/978-3-032-06924-5_4

where $m, n_1, \ldots, n_m \geq 1, n_1 + \cdots + n_m = n$, and

$$\mathcal{K}^{n_i} = \{(x_1, x_2) \in \mathbb{R} \times \mathbb{R}^{n_i - 1} \mid \|x_2\| \leq x_1\},$$

with $\|\cdot\|$ denoting the Euclidean norm and $\mathcal{K}^1$ denoting the set of nonnegative reals $\mathbb{R}_+$. In addition, we also recall the YF complementarity function:

$$\psi_{\text{YF}}(x, y) := \psi_0(\langle x, y \rangle) + \psi_{\text{FB}}(x, y), \tag{4.1}$$

where $\psi_0 : \mathbb{R} \to [0, \infty)$ is any smooth function satisfying

$$\psi_0(t) = 0 \ \ \forall t \leq 0 \quad \text{and} \quad \psi_0'(t) > 0 \ \ \forall t > 0. \tag{4.2}$$

In [10], $\psi_0(t) = \frac{1}{4}(\max\{0, t\})^4$ was considered. As shown in Sect. 3.1.4, the function ψ_{YF} is a C-function as well as a merit function, which enjoys favorable properties as what ψ_{FB} has. Moreover, ψ_{YF} possesses properties of bounded level sets and error bound.

In this section, we focus on the following equivalent reformulation of SOCCP, which arises via the merit function ψ_{YF} defined as in (4.1)–(4.2):

$$\min_{\zeta \in \mathbb{R}^n} f_{\text{YF}}(\zeta) \quad \text{where} \quad f_{\text{YF}}(\zeta) := \psi_{\text{YF}}\left(F(\zeta), \zeta\right). \tag{4.3}$$

The algorithm is described as below, where the proposed method uses $d(\zeta)$ given in Proposition 3.37, i.e.:

$$d(\zeta) := -\left(\psi_0'(\langle F(\zeta), \zeta \rangle)\zeta + \nabla_x \psi_{\text{FB}}(F(\zeta), \zeta)\right) \tag{4.4}$$

as its direction.

Algorithm 4.1 (Step 0) Choose $\zeta^0 \in \mathbb{R}^n, \varepsilon \geq 0, \sigma \in (0, 1/2), \beta \in (0, 1)$ and set $k := 0$.

(Step 1) If $f_{\text{YF}}(\zeta^k) \leq \varepsilon$, then stop.

(Step 2) Compute $d(\zeta^k) := -\left(\psi_0'(\langle F(\zeta^k), \zeta^k \rangle)\zeta^k + \nabla_x \psi_{\text{FB}}(F(\zeta^k), \zeta^k)\right)$.

(Step 3) Find a step size $t_k := \beta^{m_k}$, where m_k is the smallest nonnegative integer m satisfying Armijo's rule:

$$f_{\text{YF}}\left(\zeta^k + \beta^m d(\zeta^k)\right) \leq (1 - \sigma\beta^{2m}) f_{\text{YF}}(\zeta^k). \tag{4.5}$$

(Step 4) Set $\zeta^{k+1} := \zeta^k + t_k\, d(\zeta^k)$, $k := k + 1$ and go to Step 1.

It is worth noting that the above algorithm is ∇F-free, that is, it does not require computation of the Jacobian matrix of F. Furthermore, the computational effort per iteration is minimal, involving only a few vector multiplications. This type of algorithm has also been studied in the context of the NCP (see [11]) and the SDCP (see

[10]). A distinctive feature of such methods is that both the step size and the search direction are adaptively adjusted via Armijo's rule. In practical implementations, the parameter σ is typically chosen close to zero, while β is often selected in the interval $(\frac{1}{10}, \frac{1}{2})$, depending on the degree of confidence in the quality of the initial step size; see [12] for further discussion.

We now proceed to establish the global convergence of Algorithm 4.1. Without loss of generality, we assume $\varepsilon = 0$, so that the algorithm generates an infinite sequence ζ^k.

Proposition 4.1 *Suppose that F is monotone and the SOCCP (3.1) is strictly feasible. Then, the sequence $\{\zeta^k\}$ generated by Algorithm 4.1 has at least one accumulation point, and any accumulation point is a solution of the SOCCP (3.1).*

Proof The proof is standard and can be found in [12]. For completeness, we here present its proof by the following three steps.

(i) First, we show that, whenever ζ^k is not a solution, there exists a nonnegative integer m_k in Step 3 of Algorithm 4.1. Suppose not, then for any positive integer m, we have

$$f_{\rm YF}\left(\zeta^k + \beta^m\, d(\zeta^k)\right) - f_{\rm YF}(\zeta^k) > -\sigma\beta^{2m} f_{\rm YF}(\zeta^k)$$

where $d(\zeta)$ is described as in (4.4). Dividing by β^m on both sides and letting $m \to \infty$ yields

$$\left\langle \nabla f_{\rm YF}(\zeta^k), d(\zeta^k)\right\rangle \geq 0. \tag{4.6}$$

Since F is monotone which is equivalent to $\nabla F(\zeta)$ is positive semidefinite, the inequality (4.6) contradicts Proposition 3.37. Hence, we can find an integer m_k in Step 3.

(ii) Secondly, we show that the sequence $\{\zeta^k\}$ generated by the algorithm has at least one accumulation point. By the descent property of Algorithm 4.1, the sequence $\{f_{\rm YF}(\zeta^k)\}_{k\in N}$ is decreasing. Hence by Proposition 3.39, we obtain that $\{\zeta^k\}$ is bounded, and consequently has at least one accumulation point.

(iii) Finally, we prove that any accumulation point of $\{\zeta^k\}$ is a solution of the SOCCP (3.1). Let ζ^* be an arbitrary accumulation point of $\{\zeta^k\}_{k\in N}$. In other words, there is a subsequence $\{\zeta^k\}_{k\in K}$ converging to ζ^*, where K is a subset of N. We know $d(\cdot)$ is continuous (since ψ_0 and $\psi_{\rm FB}$ are smooth) which implies $\{d(\zeta^k)\}_{k\in K}$ converges to $d(\zeta^*)$. Next, we need to discuss two cases. First, we consider the case where there exists a constant $\bar{\beta}$ such that $\beta^{m_k} \geq \bar{\beta} > 0$ for all $k \in K$. Then, from (4.5), we have

$$f_{\rm YF}(\zeta^{k+1}) \leq (1 - \sigma\bar{\beta}^2) f_{\rm YF}(\zeta^k)$$

for all $k \in K$ and the entire sequence $\{f_{\rm YF}(\zeta^k)\}_{k\in K}$ is decreasing. Thus, we obtain $f_{\rm YF}(\zeta^*) = 0$ (by taking the limit) which says ζ^* is a solution of the SOCCP (3.1).

Now, we consider the other case where there exists a further subsequence such that $\beta^{m_k} \to 0$. Note that by Armijo's rule (4.5) in Step 3, we have

$$f_{\text{YF}}(\zeta^k + \beta^{m_k-1} d(\zeta^k)) - f_{\text{YF}}(\zeta^k) > -\sigma\beta^{2(m_k-1)} f_{\text{YF}}(\zeta^k).$$

Dividing by β^{m_k-1} both sides and passing the limit on the further subsequence, we obtain

$$\big\langle \nabla f_{\text{YF}}(\zeta^*), d(\zeta^*)\big\rangle \geq 0,$$

which yields that ζ^* is a solution of the SOCCP (3.1) by Proposition 3.37. □

Proposition 4.2 *Let F be a continuously differentiable and strongly monotone mapping. Then, the sequence $\{\zeta^k\}$ generated by Algorithm 4.1 converges to the unique solution of the SOCCP (3.1).*

Proof The proof is routine (see [13]), however, we present it for completeness. We know that the property of bounded level sets is also held when F is strongly monotone, so following the same arguments as in the proof of Proposition 4.1, we again obtain that $\{\zeta^k\}$ has at least one accumulation point and any accumulation point is a solution of the SOCCP (3.1).

On the other hand, the strong monotonicity of F implies that the SOCCP (3.1) has at most one solution. To see this, say there are two solutions $\zeta^*, \xi^* \in \mathbb{R}^n$ such that

$$\begin{cases} \langle F(\zeta^*), \zeta^* \rangle = 0, \\ F(\zeta^*) \in \mathcal{K}^n, \ \zeta^* \in \mathcal{K}^n \end{cases} \quad \text{and} \quad \begin{cases} \langle F(\xi^*), \xi^* \rangle = 0, \\ F(\xi^*) \in \mathcal{K}^n, \ \xi^* \in \mathcal{K}^n. \end{cases}$$

Since F is strongly monotone, we have $\langle F(\zeta^*) - F(\xi^*), \zeta^* - \xi^* \rangle > 0$. However,

$$\begin{aligned} & \langle F(\zeta^*) - F(\xi^*), \zeta^* - \xi^* \rangle \\ = & \langle F(\zeta^*), \zeta^* \rangle + \langle F(\xi^*), \xi^* \rangle - \langle F(\zeta^*), \xi^* \rangle - \langle F(\xi^*), \zeta^* \rangle \\ = & -\langle F(\zeta^*), \xi^* \rangle - \langle F(\xi^*), \zeta^* \rangle \\ \leq & \ 0, \end{aligned}$$

where the inequality is due to $F(\zeta^*), \zeta^*, F(\xi^*), \xi^*$ are all in $\mathcal{K}^n$. Hence, it is a contradiction and therefore there is at most one solution for the SOCCP (3.1).

From all the above, it says there is a unique solution ζ^*, so the entire sequence $\{x^k\}$ must converge to ζ^*. □

We observe that Proposition 3.39 plays a crucial role in establishing Propositions 4.1 and 4.2. Notably, the assumption of strict feasibility is essential for the validity of Proposition 3.39. For instance, if $F(\zeta) \equiv 0$, then every $\zeta \in \mathcal{K}^n$ is a solution to the SOCCP (3.1), resulting in an unbounded solution set. In what follows, we explore a further refinement by replacing the strict feasibility condition with a weaker one—namely, the assumption that F is an R_{01}-function. Under this framework, Proposition 3.39 and Proposition 4.1 are improved and generalized as Lemma 4.1 and Proposition 4.3, respectively. These results are of particular importance not only because they are novel but also because they eliminate the need for strict feasibility assumptions.

Lemma 4.1 *Let $f_{\rm YF}$ be given as in (4.3). Suppose that F is a R_{01}-function. Then the level set*

$$\mathcal{L}(\gamma) := \{\zeta \in \mathbb{R}^n \mid f_{\rm YF}(\zeta) \leq \gamma\}$$

is bounded for all $\gamma \geq 0$.

Proof We will prove this result by contradiction. Suppose there exists an unbounded sequence $\{\zeta^k\} \subset \mathcal{L}(\gamma)$ for some $\gamma \geq 0$. It can be seen that the sequence of the smaller spectral values of $\{\zeta^k\}$ and $\{F(\zeta^k)\}$ are bounded below. In fact, if not, it follows form Lemma 3.15 that $f_{\rm YF}(\zeta^k) \to \infty$, which contradicts $\{\zeta^k\} \subset \mathcal{L}(\gamma)$. Therefore, $\{(-\zeta^k)_+\}$ and $\{(-F(\zeta^k))_+\}$ are bounded above, which says the conditions of R_{01}-function given in (1.51) are satisfied. Then, by the assumption of R_{01}-function, we have

$$\liminf_{k\to\infty} \frac{\langle \zeta^k, F(\zeta^k)\rangle}{\|\zeta^k\|^2} > 0.$$

This yields $\langle \zeta^k, F(\zeta^k)\rangle \to \infty$, and hence $f_{\rm YF}(\zeta^k) \to \infty$ by definition of $f_{\rm YF}$ given as in (4.3). Thus, it is a contradiction to $\{\zeta^k\} \subset \mathcal{L}(\gamma)$. □

Proposition 4.3 *Let F be a continuously differentiable mapping. Suppose that F is R_{01}-function. Then, the sequence $\{\zeta^k\}$ generated by Algorithm 4.1 has at least one accumulation point, and any accumulation point is a solution of the SOCCP (3.1).*

Proof By applying Lemma 4.1 and following the same arguments as in Proposition 4.1, the desired results hold. □

As shown in [14, 15], the R_{01}-function condition is weaker than strong monotonicity and, in a certain sense, also weaker than the combination of monotonicity and strict feasibility. However, it remains unclear whether the R_{01}-function condition can be further relaxed to that of an R_{02}-function.

4.2 Nonsmooth Function Approach

In this section, we introduce a semismooth Newton method for solving the SOCCP. Specifically, we formulate the problem as a nonlinear least-squares problem by employing the Fischer–Burmeister function and the plus function. Consider the general SOCCP: find $\zeta \in \mathbb{R}^n$ such that

$$F(\zeta) \in \mathcal{K}, \quad G(\zeta) \in \mathcal{K}, \quad \langle F(\zeta), G(\zeta) \rangle = 0, \tag{4.7}$$

where $\langle \cdot, \cdot \rangle$ denotes the Euclidean inner product, $F : \mathbb{R}^n \to \mathbb{R}^n$ and $G : \mathbb{R}^n \to \mathbb{R}^n$ are assumed to be continuously differentiable throughout this section, and $\mathcal{K}$ is the Cartesian product of second-order cones (SOCs), i.e.:

$$\mathcal{K} = \mathcal{K}^{n_1} \times \mathcal{K}^{n_2} \times \cdots \times \mathcal{K}^{n_q},$$

where $q, n_1, \ldots, n_q \geq 1$, $n_1 + n_2 + \cdots + n_q = n$, and

$$\mathcal{K}^{n_i} := \left\{ (x_{i1}, x_{i2}) \in \mathbb{R} \times \mathbb{R}^{n_i - 1} \mid x_{i1} \geq \|x_{i2}\| \right\}.$$

In the rest of this section, corresponding to the Cartesian structure of $\mathcal{K}$, we write $F = (F_1, \ldots, F_q)$ and $G = (G_1, \ldots, G_q)$ with F_i and G_i being mappings from $\mathbb{R}^n$ to $\mathbb{R}^{n_i}$.

As mentioned in Chap. 3, the SOCCP (4.7) can be reformulated as the following system of nonsmooth equations

$$\Phi_{\rm FB}(\zeta) := \begin{pmatrix} \phi_{\rm FB}(F_1(\zeta), G_1(\zeta)) \\ \vdots \\ \phi_{\rm FB}(F_q(\zeta), G_q(\zeta)) \end{pmatrix} = 0, \tag{4.8}$$

which induces a natural merit function $\Psi_{\rm FB} : \mathbb{R}^n \to \mathbb{R}_+$ for (4.7), defined by

$$\Psi_{\rm FB}(\zeta) := \frac{1}{2}\|\Phi_{\rm FB}(\zeta)\|^2 = \sum_{i=1}^{q} \psi_{\rm FB}(F_i(\zeta), G_i(\zeta)) \tag{4.9}$$

with

$$\psi_{\rm FB}(x_i, y_i) := \frac{1}{2}\|\phi_{\rm FB}(x_i, y_i)\|^2. \tag{4.10}$$

Recently, we analyzed in [16] that, to guarantee the boundedness of the level sets of the FB merit function $\Psi_{\rm FB}$, it requires that the mapping F at least has the uniform Cartesian P-property (also see Sect. 3.1). This means that $\phi_{\rm FB}$ has some limitations in handling monotone SOCCPs.

Motivated by the work [17] for the NCP setting, we give a new reformulation for (4.7) to overcome the disadvantage of $\phi_{\rm FB}$. Let $\phi_0 : \mathbb{R}^{n_i} \times \mathbb{R}^{n_i} \to \mathbb{R}_+$ be given by

$$\phi_0(x_i, y_i) := \max\left\{0, x_i^{\rm T} y_i\right\}, \tag{4.11}$$

and define the operator $\Phi : \mathbb{R}^n \to \mathbb{R}^{n+q}$ as

$$\Phi(\zeta) := \begin{pmatrix} \rho_1 \phi_{\text{FB}}(F_1(\zeta), G_1(\zeta)) \\ \vdots \\ \rho_1 \phi_{\text{FB}}(F_q(\zeta), G_q(\zeta)) \\ \rho_2 \phi_0(F_1(\zeta), G_1(\zeta)) \\ \vdots \\ \rho_2 \phi_0(F_q(\zeta), G_q(\zeta)) \end{pmatrix}, \tag{4.12}$$

where ρ_1, ρ_2 are arbitrary but fixed constants from (0, 1) used as the weights between the first type of terms and the second one. In other words, we define Φ by appending q components to the mapping Φ_{FB}. These additional components, as will be shown later, play a crucial role in overcoming the disadvantage of Ψ_{FB} mentioned above. Noting that

$$\zeta^* \text{ solves } \Phi(\zeta) = 0 \quad \Longleftrightarrow \quad \zeta^* \text{ solves the SOCCP (4.7)},$$

we have the following nonlinear least-square reformulation for the SOCCP (4.7)

$$\min_{\zeta \in \mathbb{R}^n} \Psi(\zeta) := \frac{1}{2}\|\Phi(\zeta)\|^2 = \sum_{i=1}^{q} \psi(F_i(\zeta), G_i(\zeta)), \tag{4.13}$$

where

$$\psi(x_i, y_i) := \rho_1^2 \psi_{\text{FB}}(x_i, y_i) + \frac{1}{2}\rho_2^2 \phi_0(x_i, y_i)^2. \tag{4.14}$$

This reformulation offers several advantages. First, the function Ψ belongs to the class of merit functions f_{YF} introduced in [18], which will be shown to possess more favorable properties than Ψ_{FB}. Second, the function Φ inherits the semismoothness of Φ_{FB}, and even exhibits strong semismoothness under certain conditions. Leveraging these properties, we propose a semismooth Levenberg–Marquardt-type method for solving (4.13), and establish superlinear, or even quadratic, convergence under the assumptions of strict complementarity and a local error bound.

Lemma 4.2 *Let $\phi_0 : \mathbb{R}^n \times \mathbb{R}^n \to \mathbb{R}_+$ be defined as in (4.11). Then,*

(a) the square of ϕ_0 is continuously differentiable everywhere;
(b) ϕ_0 is strongly semismooth everywhere on $\mathbb{R}^n \times \mathbb{R}^n$;
(c) the B-subdifferential $\partial_B \phi_0(x, y)$ of ϕ_0 at any $(x, y) \in \mathbb{R}^n \times \mathbb{R}^n$ is given by

$$\partial_B \phi_0(x, y) = \left[\partial_B (x^\top y)_+ y^\top \;\; \partial_B (x^\top y)_+ x^\top\right],$$

where

$$\partial_B\left(x^\top y\right)_+ = \begin{cases} \{1\} & \text{if } x^\top y > 0, \\ \{1,0\} & \text{if } x^\top y = 0, \\ \{0\} & \text{if } x^\top y < 0. \end{cases}$$

Proof The results come from direct computation. □

Using Proposition 3.3 (b) and Lemma 4.2 (b), we readily obtain the semismoothness of Φ.

Proposition 4.4 *The operator $\Phi : \mathbb{R}^n \to \mathbb{R}^{n+q}$ defined by (4.12) is semismooth. If, in addition, F' and G' are Lipschitz continuous, then Φ is strongly semismooth.*

Proof Let Φ_i denote the i-th component function of Φ for $i = 1, 2, \ldots, 2q$, i.e., $\Phi_i(\zeta) = \phi_{\text{FB}}(F_i(\zeta), G_i(\zeta))$ for $i = 1, 2, \ldots, q$ and $\Phi_i(\zeta) = \phi_0(F_i(\zeta), G_i(\zeta))$ for $i = q+1, \ldots, 2q$. Then, the mapping Φ is (strongly) semismooth if every Φ_i is (strongly) semismooth. For $i = 1, 2, \ldots, q$, $\Phi_i : \mathbb{R}^n \to \mathbb{R}^{n_i}$ is the composite of the strongly semismooth function ϕ_{FB} and the smooth function $\zeta \mapsto (F_i(\zeta), G_i(\zeta))$, whereas $\Phi_{q+i} : \mathbb{R}^n \to \mathbb{R}$ is the composite of the strongly semismooth function ϕ_0 and the function $\zeta \mapsto (F_i(\zeta), G_i(\zeta))$. Moreover, when F' and G' are Lipschitz continuous, $\zeta \mapsto (F_i(\zeta), G_i(\zeta))$ is strongly semismooth. By [19, Theorem 19], we have that every component function of Φ is semismooth, and strongly semismooth if F' and G' are Lipschitz continuous. □

Proposition 4.5 *Let $\Phi : \mathbb{R}^n \to \mathbb{R}^{n+q}$ be defined by (4.12). Then, for any given $\zeta \in \mathbb{R}^n$,*

$$\partial_B \Phi(\zeta)^\top \subseteq \nabla F(\zeta)\big[\rho_1\,(A(\zeta) - I)\ \ \rho_2\, C(\zeta)\big] + \nabla G(\zeta)\big[\rho_1\,(B(\zeta) - I)\ \ \rho_2\, D(\zeta)\big]$$

where $C(\zeta) = \operatorname{diag}(C_1(\zeta), \ldots, C_q(\zeta))$ and $D(\zeta) = \operatorname{diag}(D_1(\zeta), \ldots, D_q(\zeta))$ with

$$C_i(\zeta) \in G_i(\zeta)\partial_B(F_i(\zeta)^\top G_i(\zeta))_+ \quad \text{and} \quad D_i(\zeta) \in F_i(\zeta)\partial_B(F_i(\zeta)^\top G_i(\zeta))_+,$$

and $A(\zeta) = \operatorname{diag}(A_1(\zeta), \ldots, A_q(\zeta))$ and $B(\zeta) = \operatorname{diag}(B_1(\zeta), \ldots, B_q(\zeta))$ with the block diagonals $A_i(\zeta), B_i(\zeta) \in \mathbb{R}^{n_i \times n_i}$ having the following representation:

(a) If $F_i(\zeta)^2 + G_i(\zeta)^2 \in \operatorname{int}\mathcal{K}^{n_i}$, then $A_i(\zeta) = L_{F_i(\zeta)} L_{z_i(\zeta)}^{-1}$ and $B_i(\zeta) = L_{G_i(\zeta)} L_{z_i(\zeta)}^{-1}$, where $z_i(\zeta) = (F_i(\zeta)^2 + G_i(\zeta)^2)^{1/2}$.
(b) If $F_i(\zeta)^2 + G_i(\zeta)^2 \in \operatorname{bd}^+\mathcal{K}^{n_i}$, then $[A_i(\zeta), G_i(\zeta)]$ belongs to the set

$$\left\{\left[\frac{1}{2\sqrt{2w_{i1}(\zeta)}} L_{F_i(\zeta)} \begin{pmatrix} 1 & \bar{w}_{i2}(\zeta)^\top \\ \bar{w}_{i2}(\zeta) & 4I - \bar{w}_{i2}(\zeta)\bar{w}_{i2}(\zeta)^\top \end{pmatrix} + \frac{1}{2} u_i \left(1, -\bar{w}_{i2}(\zeta)^\top\right),\right.\right.$$
$$\left.\frac{1}{2\sqrt{2w_{i1}(\zeta)}} L_{G_i(\zeta)} \begin{pmatrix} 1 & \bar{w}_{i2}(\zeta)^\top \\ \bar{w}_{i2}(\zeta) & 4I - \bar{w}_{i2}(\zeta)\bar{w}_{i2}(\zeta)^\top \end{pmatrix} + \frac{1}{2} v_i \left(1, -\bar{w}_{i2}(\zeta)^\top\right)\right] \Bigg|$$
$$\left. u_i = (u_{i1}, u_{i2}),\ v_i = (v_{i1}, v_{i2}) \text{ satisfy } |u_{i1}| \le \|u_{i2}\| \le 1, |v_{i1}| \le \|v_{i2}\| \le 1 \right\},$$

where $w_i(\zeta) = (w_{i1}(\zeta), w_{i2}(\zeta)) = F_i(\zeta)^2 + G_i(\zeta)^2$ *and* $\bar{w}_{i2}(\zeta) = w_{i2}(\zeta)/\|w_{i2}(\zeta)\|$.

(c) *If* $(F_i(\zeta), G_i(\zeta)) = (0, 0)$, *then* $[A_i(\zeta), B_i(\zeta)] \in \left\{[L_{\hat{u}_i}, L_{\hat{v}_i}] \mid \|\hat{u}_i\|^2 + \|\hat{v}_i\|^2 = 1\right\}$ *or*

$$\left\{\left[\frac{1}{2}\xi_i\left(1, \bar{w}_{i2}^T\right) - \frac{1}{2}u_i\left(-1, \bar{w}_{i2}^T\right) + 2L_{s_i}\begin{pmatrix}0 & 0\\ 0 & (I - \bar{w}_{i2}\bar{w}_{i2}^{\mathsf{T}})\end{pmatrix},\right.\right.$$
$$\left.\frac{1}{2}\eta_i\left(1, \bar{w}_{i2}^{\mathsf{T}}\right) - \frac{1}{2}v_i\left(-1, \bar{w}_{i2}^{\mathsf{T}}\right) + 2L_{\omega_i}\begin{pmatrix}0 & 0\\ 0 & (I - \bar{w}_{i2}\bar{w}_{i2}^{\mathsf{T}})\end{pmatrix}\right] \;\Big|$$
$\bar{w}_{i2} \in \mathbb{R}^{n_i-1}$ satisfies $\|\bar{w}_{i2}\| = 1$ and $\xi_i = (\xi_{i1}, \xi_{i2}), u_i = (u_{i1}, u_{i2}), \eta_i = (\eta_{i1}, \eta_{i2})$
$v_i = (v_{i1}, v_{i2}), s_i = (s_{i1}, s_{i2}), \omega_i = (\omega_{i1}, \omega_{i2})$ satisfy $|\xi_1| \le \|\xi_2\| \le 1$,
$$\left.|u_{i1}| \le \|u_{i2}\| \le 1, |\eta_{i1}| \le \|\eta_{i2}\| \le 1, |v_{i1}| \le \|v_{i2}\| \le 1, \|s_i\|^2 + \|\omega_i\|^2 \le 1/2\right\}.$$

Proof Let Φ_i denote the i-th component function of Φ, i.e., $\Phi_i(\zeta) = \phi_{\rm FB}(F_i(\zeta), G_i(\zeta))$ and $\Phi_{q+i}(\zeta) = \phi_0(F_i(\zeta), G_i(\zeta))$ for $i = 1, \ldots, q$. By the definition of the B-subdifferential,

$$\partial_B\Phi(\zeta)^{\mathsf{T}} \subseteq \partial_B\Phi_1(\zeta)^{\mathsf{T}} \times \partial_B\Phi_2(\zeta)^{\mathsf{T}} \times \cdots \times \partial_B\Phi_{2q}(\zeta)^{\mathsf{T}}, \tag{4.15}$$

where the latter means the set of all matrices whose $(n_{i-1} + 1)$-th to n_i-th columns belong to $\partial_B\Phi_i(\zeta)^{\mathsf{T}}$ with $n_0 = 0$, and $(n+i)$-th column belongs to $\partial_B\Phi_{q+i}(\zeta)^{\mathsf{T}}$. Notice that

$$\begin{aligned}\partial_B\Phi_i(\zeta)^{\mathsf{T}} &\subseteq \rho_1\left[\nabla F_i(\zeta)\;\; \nabla G_i(\zeta)\right]\partial_B\phi_{\rm FB}(F_i(\zeta), G_i(\zeta))^{\mathsf{T}},\\ \partial_B\Phi_{q+i}(\zeta)^{\mathsf{T}} &\subseteq \rho_2\left[\nabla F_i(\zeta)\;\; \nabla G_i(\zeta)\right]\partial_B\phi_0(F_i(\zeta), G_i(\zeta))^{\mathsf{T}}.\end{aligned} \tag{4.16}$$

Moreover, using Proposition 3.9 and Lemma 4.2 (c), each element in $\partial_B\phi_{\rm FB}(F_i(\zeta), G_i(\zeta))^{\mathsf{T}}$ and $\partial_B\phi_0(F_i(\zeta), G_i(\zeta))^{\mathsf{T}}$ has the form of $\begin{bmatrix}A_i(\zeta) - I\\ B_i(\zeta) - I\end{bmatrix}$ and $\begin{bmatrix}C_i(\zeta)\\ D_i(\zeta)\end{bmatrix}$, respectively, with $A_i(\zeta), B_i(\zeta)$ and $C_i(\zeta), D_i(\zeta)$ for $i = 1, 2, \ldots, q$ characterized as in the proposition. Therefore, combining with Eqs. (4.15)–(4.16) yields the desired result. □

To prove the fast local convergence of nonsmooth Levenberg–Marquardt methods, we need to know that under what assumptions every element $H \in \partial_B\Phi(\zeta^*)$ has full rank n, where ζ^* is an optimal solution of the SOCCP (4.7). To the end, define the index sets

$$\mathcal{I} := \left\{i \in \{1, 2, \ldots, q\} \mid F_i(\zeta^*) = 0,\; G_i(\zeta^*) \in {\rm int}\mathcal{K}^{n_i}\right\},$$
$$\mathcal{B} := \left\{i \in \{1, 2, \ldots, q\} \mid F_i(\zeta^*) \in {\rm bd}^+\mathcal{K}^{n_i},\; G_i(\zeta^*) \in {\rm bd}^+\mathcal{K}^{n_i}\right\},$$

$$\mathcal{J} := \left\{ i \in \{1, 2, \ldots, q\} \mid F_i(\zeta^*) \in \text{int}\mathcal{K}^{n_i},\ G_i(\zeta^*) = 0 \right\}. \tag{4.17}$$

If ζ^* satisfies *strict complementarity*, i.e., $F_i(\zeta^*) + G_i(\zeta^*) \in \text{int}\mathcal{K}^{n_i}$ for all i, then $\{1, 2, \ldots, q\}$ can be partitioned as $\mathcal{I} \cup \mathcal{B} \cup \mathcal{J}$. Thus, suppose that $\nabla G(\zeta^*)$ is invertible, then by rearrangement the matrix $P(\zeta^*) = \nabla G(\zeta^*)^{-1} \nabla F(\zeta^*)$ can be rewritten as

$$P(\zeta^*) = \begin{pmatrix} P(\zeta^*)_{\mathcal{I}\mathcal{I}} & P(\zeta^*)_{\mathcal{I}\mathcal{B}} & P(\zeta^*)_{\mathcal{I}\mathcal{J}} \\ P(\zeta^*)_{\mathcal{B}\mathcal{I}} & P(\zeta^*)_{\mathcal{B}\mathcal{B}} & P(\zeta^*)_{\mathcal{B}\mathcal{J}} \\ P(\zeta^*)_{\mathcal{J}\mathcal{I}} & P(\zeta^*)_{\mathcal{J}\mathcal{B}} & P(\zeta^*)_{\mathcal{J}\mathcal{J}} \end{pmatrix}.$$

Now we have the following results for the full rank of every element $H \in \partial_B \Phi(\zeta^*)$.

Proposition 4.6 *Let ζ^* be a strictly complementary solution to the SOCCP* (4.7) *and $\mathcal{I}, \mathcal{J}, \mathcal{B}$ be index sets described as in* (4.17)*. Suppose that $\nabla G(\zeta^*)$ is invertible and let $P(\zeta^*) = \nabla G(\zeta^*)^{-1} \nabla F(\zeta^*)$. If $P(\zeta^*)_{\mathcal{I}\mathcal{I}}$ is nonsingular and its Schur complement $\widehat{P}(\zeta^*)_{\mathcal{I}\mathcal{I}} := P(\zeta^*)_{\mathcal{B}\mathcal{B}} - P(\zeta^*)_{\mathcal{B}\mathcal{I}} P(\zeta^*)_{\mathcal{I}\mathcal{I}}^{-1} P(\zeta^*)_{\mathcal{I}\mathcal{B}}$, in the matrix*

$$\begin{pmatrix} P(\zeta^*)_{\mathcal{I}\mathcal{I}} & P(\zeta^*)_{\mathcal{I}\mathcal{B}} \\ P(\zeta^*)_{\mathcal{B}\mathcal{I}} & P(\zeta^*)_{\mathcal{B}\mathcal{B}} \end{pmatrix}$$

has the Cartesian P-property, then every element $H \in \partial_B \Phi(\zeta^*)$ has full column rank n.

Proof Let $H \in \partial_B \Phi(\zeta^*)$. By Proposition 4.5, we know $H = \begin{pmatrix} \rho_1 H_1 \\ \rho_2 H_2 \end{pmatrix}$ with H_1^{T} from the set $\partial_B \Phi_1(\zeta^*)^{\mathsf{T}} \times \cdots \times \partial_B \Phi_q(\zeta^*)^{\mathsf{T}}$. From Proposition 3.12, it follows that H_1^{T} is nonsingular under the given assumptions. This implies the desired result rank $(H) = n$. □

The proof of Proposition 4.6 relies on a key property of the first block H_1. However, even when H_1 is singular, the second block H_2 may still contribute to ensuring that the overall matrix H attains full column rank n.

Lemma 4.3 *Let ζ^* be a solution of (4.7) such that all elements in $\partial_B \Phi(\zeta^*)$ have full column rank. Then, there exist constants $\varepsilon > 0$ and $c > 0$ such that $\left\| (H^{\mathsf{T}} H)^{-1} \right\| \leq c$ for all $\|\zeta - \zeta^*\| < \varepsilon$ and all $H \in \partial_B \Phi(\zeta)$. Furthermore, for any given $\bar{\nu} > 0$, $H^{\mathsf{T}} H + \nu I$ are uniformly positive definite for all $\nu \in [0, \bar{\nu}]$ and $H \in \partial_B \Phi(\zeta)$ with $\|\zeta - \zeta^*\| < \varepsilon$.*

Proof The proof is similar to [20, Lemma 2.6]. For completeness, we here include it. Suppose that the claim of the lemma is not true. Then, there exists a sequence $\{\zeta^k\}$ converging to ζ^* and a corresponding sequence of matrices $\{H_k\}$ with $H_k \in \partial_B \Phi(\zeta^k)$ for all $k \in \mathbb{N}$ such that either $H_k^{\mathsf{T}} H_k$ is singular or $\left\| (H_k^{\mathsf{T}} H_k)^{-1} \right\| \to +\infty$ on a subsequence. Noting that $H_k^{\mathsf{T}} H_k$ is symmetric positive semidefinite, for the nonsingular case, we have

$$\left\|(H_k^\top H_k)^{-1}\right\| = \frac{1}{\lambda_{\min}(H_k^\top H_k)},$$

which implies the condition $\left\|(H_k^\top H_k)^{-1}\right\| \to +\infty$ is equivalent to $\lambda_{\min}(H_k^\top H_k) \to 0$. Since $\zeta^k \to \zeta^*$ and the mapping $\zeta \mapsto \partial_B \Phi(\zeta)$ is upper semicontinuous, it follows that the sequence $\{H_k\}$ is bounded, and hence it has a convergent subsequence. Let H_* be a limit of such a sequence. Then, $\lambda_{\min}(H_*^\top H_*) = 0$ by the continuity of the minimum eigenvalue. This means that $H_*^\top H_*$ is singular. However, from the fact that the mapping $\zeta \mapsto \partial_B \Phi(\zeta)$ is closed, we have $H_* \in \partial_B \Phi(\zeta^*)$, which by the given condition implies that $H_*^\top H_*$ is nonsingular. Thus, we obtain a contradiction.

By the definition of matrix norm and the result of the first part, there exist constants $\varepsilon > 0$ and $c > 0$ such that $\left[\lambda_{\min}(H^\top H + \nu I)\right]^{-1} = \left\|(H^\top H + \nu I)^{-1}\right\| \le c$ for all $\nu \in [0, \bar{\nu}]$ and $H \in \partial_B \Phi(\zeta)$ with ζ with $\|\zeta - \zeta^*\| < \varepsilon$. This implies

$$u^\top \left(H^\top H + \nu I\right) u \ge \lambda_{\min}\left(H^\top H + \nu I\right) \|u\|^2 \ge \frac{1}{c}\|u\|^2 \quad \forall\, u \in \mathbb{R}^n.$$

Therefore, the matrices $H^\top H + \nu I$ are uniformly positive definite. □

Lemma 4.4 *Let $\psi : \mathbb{R}^n \times \mathbb{R}^n \to \mathbb{R}_+$ be defined as in (4.14). Then, for any $x, y \in \mathbb{R}^n$,*

(a) $\psi(x, y) = 0 \iff \psi_{\text{FB}}(x, y) = 0 \iff x \in \mathcal{K}^n,\ y \in \mathcal{K}^n,\ \langle x, y\rangle = 0$;
(b) $\psi(x, y)$ *is continuously differentiable;*
(c) $\langle x, \nabla_x \psi(x, y)\rangle + \langle y, \nabla_y \psi(x, y)\rangle \ge 2\psi(x, y)$;
(d) $\langle \nabla_x \psi(x, y), \nabla_y \psi(x, y)\rangle \ge 0$*, and the equality holds if and only if* $\psi(x, y) = 0$;
(e) $\psi(x, y) = 0 \iff \nabla\psi(x, y) = 0 \iff \nabla_x \psi(x, y) = 0 \iff \nabla_y \psi(x, y) = 0$.

Proof Part (a) is direct by the definition of ψ, and part (b) is from Proposition 3.5 and Lemma 4.2(a). We next consider part (c). By the definition of ψ, for any $x, y \in \mathbb{R}^n$,

$$\begin{aligned}
\nabla_x \psi(x, y) &= \rho_1^2 \nabla_x \psi_{\text{FB}}(x, y) + \rho_2^2 \phi_0(x, y) y, \\
\nabla_y \psi(x, y) &= \rho_1^2 \nabla_y \psi_{\text{FB}}(x, y) + \rho_2^2 \phi_0(x, y) x.
\end{aligned} \tag{4.18}$$

From Proposition 3.6 and the definition of $\phi_0(x, y)$, it then follows that

$$\begin{aligned}
&\langle x, \nabla_x \psi(x, y)\rangle + \langle y, \nabla_y \psi(x, y)\rangle \\
&= \rho_1^2 \left[\langle x, \nabla_x \psi_{\text{FB}}(x, y)\rangle + \langle y, \nabla_y \psi_{\text{FB}}(x, y)\rangle\right] + 2\rho_2^2\, \phi_0(x, y) x^\top y \\
&= \rho_1^2 \left\|\phi_{\text{FB}}(x, y)\right\|^2 + 2\rho_2^2\, \phi_0(x, y)^2 \\
&= 2\left(\rho_1^2\, \psi_{\text{FB}}(x, y) + \frac{1}{2}\rho_2^2\, \phi_0(x, y)^2\right) + \rho_2^2\, \phi_0(x, y)^2 \\
&\ge 2\psi(x, y).
\end{aligned}$$

(d) Using the formulas in (4.18) and Proposition 3.6, it follows that

$$\begin{aligned}\langle \nabla_x \psi(x,y), \nabla_y \psi(x,y)\rangle &= \rho_1^4 \langle \nabla_x \psi_{\mathrm{FB}}(x,y), \nabla_y \psi_{\mathrm{FB}}(x,y)\rangle + \rho_2^4 x^\mathsf{T} y \phi_0(x,y)^2 \\ &\quad + \rho_1^2\rho_2^2 \phi_0(x,y)\left[\langle x, \nabla_x \psi_{\mathrm{FB}}(x,y)\rangle + \langle y, \nabla_y \psi_{\mathrm{FB}}(x,y)\rangle\right] \\ &= \rho_1^4 \langle \nabla_x \psi_{\mathrm{FB}}(x,y), \nabla_y \psi_{\mathrm{FB}}(x,y)\rangle + \rho_2^4 \phi_0(x,y)^3 \\ &\quad + 2\rho_1^2\rho_2^2 \phi_0(x,y)\psi_{\mathrm{FB}}(x,y). \end{aligned} \tag{4.19}$$

Note that for the second equality, we use the fact

$$(x^\mathsf{T} y)\,\phi_0(x,y)^2 = (x^\mathsf{T} y)\left(\max\{0, x^\mathsf{T} y\}\right)^2 = \begin{cases} (x^\mathsf{T} y)^3 & \text{if } x^\mathsf{T} y \geq 0, \\ 0 & \text{otherwise,} \end{cases}$$

which says $x^\mathsf{T} y \phi_0(x,y) = \phi_0(x,y)^3$. The first term on the right-hand side of (4.19) is nonnegative by Proposition 3.6, and the last two terms are also nonnegative. Therefore, $\langle \nabla_x \psi(x,y), \nabla_y \psi(x,y)\rangle \geq 0$, and moreover, $\langle \nabla_x \psi(x,y), \nabla_y \psi(x,y)\rangle = 0$ if and only if

$$\left\langle \nabla_x \psi_{\mathrm{FB}}(x,y), \nabla_y \psi_{\mathrm{FB}}(x,y)\right\rangle = 0 \quad \text{and} \quad \phi_0(x,y) = 0,$$

which, together with Proposition 3.6, implies the desired result.

(e) If $\psi(x,y) = 0$, then from the definition of ψ, we have $\phi_{\mathrm{FB}}(x,y) = 0$ and $\phi_0(x,y) = 0$. From Proposition 3.4, we immediately obtain $\nabla_x \psi_{\mathrm{FB}}(x,y) = \nabla_y \psi_{\mathrm{FB}}(x,y) = 0$, and consequently $\nabla_x \psi(x,y) = 0$ and $\nabla_y \psi(x,y) = 0$ by (4.18). If $\nabla \psi(x,y) = 0$, then by part (c) and the nonnegativity of ψ we get $\psi(x,y) = 0$. Thus we prove the first equivalence. For the second equivalence, it suffices to prove the sufficiency. Suppose that $\nabla_x \psi(x,y) = 0$. From part (d), we readily obtain $\psi(x,y) = 0$, which together with part (a) and (4.18) implies $\nabla \psi(x,y) = 0$. Consequently, $\nabla \psi(x,y) = 0 \Longleftrightarrow \nabla_x \psi(x,y) = 0$. Similarly, $\nabla \psi(x,y) = 0 \Longleftrightarrow \nabla_y \psi(x,y) = 0$. This implies the last equivalence. □

From Lemma 4.4(b), it is clear that the function Ψ is continuously differentiable. In addition, in light of Lemma 4.4(d), we shall prove that every stationary point of Ψ is a solution of (4.7) under mild conditions. To this end, we recall that, two matrices $M_1, M_2 \in \mathbb{R}^{n\times n}$ are called column monotone if, for any $u, v \in \mathbb{R}^n$, $M_1 u + M_2 v = 0 \Rightarrow u^\mathsf{T} v = 0$.

Proposition 4.7 *Let $\Psi : \mathbb{R}^n \to \mathbb{R}_+$ be defined by (4.13)–(4.14). Then, every stationary point of Ψ is a solution of the SOCCP (4.7) under one of the following assumptions:*

(a) $\nabla F(\zeta)$ and $-\nabla G(\zeta)$ are column monotone for any $\zeta \in \mathbb{R}^n$.
(b) For any $\zeta \in \mathbb{R}^n$, $\nabla G(\zeta)$ is invertible and $\nabla G(\zeta)^{-1}\nabla F(\zeta)$ has Cartesian P_0-property.

Proof When the assumption (a) is satisfied, using the same arguments as those of [18, Proposition 3] yields the desired result. Now suppose that the assumption (b) holds. Let $\bar{\zeta}$ be an arbitrary stationary point of Ψ and write

$$\begin{aligned}\nabla_x\psi(F(\zeta),G(\zeta)) &= \left(\nabla_{x_1}\psi(F_1(\zeta),G_1(\zeta)),\ldots,\nabla_{x_q}\psi(F_q(\zeta),G_q(\zeta))\right),\\ \nabla_y\psi(F(\zeta),G(\zeta)) &= \left(\nabla_{y_1}\psi(F_1(\zeta),G_1(\zeta)),\ldots,\nabla_{y_q}\psi(F_q(\zeta),G_q(\zeta))\right).\end{aligned}$$

Then,

$$\nabla\Psi(\bar{\zeta}) = \nabla F(\bar{\zeta})\nabla_x\psi(F(\bar{\zeta}),G(\bar{\zeta})) + \nabla G(\bar{\zeta})\nabla_y\psi(F(\bar{\zeta}),G(\bar{\zeta})) = 0,$$

which, by the invertibility of ∇G, can be rewritten as

$$\nabla G(\bar{\zeta})^{-1}\nabla F(\bar{\zeta})\nabla_x\psi(F(\bar{\zeta}),G(\bar{\zeta})) + \nabla_y\psi(F(\bar{\zeta}),G(\bar{\zeta})) = 0. \tag{4.20}$$

Suppose that $\bar{\zeta}$ is not the solution of (4.7). By Lemma 4.4(e), we necessarily have

$$\nabla_x\psi(F(\bar{\zeta}),G(\bar{\zeta})) \neq 0.$$

From the Cartesian P_0-property of $\nabla G(\bar{\zeta})^{-1}\nabla F(\bar{\zeta})$, there exists an index $\nu \in \{1,2,\ldots,q\}$ such that $\nabla_{x_\nu}\psi(F_\nu(\bar{\zeta}),G_\nu(\bar{\zeta})) \neq 0$ and

$$\left\langle\nabla_{x_\nu}\psi(F_\nu(\bar{\zeta}),G_\nu(\bar{\zeta})),\left[\nabla G(\bar{\zeta})^{-1}\nabla F(\bar{\zeta})\nabla_x\psi(F(\bar{\zeta}),G(\bar{\zeta}))\right]_\nu\right\rangle \geq 0. \tag{4.21}$$

In addition, notice that (4.20) is equivalent to

$$\left[\nabla G(\bar{\zeta})^{-1}\nabla F(\bar{\zeta})\nabla_x\psi(F(\bar{\zeta}),G(\bar{\zeta}))\right]_i + \nabla_{y_i}\psi(F_i(\bar{\zeta}),G_i(\bar{\zeta})) = 0, \quad i = 1,2,\ldots,q.$$

Making the inner product with $\nabla_{x_\nu}\psi(F(\bar{\zeta}),G(\bar{\zeta}))$ for the νth equality, we obtain

$$\begin{aligned}&\left\langle\nabla_{x_\nu}\psi(F_\nu(\bar{\zeta}),G_\nu(\bar{\zeta})),\left[\nabla G(\bar{\zeta})^{-1}\nabla F(\bar{\zeta})\nabla_x\psi(F(\bar{\zeta}),G(\bar{\zeta}))\right]_\nu\right\rangle\\ &+\langle\nabla_{x_\nu}\psi(F_\nu(\bar{\zeta}),G_\nu(\bar{\zeta})),\nabla_{y_\nu}\psi(F_\nu(\bar{\zeta}),G_\nu(\bar{\zeta}))\rangle = 0.\end{aligned}$$

The first term on the left-hand side is nonnegative by (4.21), whereas the second term is positive by Lemma 4.4(d) since ζ is not a solution of (4.7). This leads to a contradiction, and consequently $\bar{\zeta}$ must be a solution of (4.7). □

When $\nabla G(\zeta)$ is invertible for any $\zeta \in \mathbb{R}^n$, the assumption in Proposition 4.7(a) is equivalent to the positive semidefiniteness of $\nabla G(\zeta)^{-1}\nabla F(\zeta)$ at any $\zeta \in \mathbb{R}^n$, which implies the Cartesian P_0-property of $\nabla G(\zeta)^{-1}\nabla F(\zeta)$. Thus, when it reduces to the SOCCP (3.1), the assumption in Proposition 4.7(a) is stronger than the assumption in Proposition 4.7(b), which is now equivalent to the Cartesian P_0-property of F. Next we provide a condition to guarantee the boundedness of the level sets of Ψ

$$\mathcal{L}_\Psi(\gamma) := \{\zeta \in \mathbb{R}^n \mid \Psi(\zeta) \le \gamma\}$$

for all $\gamma \ge 0$. This property is important since it guarantees that the descent sequence of Ψ must have a limit point, and furthermore, the solution set of (4.7) is bounded if it is nonempty. It turns out that the following condition for F and G is sufficient.

Condition 4.1 For any sequence $\{\zeta^k\}$ satisfying $\|\zeta^k\| \to +\infty$, whenever

$$\limsup \left\|[-F(\zeta^k)]_+\right\| < +\infty \quad \text{and} \quad \limsup \left\|[-G(\zeta^k)]_+\right\| < +\infty, \tag{4.22}$$

there exists an index $\nu \in \{1, 2, \ldots, q\}$ such that $\limsup \langle F_\nu(\zeta^k), G_\nu(\zeta^k)\rangle = +\infty$.

Proposition 4.8 *If the mappings F and G satisfy Condition 4.1, then the level sets $\mathcal{L}_\Psi(\gamma)$ are bounded for all $\gamma \ge 0$.*

Proof Assume that there is a unbounded sequence $\{\zeta^k\} \subseteq \mathcal{L}_\Psi(\gamma)$ for some $\gamma \ge 0$. Since $\Psi(\zeta^k) \le \gamma$ for all k, the sequence $\{\Psi_{\text{FB}}(\zeta^k)\}$ is bounded. By Lemma 3.7,

$$\limsup \left\|[-F_i(x^k)]_+\right\| < +\infty \quad \text{and} \quad \limsup \left\|[-G_i(x^k)]_+\right\| < +\infty$$

hold for all $i \in \{1, 2, \ldots, q\}$. This shows that F and G satisfy Condition 4.1, and hence there exists an index ν such that $\limsup \langle F_\nu(\zeta^k), G_\nu(\zeta^k)\rangle = +\infty$. From the definition of Ψ, it follows that the sequence $\{\Psi(\zeta^k)\}$ is unbounded, which clearly contradicts the fact that $\{\zeta^k\} \subseteq \mathcal{L}_\Psi(\gamma)$. Thus, the proof is complete. □

Condition 4.1 is relatively mild for ensuring that Ψ has bounded level sets. As will be shown below, this condition is satisfied under various settings, including jointly monotone functions with a strictly feasible point, used in Sect. 3.1.4 for f_{LT} and f_{YF}, the jointly uniform Cartesian P-functions with a feasible point, and the joint $\tilde{R}_{01}$-functions (see Definition 1.16).

Proposition 4.9 *Condition 4.1 is satisfied if one of the following assumptions holds:*

(a) F and G are jointly monotone mappings satisfying $\lim_{\|\zeta\| \to +\infty} \|F(\zeta)\| + \|G(\zeta)\| = +\infty$, and there exists $\hat{\zeta} \in \mathbb{R}^n$ such that $F(\hat{\zeta}), G(\hat{\zeta}) \in \text{int}(\mathcal{K})$.
(b) F and G have jointly uniform Cartesian P-property, and there exists a point $\hat{\zeta} \in \mathbb{R}^n$ such that $F(\hat{\zeta}), G(\hat{\zeta}) \in \mathcal{K}$.
(c) F and G have the joint $\tilde{R}_{01}$-property.

Proof In the proof, let $\{\zeta^k\}$ be a sequence such that $\|\zeta^k\| \to +\infty$ and (4.22) holds.

(a) First, $\{\lambda_1[F(\zeta^k)]\}$ and $\{\lambda_1[G(\zeta^k)]\}$ must be bounded from below. If not, using

$$\|[-x]_+\|^2 = (\max\{0, -\lambda_1(x)\})^2 + (\max\{0, -\lambda_2(x)\})^2,$$

we obtain $\limsup \|[-F(\zeta^k)]_+\| = +\infty$ or $\limsup \|[-G(\zeta^k)]_+\| = +\infty$, which contradicts the assumption that $\{\zeta^k\}$ satisfies (4.22). Noting that $\|F(\zeta^k)\| + \|G(\zeta^k)\| \to +\infty$ and

$$\|F(\zeta^k)\| + \|G(\zeta^k)\| = \sqrt{\frac{\lambda_1^2[F(\zeta^k)] + \lambda_2^2[F(\zeta^k)]}{2}} + \sqrt{\frac{\lambda_1^2[G(\zeta^k)] + \lambda_2^2[G(\zeta^k)]}{2}},$$

the lower boundness of $\{\lambda_i[F(\zeta^k)]\}$ and $\{\lambda_i[G(\zeta^k)]\}$ for $i = 1, 2$ implies

$$\limsup \lambda_2\left[F(\zeta^k)\right] = +\infty \quad \text{or} \quad \limsup \lambda_2\left[G(\zeta^k)\right] = +\infty.$$

From the proof of Lemma 3.15 (b), it then follows that

$$\limsup\left\{\langle F(\zeta^k), G(\hat{\zeta})\rangle + \langle F(\hat{\zeta}), G(\zeta^k)\rangle\right\} = +\infty. \tag{4.23}$$

Now suppose that Condition 4.1 is not satisfied. Then, we necessarily have

$$\limsup\langle F_i(\zeta^k), G_i(\zeta^k)\rangle < +\infty \quad \text{for all } i = 1, 2, \ldots, q.$$

In addition, from the joint monotonicity of F and G, we have

$$\begin{aligned}\langle F(\zeta^k), G(\hat{\zeta})\rangle + \langle F(\hat{\zeta}), G(\zeta^k)\rangle \leq &\langle F(\zeta^k), G(\zeta^k)\rangle + \langle F(\hat{\zeta}), G(\hat{\zeta})\rangle \\ = &\sum_{i=1}^{q}\langle F_i(\zeta^k), G_i(\zeta^k)\rangle + \langle F(\hat{\zeta}), G(\hat{\zeta})\rangle.\end{aligned}$$

The last two equations imply $\limsup\{\langle F(\zeta^k), G(\hat{\zeta})\rangle + \langle F(\hat{\zeta}), G(\zeta^k)\rangle\} < +\infty$. This clearly contradicts (4.23), and consequently the desired result follows.

(b) From Definition 1.10, there exists a constant $\rho > 0$ such that

$$\begin{aligned}\rho\|\zeta^k - \hat{\zeta}\|^2 \leq &\max_{i\in\{1,\ldots,q\}}\left\{\langle F_i(\zeta^k) - F_i(\hat{\zeta}), G_i(\zeta^k) - G_i(\hat{\zeta})\rangle\right\} \\ = &\langle F_\nu(\zeta^k), G_\nu(\zeta^k)\rangle + \langle F_\nu(\hat{\zeta}), -G_\nu(\zeta^k)\rangle \\ &+ \langle -F_\nu(\zeta^k), G_\nu(\hat{\zeta})\rangle + \langle F_\nu(\hat{\zeta}), G_\nu(\hat{\zeta})\rangle \\ \leq &\langle F_\nu(\zeta^k), G_\nu(\zeta^k)\rangle + \langle F_\nu(\hat{\zeta}), [-G_\nu(\zeta^k)]_+\rangle \\ &+ \langle [-F_\nu(\zeta^k)]_+, G_\nu(\hat{\zeta})\rangle + \langle F(\hat{\zeta}), G_\nu(\hat{\zeta})\rangle,\end{aligned}$$

where ν is one of the indices for which the max is attained which we have, without loss of generality, assumed to be independent of k, and the second inequality is since

$$F_\nu(\hat{\zeta}) \in \mathcal{K}^{n_\nu}, \;\; G_\nu(\hat{\zeta}) \in \mathcal{K}^{n_\nu}, \;\; [-F_\nu(\zeta^k)]_- \in -\mathcal{K}^{n_\nu}, \;\; [-G_\nu(\zeta^k)]_- \in -\mathcal{K}^{n_\nu}.$$

Dividing the last inequality by $\|\zeta^k\|^2$ and taking the limit, it follows from (4.22) that

$$\lim_{k\to+\infty} \frac{\langle F_\nu(\zeta^k), G_\nu(\zeta^k)\rangle}{\|\zeta^k\|^2} \geq \rho > 0,$$

which immediately implies the result.

(c) Clearly, $\{\zeta^k\}$ satisfies (1.58), and the result then follows from the following implications:

$$\liminf_{k\to+\infty} \frac{\langle F(\zeta^k), G(\zeta^k)\rangle}{\|\zeta^k\|} > 0 \Longrightarrow \liminf_{k\to+\infty} \frac{\max_i\{\langle F_i(\zeta^k), G_i(\zeta^k)\rangle\}}{\|\zeta^k\|} > 0$$
$$\Longrightarrow \max_i\{\langle F_i(\zeta^k), G_i(\zeta^k)\rangle\} \to +\infty.$$

Thus, we complete the proof of this proposition. □

When $G(\zeta) \equiv \zeta$, if we replace (1.59) with $\liminf_{k\to+\infty} \langle F(\zeta^k), G(\zeta^k)\rangle/\|\zeta^k\|^2 > 0$, then Definition 1.16 indicates that F is a R_{01} function. Thus, Proposition 4.8 and Proposition 4.9 (a) show that Ψ has bounded level sets under a weaker condition than the one given by Proposition 3.39 for the class of merit functions f_{YF}. Now, we show that the function Ψ provides a global error bound for the solution of the SOCCP (4.7) under the jointly uniform Cartesian P-property of F and G. Since the jointly strong monotonicity implies the jointly uniform Cartesian P-property, the global error bound condition is weaker than the ones for Proposition 3.38 and Proposition 3.48.

Proposition 4.10 *Let ζ^* be a solution of the SOCCP (4.7). Suppose that F and G have the jointly uniform Cartesian P-property. Then, there exists a scalar $\kappa > 0$ such that*

$$\|\zeta - \zeta^*\|^2 \leq \kappa\Psi(\zeta)^{1/2} \quad \forall \zeta \in \mathbb{R}^n.$$

Proof Since F and G have the jointly uniform Cartesian P-property, there exists a scalar $\rho > 0$ such that, for any $\zeta \in \mathbb{R}^n$, there is an index $\nu \in \{1, 2, \ldots, q\}$ such that

$$\begin{aligned}
\rho\|\zeta - \zeta^*\|^2 &\leq \langle F_\nu(\zeta) - F_\nu(\zeta^*), G_\nu(\zeta) - G_\nu(\zeta^*)\rangle \\
&= \langle F_\nu(\zeta), G_\nu(\zeta)\rangle + \langle -F_\nu(\zeta), G_\nu(\zeta^*)\rangle + \langle F_\nu(\zeta^*), -G_\nu(\zeta)\rangle \\
&\leq \langle F_\nu(\zeta), G_\nu(\zeta)\rangle + \langle [-F_\nu(\zeta)]_+, G_\nu(\zeta^*)\rangle + \langle F_\nu(\zeta^*), [-G_\nu(\zeta)]_+\rangle \\
&\leq \phi_0(F_\nu(\zeta), G_\nu(\zeta)) + \|[-F_\nu(\zeta)]_+\|\|G_\nu(\zeta^*)\| + \|F_\nu(\zeta^*)\|\|[-G_\nu(\zeta)]_+\| \\
&\leq c\Big(\phi_0(F_\nu(\zeta), G_\nu(\zeta)) + \|[-F_\nu(\zeta)]_+\| + \|[-G_\nu(\zeta)]_+\|\Big) \\
&\leq c\Big(\phi_0(F_\nu(\zeta), G_\nu(\zeta)) + 4\psi_{\text{FB}}(F_\nu(\zeta), G_\nu(\zeta))^{1/2}\Big) \\
&\leq c\left(\sqrt{2}/\rho_2 + 4/\rho_1\right)\Psi(\zeta)^{1/2},
\end{aligned}$$

where $c := \max\{1, \|G_\nu(\zeta^*)\|, \|F_\nu(\zeta^*)\|\}$, the second inequality is using the fact that $G_\nu(\zeta^*) \in \mathcal{K}^{n_\nu}$ and $F_\nu(\zeta^*) \in \mathcal{K}^{n_\nu}$, and the next to last inequality is due to Lemma 3.7. Letting $\kappa := (c/\rho)(\sqrt{2}/\rho_2 + 4/\rho_1)$, we obtain the desired result. □

It is well known that the Levenberg–Marquardt method based on Equation (4.12) offers the advantage of reducing the complementarity gap $\langle x, F(x)\rangle$ for the NCP more effectively than the traditional nonsmooth method using Equation (4.8) (see [17]). This observation motivates our adoption of a Levenberg–Marquardt-type method with line search for solving the nonlinear least-squares problem (4.13). The iterative scheme is presented below.

Algorithm 4.2 (Semismooth Levenberg–Marquardt Method)

(S.0) Choose a starting point $\zeta^0 \in \mathbb{R}^n$, the parameters $\rho_1, \rho_2 \in (0, 1)$, $\eta, \beta \in (0, 1)$, and $\sigma \in (0, 1/2)$. Given a tolerance $\varepsilon \geq 0$, and set $k := 0$.

(S.1) If $\|\nabla\Psi(\zeta^k)\| \leq \varepsilon$, then stop.

(S.2) Choose $H_k \in \partial_B\Phi(\zeta^k)$ and $\nu_k > 0$. Find a solution $d^k \in \mathbb{R}^n$ of linear system

$$\left(H_k^\top H_k + \nu_k I\right) d = -\nabla\Psi(\zeta^k), \tag{4.24}$$

where $\nu_k > 0$ is the Levenberg–Marquardt parameter.

(S.3) If d^k satisfies

$$\|\Phi(\zeta^k + d^k)\| \leq \eta \|\Phi(\zeta^k)\|, \tag{4.25}$$

then $\zeta^{k+1} := \zeta^k + d^k$. Otherwise, compute $t_k = \max\{\beta^l \mid l = 0, 1, 2, \cdots\}$ such that

$$\Psi(\zeta^k + t_k d^k) \leq \Psi(\zeta^k) + \sigma t_k \nabla\Psi(\zeta^k)^\top d^k, \tag{4.26}$$

and let $\zeta^{k+1} := \zeta^k + t_k d^k$.

(S.4) Set $k := k + 1$, and go to (S.1).

Notice that the above method is different from the classical Levenberg–Marquardt method for nonlinear least-square problems in that Φ is not continuously differentiable. If $\nu_k \equiv 0$, the solution of (4.24) is exactly the solution of the linear least-square problem

$$\min_{d\in\mathbb{R}^n} \frac{1}{2} \|H_k d + \Phi(\zeta^k)\|^2,$$

since $\nabla\Psi(\zeta^k) = H_k^\top \Phi(\zeta^k)$. In this algorithm, we choose the parameter ν_k by

$$\nu_k := \min\left\{p_1, p_2\|\Phi(\zeta^k)\|^\varrho\right\}, \tag{4.27}$$

where $p_1, p_2 > 0$ are given constants and ϱ is a real number from interval $[1, 2]$. Such choice is consistent with the requirements for local superlinear (quadratic) convergence stated in Proposition 4.12 and Proposition 4.13, as well as adopted by our numerical experiments.

In the following, we examine the convergence properties of the proposed algorithm. To facilitate this analysis, we assume $\varepsilon = 0$. We begin by presenting a result on global convergence.

Proposition 4.11 *Let $\{\zeta^k\}$ be the sequence generated by Algorithm 4.2 with ν_k updated by (4.27). Then every accumulation point of $\{\zeta^k\}$ is a stationary point of Ψ.*

Proof From the steps of Algorithm 4.2, $\{\zeta^k\}$ is well defined since $\nu_k > 0$, and d^k determined by (4.24) is always a descent direction of Ψ at ζ^k. Let ζ^* be any accumulation point of $\{\zeta^k\}$ and $\{\zeta^k\}_K$ be a subsequence converging to ζ^*. Suppose that $\nabla\Psi(\zeta^*) \neq 0$. Since $\{\Psi(\zeta^k)\}$ is monotonically decreasing and bounded below, and $\{\Psi(\zeta^k)\}_K$ converges to $\Psi(\zeta^*)$, the entire sequence $\{\Psi(\zeta^k)\}$ converges to $\Psi(\zeta^*) > 0$. This implies that (4.25) holds for only finitely many $k \in K$, and the inequality (4.26) is satisfied for all sufficiently large k. Since $\Psi(\zeta^{k+1}) - \Psi(\zeta^k) \leq \sigma t_k \nabla\Psi(\zeta^k)^\top d^k \leq 0$ for all sufficiently large k, using $\Psi(\zeta^{k+1}) - \Psi(\zeta^k) \to 0$ yields

$$\left\{t_k \nabla\Psi(\zeta^k)^\top d^k\right\}_K \to 0. \tag{4.28}$$

We next prove $\left\{\nabla\Psi(\zeta^k)^\top d^k\right\}_K$ has a nonzero limit as $k \to +\infty$. By the definition of d^k,

$$\nabla\Psi(\zeta^k)^\top d^k = -\nabla\Psi(\zeta^k)^\top \left(H_k^\top H_k + \nu_k I\right)^{-1} \nabla\Psi(\zeta^k) \quad \forall k. \tag{4.29}$$

Since the B-subdifferential $\partial_B\Phi(\zeta)$ is a nonempty compact set for any $\zeta \in \mathbb{R}^n$, $\{H_k\}_K$ is bounded. Without loss of generality, assume that $\{H_k\}_K \to H_*$. Considering that the set-valued mapping $\zeta \mapsto \partial_B\Phi(\zeta)$ is closed and $\{\zeta^k\}_K \to \zeta^*$, we have $H_* \in \partial_B\Phi(\zeta^*)$. In addition, since $\Phi(\zeta^*) \neq 0$, we have $\nu_k \to \nu_*$ with $\nu_* = \min\{p_1, p_2\|\Phi(\zeta^*)\|^\varrho\} > 0$. Thus, $\{H_k^\top H_k + \nu_k I\}_{k\in K} \to H_*^\top H_* + \nu_* I \succ O$. This, together with (4.29) and the continuity of $\nabla\Psi$, implies that $\{\nabla\Psi(\zeta^k)^\top d^k\}_K$ has a nonzero limit as $k \to +\infty$. From (4.28), it then follows that $\{t_k\}_K \to 0$. Now, for all sufficiently large k, let $l_k \in \{0, 1, \ldots\}$ be the unique exponent such that $t_k = \beta^{l_k}$. Since $\{t_k\}_K \to 0$, we have $\{l_k\}_{k\in K} \to \infty$. From the Armijo line search in (S.3), for all $k \in K$ sufficiently large,

$$\frac{\Psi(\zeta^k + \beta^{l_k-1}d^k) - \Psi(\zeta^k)}{\beta^{l_k-1}} > \sigma\nabla\Psi(\zeta^k)^\top d^k.$$

Taking the limit $k \to \infty$ with $k \in K$ and using $\{l_k\}_K \to \infty$ and $\{\zeta^k\}_K \to \zeta^*$, we have $\nabla\Psi(\zeta^*)^\top d^* \geq \sigma\nabla\Psi(\zeta^*)^\top d^*$. This means $\nabla\Psi(\zeta^*)^\top d^* \geq 0$. On the other hand, we learn from (4.24) that $\{d^k\}_K \to d^*$ with d^* being the solution of

$$\left(H_*^\top H_* + \nu_* I\right) d = -\nabla\Psi(\zeta^*),$$

which implies that $\nabla\Psi(\zeta^*)^T d^* < 0$ since $\left(H_*^{\mathsf{T}} H_* + \nu_* I\right) \succ O$. Thus, we obtain a contradiction. □

Observe that the sequence $\{\zeta^k\}$ generated by Algorithm 4.2 always belongs to the level set $\mathcal{L}_\Psi(\Psi(\zeta^0))$. By Proposition 4.8 and Proposition 4.9, the existence of accumulation points of $\{\zeta^k\}$ is guaranteed by one of the assumptions of Proposition 4.9. Since, when F and G have the jointly uniform Cartesian P-property, the SOCCP (4.7) has at most one solution, $\{\zeta^k\}$ must have a unique accumulation point which is the unique solution of (4.7) if F and G satisfy the assumption (c) of Proposition 4.9. For the SOCCP (3.1), the sequence $\{\zeta^k\}$ has accumulation points and each of them is a solution under the assumption that F is monotone and (3.1) is strictly feasible.

We now establish the superlinear (or quadratic) convergence rate of Algorithm 4.2 under the assumption of strict complementarity at the solution. While this condition may appear somewhat stringent, we will subsequently relax it by employing a local error bound assumption.

Proposition 4.12 *Let $\{\zeta^k\}$ be generated by Algorithm 4.2 with ν_k given by (4.27). Suppose that ζ^* is an accumulation point of $\{\zeta^k\}$ with ζ^* being a strictly complementary solution of (4.7), and F and G at ζ^* satisfy the condition of Proposition 4.6. Then,*

(a) The entire sequence $\{\zeta^k\}$ converges to ζ^*.
(b) The full step size $t_k = 1$ is always accepted for sufficiently large k and the rate of convergence is Q-superlinear.
(c) The rate of convergence is Q-quadratic if, in addition, F' and G' are locally Lipschitz continuous around ζ^* and $\nu_k = O(\|\Phi(\zeta^k)\|)$.

Proof The proof is similar to the one given by [17]. For completeness, we include it.

(a) By the proof technique of [21, Theorem 3.1 (b)], it suffices to prove that ζ^* is an isolated solution. From Proposition 4.6 and Lemma 4.3, there exist $\varepsilon_1, \kappa_1 > 0$ such that

$$\|H(\zeta - \zeta^*)\|^2 = (\zeta - \zeta^*)H^{\mathsf{T}}H(\zeta - \zeta^*) \geq \kappa_1\|\zeta - \zeta^*\|^2$$

for all ζ satisfying $\|\zeta - \zeta^*\| < \varepsilon_1$ and all $H \in \partial_B\Phi(\zeta)$. In addition, the semismoothness of Φ implies that there exists $\varepsilon_2 > 0$ such that

$$\|\Phi(\zeta) - \Phi(\zeta^*) - H(\zeta - \zeta^*)\| \leq (\sqrt{\kappa_1}/2)\|\zeta - \zeta^*\|$$

for all $H \in \partial_B\Phi(\zeta)$ with ζ satisfying $\|\zeta - \zeta^*\| < \varepsilon_2$. Set $\varepsilon = \min\{\varepsilon_1, \varepsilon_2\}$. Then, we have

$$\|\Phi(\zeta)\| = \|H(\zeta - \zeta^*) + (\Phi(\zeta) - \Phi(\zeta^*) - H(\zeta - \zeta^*))\|$$

$$\geq \|H(\zeta - \zeta^*)\| - \|\Phi(\zeta) - \Phi(\zeta^*) - H(\zeta - \zeta^*)\|$$
$$\geq (\sqrt{\kappa_1}/2)\|\zeta - \zeta^*\|$$

for all ζ with $\|\zeta - \zeta^*\| < \varepsilon$. This means that ζ^* is an isolated solution of the SOCCP.

(b) We first prove that for all sufficiently large k,

$$\|\zeta^k + d^k - \zeta^*\| = o(\|\zeta^k - \zeta^*\|). \tag{4.30}$$

By part (a), the sequence $\{\zeta^k\}$ converges to a solution ζ^* satisfying the assumptions of Theorem 4.6. From Lemma 4.3, there exists $c > 0$ such that $\left\|\left(H_k^{\mathsf{T}} H_k + \nu_k I\right)^{-1}\right\| \leq c$ for all k. Noting that the sequence $\{H_k\}$ is bounded, there exists $c_1 > 0$ such that $\|H_k^{\mathsf{T}}\| \leq c_1$ for all k. Using Proposition 4.11 and the fact that $\Phi(\zeta^*) = 0$, we obtain

$$\begin{aligned}
\|\zeta^k + d^k - \zeta^*\| &= \|\zeta^k - (H_k^{\mathsf{T}} H_k + \nu_k I)^{-1}\nabla\Psi(\zeta^k) - \zeta^*\| \\
&\leq \|(H_k^{\mathsf{T}} H_k + \nu_k I)^{-1}\|\|\nabla\Psi(\zeta^k) - (H_k^{\mathsf{T}} H_k + \nu_k I)(\zeta^k - \zeta^*)\| \\
&\leq c\|H_k^{\mathsf{T}}\Phi(\zeta^k) - H_k^{\mathsf{T}} H_k(\zeta^k - \zeta^*) - \nu_k(\zeta^k - \zeta^*)\| \\
&= c\|H_k^{\mathsf{T}}(\Phi(\zeta^k) - \Phi(\zeta^*) - H_k(\zeta^k - \zeta^*)) - \nu_k(\zeta^k - \zeta^*)\| \\
&\leq c(c_1\|\Phi(\zeta^k) - \Phi(\zeta^*) - H_k(\zeta^k - \zeta^*)\| + \nu_k\|\zeta^k - \zeta^*\|).
\end{aligned}$$

Notice that $\Phi(\zeta^k) - \Phi(\zeta^*) - H_k(\zeta^k - \zeta^*) = o(\|\zeta^k - \zeta^*\|)$ by the semismoothness of Φ, whereas $\nu_k \to 0$ by part (a) and the continuity of Φ. Thus, the inequality implies (4.30).

To prove that the full step is eventually accepted, by (4.25), it suffices to show that

$$\lim_{k\to\infty} \frac{\Psi(\zeta^k + d^k)}{\Psi(\zeta^k)} = 0.$$

Since all elements $V \in \partial_B \Phi_{\mathrm{FB}}(\zeta^*)$ are nonsingular by Proposition 3.12, from Lemma 4.3 and the proof of part (a), there exists a constant $\alpha > 0$ such that

$$\|\Phi(\zeta^k)\| \geq \rho_1\|\Phi_{\mathrm{FB}}(\zeta^k)\| \geq \alpha\|\zeta^k - \zeta^*\|.$$

Using the locally Lipschitz continuity of Φ and (4.30) then yields

$$\frac{\|\Phi(\zeta^k + d^k)\|}{\|\Phi(\zeta^k)\|} \leq \frac{\|\Phi(\zeta^k + d^k) - \Phi(\zeta^*)\|}{\alpha\|\zeta^k - \zeta^*\|} \leq \frac{L\|\zeta^k + d^k - \zeta^*\|}{\alpha\|\zeta^k - \zeta^*\|} \to 0,$$

where $L > 0$ denotes the locally Lipschitz constant of Φ. Thus, the step size $t_k = 1$ is eventually accepted in the line search criterion, i.e., $\zeta^{k+1} = \zeta^k + d^k$ for all k large enough. Consequently, Q-suplinear convergence of $\{\zeta^k\}$ to ζ^* follows from (4.30).

(c) The proof is essentially same as for the superlinear convergence. We only note that ν_k in (4.27) satisfies $\nu_k = O(\|\Phi(\zeta^k)\|) = O(\|\zeta^k - \zeta^*\|)$ for k large enough, and

$$\Phi(\zeta^k) - \Phi(\zeta^*) - H_k(\zeta^k - \zeta^*) = O(\|\zeta^k - \zeta^*\|^2)$$

due to the strong semismoothness of Φ by Proposition 4.4. □

Assumption A. There exist constants $\kappa_2 > 0$ and $0 < \delta < 1$ such that

$$\kappa_2 \operatorname{dist}(\zeta, S^*) \leq \|\Phi(\zeta)\| \quad \forall \zeta \in \mathcal{N}(\zeta^*, \delta), \tag{4.31}$$

where S^* denotes the solution set of the SOCCP (4.7) and is assumed to be nonempty.

Lemma 4.5 *Let ζ^k be generated by Algorithm 4.2 with ν_k given by (4.27). Suppose that F' and G' are Lipschitz continuous on $\mathcal{N}(\zeta^*, \delta)$ and Assumption A holds. If $\nu_k = p_2\|\Phi(\zeta^k)\|^{\varrho}$ and $\zeta^k \in \mathcal{N}(\zeta^*, \delta/2)$, then there exists a constant $c_1 > 0$ such that $\|d^k\| \leq c_1 \operatorname{dist}(\zeta^k, S^*)$. If, in addition, $\zeta^k + d^k \in \mathcal{N}(\zeta^*, \delta/2)$, then there exists a constant $c_3 > 0$ such that*

$$\operatorname{dist}\left(\zeta^k + d^k, S^*\right) \leq c_3 \operatorname{dist}\left(\zeta^k, S^*\right)^{(\varrho+2)/2}.$$

Proof Let $\bar{\zeta}^k \in S^*$ be such that $\left\|\zeta^k - \bar{\zeta}^k\right\| = \operatorname{dist}(\zeta^k, S^*)$. Then, $\bar{\zeta}^k \in \mathcal{N}(\zeta^*, \delta)$ since

$$\left\|\bar{\zeta}^k - \zeta^*\right\| \leq \left\|\bar{\zeta}^k - \zeta^k\right\| + \left\|\zeta^k - \zeta^*\right\| \leq 2\left\|\zeta^k - \zeta^*\right\| \leq \delta.$$

Noting that Φ is Lipschitz continuous on $\mathcal{N}(\zeta^*, \delta)$, there is a constant $L_1 > 0$ such that

$$\left\|\Phi(\zeta^k)\right\| = \left\|\Phi(\zeta^k) - \Phi(\bar{\zeta}^k)\right\| \leq L_1 \left\|\zeta^k - \bar{\zeta}^k\right\|.$$

Combining with the inequality (4.31), we have

$$p_2 \kappa_2^{\varrho} \left\|\bar{\zeta}^k - \zeta^k\right\|^{\varrho} \leq \nu_k = p_2 \left\|\Phi(\zeta^k)\right\|^{\varrho} \leq p_2 L_1^{\varrho} \left\|\zeta^k - \bar{\zeta}^k\right\|^{\varrho}. \tag{4.32}$$

On the other hand, since Φ is strongly semismooth on $\mathcal{N}(\zeta^*, \delta)$ by Proposition 4.4, there exists a constant $\hat{c} > 0$ such that

$$\left\|\Phi(\zeta^k) + H_k(\bar{\zeta}^k - \zeta^k)\right\| = \left\|\Phi(\zeta^k) - \Phi(\bar{\zeta}^k) - H_k(\zeta^k - \bar{\zeta}^k)\right\| \leq \hat{c} \left\|\zeta^k - \bar{\zeta}^k\right\|^2. \tag{4.33}$$

Now, define

$$\varphi_k(d) := \left\|\Phi(\zeta^k) + H_k d\right\|^2 + \nu_k \|d\|^2. \tag{4.34}$$

Then, from (4.34), it is clear to check that d^k is a minimizer of $\varphi_k(d)$. This together with (4.33) and (4.32) yields

$$
\begin{aligned}
\|d^k\|^2 \le \frac{\varphi_k(d^k)}{\nu_k} \le \frac{\varphi_k(\bar{\zeta}^k - \zeta^k)}{\nu_k} &= \frac{\left\|\Phi(\zeta^k) + H_k(\bar{\zeta}^k - \zeta^k)\right\|^2 + \nu_k \left\|\bar{\zeta}^k - \zeta^k\right\|^2}{\nu_k} \\
&\le \hat{c}^2 p_2^{-1} \kappa_2^{-\varrho} \left\|\bar{\zeta}^k - \zeta^k\right\|^{4-\varrho} + \left\|\bar{\zeta}^k - \zeta^k\right\|^2 \\
&= \left(\hat{c}^2 p_2^{-1} \kappa_2^{-\varrho} + 1\right) \left\|\bar{\zeta}^k - \zeta^k\right\|^2,
\end{aligned}
$$

which implies the first part with $c_1 = \sqrt{\hat{c}^2 p_2^{-1} \kappa_2^{-\varrho} + 1}$. Noting that

$$
\begin{aligned}
\varphi_k(d^k) \le \varphi_k(\bar{\zeta}^k - \zeta^k) &\le \left\|\Phi(\zeta^k) + H_k(\bar{\zeta}^k - \zeta^k)\right\|^2 + \nu_k \left\|\bar{\zeta}^k - \zeta^k\right\|^2 \\
&\le \hat{c}^2 \left\|\bar{\zeta}^k - \zeta^k\right\|^4 + p_2 L_1^{\varrho} \left\|\zeta^k - \bar{\zeta}^k\right\|^{2+\varrho} \\
&\le (\hat{c}^2 + p_2 L_1^{\varrho}) \left\|\zeta^k - \bar{\zeta}^k\right\|^{2+\varrho},
\end{aligned}
$$

we have

$$
\begin{aligned}
\left\|\Phi(\zeta^k + d^k)\right\| &= \left\|\Phi(\zeta^k + d^k) - \Phi(\zeta^k) - H_k d^k + \Phi(\zeta^k) + H_k d^k\right\| \\
&\le \left\|\Phi(\zeta^k + d^k) - \Phi(\zeta^k) - H_k d^k\right\| + \sqrt{\varphi_k(d^k)} \\
&\le \hat{c}\, \|d^k\|^2 + (\hat{c}^2 + p_2 L_1^{\varrho})^{1/2} \left\|\zeta^k - \bar{\zeta}^k\right\|^{(\varrho+2)/2} \\
&\le \hat{c} \left(\hat{c}^2 p_2^{-1} \kappa_2^{-\varrho} + 1\right) \left\|\bar{\zeta}^k - \zeta^k\right\|^2 + (\hat{c}^2 + p_2 L_1^{\varrho})^{1/2} \left\|\zeta^k - \bar{\zeta}^k\right\|^{(\varrho+2)/2} \\
&\le c_2 \left\|\zeta^k - \bar{\zeta}^k\right\|^{(\varrho+2)/2}
\end{aligned}
$$

with $c_2 = \hat{c}(\hat{c}^2 p_2^{-1} \kappa_2^{-\varrho} + 1) + (\hat{c}^2 + p_2 L_1^{\varrho})^{1/2}$. Consequently,

$$
\operatorname{dist}(\zeta^k + d^k, S^*) \le \frac{1}{\kappa_2} \left\|\Phi(\zeta^k + d^k)\right\| \le \frac{c_2}{\kappa_2} \left\|\zeta^k - \bar{\zeta}^k\right\|^{(\varrho+2)/2} = c_3 \operatorname{dist}\left(\zeta^k, S^*\right)^{(\varrho+2)/2}.
$$

Thus, we complete the proof of the second part. □

Invoking Lemma 4.5 and following arguments similar to those in [22, Theorem 2.1] and [23, Theorem 3.1], we derive the quadratic convergence rate of Algorithm 4.2 under Assumption A.

Proposition 4.13 *Let $\{\zeta^k\}$ be generated by Algorithm 4.2 with ν_k given by (4.27), and ζ^* be an accumulation point of $\{\zeta^k\}$. If ζ^* is a solution of (4.7), then the sequence $\{\zeta^k\}$ converges to ζ^* superlinearly, and, moreover, quadratically when $\varrho = 2$, provided that F' and G' are locally Lipschitz continuous and Assumption A holds.*

It remains unclear whether Assumption A is indeed weaker than the strict complementarity condition at the solution. While the assumptions in Proposition 4.13 are weaker than those in Proposition 4.12, the latter ensures that every element of $\partial_B \Phi(\zeta^*)$ is nonsingular, thereby guaranteeing that $\|\Phi(\zeta)\|$ provides a local error bound in a neighborhood of the solution ζ^*. In contrast, as noted in [23], the assumptions in Proposition 4.13 do not imply the nonsingularity of each element in $\partial_B \Phi(\zeta^*)$. From the proof of Lemma 4.5, we find that the condition (4.31) cannot be weakened to

$$\kappa_2 \text{dist}(\zeta, S^*) \leq \|\Phi(\zeta)\|^{1/2} \quad \forall \zeta \in \mathcal{N}(\zeta^*, \delta),$$

in order to guarantee the superlinear (or quadratic) convergence of Algorithm 4.2, and therefore the global error bound result of Proposition 4.10 may not be applied for it. If let $\Psi(\zeta) = \|\Phi(\zeta)\|^4/4$ instead of $\Psi(\zeta) = \|\Phi(\zeta)\|^2/2$, then Assumption A holds automatically under the jointly uniform Cartesian P-property of F and G, but this will bring difficulty to numerical implementation due to the bad scaling of Ψ. Thus, it is worthwhile to study what conditions of F and G are sufficient for Assumption A to hold.

Numerical experiments and performance results for Algorithm 4.2 can be found in [24]. In general, it has been observed that the Levenberg–Marquardt semismooth method outperforms the Fischer–Burmeister semismooth method primarily on more challenging problem instances.

4.3 Smoothing Function Approach

In this section, we present the smoothing function approach for solving the mixed complementarity problem (MCP). The MCP arises in a wide range of applications, including economics, engineering, and operations research [25–28], and has garnered considerable attention over the past few decades [17, 29–33]. A curated collection of nonlinear mixed complementarity problems, known as MCPLIB, is available in [34].

Given a mapping $F : [l, u] \to \mathbb{R}^n$ with $F = (F_1, \ldots, F_n)^\top$, where $l = (l_1, \ldots, l_n)^\top$ and $u = (u_1, \ldots, u_n)^\top$ with $l_i \in \mathbb{R} \cup \{-\infty\}$ and $u_i \in \mathbb{R} \cup \{+\infty\}$ being given lower and upper bounds satisfying $l_i < u_i$ for $i = 1, 2, \ldots, n$. The MCP is to find a vector $x^* \in [l, u]$ such that each component x_i^* satisfies exactly one of the following implications:

$$\begin{aligned} x_i^* = l_i &\Longrightarrow F_i(x^*) \geq 0, \\ x_i^* \in (l_i, u_i) &\Longrightarrow F_i(x^*) = 0, \\ x_i^* = u_i &\Longrightarrow F_i(x^*) \leq 0. \end{aligned} \tag{4.35}$$

It is not hard to see that, when $l_i = -\infty$ and $u_i = +\infty$ for all $i = 1, 2, \ldots, n$, the MCP (4.35) is equivalent to solving the nonlinear system of equations:

$$F(x) = 0,$$

whereas when $l_i = 0$ and $u_i = +\infty$ for all $i = 1, 2, \ldots, n$, it reduces to the NCP, which is to find a point $x \in \mathbb{R}^n$ such that

$$x \geq 0, \quad F(x) \geq 0, \quad \langle x, F(x) \rangle = 0.$$

In fact, from [35, Theorem 2], the MCP (4.35) is also equivalent to the famous variational inequality problem (VIP) which is to find a vector $x^* \in [l, u]$ such that

$$\langle F(x^*), x - x^* \rangle \geq 0, \quad \forall x \in [l, u].$$

In the rest of this section, we assume the mapping F to be continuously differentiable.

Lemma 4.6 *Let $\phi_{\text{FB}}^p : \mathbb{R} \times \mathbb{R} \to \mathbb{R}$ be defined by (2.14). Then, the following limits hold.*

(a) $\lim\limits_{l_i \to -\infty} \phi_{\text{FB}}^p \left(x_i - l_i, \phi_{\text{FB}}^p(u_i - x_i, -F_i(x))\right) = -\phi_{\text{FB}}^p(u_i - x_i, -F_i(x)).$

(b) $\lim\limits_{u_i \to \infty} \phi_{\text{FB}}^p \left(x_i - l_i, \phi_{\text{FB}}^p(u_i - x_i, -F_i(x))\right) = \phi_{\text{FB}}^p(x_i - l_i, F_i(x)).$

(c) $\lim\limits_{l_i \to -\infty} \lim\limits_{u_i \to \infty} \phi_{\text{FB}}^p \left(x_i - l_i, \phi_{\text{FB}}^p(u_i - x_i, -F_i(x))\right) = -F_i(x).$

Proof Let $\{a^k\} \subseteq \mathbb{R}$ be any sequence converging to $+\infty$ as $k \to \infty$ and $b \in \mathbb{R}$ be any fixed real number. We will prove $\lim\limits_{k \to \infty} \phi_{\text{FB}}^p(a^k, b) = -b$, and part (a) then follows by continuity arguments. Without loss of generality, assume that $a^k > 0$ for each k. Then,

$$\begin{aligned}
\phi_{\text{FB}}^p(a^k, b) &= a^k \left(1 + (|b|/a^k)^p\right)^{1/p} - a^k - b \\
&= a^k \Bigg[1 + \frac{1}{p}\left(\frac{|b|}{a^k}\right)^p + \frac{1-p}{2p^2}\left(\frac{|b|}{a^k}\right)^{2p} + \cdots + \\
&\qquad \frac{(1-p)\cdots(1-pn+p)}{n!p^n}\left(\frac{|b|}{a^k}\right)^{np} + o\left(\left(\frac{|b|}{a^k}\right)^{pn}\right)\Bigg] - a^k - b \\
&= \frac{1}{p}\frac{|b|^p}{(a^k)^{p-1}} + \frac{1-p}{2p^2}\frac{|b|^{2p}}{(a^k)^{2p-1}} + \cdots + \frac{(1-p)\cdots(1-pn+p)}{n!p^n}\frac{|b|^{np}}{(a^k)^{np-1}} \\
&\quad + \frac{(a^k)|b|^{np}}{(a^k)^{np}}\frac{o\left(|b|/a^k\right)^{pn}}{\left(|b|/a^k\right)^{pn}} - b
\end{aligned}$$

where the second equality is using the Taylor expansion of the function $(1+t)^{1/p}$ and the notation $o(t)$ means $\lim_{t \to 0} o(t)/t = 0$. Since $a^k \to +\infty$ as $k \to \infty$, we have $\dfrac{|b|^{np}}{(a^k)^{np-1}} \to 0$ for all n. This together with the last equation implies $\lim_{k \to \infty} \phi_p(a^k, b) = -b$. This proves part (a). Parts (b) and (c) are direct by part (a) and the continuity of ϕ_{FB}^p. □

Below, we summarize the monotonicity properties of two scalar-valued functions that will be used in the subsequent section. As the proofs are straightforward, they are omitted here.

Lemma 4.7 *For any fixed* $0 \le \mu_1 < \mu_2$, *the following functions*

$$f_1(t) := (t+\mu_1)^{-\frac{p-1}{p}} - (t+\mu_2)^{-\frac{p-1}{p}} \quad (t > 0)$$

and

$$f_2(t) := (t+\mu_2)^{\frac{p-1}{p}} - (t+\mu_1)^{\frac{p-1}{p}} \quad (t \ge 0)$$

are decreasing on $(0, +\infty)$, *and furthermore,* $f_2(t) \le f_2(0) = \mu_2^{(p-1)/p} - \mu_1^{(p-1)/p}$.

For convenience, we adopt the following notations of index sets:

$$\begin{aligned} I_l &:= \{i \in \{1,2,\ldots,n\} \mid -\infty < l_i < u_i = +\infty\}, \\ I_u &:= \{i \in \{1,2,\ldots,n\} \mid -\infty = l_i < u_i < +\infty\}, \\ I_{lu} &:= \{i \in \{1,2,\ldots,n\} \mid -\infty < l_i < u_i < +\infty\}, \\ I_f &:= \{i \in \{1,2,\ldots,n\} \mid -\infty = l_i < u_i = +\infty\}. \end{aligned} \tag{4.36}$$

With the generalized FB function *phifbp*, we define an operator $\Phi^p_{\text{FB}} : \mathbb{R}^n \to \mathbb{R}^n$ componentwise as

$$(\Phi^p_{\text{FB}})^i(x) := \begin{cases} \phi^p_{\text{FB}}(x_i - l_i, F_i(x)) & \text{if } i \in I_l, \\ -\phi^p_{\text{FB}}(u_i - x_i, -F_i(x)) & \text{if } i \in I_u, \\ \phi^p_{\text{FB}}(x_i - l_i, \phi^p_{\text{FB}}(u_i - x_i, -F_i(x))) & \text{if } i \in I_{lu}, \\ -F_i(x) & \text{if } i \in I_f, \end{cases} \tag{4.37}$$

where the minus sign for $i \in I_u$ and $i \in I_f$ is motivated by Lemma 4.6. In fact, all results of this paper would be true without the minus sign. Using the fact that ϕ^p_{FB} is an NCP function, it is not difficult to verify that the following result holds.

Proposition 4.14 *Let* $\Phi^p_{\text{FB}} : \mathbb{R}^n \to \mathbb{R}^n$ *be defined as in* (4.36)–(4.37). *Then,* $x^* \in \mathbb{R}^n$ *is a solution to the MCP (4.35) if and only if* x^* *solves the nonlinear system of equations* $\Phi^p_{\text{FB}}(x) = 0$.

We point out that, unlike for the NCP, when writing the generalized FB function ϕ^p_{FB} as $\phi^p_{\text{FB}}(a,b) = (a+b) - \|(a,b)\|_p$, the conclusion of Proposition 4.14 does not necessarily hold since, if $I_l = \{1,2,\ldots,n\}$, then $\bar{x} = l$ satisfies $\Phi^p_{\text{FB}}(\bar{x}) = 0$, but $F(\bar{x}) \ge 0$ does not necessarily hold. Similar phenomenon also appears when replacing ϕ^p_{FB} by the ϕ_{NR} function.

Since ϕ^p_{FB} is not differentiable at the origin, the system $\Phi^p_{\text{FB}}(x) = 0$ is nonsmooth. In this paper, we will find a solution of nonsmooth system $\Phi^p_{\text{FB}}(x) = 0$ by solving a sequence of smooth approximations $\Psi^p_{\text{FB}}(x, \varepsilon) = 0$, where $\varepsilon > 0$ is a smoothing parameter and the operator $\Psi^p_{\text{FB}} : \mathbb{R}^n \times \mathbb{R}_{++} \to \mathbb{R}^n$ is defined componentwise as

$$(\Psi^p_{\text{FB}})^i(x,\varepsilon) := \begin{cases} \psi^p_{\text{FB}}(x_i - l_i, F_i(x), \varepsilon) & \text{if } i \in I_l, \\ -\psi^p_{\text{FB}}(u_i - x_i, -F_i(x), \varepsilon) & \text{if } i \in I_u, \\ \psi^p_{\text{FB}}\left(x_i - l_i, \psi^p_{\text{FB}}(u_i - x_i, -F_i(x), \varepsilon), \varepsilon\right) & \text{if } i \in I_{lu}, \\ -F_i(x) & \text{if } i \in I_f, \end{cases} \tag{4.38}$$

with

$$\psi^p_{\text{FB}}(a,b,\varepsilon) := \sqrt[p]{|a|^p + |b|^p + \varepsilon^p} - (a+b). \tag{4.39}$$

In the following, we focus on the favorable properties of the smoothing function ψ^p_{FB} and the associated operator Ψ^p_{FB}. We begin by presenting the key properties of ψ^p_{FB}.

Lemma 4.8 *Let $\psi^p_{\text{FB}} : \mathbb{R}^3 \to \mathbb{R}$ be defined by (4.39). Then, the following result holds.*

(a) For any fixed $\varepsilon > 0$, $\psi^p_{\text{FB}}(a,b,\varepsilon)$ is continuously differentiable at all $(a,b) \in \mathbb{R}^2$ with

$$-2 < \frac{\partial \psi^p_{\text{FB}}(a,b,\varepsilon)}{\partial a} < 0, \quad -2 < \frac{\partial \psi^p_{\text{FB}}(a,b,\varepsilon)}{\partial b} < 0. \tag{4.40}$$

(b) For any fixed $(a,b) \in \mathbb{R}^2$, $\psi^p_{\text{FB}}(a,b,\varepsilon)$ is continuously differentiable, strictly increasing and convex with respect to $\varepsilon > 0$. Moreover, for any $0 < \varepsilon_1 \le \varepsilon_2$,

$$0 \le \psi^p_{\text{FB}}(a,b,\varepsilon_2) - \psi^p_{\text{FB}}(a,b,\varepsilon_1) \le (\varepsilon_2 - \varepsilon_1). \tag{4.41}$$

In particular, $|\psi^p_{\text{FB}}(a,b,\varepsilon) - \phi^p_{\text{FB}}(a,b)| \le \varepsilon$ for all $\varepsilon \ge 0$.

(c) For any fixed $(a,b) \in \mathbb{R}^2$, let $(\psi^p_{\text{FB}})^0(a,b) := \left(\lim_{\varepsilon \downarrow 0} \frac{\partial \psi^p_{\text{FB}}(a,b,\varepsilon)}{\partial a}, \lim_{\varepsilon \downarrow 0} \frac{\partial \psi^p_{\text{FB}}(a,b,\varepsilon)}{\partial b}\right)$. Then,

$$\lim_{h=(h_1,h_2)\to(0,0)} \frac{\phi_p(a+h_1, b+h_2) - \phi^p_{\text{FB}}(a,b) - (\psi^p_{\text{FB}})^0(a+h_1, b+h_2)^{\mathsf{T}} h}{\|h\|} = 0.$$

(d) For any given $\varepsilon > 0$, if $p \ge 2$, then $\psi^p_{\text{FB}}(a,b,\varepsilon) = 0 \Longrightarrow a > 0,\ b > 0,\ 2ab \le \varepsilon^2$, and whenever $p > 1$, $\psi^p_{\text{FB}}(a,b,\varepsilon) = 0 \Longrightarrow a > 0,\ b > 0,\ \min\{a,b\} \le \dfrac{\varepsilon}{\sqrt[p]{2^p - 2}}$.

Proof (a) Using an elementary calculation, we immediately obtain that

$$\begin{aligned} \frac{\partial \psi^p_{\text{FB}}(a,b,\varepsilon)}{\partial a} &= \frac{\text{sgn}(a)|a|^{p-1}}{\left(\sqrt[p]{|a|^p + |b|^p + \varepsilon^p}\right)^{p-1}} - 1, \\ \frac{\partial \psi^p_{\text{FB}}(a,b,\varepsilon)}{\partial b} &= \frac{\text{sgn}(b)|b|^{p-1}}{\left(\sqrt[p]{|a|^p + |b|^p + \varepsilon^p}\right)^{p-1}} - 1. \end{aligned} \tag{4.42}$$

For any fixed $\varepsilon > 0$, since $\dfrac{\partial \psi^p_{\text{FB}}(a, b, \varepsilon)}{\partial a}$ and $\dfrac{\partial \psi^p_{\text{FB}}(a, b, \varepsilon)}{\partial b}$ are continuous at all $(a, b) \in \mathbb{R}^2$, it follows that $\psi^p_{\text{FB}}(a, b, \varepsilon)$ is continuously differentiable at all $(a, b) \in \mathbb{R}^2$. Noting that

$$\left|\frac{\operatorname{sgn}(a)|a|^{p-1}}{\left(\sqrt[p]{|a|^p + |b|^p + \varepsilon^p}\,\right)^{p-1}}\right| < 1 \quad \text{and} \quad \left|\frac{\operatorname{sgn}(b)|b|^{p-1}}{\left(\sqrt[p]{|a|^p + |b|^p + \varepsilon^p}\,\right)^{p-1}}\right| < 1,$$

we readily achieve the inequality (4.40).

(b) For any $\varepsilon > 0$, an elementary calculation yields

$$\frac{\partial \psi^p_{\text{FB}}(a, b, \varepsilon)}{\partial \varepsilon} = \frac{\varepsilon^{p-1}}{\left(\sqrt[p]{|a|^p + |b|^p + \varepsilon^p}\,\right)^{p-1}} > 0,$$

$$\frac{\partial^2 \psi^p_{\text{FB}}(a, b, \varepsilon)}{\partial \varepsilon^2} = \frac{(p-1)\varepsilon^{p-2}}{\left(\sqrt[p]{|a|^p + |b|^p + \varepsilon^p}\right)^{p-1}} \left(1 - \frac{\varepsilon^p}{|a|^p + |b|^p + \varepsilon^p}\right) \geq 0.$$

Therefore, for any fixed $(a, b) \in \mathbb{R}^2$, $\psi^p_{\text{FB}}(a, b, \varepsilon)$ is continuously differentiable, strictly increasing and convex with respect to $\varepsilon > 0$. By the Mean-Value Theorem, for any $0 < \varepsilon_1 \leq \varepsilon_2$, there exists some $\varepsilon_0 \in (\varepsilon_1, \varepsilon_2)$ such that

$$\psi^p_{\text{FB}}(a, b, \varepsilon_2) - \psi^p_{\text{FB}}(a, b, \varepsilon_1) = \frac{\partial \psi^p_{\text{FB}}}{\partial \varepsilon}(a, b, \varepsilon_0)(\varepsilon_2 - \varepsilon_1).$$

Since $\frac{\partial \psi^p_{\text{FB}}}{\partial \varepsilon}(a, b, \varepsilon_0) \leq 1$ by the proof of part (a), inequality (4.41) holds for all $0 < \varepsilon_1 \leq \varepsilon_2$. Letting $\varepsilon_1 \downarrow 0$, the desired result then follows.

(c) Using the formula (4.42), it is easy to calculate that

$$\lim_{\varepsilon \downarrow 0} \frac{\partial \psi^p_{\text{FB}}(a, b, \varepsilon)}{\partial a} = \begin{cases} \dfrac{\operatorname{sgn}(a)|a|^{p-1}}{\left(\sqrt[p]{|a|^p + |b|^p}\,\right)^{p-1}} - 1 & \text{if } (a, b) \neq (0, 0), \\ -1 & \text{if } (a, b) = (0, 0); \end{cases}$$

$$\lim_{\varepsilon \downarrow 0} \frac{\partial \psi^p_{\text{FB}}(a, b, \varepsilon)}{\partial b} = \begin{cases} \dfrac{\operatorname{sgn}(b)|b|^{p-1}}{\left(\sqrt[p]{|a|^p + |b|^p}\,\right)^{p-1}} - 1 & \text{if } (a, b) \neq (0, 0), \\ -1 & \text{if } (a, b) = (0, 0). \end{cases}$$

From this, we see that $(\psi^p_{\text{FB}})^0(a, b) = \left(\frac{\partial \phi_p(a,b)}{\partial a}, \frac{\partial \phi^p_{\text{FB}}(a,b)}{\partial b}\right)$ at $(a, b) \neq (0, 0)$. Therefore, we only need to check the case $(a, b) = (0, 0)$. The desired result follows by

$$\phi^p_{\text{FB}}(h_1, h_2) - \phi_p(0, 0) - \psi^0_p(h_1, h_2)^{\mathsf{T}} h$$

$$
\begin{aligned}
&= \sqrt[p]{|h_1|^p + |h_2|^p} - \frac{|h_1|^p + |h_2|^p}{(\sqrt[p]{|h_1|^p + |h_2|^p})^{p-1}} \\
&= \sqrt[p]{|h_1|^p + |h_2|^p} - \sqrt[p]{|h_1|^p + |h_2|^p} \\
&= 0.
\end{aligned}
$$

(d) From the definition of $\psi^p_{\rm FB}(a, b, \varepsilon)$, clearly, $\psi^p_{\rm FB}(a, b, \varepsilon) = 0$ implies $a + b \geq 0$, and hence $a \geq 0$ or $b \geq 0$. Note that, whenever $a \geq 0, b \leq 0$ or $a \leq 0, b \geq 0$, there holds

$$\sqrt[p]{|a|^p + |b|^p + \varepsilon^p} > \sqrt[p]{|a|^p + |b|^p} \geq \max\{|a|, |b|\} \geq a + b,$$

i.e., $\psi^p_{\rm FB}(a, b, \varepsilon) > 0$. Hence, for any given $\varepsilon > 0$, $\psi^p_{\rm FB}(a, b, \varepsilon) = 0$ implies $a > 0$ and $b > 0$.

(i) If $p \geq 2$, using the nonincreasing of p-norm with respect to p leads to

$$
\begin{aligned}
\psi_p(a, b, \varepsilon) = 0 &\Longleftrightarrow a + b = \sqrt[p]{|a|^p + |b|^p + \varepsilon^p} \leq \sqrt{|a|^2 + |b|^2 + \varepsilon^2} \\
&\Longrightarrow (a + b)^2 \leq a^2 + b^2 + \varepsilon^2 \Longrightarrow 2ab \leq \varepsilon^2.
\end{aligned}
$$

(ii) For $p > 1$, without loss of generality, we assume $0 < a \leq b$. For any fixed $a \geq 0$, consider $f(t) = (t + a)^p - t^p - a^p - \varepsilon^p$ $(t \geq 0)$. It is easy to verify that the function f is strictly increasing on $[0, +\infty)$. Since $\psi^p_{\rm FB}(a, b, \varepsilon) = 0$, we have $f(b) = 0$ which says $f(a) = (2^p - 2)a^p - \varepsilon^p \leq f(b) = 0$. From this inequality, we obtain $\min\{a, b\} = a \leq \dfrac{\varepsilon}{\sqrt[p]{2^p - 2}}$. □

Proposition 4.15 *Let $\Psi^p_{\rm FB}$ be defined by (4.38). Then, the following results hold.*

(a) For any fixed $\varepsilon > 0$, $\Psi^p_{\rm FB}(x, \varepsilon)$ is continuously differentiable on $\mathbb{R}^n$ with

$$\nabla_x \Psi^p_{\rm FB}(x, \varepsilon) = D_a(x, \varepsilon) + \nabla F(x) D_b(x, \varepsilon),$$

where $D_a(x, \varepsilon)$ and $D_b(x, \varepsilon)$ are $n \times n$ diagonal matrices with the diagonal elements $(D_a)_{ii}(x, \varepsilon)$ and $(D_b)_{ii}(x, \varepsilon)$ defined as follows:

(a1) For $i \in I_l$,

$$
\begin{aligned}
(D_a)_{ii}(x, \varepsilon) &= \frac{\text{sgn}(x_i - l_i)|x_i - l_i|^{p-1}}{\|(x_i - l_i, F_i(x), \varepsilon)\|_p^{p-1}} - 1, \\
(D_b)_{ii}(x, \varepsilon) &= \frac{\text{sgn}(F_i(x))|F_i(x)|^{p-1}}{\|(x_i - l_i, F_i(x), \varepsilon)\|_p^{p-1}} - 1.
\end{aligned}
$$

(a2) For $i \in I_u$,

$$(D_a)_{ii}(x,\varepsilon) = \frac{\mathrm{sgn}(u_i - x_i)|u_i - x_i|^{p-1}}{\|(u_i - x_i, F_i(x), \varepsilon)\|_p^{p-1}} - 1,$$

$$(D_b)_{ii}(x,\varepsilon) = \frac{-\mathrm{sgn}(F_i(x))|F_i(x)|^{p-1}}{\|(u_i - x_i, F_i(x), \varepsilon)\|_p^{p-1}} - 1.$$

(a3) For $i \in I_{lu}$,

$$(D_a)_{ii}(x,\varepsilon) = a_i(x,\varepsilon) + b_i(x,\varepsilon)c_i(x,\varepsilon) \text{ and } (D_b)_{ii}(x,\varepsilon) = b_i(x,\varepsilon)d_i(x,\varepsilon)$$

with

$$a_i(x,\varepsilon) = \frac{\mathrm{sgn}(x_i - l_i)|x_i - l_i|^{p-1}}{\left\|(x_i - l_i, \psi^p_{\text{FB}}(u_i - x_i, -F_i(x), \varepsilon), \varepsilon)\right\|_p^{p-1}} - 1,$$

$$b_i(x,\varepsilon) = \frac{\mathrm{sgn}(\psi^p_{\text{FB}}(u_i - x_i, -F_i(x), \varepsilon))|\psi^p_{\text{FB}}(u_i - x_i, -F_i(x), \varepsilon)|^{p-1}}{\left\|(x_i - l_i, \psi^p_{\text{FB}}(u_i - x_i, -F_i(x), \varepsilon), \varepsilon)\right\|_p^{p-1}} - 1,$$

$$c_i(x,\varepsilon) = -\frac{\mathrm{sgn}(u_i - x_i)|u_i - x_i|^{p-1}}{\|(u_i - x_i, F_i(x), \varepsilon)\|_p^{p-1}} + 1,$$

$$d_i(x,\varepsilon) = \frac{\mathrm{sgn}(F_i(x))|F_i(x)|^{p-1}}{\|(u_i - x_i, F_i(x), \varepsilon)\|_p^{p-1}} + 1.$$

(a4) For $i \in I_f$, $(D_a)_{ii}(x,\varepsilon) = 0$ and $(D_b)_{ii}(x,\varepsilon) = -1$.

Moreover, $-2 < (D_a)_{ii}(x,\varepsilon) < 0$ and $-2 < (D_b)_{ii}(x,\varepsilon) < 0$ or all $i \in I_l \cup I_u$, and $-6 < (D_a)_{ii}(x,\varepsilon) < 0$ and $-4 < (D_b)_{ii}(x,\varepsilon) < 0$ for $i \in I_{lu}$.

(b) For any given $\varepsilon_1 > 0$ and $\varepsilon_2 > 0$, we have

$$\|\Psi^p_{\text{FB}}(x,\varepsilon_2) - \Psi^p_{\text{FB}}(x,\varepsilon_1)\| \le \sqrt{n}\left(\sqrt[p]{2} + 1\right)|\varepsilon_2 - \varepsilon_1|, \quad \forall x \in \mathbb{R}^n.$$

Particularly, for any given $\varepsilon > 0$,

$$\|\Psi^p_{\text{FB}}(x,\varepsilon) - \Phi^p_{\text{FB}}(x)\| \le \sqrt{n}\left(\sqrt[p]{2} + 1\right)\varepsilon, \quad \forall x \in \mathbb{R}^n.$$

Proof This is a direct consequence of Lemma 4.8 and the expression of Ψ^p_{FB}. □

The Jacobian consistency property is fundamental to the analysis of local fast convergence in smoothing algorithms [36]. To establish that the smoothing operator Ψ^p_{FB} satisfies this property, we first present a characterization of the generalized Jacobian $\partial_C \Phi^p_{\text{FB}}(x)$, which follows directly from Proposition 2.1(f).

Proposition 4.16 *For any given* $x \in \mathbb{R}^n$, $\partial_C \Phi^p_{\text{FB}}(x)^{\mathsf{T}} = \{D_a(x) + \nabla F(x) D_b(x)\}$, *where* $D_a(x)$, $D_b(x)$ *are* $n \times n$ *diagonal matrices whose diagonal elements are given below:*

(a) For $i \in I_l$, *if* $(x_i - l_i, F_i(x)) \neq (0, 0)$, *then*

$$(D_a)_{ii}(x) = \frac{\operatorname{sgn}(x_i - l_i) \cdot |x_i - l_i|^{p-1}}{\|(x_i - l_i, F_i(x))\|_p^{p-1}} - 1,$$
$$(D_b)_{ii}(x) = \frac{\operatorname{sgn}(F_i(x)) \cdot |F_i(x)|^{p-1}}{\|(x_i - l_i, F_i(x))\|_p^{p-1}} - 1,$$

and otherwise

$$((D_a)_{ii}(x), (D_b)_{ii}(x)) \in \left\{(\xi - 1, \zeta - 1) \in \mathbb{R}^2 \mid |\xi|^{\frac{p}{p-1}} + |\zeta|^{\frac{p}{p-1}} \leq 1\right\}.$$

(b) For $i \in I_u$, *if* $(u_i - x_i, -F_i(x)) \neq (0, 0)$, *then*

$$(D_a)_{ii}(x) = \frac{\operatorname{sgn}(u_i - x_i) \cdot |u_i - x_i|^{p-1}}{\|(u_i - x_i, -F_i(x))\|_p^{p-1}} - 1,$$
$$(D_b)_{ii}(x) = -\frac{\operatorname{sgn}(F_i(x)) \cdot |F_i(x)|^{p-1}}{\|(u_i - x_i, -F_i(x))\|_p^{p-1}} - 1,$$

and otherwise

$$((D_a)_{ii}(x), (D_b)_{ii}(x)) \in \left\{(\xi - 1, \zeta - 1) \in \mathbb{R}^2 \mid |\xi|^{\frac{p}{p-1}} + |\zeta|^{\frac{p}{p-1}} \leq 1\right\}.$$

(c) For $i \in I_{lu}$, $(D_a)_{ii}(x) = a_i(x) + b_i(x)c_i(x)$ *and* $(D_b)_{ii}(x) = b_i(x)d_i(x)$ *where, if* $(x_i - l_i, \phi^p_{\text{FB}}(u_i - x_i, -F_i(x))) \neq (0, 0)$, *then*

$$a_i(x) = \frac{\operatorname{sgn}(x_i - l_i) \cdot |x_i - l_i|^{p-1}}{\left\|\left(x_i - l_i, \phi^p_{\text{FB}}(u_i - x_i, -F_i(x))\right)\right\|_p^{p-1}} - 1,$$
$$b_i(x) = \frac{\operatorname{sgn}\left(\phi^p_{\text{FB}}(u_i - x_i, -F_i(x))\right) \cdot \left|\phi^p_{\text{FB}}(u_i - x_i, -F_i(x))\right|^{p-1}}{\left\|\left(x_i - l_i, \phi^p_{\text{FB}}(u_i - x_i, -F_i(x))\right)\right\|_p^{p-1}} - 1,$$

and otherwise

$$(a_i(x), b_i(x)) \in \left\{(\xi - 1, \zeta - 1) \in \mathbb{R}^2 \mid |\xi|^{\frac{p}{p-1}} + |\zeta|^{\frac{p}{p-1}} \leq 1\right\};$$

and if $(u_i - x_i, -F_i(x)) \neq (0, 0)$, *then*

$$c_i(x) = \frac{-\text{sgn}(u_i - x_i) \cdot |u_i - x_i|^{p-1}}{\|(u_i - x_i, -F_i(x))\|_p^{p-1}} + 1,$$
$$d_i(x) = \frac{\text{sgn}\,(F_i(x)) \cdot |F_i(x)|^{p-1}}{\|(u_i - x_i, -F_i(x))\|_p^{p-1}} + 1,$$

and otherwise

$$(c_i(x), d_i(x)) \in \left\{(\xi + 1, \zeta + 1) \in \mathbb{R}^2 \mid |\xi|^{\frac{p}{p-1}} + |\zeta|^{\frac{p}{p-1}} \leq 1\right\}.$$

(d) For $i \in I_f$, $(D_a)_{ii}(x) = 0$ and $(D_b)_{ii}(x) = -1$.

Proposition 4.17 *Let Ψ^p_{FB} be defined by (4.38). Then, for any fixed $x \in \mathbb{R}^n$,*

$$\lim_{\varepsilon \downarrow 0} \text{dist}\left(\nabla_x \Psi^p_{\text{FB}}(x, \varepsilon)^{\mathsf{T}}, \partial_C \Phi^p_{\text{FB}}(x)\right) = 0.$$

Proof For the sake of notation, for any given $x \in \mathbb{R}^n$, we define the index sets:

$$\begin{aligned}
\beta_1(x) &:= \{i \in I_l \mid (x_i - l_i, F_i(x)) = (0, 0)\}, \quad \bar{\beta}_1(x) := I_l \setminus \beta_1(x),\\
\beta_2(x) &:= \{i \in I_u \mid (u_i - x_i, F_i(x)) = (0, 0)\}, \quad \bar{\beta}_2(x) := I_u \setminus \beta_2(x),\\
\beta_3(x) &:= \{i \in I_{lu} \mid \left(x_i - l_i, \phi_p(u_i - x_i, -F_i(x))\right) = (0, 0)\}, \quad \bar{\beta}_3(x) := I_{lu} \setminus \beta_3(x),\\
\beta_4(x) &:= \{i \in \bar{\beta}_3(x) \mid (u_i - x_i, F_i(x)) = (0, 0)\}, \quad \bar{\beta}_4(x) := \bar{\beta}_3(x) \setminus \beta_4(x).
\end{aligned} \tag{4.43}$$

We proceed the arguments by the cases $i \in I_l \cup I_u$, $i \in I_{lu}$, and $i \in I_f$, respectively.
Case 1: $i \in I_l \cup I_u$. When $i \in \beta_1(x) \cup \beta_2(x)$, it is easy to see that

$$(D_a)_{ii}(x, \varepsilon) = -1 \quad \text{and} \quad (D_b)_{ii}(x, \varepsilon) = -1.$$

By Proposition 4.15 (a1) and (a2), $\nabla_x (\Psi^p_{\text{FB}})^i(x, \varepsilon)^{\mathsf{T}} = -e_i^{\mathsf{T}} - F_i'(x)$ for all $\varepsilon > 0$. Since

$$(-1, -1) \in \left\{(\xi - 1, \zeta - 1) \in \mathbb{R}^2 \,\middle|\, |\xi|^{\frac{p}{p-1}} + |\zeta|^{\frac{p}{p-1}} \leq 1\right\},$$

from Proposition 4.16(a) and Proposition 4.16(b), we obtain $\nabla_x (\Psi^p_{\text{FB}})^i(x, \varepsilon)^{\mathsf{T}} \in \partial_C (\Phi^p_{\text{FB}})^i(x)$. When $i \in \bar{\beta}_1(x) \cup \bar{\beta}_2(x)$,

$$\lim_{\varepsilon \downarrow 0} (D_a)_{ii}(x, \varepsilon) = (D_a)_{ii}(x) \quad \text{and} \quad \lim_{\varepsilon \downarrow 0} (D_b)_{ii}(x, \varepsilon) = (D_b)_{ii}(x),$$

which together with Proposition 4.15 (a1) and (a2) implies

$$\lim_{\varepsilon \downarrow 0} \nabla_x (\Psi^p_{\text{FB}})^i(x, \varepsilon)^{\mathsf{T}} = (D_a)_{ii}(x) e_i^T + (D_b)_{ii}(x) F_i'(x) \in \partial_C \Phi_{p,i}(x).$$

Since $I_l \cup I_u = \beta_1(x) \cup \beta_2(x) \cup \bar{\beta}_1(x) \cup \bar{\beta}_2(x)$, the last two subcases show that

$$\lim_{\varepsilon \downarrow 0} \nabla_x (\Psi^p_{\text{FB}})^i(x,\varepsilon)^\mathsf{T} \in \partial_C (\Phi^p_{\text{FB}})^i(x), \quad \forall\, i \in I_l \cup I_u. \tag{4.44}$$

Case 2: $i \in I_{lu}$. When $i \in \beta_3(x)$, we have $x_i - l_i = 0$, $\phi^p_{\text{FB}}(u_i - x_i, -F_i(x)) = 0$, $u_i - x_i > 0$ and $F_i(x) = 0$. Hence, $c_i(x) = 0$ and $d_i(x) = 1$. From Proposition 4.16(c), it follows that

$$\partial_C (\Phi^p_{\text{FB}})^i(x) = \left\{ a_i(x) e_i^\mathsf{T} + b_i(x) F_i'(x) \right\}$$

with

$$(a_i(x), b_i(x)) \in \left\{ (\xi - 1, \zeta - 1) \in \mathbb{R}^2 \;\middle|\; |\xi|^{\frac{p}{p-1}} + |\zeta|^{\frac{p}{p-1}} \le 1 \right\}.$$

On the other hand, since $a_i(x,\varepsilon) = -1$, $d_i(x,\varepsilon) = 1$ and

$$b_i(x,\varepsilon) = \frac{|\psi_p(u_i - x_i, -F_i(x), \varepsilon)|^{p-1}}{\left(|\psi_p(u_i - x_i, -F_i(x), \varepsilon)|^p + \varepsilon^p\right)^{\frac{p-1}{p}}} - 1,$$

$$c_i(x,\varepsilon) = 1 - \frac{|u_i - x_i|^{p-1}}{(|u_i - x_i|^p + \varepsilon^p)^{(p-1)/p}},$$

from Proposition 4.15(a3), it follows that

$$\nabla_x (\Psi^p_{\text{FB}})^i(x,\varepsilon)^\mathsf{T} = (-1 + b_i(x,\varepsilon) c_i(x,\varepsilon))\, e_i^\mathsf{T} + b_i(x,\varepsilon) F_i'(x).$$

Taking

$$\xi = 0 \ \text{ and } \ \zeta = \frac{|\psi_p(u_i - x_i, -F_i(x), \varepsilon)|^{p-1}}{\left(|\psi_p(u_i - x_i, -F_i(x), \varepsilon)|^p + \varepsilon^p\right)^{\frac{p-1}{p}}},$$

it is not hard to verify that $|\xi|^{\frac{p}{p-1}} + |\zeta|^{\frac{p}{p-1}} \le 1$, and consequently

$$-\, e_i^\mathsf{T} + b_i(x,\varepsilon) F_i'(x) \in \partial_C (\Phi^p_{\text{FB}})^i(x).$$

Noting that

$$\lim_{\varepsilon \downarrow 0} \left\| \nabla_x (\Psi^p_{\text{FB}})^i(x,\varepsilon)^\mathsf{T} - \left(-e_i^\mathsf{T} + b_i(x,\varepsilon) F_i'(x)\right) \right\| = \lim_{\varepsilon \downarrow 0} \|b_i(x,\varepsilon) c_i(x,\varepsilon) e_i^\mathsf{T}\| = 0,$$

it then follows that

$$\lim_{\varepsilon \downarrow 0} \operatorname{dist}\left(\nabla_x (\Psi^p_{\text{FB}})^i(x,\varepsilon)^\mathsf{T}, \partial_C (\Phi^p_{\text{FB}})^i(x) \right) = 0, \quad i \in \beta_3(x).$$

When $i \in \bar{\beta}_3(x)$, we have $\lim_{\varepsilon \downarrow 0} a_i(x,\varepsilon) = a_i(x)$ and $\lim_{\varepsilon \downarrow 0} b_i(x,\varepsilon) = b_i(x)$. Also,

$$c_i(x,\varepsilon) = 1, \quad d_i(x,\varepsilon) = 1 \quad \text{for } i \in \beta_4(x)$$

and

$$\lim_{\varepsilon\downarrow 0} c_i(x,\varepsilon) = c_i(x), \quad \lim_{\varepsilon\downarrow 0} d_i(x,\varepsilon) = d_i(x) \quad \text{for } i \in \bar{\beta}_4(x).$$

Using Proposition 4.16(c) and noting that

$$(1,1) \in \left\{(\xi+1, \zeta+1) \in \mathbb{R}^2 \,\middle|\, |\xi|^{\frac{p}{p-1}} + |\zeta|^{\frac{p}{p-1}} \le 1\right\},$$

we obtain $\lim_{\varepsilon\downarrow 0} \nabla_x (\Psi_{\text{FB}}^p)^i(x,\varepsilon)^{\mathsf{T}} \in \partial_C(\Phi_{\text{FB}}^p)^i(x)$ for $i \in \bar{\beta}_3(x)$. Along with the above discussions,

$$\lim_{\varepsilon\downarrow 0} \nabla_x (\Psi_{\text{FB}}^p)^i(x,\varepsilon)^{\mathsf{T}} \in \partial_C(\Phi_{\text{FB}}^p)^i(x) \quad \text{for } i \in I_{lu}. \tag{4.45}$$

<u>Case 3</u>: $i \in I_f$. By Proposition 4.15 (a4) and Proposition 4.16(d), it is obvious that

$$\lim_{\varepsilon\downarrow 0} \nabla_x (\Psi_{\text{FB}}^p)^i(x,\varepsilon)^{\mathsf{T}} \in \partial_C(\Phi_{\text{FB}}^p)^i(x) \quad \text{for } i \in I_f. \tag{4.46}$$

Now the desired result follows from (4.44), (4.45), (4.46), and $\{1,2,\ldots,n\} = I_f \cup I_l \cup I_u \cup I_{lu}$. □

Proposition 4.17 implies that for any $\delta > 0$, there exists an $\varepsilon(x,\delta) > 0$ such that

$$\operatorname{dist}\left(\nabla_x \Psi_{\text{FB}}^p(x,\varepsilon)^{\mathsf{T}},\ \partial_C \Phi_{\text{FB}}^p(x)\right) \le \delta \quad \text{for all } 0 < \varepsilon \le \varepsilon(x,\delta).$$

The following lemma offers a way to choose such $\varepsilon(x,\delta)$.

Lemma 4.9 *Let Ψ_{FB}^p be defined by (4.38). Suppose that x is not a solution of (4.35). Let*

$$\alpha(x) := \min\{\alpha_1(x), \alpha_2(x), \alpha_3(x)\} > 0, \quad \gamma(x) := \max\{\gamma_1(x), \gamma_2(x), \gamma_3(x)\} \ge 0$$

with

$$\begin{aligned}
\alpha_1(x) &:= \min_{i\in\bar{\beta}_1(x)} |x_i - l_i|^p + |F_i(x)|^p, \\
\alpha_2(x) &:= \min_{i\in\bar{\beta}_2(x)\cup\bar{\beta}_4(x)} |u_i - x_i|^p + |F_i(x)|^p, \\
\alpha_3(x) &:= \min_{i\in\bar{\beta}_4(x)\cup\{i\,|\,|x_i-l_i|\neq 0\}} |x_i - l_i|^p + |\phi_{\text{FB}}^p(u_i - x_i, -F_i(x))|^p \\
\gamma_1(x) &:= \max_{i\in\bar{\beta}_1(x)} \left\|\operatorname{sgn}(x_i - l_i)|x_i - l_i|^{p-1} e_i + \operatorname{sgn}(F_i(x))|F_i(x)|^{p-1}\nabla F_i(x)\right\| \\
\gamma_2(x) &:= \max_{i\in\bar{\beta}_2(x)} \left\|\operatorname{sgn}(F_i(x))|F_i(x)|^{p-1}\nabla F_i(x) - \operatorname{sgn}(u_i - x_i)|u_i - x_i|^{p-1} e_i\right\|
\end{aligned}$$

$$\gamma_3(x) := \max_{i \in \bar{\beta}_4(x)} |u_i - x_i|^{p-1} + |F_i(x)|^{p-1}.$$

Then, for any $\delta > 0$, there exists an $\varepsilon(x, \delta) > 0$ such that

$$\operatorname{dist}\left(\nabla_x \Psi^p_{\text{FB}}(x, \varepsilon)^\top, \partial_C \Phi^p_{\text{FB}}(x)\right) \le \delta \quad \text{for all } 0 < \varepsilon \le \varepsilon(x, \delta),$$

where

$$\varepsilon(x, \delta) := \min\left\{\varepsilon_0(x, \delta), \varepsilon_1(x, \delta), \varepsilon_2(x, \delta), \varepsilon_3(x, \delta), \left(\frac{\delta}{\sqrt{n}M(x)}\right)^{\frac{p-1}{p}}\right\}$$

with

$$\varepsilon_0(x, \delta) := \min_{i \in \beta_3(x)} \left[\frac{|u_i - x_i|^{p-1}}{(1 - \delta/\sqrt{n})^{\frac{p}{p-1}}} - |u_i - x_i|^p\right]^{1/p}, \quad \varepsilon_2(x, \delta) := \min_{i \in \beta_4(x)} \frac{1}{2}|x_i - l_i|,$$

$$\varepsilon_1(x, \delta) := \begin{cases} 1 & \text{if } (\frac{\sqrt{n}\gamma(x)}{\delta})^{\frac{p}{p-1}} - \alpha(x) \le 0, \\ \alpha(x)^{2/p}\left((\frac{\sqrt{n}\gamma(x)}{\delta})^{p/(p-1)} - \alpha(x)\right)^{-1/p} & \text{otherwise,} \end{cases}$$

$$\varepsilon_3(x, \delta) := \begin{cases} 1 & \text{if } \phi^p_{\text{FB}}(u_i - x_i, -F_i(x)) \ge 0, \\ \frac{1}{2}\left[(u_i - x_i - F_i(x))^p - |u_i - x_i|^p - |F_i(x)|^p\right]^{1/p} & \text{otherwise.} \end{cases}$$

Proof From Eq. (4.43), clearly, the index set $\{1, 2, \ldots, n\}$ can be partitioned as

$$I_f \cup \beta_1(x) \cup \bar{\beta}_1(x) \cup \beta_2(x) \cup \bar{\beta}_2(x) \cup \beta_3(x) \cup \beta_4(x) \cup \bar{\beta}_4(x).$$

In view of this, we proceed the proof by the following several cases.

Case 1: $i \in I_f$. From Propositions 4.15 (a4) and 4.16 (d), we have

$$\nabla_x (\Psi^p_{\text{FB}})^i(x, \varepsilon)^\top = -F_i'(x) \quad \text{and} \quad \partial_C (\Phi^p_{\text{FB}})^i(x) = -F_i'(x),$$

which implies

$$\operatorname{dist}\left(\nabla_x (\Psi^p_{\text{FB}})^i(x, \varepsilon)^\top, \partial_C (\Phi^p_{\text{FB}})^i(x)\right) = 0 \quad \text{for all } \varepsilon > 0.$$

Case 2: $i \in \beta_1(x) \cup \beta_2(x)$. From Proposition 4.15 (a1) and (a2), it follows that

$$\nabla_x (\Psi^p_{\text{FB}})^i(x, \varepsilon)^\top = -e_i^\top - F_i'(x).$$

In addition, by Proposition 4.16 (a) and (b), we have $\nabla_x (\Psi^p_{\text{FB}})^i(x, \varepsilon)^\top \in \partial(\Phi^p_{\text{FB}})^i(x)$ since

$$(-1, -1) \in \left\{(\xi - 1, \zeta - 1) \in \mathbb{R}^2 \mid |\xi|^{\frac{p}{p-1}} + |\zeta|^{\frac{p}{p-1}} \le 1\right\}.$$

Therefore, we have

$$\text{dist}\left(\nabla_x(\Psi^p_{\text{FB}})^i(x,\varepsilon)^\top, \partial_C(\Phi^p_{\text{FB}})^i(x)\right) = 0 \quad \text{for all } \varepsilon > 0.$$

Case 3: $i \in \beta_3(x)$. Under this case, $x_i - l_i = 0, \phi_p(u_i - x_i, -F_i(x)) = 0, u_i - x_i > 0$ and $F_i(x) = 0$. Hence, $c_i(x) = 0$ and $d_i(x) = 1$. From Proposition 4.16(c), it follows that

$$\partial_C(\Phi^p_{\text{FB}})^i(x) = \{a_i(x)e_i^\top + b_i(x)F_i'(x)\}$$

with

$$(a_i(x), b_i(x)) \in \left\{(\xi - 1, \zeta - 1) \in \mathbb{R}^2 \mid |\xi|^{\frac{p}{p-1}} + |\zeta|^{\frac{p}{p-1}} \leq 1\right\}.$$

On the other hand, since $a_i(x,\varepsilon) = -1, d_i(x,\varepsilon) = 1$ and

$$b_i(x,\varepsilon) = \frac{|\psi^p_{\text{FB}}(u_i - x_i, -F_i(x), \varepsilon)|^{p-1}}{\left(|\psi^p_{\text{FB}}(u_i - x_i, -F_i(x), \varepsilon)|^p + \varepsilon^p\right)^{\frac{p-1}{p}}} - 1,$$

$$c_i(x,\varepsilon) = 1 - \frac{|u_i - x_i|^{p-1}}{(|u_i - x_i|^p + \varepsilon^p)^{(p-1)/p}},$$

from Proposition 4.15 (a3) it follows that

$$\nabla_x(\Psi^p_{\text{FB}})^i(x,\varepsilon)^\top = (-1 + b_i(x,\varepsilon)c_i(x,\varepsilon))e_i^\top + b_i(x,\varepsilon)F_i'(x).$$

Taking

$$\xi = 0 \;\text{ and }\; \zeta = \frac{|\psi^p_{\text{FB}}(u_i - x_i, -F_i(x), \varepsilon)|^{p-1}}{\left(|\psi^p_{\text{FB}}(u_i - x_i, -F_i(x), \varepsilon)|^p + \varepsilon^p\right)^{\frac{p-1}{p}}},$$

we can verify that $|\xi|^{\frac{p}{p-1}} + |\zeta|^{\frac{p}{p-1}} \leq 1$, and consequently $-e_i^\top + b_i(x,\varepsilon)F_i'(x) \in \partial_C(\Phi^p_{\text{FB}})^i(x)$. Using the definition of $\varepsilon_0(x,\delta)$, it is easy to verify that, for all $\varepsilon \leq \varepsilon_0(x,\delta)$,

$$\left\|\nabla_x(\Psi^p_{\text{FB}})^i(x,\varepsilon)^\top - \left(-e_i^\top + b_i(x,\varepsilon)F_i'(x)\right)\right\| = \|b_i(x,\varepsilon)c_i(x,\varepsilon)e_i^\top\| \leq |c_i(x,\varepsilon)| \leq \frac{\delta}{\sqrt{n}}.$$

Therefore, for all $0 < \varepsilon \leq \varepsilon_0(x,\delta)$,

$$\text{dist}\left(\nabla_x(\Psi^p_{\text{FB}})^i(x,\varepsilon)^\top, \partial_C(\Phi^p_{\text{FB}})^i(x)\right) \leq \frac{\delta}{\sqrt{n}}.$$

Case 4: $i \in \bar{\beta}_1(x)$. From Propositions 4.16 (a) and 4.15 (a1), it follows that

$$\begin{aligned} &\text{dist}\left(\nabla_x(\Psi^p_{\text{FB}})^i(x,\varepsilon)^\top, \partial_C(\Phi^p_{\text{FB}})^i(x)\right) \\ &= \left\|\nabla_x(\Psi^p_{\text{FB}})^i(x,\varepsilon)^\top - \nabla(\Phi^p_{\text{FB}})^i(x)^\top\right\| \end{aligned} \tag{4.47}$$

$$
\begin{aligned}
&= \left(\frac{1}{\|(x_i - l_i, F_i(x))\|_p^{p-1}} - \frac{1}{\|(x_i - l_i, F_i(x), \varepsilon)\|_p^{p-1}} \right) \\
&\quad \left\| \operatorname{sgn}(x_i - l_i)|x_i - l_i|^{p-1} e_i + \operatorname{sgn}(F_i(x))|F_i(x)|^{p-1} \nabla F_i(x) \right\| \\
&\le \left(\alpha_1(x)^{\frac{1-p}{p}} - (\alpha_1(x) + \varepsilon^p)^{\frac{1-p}{p}} \right) \gamma_1(x) \\
&\le \left(\alpha(x)^{\frac{1-p}{p}} - (\alpha(x) + \varepsilon^p)^{\frac{1-p}{p}} \right) \gamma(x) \\
&= \frac{(\alpha(x) + \varepsilon^p)^{\frac{p-1}{p}} - \alpha(x)^{\frac{p-1}{p}}}{[\alpha(x)(\alpha(x) + \varepsilon^p)]^{\frac{p-1}{p}}} \gamma(x) \\
&\le \frac{\varepsilon^{p-1}}{[\alpha(x)(\alpha(x) + \varepsilon^p)]^{\frac{p-1}{p}}} \gamma(x),
\end{aligned}
\tag{4.48}
$$

where the inequalities are using Lemma 4.7 and the definition of $\alpha(x)$ and $\gamma(x)$. Now using Equation (4.47), it is not hard to verify that for all $0 < \varepsilon \le \varepsilon_1(x, \delta)$

$$
\operatorname{dist}\left(\nabla_x (\Psi_{\rm FB}^p)^i (x, \varepsilon)^{\mathsf{T}}, \partial_C (\Phi_{\rm FB}^p)^i (x) \right) \le \frac{\delta}{\sqrt{n}}. \tag{4.49}
$$

Indeed, if $\gamma(x) = 0$, this inequality obviously holds for all $\varepsilon > 0$. Suppose that $\gamma(x) > 0$. Then, a simple calculation shows that

$$
\frac{\varepsilon^{p-1}\gamma(x)}{[\alpha(x)(\alpha(x) + \varepsilon^p)]^{\frac{p-1}{p}}} \le \frac{\delta}{\sqrt{n}} \iff \alpha(x)^2 \ge \varepsilon^p \left(\left(\frac{\sqrt{n}\gamma(x)}{\delta} \right)^{p/(p-1)} - \alpha(x) \right).
$$

Clearly, the inequality on the right-hand side holds for all $0 < \varepsilon \le \varepsilon_1(x, \delta)$. Consequently, the result in (4.49) follows from the above equivalence and (4.47).

Case 5: $i \in \bar{\beta}_2(x)$. From Propositions 4.16(b) and 4.15 (a2), it follows that

$$
\begin{aligned}
&\operatorname{dist}\left(\nabla_x (\Psi_{\rm FB}^p)^i (x, \varepsilon)^{\mathsf{T}}, \partial_C (\Phi_{\rm FB}^p)^i (x) \right) \\
&= \left(\frac{1}{\|(u_i - x_i, F_i(x))\|_p^{p-1}} - \frac{1}{\|(u_i - x_i, F_i(x), \varepsilon)\|_p^{p-1}} \right) \\
&\quad \left\| \operatorname{sgn}(F_i(x))|F_i(x)|^{p-1} \nabla F_i(x) - \operatorname{sgn}(u_i - x_i)|u_i - x_i|^{p-1} e_i \right\| \\
&\le \left(\alpha_2(x)^{\frac{1-p}{p}} - (\alpha_2(x) + \varepsilon^p)^{\frac{1-p}{p}} \right) \gamma_2(x) \\
&\le \left(\alpha(x)^{\frac{1-p}{p}} - (\alpha(x) + \varepsilon^p)^{\frac{1-p}{p}} \right) \gamma(x),
\end{aligned}
$$

where the inequalities are using Lemma 4.7 and the definition of $\alpha(x)$ and $\gamma(x)$. Using the same arguments as Case 4, we can prove that for all $0 < \varepsilon \le \varepsilon_1(x, \delta)$,

$$\operatorname{dist}\left(\nabla_x(\Psi^p_{_{\rm FB}})^i(x,\varepsilon)^{\mathsf T}, \partial_C(\Phi^p_{_{\rm FB}})^i(x)\right) \le \frac{\delta}{\sqrt{n}}.$$

<u>Case 6:</u> $i \in \beta_4(x)$. Since $(u_i - x_i, -F_i(x)) = (0,0)$, we necessarily have

$$x_i - l_i > 0,\ \ \phi^p_{_{\rm FB}}(u_i - x_i, -F_i(x)) = 0 \ \text{ and } \ \psi^p_{_{\rm FB}}(u_i - x_i, -F_i(x), \varepsilon) = \varepsilon,$$

which in turn implies $a_i(x) = 0$ and $b_i(x) = -1$. By Proposition 4.16 (c),

$$\partial_C(\Phi^p_{_{\rm FB}})^i(x) = \left\{-c_i(x)e_i^{\mathsf T} - d_i(x)F_i'(x)\right\}$$

with

$$(c_i(x), d_i(x)) \in \left\{(\xi+1, \zeta+1) \in \mathbb{R}^2 \mid |\xi|^{\frac{p}{p-1}} + |\zeta|^{\frac{p}{p-1}} \le 1\right\}.$$

In addition, we notice that under this case $c_i(x,\varepsilon) = 1$, $d_i(x,\varepsilon) = 1$ and

$$a_i(x,\varepsilon) = \frac{|x_i - l_i|^{p-1}}{\left(\sqrt[p]{|x_i - l_i|^p + 2\varepsilon^p}\,\right)^{p-1}} - 1,\ \ b_i(x,\varepsilon) = \frac{\varepsilon^{p-1}}{\left(\sqrt[p]{|x_i - l_i|^p + 2\varepsilon^p}\,\right)^{p-1}} - 1.$$

Therefore, from Proposition 4.15(a3), it follows that

$$\begin{aligned}
\nabla_x(\Psi^p_{_{\rm FB}})^i(x,\varepsilon)^{\mathsf T} &= (a_i(x,\varepsilon) + b_i(x,\varepsilon))e_i^{\mathsf T} + b_i(x,\varepsilon)d_i(x,\varepsilon)F_i'(x) \\
&= -\left(1 - \frac{|x_i - l_i|^{p-1} + \varepsilon^{p-1}}{\left(\sqrt[p]{|x_i - l_i|^p + 2\varepsilon^p}\,\right)^{p-1}} + 1\right) e_i^{\mathsf T} \\
&\quad - \left(-\frac{\varepsilon^{p-1}}{\left(\sqrt[p]{|x_i - l_i|^p + 2\varepsilon^p}\,\right)^{p-1}} + 1\right) F_i'(x).
\end{aligned}$$

We next want to prove that for any $0 < \varepsilon \le \varepsilon_2(x,\delta)$,

$$\left|1 - \frac{|x_i - l_i|^{p-1} + \varepsilon^{p-1}}{\left(\sqrt[p]{|x_i - l_i|^p + 2\varepsilon^p}\,\right)^{p-1}}\right|^{\frac{p}{p-1}} + \left|\frac{\varepsilon^{p-1}}{\left(\sqrt[p]{|x_i - l_i|^p + 2\varepsilon^p}\,\right)^{p-1}}\right|^{\frac{p}{p-1}} \le 1, \quad (4.50)$$

and consequently $\nabla_x(\Psi^p_{_{\rm FB}})^i(x,\varepsilon)^{\mathsf T} \in \partial_C(\Phi^p_{_{\rm FB}})^i(x)$. It is easily verified that the function

$$h_1(\varepsilon) = \frac{|x_i - l_i|^{p-1} + \varepsilon^{p-1}}{(\sqrt[p]{|x_i - l_i|^p + 2\varepsilon^p}\,)^{p-1}}$$

is increasing on $[0, \varepsilon_2(x,\delta)]$. Since $h_1(0) = 1$, we have $h_1(\varepsilon) \ge 1$ on $[0, \varepsilon_2(x,\delta)]$. Therefore,

$$\left|1-\frac{|x_i-l_i|^{p-1}+\varepsilon^{p-1}}{\left(\sqrt[p]{|x_i-l_i|^p+2\varepsilon^p}\right)^{p-1}}\right|^{\frac{p}{p-1}}+\left|\frac{\varepsilon^{p-1}}{\left(\sqrt[p]{|x_i-l_i|^p+2\varepsilon^p}\right)^{p-1}}\right|^{\frac{p}{p-1}}$$
$$=\left(\frac{|x_i-l_i|^{p-1}+\varepsilon^{p-1}}{\left(\sqrt[p]{|x_i-l_i|^p+2\varepsilon^p}\right)^{p-1}}-1\right)^{\frac{p}{p-1}}+\frac{\varepsilon^p}{|x_i-l_i|^p+2\varepsilon^p}$$
$$:=h_2(\varepsilon).$$

We can verify that $h_2(\varepsilon)$ is strictly increasing on $[0,\varepsilon_2(x,\delta)]$ and

$$h_2(\varepsilon_2(x,\delta))=h_2(|x_i-l_i|/2)\le\left[\left(1+\frac{1}{2^{p-1}}\right)^{1/p}-1\right]^{\frac{p}{p-1}}+1/2$$
$$\le\left[1+(1/2)^{\frac{p-1}{p}}-1\right]^{\frac{p}{p-1}}+1/2\le 1,$$

where the second inequality is since $(1+t)^{1/p}\le 1+t^{1/p}$ for $t\ge 0$. The last two equations imply that the inequality (4.50) holds. Consequently,

$$\operatorname{dist}\left(\nabla_x(\Psi^p_{\text{FB}})^i(x,\varepsilon)^{\mathsf T},\partial_C(\Phi^p_{\text{FB}})^i(x)\right)\le\frac{\delta}{\sqrt{n}}\quad\text{for all } 0<\varepsilon\le\varepsilon_2(x,\delta).$$

Case 7: $i\in\bar{\beta}_4(x)$. Since $(u_i-x_i,-F_i(x))\ne(0,0)$, by Propositions 4.16(c) and 4.15(a3),

$$\begin{aligned}
&\operatorname{dist}\left(\nabla_x(\Psi^p_{\text{FB}})^i(x,\varepsilon)^{\mathsf T},\partial_C(\Phi^p_{\text{FB}})^i(x)\right)\\
&=\|(a_i(x,\varepsilon)-a_i(x))\,e_i+(b_i(x,\varepsilon)c_i(x,\varepsilon)-b_i(x)c_i(x))\,e_i\\
&\quad+(b_i(x,\varepsilon)d_i(x,\varepsilon)-b_i(x)c_i(x))\,\nabla F_i(x)\|\\
&=\|(a_i(x,\varepsilon)-a_i(x))\,e_i+(b_i(x,\varepsilon)-b_i(x))\,c_i(x)e_i+(b_i(x,\varepsilon)-b_i(x))\,d_i(x)\nabla F_i(x)\\
&\quad+b_i(x,\varepsilon)\,(c_i(x,\varepsilon)-c_i(x))\,e_i+b_i(x,\varepsilon)\,(d_i(x,\varepsilon)-d_i(x))\,\nabla F_i(x)\|\,.
\end{aligned}\tag{4.51}$$

In what follows, we will successively estimate the value of $|a_i(x,\varepsilon)-a_i(x)|$, $|b_i(x,\varepsilon)-b_i(x)|$, $|c_i(x,\varepsilon)-c_i(x)|$ and $|d_i(x,\varepsilon)-d_i(x)|$ for $0<\varepsilon<\varepsilon_3(x,\delta)$. Note that $\psi^p_{\text{FB}}(u_i-x_i,-F_i(x),\varepsilon)$ and $\phi^p_{\text{FB}}(u_i-x_i,-F_i(x))$ have the same sign for all $0<\varepsilon\le\varepsilon_3(x,\delta)$. Indeed, if $\phi^p_{\text{FB}}(u_i-x_i,-F_i(x))\ge 0$, then $\psi^p_{\text{FB}}(u_i-x_i,-F_i(x),\varepsilon)>0$ clearly holds. Otherwise, since

$$\begin{aligned}
&\psi^p_{\text{FB}}(u_i-x_i,-F_i(x),\varepsilon)<0\\
\Longleftrightarrow\ &|u_i-x_i|^p+|F_i(x)|^p+\varepsilon^p<(u_i-x_i-F_i(x))^p,\\
\Longleftrightarrow\ &\varepsilon<\left((u_i-x_i-F_i(x))^p-|u_i-x_i|^p-|F_i(x)|^p\right)^{1/p},
\end{aligned}$$

the definition of $\varepsilon_3(x,\delta)$ implies $\psi^p_{\text{FB}}(u_i-x_i,-F_i(x),\varepsilon)<0$ for all $0<\varepsilon\le\varepsilon_3(x,\delta)$.

Step1: To estimate $|a_i(x,\varepsilon)-a_i(x)|$. For $0<\varepsilon\le\varepsilon_3(x,\delta)$, we first estimate

$$r(x,\varepsilon):=\left|\frac{1}{\|(x_i-l_i,\psi_{\rm FB}^p(u_i-x_i,-F_i(x),\varepsilon),\varepsilon)\|_p^{p-1}}-\frac{1}{\|(x_i-l_i,\phi_{\rm FB}^p(u_i-x_i,-F_i(x)))\|_p^{p-1}}\right|.$$

Let

$$g_1(\varepsilon):=|\psi_{\rm FB}^p(u_i-x_i,-F_i(x),\varepsilon)|^p-|\phi_{\rm FB}^p(u_i-x_i,-F_i(x))|^p$$

and

$$\Delta(\varepsilon):=\psi_{\rm FB}^p(u_i-x_i,-F_i(x),\varepsilon)-\phi_{\rm FB}^p(u_i-x_i,-F_i(x))$$

for $0<\varepsilon\le\varepsilon_3(x,\delta)$. If $\phi_p(u_i-x_i,-F_i(x))\ge 0$, then $\psi_{\rm FB}^p(u_i-x_i,-F_i(x),\varepsilon)>0$, and hence $g_1(\varepsilon)>0$. In addition, applying the Mean-Value Theorem and Lemma 4.8(c), we have

$$\begin{aligned}
g_1(\varepsilon)&=\psi_{\rm FB}^p(u_i-x_i,-F_i(x),\varepsilon)^p-\phi_{\rm FB}^p(u_i-x_i,-F_i(x))^p\\
&=p\left[\phi_{\rm FB}^p(u_i-x_i,-F_i(x))+t_1\Delta(\varepsilon)\right]^{p-1}\Delta(\varepsilon)\quad\text{for some } t_1\in(0,1)\\
&\le p\left[\phi_{\rm FB}^p(u_i-x_i,-F_i(x))+\varepsilon_3(x,\delta)\right]^{p-1}\varepsilon.
\end{aligned}\tag{4.52}$$

Under this case, taking into account the definition of $\alpha_3(x)$ and $a(x)$, we have

$$\begin{aligned}
r(x,\varepsilon)&=\frac{1}{\left\|(x_i-l_i,\phi_{\rm FB}^p(u_i-x_i,-F_i(x)))\right\|_p^{p-1}}-\frac{1}{\left\|(x_i-l_i,\psi_{\rm FB}^p(u_i-x_i,-F_i(x),\varepsilon),\varepsilon)\right\|_p^{p-1}}\\
&\le\left[\alpha_3(x)^{-\frac{p-1}{p}}-(\alpha_3(x)+g_1(\varepsilon)+\varepsilon^p)^{-\frac{p-1}{p}}\right]\\
&\le\left[\alpha(x)^{-\frac{p-1}{p}}-(\alpha(x)+g_1(\varepsilon)+\varepsilon^p)^{-\frac{p-1}{p}}\right]\\
&=\frac{(\alpha(x)+g_1(\varepsilon)+\varepsilon^p)^{\frac{p-1}{p}}-\alpha(x)^{\frac{p-1}{p}}}{[\alpha(x)(\alpha(x)+g_1(\varepsilon)+\varepsilon^p)]^{\frac{p-1}{p}}}\le\frac{(g_1(\varepsilon)+\varepsilon^p)^{\frac{p-1}{p}}}{\alpha(x)^{\frac{2(p-1)}{p}}}\le M_1(x)\varepsilon^{\frac{p-1}{p}}
\end{aligned}$$

where the first three inequalities are due to Lemma 4.7, the last one is by (4.52), and

$$M_1(x):=\left[\frac{p\left[\phi_{\rm FB}^p(u_i-x_i,-F_i(x))+\varepsilon_3(x,\delta)\right]^{p-1}+\varepsilon_3(x,\delta)^{p-1}}{\alpha(x)^{2/p}}\right]^{p-1}.$$

If $\phi_{\rm FB}^p(u_i-x_i,-F_i(x))<0$, then $\psi_{\rm FB}^p(u_i-x_i,-F_i(x),\varepsilon)<0$, and hence $g_1(\varepsilon)<0$. Now,

$$r(x,\varepsilon)\le\frac{1}{\left\|(x_i-l_i,\psi_{\rm FB}^p(u_i-x_i,-F_i(x),\varepsilon),\varepsilon)\right\|_p^{p-1}}-\frac{1}{\left\|(x_i-l_i,\phi_{\rm FB}^p(u_i-x_i,-F_i(x)),\varepsilon)\right\|_p^{p-1}}$$

$$
\begin{aligned}
&+\left|\frac{1}{\left\|(x_i-l_i,\phi^p_{\rm FB}(u_i-x_i,-F_i(x)),\varepsilon)\right\|_p^{p-1}}-\frac{1}{\left\|(x_i-l_i,\phi^p_{\rm FB}(u_i-x_i,-F_i(x)))\right\|_p^{p-1}}\right|\\
&\le\frac{\left[|\phi^p_{\rm FB}(u_i-x_i,-F_i(x))|^{p-1}-|\psi^p_{\rm FB}(u_i-x_i,-F_i(x),\varepsilon)|^{p-1}\right]}{\left\|(x_i-l_i,\psi^p_{\rm FB}(u_i-x_i,-F_i(x),\varepsilon),\varepsilon)\right\|_p^{p-1}\left\|(x_i-l_i,\phi^p_{\rm FB}(u_i-x_i,-F_i(x)),\varepsilon)\right\|_p^{p-1}}\\
&\quad+\left[\alpha_3(x)^{-\frac{p-1}{p}}-(\alpha_3(x)+\varepsilon^p)^{-\frac{p-1}{p}}\right]\\
&\le\frac{\left[|\phi^p_{\rm FB}(u_i-x_i,-F_i(x))|^{p-1}-|\psi^p_{\rm FB}(u_i-x_i,-F_i(x),\varepsilon)|^{p-1}\right]}{\left\|(x_i-l_i,\psi^p_{\rm FB}(u_i-x_i,-F_i(x),\varepsilon),\varepsilon)\right\|_p^{2p-2}}\\
&\quad+\left[\alpha_3(x)^{-\frac{p-1}{p}}-(\alpha_3(x)+\varepsilon^p)^{-\frac{p-1}{p}}\right].
\end{aligned}\tag{4.53}
$$

Notice that

$$
\begin{aligned}
&|\phi^p_{\rm FB}(u_i-x_i,-F_i(x))|^{p-1}-|\psi_p(u_i-x_i,-F_i(x),\varepsilon)|^{p-1}\\
&=\left[-\phi^p_{\rm FB}(u_i-x_i,-F_i(x))\right]^{p-1}-\left[-\psi^p_{\rm FB}(u_i-x_i,-F_i(x),\varepsilon)\right]^{p-1}\\
&=(p-1)\left[-\psi^p_{\rm FB}(u_i-x_i,-F_i(x),\varepsilon)+t_2\Delta(\varepsilon)\right]^{p-2}\Delta(\varepsilon)\ \text{ for some } t_2\in(0,1)\\
&\le\begin{cases}(p-1)\left[-\phi^p_{\rm FB}(u_i-x_i,-F_i(x))\right]^{p-2}\varepsilon & \text{if } p\ge 2;\\ (p-1)\left[-\psi^p_{\rm FB}(u_i-x_i,-F_i(x),\varepsilon_3(x,\delta))\right]^{p-2}\varepsilon & \text{if } 1<p<2,\end{cases}
\end{aligned}
$$

and

$$
\left\|(x_i-l_i,\psi^p_{\rm FB}(u_i-x_i,-F_i(x),\varepsilon),\varepsilon)\right\|_p\ge\left\|(x_i-l_i,\psi^p_{\rm FB}(u_i-x_i,-F_i(x),\varepsilon_3(x,\delta)))\right\|_p.
$$

Therefore,

$$
\begin{aligned}
&\frac{\left[|\phi^p_{\rm FB}(u_i-x_i,-F_i(x))|^{p-1}-|\psi^p_{\rm FB}(u_i-x_i,-F_i(x),\varepsilon)|^{p-1}\right]}{\left\|(x_i-l_i,\psi^p_{\rm FB}(u_i-x_i,-F_i(x),\varepsilon),\varepsilon)\right\|_p^{2p-2}}\\
&\le(p-1)\frac{\left[-\phi^p_{\rm FB}(u_i-x_i,-F_i(x))\right]^{p-2}\varepsilon+\left[-\psi^p_{\rm FB}(u_i-x_i,-F_i(x),\varepsilon_3(x,\delta))\right]^{p-2}\varepsilon}{\left\|(x_i-l_i,\psi^p_{\rm FB}(u_i-x_i,-F_i(x),\varepsilon_3(x,\delta))\right\|_p^{2p-2}}\\
&\le(p-1)\left(\frac{\left[-\phi^p_{\rm FB}(u_i-x_i,-F_i(x))\right]^{p-2}}{\left\|(x_i-l_i,\psi^p_{\rm FB}(u_i-x_i,-F_i(x),\varepsilon_3(x,\delta)))\right\|_p^{p-2}}+1\right)\varepsilon\\
&:=M_2(x)\varepsilon.
\end{aligned}
$$

This together with (4.53) and Lemma 4.7 yields

$$\begin{aligned}
r(x,\varepsilon) \leq & M_2(x)\varepsilon + \frac{(\alpha(x)+\varepsilon^p)^{\frac{p-1}{p}} - \alpha(x)^{\frac{p-1}{p}}}{[\alpha(x)(\alpha(x)+\varepsilon^p)]^{\frac{p-1}{p}}} \\
\leq & M_2(x)\varepsilon + \frac{\varepsilon^{p-1}}{\alpha(x)^{\frac{2(p-1)}{p}}} \\
\leq & \begin{cases} M_3(x)\varepsilon & \text{if } p \geq 2 \\ M_3(x)\varepsilon^{p-1} & \text{if } 1 < p < 2 \end{cases}
\end{aligned}$$

where

$$M_3(x) := \begin{cases} M_2(x) + \dfrac{\varepsilon_3(x,\delta)^{p-2}}{\alpha(x)^{\frac{2p-2}{p}}} & \text{if } p \geq 2; \\ M_2(x)\varepsilon_3(x,\delta)^{2-p} + \alpha(x)^{\frac{2-2p}{p}} & \text{if } 1 < p < 2. \end{cases}$$

Summing up the above discussions, it then follows that

$$\begin{aligned}
r(x,\varepsilon) \leq & \begin{cases} \max\left\{M_1(x), M_3(x)\varepsilon_3(x,\delta)^{1/p}\right\}\varepsilon^{\frac{p-1}{p}} & \text{if } p \geq 2; \\ \max\left\{M_1(x), M_3(x)\varepsilon_3(x,\delta)^{(p+\frac{1}{p}-2)}\right\}\varepsilon^{\frac{p-1}{p}} & \text{if } 1 < p < 2, \end{cases} \\
\leq & M_4(x)\varepsilon^{\frac{p-1}{p}},
\end{aligned}$$

where

$$M_4(x) := \max\left\{M_1(x), M_3(x)\varepsilon_3(x,\delta)^{1/p}, M_3(x)\varepsilon_3(x,\delta)^{(p+1/p-2)}\right\}.$$

Consequently, we establish

$$|a_i(x,\varepsilon) - a_i(x)| = r(x,\varepsilon)|x_i - l_i|^{p-1} \leq M_4(x)|x_i - l_i|^{p-1}\varepsilon^{\frac{p-1}{p}}.$$

Step 2: To estimate $|b_i(x,\varepsilon) - b_i(x)|$. From the expressions of $b_i(x,\varepsilon)$ and $b_i(x)$,

$$\begin{aligned}
|b_i(x,\varepsilon) - b_i(x)| = & \left| \frac{\operatorname{sgn}(\psi_{\text{FB}}^p(u_i - x_i, -F_i(x), \varepsilon))|\psi_{\text{FB}}^p(u_i - x_i, -F_i(x), \varepsilon)|^{p-1}}{\left\|\left(x_i - l_i, \psi_p(u_i - x_i, -F_i(x), \varepsilon), \varepsilon\right)\right\|_p^{p-1}} \right. \\
& - \frac{\operatorname{sgn}(\phi_{\text{FB}}^p(u_i - x_i, -F_i(x)))|\phi_{\text{FB}}^p(u_i - x_i, -F_i(x))|^{p-1}}{\left\|\left(x_i - l_i, \psi_p(u_i - x_i, -F_i(x), \varepsilon), \varepsilon\right)\right\|_p^{p-1}} \\
& + \frac{\operatorname{sgn}(\phi_{\text{FB}}^p(u_i - x_i, -F_i(x)))|\phi_{\text{FB}}^p(u_i - x_i, -F_i(x))|^{p-1}}{\left\|\left(x_i - l_i, \psi_p(u_i - x_i, -F_i(x), \varepsilon), \varepsilon\right)\right\|_p^{p-1}} \\
& \left. - \frac{\operatorname{sgn}(\phi_{\text{FB}}^p(u_i - x_i, -F_i(x)))|\phi_{\text{FB}}^p(u_i - x_i, -F_i(x))|^{p-1}}{\left\|\left(x_i - l_i, \phi_{\text{FB}}^p(u_i - x_i, -F_i(x))\right)\right\|_p^{p-1}} \right|
\end{aligned}$$

$$\le \frac{g_2(\varepsilon)}{\left\|(x_i - l_i, \psi^p_{\text{FB}}(u_i - x_i, -F_i(x), \varepsilon), \varepsilon)\right\|_p^{p-1}} + r(x, \varepsilon)|\phi^p_{\text{FB}}(u_i - x_i, -F_i(x))|^{p-1}, \tag{4.54}$$

where $r(x, \varepsilon)$ is same as above, and $g_2(\varepsilon)$ is defined by

$$\begin{aligned} g_2(\varepsilon) := \big|&\text{sgn}(\psi^p_{\text{FB}}(u_i - x_i, -F_i(x), \varepsilon))|\psi^p_{\text{FB}}(u_i - x_i, -F_i(x), \varepsilon)|^{p-1} \\ &-\text{sgn}(\phi^p_{\text{FB}}(u_i - x_i, -F_i(x)))|\phi^p_{\text{FB}}(u_i - x_i, -F_i(x))|^{p-1}\big|. \end{aligned}$$

If $\phi^p_{\text{FB}}(u_i - x_i, -F_i(x)) \ge 0$, then $\psi^p_{\text{FB}}(u_i - x_i, -F_i(x), \varepsilon) > 0$, and therefore

$$\begin{aligned} g_2(\varepsilon) &= \psi^p_{\text{FB}}(u_i - x_i, -F_i(x), \varepsilon)^{p-1} - \phi^p_{\text{FB}}(u_i - x_i, -F_i(x))^{p-1} \\ &= (p-1)\left[\phi^p_{\text{FB}}(u_i - x_i, -F_i(x)) + t_3\Delta(\varepsilon)\right]^{p-2}\Delta(\varepsilon) \ \text{ for some } t_3 \in (0, 1) \\ &\le \begin{cases} (p-1)\left[\phi^p_{\text{FB}}(u_i - x_i, -F_i(x)) + \varepsilon_3(x, \delta)\right]^{p-2}\varepsilon & \text{if } p \ge 2; \\ (p-1)\left[\phi^p_{\text{FB}}(u_i - x_i, -F_i(x))\right]^{p-2}\varepsilon & \text{if } 1 < p < 2. \end{cases} \end{aligned}$$

If $\phi^p_{\text{FB}}(u_i - x_i, -F_i(x)) < 0$, then $\psi^p_{\text{FB}}(u_i - x_i, -F_i(x), \varepsilon) < 0$ and

$$|\psi^p_{\text{FB}}(u_i - x_i, -F_i(x), \varepsilon)|^{p-1} < |\phi^p_{\text{FB}}(u_i - x_i, -F_i(x))|^{p-1} \quad \text{for} \quad 0 < \varepsilon \le \varepsilon_3(x, \delta).$$

Consequently,

$$\begin{aligned} g_2(\varepsilon) &= \left[-\phi^p_{\text{FB}}(u_i - x_i, -F_i(x))\right]^{p-1} - \left[-\psi^p_{\text{FB}}(u_i - x_i, -F_i(x), \varepsilon)\right]^{p-1} \\ &= (p-1)\left[-\psi^p_{\text{FB}}(u_i - x_i, -F_i(x), \varepsilon) + t_4\Delta(\varepsilon)\right]^{p-2}\Delta(\varepsilon) \ \text{ for some } t_4 \in (0, 1) \\ &\le \begin{cases} (p-1)\left[-\phi^p_{\text{FB}}(u_i - x_i, -F_i(x))\right]^{p-2}\varepsilon & \text{if } p \ge 2; \\ (p-1)\left[-\psi^p_{\text{FB}}(u_i - x_i, -F_i(x), \varepsilon_3(x, \delta))\right]^{p-2}\varepsilon & \text{if } 1 < p < 2. \end{cases} \end{aligned}$$

In addition, if $\phi^p_{\text{FB}}(u_i - x_i, -F_i(x)) \ge 0$, then

$$\left\|(x_i - l_i, \psi^p_{\text{FB}}(u_i - x_i, -F_i(x), \varepsilon), \varepsilon)\right\|_p^{p-1} > \left\|(x_i - l_i, \phi^p_{\text{FB}}(u_i - x_i, -F_i(x)))\right\|_p^{p-1},$$

whereas if $\phi^p_{\text{FB}}(u_i - x_i, -F_i(x)) < 0$, then for all $0 < \varepsilon \le \varepsilon_3(x, \delta)$,

$$\begin{aligned} \left\|(x_i - l_i, \psi^p_{\text{FB}}(u_i - x_i, -F_i(x), \varepsilon), \varepsilon)\right\|_p^{p-1} &\ge |\psi^p_{\text{FB}}(u_i - x_i, -F_i(x), \varepsilon)|^{p-1} \\ &\ge |\psi^p_{\text{FB}}(u_i - x_i, -F_i(x), \varepsilon_3(x, \delta))|^{p-1}. \end{aligned}$$

The align discussions show that for all $0 < \varepsilon \le \varepsilon_3(x, \delta)$, we have

$$\frac{g_2(\varepsilon)}{\left\|(x_i - l_i, \psi^p_{\text{FB}}(u_i - x_i, -F_i(x), \varepsilon), \varepsilon)\right\|_p^{p-1}} \le (p-1)M_5(x)\varepsilon,$$

where

$$M_5(x) := \begin{cases} \dfrac{\left[|\phi^p_{\text{FB}}(u_i - x_i, -F_i(x))| + \varepsilon_3(x,\delta)\right]^{p-2}}{\left\|(x_i - l_i, \phi^p_{\text{FB}}(u_i - x_i, -F_i(x)))\right\|_p^{p-1}} & \text{if } p \ge 2, \\ \dfrac{\max\{|\phi^p_{\text{FB}}(u_i - x_i, -F_i(x))|^{p-2}, |\psi^p_{\text{FB}}(u_i - x_i, -F_i(x), \varepsilon_3(x,\delta))|^{p-2}\}}{|\psi^p_{\text{FB}}(u_i - x_i, -F_i(x), \varepsilon_3(x,\delta))|^{p-1}} & \text{if } 1 < p < 2. \end{cases}$$

This along with (4.54) and the result of Step 1 gives $|b_i(x,\varepsilon) - b_i(x)| \le M_6(x)\varepsilon^{\frac{p-1}{p}}$ where

$$M_6(x) := (p-1)M_5(x)\varepsilon_3(x,\delta)^{1/p} + M_4(x)|\phi^p_{\text{FB}}(u_i - x_i, -F_i(x))|^{p-1}. \quad (4.55)$$

Step 3: To estimate $|c_i(x,\varepsilon) - c_i(x)|$ and $|d_i(x,\varepsilon) - d_i(x)|$. Using Lemma 4.7 gives

$$\begin{aligned} |c_i(x,\varepsilon) - c_i(x)| &= \left| \frac{\text{sgn}(u_i - x_i)|u_i - x_i|^{p-1}}{\|(u_i - x_i, -F_i(x), \varepsilon)\|_p^{p-1}} - \frac{\text{sgn}(u_i - x_i)|u_i - x_i|^{p-1}}{\|(u_i - x_i, -F_i(x)\|_p^{p-1}} \right| \\ &= \frac{|u_i - x_i|^{p-1}}{\|(u_i - x_i, -F_i(x)\|_p^{p-1}} - \frac{|u_i - x_i|^{p-1}}{\|(u_i - x_i, -F_i(x), \varepsilon)\|_p^{p-1}} \\ &\le \left[\alpha_2(x)^{\frac{1-p}{p}} - \left(\alpha_2(x) + \varepsilon^p\right)^{\frac{1-p}{p}}\right] |u_i - x_i|^{p-1} \\ &\le \left[\alpha(x)^{\frac{1-p}{p}} - \left(\alpha(x) + \varepsilon^p\right)^{\frac{1-p}{p}}\right] |u_i - x_i|^{p-1} \\ &= \frac{(\alpha(x) + \varepsilon^p)^{\frac{p-1}{p}} - \alpha(x)^{\frac{p-1}{p}}}{[\alpha(x)(\alpha(x) + \varepsilon)]^{\frac{p-1}{p}}} |u_i - x_i|^{p-1} \\ &\le \frac{|u_i - x_i|^{p-1}\varepsilon^{p-1}}{\alpha(x)^{\frac{2p-2}{p}}}. \end{aligned}$$

Using the similar arguments, we also have $|d_i(x,\varepsilon) - d_i(x)| \le \frac{|F_i(x)|^{p-1}\varepsilon^{p-1}}{\alpha(x)^{\frac{2p-2}{p}}}$.

Now using (4.51) and the results of the above three steps, and noting that $|b_i(x,\varepsilon)| \le 1$, $|d_i(x)| \le 1$ and $|c_i(x)| \le 1$, it follows that for all $0 < \varepsilon \le \varepsilon_3(x,\delta)$,

$$\begin{aligned} &\text{dist}\left(\nabla_x(\Psi^p_{\text{FB}})^i(x,\varepsilon)^{\mathsf{T}}, \partial_C(\Phi^p_{\text{FB}})^i(x)\right) \\ &\le M_4(x)|x_i - l_i|\varepsilon^{\frac{p-1}{p}} + M_6(x)\left(1 + \|\nabla F_i(x)\|\right)\varepsilon^{\frac{p-1}{p}} \\ &+ \frac{|u_i - x_i|^{p-1}\varepsilon^{p-1}}{\alpha(x)^{\frac{2p-2}{p}}} + \frac{|F_i(x)|^{p-1}\varepsilon^{p-1}}{\alpha(x)^{\frac{2p-2}{p}}} \\ &\le M(x)\varepsilon^{\frac{p-1}{p}} \end{aligned}$$

where

$$M(x) := M_4(x)|x_i - l_i| + M_6(x)\,(1 + \|\nabla F_i(x)\|) + \frac{\gamma_3(x)}{\alpha(x)^{\frac{2p-2}{p}}}.$$

Therefore, when $i \in \bar{\beta}_4(x)$, we have

$$\operatorname{dist}\left(\nabla_x(\Psi^p_{\text{FB}})^i(x,\varepsilon)^{\mathsf{T}}, \partial_C(\Phi^p_{\text{FB}})^i(x)\right) \le \frac{\delta}{\sqrt{n}} \quad \text{for all } 0 < \varepsilon \le \left(\frac{\delta}{\sqrt{n}M(x)}\right)^{\frac{p-1}{p}}.$$

Based on the analysis of the preceding seven cases and the definition of $\varepsilon(x,\delta)$, the desired result follows. □

We are now prepared to describe the iterative steps of the smoothing algorithm, which is based on the smooth approximation $\Psi^p_{\text{FB}}(x,\varepsilon) = 0$ of the original equation $\Phi^p_{\text{FB}}(x) = 0$. In addition, we will present both global and local convergence results for the algorithm. To this end, we first introduce the following merit functions:

$$\Theta_p(x) := \frac{1}{2}\|\Phi^p_{\text{FB}}(x)\|^2$$

and

$$H_p(x,\varepsilon) := \frac{1}{2}\|\Psi^p_{\text{FB}}(x,\varepsilon)\|^2.$$

Algorithm 4.3 (Smoothing Algorithm)

(S.0) Given a starting point $x^0 \in \mathbb{R}^n$, the parameters $\rho, \alpha, \eta \in (0,1)$ and $\nu \in (0,+\infty)$. Choose $\sigma \in (0,(1-\alpha)/2)$. Let $\beta_0 = \|\Phi^p_{\text{FB}}(x^0)\|$ and $\varepsilon_0 := \frac{\alpha}{2\sqrt{n}}$. Set $k := 0$.

(S.1) Solve the following linear system of equations

$$\Phi^p_{\text{FB}}(x^k) + \Psi^p_{\text{FB}}(x^k,\varepsilon^k)d = 0,$$

and denote its solution by d^k.

(S.2) Let m_k be the smallest nonnegative integer m such that

$$H_p(x^k + \rho^m d^k, \varepsilon^k) - H_p(x^k,\varepsilon^k) \le -2\sigma\rho^m\Theta_p(x^k).$$

Set $t_k := \rho^{m_k}$ and $x^{k+1} := x^k + t_k d^k$.

(S.3) If $\|\Phi^p_{\text{FB}}(x^{k+1})\| = 0$, then terminate. If

$$0 < \|\Phi^p_{\text{FB}}(x^{k+1})\| \le \max\left\{\eta\beta_k, \alpha^{-1}\|\Phi^p_{\text{FB}}(x^{k+1}) - \Psi^p_{\text{FB}}(x^{k+1},\varepsilon^k)\|\right\}, \quad (4.56)$$

let $\beta_{k+1} = \|\Phi^p_{\text{FB}}(x^{k+1})\|$ and choose an ε_{k+1} satisfying

$$0 < \varepsilon_{k+1} \le \min\left\{\frac{\alpha\beta_{k+1}}{2\sqrt{n}}, \frac{\varepsilon_k}{2}\right\} \tag{4.57}$$

and

$$\operatorname{dist}\left(\nabla_x \Psi^p_{\text{FB}}(x^{k+1}, \varepsilon^{k+1}), \partial_C \Phi^p_{\text{FB}}(x^{k+1})\right) \le \beta_{k+1}\nu. \tag{4.58}$$

If $\|\Phi^p_{\text{FB}}(x^{k+1})\| > 0$ but (4.56) does not hold, then let $\beta_{k+1} = \beta_k$ and $\varepsilon_{k+1} = \varepsilon_k$.
(S.4) Set $k := k+1$, and go to (S.1).

In Algorithm 4.3, the parameter σ, chosen from the interval $\left(0, \frac{(1-\alpha)}{2}\right)$, serves two purposes: it ensures the existence of m_k in step (S.2) and facilitates the superlinear convergence analysis of the algorithm. The initial values β_0 and ε_0 are set to $\|\Phi_{\text{FB}}(x^0)\|$ and $\frac{\alpha}{2\sqrt{n}}$, respectively, primarily for the purpose of establishing global convergence. These parameter choices are also adopted in the numerical experiments. Algorithm 4.3 is computationally efficient, requiring only the solution of a linear system at each iteration. Owing to the Jacobian consistency property of the operator Ψ^p_{FB}, it is always possible to find an $\varepsilon_{k+1} > 0$ that satisfies conditions (4.57) and (4.58), by construction. Moreover, Lemma 4.9 provides explicit guidance on how to select such an $\varepsilon_{k+1} > 0$ for the MCP (4.35).

Lemma 4.10 *For any fixed $\varepsilon > 0$, the Jacobian matrix of Ψ^p_{FB} at any $x \in \mathbb{R}^n$ is nonsingular if F is a P_0-function and the submatrix $[F'(x)]_{I_f I_f}$ is nonsingular. Particularly, if $I_f = \emptyset$, the Jacobian matrix of Ψ^p_{FB} at any $x \in \mathbb{R}^n$ is nonsingular if and only if F is a P_0-function.*

Proof For any given $\varepsilon > 0$, the Jacobian matrix of Ψ^p_{FB} at any $x \in \mathbb{R}^n$ is given by

$$\nabla_x \Psi^p_{\text{FB}}(x, \varepsilon)^{\mathsf{T}} = D_a(x, \varepsilon) + D_b(x, \varepsilon)F'(x)$$

where $D_a(x, \varepsilon)$ and $D_b(x, \varepsilon)$ are $n \times n$ diagonal matrices whose diagonal elements $(D_a)_{ii}(x, \varepsilon)$ and $(D_b)_{ii}(x, \varepsilon)$ are negative for $i \in I_l \cup I_u \cup I_{lu}$, and $(D_a)_{ii}(x, \varepsilon) = 0$, $(D_b)_{ii}(x, \varepsilon) = -1$ for $i \in I_f$. Now suppose that $\nabla_x \Psi^p_{\text{FB}}(x, \varepsilon)^{\mathsf{T}} z = 0$. Then,

$$z_i = -\frac{(D_b)_{ii}(x, \varepsilon)}{(D_a)_{ii}(x, \varepsilon)}\left(F'(x)z\right)_i, \quad \text{for } i \in I_l \cup I_u \cup I_{lu}$$

and

$$\left(F'(x)z\right)_i = 0, \quad \text{for } i \in I_f. \tag{4.59}$$

Since F is a continuously differentiable P_0-function, $F'(x)$ is a P_0-matrix. From Lemma 1.5, we obtain $z_i = 0$ for $i \in I_l \cup I_u \cup I_{lu}$. Substituting this into (4.59), we obtain

$$\left[F'(x)_{I_f I_f}\right] z_{I_f} = 0,$$

where z_{I_f} is a vector consisting of z_i with $i \in I_f$. This along with the nonsingularity of $[F'(x)]_{I_f I_f}$ implies $z_i = 0$ for $i \in I_f$. Thus, we prove $z = 0$, and consequently the first part of the conclusions follows. The second part is implied by the above arguments. □

Remark 4.1 We want to point out when $p \to +\infty$, the diagonal elements $(D_a)_{ii}(x, \varepsilon)$ and $(D_b)_{ii}(x, \varepsilon)$ for $i \in I_l \cup I_u \cup I_{lu}$ will tend to 0, though $(D_a)_{ii}(x, \varepsilon) + (D_b)_{ii}(x, \varepsilon) < 0$. This implies that for a larger p the nonsingularity of $\nabla \Psi^p_{\text{FB}}(x, \varepsilon)$ actually requires stronger conditions than those given by Lemma 4.10.

By Lemma 4.10 and [36, Lemma 3.1], Algorithm 4.3 is well defined under the assumptions that F is a P_0-function and the submatrix $[F'(x)]_{I_f I_f}$ is nonsingular. The following lemma provides a sufficient condition to ensure that the merit function $\Theta_p(x)$ possesses bounded level sets.

Lemma 4.11 *The level sets* $\mathcal{L}(\gamma) := \left\{x \in \mathbb{R}^n \mid \|\Phi^p_{\text{FB}}(x)\| \le \gamma\right\}$ *are bounded for all* $\gamma > 0$ *if one of the following two conditions is satisfied:*

(a) l and u are both bounded.
(b) F is a uniform P-function.

Proof Under the condition (a), we have $\{1, 2, \ldots, n\} = I_{lu}$. The result is clear by the definition of Φ^p_{FB} and Lemma 2.1(d). Next we prove the boundedness of $\mathcal{L}(\gamma)$ under the condition (b). Suppose that there exists some $\gamma > 0$ such that $\mathcal{L}(\gamma)$ is unbounded, i.e., there exists a sequence $\{x^k\} \subseteq \mathcal{L}(\gamma)$ such that $\|x^k\| \to \infty$. Define the index set

$$J := \left\{i \in \{1, 2, \ldots, n\} \mid \{x_i^k\} \text{ is unbounded}\right\}.$$

Then $J \neq \emptyset$. We choose a bounded sequence y^k with

$$y_i^k = \begin{cases} 0 & \text{if } i \in J, \\ x_i^k & \text{otherwise.} \end{cases}$$

Since F is a uniform P-function, there is a constant $\mu > 0$ such that

$$\begin{aligned} \mu \|x^k - y^k\|^2 &\le \max_{1 \le i \le n} (x_i^k - y_i^k)(F_i(x^k) - F_i(y^k)) \\ &= \max_{i \in J} (x_i^k)(F_i(x^k) - F_i(y^k)) \\ &\le |x_{j_0}^k| |F_{j_0}(x^k) - F_{j_0}(y^k)| \end{aligned}$$

where j_0 is an index from $\{1, 2, \ldots, n\}$ for which the maximum is attained, and without loss of generality it is assumed to be independent of k. Clearly, $j_0 \in J$, which means that $\{x^k_{j_0}\}$ is unbounded. Consequently, there exists a subsequence, assumed to be $\{x^k_{j_0}\}$ without loss of generality, such that $|x^k_{j_0}| \to \infty$. Notice that

$$\|x^k - y^k\|^2 \geq |x^k_{j_0} - y^k_{j_0}|^2 = |x^k_{j_0}|^2 \quad \text{for each } k.$$

Therefore, $\mu |x^k_{j_0}|^2 \leq |x^k_{j_0}||F_{j_0}(x^k) - F_{j_0}(y^k)|$ and

$$\mu |x^k_{j_0}| \leq |F_{j_0}(x^k) - F_{j_0}(y^k)| \leq |F_{j_0}(x^k)| + |F_{j_0}(y^k)|,$$

which in turn implies $|F_{j_0}(x^k)| \to \infty$ as $|x^k_{j_0}| \to \infty$. Thus, we prove that

$$|x^k_{j_0}| \to +\infty \quad \text{and} \quad |F_{j_0}(x^k)| \to +\infty. \tag{4.60}$$

On the other hand, we notice that (4.60) implies

$$\left|x^k_{j_0} - l_i\right| \to +\infty \quad \text{and} \quad \left|F_{j_0}(x^k)\right| \to +\infty.$$

Combining the last two equations with Lemma 2.1 (d), we have $\left|(\Phi^p_{\text{FB}})^{j_0}(x^k)\right| \to +\infty$ from the definition of Φ^p_{FB}. This contradicts the fact that $\{x^k\} \subseteq \mathcal{L}(\gamma)$. □

Proposition 4.18 *Suppose that F is a uniform P-function. Then, the iteration sequence $\{x^k\}$ generated by Algorithm 4.3 is well defined and converges to the unique solution x^* of the MCP (4.35) superlinearly. Furthermore, if F' is locally Lipschitz continuous around x^*, then the convergence rate is quadratic.*

Proof Using Lemma 4.10 and Lemma 4.11, following the same arguments as in [36], the desired global and local convergence results are obtained. □

Detailed numerical experiments related to Algorithm 4.3 are reported in [37]. In general, it has been observed that using a smaller value of p, particularly with $p < 2$, tends to yield greater robustness compared to larger values of p. However, as p approaches 1, the algorithm typically requires more iterations to converge. Based on these findings, choosing p within the range $[1.1, 2]$ is generally recommended for a good balance between robustness and efficiency. Additionally, the parameter α also influences the numerical performance of Algorithm 4.3; empirical results suggest that selecting α from the interval $[0.3, 0.7]$ tends to produce favorable outcomes.

4.4 Regularization Approach

It is well known that the regularization approach is developed to address ill-posed problems by replacing the original problem with a sequence of well-posed approximations, whose solutions converge to that of the original problem; see [38–40] and references therein. In this section, we present a regularization framework for solving complementarity problems. Specifically, we consider two distinct settings: (i) applying the regularization approach to the NCP using ϕ^p_{FB} and (ii) applying it to the SOCCP using ϕ_{FB}.

In the context of nonlinear complementarity problems (2.1), if we consider the so-called *Tikhonov regularization*, this scheme consists of solving a sequence of nonlinear complementarity problems NCP(F_ε):

$$x \geq 0, \quad F_\varepsilon(x) \geq 0, \quad \langle x, F_\varepsilon(x)\rangle = 0, \tag{4.61}$$

where $\varepsilon > 0$ is a parameter tending to zero and F_ε is given by

$$F_\varepsilon(x) \ := \ F(x) + \varepsilon x. \tag{4.62}$$

Let $F_{\varepsilon,i}(x)$ denote the i-th component of $F_\varepsilon(x)$ given as in (4.62) and define the map $\Phi_{p,\varepsilon} : \mathbb{R}^n \to \mathbb{R}^n$ by

$$\Phi_{p,\varepsilon}(x) := \begin{pmatrix} \phi^p_{\text{FB}}(x_1, F_{\varepsilon,1}(x)) \\ \cdot \\ \cdot \\ \cdot \\ \phi^p_{\text{FB}}(x_n, F_{\varepsilon,n}(x)) \end{pmatrix}. \tag{4.63}$$

Then the regularized problem NCP(F_ε) for any given $\varepsilon > 0$ (described as in (4.61)) can be reformulated as

$$\Phi_{p,\varepsilon}(x) = 0,$$

which leads to a merit function $\Psi_{p,\varepsilon} : \mathbb{R}^n \to \mathbb{R}_+$ for the NCP(F_ε):

$$\Psi_{p,\varepsilon}(x) := \frac{1}{2}\|\Phi_{p,\varepsilon}(x)\|^2 = \frac{1}{2}\sum_{i=1}^{n} \phi^p_{\text{FB}}(x_i, F_{\varepsilon,i}(x))^2. \tag{4.64}$$

Therefore, the original NCP (2.1) is effectively equivalent to solving a sequence of nonsmooth systems of equations $\Phi_{p,\varepsilon}(x) = 0$ as $\varepsilon \to 0$. In this context, the parameter ε serves a role analogous to that of smoothing parameter in traditional smoothing methods for the NCP, with the key distinction that ε is applied to the mapping F, rather than to the NCP function ϕ^p_{FB}.

As will be shown later, the sequence of subproblems $\Phi_{p,\varepsilon}(x) = 0$, with $\varepsilon \to 0$, will be approximately solved using a generalized Newton method applied to an augmented system of equations that is equivalent to the original NCP. Specifically, we let $z := (\varepsilon, x) \in \mathbb{R}_+ \times \mathbb{R}^n$ by viewing ε as a variable, and define the mapping $H_p : \mathbb{R}_+ \times \mathbb{R}^n \to \mathbb{R}^{n+1}$ by

$$H_p(z) := \begin{bmatrix} \varepsilon \\ \phi^p_{\text{FB}}(x_1, F_{\varepsilon,1}(x)) \\ \vdots \\ \phi^p_{\text{FB}}(x_n, F_{\varepsilon,n}(x)) \end{bmatrix}. \tag{4.65}$$

Notice that if the function $\Phi_{p,\varepsilon}(x)$ defined by (4.63) is viewed as a function of ε and x, then we may denote it as $\Phi^p_{\text{FB}}(z) := \Phi^p_{\text{FB}}(\varepsilon, x) = \Phi_{p,\varepsilon}(x)$. Hence, (4.65) is the same as

$$H_p(z) = \begin{bmatrix} \varepsilon \\ \Phi^p_{\text{FB}}(z) \end{bmatrix}.$$

It is easily verified that the NCP is equivalent to the augmented system of equations

$$H_p(z) = H_p(\varepsilon, x) = 0, \tag{4.66}$$

which naturally induces a merit function $G_p : \mathbb{R}^{n+1} \to \mathbb{R}_+$ given by

$$G_p(z) = \frac{1}{2}\|H_p(z)\|^2 = \frac{1}{2}\left(\varepsilon^2 + \|\Phi_{p,\varepsilon}(x)\|^2\right) = \frac{1}{2}\varepsilon^2 + \Psi^p_{\text{FB}}(z). \tag{4.67}$$

The function H_p is locally Lipschitz continuous, owing to the Lipschitz continuity of ϕ^p_{FB}; see Proposition 2.1(e). Moreover, H_p is semismooth. Based on this, we employ the generalized Newton method developed in [20, 41] to solve (4.66), thereby formulating a regularized semismooth Newton-type algorithm in which each iteration approximately solves the regularized problem NCP(F_ε). Compared with the semismooth Newton method based on (2.16), this approach offers a notable advantage in handling P_0-NCPs, as the associated merit function $\Psi_{p,\varepsilon}(x)$ possesses bounded level sets for such problems.

Proposition 4.19 *The mapping $H_p : \mathbb{R}_+ \times \mathbb{R}^n \to \mathbb{R}^n$ defined as in (4.65) is semismooth. Moreover, it is strongly semismooth if F' is locally Lipschitz continuous.*

Proof Since a function is (strongly) semismooth if and only if its component functions are (strongly) semismooth, to prove that H_p is (strongly) semismooth we only need to prove that $H_{p,i},\ i = 1, 2, \cdots, n+1$ are (strongly) semismooth. Apparently, $H_{p,1}$ is strongly semismooth by formula (1.42) since $H_{p,1}(z) = \varepsilon$. For $H_{p,i},\ i = 2, 3, \cdots, n+1$, since ϕ_p is strongly semismooth by Proposition 2.16 and the composite of two (strongly) semismooth functions is (strongly) semismooth by [19, Theorem 19], we conclude that $H_{p,i},\ i = 2, 3, \cdots, n+1$ are semismooth. If F' is locally Lipschitz continuous, then F_ε is strongly semismooth, and, consequently, $H_{p,i},\ i = 2, 3, \cdots, n+1$ are strongly semismooth. □

We next give the estimation of the generalized Jacobian of H_p by Proposition 2.1(f).

Proposition 4.20 *Let $H_p : \mathbb{R}_+ \times \mathbb{R}^n \to \mathbb{R}^n$ be defined as in (4.65). For any $z = (\varepsilon, x) \in \mathbb{R}_+ \times \mathbb{R}^n$, we have*

$$\left(\partial H_p(z)\right)^{\mathsf{T}} \subseteq \begin{bmatrix} 1 & x^{\mathsf{T}} B(z) \\ 0 & (A(z) - I) + (\nabla F(x) + \varepsilon I)(B(z) - I) \end{bmatrix}, \tag{4.68}$$

where $A(z)$ and $B(z)$ are possibly multi-valued $n \times n$ diagonal matrices with i-th diagonal elements $A_{ii}(z)$ and $B_{ii}(z)$ given by

$$A_{ii}(z) = \frac{\operatorname{sgn}(x_i) \cdot |x_i|^{p-1}}{\|(x_i, F_{\varepsilon,i}(x))\|_p^{p-1}}, \quad B_{ii}(z) = \frac{\operatorname{sgn}(F_{\varepsilon,i}(x)) \cdot |F_{\varepsilon,i}(x)|^{p-1}}{\|(x_i, F_{\varepsilon,i}(x))\|_p^{p-1}}$$

if $(x_i, F_{\varepsilon,i}(x)) \neq (0, 0)$, and otherwise given by

$$A_{ii}(z) = \xi_i, \quad B_{ii}(z) = \zeta_i \quad \text{for any } (\xi_i, \zeta_i) \text{ such that } |\xi_i|^{\frac{p}{p-1}} + |\zeta_i|^{\frac{p}{p-1}} \leq 1.$$

Proof According to the known rules on the evaluation of the generalized Jacobian (see [42, Proposition 2.6.2(e)]), we have

$$\partial H_p(z)^{\mathsf{T}} \subseteq \partial H_{p,1}(z) \times \partial H_{p,2}(z) \times \cdots \times \partial H_{p,n+1}(z)$$

where the right-hand side denotes a set of matrices whose i-th column belongs to $\partial H_{p,i}(z)$, and $H_{p,i}$ is the i-th component function of H_p. Clearly,

$$\partial H_{p,1}(z) = \begin{pmatrix} 1 \\ 0 \end{pmatrix} \in \mathbb{R}^{n+1}.$$

For $j = 2, 3, \cdots, n+1$, letting $i = j - 1$ and applying Proposition 2.1(f) yield

$$\begin{aligned} \partial H_{p,j}(z) = {} & \left(\frac{\operatorname{sgn}(x_i) \cdot |x_i|^{p-1}}{\|(x_i, F_{\varepsilon,i}(x))\|_p^{p-1}} - 1 \right) \begin{pmatrix} 0 \\ e_i \end{pmatrix} \\ & + \begin{pmatrix} x_i \\ \nabla F_i(x) + \varepsilon e_i \end{pmatrix} \left(\frac{\operatorname{sgn}(F_{\varepsilon,i}(x)) \cdot |F_{\varepsilon,i}(x)|^{p-1}}{\|(x_i, F_{\varepsilon,i}(x))\|_p^{p-1}} - 1 \right) \end{aligned}$$

if $(x_i, F_{\varepsilon,i}(x)) \neq (0, 0)$, and otherwise

$$\partial H_{p,j}(z) = (\xi_i - 1) \begin{pmatrix} 0 \\ e_i \end{pmatrix} + \begin{pmatrix} x_i \\ \nabla F_i(x) + \varepsilon e_i \end{pmatrix} (\zeta_i - 1)$$

with $|\xi_i|^{\frac{p}{p-1}} + |\zeta_i|^{\frac{p}{p-1}} \leq 1$, where e_i denotes the vector whose i-th element is zero and other elements are 1. From these equalities, the conclusion easily follows. □

Utilizing the estimation of $\partial H_p(z)$ given in (4.68), we now present a sufficient condition to ensure the nonsingularity of all generalized Jacobians of H_p at a solution z^* of (4.66). This result is crucial for establishing the superlinear (or quadratic) convergence of the semismooth Newton method; see [43]. Let $z^* = (\varepsilon^*, x^*) \in \mathbb{R}_+ \times \mathbb{R}^n$ be a solution of (4.66). Clearly, $\varepsilon^* = 0$ and x^* is a solution of the NCP. For the sake of notation, let

$$
\begin{aligned}
\mathcal{I} &:= \left\{ i \in \{1, 2, \cdots, n\} \,|\, x_i^* > 0,\ F_i(x^*) = 0 \right\}, \\
\mathcal{J} &:= \left\{ i \in \{1, 2, \cdots, n\} \,|\, x_i^* = 0,\ F_i(x^*) = 0 \right\}, \\
\mathcal{K} &:= \left\{ i \in \{1, 2, \cdots, n\} \,|\, x_i^* = 0,\ F_i(x^*) > 0 \right\}.
\end{aligned}
$$

By rearrangement we assume that $\nabla F(x^*)$ can be written as

$$
\nabla F(x^*) = \begin{bmatrix} \nabla F_{\mathcal{I}\mathcal{I}}(x^*) & \nabla F_{\mathcal{I}\mathcal{J}}(x^*) & \nabla F_{\mathcal{I}\mathcal{K}}(x^*) \\ \nabla F_{\mathcal{J}\mathcal{I}}(x^*) & \nabla F_{\mathcal{J}\mathcal{J}}(x^*) & \nabla F_{\mathcal{J}\mathcal{K}}(x^*) \\ \nabla F_{\mathcal{K}\mathcal{I}}(x^*) & \nabla F_{\mathcal{K}\mathcal{J}}(x^*) & \nabla F_{\mathcal{K}\mathcal{K}}(x^*) \end{bmatrix}. \tag{4.69}
$$

The NCP is called R-regular at x^* if $\nabla F_{\mathcal{I}\mathcal{I}}(x^*)$ is nonsingular and its Schur complement in the matrix $\begin{bmatrix} \nabla F_{\mathcal{I}\mathcal{I}}(x^*) & \nabla F_{\mathcal{I}\mathcal{J}}(x^*) \\ \nabla F_{\mathcal{J}\mathcal{I}}(x^*) & \nabla F_{\mathcal{J}\mathcal{J}}(x^*) \end{bmatrix}$ is a P-matrix.

Proposition 4.21 *Suppose that $z^* = (\varepsilon^*, x^*) \in \mathbb{R}_+ \times \mathbb{R}^n$ be a solution of (4.66) and the NCP is R-regular at x^*, then all $V \in \partial H_p(z^*)$ are nonsingular.*

Proof From Proposition 4.20, it is easy to see that for any $V \in \partial H_p(z^*)^T$, there exists a vector $u(z^*) \in \mathbb{R}^n$ and a matrix $W(z^*) \in \mathbb{R}^{n\times n}$ such that

$$
V = \begin{bmatrix} 1 & u(z^*)^{\mathrm{T}} \\ 0 & W(z^*) \end{bmatrix},
$$

where

$$
W(z^*) = (A(z^*) - I) + (\nabla F(x^*) + \varepsilon^* I)(B(z^*) - I)
$$

with $A(z^*)$ and $B(z^*)$ characterized as in Proposition 4.20. Therefore, proving that V is nonsingular is equivalent to arguing that $W(z^*)$ is nonsingular. Using the expression of $\nabla F(x^*)$ in (4.69) and noting that $\varepsilon^* = 0$, we can rewrite $W(z^*)$ in the partitioned form:

$$
W(z^*) = \begin{bmatrix} -\nabla F_{\mathcal{I}\mathcal{I}} & \nabla F_{\mathcal{I}\mathcal{J}}(B_{\mathcal{J}\mathcal{J}} - I_{\mathcal{J}\mathcal{J}}) & 0_{\mathcal{I}\mathcal{K}} \\ -\nabla F_{\mathcal{J}\mathcal{I}} & \nabla F_{\mathcal{I}\mathcal{J}}(B_{\mathcal{J}\mathcal{J}} - I_{\mathcal{J}\mathcal{J}}) + (A_{\mathcal{J}\mathcal{J}} - I_{\mathcal{J}\mathcal{J}}) & 0_{\mathcal{J}\mathcal{K}} \\ -\nabla F_{\mathcal{K}\mathcal{I}} & \nabla F_{\mathcal{K}\mathcal{J}}(B_{\mathcal{J}\mathcal{J}} - I_{\mathcal{J}\mathcal{J}}) & -I_{\mathcal{K}\mathcal{K}} \end{bmatrix},
$$

where for convenience we dispense with the notations z^* and x^*. The rest of the proof is identical to that of [43, Proposition 3.2]. □

Proposition 4.22 *For any $\varepsilon \geq 0$, the function $\Psi_{p,\varepsilon}$ defined by (4.64) is continuously differentiable everywhere, and consequently G_p defined as in (4.67) is continuously differentiable everywhere and $\nabla G_p(z) = V^{\mathsf{T}} H_p(z)$ for any $V \in \partial H_p(z)$.*

Proof By applying [44, Proposition 3.2(c)] and Theorem 2.6.6 from [42], we immediately arrive at the conclusion. □

Proposition 4.23 *Suppose that F is a P_0-function and $\hat{\varepsilon}$, $\tilde{\varepsilon}$ are two given positive numbers such that $\hat{\varepsilon} < \tilde{\varepsilon}$. Then, the merit function G_p defined as in (4.67) has the property:*

$$\lim_{k \to +\infty} G_p(z^k) = +\infty$$

for any sequence $\{z^k = (\varepsilon^k, x^k)\}$ such that $\varepsilon^k \in [\hat{\varepsilon}, \tilde{\varepsilon}]$ and $\|x^k\| \to +\infty$.

Proof We prove this by contradiction which is a standard and common technique. Suppose $\lim_{k\to+\infty} G_p(z^k) \neq +\infty$. Then from (4.67) and (4.64) it follows that there exists an unbounded sequence $\{x^k\}$ such that $\{\Psi_{p,\varepsilon^k}(x^k)\}$ is bounded. Let

$$J := \left\{ i \in \{1, 2, \cdots, n\} \mid \{x_i^k\} \text{ is unbounded} \right\}.$$

Since $\{x^k\}$ is unbounded, we have $J \neq \emptyset$. Without loss of generality, we assume that $\{|x_j^k|\} \to \infty$ for any $j \in J$. Now, we define a bounded sequence by

$$y_i^k := \begin{cases} 0 & \text{if } i \in J, \\ x_i^k & \text{if } i \notin J. \end{cases}$$

From the definition of $\{y^k\}$ and F being a P_0-function, we have

$$\begin{aligned} 0 &\leq \max_{\substack{1 \leq i \leq n \\ x_i^k \neq y_i^k}} (x_i^k - y_i^k)(F_i(x^k) - F_i(y^k)) \\ &= \max_{i \in J} x_i^k \cdot (F_i(x^k) - F_i(y^k)) \\ &= x_{j_0}^k \cdot (F_{j_0}(x^k) - F_{j_0}(y^k)), \end{aligned} \tag{4.70}$$

where j_0 is one of the indices for which the max is attained. Since $j_0 \in J$, we have that $\{|x_{j_0}^k|\} \to +\infty$ as $k \to +\infty$. If $x_{j_0}^k \to -\infty$ as $k \to +\infty$, using Proposition (2.8) immediately yields that $\phi_{\text{FB}}^p(x_{j_0}^k, F_{\varepsilon^k, j_0}(x^k)) \to +\infty$. If $x_{j_0}^k \to +\infty$ as $k \to +\infty$, noting that $F_{j_0}(y^k)$ is bounded by the continuity of F_{j_0}, we have from (4.70) that $F_{j_0}(x^k)$ does not tend to $-\infty$, which in turn implies that $\{F_{j_0}(x^k) + \varepsilon^k x_{j_0}^k\} \to +\infty$. From Proposition (2.8) where $\{x_{j_0}^k\} \to +\infty$ and $\{F_{j_0}(x^k) + \varepsilon^k x_{j_0}^k\} \to +\infty$, we also obtain that $\phi_{\text{FB}}^p(x_{j_0}^k, F_{\varepsilon^k, j_0}(x^k)) \to +\infty$. Thus, both cases yield $\phi_{\text{FB}}^p(x_{j_0}^k, F_{\varepsilon^k, j_0}(x^k)) \to +\infty$ which is a contradiction to the boundedness of $\{\Psi_{p,\varepsilon^k}(x^k)\}$. Consequently, we prove that $\lim_{k\to+\infty} G_p(z^k) = +\infty$. □

Remark 4.2 Proposition 4.23 establishes that $\Psi_{p,\varepsilon}$ possesses bounded level sets under the assumption that F is a P_0-function. In contrast, as shown in [44, Proposition 3.5], a stronger condition, namely, that F is a uniform P-function, is required to ensure the boundedness of the level sets of Ψ^p_{FB}.

We now present two results that will be instrumental in analyzing the global convergence of the algorithm in the subsequent section. The first is adapted from [38, Theorem 5.4], while the second follows from Lemma 2.3 and employs arguments similar to those used in [40, Proposition 2.2].

Proposition 4.24 *Suppose that F is a P_0-function and the solution set S^* of the NCP is nonempty and bounded. Suppose that $\{\varepsilon^k\}$ and $\{x^k\}$ are two infinite sequences such that for each $k \geq 0$, $\varepsilon^k > 0$, $\eta^k \geq 0$ satisfying $\lim_{k\to+\infty} \varepsilon^k = 0$, $\lim_{k\to+\infty} \eta^k = 0$. For each $k \geq 0$, let $x^k \in \mathbb{R}^n$ satisfy $\left\|\Phi^p_{\text{FB}}(\varepsilon^k, x^k)\right\| \leq \eta^k$. Then, $\{x^k\}$ remains bounded and every accumulation point of $\{x^k\}$ is a solution of the NCP.*

Proposition 4.25 *Suppose that F is a monotone function and the solution set S^* of the NCP is nonempty. Suppose that $\{\varepsilon^k\}$ and $\{x^k\}$ are two infinite sequences such that for each $k \geq 0$, $\varepsilon^k > 0$, $\eta^k \geq 0$, $\eta^k \geq C\varepsilon^k$, and $\lim_{k\to+\infty} \varepsilon^k = 0$, where $C > 0$ is a constant. For each $k \geq 0$, let $x^k \in \mathbb{R}^n$ satisfy $\left\|\Phi^p_{\text{FB}}(\varepsilon^k, x^k)\right\| \leq \eta^k$. Suppose that $x^* = \operatorname{argmin}_{x\in S^*} \|x\|$ and F is Lipschitz continuous. Then, $\{x^k\}$ remains bounded and every accumulation point of $\{x^k\}$ is a solution of the NCP.*

We are now ready to describe the specific algorithm, adopting notation that closely follows that of [40]. Choose $\bar{\varepsilon} \in (0, +\infty)$ and $\gamma \in (0, 1)$ such that $\gamma\bar{\varepsilon} < 1$. Let $t \in [1/2, 1]$ and $\bar{z} := (\bar{\varepsilon}, 0) \in \mathbb{R}_{++} \times \mathbb{R}^n$. We define $\beta : \mathbb{R}_+ \times \mathbb{R}^n \to \mathbb{R}_+$ by

$$\beta(z) \;:=\; \gamma \min\left\{1, G_p(z)^t\right\} \tag{4.71}$$

and denote

$$\Omega := \left\{z = (\varepsilon, x) \in \mathbb{R}_+ \times \mathbb{R}^n \mid \varepsilon \geq \beta(z)\bar{\varepsilon}\right\}. \tag{4.72}$$

Note that $\beta(z) \leq \gamma$ for any $z \in \mathbb{R}_+ \times \mathbb{R}^n$ by (4.71). Hence, $(\bar{\varepsilon}, x) \in \Omega$ for any $x \in \mathbb{R}^n$. In addition, by the definition of $\beta(z)$, it is easily verified that the following relation holds.

Proposition 4.26 *Let H_p and β be defined as in (4.65) and (4.71), respectively. Then,*

$$H_p(z) = 0 \iff \beta(z) = 0 \iff H_p(z) = \beta(z)\bar{z}.$$

Algorithm 4.4 (The Regularization Newton Algorithm)

(Step 0) Given any $p > 1$ and choose constants $\delta \in (0, 1)$, $t \in [1/2, 1]$ and $\sigma \in (1, 1/2)$. Let $\varepsilon^0 := \bar{\varepsilon}$ and $x^0 \in \mathbb{R}^n$ be an arbitrary point. Set $k := 0$.

(Step 1) If $H_p(z^k) = 0$, then stop. Otherwise, let

$$\beta_k := \beta(z^k) = \gamma \min\left\{1, G_p(z^k)^t\right\}.$$

(Step 2) Choose $V_k \in \partial H_p(z^k)$ and compute $\Delta z^k = (\Delta\varepsilon^k, \Delta x^k) \in \mathbb{R} \times \mathbb{R}^n$ by

$$H_p(z^k) + V_k \Delta z^k = \beta_k \bar{z}. \tag{4.73}$$

(Step 3) Let l_k be the smallest nonnegative integer l such that

$$G_p(z^k + \delta^l \Delta z^k) \le \left[1 - 2\sigma(1 - \gamma\bar{\varepsilon})\delta^l\right] G_p(z^k).$$

Set $z^{k+1} := z^k + \delta^{l_k} \Delta z^k$.

(Step 4) Set $k := k + 1$ and go to Step 1.

From Proposition 4.20, we know that for any $V \in \partial H_p(z)$ with $z = (\varepsilon, x) \in \mathbb{R}_{++} \times \mathbb{R}^n$, there exists a $W = (u(z)\; W(z)) \in \partial \Phi^p_{\text{FB}}(z)$ with $u(z) \in \mathbb{R}^n$ and $W(z) \in \mathbb{R}^{n\times n}$ such that

$$V = \begin{bmatrix} 1 & 0 \\ u(z) & W(z) \end{bmatrix}.$$

Suppose that F is a P_0-function. Then, by Proposition 1.6(a), $F'_\varepsilon(x)$ is a P-matrix. Hence, for any $x \in \mathbb{R}^n$ and $\varepsilon > 0$, $W(z)$ is nonsingular by the proof of [45, Proposition 3.2]. It thus follows that all $V \in \partial H_p(z)$ with $z = (\varepsilon, x) \in \mathbb{R}_{++} \times \mathbb{R}^n$ are nonsingular. Therefore, the Newton step in (4.73) is well defined, and moreover from (4.73), for any $k \ge 0$ and $\varepsilon^k > 0$, there exists a $W_k \in \partial \Phi^p_{\text{FB}}(z^k)$ such that

$$(\nabla \Psi^p_{\text{FB}}(z^k))^{\mathsf{T}} \Delta z^k = \Phi^p_{\text{FB}}(z^k)^{\mathsf{T}} W_k \Delta z^k = -\Phi^p_{\text{FB}}(z^k)^{\mathsf{T}} \Phi^p_{\text{FB}}(z^k) = -2\Psi^p_{\text{FB}}(z^k). \tag{4.74}$$

Using the equality and Proposition 4.26, we next show that Algorithm 4.4 is well defined.

Proposition 4.27 *Suppose that F is a P_0-function and $z^k = (\varepsilon^k, x^k) \in \mathbb{R}_{++} \times \mathbb{R}^n$ for $k \ge 0$. Then, $z^{k+1} \in \mathbb{R}_{++} \times \mathbb{R}^n$ and Algorithm 4.4 is well defined.*

Proof Since $\varepsilon^k > 0$, from the definition of $\beta(z)$ it follows that $\beta_k = \beta(z^k) > 0$. From the first component in the relation (4.73) in Algorithm 4.4, we have

$$\varepsilon^k + \Delta\varepsilon^k = \beta_k \bar{\varepsilon} \implies \Delta\varepsilon^k = -\varepsilon^k + \beta_k \bar{\varepsilon}.$$

Then, for any $\alpha \in [0, 1]$, there has

$$\varepsilon^k + \alpha \Delta\varepsilon^k = (1 - \alpha)\varepsilon^k + \alpha\beta_k\bar{\varepsilon} > 0. \tag{4.75}$$

Thus, combining the fact that $\beta(z) \le \gamma G_p(z)^{1/2}$ with (4.73) and (4.75) yields

$$\begin{aligned}
\left(\varepsilon^k + \alpha\Delta\varepsilon^k\right)^2 &= \left[(1-\alpha)\varepsilon^k + \alpha\beta_k\bar{\varepsilon}\right]^2 \\
&= (1-\alpha)^2(\varepsilon^k)^2 + 2(1-\alpha)\alpha\beta_k\varepsilon^k\bar{\varepsilon} + \alpha^2\beta_k^2\bar{\varepsilon}^2 \\
&\le (1-\alpha)^2(\varepsilon^k)^2 + 2\alpha\beta_k\varepsilon^k\bar{\varepsilon} + O(\alpha^2) \\
&\le (1-\alpha)^2(\varepsilon^k)^2 + 2\alpha\gamma G_p(z^k)^{1/2}\|H_p(z^k)\|\bar{\varepsilon} + O(\alpha^2) \\
&= (1-2\alpha)(\varepsilon^k)^2 + 2\sqrt{2}\alpha\gamma\bar{\varepsilon}G_p(z^k) + O(\alpha^2). \qquad (4.76)
\end{aligned}$$

Now, we define

$$\theta(\alpha) := \Psi^p_{\text{FB}}(z^k + \alpha\Delta z^k) - \Psi^p_{\text{FB}}(z^k) - \alpha(\nabla\Psi^p_{\text{FB}}(z^k))^\top\Delta z^k.$$

Since Ψ^p_{FB} is continuously differentiable at any $z^k \in \mathbb{R}_{++} \times \mathbb{R}^n$ by Proposition 4.22, we obtain $\theta(\alpha) = o(\alpha)$. On the other hand, from (4.73) and (4.74), it follows that

$$\begin{aligned}
\frac{1}{2}\left\|\Phi^p_{\text{FB}}(z^k + \alpha\Delta z^k)\right\|^2 &= \Psi^p_{\text{FB}}(z^k + \alpha\Delta z^k) \\
&= \Psi^p_{\text{FB}}(z^k) + \alpha(\nabla\Psi^p_{\text{FB}}(z^k))^\top\Delta z^k + \theta(\alpha) \qquad (4.77) \\
&= \Psi^p_{\text{FB}}(z^k) - 2\alpha\Psi^p_{\text{FB}}(z^k) + o(\alpha) \\
&= (1-2\alpha)\Psi^p_{\text{FB}}(z^k) + o(\alpha)
\end{aligned}$$

for any $\alpha \in [0, 1]$. Therefore, using Eqs. (4.76) and (4.77), we obtain

$$\begin{aligned}
G_p(z^k + \alpha\Delta z^k) &= \frac{1}{2}\left\|H_p(z^k + \alpha\Delta z^k)\right\|^2 \\
&= \frac{1}{2}(\varepsilon^k + \alpha\Delta\varepsilon^k)^2 + \frac{1}{2}\|\Phi^p_{\text{FB}}(z^k + \alpha\Delta z^k)\|^2 \\
&\le \frac{1}{2}(1-2\alpha)(\varepsilon^k)^2 + \sqrt{2}\alpha\gamma\bar{\varepsilon}G_p(z^k) + (1-2\alpha)\Psi^p_{\text{FB}}(z^k) + o(\alpha) \\
&\le (1-2\alpha)G_p(z^k) + 2\alpha\gamma\bar{\varepsilon}G_p(z^k) + o(\alpha) \\
&= \left[1 - 2(1-\gamma\bar{\varepsilon})\alpha\right]G_p(z^k) + o(\alpha) \qquad (4.78)
\end{aligned}$$

for any $\alpha \in [0, 1]$. The inequality (4.78) implies that there exists $\bar{\alpha} \in (0, 1]$ such that

$$G_p(z^k + \alpha\Delta z^k) \le \left[1 - 2\sigma(1-\gamma\bar{\varepsilon})\alpha\right]G_p(z^k) \quad \forall\alpha \in [0, \bar{\alpha}],$$

which indicates that Algorithm 4.4 is well defined. □

Proposition 4.28 *Let Ω be defined as in* (4.72). *Suppose that F is a P_0-function. For each $k \ge 0$, if $\varepsilon^k > 0$ and $z^k \in \Omega$, then for any $\alpha \in [0, 1]$ such that*

$$G_p\left(z^k+\alpha\Delta z^k\right)\le\left[1-2\sigma(1-\gamma\bar{\varepsilon})\alpha\right]G_p(z^k), \tag{4.79}$$

there holds that $z^k+\alpha\Delta z^k\in\Omega$.

Proof We prove this proposition by considering the following two cases:

Case (i): $G_p(z^k)>1$. Then $\beta_k=\gamma$. From $z^k\in\Omega$ and $\beta(z)=\gamma\min\{1,G_p(z)^t\}\le\gamma$ for any $z\in\mathbb{R}_+\times\mathbb{R}^n$, it follows that for any $\alpha\in[0,1]$,

$$\begin{aligned}(\varepsilon^k+\alpha\Delta\varepsilon^k)-\beta(z^k+\alpha\Delta z^k)\bar{\varepsilon}&\ge(1-\alpha)\varepsilon^k+\alpha\beta_k\bar{\varepsilon}-\gamma\bar{\varepsilon}\\&\ge(1-\alpha)\beta_k\bar{\varepsilon}+\alpha\beta_k\bar{\varepsilon}-\gamma\bar{\varepsilon}\\&=0.\end{aligned} \tag{4.80}$$

Case (ii): $G_p(z^k)\le 1$. Then, for any $\alpha\in[0,1]$ satisfying (4.79), we have

$$G_p(z^k+\alpha\Delta z^k)\le\left[1-2\sigma(1-\gamma\bar{\varepsilon})\alpha\right]G_p(z^k)\le 1. \tag{4.81}$$

Therefore, for any $\alpha\in[0,1]$ satisfying (4.79),

$$\beta(z^k+\alpha\Delta z^k)=\gamma G_p(z^k+\alpha\Delta z^k)^t.$$

Using the fact that $z^k\in\Omega$ and the first inequality in (4.81), we then obtain that for any $\alpha\in[0,1]$ satisfying (4.79),

$$\begin{aligned}&(\varepsilon^k+\alpha\Delta\varepsilon^k)-\beta(z^k+\alpha\Delta z^k)\bar{\varepsilon}\\&\ge(1-\alpha)\varepsilon^k+\alpha\beta_k\bar{\varepsilon}-\gamma G_p(z^k+\alpha\Delta z^k)^t\bar{\varepsilon}\\&\ge(1-\alpha)\beta_k\bar{\varepsilon}+\alpha\beta_k\bar{\varepsilon}-\gamma[1-2\sigma(1-\gamma\bar{\varepsilon})\alpha]^tG_p(z^k)^t\bar{\varepsilon}\\&=\beta_k\bar{\varepsilon}-\gamma[1-2\sigma(1-\gamma\bar{\varepsilon})\alpha]^tG_p(z^k)^t\bar{\varepsilon}\\&=\gamma\,G_p(z^k)^t\bar{\varepsilon}-\gamma\left[1-2\sigma(1-\gamma\bar{\varepsilon})\alpha\right]^tG_p(z^k)^t\bar{\varepsilon}\\&=\gamma\left\{1-\left[1-2\sigma(1-\gamma\bar{\varepsilon})\alpha\right]^t\right\}G_p(z^k)^t\bar{\varepsilon}\\&\ge 0.\end{aligned} \tag{4.82}$$

Combining (4.80) and (4.82) immediately yields the desired result. □

Proposition 4.29 *Suppose that* F *is a* P_0*-function. Then, Algorithm 4.4 generates an infinite sequence* $\{z^k\}$ *with* $z^k\in\Omega$ *for all* k *and*

$$0<\varepsilon^{k+1}\le\varepsilon^k\le\bar{\varepsilon}\quad\text{for all }k. \tag{4.83}$$

Proof Since $z^0=(\bar{\varepsilon},x^0)\in\Omega$, the first part of the conclusions follows by repeatedly resorting to Propositions 4.27 and 4.28. We next concentrate on the proof of (4.83). First, $\varepsilon^0 = \bar{\varepsilon} > 0$. From the design of Algorithm 4.4 and the fact that $\beta(z) = \gamma \min\{1, G_p(z)^t\} \le \gamma$ for any $z \in \mathbb{R}_+ \times \mathbb{R}^n$, it then follows that

$$\varepsilon^1 = (1 - \delta^{l_0})\varepsilon^0 + \delta^{l_0}\beta(z^0)\bar{\varepsilon} \le (1 - \delta^{l_0})\bar{\varepsilon} + \delta^{l_0}\gamma\bar{\varepsilon} \le \bar{\varepsilon}.$$

Hence (4.83) holds for $k = 0$. Suppose that (4.83) holds for $k = i - 1$. We next prove that (4.83) holds for $k = i$. From the design of Algorithm 4.4, we have

$$\varepsilon^{i+1} = (1 - \delta^{l_i})\varepsilon^i + \delta^{l_i}\beta(z^i)\bar{\varepsilon}.$$

Noting that $\varepsilon^i \ge \beta(z^i)\bar{\varepsilon}$ since $z^i \in \Omega$, we then obtain

$$\varepsilon^{i+1} \le (1 - \delta^{l_i})\varepsilon^i + \delta^{l_i}\varepsilon^i = \varepsilon^i$$

and

$$\varepsilon^{i+1} \ge (1 - \delta^{l_i})\beta(z^i)\bar{\varepsilon} + \delta^{l_i}\beta(z^i)\bar{\varepsilon} = \beta(z^i)\bar{\varepsilon} > 0.$$

Therefore, (4.83) holds for $k = i$. Thus, the proof is complete. □

Now, by applying Propositions 4.23âĽ“-4.25 and Proposition 4.29, and following the same line of reasoning as in [40], we derive the following global convergence results for Algorithm 4.4.

Proposition 4.30 *Suppose that F is a P_0-function and the solution set S^* of the NCP is nonempty and bounded. Then the infinite sequence $\{z^k\}$ generated by Algorithm 4.4 is bounded and any accumulation point of $\{z^k\}$ is a solution of $H(z) = 0$.*

Proposition 4.31 *Suppose that F is a monotone function and in Algorithm 4.4 the parameter $t = \frac{1}{2}$. Then, if the iteration sequence $\{z^k\}$ is bounded, then the solution set S^* of the NCP is nonempty. Conversely, if the solution set S^* of the NCP is nonempty and F is Lipschitz continuous, then the infinite sequence $\{z^k\}$ generated by Algorithm 4.1 is bounded and any accumulation point of $\{z^k\}$ is a solution of $H(z) = 0$.*

In addition, by applying Proposition 4.19 and following a proof similar to that of [40, Theorem 5.1], we establish the local superlinear (or quadratic) convergence results for Algorithm 4.4.

Proposition 4.32 *Suppose that F is a P_0-function and the solution set S^* of the NCP is nonempty and bounded. Suppose that $z^* := (\varepsilon^*, x^*) \in \mathbb{R} \times \mathbb{R}^n$ is an accumulation point of the infinite sequence $\{z^k\}$ generated by Algorithm 4.4 and all $V \in \partial H_p(z^*)$ are nonsingular. Then, the whole sequence $\{z^k\}$ converges to z^* with*

$$\|z^{k+1} - z^*\| = o(\|z^k - z^*\|), \quad \varepsilon^{k+1} = o(\varepsilon^k).$$

Furthermore, if F' is locally Lipschitz continuous around x^*, then

$$\|z^{k+1} - z^*\| = O(\|z^k - z^*\|^2), \quad \varepsilon^{k+1} = O(\varepsilon^k)^2.$$

Moreover, Proposition 4.21 implies that all conclusions of Proposition 4.32 remain valid if the assumption that all $V \in \partial H_p(z^*)$ are nonsingular is replaced by the condition that the nonlinear complementarity problem is R-regular at x^*. Numerical performance of Algorithm 4.4 is reported in [46], where the results indicate that choosing a smaller value of p, particularly within the range $p \in [1.1, 2]$, generally yields superior numerical performance. In this regard, the generalized Fischer–Burmeister functions ϕ^p_{FB} with $p \in [1.1, 2)$ can serve as effective alternatives to the classical squared Fischer–Burmeister function ϕ^2_{FB}.

Next, we demonstrate the regularization approach by using the Fischer–Burmeister SOC complementarity function for solving the SOCCP (3.1). Again, in the context of SOCCPs, the regularization scheme consists in solving a sequence of SOCCP(F_ε):

$$\zeta \in \mathcal{K}, \quad F_\varepsilon(\zeta) \in \mathcal{K}, \quad \langle \zeta, F_\varepsilon(\zeta)\rangle = 0, \tag{4.84}$$

where ε is a positive parameter tending to zero and $F_\varepsilon : \mathbb{R}^n \to \mathbb{R}^n$ is given by

$$F_\varepsilon(\zeta) := F(\zeta) + \varepsilon\zeta. \tag{4.85}$$

For convenience, we continue to use $x(\varepsilon)$ in place of $\zeta(\varepsilon)$ to denote the solution trajectory. Specifically, drawing parallels to the classical results of regularization methods in convex optimization, we aim to extend, as broadly as possible, the following results to the general class of SOCCPs in which F possesses the Cartesian P_0-property:

(a) The regularized problem SOCCP(F_ε) has a unique solution for every $\varepsilon > 0$.
(b) The trajectory $x(\varepsilon)$ is continuous for $\varepsilon > 0$.
(c) For $\varepsilon \to 0$, the trajectory $x(\varepsilon)$ converges to the least l_2-norm solution of SOCCP(F) if the SOCCP(F) has a nonempty solution set, and otherwise it diverges.

We now proceed to show that the regularized problem SOCCP(F_ε) (4.84) admits a unique solution $x(\varepsilon)$ for every $\varepsilon > 0$, under the assumption that F satisfies the Cartesian P_0-property and the following condition:

Condition 4.2 For any sequence $\{x^k\} \subseteq \mathbb{R}^n$, when there exists $i \in \{1, 2, \ldots, m\}$ such that the sequences $\{\|x_i^k\|\}$ and $\left\{\frac{\|F_i(x^k)\|}{\|x_i^k\|}\right\}$ are unbounded, there holds

$$\limsup_{k\to+\infty} \left\langle \frac{x_i^k}{\|x_i^k\|}, \frac{F_i(x^k)}{\|F_i(x^k)\|} \right\rangle > 0.$$

Analogously, for the SOCCP(F_ε), we define the operator $\Phi_\varepsilon : \mathbb{R}^n \to \mathbb{R}^n$ by

$$\Phi_\varepsilon(x) := \begin{pmatrix} \phi_{\text{FB}}(x_1, F_{\varepsilon,1}(x)) \\ \vdots \\ \phi_{\text{FB}}(x_m, F_{\varepsilon,m}(x)) \end{pmatrix}, \tag{4.86}$$

where $F_{\varepsilon,i} : \mathbb{R}^n \to \mathbb{R}^{n_i}$ denotes the i-th subvector of F_ε. The natural merit function $\Psi_\varepsilon : \mathbb{R}^n \to \mathbb{R}_+$ corresponding to Φ_ε given as (4.86) is then described by

$$\Psi_\varepsilon(x) := \frac{1}{2}\|\Phi_\varepsilon(x)\|^2 = \frac{1}{2}\sum_{i=1}^{m}\left\|\phi_{\text{FB}}(x_i, F_{\varepsilon,i}(x))\right\|^2. \tag{4.87}$$

Proposition 4.33 *Suppose that $F : \mathbb{R}^n \to \mathbb{R}^n$ has the Cartesian P_0-property and satisfies Condition 4.2. Then, the function Ψ_ε given as (4.87) for any $\varepsilon > 0$ is coercive, i.e.:*

$$\lim_{\|x^k\|\to\infty} \Psi_\varepsilon(x^k) = +\infty.$$

Proof Suppose by contradiction that the conclusion does not hold. Then, we can find an unbounded sequence $\{x^k\} \subseteq \mathbb{R}^n$ with $x^k = (x_1^k, \ldots, x_m^k)$ and $x_i^k \in \mathbb{R}^{n_i}$ such that the sequence $\{\Psi_\varepsilon(x^k)\}$ is bounded. Define the index set

$$J := \left\{ i \in \{1, 2, \ldots, m\} \;\middle|\; \{\|x_i^k\|\} \text{ is unbounded} \right\}.$$

Since $\{x^k\}$ is unbounded, $J \neq \emptyset$. Subsequencing if necessary, we assume without loss of generality that $\{\|x_i^k\|\} \to +\infty$ for all $i \in J$. For each $i \in J$, we define

$$J_i := \left\{ \nu \in \{1, 2, \ldots, n_i\} \;\middle|\; \{|x_{i\nu}^k|\} \text{ is unbounded} \right\}.$$

Let $\{y^k\}$ be a bounded sequence with $y^k = (y_1^k, \ldots, y_m^k)$ and $y_i^k \in \mathbb{R}^{n_i}$ defined as follows:

$$y_{i\nu}^k = \begin{cases} 0 & \text{if } i \in J \text{ and } \nu \in J_i; \\ x_{i\nu}^k & \text{otherwise.} \end{cases}$$

From the definition of $\{y^k\}$ and the Cartesian P_0-property of F, it follows that

$$\begin{aligned} 0 &\le \max_{1\le l\le m} \left\langle x_l^k - y_l^k,\; F_l(x^k) - F_l(y^k) \right\rangle \\ &= \left\langle x_i^k - y_i^k,\; F_i(x^k) - F_i(y^k) \right\rangle \\ &\le n_i \max_{\nu\in J_i} x_{i\nu}^k \left[F_{i\nu}(x^k) - F_{i\nu}(y^k) \right] \\ &= n_i x_{ij}^k \left[F_{ij}(x^k) - F_{ij}(y^k) \right], \end{aligned} \tag{4.88}$$

where i is an index from J for which the first maximum is attained, and j is an index from J_i for which the second maximum is attained. Without loss of generality, we assume that i and j are independent of k. Since $i \in J$ and $j \in J_i$,

$$|x_{ij}^k| \rightarrow +\infty. \tag{4.89}$$

We now consider the two cases where $x_{ij}^k \rightarrow +\infty$ and $x_{ij}^k \rightarrow -\infty$, respectively.

Case (1): $x_{ij}^k \rightarrow +\infty$. In this case, since $F_{ij}(y^k)$ is bounded by the continuity of $F_{ij}(y)$, the inequality (4.88) implies that $F_{ij}(x^k)$ does not tend to $-\infty$. This in turn implies that

$$\{F_{ij}(x^k) + \varepsilon x_{ij}^k\} \rightarrow +\infty.$$

Case (2): $x_{ij}^k \rightarrow -\infty$. Now, using the inequality (4.88) and the boundedness of $F_{ij}(y^k)$ immediately yields that $F_{ij}(x^k)$ does not tend to $+\infty$. This in turn implies that

$$\{F_{ij}(x^k) + \varepsilon x_{ij}^k\} \rightarrow -\infty. \tag{4.90}$$

From Equations (4.89)–(4.90) and the definition of $F_{\varepsilon,i}(x)$, we thus obtain

$$\|x_i^k\| \rightarrow +\infty, \quad \|F_{\varepsilon,i}(x^k)\| \rightarrow +\infty. \tag{4.91}$$

If $\lambda_1(x_i^k) \rightarrow -\infty$ or $\lambda_1[F_{\varepsilon,i}(x^k)] \rightarrow -\infty$, then from Lemma 3.15(a) we readily obtain that $\|\phi_{\text{FB}}(x_i^k, F_{\varepsilon,i}(x^k))\| \rightarrow +\infty$. Otherwise, Eq. (4.91) implies that $\{x_i^k\}$ and $\{F_{\varepsilon,i}(x^k)\}$ are bounded below: but $\lambda_2(x_i^k) \rightarrow +\infty$ and $\lambda_2[F_{\varepsilon,i}(x^k)] \rightarrow +\infty$. We next prove that

$$\lim_{k\rightarrow+\infty} \frac{x_i^k}{\|x_i^k\|} \circ \frac{F_{\varepsilon,i}(x^k)}{\|F_{\varepsilon,i}(x^k)\|} \nrightarrow 0, \tag{4.92}$$

and consequently from Lemma 3.15(b) it follows that $\|\phi_{\text{FB}}(x_i^k, F_{\varepsilon,i}(x^k))\| \rightarrow +\infty$. From the first two equations of (4.88) and the boundedness of $\{y^k\}$ and $\{F_i(y^k)\}$, it is not hard to verify that $\left\langle \frac{x_i^k}{\|x_i^k\|}, \frac{F_i(x^k)}{\|F_{\varepsilon,i}(x^k)\|} \right\rangle \geq 0$ for all sufficiently large k. Notice that

$$\left\langle \frac{x_i^k}{\|x_i^k\|}, \frac{F_{\varepsilon,i}(x^k)}{\|F_{\varepsilon,i}(x^k)\|} \right\rangle = \left\langle \frac{x_i^k}{\|x_i^k\|}, \frac{F_i(x^k)}{\|F_{\varepsilon,i}(x^k)\|} \right\rangle + \frac{\varepsilon\|x_i^k\|}{\|F_{\varepsilon,i}(x^k)\|}, \quad \forall\, k. \tag{4.93}$$

Therefore, if the sequence $\left\{\frac{\|F_i(x^k)\|}{\|x_i^k\|}\right\}$ is bounded, then equality (4.93) implies that

$$\limsup_{k\rightarrow+\infty} \left\langle \frac{x_i^k}{\|x_i^k\|}, \frac{F_{\varepsilon,i}(x^k)}{\|F_{\varepsilon,i}(x^k)\|} \right\rangle > 0. \tag{4.94}$$

If the sequence $\left\{\frac{\|F_i(x^k)\|}{\|x_i^k\|}\right\}$ is unbounded, then using Condition A and equality (4.93), it is easy to verify that (4.94) also holds. Clearly, Equation (4.94) implies (4.92), and we thus achieve $\|\phi_{\text{FB}}(x_i^k, F_{\varepsilon,i}(x^k))\| \to +\infty$. This contradicts the boundedness of $\{\Psi_\varepsilon(x^k)\}$. □

Notice that Proposition 4.33 is equivalent to saying that the level set

$$\mathcal{L}_\gamma(x) := \{x \in \mathbb{R}^n \mid \Psi_\varepsilon(x) \le \gamma\}$$

is bounded for every $\gamma \ge 0$.

Proposition 4.34 *Assume that the mapping $F : \mathbb{R}^n \to \mathbb{R}^n$ has the Cartesian P_0-property and satisfies Condition 4.2. Then, for every $\varepsilon > 0$ the problem SOCCP(F_ε) has a unique bounded solution $x(\varepsilon)$.*

Proof Let $\varepsilon > 0$. Then, the mapping F_ε has the Cartesian P-property by Proposition 1.6(b). This means that the regularized problem SOCCP(F_ε) has at most one solution. Now let us prove the fact by contradiction. Suppose that $x(\varepsilon)$ and $\hat{x}(\varepsilon)$ are two different solutions of the SOCCP(F_ε). Then, from the Cartesian P-property of F_ε, there exists an index $i \in \{1, 2, \ldots, m\}$ such that

$$\begin{aligned} 0 &< \langle x_i(\varepsilon) - \hat{x}_i(\varepsilon),\ F_{\varepsilon,i}(x(\varepsilon)) - F_{\varepsilon,i}(\hat{x}(\varepsilon))\rangle \\ &= \langle x_i(\varepsilon),\ F_{\varepsilon,i}(x(\varepsilon))\rangle - \langle x_i(\varepsilon),\ F_{\varepsilon,i}(\hat{x}(\varepsilon))\rangle \\ &\quad - \langle \hat{x}_i(\varepsilon),\ F_{\varepsilon,i}(x(\varepsilon))\rangle + \langle \hat{x}_i(\varepsilon),\ F_{\varepsilon,i}(\hat{x}(\varepsilon))\rangle \\ &= -\langle x_i(\varepsilon),\ F_{\varepsilon,i}(\hat{x}(\varepsilon))\rangle - \langle \hat{x}_i(\varepsilon),\ F_{\varepsilon,i}(x(\varepsilon))\rangle, \end{aligned} \tag{4.95}$$

where the last equality is due to $\langle x_i(\varepsilon), F_{\varepsilon,i}(x(\varepsilon))\rangle = 0$ and $\langle \hat{x}_i(\varepsilon), F_{\varepsilon,i}(\hat{x}(\varepsilon))\rangle = 0$. Note that the two terms on the right-hand side of (4.95) are non-positive since $x_i(\varepsilon), \hat{x}_i(\varepsilon) \in \mathcal{K}^{n_i}$ and $F_{\varepsilon,i}(x(\varepsilon)), F_{\varepsilon,i}(\hat{x}(\varepsilon)) \in \mathcal{K}^{n_i}$. Then we obtain a contradiction with the inequality (4.95).

To prove the existence of a solution, let $x^0 \in \mathbb{R}^n$ be an arbitrary point and define $\gamma := \Psi_\varepsilon(x^0)$. By Proposition 4.33, the corresponding level set $\mathcal{L}_\gamma(x)$ is nonempty and compact. Therefore, the continuous function $\Psi_\varepsilon(x)$ attains a global minimum $x(\varepsilon)$ on $\mathcal{L}_\gamma(x)$ which, by the definition of level sets, is also a global minimum of $\Psi_\varepsilon(x)$ on $\mathbb{R}^n$. Therefore, $x(\varepsilon)$ is a stationary point of $\Psi_\varepsilon(x)$. Since the mapping F_ε has the Cartesian P-property, we have from Proposition 3.13 that $x(\varepsilon)$ is a solution of the regularized problem SOCCP(F_ε). Furthermore, this solution is bounded. Combining with the discussions above, we thus complete the proof. □

From Proposition 4.34, we learn that the regularized problem SOCCP(F_ε) for every $\varepsilon > 0$ has a unique solution $x(\varepsilon)$ when the mapping F has the Cartesian P_0-property and satisfies Condition A. Thus, as the parameter ε tends to 0, the solution of the regularized problem SOCCP(F_ε) generates a solution path $\mathcal{P} := \{x(\varepsilon) \mid \varepsilon >$

$0\}$. The aim of the subsequent work is to study the properties of the trajectory $\mathcal{P}$. Specifically, we prove that the path $\mathcal{P}$ is bounded as $\varepsilon \to 0$ if F has the uniform Cartesian P-property, but the bound is dependent on the constant ρ involved in the uniform Cartesian P-property. We also illustrate that in this case the path $\mathcal{P}$ is not locally Lipschitz continuous as $\varepsilon \to 0$. Then, for the case that F has the Cartesian P_0-property and satisfies Condition A, we provide the condition to guarantee that $x(\varepsilon)$ remains bounded as $\varepsilon \to 0$. The reason why we are interested in the boundedness of $x(\varepsilon)$ is due to the following evident result.

Proposition 4.35 *Let $\{\varepsilon_k\}$ be a sequence of positive values converging to* 0. *If $\{x(\varepsilon_k)\}$ converges to a point $\bar{x}$, then $\bar{x}$ solves the SOCCP(F).*

The following proposition establishes that the solution $x(\varepsilon)$ of SOCCP(F_ε) remains bounded for all $\varepsilon \geq 0$, provided that F possesses the uniform Cartesian P-property. However, the bound on $x(\varepsilon)$ depends on the constant ρ associated with this property.

Proposition 4.36 *Suppose that the mapping F has the uniform Cartesian P-property. Then, for any $\varepsilon \geq 0$, we have*

$$\|x(\varepsilon)\| \leq \rho^{-1} \, \|[-F(0)]_+\| \,, \tag{4.96}$$

where $\rho > 0$ is the constant involved in the uniform Cartesian P-property.

Proof Since the uniform Cartesian P-property implies the Cartesian R_{02}-property and the P-property, from [15, Theorem 3.1] and the proof of Proposition 4.40(b) in the sequel, it follows that $x(\varepsilon)$ exists for any $\varepsilon \geq 0$. If $x(\varepsilon) \equiv 0$ for any $\varepsilon \geq 0$, then the inequality (4.96) is direct. Suppose that $x(\varepsilon) \neq 0$ for some $\varepsilon \geq 0$. Since $x(\varepsilon)$ is the solution of the SOCCP(F_ε), it follows that

$$x_i(\varepsilon) \in \mathcal{K}^{n_i}, \quad F_{\varepsilon,i}(x(\varepsilon)) \in \mathcal{K}^{n_i} \;\text{ and }\; \big\langle x_i(\varepsilon),\; F_{\varepsilon,i}(x(\varepsilon))\big\rangle = 0, \quad i = 1, 2, \ldots, m.$$

Using the fact and the uniform Cartesian P-property of F, we have that

$$\begin{aligned}
\rho\|x(\varepsilon)\|^2 &\leq \max_{1\leq i\leq m} \langle x_i(\varepsilon),\; F_i(x(\varepsilon)) - F_i(0)\rangle \\
&= \max_{1\leq i\leq m} \langle x_i(\varepsilon),\; -\varepsilon x_i(\varepsilon) - F_i(0)\rangle \\
&\leq \max_{1\leq i\leq m} \langle x_i(\varepsilon),\; -F_i(0)\rangle \\
&\leq \max_{1\leq i\leq m} \langle x_i(\varepsilon),\; [-F_i(0)]_+\rangle \\
&\leq \|x(\varepsilon)\| \; \|[-F_i(0)]_+\| \,,
\end{aligned}$$

where the third inequality is since $x_i(\varepsilon) \in \mathcal{K}^{n_i}$, $-F_i(0) = [-F_i(0)]_+ + [-F_i(0)]_-$, and $[-F_i(0)]_- \in -\mathcal{K}^{n_i}$. This leads to the desired result. $\square$

Remark 4.3 (a) From Proposition 4.36, when F has the uniform Cartesian P-property, the SOCCP(F) has a unique bounded solution. Furthermore, if $F(0) \in \mathcal{K}$, the regularized problem SOCCP(F_ε) for every $\varepsilon \geq 0$ has the unique solution $x(\varepsilon) = 0$.

(b) When F is an affine function $Mx + q$ with $M \in \mathbb{R}^{n\times n}$ and $q \in \mathbb{R}^n$, the assumption of Proposition 4.36 is equivalent to requiring that M has the Cartesian P-property.

Proposition 4.37 *Suppose that the mapping F has the uniform Cartesian P-property. Then, for any $\varepsilon_1, \varepsilon_2 \geq 0$, there holds that*

$$\|x(\varepsilon_1) - x(\varepsilon_2)\| \leq \rho^{-1}\|\varepsilon_1 x(\varepsilon_1) - \varepsilon_2 x(\varepsilon_2)\|,$$

where $\rho > 0$ is the constant same as Proposition 4.36.

Proof Without loss of generality, we assume that $\varepsilon_1 \neq \varepsilon_2$. Let

$$y(\varepsilon_1) := F_{\varepsilon_1}(x(\varepsilon_1)), \quad y(\varepsilon_2) := F_{\varepsilon_2}(x(\varepsilon_2)).$$

Since $x(\varepsilon_1)$ and $x(\varepsilon_2)$ are the solution of the problem SOCCP(F_{ε_1}) and SOCCP(F_{ε_2}), respectively, we have $x_i(\varepsilon_1), y_i(\varepsilon_1) \in \mathcal{K}^{n_i}$ with $\langle x_i(\varepsilon_1), y_i(\varepsilon_1)\rangle = 0$ and $x_i(\varepsilon_2), y_i(\varepsilon_2) \in \mathcal{K}^{n_i}$ with $\langle x_i(\varepsilon_2), y_i(\varepsilon_2)\rangle = 0$ for all $i = 1, 2, \ldots, m$. From this, it then follows that

$$\begin{aligned}
&\langle x_i(\varepsilon_1) - x_i(\varepsilon_2),\ F_i(x(\varepsilon_1)) - F_i(x(\varepsilon_2))\rangle \\
&= \langle x_i(\varepsilon_1) - x_i(\varepsilon_2),\ y_i(\varepsilon_1) - \varepsilon_1 x_i(\varepsilon_1) - y_i(\varepsilon_2) + \varepsilon_2 x_i(\varepsilon_2)\rangle \\
&= -\langle x_i(\varepsilon_1), y_i(\varepsilon_2)\rangle - \langle x_i(\varepsilon_2), y_i(\varepsilon_1)\rangle + \langle x_i(\varepsilon_1) - x_i(\varepsilon_2),\ \varepsilon_2 x_i(\varepsilon_2) - \varepsilon_1 x_i(\varepsilon_1)\rangle \\
&\leq \langle x_i(\varepsilon_1) - x_i(\varepsilon_2),\ \varepsilon_2 x_i(\varepsilon_2) - \varepsilon_1 x_i(\varepsilon_1)\rangle
\end{aligned}$$

where the second inequality holds since $-\langle x_i(\varepsilon_1), y_i(\varepsilon_2)\rangle \leq 0$ and $-\langle x_i(\varepsilon_2), y_i(\varepsilon_1)\rangle \leq 0$. Using the last inequality and the uniform Cartesian P-property of F, we have that

$$\begin{aligned}
\rho\|x(\varepsilon_1) - x(\varepsilon_2)\|^2 &\leq \max_{1\leq i\leq m} \langle x_i(\varepsilon_1) - x_i(\varepsilon_2),\ F_i(x(\varepsilon_1)) - F_i(x(\varepsilon_2))\rangle \\
&\leq \max_{1\leq i\leq m} \langle x_i(\varepsilon_1) - x_i(\varepsilon_2),\ \varepsilon_2 x_i(\varepsilon_2) - \varepsilon_1 x_i(\varepsilon_1)\rangle, \\
&\leq \max_{1\leq i\leq m} \|x_i(\varepsilon_1) - x_i(\varepsilon_2)\|\ \|\varepsilon_2 x_i(\varepsilon_2) - \varepsilon_1 x_i(\varepsilon_1)\| \\
&\leq \|x(\varepsilon_1) - x(\varepsilon_2)\|\ \|\varepsilon_2 x(\varepsilon_2) - \varepsilon_1 x(\varepsilon_1)\|,
\end{aligned}$$

which immediately implies the desired result. Thus, we complete the proof. □

Propositions 4.36 and 4.37 characterize certain properties of the solution path $\mathcal{P}$ as $\varepsilon \to 0$, under the uniform Cartesian P-property of F. However, these results do not guarantee the local Lipschitz continuity of $\mathcal{P}$ in the limit $\varepsilon \to 0$. The following counterexample illustrates this limitation.

Example 4.1 Let $m = 2$ and $n_1 = n_2 = 2$. Let F be given by $F(x) = Mx + q$, where

$$M = \begin{pmatrix} 1 & 0 & 0 & 0 \\ 0 & 1 & 0 & 0 \\ 0 & 0 & 2 & 0 \\ 0 & 0 & 0 & 2 \end{pmatrix}, \quad q = \begin{pmatrix} -\frac{1+\varepsilon}{\varepsilon} \\ 0 \\ 0 \\ 0 \end{pmatrix} \quad \text{for any given } \varepsilon > 0.$$

Since the matrix M has the Cartesian P-property, the mapping F has the uniform Cartesian P-property. For the SOCCP(F_ε), i.e., to find x such that

$$x \in \mathcal{K}^2 \times \mathcal{K}^2, \quad F_\varepsilon(x) \in \mathcal{K}^2 \times \mathcal{K}^2, \quad \langle x, F_\varepsilon(x) \rangle = 0,$$

we can verify that $x(\varepsilon) = (1/\varepsilon, 0, 0, 0)^{\mathsf{T}}$ is the unique solution. Obviously, $x(\varepsilon)$ is not locally Lipschitz continuous as $\varepsilon \to 0$. Furthermore, $x(\varepsilon)$ even has no bound since the constant ρ involved in the uniform Cartesian P-property of F approaches to 0.

We now focus on the case where F satisfies the Cartesian P_0-property and Condition A. Unlike the NCP setting, we are currently unable to establish the continuity of the mapping $\varepsilon \to x(\varepsilon)$ at any $\varepsilon > 0$. The primary difficulty lies in the lack of an analogue to [47, Theorem 3.1] under the Cartesian P-property of F. Although the Cartesian P-property is preserved by every principal block of $\nabla F(x)$, and by the Schur complement of a matrix possessing this property (as shown in Proposition 1.4), these facts alone are insufficient to derive the desired continuity result. In this case, we state the following result without proof.

Proposition 4.38 *Suppose that the mapping F has the Cartesian P_0-property and satisfies Condition 4.2. If the solution set $\mathcal{S}^*$ of the SOCCP(F) is nonempty and bounded, then the path $\mathcal{P}_{\bar\varepsilon} = \{x(\varepsilon) \mid \varepsilon \in (0, \bar\varepsilon\,]\}$ is bounded for any $\bar\varepsilon > 0$ and*

$$\lim_{\varepsilon \downarrow 0} \operatorname{dist}\left(x(\varepsilon) | \mathcal{S}^*\right) = 0.$$

Proposition 4.39 *Suppose that the mapping F has the Cartesian P_0-property and satisfies Condition 4.2. If the SOCCP(F) has a unique solution $\bar{x}$, then $\lim_{\varepsilon \downarrow 0} x(\varepsilon) = \bar{x}$.*

Proof This is an immediate consequence of Proposition 4.38. □

As demonstrated in [38, Example 4.6], the boundedness assumption of S^* is essential, removing it may result in the loss of boundedness of the solution path $\mathcal{P}_{\bar\varepsilon}$. To this end, we now present several conditions that ensure the nonemptiness and boundedness of S^*.

Proposition 4.40 *The SOCCP(F) has a nonempty and bounded solution set S^* if one of the following conditions holds:*

(a) The mapping F is monotone, and the SOCCP(F) is strictly feasible, i.e., there exists a point $\bar{x} \in \mathbb{R}^n$ satisfying $\bar{x}$, $F(\bar{x}) \in \operatorname{int}(\mathcal{K})$.
(b) The mapping F has the P_0-property and the Cartesian R_{02}-property.

Proof (a) Since $F(x)$ is monotone and $\nabla F(x)$ is positive semidefinite, the result directly follows from Proposition 3.50.

(b) We prove that a stronger result holds for this case, i.e., the following SOCCP(F, q)

$$x \in \mathcal{K}, \quad F(x) + q \in \mathcal{K}, \quad \langle x, F(x) + q \rangle = 0 \tag{4.97}$$

has a nonempty and bounded solution set for all $q \in \mathbb{R}^n$. By [15, Theorem 3.1], we only need to prove that for any $\Delta > 0$, the following set

$$\{x : \ x \text{ solves (4.97) with } \|q\| \leq \Delta\} \tag{4.98}$$

is bounded. Suppose that the set is not bounded. Then there exists a sequence $\{q^k\}$ with $\|q^k\| \leq \Delta$ and a sequence $\{x^k\}$ with $\|x^k\| \to +\infty$ such that for any k,

$$x^k \in \mathcal{K}, \quad y^k = F(x^k) + q^k \in \mathcal{K} \quad \text{and} \quad x^k \circ y^k = 0. \tag{4.99}$$

This is equivalent to saying that for any k,

$$\frac{1}{2} \sum_{i=1}^{m} \left\| \phi_{\text{FB}}(x_i^k, y_i^k) \right\|^2 = 0.$$

Using Lemma 3.7 and the boundedness of q^k, we then obtain

$$\|x^k\| \to +\infty, \quad \lim_{k \to +\infty} \frac{[-x^k]_+}{\|x^k\|} \to 0, \quad \lim_{k \to +\infty} \frac{[-y^k]_+}{\|x^k\|} \to 0, \text{ and } \lim_{k \to +\infty} \frac{\|[q^k]_+\|}{\|x^k\|} \to 0. \tag{4.100}$$

Noting that

$$\left\| [q^k]_+ \right\| = \left\| [y^k - F(x^k)]_+ \right\| \geq \left\| [-F(x^k)]_+ \right\|,$$

where the inequality is due to Lemma 1.1(c), we have from the last term in (4.100) that

$$\lim_{k \to +\infty} \frac{\left\| [-F(x^k)]_+ \right\|}{\|x^k\|} \to 0.$$

This together with the first two terms in (4.100) shows that $\{x^k\}$ satisfies the condition (1.49). By the Cartesian R_{02}-property of F, there exists a $\nu \in \{1, 2, \ldots, m\}$ such that

$$\liminf_{k\to+\infty} \frac{\lambda_2[x_\nu^k \circ F_\nu(x^k)]}{\|x^k\|^2} > 0.$$

However, from Equation (4.99) and the boundedness of q^k, we have

$$\frac{\lambda_2[x_\nu^k \circ F_\nu(x^k)]}{\|x^k\|^2} = \frac{\lambda_2[-x_\nu^k \circ q_\nu^k]}{\|x^k\|^2} \to 0.$$

This leads to a contradiction. Consequently, the set defined by (4.98) is bounded. □

It is worth noting that the Cartesian R_{02}-property is implied by the R_0-property as discussed in [15]. Therefore, Proposition 4.40(b) offers a weaker condition under which the solution set S^* is guaranteed to be nonempty and bounded. Combining Proposition 4.34, Proposition 4.40(b), and Proposition 1.7(b), we obtain the following result.

Proposition 4.41 *Suppose that F has the Cartesian P_0-property and the Cartesian R_{02}-property and satisfies Condition 4.2. Then, the path $\mathcal{P}_{\bar\varepsilon} = \{x(\varepsilon) \mid \varepsilon \in (0, \bar\varepsilon\,]\}$ is bounded for any $\bar\varepsilon > 0$ and $\lim_{\varepsilon\downarrow 0} \operatorname{dist}(x(\varepsilon) \mid S^*) = 0$.*

The preceding discussion demonstrates that the original SOCCP(F) can, in principle, be solved by computing the exact solutions to a sequence of regularized problems SOCCP(F_ε). However, in practical settings, it is often infeasible to solve each SOCCP(F_ε) exactly for every $\varepsilon > 0$. To address this, we propose an inexact regularization algorithm that allows for approximate solutions to the subproblems, while still preserving all the convergence properties of its exact counterpart.

Algorithm 4.5 (Inexact Regularization Method)

(S.0) Choose $\varepsilon_0 > 0$ and $\tau_0 > 0$, and set $k := 0$.
(S.1) Compute an approximate solution x^k of SOCCP (F_ε) such that

$$\Psi_\varepsilon(x^k) \le \tau_k.$$

(S.2) Terminate the iteration if a suitable criterion is satisfied.
(S.3) Choose $\varepsilon_{k+1} > 0$ and $\tau_{k+1} > 0$, set $k := k + 1$, and go to (S.1).

Clearly, if we take $\tau_k = 0$ at each iteration, then $x^k = x(\varepsilon_k)$. In addition, we note that the point x^k can be easily obtained by applying any effective gradient-type unconstrained optimization algorithm to the minimization problem

$$\min_{x \in \mathbb{R}^n} \Psi_\varepsilon(x), \tag{4.101}$$

because the objective function $\Psi_\varepsilon(x)$ in (4.101) is continuously differentiable everywhere and has bounded level sets for those SOCCPs with F having the Cartesian P_0-property and satisfying Condition 4.2. In our numerical experiments, we adopt the BFGS algorithm to compute x^k.

Lemma 4.12 *Suppose that $f : \mathbb{R}^n \to \mathbb{R}$ is smooth and coercive. Let $C \subseteq \mathbb{R}^n$ be a nonempty compact set and denote $\bar{c}$ by the least value of f on the boundary of C, i.e., $\bar{c} := \min_{x \in \partial C} f(x)$. If there have two points $a \in C$ and $b \notin C$ such that $f(a) < \bar{c}$ and $f(b) < \bar{c}$, then there exists a point $z \in \mathbb{R}^n$ such that $\nabla f(z) = 0$ and $f(z) \geq \bar{c}$.*

Proof This is the well-known Mountain Pass Theorem [48], which will be employed in the convergence analysis of Algorithm 4.5. □

We now proceed to establish the convergence results of Algorithm 4.5. To this end, we assume that the algorithm generates an infinite sequence–that is, the termination criterion in step (S.2) is never triggered.

Proposition 4.42 *Let F be the mapping having the Cartesian P_0-property and satisfying Condition 4.2. Assume that the solution set $\mathcal{S}^*$ of the SOCCP(F) is nonempty and bounded. If $\varepsilon_k \to 0$ and $\tau_k \to 0$, then the sequence $\{x^k\}$ generated by Algorithm 4.5 remains bounded, and every accumulation point of $\{x^k\}$ is a solution of the SOCCP(F).*

Proof Suppose that the sequence $\{x^k\}$ is unbounded. Then, passing to a subsequence if necessary, we assume that $\{\|x^k\|\} \to +\infty$. This together with the boundedness of $\mathcal{S}^*$ means that there exists a compact set $C \subseteq \mathbb{R}^n$ with $\mathcal{S}^* \subset \text{int}C$ and $x^k \notin C$ for sufficiently large k. Let $x^* \in \mathcal{S}^*$ be a solution of the SOCCP(F). Then, we have

$$\Psi_{\text{FB}}(x^*) = 0 \quad \text{and} \quad \bar{c} := \min_{x \in \partial C} \Psi_{\text{FB}}(x) > 0. \tag{4.102}$$

Let $\delta := \bar{c}/4$. Notice that $\Psi_\varepsilon(x)$ viewed as the function of x and ε is continuous on the compact set $C \times [0, \tilde{\varepsilon}]$, and so is uniformly continuous on $C \times [0, \tilde{\varepsilon}]$. Hence, there exists an $\tilde{\varepsilon} > 0$ such that for all $x \in C$ and $\varepsilon \in [0, \tilde{\varepsilon}]$

$$|\Psi_\varepsilon(x) - \Psi_{\text{FB}}(x)| \leq \delta. \tag{4.103}$$

Combining (4.103) with (4.102), we have that for all sufficiently large k,

$$\Psi_{\varepsilon_k}(x^*) \leq \frac{1}{4}\bar{c} \tag{4.104}$$

and

$$c := \min_{x \in \partial C} \Psi_{\varepsilon_k}(x) \geq \frac{3}{4}\bar{c}. \tag{4.105}$$

On the other hand, $\Psi_{\varepsilon_k}(x^k) \leq \tau_k$ by Algorithm 4.5 and $\tau_k \to 0$, which means

$$\Psi_{\varepsilon_k}(x^k) \leq \frac{1}{4}\bar{c} \tag{4.106}$$

for all k large enough. Now using (4.104), (4.105), (4.106) and setting $a = x^*$ and $b = x^k$ in Lemma 4.12, there exists a vector $\hat{x} \in \mathbb{R}^n$ such that

$$\nabla \Psi_{\varepsilon_k}(\hat{x}) = 0 \quad \text{and} \quad \Psi_{\varepsilon_k}(\hat{x}) \geq \frac{1}{4}\bar{c} > 0.$$

This says that $\hat{x}$ is a stationary point of $\Psi_{\varepsilon_k}(x)$, but not a solution of the SOCCP(F_{ε_k}). However, by Proposition 3.13, we know that any stationary point of $\Psi_{\varepsilon_k}(x)$ is a solution of the SOCCP(F_{ε_k}). Thus, we obtain a contradiction. □

Clearly, Proposition 4.38 follows directly from Proposition 4.42 by setting $\tau_k = 0$ for all k. Moreover, Proposition 4.39 and Proposition 4.41 can be readily extended to the inexact framework.

Proposition 4.43 *Suppose that the mapping F has the Cartesian P_0-property and satisfies Condition 4.2. Let $\{x^k\}$ be the sequence generated by Algorithm 4.5. If $\varepsilon_k \to 0$ and $\tau_k \to 0$, and the SOCCP(F) has a unique solution $\bar{x}$, then we have $\lim_{\varepsilon_k \to 0} x^k = \bar{x}$.*

Proposition 4.44 *Suppose that F has the Cartesian P_0-property and the Cartesian R_{02}-property and satisfies Condition 4.2. Let $\{x^k\}$ be the sequence generated by Algorithm 4.5. If $\varepsilon_k \to 0$ and $\tau_k \to 0$, then $\{x^k\}$ is bounded and its every accumulation point is a solution of the SOCCP(F).*

Proposition 4.45 *Suppose that F is a monotone mapping satisfying Condition 4.2 and the SOCCP(F) is strictly feasible. Let $\{x^k\}$ be the sequence generated by Algorithm 4.5. If $\varepsilon_k \to 0$ and $\tau_k \to 0$, then $\{x^k\}$ is bounded and every accumulation point is a solution of the SOCCP(F).*

Proof Applying Proposition 4.40(a), the desired result follows. □

Detailed numerical performance of Algorithm 4.5 is reported in [49]. We highlight several aspects regarding its implementation. To assess the effectiveness of the regularization method, we first applied the inexact regularization algorithm to a class of monotone SOCCPs arising as the KKT optimality conditions of linear SOCPs from the DIMACS Implementation Challenge library [50]. In addition, we tested the method on a class of SOCCPs where the mapping F satisfies the Cartesian P_0-property. Since suitable benchmark examples are not readily available in the literature, we considered the case $F = Mx + q$, where $M \in \mathbb{R}^{n\times n}$ and $q = (q_1, \ldots, q_m)$ is generated randomly, with M constructed to satisfy the Cartesian P_0-property.

Several open questions merit further investigation in future work. First, for monotone SOCCPs, it remains to establish sufficient conditions under which the solution path $x(\varepsilon)$ is continuous, and to determine whether the trajectory $x(\varepsilon)$ converges to the least l_2-norm solution of the SOCCP(F) when the solution set is nonempty and bounded. Second, for SOCCPs with the Cartesian P_0-property, it is of interest to identify appropriate conditions that guarantee the continuity of the solution path $x(\varepsilon)$, and to examine the convergence behavior of the trajectory $x(\varepsilon)$ under boundedness.

References

1. J.-S. Chen, Two classes of merit functions for the second-order cone complementarity problem. Math. Methods Oper. Res. **64**(3), 495–519 (2006)
2. J.-S. Chen, S.-H. Pan, A survey on SOC complementarity functions and solution methods for SOCPs and SOCCPs. Pacific J. Optim. **8**(1), 33–74 (2012)
3. X. Chen, Smoothing methods for complementarity problems and their applications: a survery. J. Oper. Res. **43**, 32–47 (2000)
4. X. Chen, L. Qi, A parameterized Newton method and a Broyden-like method for solving nonsmooth equations. Comput. Optim. Appl. **3**, 157–179 (1994)
5. M. Doljansky, M. Teboulle, An interior proximal algorithm and the exponential multiplier method for semidefinite programming. SIAM J. Optim. **9**, 1–13 (1998)
6. H. Jiang, D. Ralph, Global and local superlinear convergence analysis of Newton-type methods for semismooth equations with smooth least squares, in *Reformulation—Nonsmooth, Piecewise Smooth, Semismooth and Smoothing Methods*. ed. by M. Fukushima, L. Qi (Kluwer Academic Publishers, Boston, 1999), pp.181–209
7. C. Kanzow, Some non-interior continuation method for linear complementarity problems. SIAM J. Matrix Anal. Appl. **17**, 851–868 (1996)
8. N. Yamashita, M. Fukushima, Modified Newton methods for solving semismooth reformulations of monotone complementarity problems. Math. Program. **76**, 469–491 (1997)
9. A. Yoshise, Interior point trajectories and a homogenous model for nonlinear complementarity problems over symmetric cones. SIAM J. Optim. **17**, 1129–1153 (2006)
10. N. Yamashita, M. Fukushima, A new merit function and a descent method for semidefinite complementarity problems, in *Reformulation—Nonsmooth, Piecewise Smooth, Semismooth and Smoothing Methods*. ed. by M. Fukushima, L. Qi (Kluwer Academic Publishers, Boston, 1999), pp.405–420
11. C. Geiger, C. Kanzow, On the resolution of monotone complementarity problems. *Computational Optimization and Applications*, vol. 5, pp. 155–173
12. D.P. Bertsekas, *Nonlinear Programming*, 2nd edn. (Athena Scientific, MA, 1999)

13. F. Facchinei, J.-S. Pang, *Finite-Dimensional Variational Inequalities and Complementarity Problems* (Springer, New York, 2003), pp.1–2
14. Y. Liu, L. Zhang, Y. Wang, Some propeties of a class of merit functions for symmetric cone complementarity problems. Asia-Pacific J. Oper. Res. **23**, 473–496 (2006)
15. J. Tao, M.S. Gowda, Some P-properties for nonlinear transformations on Euclidean Jordan algbras. Math. Oper. Res. **4**, 985–1004 (2005)
16. S.-H. Pan, J.-S. Chen, A damped Gauss-Newton method for the second-order cone complementarity problem. Appl. Math. Optim. **59**, 293–318 (2009)
17. C. Kanzow, S. Petra, On a semismooth least squares formulation of mixed complementarity problems with gap reduction. Optim. Methods Softw. **19**, 507–525 (2004)
18. J.-S. Chen, P. Tseng, An unconstrained smooth minimization reformulation of the second-order cone complementarity problem. Math. Program. **104**, 293–327 (2005)
19. A. Fischer, Solution of monotone complementarity problems with locally Lipschitzian functions. Math. Program. **76**, 513–532 (1997)
20. L. Qi, Convergence analysis of some algorithms for solving nonsmooth equations. Math. Oper. Res. **18**, 227–244 (1993)
21. T. De Luca, F. Facchinei, C. Kanzow, A semismooth equation approach to the solution of nonlinear complementarity problems. Math. Program. **75**, 407–439 (1996)
22. J.Y. Fan, Y.X. Yuan, On the quadratic convergence of the Levenberg-Marquardt method without nonsingularity assumption. Computing **74**, 23–39 (2005)
23. N. Yamashita, M. Fukushima, On the rate of convergence of the Levenberg-Marquardt method. Computing **15**, 237–249 (2001)
24. S.-H. Pan, J.-S. Chen, A least-square semismooth Newton method for the second-order cone complementarity problem. Optim. Methods Softw. **26**(1), 1–22 (2011)
25. R.W. Cottle, J.-S. Pang, R.-E. Stone, *The Linear Complementarity Problem* (Academic Press, New York, 1992)
26. M.C. Ferris, J.-S. Pang, Engineering and economic applications of complementarity problems. SIAM Rev. **39**, 669–713 (1997)
27. M. Ferris, K. Sinapiromsaran, Formulating and solving nonlinear programs as mixed complementarity problems, Optimization in *Lecture Notes in Economics and Mathematical Systems*, ed. by V.J. Strodiot, P.T. Nguyen, vol. 481 (2000)
28. P. Harker, J.-S. Pang, Finite-dimensional variational inequality and nonlinear complementarity problems: a survey of theory, algorithms and applications. Math. Program. **48**, 161–220 (1990)
29. S.C. Billups, S.P. Dirkse, M.C. Soares, A comparison of algorithms for large scale mixed complementarity problems. Comput. Optim. Appl. **7**, 3–25 (1997)
30. S.C. Billups, M.C. Soares, QPCOMP: a quadratic programming based solver for mixed complementarity problems. Math. Program. **76**, 533–562 (1997)
31. M. Ferris, C. Kanzow, T. Munson, Feasible descent algorithms for mixed complementarity problems. Math. Program. **86**, 475–497 (1999)
32. C. Kanzow, Global convergence properties of some iterative methods for linear complementarity problems. SIAM J. Optim. **6**, 326–341 (1996)
33. C. Kanzow, S. Petra, Projected filter trust region methods for a semismooth least squares formulation of mixed complementarity problems. Optim. Methods Softw. **22**, 713–735 (2007)
34. S. Dirkse, M. Ferris, MCPLIB: a collection of nonlinear mixed complementarity problems. Optim. Methods Softw. **5**, 319–345 (1995)
35. S. Dirkse, *Robusr Solution of Mixed Complementarity Problems*, Ph.D. thesis, Computer Sciences Department, University of Wisconsin (1994)
36. X. Chen, L. Qi, D. Sun, Global and superlinear convergence of the smoothing Newton method and its application to general box constrained variational inequalities. Math. Comput. **67**, 519–540 (1998)
37. J.-S. Chen, S.-H. Pan, T.-C. Lin, A smoothing Newton method based on the generalized Fischer-Burmeister function for MCPs. Nonlinear Anal. Theory Methods Appl. **72**(9–10), 3739–3758 (2010)

38. F. Facchinei, C. Kanzow, Beyond monotonicity in regularization methods for nonlinear complementarity problems. SIAM J. Control Optim. **37**, 1150–1161 (1999)
39. Z.-H. Huang, Locating a maximally complementary solutions of the monotone NCP by using non-interior-point smoothing algorithms. Math. Methods Oper. Res. **61**, 41–55 (2005)
40. D. Sun, A regularization Newton method for solving nonlinear complementarity problems. Appl. Math. Optim. **40**, 315–339 (1999)
41. L. Qi, J. Sun, A nonsmooth version of Newton's method. Math. Program. **58**, 353–367 (1993)
42. F.H. Clarke, *Optimization and Nonsmooth Analysis* (Wiley, New York, 1983)
43. F. Facchinei, J. Soares, A new merit function for nonlinear complementarity problems and a related algorithm. SIAM J. Optim. **7**, 225–247 (1997)
44. J.-S. Chen, S.-H. Pan, A family of NCP functions and a descent method for the nonlinear complementarity problem. Comput. Optim. Appl. **40**, 389–404 (2008)
45. H. Jiang, L. Qi, A new nonsmooth equation approach to nonlinear complementarity problems. SIAM J. Control. Optim. **35**, 178–193 (1997)
46. J.-S. Chen, S.-H. Pan, A regularization semismooth Newton method based on the generalized Fischer-Burmeister function for P_0-NCPs. J. Comput. Appl. Math. **220**(1–2), 464–479 (2008)
47. J. Kyparisis, Uniqueness and differentiability of solutions of parametric nonlinear complementarity problems. Math. Program. Study **36**, 105–113 (1986)
48. R.S. Palais, C.-L. Terng, *Critical Point Theorem and Submanifold Geometry*. Lecture Notes in Mathematics, vol. 1353. (Springer, Berlin, 1988)
49. S.-H. Pan, J.-S. Chen, A regularization method for the second-order cone complementarity problems with the Cartesian P_0-property. Nonlinear Anal. Theory Methods Appl. **70**(4), 1475–1491 (2009)
50. G. Pataki, S. Schmieta, The DIMACS library of semidefinite-quadratic-linear programs, *Preliminary draft, Computational Optimization Research Center* (Columbia University, New York). http://dimacs.rutgers.edu/Challenges/Seventh Instances/

Chapter 5
Dynamical Methods Using Complementarity Functions

Abstract In this chapter, we explore the applications of complementarity functions within neural network methods. In particular, we present two classes of target problems, nonlinear complementarity problems and optimization problems involving second-order cones (SOCs), and demonstrate how they can be addressed using neural networks in conjunction with complementarity functions.

Neural network approaches to optimization were first introduced in the 1980s by Hopfield and Tank [1, 2]. Since then, they have been successfully applied to a wide range of optimization problems, including linear and nonlinear programming, variational inequalities, and both linear and nonlinear complementarity problems; see [3–15]. Moreover, neural networks have also found applications in solving real-world problems across various domains, as discussed in [16–18]. The central idea behind neural network approaches to optimization is to construct a nonnegative energy function and to design a dynamic system—typically modeled by a first-order ordinary differential equation (ODE), whose evolution simulates the behavior of an artificial neural network. The system is expected to converge to a steady state (equilibrium point), which corresponds to a solution of the underlying optimization problem. Additionally, these neural networks are hardware implementable and can be realized using integrated circuit technologies.

In essence, neural networks serve as ODE-based models whose trajectories represent the solution paths of the target problems. Unlike traditional optimization algorithms, the stability of these systems is interpreted as the analog of convergence and convergence rate. To set the stage for subsequent discussions, we begin by reviewing a few fundamental concepts related to trajectories and stability, which are standard in the theory of ordinary differential equations; see, for instance, [19]. Consider the first order differential equations (ODE):

J. Chen, *Complementarity Functions in Optimization*, Springer Optimization and Its Applications 228, https://doi.org/10.1007/978-3-032-06924-5_5

$$\dot{x}(t) = H(x(t)), \quad x(t_0) = x_0 \in \mathbb{R}^n \tag{5.1}$$

where $H : \mathbb{R}^n \to \mathbb{R}^n$ is a mapping.

Definition 5.1 A point $x^* = x(t^*)$ is called an equilibrium point or a steady state of the dynamic system (5.1) if $H(x^*) = 0$. If there is a neighborhood $\Omega^* \subseteq \mathbb{R}^n$ of x^* such that $H(x^*) = 0$ and $H(x) \neq 0 \; \forall x \in \Omega^* \backslash \{x^*\}$, then x^* is called an isolated equilibrium point.

Lemma 5.1 *Assume that $H : \mathbb{R}^n \to \mathbb{R}^n$ is a continuous mapping. Then, for any $t_0 \geq 0$ and $x_0 \in \mathbb{R}^n$, there exists a local solution $x(t)$ for (5.1) with $t \in [t_0, \tau)$ for some $\tau > t_0$. If, in addition, H is locally Lipschitz continuous at x_0, then the solution is unique; if H is Lipschitz continuous in $\mathbb{R}^n$, then τ can be extended to ∞.*

If a local solution defined on $[t_0, \tau)$ cannot be extended to a local solution on a larger interval $[t_0, \tau_1)$, $\tau_1 > \tau$, then it is called a maximal solution, and the interval $[t_0, \tau)$ is the maximal interval of existence. Clearly, any local solution has an extension to a maximal one. We denote $[t_0, \tau(x_0))$ by the maximal interval of existence associated with x_0.

Lemma 5.2 *Assume that $H : \mathbb{R}^n \to \mathbb{R}^n$ is continuous. If $x(t)$ with $t \in [t_0, \tau(x_0))$ is a maximal solution and $\tau(x_0) < \infty$, then $\lim_{t \uparrow \tau(x_0)} \|x(t)\| = \infty$.*

Definition 5.2 (*Stability in the sense of Lyapunov*) Let $x(t)$ be a solution for (5.1). An isolated equilibrium point x^* is Lyapunov stable if for any $x_0 = x(t_0)$ and any $\varepsilon > 0$, there exists a $\delta > 0$ such that $\|x(t) - x^*\| < \varepsilon$ for all $t \geq t_0$ and $\|x(t_0) - x^*\| < \delta$.

Definition 5.3 (*Asymptotic stability*) An isolated equilibrium point x^* is said to be asymptotically stable if in addition to being Lyapunov stable, it has the property that $x(t) \to x^*$ as $t \to \infty$ for all $\|x(t_0) - x^*\| < \delta$.

Definition 5.4 (*Lyapunov function*) Let $\Omega \subseteq \mathbb{R}^n$ be an open neighborhood of $\bar{x}$. A continuously differentiable function $W : \mathbb{R}^n \to \mathbb{R}$ is said to be a Lyapunov function at the state $\bar{x}$ over the set Ω for Eq. (5.1) if

$$\begin{cases} W(\bar{x}) = 0, \quad W(x) > 0, \quad \forall x \in \Omega \backslash \{\bar{x}\}. \\ \dfrac{dW(x(t))}{dt} = \nabla W(x(t))^{\mathsf{T}} H(x(t)) \leq 0, \quad \forall x \in \Omega. \end{cases} \tag{5.2}$$

Lemma 5.3 *(a) An isolated equilibrium point x^* is Lyapunov stable if there exists a Lyapunov function over some neighborhood Ω^* of x^*.*
(b) An isolated equilibrium point x^* is asymptotically stable if there is a Lyapunov function over some neighborhood Ω^* of x^* such that $\dfrac{dW(x(t))}{dt} < 0$ for all $x \in \Omega^* \backslash \{x^*\}$.

Definition 5.5 (*Exponential stability*) An isolated equilibrium point x^* is exponentially stable if there exists a $\delta > 0$ such that arbitrary point $x(t)$ of (5.1) with the initial condition $x(t_0) = x_0$ and $\|x(t_0) - x^*\| < \delta$ is well-defined on $[0, +\infty)$ and satisfies

$$\|x(t) - x^*\|_2 \leq ce^{-\omega t}\|x(t_0) - x^*\| \quad \forall t \geq t_0,$$

where $c > 0$ and $\omega > 0$ are constants independent of the initial point.

5.1 Neural Networks for NCP

5.1.1 Neural Network Using ψ^p_{FB} for NCP

In this section, we focus on a neural network approach to the NCP (2.1), utilizing $\Psi^p_{\text{FB}}(x)$ as the energy function. As discussed in Chap. 2, the NCP can be reformulated as the following unconstrained smooth minimization problem:

$$\min_{x \in \mathbb{R}^n} \Psi^p_{\text{FB}}(x) = \frac{1}{2}\|\Phi^p_{\text{FB}}(x)\|^2.$$

Accordingly, it is natural to adopt the following steepest descent-based neural network model for the NCP:

$$\frac{dx(t)}{dt} = -\rho \nabla \Psi^p_{\text{FB}}(x(t)), \quad x(0) = x_0, \tag{5.3}$$

where $\rho > 0$ is a scaling factor. Most neural network models in the existing literature are projection-based and rely on alternative NCP functions, such as the natural residual function (e.g., [8, 13]) or the regularized gap function (e.g., [3]). More recently, neural networks based on the Fischer-Burmeister (FB) function have been developed for linear and quadratic programming, as well as for nonlinear complementarity problems [4, 10]. The model considered here is based on the generalized FB function, thereby extending the approaches found in [4, 10].

We shall demonstrate that the neural network defined in (5.3) possesses desirable stability properties; it is Lyapunov stable, asymptotically stable, and exponentially stable. Furthermore, as observed in [20], the parameter p significantly influences the numerical performance of certain descent-type methods. Specifically, larger values of p tend to improve convergence rates, while smaller values promote better global convergence. In addition, we investigate whether similar phenomena arise in the context of our neural network model.

From Proposition 2.2(c), we know that the function Ψ^p_{FB} is continuously differentiable everywhere with

$$\nabla\Psi^p_{\text{FB}}(x) = V^\top \Phi^p_{\text{FB}}(x) \quad \text{for any } V \in \partial\Phi^p_{\text{FB}}(x) \tag{5.4}$$

or

$$\nabla\Psi^p_{\text{FB}}(x) = \nabla_a \psi^p_{\text{FB}}(x, F(x)) + \nabla F(x)\nabla_b \psi^p_{\text{FB}}(x, F(x)) \tag{5.5}$$

with

$$\begin{aligned}
\nabla_a \psi^p_{\text{FB}}(x, F(x)) &:= \left[\nabla_a \psi^p_{\text{FB}}(x_1, F_1(x)), \ldots, \nabla_a \psi^p_{\text{FB}}(x_n, F_n(x))\right]^\top, \\
\nabla_b \psi^p_{\text{FB}}(x, F(x)) &:= \left[\nabla_b \psi^p_{\text{FB}}(x_1, F_1(x)), \ldots, \nabla_b \psi^p_{\text{FB}}(x_n, F_n(x))\right]^\top.
\end{aligned}$$

In view of the above, there have two ways to compute $\nabla\Psi^p_{\text{FB}}(x)$, which is needed in the network (5.3). One is to use formula (5.4), for which we give an algorithm (see Algorithm 5.1 below), to evaluate an element $V \in \partial\Phi^p_{\text{FB}}(x)$, see Proposition 2.1 for $\partial\phi^p_{\text{FB}}(x_i, F_i(x))$. The other is to adopt formula (5.5).

Algorithm 5.1 (The procedure to evaluate an element $V \in \partial\Phi^p_{\text{FB}}(x)$)

(S.0) Let $x \in \mathbb{R}^n$ be given, and let V_i denote the i-th row of a matrix $V \in \mathbb{R}^{n\times n}$.
(S.1) Set $I(x) := \{i \in \{1, 2, \ldots, n\} \mid x_i = F_i(x) = 0\}$.
(S.2) Set $z \in \mathbb{R}^n$ such that $z_i = 0$ for $i \notin I(x)$, and $z_i = 1$ for $i \in I(x)$.
(S.3) For $i \in I(x)$, let $u_i = \left[|z_i|^{\frac{p}{p-1}} + |\nabla F_i(x)^\top z|^{\frac{p}{p-1}}\right]^{\frac{p-1}{p}}$, and

$$V_i = \left(\frac{z_i}{u_i} - 1\right) e_i^\top + \left(\frac{\nabla F_i(x)^\top z}{u_i} - 1\right) \nabla F_i(x)^\top.$$

(S.4) For $i \notin I(x)$, set

$$V_i = \left(\frac{\text{sgn}(x_i) \cdot |x_i|^{p-1}}{\|(x_i, F_i(x))\|_p^{p-1}} - 1\right) e_i^\top + \left(\frac{\text{sgn}(F_i(x)) \cdot |F_i(x)|^{p-1}}{\|(x_i, F_i(x))\|_p^{p-1}} - 1\right) \nabla F_i(x)^\top.$$

The procedure outlined above represents the conventional approach to computing $\nabla\Psi^p_{\text{FB}}(x(t))$. For instance, the neural network model in [10] employs equation (5.4) along with a similar algorithm to evaluate an element of $V \in \partial\Phi_{\text{FB}}(x)$. In contrast, we

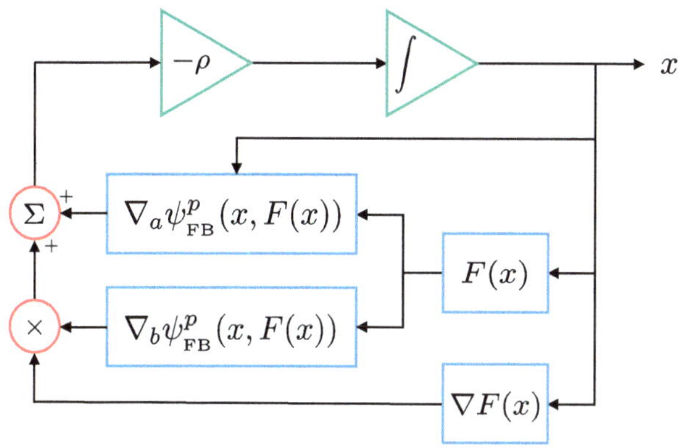

Fig. 5.1 A simplified block diagram for the neural network (5.3)

propose a simpler and more efficient method for computing $\nabla \Psi^p_{FB}(x(t))$: specifically, by using the formula given in (5.5) rather than (5.4). This alternative formulation not only simplifies computation but also provides valuable insight into how the neural network (5.3) can be implemented in hardware. See Fig. 5.1 below for an illustration.

We now assert that Ψ^p_{FB} serves as a global error bound for the solution of the NCP. This result is of fundamental importance, as it will be used to analyze the influence of the parameter p on the convergence rate of the trajectory $x(t)$ generated by the neural network (5.3).

Proposition 5.1 *Suppose F is a uniform P-function with modulus $\kappa > 0$ and Lipschitz continuous with constant $L > 0$. Then, the NCP has a unique solution x^*, and*

$$\|x - x^*\|^2 \leq \frac{4L^2}{\kappa^2(2 - 2^{1/p})^2} \Psi^p_{FB}(x) \quad \forall x \in \mathbb{R}^n.$$

Proof Since F is a uniform P-function, by Proposition 2.5, there exists a global minimizer of $\Psi^p_{FB}(x)$ which says the NCP has a solution. Assume that the NCP has two different solutions x^* and y^*, then by Definition 1.7(f) we have

$$\begin{aligned}
\kappa\|x^* - y^*\|^2 &\leq \max_{1 \leq i \leq m} (x_i^* - y_i^*)(F_i(x^*) - F_i(y^*)) \\
&= \max_{1 \leq i \leq m} \Big\{ - x_i^* F_i(y^*) - y_i^* F_i(x^*) \Big\} \leq 0
\end{aligned}$$

where the equality is due to the fact that $x_i^* F_i(x^*) = y_i^* F_i(y^*) = 0$ for $i = 1, 2, \ldots, n$ (note that x^* and y^* are the solutions to the NCP), and the last inequality holds since $x^*, y^* \geq 0$ and $F(x^*), F(y^*) \geq 0$. This leads to a contradiction. Hence, the NCP has a unique solution.

For any $x \in \mathbb{R}^n$, let $r(x) := (r_1(x), \ldots, r_n(x))^\mathsf{T}$ with $r_i(x) = \min\{x_i, F_i(x)\}$ for $i = 1, \ldots, n$. Since F is Lipschitz continuous with constant $L > 0$, by [21, Lemma 7.4] we have

$$(x_i - x_i^*)(F_i(x) - F_i(x^*)) \leq 2L|r_i(x)|\|x - x^*\|,$$

for all $x \in \mathbb{R}^n$ and $i = 1, 2, \ldots, n$. On the other hand, since F is a uniform P-function with modulus $\kappa > 0$, from Definition 1.7(f) it follows that

$$\kappa\|x - x^*\|^2 \leq \max_{1\leq i\leq n}(x_i - x_i^*)(F_i(x) - F_i(x^*))$$

for any $x \in \mathbb{R}^n$. Combining the last two equations yields

$$\|x - x^*\| \leq (2L/\kappa) \max_{1\leq i\leq n} |r_i(x)| \quad \forall x \in \mathbb{R}^n.$$

This together with Lemma 2.3 implies

$$\|x - x^*\| \leq \frac{2L}{\kappa(2 - 2^{1/p})} \max_{1\leq i\leq n} |\phi^p_{\text{FB}}(x_i, F_i(x))| \leq \frac{2L}{\kappa(2 - 2^{1/p})} \|\Phi^p_{\text{FB}}(x)\|.$$

Consequently, we obtain the desired result. □

Next, we turn to the convergence and stability properties of the neural network (5.3). Our analysis focuses on the behavior of the solution trajectory, including its existence and convergence, and we establish three types of stability for an isolated equilibrium point. We begin by stating the relationship between an equilibrium point of (5.3) and a solution to the NCP.

Proposition 5.2 *(a) Every solution to the NCP is an equilibrium point of the neural network (5.3).*

(b) If F is an P_0-function, then every equilibrium point of (5.3) is a solution to the NCP.

Proof (a) Suppose that x is a solution to the NCP. Then, from Proposition 2.3, it is clear that $\Phi^p_{\text{FB}}(x) = 0$. Using Proposition 2.2(e) and (5.5), we then have $\nabla\Psi^p_{\text{FB}}(x) = 0$. This, by Definition 5.1, shows that x is an equilibrium point of (5.3).

(b) This is a direct consequence of Proposition 2.4. □

Lemma 5.4 *Let $\Psi^p_{\text{FB}} : \mathbb{R}^n \to \mathbb{R}_+$ be given by (2.17). Then, the function $\Psi^p_{\text{FB}}(x(t))$ is nonincreasing with respect to t.*

Proof By the definition of $\Psi^p_{\text{FB}}(x)$ and (5.3), it is not difficult to compute

$$\begin{aligned}\frac{d\Psi^p_{\text{FB}}(x(t))}{dt} &= \nabla\Psi^p_{\text{FB}}(x(t))^\mathsf{T}\frac{dx(t)}{dt} \\ &= \nabla\Psi^p_{\text{FB}}(x(t))^\mathsf{T}\left(-\rho\nabla\Psi^p_{\text{FB}}(x(t))\right) \\ &= -\rho\|\nabla\Psi^p_{\text{FB}}(x(t))\|^2 \leq 0.\end{aligned} \tag{5.6}$$

Therefore, $\Psi^p_{\rm FB}(x(t))$ is a monotonically decreasing function with respect to t. □

Proposition 5.3 *For any fixed $p \ge 2$, the following hold.*

(a) For any initial state $x_0 = x(t_0)$, there exists exactly one maximal solution $x(t)$ with $t \in [t_0, \tau(x_0))$ for the neural network (5.3).
(b) If the level set $\mathcal{L}(x_0) = \{x \in \mathbb{R}^n \mid \Psi^p_{\rm FB}(x) \le \Psi^p_{\rm FB}(x_0)\}$ is bounded or F is Lipschitz continuous, then $\tau(x_0) = +\infty$.

Proof (a) Since F is continuously differentiable, $\nabla F(x)$ is continuous, and therefore, $\nabla F(x)$ is bounded on a local compact neighborhood of x. On the other hand, $\nabla_a \psi^p_{\rm FB}$ and $\nabla_b \psi^p_{\rm FB}$ are Lipschitz continuous by Lemma 2.5. These two facts together with formula (5.5) show that $\nabla \Psi^p_{\rm FB}(x)$ is locally Lipschitz continuous. Thus, applying Lemma 5.1 leads to the desired result.

(b) We proceed the arguments by the two cases as shown below.

Case (i): The level set $\mathcal{L}(x_0)$ is bounded. We prove the result by contradiction. Suppose $\tau(x_0) < \infty$. Then, by Lemma 5.2, $\lim_{t \uparrow \tau(x_0)} \|x(t)\| = \infty$. Let $\mathcal{L}^c(x_0) := \mathbb{R}^n \backslash \mathcal{L}(x_0)$ and

$$\tau_0 := \inf\{s \ge 0 \mid s < \tau(x_0), x(s) \in \mathcal{L}^c(x_0)\} < \infty.$$

We know that $x(\tau_0)$ lies on the boundary of $\mathcal{L}(x_0)$ and $\mathcal{L}^c(x_0)$. Moreover, $\mathcal{L}(x_0)$ is compact since it is bounded by assumption and it is also closed because of the continuity of $\Psi^p_{\rm FB}(x)$. Therefore, we have $x(\tau_0) \in \mathcal{L}(x_0)$ and $\tau_0 < \tau(x_0)$, implying that

$$\Psi^p_{\rm FB}(x(s)) > \Psi^p_{\rm FB}(x_0) > \Psi^p_{\rm FB}(x(\tau_0)) \quad \text{for some } s \in (\tau_0, \tau(x_0)). \tag{5.7}$$

However, Lemma 5.4 says that $\Psi^p_{\rm FB}(x(\cdot))$ is nonincreasing on $[t_0, \tau(x_0))$, which contradicts (5.7). This completes the proof of Case (i).

Case (ii): F is Lipschitz continuous. From the proof of part (a), we know that $\nabla \Psi^p_{\rm FB}(x)$ is Lipschitz continuous. Thus, by Lemma 5.1, we have $\tau(x_0) = \infty$. □

Proposition 5.4 *(a) Let $x(t)$ with $t \in [t_0, \tau(x_0))$ be the unique maximal solution to the neural network (5.3). If $\tau(x_0) = \infty$ and $\{x(t)\}$ is bounded, then* $\lim_{t \to \infty} \nabla \Psi^p_{\rm FB}(x(t)) = 0$.
(b) If F is strongly monotone or a uniform P-function, then $\mathcal{L}(x_0)$ is bounded and every accumulation point of the trajectory $x(t)$ is a solution to the NCP.

Proof With Proposition 2.4, Lemma 5.4, and Proposition 5.3 in place, the subsequent arguments follow directly from those in [10, Corollary 4.3]. Therefore, we omit the details here. □

From Proposition 5.2(a), every solution x^* to the NCP corresponds to an equilibrium point of the neural network (5.3). Moreover, if x^* is an isolated equilibrium point of (5.3), then it can be shown that x^* is not only Lyapunov stable but also asymptotically stable.

Proposition 5.5 *Let x^* be an isolated equilibrium point of the neural network (5.3). Then, x^* is Lyapunov stable for the neural network (5.3), and furthermore, it is asymptotically stable.*

Proof Since x^* is a solution to the NCP, $\Psi^p_{\text{FB}}(x^*) = 0$. In addition, since x^* is an isolated equilibrium point of (5.3), there exists a neighborhood $\Omega^* \subseteq \mathbb{R}^n$ of x^* such that

$$\nabla\Psi^p_{\text{FB}}(x^*) = 0, \quad \text{and} \quad \nabla\Psi^p_{\text{FB}}(x) \neq 0 \quad \forall x \in \Omega^* \backslash \{x^*\}.$$

Next, we argue that $\Psi^p_{\text{FB}}(x)$ is indeed a Lyapunov function at x^* over the set Ω^* for (5.3) by showing that the conditions in (5.2) are satisfied. First, notice that $\Psi^p_{\text{FB}}(x) \geq 0$. Suppose that there is an $\bar{x} \in \Omega^* \backslash \{x^*\}$ such that $\Psi^p_{\text{FB}}(\bar{x}) = 0$. Then, by formula (5.5) and Proposition 2.2(e), we have $\nabla\Psi(\bar{x}) = 0$, i.e., $\bar{x}$ is also an equilibrium point of (5.3), which clearly contradicts the assumption that x^* is an isolated equilibrium point in Ω^*. Thus, we prove that $\Psi^p_{\text{FB}}(x) > 0$ for any $x \in \Omega^* \backslash \{x^*\}$. This together with (5.6) shows that the conditions in (5.2) are satisfied, and hence $\Psi^p_{\text{FB}}(x)$ is a Lyapunov function at x^* over the set Ω^* for (5.3). Therefore, x^* is Lyapunov stable by Lemma 5.3(a).

Now, we show that x^* is asymptotically stable. Since x^* is isolated, from (5.6) we have

$$\frac{d\Psi^p_{\text{FB}}(x(t))}{dt} < 0, \quad \forall\, x(t) \in \Omega^* \backslash \{x^*\}.$$

This, by Lemma 5.3(b), implies that x^* is asymptotically stable. □

Proposition 5.6 *If x^* is a regular solution of the NCP, then it is exponentially stable.*

Proof Recall that x^* is a regular solution to the NCP if every element $V \in \partial\Phi^p_{\text{FB}}(x^*)$ is nonsingular. Then, using the same arguments, we can verify that the neural network (5.3) is also exponentially stable if x^* is a regular solution to the NCP. □

To conclude this section, we provide further elaboration on the various notions, conditions, and related numerical issues.

1. Using arguments similar to those used in [22, Proposition 3.2], we can prove that x^* is regular if $\nabla F_{\alpha\alpha}$ is nonsingular and the Schur complement of $\nabla F_{\alpha\alpha}$ in

$$\begin{bmatrix} \nabla F_{\alpha\alpha}(x^*) & \nabla F_{\alpha\beta}(x^*) \\ \nabla F_{\beta\alpha}(x^*) & \nabla F_{\beta\beta}(x^*) \end{bmatrix}$$

is an P-matrix, where $\alpha := \{i \mid x_i^* > 0\}$ and $\beta := \{i \mid x_i^* = F_i(x^*) = 0\}$. Clearly, if ∇F is positive definite, then the conditions hold true.

2. From Definition 5.5, if an isolated equilibrium point x^* is exponentially stable, then there exists a $\delta > 0$ such that $x(t)$ with $x_0 = (t_0)$, and $\|x(t_0) - x^*\| < \delta$ satisfies
$$\|x(t) - x^*\| \le ce^{-\omega t}\|x(t_0) - x^*\| \quad \forall t \ge t_0,$$
which together with Proposition 5.1 implies that
$$\|x(t) - x^*\| \le \frac{2cL}{\kappa(2 - 2^{1/p})}\sqrt{\Psi^p_{\text{FB}}(x_0)}e^{-\omega t} \quad \forall t \ge t_0. \tag{5.8}$$
Since the strong monotonicity of F implies that F is a uniform P-function and that ∇F is positive definite, from (5.8), we obtain that the neural network (5.3) can yield a trajectory with an exponential convergence rate under the condition that F is strongly monotone and Lipschitz continuous.
3. From Eq. (5.8), we observe that as the parameter p increases, the coefficient of the exponential term $e^{-\omega t}$ on the right-hand side decreases. This indicates that a larger p leads to a faster convergence rate, a conclusion consistent with the findings of [20] for descent-type methods based on Ψ^p_{FB}. Furthermore, (5.8) also reveals that the energy of the initial state, $\Psi^p_{\text{FB}}(x_0)$, affects the convergence behavior. Specifically, a higher initial energy tends to result in a slower convergence rate.
4. For detailed numerical simulations, please see [23].

5.1.2 Neural Network Using ϕ^p_{NR}, $\phi^p_{\text{S-NR}}$, and $\psi^p_{\text{S-NR}}$ for NCP

Analogous to what we do in Sect. 5.1.1, we consider the steepest descent-based neural network :
$$\frac{dx(t)}{dt} = -\rho\nabla\Psi(x(t)), \quad x(t_0) = x^0, \tag{5.9}$$
where $\rho > 0$ is a time-scaling factor. Here, we will employ different types of NCP functions to work along with dynamical system (5.9). To this end, given an NCP function $\phi \in \{\phi^p_{\text{NR}}, \phi^p_{\text{S-NR}}, \psi^p_{\text{S-NR}}\}$, we denote
$$\psi(a, b) := \frac{1}{2}|\phi(a, b)|^2. \tag{5.10}$$
Moreover, let $\Phi : \mathbb{R}^n \to \mathbb{R}^n$ be defined by
$$\Phi(x) = \begin{pmatrix} \phi(x_1, F_1(x)) \\ \vdots \\ \phi(x_n, F_n(x)) \end{pmatrix} \tag{5.11}$$

and $\Psi : \mathbb{R}^n \to \mathbb{R}_+$ be given by

$$\Psi(x) = \frac{1}{2}\|\Phi(x)\|^2. \tag{5.12}$$

To proceed, we first summarize several key lemmas and important properties of Ψ, as defined in (5.12), for general NCP functions. These results can be found in Chap. 2.

Lemma 5.5 *Let F be locally Lipschitzian. If all $V \in \partial F(x)$ are nonsingular, then there is a neighborhood $N(x)$ of x and a constant C such that for any $y \in N(x)$ and any $V \in \partial F(y)$, V is nonsingular and $\|V^{-1}\| \leq C$*

Proof Please see [24, Propositions 3.1]. □

Proposition 5.7 *Let $\Psi : \mathbb{R}^n \to \mathbb{R}_+$ be defined as in (5.12) with ϕ being any NCP function and ψ being given as in (5.10). Suppose that F is continuously differentiable. Then, the following hold.*

(a) $\Psi(x) \geq 0$ for all $x \in \mathbb{R}^n$. If the NCP (2.1) has a solution, x is a global minimizer of $\Psi(x)$ if and only if x solves the NCP.
(b) $\Psi(x(t))$ is a nonincreasing function of t, where $x(t)$ is a solution of (5.9).
(c) Let $x \in \mathbb{R}^n$, and suppose that ϕ is differentiable at $(x_i, F_i(x))$ for each $i = 1, \ldots, n$. Then,

$$\nabla\Psi(x) = \nabla_a\psi(x, F(x)) + \nabla F(x)\nabla_b\psi(x, F(x)) \tag{5.13}$$

where

$$\begin{aligned}\nabla_a\psi(x, F(x)) &:= [\nabla_a\psi(x_1, F_1(x)), \ldots, \nabla_a\psi(x_n, F_n(x))]^{\mathsf{T}},\\ \nabla_b\psi(x, F(x)) &:= [\nabla_b\psi(x_1, F_1(x)), \ldots, \nabla_b\psi(x_n, F_n(x))]^{\mathsf{T}}.\end{aligned}$$

(d) Let x be a solution to the NCP such that ϕ is differentiable at $(x_i, F_i(x))$ for each $i = 1, \ldots, n$. Then, x is a stationary point of Ψ.
(e) Every accumulation point of a solution $x(t)$ of neural network (5.9) is an equilibrium point.

Proof (a) It is clear that $\Psi \geq 0$. Notice that $\Psi(x) = 0$ if and only if $\Phi(x) = 0$, which occurs if and only if $\phi(x_i, F_i(x)) = 0$ for all i. Since ϕ is an NCP-function, this is equivalent to having $x_i \geq 0$, $F_i(x) \geq 0$ and $x_i F_i(x) = 0$. Thus, $\Psi(x) = 0$ if and only if $x \geq 0$, $F(x) \geq 0$ and $\langle x, F(x)\rangle = 0$. This proves part (a).

(b) The desired result follows from

$$\frac{d\Psi(x(t))}{dt} = \nabla\Psi(x(t))^{\mathsf{T}}\frac{dx}{dt}$$

$$= \nabla\Psi(x(t))^T(-\rho\nabla\Psi(x(t)))$$
$$= -\rho\|\nabla\Psi(x(t))\|^2 \le 0$$

for all solutions $x(t)$.

(c) The formula is clear from chain rule.

(d) First, note that from Eq. (5.10), we have $\nabla\psi(a,b) = \phi(a,b)\cdot\nabla\phi(a,b)$. Thus, if x is a solution to the NCP, it gives $\nabla\psi(x_i, F_i(x)) = 0$ for all $i = 1,\dots,n$. Then, it follows from formula (5.13) in part(c) that $\nabla\Psi(x) = 0$. That is, x is a stationary point of Ψ.

(e) Please see page 232 in [25] for a proof. □

As mentioned, we adopt the neural network (5.9) with $\Psi(x) = \frac{1}{2}\|\Phi(x)\|^2$, where Φ is given by (5.11) with $\phi \in \{\phi^p_{\text{NR}}, \phi^p_{\text{S-NR}}, \psi^p_{\text{S-NR}}\}$. The function Φ corresponding to ϕ^p_{NR}, $\phi^p_{\text{S-NR}}$ and $\psi^p_{\text{S-NR}}$ is denoted, respectively, by Φ^p_{NR}, $\Phi^p_{\text{S1-NR}}$ and $\Phi^p_{\text{S2-NR}}$. Their corresponding merit functions will be denoted by Ψ^p_{NR}, $\Psi^p_{\text{S1-NR}}$ and $\Psi^p_{\text{S2-NR}}$, respectively. We note that by formula (5.13) and the differentiability of $\Psi \in \{\Psi^p_{\text{NR}}, \Psi^p_{\text{S1-NR}}, \Psi^p_{\text{S2-NR}}\}$, the neural network (5.9) can be implemented on hardware as in Fig. 5.1.

Proposition 5.8 *Let $p > 1$ be an odd integer. Then, the following hold.*

(a) Ψ^p_{NR} and $\Psi^p_{\text{S2-NR}}$ are both continuously differentiable on $\mathbb{R}^n$.
(b) $\Psi^p_{\text{S1-NR}}$ is continuously differentiable on the open set $\Omega = \{x \in \mathbb{R}^n \mid x_i \neq F_i(x), \forall i = 1, 2, \cdots, n\}$.

Consequently, the neural network (5.9) with Ψ^p_{NR} or $\Psi^p_{\text{S2-NR}}$ has a unique solution for all $x^0 \in \mathbb{R}^n$. The neural network (5.9) with $\Psi^p_{\text{S1-NR}}$ has a unique solution for all $x^0 \in \Omega$.

Proof Parts (a) and (b) follow directly from Propositions 5.7(c), 2.23, 2.27, and 2.33. The existence and uniqueness of the solutions are guaranteed by Lemma 5.1, given the continuous differentiability of F, as well as that of Ψ^p_{NR}, $\Psi^p_{\text{S1-NR}}$ (on Ω), and $\Psi^p_{\text{S2-NR}}$. □

As noted in Proposition 5.8(b), we restrict our consideration of the neural network (5.9) with $\Psi = \Psi^p_{\text{S1-NR}}$ to the domain Ω, treating it as a dynamical system defined on this set. Our next objective is to identify conditions under which the equilibrium points of (5.9) coincide with the global minimizers of Ψ. When the underlying NCP function satisfies the following properties:

(P1) $\nabla_a\psi(a,b)\cdot\nabla_b\psi(a,b) \ge 0$ for all $(a,b) \in \mathbb{R}^2$; and
(P2) for all $(a,b) \in \mathbb{R}^2$, $\nabla_a\psi(a,b) = 0 \Longleftrightarrow \nabla_b\psi(a,b) = 0 \Longleftrightarrow \phi(a,b) = 0$,

an equilibrium point of the neural network corresponds to a global minimizer of Ψ, provided that F is a P_0-function. However, as discussed in Sect. 2.2, the functions ϕ^p_{NR}, $\phi^p_{\text{S-NR}}$, and $\psi^p_{\text{S-NR}}$ satisfy these properties only on a proper subset of $\mathbb{R}^n$. Consequently, we seek alternative conditions to achieve the desired characterization. We begin by examining the merit function Ψ^p_{NR}.

Proposition 5.9 *If F is strongly monotone with modulus $\mu > 1$, then every stationary point of Ψ^p_{NR} is a global minimizer.*

Proof Let x^* be a stationary point of Ψ^p_{NR}, that is, $\nabla\Psi^p_{\text{NR}}(x^*) = 0$. For convenience, we denote by $A(x^*)$ and $B(x^*)$ the diagonal matrices such that for each $i = 1, \ldots, n$,

$$A_{ii}(x^*) = (x_i^*)^{p-1} \quad \text{and} \quad B_{ii}(x) = (x_i^* - F_i(x^*))^{p-2}(x_i^* - F_i(x^*))_+.$$

Then, by formula (5.13) and Proposition 2.23, we have

$$p[A(x^*) - B(x^*)]\Phi^p_{\text{NR}}(x^*) + p\nabla F(x^*)B(x^*)\Phi^p_{\text{NR}}(x^*) = 0,$$

which yields

$$A(x^*)\Phi^p_{\text{NR}}(x^*) + (\nabla F(x^*) - I)B(x^*)\Phi^p_{\text{NR}}(x^*) = 0. \tag{5.14}$$

Analogous to the technique in [26], pre-multiplying both sides of (5.14) by $(B(x^*)\Phi^p_{\text{NR}}(x^*))^{\mathsf{T}}$ leads to

$$\Phi^p_{\text{NR}}(x^*)^{\mathsf{T}}\big[B(x^*)A(x^*)\big]\Phi^p_{\text{NR}}(x^*) + \big(B(x^*)\Phi^p_{\text{NR}}(x^*)\big)^{\mathsf{T}}(\nabla F(x^*) - I)B(x^*)\Phi^p_{\text{NR}}(x^*) = 0. \tag{5.15}$$

Since p is odd integer, we have $A(x^*) \geq 0$ and $B(x^*) \geq 0$; and hence,

$$\Phi^p_{\text{NR}}(x^*)^{\mathsf{T}}[B(x^*)A(x^*)]\Phi^p_{\text{NR}}(x^*) \geq 0.$$

On the other hand, since F is strongly monotone with modulus $\mu > 1$, defining $G(x) := F(x) - x$ gives

$$\begin{aligned}
\langle x - y, G(x) - G(y)\rangle &= \langle x - y, F(x) - x - F(y) + y\rangle \\
&= \langle x - y, F(x) - F(y)\rangle - \|x - y\|^2 \\
&\geq (\mu - 1)\|x - y\|^2 \\
&> 0,
\end{aligned}$$

for all $x, y \in \mathbb{R}^n$. Note then that $\nabla G(x) = \nabla F(x) - I$ is positive definite. Consequently, each term of the left-hand side of (5.15) is non-negative. With $(\nabla F(x^*) - I)$ being positive definite, it yields $B(x^*)\Phi^p_{\text{NR}}(x^*) = 0$. In addition, from (5.14), we have $A(x^*)\Phi^p_{\text{NR}}(x^*) = 0$. To sum up, we have proved that $A_{ii}(x^*)\phi^p_{\text{NR}}(x_i^*, F_i(x^*)) = 0$ and $B_{ii}(x^*)\phi^p_{\text{NR}}(x_i^*, F_i(x^*)) = 0$ for all i.

Now, if $\phi^p_{\text{NR}}(x_i^*, F_i(x^*)) \neq 0$ for some i, then we must have $A_{ii}(x^*) = B_{ii}(x^*) = 0$. Thus, $(x_i^*)^{p-1} = 0$ (i.e., $x_i^* = 0$), and $x_i^* \leq F_i(x^*)$. Since ϕ^p_{NR} is an NCP-function, the latter implies that $\phi^p_{\text{NR}}(x_i, F_i(x^*)) = 0$. Hence, $\phi^p_{\text{NR}}(x_i, F_i(x^*)) = 0$ for all i, that is, x^* is a global minimizer of Ψ^p_{NR}. This completes the proof. □

The following proposition establishes a weaker condition on F under which any stationary point of Ψ^p_{NR} is guaranteed to be a global minimizer.

Proposition 5.10 *If $(\nabla F - I)$ is a P-matrix, then every stationary point of Ψ^p_{NR} is a global minimizer.*

Proof Suppose that $\nabla \Psi^p_{\text{NR}}(x^*) = 0$. If $B(x^*)\Phi^p_{\text{NR}}(x^*) = 0$, then $A(x^*)\Phi^p_{\text{NR}}(x^*) = 0$ by Eq. (5.14). As in the preceding proof, we obtain $\Phi^p_{\text{NR}}(x^*) = 0$, and hence we are done. It remains to consider another case that $B(x^*)\Phi^p_{\text{NR}}(x^*) \neq 0$. Note that

$$
\begin{aligned}
&(B(x^*)\Phi^p_{\text{NR}}(x^*))_i \\
&= (x_i^* - F_i(x^*))^{p-2}(x_i^* - F_i(x^*))_+ \phi^p_{\text{NR}}(x_i^*, F_i(x^*)) \\
&= \begin{cases} 0 & \text{if } x_i^* \leq F_i(x^*) \text{ or } x_i^* > F_i(x^*) = 0, \\ (x_i^* - F_i(x^*))^{p-1}\phi^p_{\text{NR}}(x_i^*, F_i(x^*)) & \text{if } x_i^* > F_i(x^*) \text{ and } F_i(x^*) \neq 0. \end{cases}
\end{aligned}
$$

Thus, the nonzero entries of $B(x^*)\Phi^p_{\text{NR}}(x^*)$ appear at indices i where $x_i^* > F_i(x^*)$ and $F_i(x^*) \neq 0$. To proceed, we denote

$$
\begin{aligned}
I_1 &= \{i \mid x_i^* \neq 0 \text{ and } (B(x^*)\Phi^p_{\text{NR}}(x^*))_i \neq 0\}, \\
I_2 &= \{i \mid x_i^* = 0 \text{ and } (B(x^*)\Phi^p_{\text{NR}}(x^*))_i \neq 0\}.
\end{aligned}
$$

With these notations, we observe the following facts.

(i) For $i \in I_1$, since p is odd, it is clear that the i-th entry of $A(x^*)\Phi^p_{\text{NR}}(x^*)$ and $B(x^*)\Phi^p_{\text{NR}}(x^*))$ are both nonzero and have the same sign.
(ii) For $i \in I_2$, then $(B(x^*)\Phi^p_{\text{NR}}(x^*))_i \neq 0$ and $(A(x^*)\Phi^p_{\text{NR}}(x^*))_i = 0$.

Because $(\nabla F - I)$ is a P-matrix, it follows from Lemma 1.5 that there exists an index j such that

$$
(B(x^*)\Phi^p_{\text{NR}}(x^*))_j[(\nabla F(x^*) - I)(B(x^*)\Phi^p_{\text{NR}}(x^*))]_j > 0.
$$

This says that $(B(x^*)\Phi^p_{\text{NR}}(x^*))_j \neq 0$ and therefore $j \in I_1 \cup I_2$. Note that by (i) above, $(A(x^*)\Phi^p_{\text{NR}}(x^*))_i$ and $(B(x^*)\Phi^p_{\text{NR}}(x^*))_i$ have the same sign if $j \in I_1$ which will contradict Eq. (5.14). On the other hand, if $j \in I_2$, we have from fact (ii) that $(A(x^*)\Phi^p_{\text{NR}}(x^*))_j = 0$. However, we also have that $[(\nabla F(x^*) - I)(B(x^*)\Phi^p_{\text{NR}}(x^*))]_j \neq 0$. This certainly violates Eq. (5.14). Thus, we conclude that $B(x^*)\Phi^p_{\text{NR}}(x^*) = 0$, and hence $\Phi^p_{\text{NR}}(x^*) = 0$. Then, the proof is complete. □

Remark 5.1 In fact, if the function F is nonnegative (or if we at least have $F(x^*) \geq 0$ for an equilibrium point x^*), then case (ii) in the above proof cannot happen. Thus, the above result is valid even when $(\nabla F - I)$ is a P_0-matrix by Lemma 1.5.

As shown in Proposition 2.27 and Proposition 2.33, the structures of $\nabla \Phi^p_{\text{S1-NR}}$ and $\nabla \Phi^p_{\text{S2-NR}}$, corresponding to the NCP functions $\phi^p_{\text{S-NR}}$ and $\psi^p_{\text{S-NR}}$, are inherently complex due to the piecewise nature of these functions. This complexity presents significant challenges in identifying conditions on F that ensure a stationary point of $\Psi^p_{\text{S1-NR}}$ or $\Psi^p_{\text{S2-NR}}$ is also a global minimizer. Nevertheless, under the assumption that F is a nonnegative function, we can establish the following result.

Proposition 5.11 *Suppose that F is a nonnegative P_0-function and $x^* \geq 0$. If x^* is a stationary point of $\Psi^p_{\text{S1-NR}}$ or $\Psi^p_{\text{S2-NR}}$, then it is a global minimizer.*

Proof If we can show that the aforementioned properties (P1) and (P2) hold for $\phi^p_{\text{S-NR}}$ and $\psi^p_{\text{S-NR}}$ on the nonnegative quadrant $\mathbb{R}^2_+$, then we can proceed as in the proof of [27, Proposition 3.4]. Thus, it is enough to show that (P1) and (P2) hold on $\mathbb{R}^2_+$. Indeed, they clearly follow from Propositions 2.28 to 2.35.

For completeness, we include the detailed arguments here. To simplify our notations, we denote $\phi_1 = \phi^p_{\text{S-NR}}$, $\phi_2 = \psi^p_{\text{S-NR}}$, and $\psi_i = \frac{1}{2}|\phi_i|^2$ $(i = 1, 2)$. Note that the domain of $\nabla\Psi^p_{\text{S1-NR}}$ is $\{x \mid x_i \neq F_i(x) \text{ or } x_i = F_i(x) = 0\}$. Thus, for ψ_1, it suffices to check that it has properties (P1) and (P2) only on the set $\{(a, b) \in \mathbb{R}^2_+ \mid a \neq b \text{ or } a = b = 0\}$.

To proceed, we observe that

$$\nabla_a\psi_i(a, b) = \phi_i(a, b)\nabla_a\phi_i(a, b) \quad \text{and} \quad \nabla_b\psi_i(a, b) = \phi_i(a, b)\nabla_b\phi_i(a, b),$$

which imply

$$\nabla_a\psi_i(a, b) \cdot \nabla_b\psi_i(a, b) = (\phi_i(a, b))^2 \cdot \nabla_a\phi_i(a, b) \cdot \nabla_b\phi_i(a, b), \quad i = 1, 2.$$

If $a \geq b = 0$ or $b \geq a = 0$, then $\phi_i(a, b) = 0$; and thus, the above product is zero. Otherwise, the above product is positive by Proposition 2.24. This asserts (P1).

To show (P2), note that it is obvious that $\nabla_a\psi_i(a, b) = \nabla_b\psi_i(a, b) = 0$ if $\phi_i(a, b) = 0$ for $i = 1, 2$.

To show the converse, it is enough to argue that if $\nabla_a\phi_i(a, b) = 0$ or $\nabla_b\phi_i(a, b) = 0$, then $\phi_i(a, b) = 0$. First, we analyze the case for ϕ_1. Suppose that $\nabla_a\phi_1(a, b) = 0$. From Proposition 2.27, we know

$$\frac{1}{p}\nabla_a\phi_1(a, b) = \begin{cases} a^{p-1} - (a - b)^{p-1} & \text{if } a > b \\ 0 & \text{if } a = b = 0 \\ (b - a)^{p-1} & \text{if } a < b \end{cases} \tag{5.16}$$

For $a = b = 0$, then $\phi_1(a, b) = 0$. For $a > b$, then $a = |a - b| = a - b$ since p is an odd integer. Thus, $b = 0$ and because $a > b$, we obtain $\phi_1(a, b) = 0$. For $a < b$, we have from (5.16) that $(b - a)^{p-1} = 0$, which is impossible. This proves that $\nabla_a\phi_1(a, b) = 0$ implies that $\phi_1(a, b) = 0$. Similarly, we can show that $\nabla_b\phi_1(a, b) = 0$ implies that $\phi_1(a, b) = 0$. This asserts (P2) for the function ψ_1.

Analogously, for ψ_2, assume that $\nabla_a \phi_2(a, b) = 0$. From Proposition 2.33, we have

$$\frac{1}{p}\nabla_a \phi_2(a, b) = \begin{cases} a^{p-1}b^p - (a-b)^{p-1}b^p & \text{if } a > b, \\ a^{2p-1} & \text{if } a = b, \\ a^{p-1}b^p - (b-a)^p a^{p-1} + (b-a)^{p-1}a^p & \text{if } a < b. \end{cases}$$

For $a = b$, then $a^{2p-1} = 0$, and hence $a = 0$ and $\phi_2(a, b) = 0$. For $a > b$, then $a^{p-1}b^p - (a-b)^{p-1}b^p = 0$. For $b = 0$, we obtain $\phi_2(a, b) = 0$ by using $a > b$. Otherwise, $a^{p-1} - (a-b)^{p-1} = 0$. Because p is odd and $a > b$, we have $a = |a - b| = a - b$. consequently, $b = 0$ and $\phi_2(a, b) = 0$. For $a < b$, then we have from the above formula for $\nabla_a \phi_2$ that $a^{p-1}b^p - (b-a)^p a^{p-1} + (b-a)^{p-1}a^p = 0$. For $a = 0$, then $\phi_2(a, b) = 0$ due to $a < b$. Otherwise, $a > 0$ and

$$\begin{aligned} 0 &= b^p - (b-a)^p + (b-a)^{p-1}a \\ &= b^p - (b-a)^{p-1}(b-2a) \\ &= (a+k)^p - k^{p-1}(k-a) \quad \text{where } k = b - a > 0 \\ &= \sum_{i=0}^{p-1} \binom{p}{i} a^{p-i}k^i + ak^{p-1} \\ &> 0 \end{aligned}$$

which is a contradiction. To sum up, we have shown that $\nabla_a \phi_2(a, b) = 0$ implies that $\phi_2(a, b) = 0$. Similarly, it can be verified $\phi_2(a, b) = 0$ provided $\nabla_b \phi_2(a, b) = 0$. Thus, ψ_2 possesses the property (P2). This completes the proof. □

We now examine the properties of the neural network (5.9) concerning the behavior of its solution trajectories. The following results follow directly from Propositions 5.7(a), 5.7(d), 5.10, and 5.11.

Proposition 5.12 *Consider the neural network* (5.9) *with* $\Psi \in \{\Psi^p_{\text{NR}}, \Psi^p_{\text{S1-NR}}, \Psi^p_{\text{S2-NR}}\}$.

(a) *Every solution of the NCP is an equilibrium point.*
(b) *If* $(\nabla F - I)$ *is a* P*-matrix, then every equilibrium point of* (5.9) *with* $\Psi = \Psi^p_{\text{NR}}$ *solves the NCP.*
(c) *If* F *is a nonnegative* P_0*-function, every equilibrium point* $x^* \geq 0$ *of* (5.9) *with* $\Psi \in \{\Psi^p_{\text{S1-NR}}, \Psi^p_{\text{S2-NR}}\}$ *solves the NCP.*

Proposition 5.13 *Let* F *be a uniform* P*-function and let* $\Psi \in \{\Psi^p_{\text{NR}}, \Psi^p_{\text{S1-NR}}, \Psi^p_{\text{S2-NR}}\}$.

(a) *The level sets* $\mathcal{L}(\Psi, \gamma) := \{x \in \mathbb{R}^n \mid \Psi(x) \leq \gamma\}$ *of* Ψ *are bounded for any* $\gamma \geq 0$. *Consequently, the trajectory* $x(t)$ *through any initial condition* $x^0 \in \mathbb{R}^n$ *is defined for all* $t \geq 0$.
(b) *The trajectory* $x(t)$ *of the neural network* (5.9) *through any* $x^0 \in \mathbb{R}^n$ *converges to an equilibrium point.*

Proof (a) Suppose otherwise. Then, there exists a sequence $\{x^k\}_{k=1}^{\infty} \subseteq \mathcal{L}(\Psi, \gamma)$ such that $\|x^k\| \to \infty$ as $k \to \infty$. A similar argument as in [22] shows that there exists an index i such that $|x_i^k| \to \infty$ and $|F_i(x^k)| \to \infty$ as $k \to \infty$. By Proposition 2.36, we have $|\phi(x_i^k, F_i(x^k))| \to \infty$, where $\phi \in \{\phi_{\text{NR}}^p, \phi_{\text{S-NR}}^p, \psi_{\text{S-NR}}^p\}$. But, this is impossible since $\Psi(x^k) \le \gamma$ for all k. Thus, the level set $\mathcal{L}(\Psi, \gamma)$ is bounded. The remaining part of the theorem can be proved similar to Proposition 4.2(b) in [23].

(b) From part(a), the level sets of Ψ are compact and so by LaSalle's Invariance Principle [28], we reach the desired conclusion. □

Proposition 5.14 *Suppose x^* is an isolated equilibrium point of the neural network* (5.9). *Then, x^* is asymptotically stable provided that either*

(i) $\Psi = \Psi_{\text{NR}}^p$ *and* $(\nabla F - I)$ *is a P-matrix; or*
(ii) $\Psi \in \{\Psi_{\text{S1-NR}}^p, \Psi_{\text{S2-NR}}^p\}$, *$F$ is a nonnegative P_0-function, and the equilibrium point is nonnegative.*

Proof Let x^* be an isolated equilibrium point of (5.9). Then, it has a neighborhood O such that

$$\nabla\Psi(x^*) = 0 \quad \text{and} \quad \nabla\Psi(x) \neq 0 \text{ for all } x \in O\backslash\{x^*\}.$$

We claim that Ψ is a Lyapunov function at x^* over Ω. To proceed, we note first that $\Psi(x) \ge 0$. By Propositions 5.12(b) and 5.12(c), $\Psi(x^*) = 0$. Further, if $\Psi(x) = 0$ for some $x \in O\backslash\{x^*\}$, then x solves the NCP and by Proposition 5.12(a), it is an equilibrium point. This contradicts the isolation of x^*. Thus, $\Psi(x) > 0$ for all $x \in O\backslash\{x^*\}$. Finally, it is clear that

$$\frac{d\Psi(x(t))}{dt} = -\rho\|\nabla\Psi(x(t))\|^2 < 0$$

over the set $O\backslash\{x^*\}$. Then, applying Lemma 5.3 yields that x^* is asymptotically stable. □

Proposition 5.15 *Consider the neural network* (5.9) *with* $\Psi \in \{\Psi_{\text{NR}}^p, \Psi_{\text{S1-NR}}^p, \Psi_{\text{S2-NR}}^p\}$. *If $\nabla\Phi(x^*)$ is nonsingular for some isolated equilibrium point x^*, then x^* solves the NCP and x^* is exponentially stable.*

Proof Let x^* be an equilibrium point such that $\nabla\Phi(x^*)$ is nonsingular. Note that $\nabla\Psi(x^*) = \nabla\Phi(x^*)\Phi(x^*)$, and so $\nabla\Psi(x^*) = 0$ implies that $\Phi(x^*) = 0$. This proves the first claim of this proposition. Further, using Ψ as a Lyapunov function as in the preceding theorem, x^* is asymptotically stable.

Note that since Φ is differentiable at x^*, we have

$$\Phi(x) = \nabla\Phi(x)^{\mathsf{T}}(x - x^*) + o(\|x - x^*\|) \quad \text{as } x \to x^* \tag{5.17}$$

By Lemma 5.5, there exists $\delta > 0$ and a constant C such that $\nabla\Phi(x)$ is nonsingular for all x with $\|x - x^*\| < \delta$, and $\|\nabla\Phi(x)^{-1}\| \le C$. Then, it gives

$$\kappa\|y\|^2 \leq \|\nabla\Phi(x)y\|^2 \tag{5.18}$$

for any x in the δ-neighborhood (call it N_δ) and any $y \in \mathbb{R}^n$, where $\kappa = 1/C^2$.

Let $\varepsilon < 2\rho\kappa$. Since x^* is asymptotically stable, we may choose δ small enough so that $o(\|x - x^*\|^2) < \varepsilon\|x - x^*\|^2$ and $x(t) \to x^*$ as $t \to \infty$ for any initial condition $x(0) \in N_\delta$. Now, define $g : [0, \infty) \to \mathbb{R}$ by

$$g(t) := \|x(t) - x^*\|^2$$

where $x(t)$ is the unique solution through $x(0) \in N_\delta$. Using Eqs. (5.17) and (5.18), we obtain

$$\begin{aligned}
\frac{dg(t)}{dt} &= 2(x(t) - x^*)^\top \frac{dx(t)}{dt} \\
&= -2\rho(x(t) - x^*)^\top \nabla\Psi(x(t)) \\
&= -2\rho(x(t) - x^*)^\top \nabla\Phi(x(t))\Phi(x(t)) \\
&= -2\rho(x(t) - x^*)^\top \nabla\Phi(x(t))\nabla\Phi(x)^\top (x(t) - x^*) + o(\|x(t) - x^*\|^2) \\
&\leq (-2\rho\kappa + \varepsilon)\|x(t) - x^*\|^2 \\
&= (-2\rho\kappa + \varepsilon)g(t).
\end{aligned}$$

Then, it follows that $g(t) \leq e^{(-2\rho\kappa+\varepsilon)t} g(0)$, which says

$$\|x(t) - x^*\| \leq e^{(-\rho\kappa+\varepsilon/2)t} \|x(0) - x^*\|,$$

where $-\rho\kappa + \varepsilon/2 < 0$. This proves that x^* is exponentially stable. □

Detailed simulations involving the neural network (5.9) with $\Psi \in \{\Psi^p_{\rm NR}, \Psi^p_{\rm S1-NR}, \Psi^p_{\rm S2-NR}\}$ are provided in [29]. In addition, a variety of comparative analyses are presented, including convergence rate comparisons across different values of p, as well as performance comparisons between these networks and those based on $\phi_{\rm FB}$ and $\phi^p_{\rm FB}$.

5.1.3 Neural Network Using $\widetilde{\phi}^p_{\rm NR}$, $\widetilde{\phi}^p_{\rm S-NR}$, and $\widetilde{\psi}^p_{\rm S-NR}$ for NCP

Following the same idea in Sects. 5.1.1 and 5.1.2, the neural network considered for solving the NCP is the gradient dynamical system

$$\frac{dx}{dt} = -\rho\nabla\Psi_{\rm F}(x(t)), \quad x(0) = x^0, \tag{5.19}$$

which is based on the unconstrained minimization problem $\min_{x\in\mathbb{R}^n} \Psi_{\rm F}(x)$, where

$$\Psi_{\rm F}(x) = \frac{1}{2}\|\Phi_{\rm F}(x)\|^2 = \frac{1}{2}\sum_{j=1}^{n} \phi(x_j, F_j(x))^2. \tag{5.20}$$

In this section, we will employ three functions $\widetilde{\phi}^p_{\rm NR}$, $\widetilde{\phi}^p_{\rm S-NR}$ and $\widetilde{\psi}^p_{\rm S-NR}$ for the ϕ function in (5.20). Note that p could be any positive real number and the case when p is an odd integer greater than 1, the neural network (5.19) reduces to the neural network (5.9) studied in previous section.

Proposition 5.16 *Let $p > 1$ and consider* (5.20). *Then the following hold:*

(a) If $(\nabla F - I)$ is a P-matrix, then every stationary point of $\widetilde{\Psi}^p_{\rm NR}$ is a global minimizer.
(b) If $F(x^*) \geq 0$, $(\nabla F(x^*) - I)$ is a P_0-matrix and x^* is a stationary point of $\widetilde{\Psi}^p_{\rm NR}$, then x^* is a global minimizer of $\widetilde{\Psi}^p_{\rm NR}$.
(c) Suppose that $x^* \in \Omega_{\rm F}$ and $\nabla F(x^*)$ is a P_0-matrix. If $x^*$is a stationary point of $\widetilde{\Psi}^p_{\rm S1-NR}$ or $\widetilde{\Psi}^p_{\rm S2-NR}$, then x^* is a global minimizer.

Proof To prove (a) and (b), we define two diagonal matrices $A(x^*)$ and $B(x^*)$ where

$$A_{ii}(x^*) = |x_i^*|^{p-1} \quad \text{and} \quad B_{ii}(x^*) = (x_i^* - F_i(x^*))\text{sgn}(x_i^* - F_i(x^*))_+,$$

where x^* is an equilibrium point of (5.19) with $\Psi_{\rm F} = \widetilde{\Psi}^p_{\rm NR}$. Then, analogous arguments as in the proof of Proposition 5.10 lead to the desired conclusion. To prove (c), we proceed as in the proof of Proposition 5.11. That is, we verify the following properties:

(P1) $\forall (a, b) \in \mathbb{R}^2_+$, we have $\nabla_a\psi(a, b) \cdot \nabla_b\psi(a, b) \geq 0$; and
(P2) $\forall (a, b) \in \mathbb{R}^2_+$, we have $\nabla_a\psi(a, b) = 0 \Longleftrightarrow \nabla_b\psi(a, b) = 0 \Longleftrightarrow \phi(a, b) = 0$,

where $\psi := \frac{1}{2}\phi^2$ and $\phi \in \{\widetilde{\phi}^p_{\rm S-NR}, \widetilde{\psi}^p_{\rm S-NR}\}$. Property (P1) can be easily verified. To show (P2), we only need to show that given $a, b \geq 0$, the following hold:

(i) $\nabla_a\widetilde{\phi}^p_{\rm S-NR}(a, b) = 0$ implies $\widetilde{\phi}^p_{\rm S-NR}(a, b) = 0$; and
(ii) $\nabla_a\widetilde{\psi}^p_{\rm S-NR}(a, b) = 0$ implies $\widetilde{\psi}^p_{\rm S-NR}(a, b) = 0$.

We first prove (i). If $\nabla_a\widetilde{\phi}^p_{\rm S-NR}(a, b) = 0$, then we see from Proposition 2.38(b) that we must have $a > b$ or $a = b = 0$. Otherwise, $\nabla_a\widetilde{\phi}^p_{\rm S-NR}(a, b) = p(b - a)^{p-1}$ would be positive. If $a = b = 0$, then $\widetilde{\phi}^p_{\rm S-NR}(a, b) = 0$ as desired. If $a > b$, then $0 = \frac{1}{p}\nabla_a\widetilde{\phi}^p_{\rm S-NR}(a, b) = a^{p-1} - (a - b)^{p-1}$. Since $t \mapsto t^{p-1}$ is strictly increasing on $[0, \infty)$, then $a = a - b$, i.e., $b = 0$. Then, $\widetilde{\phi}^p_{\rm S-NR}(a, b) = 0$ since $a > b = 0$ and $\widetilde{\phi}^p_{\rm S-NR}$ is an NCP function. To prove (ii), assume that $\nabla_a\widetilde{\psi}^p_{\rm S-NR}(a, b) = 0$. From Proposition 2.38(c), we must have

$$0 = \frac{1}{p}\nabla_a\widetilde{\psi}^p_{\rm S-NR}(a, b) = \begin{cases} a^{p-1}b^p - (a - b)^{p-1}b^p & \text{if } a \geq b, \\ a^{p-1}b^p - (b - a)^p a^{p-1} + (b - a)^{p-1}a^p & \text{if } a < b. \end{cases}$$

If $a \geq b$, then we can proceed as in Proposition 5.11. If $a < b$, then

$$0 = a^{p-1}b^p - (b-a)^p a^{p-1} + (b-a)^{p-1}a^p = a^{p-1}(b^p - (b-a)^p + (b-a)^{p-1}a). \tag{5.21}$$

From here, we conclude that $a = 0$. Otherwise, we must have $b^p > (b-a)^p$ and so $b^p - (b-a)^p + (b-a)^{p-1}a > (b-a)^{p-1}a > 0$. This contradicts (5.21). Hence, $a = 0$ and since $b > a = 0$, we obtain that $\widetilde{\psi}^p_{\text{S-NR}}(a, b) = 0$ by definition of an NCP function. □

In light of the above proposition, we now present analogous stability results to those established in Sect. 5.1.2. Due to the close similarity in the underlying arguments, the proofs are omitted.

Proposition 5.17 *Let x^* be an equilibrium point of dynamical system* (5.19).

(a) If $\Psi_{\text{F}} \in \{\widetilde{\Psi}^p_{\text{NR}}, \widetilde{\Psi}^p_{\text{S1-NR}}, \widetilde{\Psi}^p_{\text{S2-NR}}\}$ and F is a uniformly P-function, then the solution to (5.19) *through any $x^0 \in \mathbb{R}^n$ converges to x^*.*

(b) If $\Psi_{\text{F}} = \widetilde{\Psi}^p_{\text{NR}}$, then $x^* \in \text{SOL}(F)$ provided that $(\nabla F - I)$ is a P-matrix . If x^* is isolated, then it is asymptotically stable.

(c) If $x^* \in \Omega_F$ and $\Psi_{\text{F}} = \widetilde{\Psi}^p_{\text{S1-NR}}$ or $\Psi_{\text{F}} = \widetilde{\Psi}^p_{\text{S2-NR}}$, then $x^* \in \text{SOL}(F)$ provided that F is a P_0-function . If x^* is isolated, then it is asymptotically stable.

(d) If $\nabla \Phi_{\text{F}}(x^*)$ is nonsingular, where $\phi \in \{\widetilde{\phi}^p_{\text{NR}}, \widetilde{\phi}^p_{\text{S-NR}}, \widetilde{\psi}^p_{\text{S-NR}}\}$, and x^* is isolated, then $x^* \in \text{SOL}(F)$ and x^* is exponentially stable.

The parameter p plays a crucial role in determining the convergence rate of the neural network. For discrete-type families, numerical experiments conducted in [29] on a selected set of test problems revealed that smaller values of $p \in \{3, 5, 7, \ldots\}$ often result in faster convergence. However, there is currently no theoretical justification for this phenomenon. In fact, as we shall observe later, the convergence behavior can vary significantly with different choices of p. Specifically, a smaller value of p does not always guarantee faster convergence; in some instances, higher values of p yield superior performance.

The numerical results presented in the later sections indicate that no simple or uniform relationship can be established between the performance of the neural network (5.19) and the parameter p, particularly when $\Psi_{\text{F}} \in \{\widetilde{\Psi}^p_{\text{NR}}, \widetilde{\Psi}^p_{\text{S1-NR}}, \widetilde{\Psi}^p_{\text{S2-NR}}\}$. Moreover, the suggest that the initial conditions have a significant influence on both the convergence behavior and the sensitivity of the network to the choice of p. To better understand these phenomena, we establish the following theorem. The first part of the proof derives an error bound for the NCP(F) (see Eq. (5.24)), assuming that F is a locally Lipschitz uniformly P-function. The derivation technique follows a similar line of reasoning to that employed in [30, Proposition 6.3.1].

Proposition 5.18 *Consider the neural network* (5.19) *with $\Psi_{\text{F}} = \widetilde{\Psi}^p_{\text{S1-NR}}$ for a given $p > 1$. Suppose that $x^* \in \text{SOL}(F)$ is exponentially stable and F is a uniformly P-function that is locally Lipschitz continuous. Then there exist positive constants K, ω and δ such that for all $t \geq 0$, we have*

$$\|x(t) - x^*\| \leq K \left(\frac{p+1}{p} \sqrt{2\widetilde{\Psi}^p_{\text{S1-NR}}(x^0)} \right)^{\frac{1}{p}} e^{-\omega t} \quad \forall x^0 \in \Omega_{\text{F}} \cap N_\delta(x^*),$$

where $N_\delta(x^*) = \{y : \|y - x^*\| < \delta\}$.

Proof Suppose F is uniformly P with modulus $\kappa > 0$. Given $x \in \mathbb{R}^n$, let $j \in \{1, \ldots, n\}$ such that

$$(x_j - x_j^*)(F_j(x) - F_j(x^*)) \geq (x_i - x_i^*)(F_i(x) - F_i(x^*)) \quad \forall i = 1, \ldots, n.$$

Then

$$\kappa\|x - x^*\|^2 \leq (x_j - x_j^*)(F_j(x) - F_j(x^*)) = -x_j F_j(x^*) - (x_j^* - x_j)F_j(x). \tag{5.22}$$

Meanwhile, note that $(s - t_+)(t_+ - t) \geq 0$ for any $s \geq 0$ and $t \in \mathbb{R}$. Since min $\{x_j, F_j(x)\} = x_j - (x_j - F_j(x))_+$, then taking $s = x_j^* \geq 0$ and $t = x_j - F_j(x)$, we have

$$(x_j^* - x_j + \min\{x_j, F_j(x)\})(F_j(x) - \min\{x_j, F_j(x)\}) \geq 0$$

which implies that

$$(x_j^* - x_j)F_j(x) \geq (x_j^* - x_j)\min\{x_j, F_j(x)\} - F_j(x)\min\{x_j, F_j(x)\}. \tag{5.23}$$

Since $x_j \geq \min\{x_j, F_j(x)\}$ and $F_j(x^*) \geq 0$, we have from inequalities (5.22) and (5.23) that

$$\begin{aligned} \kappa\|x - x^*\|^2 &\leq [(F_j(x) - F_j(x^*)) - (x_j^* - x_j)]\min\{x_j, F_j(x)\} \\ &\leq (\|F(x) - F(x^*)\| + \|x - x^*\|)|\min\{x_j, F_j(x)\}| \end{aligned}$$

Since F is locally Lipschitz, we conclude that given any $x \in \mathbb{R}^n$ in some neighborhood of x^*, there exists an index $j = j(x)$ and $L > 0$ such that

$$\kappa\|x - x^*\|^2 \leq (1 + L) \cdot |\min\{x_j, F_j(x)\}| \cdot \|x - x^*\|. \tag{5.24}$$

Now, let $x^0 \in \Omega_{\text{F}}$. We have from part (a) of the proof of Proposition 2.36 and using Lemma 2.10 that $\widetilde{\phi}^p_{\text{S-NR}}(a, b) \geq \frac{p}{p+1}(\min\{a, b\})^p$ for any $a, b \geq 0$. By (5.24), there exists $j = j(x^0) \in \{1, \ldots, n\}$ such that

$$\kappa\|x^0 - x^*\| \leq (1 + L) \cdot \left[\frac{p+1}{p} \widetilde{\phi}^p_{\text{S-NR}}(x_j^0, F_j(x^0)) \right]^{\frac{1}{p}}. \tag{5.25}$$

Since x^* is exponentially stable, there exist positive constants δ, c and ω such that for any $t \geq 0$, $\|x(t) - x^*\| \leq ce^{-\omega t}\|x^0 - x^*\|$ for all $x^0 \in N_\delta(x^*)$. This, together with inequality (5.25), gives the desired result with $K := \dfrac{c}{\kappa}(1 + L)$. □

Proposition 5.19 *Consider the neural network* (5.19) *for a given* $p > 1$*, and let* $x^* \in \mathrm{SOL}(F)$ *be exponentially stable. Suppose that* F *is a uniformly* P*-function and locally Lipschitz continuous. Then*

(a) If $\Psi_{\mathrm{F}} = \widetilde{\Psi}^p_{\mathrm{NR}}$*, there exist positive constants* K*,* ω *and* δ *such that for all* $t \geq 0$*, we have*

$$\|x(t) - x^*\| \leq K \left(\frac{p+1}{p} \sqrt{2\widetilde{\Psi}^p_{\mathrm{NR}}(x^0)} \right)^{\frac{1}{p}} e^{-\omega t} \qquad \forall x^0 \in \Omega_{\mathrm{F}} \cap N_\delta(x^*).$$

(b) If $\Psi_{\mathrm{F}} = \widetilde{\Psi}^p_{\mathrm{S2-NR}}$*, there exist positive constants* K*,* ω *and* δ *such that for all* $t \geq 0$*, we have*

$$\|x(t) - x^*\| \leq K \left(\frac{p+1}{p} \sqrt{2\widetilde{\Psi}^p_{\mathrm{S2-NR}}(x^0)} \right)^{\frac{1}{2p}} e^{-\omega t} \qquad \forall x^0 \in \Omega_{\mathrm{F}} \cap N_\delta(x^*).$$

Proof For $a \geq b \geq 0$, then $\widetilde{\phi}^p_{\mathrm{NR}}(a, b) = \widetilde{\phi}^p_{\mathrm{S-NR}}(a, b) \geq \frac{p}{p+1} b^p$ as in part (a) of the proof of Proposition 2.36. When $0 \leq a < b$, we have $\widetilde{\phi}^p_{\mathrm{NR}}(a, b) = a^p \geq \frac{p}{p+1} a^p$. It follows that $\widetilde{\phi}^p_{\mathrm{NR}}(a, b) \geq \frac{p}{p+1}(\min\{a, b\})^p$. On the other hand, using the identity (2.49) and the fact that $\widetilde{\phi}^p_{\mathrm{S-NR}}(a, b) \geq \frac{p}{p+1}(\min\{a, b\})^p$ for any $a, b \geq 0$, we derive that $\widetilde{\psi}^p_{\mathrm{S-NR}}(a, b) \geq \frac{p}{p+1}(\min\{a, b\})^{2p}$. Using these identities and the same arguments as in Proposition 5.18, we get the desired inequalities. □

As mentioned in the discussion before Proposition 5.18, there is no simple relation describing the influence of p. To see this clearly, consider the function $\widetilde{\phi}^p_{\mathrm{S-NR}}$. From the proof of Proposition 5.18, there exists an index $j = j(x^0)$ given any $x^0 \in \Omega_{\mathrm{F}}$ close enough to x^* such that

$$\|x(t) - x^*\| \leq \frac{c(1 + L)}{\kappa} \left[\frac{p+1}{p} \widetilde{\phi}^p_{\mathrm{S-NR}}(x^0_j, F_j(x^0)) \right]^{\frac{1}{p}} e^{-\omega t}, \qquad \forall t \geq 0. \qquad (5.26)$$

For a fixed $x^0 \in \Omega_{\mathrm{F}} \cap N_\delta(x^*)$, we define the function

$$g_{a,b}(p) := \left[\frac{p+1}{p} \widetilde{\phi}^p_{\mathrm{S-NR}}(a, b) \right]^{\frac{1}{p}},$$

where $a = x^0_j$ and $b = F_j(x^0)$ and $p > 1$. Without loss of generality, by taking into account the symmetry of $\widetilde{\phi}^p_{\mathrm{S-NR}}$, we may suppose that $a \geq b$. Then

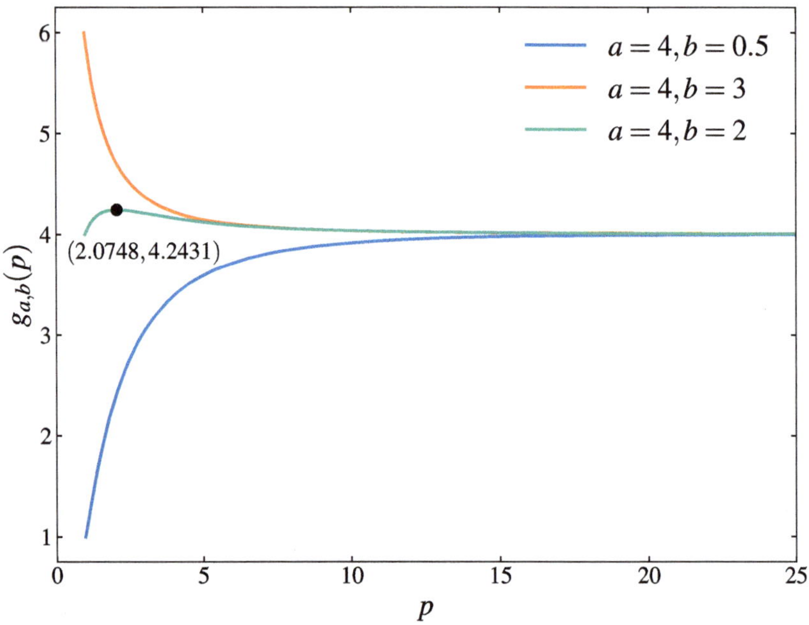

Fig. 5.2 Graph of upper bound for the error term $\|x(t) - x^*\|$ for some values of a and b with $a, b \geq 0$ and $a > b$

$$g_{a,b}(p) = \left[\frac{p+1}{p}(a^p - (a-b)^p)\right]^{\frac{1}{p}}.$$

Note that $M := \lim_{p\to\infty} g_{a,b}(p) = a$. As we shall see in the following example, the function $g_{a,b}$ is not necessarily monotonic, and the values of a and b have a significant effect on the behavior of $g_{a,b}$.

Example 5.1 In Fig. 5.2, we see that $g_{a,b}(p)$ increases for increasing values of p for $(a, b) = (4, 0.5)$ on the interval $(1, 25]$. In view of the error bound (5.26), this indicates that lower values of $p \in (1, 25]$ will provide faster convergence rate. We shall note that $g_{4,0.5}$ does not continue to increase on $[25, \infty)$. In particular, it is increasing from $p = 1$ to $p \approx 34.4458$, then decreases afterwards (see Fig. 5.3). On the other hand, Fig. 5.2 suggests that for $(a, b) = (4, 3)$, higher values of p result to faster convergence rate. Finally, the nonmonotonic graph depicted in Fig. 5.2 for $(a, b) = (4, 2)$ indicates different convergence behaviors for values of p on different intervals. However, observe too that the values of $g_{4,2}(p)$ are close to the limit value $M = 4$ when p belongs to some interval $(1, 1 + \varepsilon)$, for some small $\varepsilon > 0$.

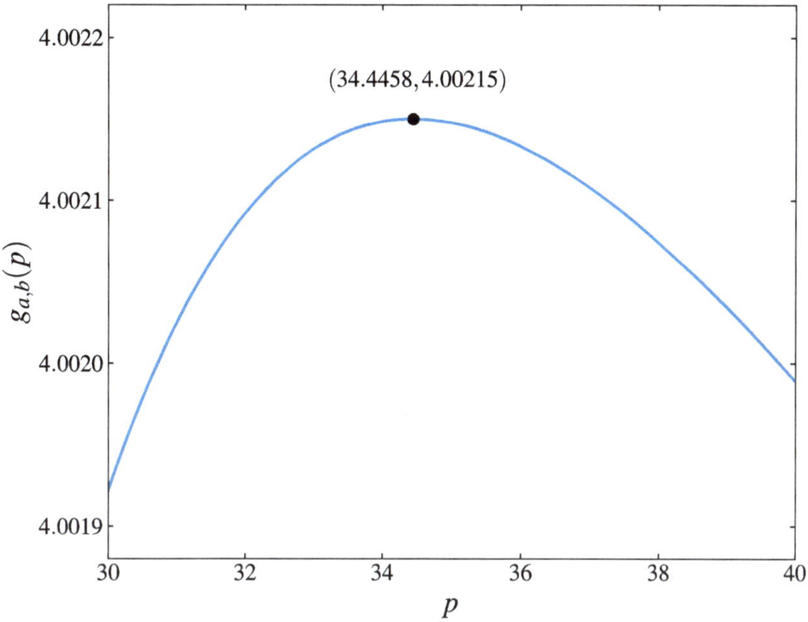

Fig. 5.3 Graph of $g_{4,0.5}(p)$ on the interval [30, 40]

Remark 5.2

(a) The preceding example clearly illustrates that the effect of varying the parameter p on the upper bound is highly sensitive to the choice of initial condition for the neural network (5.19). Nonetheless, it is important to note that the function $g_{a,b}(p)$ exhibits minimal variation for large values of p. As a result, we expect that the convergence behavior of the neural network remains largely unaffected as p becomes large.

(b) We remark that these observed behaviors hold under the hypotheses of Propositions 5.18 and 5.19, which include very strong assumptions on F and on the equilibrium point x^*. Hence, we expect more varying convergence behaviors for other classes of functions F.

(c) Finally, note that for the generalized FB function ϕ^p_{FB} we may define a similar upper bound function $h_{a,b}$ (see [23]) as

$$h_{a,b}(p) = \frac{|\phi^p_{\text{FB}}(a,b)|}{2 - 2^{1/p}}, \quad p > 1.$$

In contrast to the function $g_{a,b}$, the function $h_{a,b}$, as defined above, can be verified to be strictly monotonically decreasing. Consistent with this observation, it was reported in [23] that neural network approaches employing ϕ^p_{FB} tend to achieve faster convergence rates when larger values of p are used.

The preceding example and accompanying remarks highlight the complex and nuanced role of the parameter p, a phenomenon we will further illustrate through numerical examples in the next section. Precisely characterizing the effect of p on the convergence behavior of the ODE trajectories remains an open question. Nevertheless, we have provided theoretical justification for the observed non-monotonic relationship between p and convergence rates when employing dynamical systems based on $\widetilde{\phi}^p_{\mathrm{NR}}$, $\widetilde{\phi}^p_{\mathrm{S-NR}}$ and $\widetilde{\psi}^p_{\mathrm{S-NR}}$. For further details and simulation results, we refer the reader to [31].

5.2 Neural Networks for Optimization Problems Involving SOC

In this section, we explore neural network methods for solving optimization problems involving second-order cones (SOCs), including the standard second-order cone programming (SOCP) problem, a more general class of SOCPs, and second-order cone constrained variational inequality (SOCCVI) problems. To construct the neural networks for these problems, we utilize certain C-functions introduced in Chap. 3. The section is organized into three subsections, each devoted to one of the aforementioned problem classes: the standard SOCP, the generalized SOCP, and the SOCCVI, respectively.

5.2.1 *Neural Networks Using ϕ_{FB} and Projection for Standard SOCP*

The target problem that we will tackle is the second-order cone program in the form of

$$\begin{array}{ll} \min & f(x) \\ \text{s.t.} & Ax = b, \; x \in \mathcal{K}. \end{array} \tag{5.27}$$

Here $f : \mathbb{R}^n \to \mathbb{R}$ is a nonlinear continuously differentiable function, $A \in \mathbb{R}^{m \times n}$ is a full row rank matrix, $b \in \mathbb{R}^m$ is a vector, and $\mathcal{K}$ is a Cartesian product of second-order cones (or Lorentz cones) given as in (3.2). The KKT optimality conditions for (5.27) are given by

$$\begin{cases} \nabla f(x) - A^{\mathsf{T}} y - \lambda = 0, \\ x^{\mathsf{T}} \lambda = 0, \; x \in \mathcal{K}, \; \lambda \in \mathcal{K}, \\ Ax = b, \end{cases} \tag{5.28}$$

where $y \in \mathbb{R}^m$ and $\lambda \in \mathbb{R}^n$. When f is convex, these conditions are sufficient for optimality. It also can be written as

$$\begin{cases} x^{\mathsf{T}}\left(\nabla f(x) - A^{\mathsf{T}}y\right) = 0,\ x \in \mathcal{K},\ \nabla f(x) - A^{\mathsf{T}}y \in \mathcal{K}, \\ Ax = b. \end{cases} \tag{5.29}$$

By solving the system (5.29), one can obtain a primal-dual optimal solution to the SOCP (5.27). It is worth noting that the system (5.29) involves a second-order cone complementarity problem (SOCCP). To solve it efficiently, we propose neural network approaches based on the Fischer-Burmeister function ϕ_{FB} and the natural residual function ϕ_{NR}, as described below.

In [32], the system (5.29) is shown to be equivalent to an unconstrained smooth minimization problem via the merit function approach, described by

$$\min E(x, y) = \Psi_{\text{FB}}(x, \nabla f(x) - A^{\mathsf{T}}y) + \frac{1}{2}\|Ax - b\|^2, \tag{5.30}$$

where $E(x, y)$ is a merit function, $\Psi_{\text{FB}}(x, y) = \frac{1}{2}\sum_{i=1}^{N} \|\phi_{\text{FB}}(x_i, y_i)\|^2$, $x = (x_1, \cdots, x_N)^{\mathsf{T}}$, $y = (y_1, \cdots, y_N)^{\mathsf{T}} \in \mathbb{R}^{n_1} \times \cdots \times \mathbb{R}^{n_N}$, and ϕ_{FB} is the Fischer-Burmeister function given by (3.10). Based on the gradient of the objective $E(x, y)$ in minimization problem (5.30), we propose the first neural network for solving the SOCP, with the following dynamic equation

$$\frac{d}{dt}\begin{pmatrix} x \\ y \end{pmatrix} = \rho \begin{pmatrix} -\nabla_x E(x, y) \\ -\nabla_y E(x, y) \end{pmatrix}, \tag{5.31}$$

where ρ is a positive scaling factor and

$$\begin{cases} \nabla_x E(x, y) = \nabla_x \Psi_{\text{FB}}(x, \nabla f(x) - A^{\mathsf{T}}y) + \nabla^2 f(x) \cdot \nabla_y \Psi_{\text{FB}}(x, \nabla f(x) - A^{\mathsf{T}}y) \\ \qquad\qquad\quad + A^{\mathsf{T}}(Ax - b), \\ \nabla_y E(x, y) = -A \cdot \nabla_y \Psi_{\text{FB}}(x, \nabla f(x) - A^{\mathsf{T}}y). \end{cases} \tag{5.32}$$

The neural network realization of the proposed model requires $n + m$ integrators, n processors for computing $\nabla f(x)$, n^2 processors for $\nabla^2 f(x)$, n processors for $\nabla_x \Psi_{\text{FB}}$, m processors for $\nabla_y \Psi_{\text{FB}}$, along with $4mn$ connection weights and several summing units. Moreover, as shown in the analysis below, the neural network (5.31) is asymptotically stable.

Proposition 5.20 *If $u^* = (x^*, y^*)$ is an isolated equilibrium point of neural network (5.31), then $u^* = (x^*, y^*)$ is asymptotically stable for (5.31).*

Proof We assume that $u^* = (x^*, y^*)$ is an isolated equilibrium point of neural network (5.31) over a neighborhood $\Omega_* \subseteq \mathbb{R}^n$ of u^* such that $\nabla E(x^*, y^*) = 0$ and $\nabla E(x, y) \neq 0, \forall (x, y) \in \Omega_* \setminus \{(x^*, y^*)\}$. First we show that $E(x, y)$ is a Lyapunov function for u^* at Ω_*. Since

$$\nabla_y E(x^*, y^*) = -A \cdot \nabla_y \Psi_{\text{FB}}(x^*, \nabla f(x^*) - A^{\mathsf{T}}y^*) = 0,$$

from Propositions 3.2 to 3.4, we have

$$\nabla_x \Psi_{\text{FB}}(x^*, \nabla f(x^*) - A^\mathsf{T} y^*) = \nabla_y \Psi_{\text{FB}}(x^*, \nabla f(x^*) - A^\mathsf{T} y^*) = 0.$$

Moreover, from Propositions 3.6(b) and 3.2, this says

$$\Psi_{\text{FB}}(x^*, \nabla f(x^*) - A^\mathsf{T} y^*) = 0.$$

Then from Eq. (5.32),

$$\begin{aligned}\nabla_x E(x^*, y^*) = {} & \nabla_x \Psi_{\text{FB}}(x^*, \nabla f(x^*) - A^\mathsf{T} y^*) \\ & + \nabla^2 f(x^*) \cdot \nabla_y \Psi_{\text{FB}}(x^*, \nabla f(x^*) - A^\mathsf{T} y^*) + A^\mathsf{T}(Ax^* - b) = 0,\end{aligned}$$

which implies that $A^\mathsf{T}(Ax^* - b) = 0$. Because $A \in \mathbb{R}^{m \times n}$ is a full row rank matrix, we must have $Ax^* - b = 0$, which yields

$$E(x^*, y^*) = \Psi_{\text{FB}}(x^*, \nabla f(x^*) - A^\mathsf{T} y^*) + \frac{1}{2}\|Ax^* - b\|^2 = 0.$$

Next, we claim that $E(x, y) > 0$, $\forall (x, y) \in \Omega_* \setminus \{(x^*, y^*)\}$. If not, there is an $(x, y) \in \Omega_* \setminus \{(x^*, y^*)\}$ such that $E(x, y) = 0$, this says that $\Psi_{\text{FB}}(x, \nabla f(x) - A^\mathsf{T} y) = 0$ and $Ax = b$, then $\nabla_x E(x, y) = 0$ and $\nabla_y E(x, y) = 0$. Hence, (x, y) is an equilibrium point of neural network (5.31), contradicting with that $u^* = (x^*, y^*)$ is an isolate equilibrium point. Finally,

$$\begin{aligned}&\frac{dE(x(t), y(t))}{dt} \\ &= \left[\nabla_{(x(t),y(t))} E(x(t), y(t))\right]^\mathsf{T} \left(-\rho \nabla_{(x(t),y(t))} E(x(t), y(t))\right) \\ &= -\rho \left\| \nabla_{(x(t),y(t))} E(x(t), y(t)) \right\|^2 \\ &\le 0.\end{aligned}$$

Therefore, the function $E(x, y)$ is a Lyapunov function for neural network (5.31) over the set Ω_*. Since $u^* = (x^*, y^*)$ is an isolated equilibrium point of neural network (5.31), we have

$$\frac{dE(x(t), y(t))}{dt} < 0, \quad \forall (x(t), y(t)) \in \Omega_* \setminus \{(x^*, y^*)\}.$$

Thus, u^* is asymptotically stable for neural network (5.31). □

Next, we consider an alternative neural network model based on the cone projection function to solve the system (5.29) for obtaining the SOCP solution. We also examine the stability of this network. In fact, as shown in (1.11), the projection onto the second-order cone $\mathcal{K}^n$ admits a closed-form expression given by

$$\Pi_{\mathcal{K}^n}(z) = [\lambda_1(z)]_+ u_z^{(1)} + [\lambda_2(z)]_+ u_z^{(2)}, \tag{5.33}$$

where $[\,\cdot\,]_+$ means the scalar projection, $\lambda_1(z)$, $\lambda_2(z)$ and $u_z^{(1)}$, $u_z^{(2)}$ are the spectral values and the associated spectral vectors of $z = (z_1, z_2) \in \mathbb{R} \times \mathbb{R}^{n-1}$, respectively, given by

$$\begin{cases} \lambda_i(z) = z_1 + (-1)^i \|z_2\|, \\ u_z^{(i)} = \dfrac{1}{2}\left(1, (-1)^i \dfrac{z_2}{\|z_2\|}\right), \end{cases}$$

for $i = 1, 2$. Moreover using Proposition 1.3, the system (5.29) can be equivalently written as

$$\begin{cases} \phi_{\text{NR}}(x, \nabla f(x) - A^\top y) = 0, \\ Ax - b = 0, \end{cases} \iff \begin{cases} x - \Pi_{\mathcal{K}}(x - \nabla f(x) + A^\top y) = 0, \\ Ax - b = 0, \end{cases} \tag{5.34}$$

where $x = (x_1, \cdots, x_N)^\top \in \mathbb{R}^{n_1} \times \cdots \times \mathbb{R}^{n_N}$ with $x_i = (x_{i1}, x_{i2}, \cdots, x_{in_i})^\top$, $i = 1, \cdots, N$, and $\Pi_{\mathcal{K}}(x) = \left[\Pi_{\mathcal{K}^{n_1}}(x_1), \cdots, \Pi_{\mathcal{K}^{n_N}}(x_N)\right]^\top$. Based on the equivalent formulation in (5.34) and employing the similar idea as mentioned earlier, we consider the second neural network for solving the SOCP, with the following dynamic equations:

$$\frac{d}{dt}\begin{pmatrix} x \\ y \end{pmatrix} = \rho \begin{pmatrix} -x + \Pi_{\mathcal{K}}(x - \nabla f(x) + A^\top y) \\ -Ax + b \end{pmatrix}, \tag{5.35}$$

where ρ is a positive scaling factor.

The dynamic equations can be implemented using a recurrent neural network incorporating the cone projection function, as illustrated in Fig. 5.4. The neural network realization requires $n + m$ integrators, n processors for computing $\nabla f(x)$, N processors for cone projection mapping $\Pi_{\mathcal{K}}$, $2mn$ connection weights, and several summing units. Compared to the first neural network described in (5.31), the second neural network (5.35) does not require the computation of $\nabla^2 f(x)$, thereby reducing the overall model complexity.

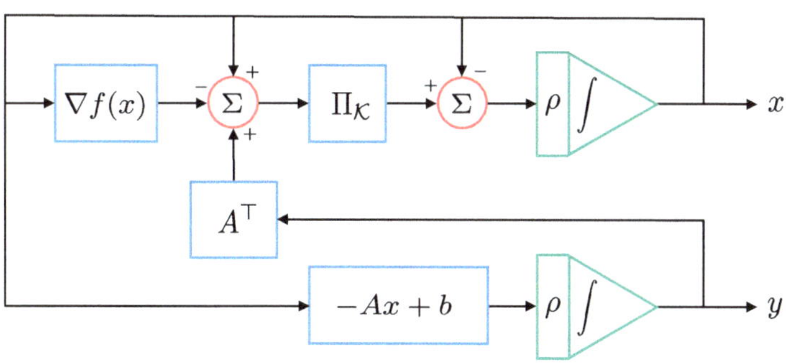

Fig. 5.4 Block diagram of the proposed neural network with projection function

To analyze the stability of the neural network defined by (5.35), we begin by presenting three lemmas and a key proposition.

Lemma 5.6 *Let $F(u)$ be defined as*

$$F(u) := F(x, y) := \begin{bmatrix} -x + \Pi_{\mathcal{K}}(x - \nabla f(x) + A^{\mathsf{T}} y) \\ -Ax + b \end{bmatrix}. \tag{5.36}$$

Then, $F(u)$ is semi-smooth. Moreover, $F(u)$ is strongly semi-smooth if $\nabla^2 f(x)$ is locally Lipschitz continuous.

Proof This is an immediate consequence of [33, Theorem 1]. □

Proposition 5.21 *For any initial point $u_0 = (x_0, y_0)$ where $x_0 := x(t_0) \in \mathcal{K}$, there exists a unique solution $u(t) = (x(t), y(t))$ for neural network (5.35). Moreover, $x(t) \in \mathcal{K}$.*

Proof For simplicity, we assume $\mathcal{K} = \mathcal{K}^n$. The analysis can be carried over to the general case where $\mathcal{K}$ is the Cartesian product of second-order cones. From Lemma 5.6, $F(u) := F(x, y)$ is semi-smooth and Lipschitz continuous. Thus, there exists a unique solution $u(t) = (x(t), y(t))$ for neural network (5.35). Therefore, it remains to show that $x(t) \in \mathcal{K}^n$. For convenience, we denote $x(t) := (x_1(t), x_2(t)) \in \mathbb{R} \times \mathbb{R}^{n-1}$. To complete the proof, we need to verify two things: (i) $x_1(t) \geq 0$ and (ii) $\|x_2(t)\| \leq x_1(t)$. First, from (5.35), we have

$$\frac{dx}{dt} + \rho x(t) = \rho \Pi_{\mathcal{K}^n}(x - \nabla f(x) + A^{\mathsf{T}} y).$$

The solution of the first-order ordinary differential equation above is

$$x(t) = e^{-\rho(t-t_0)} x(t_0) + \rho e^{-\rho t} \int_{t_0}^{t} e^{\rho s} \Pi_{\mathcal{K}^n}(x - \nabla f(x) + A^{\mathsf{T}} y) ds.$$

If we let $x(t_0) := (x_1(t_0), x_2(t_0)) \in \mathbb{R} \times \mathbb{R}^{n-1}$ and denote $z(t) := (z_1(t_0), z_2(t_0))$ as the term $\Pi_{\mathcal{K}^n}(x - (\nabla f(x) - A^{\mathsf{T}} y))$, which leads to

$$\begin{aligned} x_1(t) &= e^{-\rho(t-t_0)} x_1(t_0) + \rho e^{-\rho t} \int_{t_0}^{t} e^{\rho s} z_1(s) ds, \\ x_2(t) &= e^{-\rho(t-t_0)} x_2(t_0) + \rho e^{-\rho t} \int_{t_0}^{t} e^{\rho s} z_2(s) ds. \end{aligned}$$

Due to both $x_0(t)$ and $z(t)$ belong to $\mathcal{K}^n$, there have $x_1(t_0) \geq 0$, $\|x_2(t_0)\| \leq x_1(t_0)$ and $z_1(t) \geq 0$, $\|z_2(t)\| \leq z_1(t)$. Therefore, $x_1(t) \geq 0$ since both terms in the right-hand side are nonnegative. In addition,

$$
\begin{aligned}
\|x_2(t)\| &\le e^{-\rho(t-t_0)}\|x_2(t_0)\| + \rho e^{-\rho t}\int_{t_0}^{t} e^{\rho s}\|z_2(s)\|ds \\
&\le e^{-\rho(t-t_0)}x_1(t_0) + \rho e^{-\rho t}\int_{t_0}^{t} e^{\rho s}z_1(s)ds \\
&= x_1(t),
\end{aligned}
$$

which implies that $x(t) \in \mathcal{K}^n$. □

Lemma 5.7 *Let $H(u)$ be defined as*

$$
H(u) := H(x, y) := \begin{bmatrix} \nabla f(x) - A^{\mathsf{T}}y \\ Ax - b \end{bmatrix}. \tag{5.37}
$$

Then, H is a monotone function if f is a convex function. Moreover, $\nabla H(u)$ is positive semi-definite if and only if $\nabla^2 f(x)$ is positive semi-definite.

Proof Let $u = (x, y)$ and $\tilde{u} = (\tilde{x}, \tilde{y})$. Then, the monotonicity of H holds since

$$
\begin{aligned}
&(u - \tilde{u})^{\mathsf{T}}(H(u) - H(\tilde{u})) \\
&= (x - \tilde{x})^{\mathsf{T}}(\nabla f(x) - \nabla f(\tilde{x})) - (x - \tilde{x})^{\mathsf{T}}(A^{\mathsf{T}}(y - \tilde{y})) + (y - \tilde{y})^{\mathsf{T}}(A(x - \tilde{x})) \\
&= (x - \tilde{x})^{\mathsf{T}}(\nabla f(x) - \nabla f(\tilde{x})) \\
&\ge 0,
\end{aligned}
$$

where the last inequality is due to the convexity of $f(x)$, see [34, Theorem 3.4.5]. Furthermore, we observe that

$$
\nabla H(u) = \begin{bmatrix} \nabla^2 f(x) & -A^{\mathsf{T}} \\ A & 0 \end{bmatrix}.
$$

Thus, we have

$$
\begin{aligned}
&u^{\mathsf{T}}\nabla H(u)u \\
&= \begin{bmatrix} x^{\mathsf{T}} & y^{\mathsf{T}} \end{bmatrix}\begin{bmatrix} \nabla^2 f(x) & -A^{\mathsf{T}} \\ A & 0 \end{bmatrix}\begin{bmatrix} x \\ y \end{bmatrix} \\
&= x^{\mathsf{T}}\nabla^2 f(x)x,
\end{aligned}
$$

which indicates that the positive semi-definiteness of $\nabla H(u)$ is equivalent to the positive semi-definiteness of $\nabla^2 f(x)$. □

Lemma 5.8 *Let $F(u)$, $H(u)$ be defined as in (5.36) and (5.37), respectively. Also, let $u^* = (x^*, y^*)$ be an equilibrium point of neural network (5.35) with x^* being an optimal solution of SOCP. Then, the following inequalities hold:*

$$
(F(u) + u - u^*)^{\mathsf{T}}(-F(u) - H(u)) \ge 0. \tag{5.38}
$$

Proof First, we denote $\lambda := \nabla f(x) - A^{\mathsf{T}} y$. Then, we obtain

$$\begin{aligned}
&(F(u) + u - u^*)^{\mathsf{T}}(-F(u) - H(u)) \\
&= \begin{bmatrix} -x + \Pi_{\mathcal{K}}(x-\lambda) + (x - x^*) \\ (-Ax+b) + (y - y^*) \end{bmatrix}^{\mathsf{T}} \begin{bmatrix} x - \Pi_{\mathcal{K}}(x-\lambda) - \lambda \\ (Ax - b) - (Ax - b) \end{bmatrix} \\
&= \begin{bmatrix} -x^* + \Pi_{\mathcal{K}}(x-\lambda) \\ (-Ax+b) + (y - y^*) \end{bmatrix}^{\mathsf{T}} \begin{bmatrix} (x-\lambda) - \Pi_{\mathcal{K}}(x-\lambda) \\ 0 \end{bmatrix} \\
&= -\left(x^* - \Pi_{\mathcal{K}}(x-\lambda)\right)^{\mathsf{T}} \left((x-\lambda) - \Pi_{\mathcal{K}}(x-\lambda)\right).
\end{aligned}$$

Since $x^* \in \mathcal{K}$, applying Lemma 1.1(b) gives

$$\left(x^* - \Pi_{\mathcal{K}}(x-\lambda)\right)^{\mathsf{T}} \left((x-\lambda) - \Pi_{\mathcal{K}}(x-\lambda)\right) \leq 0.$$

Thus, inequality (5.38) is proved. □

We now investigate the stability and convergence properties of the neural network (5.35). We begin by analyzing the behavior of its solution trajectories, including their existence and convergence. Subsequently, we establish two forms of stability for an isolated equilibrium point. In particular, it is known that every solution u^* to the SOCP corresponds to an equilibrium point of the neural network (5.35). Moreover, if u^* is an isolated equilibrium point, we show that it is Lyapunov stable.

Proposition 5.22 *If f is convex and twice differentiable, then the solution of neural network (5.35), with initial point $u_0 = (x_0, y_0)$ where $x_0 \in \mathcal{K}$, is Lyapunov stable. Moreover, the solution trajectory of neural network (5.35) is extendable to the global existence.*

Proof Again, for simplicity, we assume $\mathcal{K} = \mathcal{K}^n$. From Proposition 5.21, there exists a unique solution $u(t) = (x(t), y(t))$ for neural network (5.35) and $x(t) \in \mathcal{K}^n$. Let $u^* = (x^*, y^*)$ be an equilibrium point of neural network (5.35) with x^* being an optimal solution of SOCP. We define a Lyapunov function as below:

$$E(u) := E(x, y) := -H(u)^{\mathsf{T}} F(u) - \frac{1}{2}\|F(u)\|^2 + \frac{1}{2}\|u - u^*\|^2, \tag{5.39}$$

where $F(u)$ and $H(u)$ are given as in (5.36) and (5.37), respectively. From [35, Theorem 3.2], we know that E is continuously differentiable with

$$\nabla E(u) = H(u) - \left[\nabla H(u) - I\right] F(u) + (u - u^*).$$

It is also trivial that $E(u^*) = 0$. Then, we have

$$\begin{aligned}
\frac{dE(u(t))}{dt} &= \nabla E(u(t))^{\mathsf{T}} \frac{du}{dt} \\
&= \left\{H(u) - \left[\nabla H(u) - I\right] F(u) + (u - u^*)\right\}^{\mathsf{T}} \rho F(u) \\
&= \rho \left\{\left[H(u) + (u - u^*)\right]^{\mathsf{T}} F(u) + \|F(u)\|^2 - F(u)^{\mathsf{T}} \nabla H(u) F(u)\right\}.
\end{aligned}$$

Hence, inequality (5.38) in Lemma 5.8 implies

$$\left(H(u)+u-u^*\right)^{\mathsf{T}} F(u) \leq -H(u)^{\mathsf{T}}(u-u^*) - \|F(u)\|^2,$$

which yields

$$\begin{aligned}
&\frac{dE(u(t))}{dt}\\
&\leq \rho\left\{-H(u)^{\mathsf{T}}(u-u^*) - F(u)^{\mathsf{T}}\nabla H(u)F(u)\right\}\\
&= \rho\left\{-H(u^*)^{\mathsf{T}}(u-u^*) - (H(u)-H(u^*))^{\mathsf{T}}(u-u^*) - F(u)^{\mathsf{T}}\nabla H(u)F(u)\right\}.
\end{aligned} \tag{5.40}$$

On the other hand, we know that

$$\begin{aligned}
&(F(u^*)+u^*-u)^{\mathsf{T}}(-F(u^*)-H(u^*))\\
&= -\left(x-\Pi_{\mathcal{K}}(x^*-\lambda^*)\right)^{\mathsf{T}}\left((x^*-\lambda^*)-\Pi_{\mathcal{K}}(x^*-\lambda^*)\right).
\end{aligned}$$

Since $x \in \mathcal{K}^n$, applying Lemma 1.1(d) gives

$$\left(x-\Pi_{\mathcal{K}}(x^*-\lambda^*)\right)^{\mathsf{T}}\left((x^*-\lambda^*)-\Pi_{\mathcal{K}}(x^*-\lambda^*)\right) \leq 0.$$

Thus, we have $(F(u^*)+u^*-u)^{\mathsf{T}}(-F(u^*)-H(u^*)) \geq 0$. Note that $F(u^*)=0$, we therefore obtain $-H(u^*)^{\mathsf{T}}(u-u^*)^{\mathsf{T}} \leq 0$. Also the monotonicity of H implies $-(H(u)-H(u^*))^{\mathsf{T}}(u-u^*) \leq 0$. In addition, f is convex and twice differentiable if and only if $\nabla^2 f(x)$ is positive semidefinite and hence ∇H is positive semidefinite by Lemma 5.7, i.e., the second term $-F(u)^{\mathsf{T}}\nabla H(u)F(u) \leq 0$. The above discussions lead to $dE(u(t))/dt \leq 0$.

To establish that $E(u)$ serves as a Lyapunov function and that u^* is Lyapunov stable, it suffices to show the following inequality:

$$-H(u)^{\mathsf{T}} F(u) \geq \|F(u)\|^2. \tag{5.41}$$

To see this, we first observe that

$$\|F(u)\|^2 + H(u)^{\mathsf{T}} F(u) = \left(x-\Pi_{\mathcal{K}}(x-\lambda)\right)^{\mathsf{T}}\left((x-\lambda)-\Pi_{\mathcal{K}}(x-\lambda)\right).$$

Since $x \in \mathcal{K}$, applying Lemma 1.1(d) again, there holds

$$\left(x-\Pi_{\mathcal{K}}(x-\lambda)\right)^{\mathsf{T}}\left((x-\lambda)-\Pi_{\mathcal{K}}(x-\lambda)\right) \leq 0,$$

which yields the desired inequality (5.41). By combining Eq. (5.39) and inequality (5.41), we have

$$E(u) \geq \frac{1}{2}\|F(u)\|^2 + \frac{1}{2}\|u-u^*\|^2,$$

which says $E(u) > 0$ if $u \neq u^*$. Hence $E(u)$ is indeed a Lyapunov function and u^* is Lyapunov stable. Moreover, it holds that

$$E(u_0) \geq E(u) \geq \frac{1}{2}\|u - u^*\|^2 \quad \text{for} \quad t \geq t_0, \tag{5.42}$$

which means the solution trajectory $u(t)$ is bounded. Hence, it can be extended to global existence. □

Proposition 5.23 *Let $u^* = (x^*, y^*)$ be an equilibrium point of (5.35) with x^* being an optimal solution of SOCP. If f is twice differentiable and $\nabla^2 f(x)$ is positive definite, the solution of neural network (5.35), with initial point $u_0 = (x_0, y_0)$ where $x_0 \in \mathcal{K}$, is globally convergent to u^* and has finite convergence time.*

Proof From (5.42), it is clear that the level set

$$\mathcal{L}(u_0) := \{u \mid E(u) \leq E(u_0)\}$$

is bounded. Then, the Invariant Set Theorem [36] implies the solution trajectory $u(t)$ converges to θ as $t \to \infty$ where θ is the largest invariant set in

$$S = \left\{ u \in \mathcal{L}(u_0) \,\middle|\, \frac{dE(u(t))}{dt} = 0 \right\}.$$

We will show that $du/dt = 0$ if and only if $dE(u(t))/dt = 0$ which yields that $u(t)$ converges globally to the equilibrium point $u^* = (x^*, y^*)$. Suppose $du/dt = 0$, then it is clear that $dE(u(t))/dt = \nabla E(u)^{\mathsf{T}}(du/dt) = 0$. Let $\hat{u} = (\hat{x}, \hat{y}) \in S$ which says $dE(\hat{u}(t))/dt = 0$. From (5.40), we know that

$$\frac{dE(\hat{u}(t))}{dt} \leq \rho\left\{-(H(\hat{u}) - H(u^*))^{\mathsf{T}}(\hat{u} - u^*) - F(\hat{u})^{\mathsf{T}}\nabla H(\hat{u})F(\hat{u})\right\}.$$

Both terms inside the big parenthesis are nonpositive as shown in Lemma 5.7, so $(H(\hat{u}) - H(u^*))^{\mathsf{T}}(\hat{u} - u^*) = 0$, $F(\hat{u})^{\mathsf{T}}\nabla H(\hat{u})F(\hat{u}) = 0$, and

$$\begin{aligned}
&F(\hat{u})^{\mathsf{T}}\nabla H(\hat{u})F(\hat{u}) \\
&= \{-\hat{x} + \Pi_{\mathcal{K}}(\hat{x} - \nabla f(\hat{x}) + A^{\mathsf{T}}\hat{y})\}^{\mathsf{T}}\nabla^2 f(\hat{x})\{-\hat{x} + \Pi_{\mathcal{K}}(\hat{x} - \nabla f(\hat{x}) + A^{\mathsf{T}}\hat{y})\} \\
&= 0.
\end{aligned}$$

The condition of $\nabla^2 f(\hat{x})$ being positive definite leads to

$$-\hat{x} + \Pi_{\mathcal{K}}(\hat{x} - \nabla f(\hat{x}) + A^{\mathsf{T}}\hat{y}) = 0,$$

which is equivalent to $d\hat{x}/dt = 0$. On the other hand, similar to the arguments in Lemma 5.7, we have

$$\begin{aligned}
&(\hat{u} - u^*)^{\mathsf{T}}(H(\hat{u}) - H(u^*)) \\
&= (\hat{x} - x^*)^{\mathsf{T}}(\nabla f(\hat{x}) - \nabla f(x^*)) \\
&= (\hat{x} - x^*)^{\mathsf{T}}\nabla^2 f(x_s)(\hat{x} - x^*) \\
&= 0,
\end{aligned}$$

where $x_s \in [x^*, \hat{x}]$. Again, the condition of $\nabla^2 f(x_s)$ being positive definite yields $\hat{x} = x^*$. Hence $d\hat{y}/dt = 0$ and therefore $d\hat{u}(t)/dt = 0$. From above, $u(t)$ converges globally to the equilibrium point $u^* = (x^*, y^*)$. Moreover, with Proposition 5.22 and following the same arguments as in [13, Theorem 2], the neural network (5.35) has finite convergence time. □

It is worth noting that the neural network employing the cone projection $\Pi_{\mathcal{K}}$ is equivalent to the one based on the natural residual function ϕ_{NR}, as shown in (5.34). In other words, this section presents neural network approaches for solving SOCPs using both the Fischer-Burmeister function ϕ_{FB} and the natural residual function ϕ_{NR}. For details on simulation results, we refer the reader to [37]. Furthermore, these functions can also be used to solve second-order cone constrained variational inequality (SOCCVI) problems; see [38].

Since the KKT conditions of SOCPs can be reformulated as a variational inequality problem, the framework in [38] addresses a broader class of optimization problems. In general, the neural networks considered therein differ from those studied in this section. Specifically, the FB-based method in [38] utilizes a smoothed version of the Fischer-Burmeister function, whereas the approach discussed here is based on the standard (non-smoothed) FB function. Similarly, the cone projection method in [38] is derived from a Lagrangian formulation which, even when specialized to SOCPs, is distinct from the model explored here. Owing to these fundamental differences, the assumptions required to establish stability also differ. These distinctions will be elaborated upon in Sect. 5.2.3.

5.2.2 Neural Networks for General SOCCP

We now turn our attention to a more general class of SOCPs beyond the standard formulation (5.27), which was examined in Sect. 5.2.1. Specifically, we aim to find a solution to the following nonlinear convex optimization problem subject to second-order cone constraints:

$$\begin{aligned}
\min\ & f(x) \\
\text{s.t.}\ & Ax = b \\
& -g(x) \in \mathcal{K}
\end{aligned} \tag{5.43}$$

where $A \in \mathbb{R}^{m\times n}$ has full row rank, $b \in \mathbb{R}^m$, $f : \mathbb{R}^n \to \mathbb{R}$, $g = [g_1, \cdots, g_l]^{\mathsf{T}} : \mathbb{R}^n \to \mathbb{R}^l$ with f and g_i's being two order continuous differentiable and convex on $\mathbb{R}^n$, and $\mathcal{K}$ is a Cartesian product of second-order cones (also called Lorentz cones), expressed as

$$\mathcal{K} = \mathcal{K}^{n_1} \times \mathcal{K}^{n_2} \times \cdots \times \mathcal{K}^{n_N}$$

with $N, n_1, \cdots, n_N \geq 1, n_1 + \cdots + n_N = l$ and

$$\mathcal{K}^{n_i} := \left\{ (x_{i1}, x_{i2}, \cdots, x_{in_i})^\mathsf{T} \in \mathbb{R}^{n_i} \mid \|(x_{i2}, \cdots, x_{in_i})\| \leq x_{i1} \right\}.$$

Compared with (5.27), we see that the constraint $-g(x) \in \mathcal{K}$ in (5.43) extends the one $x \in \mathcal{K}$ in (5.27). In fact, the problem (5.43) is equivalent to the following variational inequality problem, which is to find $x \in D$ satisfying

$$\langle \nabla f(x), y - x \rangle \geq 0 \quad \forall y \in D,$$

where $D = \{x \in \mathbb{R}^n \mid Ax = b, \ -g(x) \in \mathcal{K}\}$. Many problems in the engineering, transportation science, and economics communities can be solved by transforming the original problems into the mentioned convex optimization problems or variational inequality problems, see [3, 30, 39–41]. Similarly, we first look into the KKT conditions of the problem (5.43), which are presented as below:

$$\begin{cases} \nabla f(x) - A^\mathsf{T} y + \nabla g(x) z = 0, \\ z \in \mathcal{K}, \ -g(x) \in \mathcal{K}, \ z^\mathsf{T} g(x) = 0, \\ Ax - b = 0, \end{cases} \tag{5.44}$$

where $y \in \mathbb{R}^m$, $\nabla g(x)$ denotes the gradient matrix of g. According to the KKT condition, it is well known that if the problem (5.43) satisfies Slater's condition, which means there exists a strictly feasible point for (5.43), i.e., there exists an $x \in \mathbb{R}^n$ such that $-g(x) \in \text{int}(\mathcal{K})$ and $Ax = b$. Then x^* is a solution of the problem (5.43) if and only if there exist y^*, z^* such that (x^*, y^*, z^*) satisfies the KKT conditions (5.44). Hence, we assume that the problem (5.43) satisfies Slater's condition in this section.

In view of the projection mapping onto SOC given as in (5.33) and the non-differentiability of ϕ_{NR}, we consider a class of smoothed NR complementarity function. To this end, we employ a continuously differentiable convex function $\hat{g} : \mathbb{R} \to \mathbb{R}$ such that

$$\lim_{a \to -\infty} \hat{g}(a) = 0, \quad \lim_{a \to \infty} (\hat{g}(a) - a) = 0, \quad \text{and} \quad 0 < \hat{g}'(a) < 1. \tag{5.45}$$

What kind of functions satisfies the condition (5.45)? Here we present two examples:

$$\hat{g}(a) = \frac{\sqrt{a^2 + 4} + a}{2} \quad \text{and} \quad \hat{g}(a) = \ln(e^a + 1).$$

Suppose $z = \lambda_1 u_z^{(1)} + \lambda_2 u_z^{(2)}$, where λ_i and u_z^i for $i = 1, 2$ are the spectral values and spectral vectors of z, respectively. By applying the function $\hat{g}(\cdot)$, we define the following function:

$$P_\mu(z) := \mu \hat{g}\left(\frac{\lambda_1}{\mu}\right) u_z^{(1)} + \mu \hat{g}\left(\frac{\lambda_2}{\mu}\right) u_z^{(2)}. \tag{5.46}$$

Fukushima, Luo, and Tseng [42] show that P_μ is smooth for any $\mu > 0$; moreover P_μ is a smoothing function of the projection $P_{\mathcal{K}}$, i.e., $\lim_{\mu \downarrow 0} P_\mu = P_{\mathcal{K}}$. Hence, a smoothed NR complementarity function is given in the form of

$$\phi_\mu(x, y) := x - P_\mu(x - y).$$

In particular, from [42, Proposition 5.1], there exists a positive constant $\gamma > 0$ such that

$$\|\phi_\mu(x, y) - \phi_{\text{NR}}(x, y)\| \leq \gamma\mu$$

for any $\mu > 0$ and $(x, y) \in \mathbb{R}^n \times \mathbb{R}^n$.

Now we look into the KKT conditions (5.44) again. Let

$$L(x, y, z) = \nabla f(x) - A^{\mathsf{T}} y + \nabla g(x) z, \quad H(u) := \begin{bmatrix} \mu \\ Ax - b \\ L(x, y, z) \\ \phi_\mu(z, -g(x)) \end{bmatrix}$$

and

$$\begin{aligned} \Psi_\mu(u) &:= \frac{1}{2}\|H(u)\|^2 \\ &= \frac{1}{2}\|\phi_\mu(z, -g(x))\|^2 + \frac{1}{2}\|L(x, y, z)\|^2 + \frac{1}{2}\|Ax - b\|^2 + \frac{1}{2}\mu^2, \end{aligned}$$

where $u = (\mu, x, y, z) \in \mathbb{R}_+ \times \mathbb{R}^n \times \mathbb{R}^m \times \mathbb{R}^l$. It is well known that $\Psi_\mu(u)$ serves as a smoothing approximation of the merit function Ψ_{NR}. This implies that the KKT conditions (5.44) can be reformulated, via the merit function approach, as the following unconstrained minimization problem:

$$\min \Psi_\mu(u) := \frac{1}{2}\|H(u)\|^2. \tag{5.47}$$

Proposition 5.24 *(a) Let P_μ be defined by (5.46). Then, $\nabla P_\mu(z)$ and $I - \nabla P_\mu(z)$ are positive definite for any $\mu > 0$ and $z \in \mathbb{R}^l$.*

(b) Let Ψ_μ be defined as in (5.47). Then, the smoothed merit function Ψ_μ is continuously differentiable everywhere with $\nabla \Psi_\mu(u) = \nabla H(u) H(u)$ where

$$\nabla H(u) = \begin{bmatrix} 1 & 0 & 0 & -\left(\frac{\partial P_\mu(z+g(x))}{\partial \mu}\right)^{\mathsf{T}} \\ 0 & A^{\mathsf{T}} & \nabla^2 f(x) + \nabla^2 g_1(x) + \cdots + \nabla^2 g_l(x) & -\nabla_x P_\mu(z+g(x)) \\ 0 & 0 & -A & 0 \\ 0 & 0 & \nabla g(x)^{\mathsf{T}} & I - \nabla_z P_\mu(z+g(x)) \end{bmatrix}.$$

Proof For the function $P_\mu(z)$ defined as in (5.46), the gradient matrix of $P_\mu(z)$ is described as below.

$$\nabla P_\mu(z) = \begin{cases} \hat{g}'(\frac{z_1}{\mu})I & \text{if } z_2 = 0; \\ \begin{bmatrix} b_\mu & \frac{c_\mu z_2^{\mathsf{T}}}{\|z_2\|} \\ \frac{c_\mu z_2}{\|z_2\|} & a_\mu I + (b_\mu - a_\mu)\frac{z_2 z_2^{\mathsf{T}}}{\|z_2\|^2} \end{bmatrix} & \text{if } z_2 \neq 0, \end{cases}$$

where

$$a_\mu = \frac{\hat{g}(\frac{\lambda_2}{\mu}) - \hat{g}(\frac{\lambda_1}{\mu})}{\frac{\lambda_2}{\mu} - \frac{\lambda_1}{\mu}},$$

$$b_\mu = \frac{1}{2}\left(\hat{g}'(\frac{\lambda_2}{\mu}) + \hat{g}'(\frac{\lambda_1}{\mu})\right),$$

$$c_\mu = \frac{1}{2}\left(\hat{g}'(\frac{\lambda_2}{\mu}) - \hat{g}'(\frac{\lambda_1}{\mu})\right),$$

and I denotes the identity matrix. Form the proof of [43, Proposition 3.1], it is clear that $\nabla P_\mu(z)$ and $I - \nabla P_\mu(z)$ are positive definite for any $\mu > 0$ and $z \in \mathbb{R}^l$. With the help of the definition of the smoothed merit function Ψ_μ, part(b) easily follows from the chain rule. □

Following the core principles for constructing artificial neural networks (see [44] for details), we formulate a specific first-order ordinary differential equation to define an artificial neural network model. In particular, based on the gradient of the objective function Ψ_μ in the minimization problem (5.47), we consider the following neural network model for solving the KKT system (5.44) associated with the nonlinear SOCP (5.43):

$$\frac{du(t)}{dt} = -\rho \nabla \Psi_\mu(u), \quad u(t_0) = u_0, \tag{5.48}$$

where $\rho > 0$ is a time scaling factor. In fact, if $\tau = \rho t$, then $\frac{du(t)}{dt} = \rho \frac{du(\tau)}{d\tau}$. Hence, it follows from (5.48) that $\frac{du(\tau)}{d\tau} = -\nabla \Psi_\mu(u)$. In view of this, for simplicity and convenience, we set $\rho = 1$. Indeed, the dynamical system (5.48) can be realized by an architecture with the cone projection function shown in Fig. 5.5. Moreover, the architecture of this artificial neural network is categorized as a "recurrent" neural network according to the classifications of artificial neural networks as in [44, Sect. 2.3.1]. The circuit for (5.48) requires $n + m + l + 1$ integrators, n processors

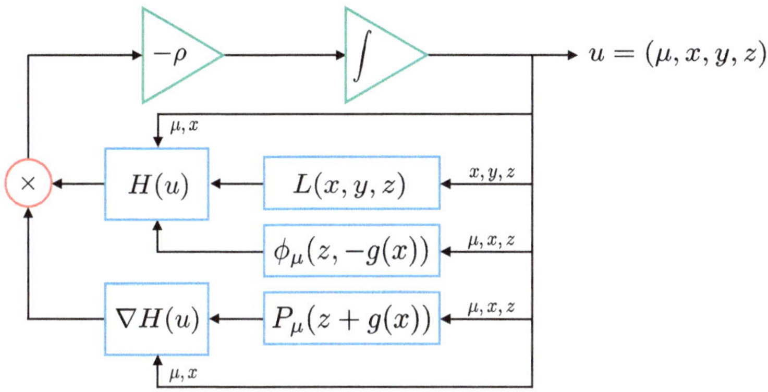

Fig. 5.5 Block diagram of the proposed neural network with smoothed NR function

for $\nabla f(x)$, l processors for $g(x)$, ln processors for $\nabla g(x)$, $(l+1)^2 n$ processors for $\nabla^2 f(x) + \sum_{i=1}^{l} \nabla^2 g_i(x)$, 1 processor for ϕ_μ, 1 processor for $\dfrac{\partial P_\mu}{\partial \mu}$, n processors for $\nabla_x P_\mu$, l processors for $\nabla_z P_\mu$, $n^2 + 4mn + 3ln + l^2 + l$ connection weights and a few summers.

To analyze the stability of the proposed neural network (5.48) for solving the problem (5.43), we begin by introducing an assumption that will be essential for the subsequent analysis.

Assumption 5.1 (a) The problem (5.43) satisfies Slater's condition.
(b) The matrix $\nabla^2 f(x) + \nabla^2 g_1(x) + \cdots + \nabla^2 g_l(x)$ is positive definite for each x.

We briefly comment on Assumption 5.1(a) and (b). Assumption 5.1(a) corresponds to Slater's condition, a classical and widely adopted regularity condition in the field of optimization. Although Assumption 5.1(b) may appear stringent at first glance, it is readily satisfied in many practical cases. Specifically, if the objective function f and the constraint functions g_i are twice continuously differentiable and convex on $\mathbb{R}^n$, then the assumption holds provided that at least one of these functions is strictly convex.

Lemma 5.9 *(a) For any u, we have*

$$\|H(u) - H(u^*) - V(u - u^*)\| = o(\|u - u^*\|) \text{ for } u \to u^* \text{ and } V \in \partial H(u)$$

where $\partial H(u)$ denotes the Clarke generalized Jacobian at u.
(b) Under Assumption 5.1, $\nabla H(u)^{\mathrm{T}}$ is nonsingular for any $u = (\mu, x, y, z) \in \mathbb{R}_{++} \times \mathbb{R}^n \times \mathbb{R}^m \times \mathbb{R}^l$, where $\mathbb{R}_{++}$ denotes the set $\{\mu \mid \mu > 0\}$.

(c) *Under Assumption 5.1 and $V \in \partial P_0(w)$ being a positive definite matrix where $\partial P_0(w)$ denotes the Clarke generalized Jacobian of the project function P at w, there has*

$$T \in \partial H(u) = \left\{ \begin{bmatrix} 1 & 0 & 0 & -\left(\frac{\partial P_\mu(z+g(x))}{\partial \mu}\right)^{\mathsf{T}}|_{\mu=0} \\ 0 & A^{\mathsf{T}} & \nabla^2 f(x) + \nabla^2 g_1(x) + \cdots + \nabla^2 g_l(x) & -V^{\mathsf{T}} \nabla g(x) \\ 0 & 0 & -A & 0 \\ 0 & 0 & \nabla g(x)^{\mathsf{T}} & I - V \end{bmatrix} \middle| V \in \partial P_0(W) \right\}$$

is nonsingular for any $u = (0, x, y, z) \in \{0\} \times \mathbb{R}^n \times \mathbb{R}^m \times \mathbb{R}^l$.

(d) *$\Psi_\mu(u(t))$ is nonincreasing with respect to t.*

Proof (a) This result follows directly from the definition of semismoothness of H, see [45] for more details.

(b) From the expression of $\nabla H(u)$ in Proposition 5.24, it follows that $\nabla H(u)^{\mathsf{T}}$ is nonsingular if and only if the following matrix

$$M := \begin{bmatrix} A & 0 & 0 \\ \nabla^2 f(x) + \nabla^2 g_1(x) + \cdots + \nabla^2 g_l(x) & -A^{\mathsf{T}} & \nabla g(x) \\ -\nabla_x P_\mu(z + g(x))^{\mathsf{T}} & 0 & (I - \nabla_z P_\mu(z + g(x)))^{\mathsf{T}} \end{bmatrix}$$

is nonsingular. Suppose $v = (x, y, z) \in \mathbb{R}^n \times \mathbb{R}^m \times \mathbb{R}^l$. To show the nonsingularity of M, it is enough to prove that

$$Mv = 0 \implies x = 0, \quad y = 0 \quad \text{and} \quad z = 0.$$

Because $-\nabla_x P_\mu(z + g(x))^{\mathsf{T}} = -\nabla P_\mu(w)^{\mathsf{T}} \nabla g(x)^{\mathsf{T}}$, where $w = z + g(x) \in \mathbb{R}^l$, from $Mv = 0$, we have

$$Ax = 0, \quad \left(\nabla^2 f(x) + \nabla^2 g_1(x) + \cdots + \nabla^2 g_l(x)\right) x - A^{\mathsf{T}} y + \nabla g(x) z = 0 \tag{5.49}$$

and

$$-\nabla P_\mu(w)^{\mathsf{T}} \nabla g(x)^{\mathsf{T}} x + (I - \nabla P_\mu(w))^{\mathsf{T}} z = 0. \tag{5.50}$$

From (5.49), it follows that

$$x^{\mathsf{T}} \left(\nabla^2 f(x) + \nabla^2 g_1(x) + \cdots + \nabla^2 g_l(x)\right) x + \left(\nabla g(x)^T x\right)^{\mathsf{T}} z = 0. \tag{5.51}$$

Moveover, Eq. (5.50) and Proposition 5.24 yield

$$\nabla g(x)^{\mathsf{T}} x = (\nabla P_\mu(w)^{\mathsf{T}})^{-1} (I - \nabla P_\mu(w))^{\mathsf{T}} z. \tag{5.52}$$

Combining (5.51)–(5.52) and Proposition 5.24, under the condition of Assumption 5.1, it is not hard to obtain that $x = 0$ and $z = 0$. By looking at Eq. (5.49) again, since A is full row rank, we have $y = 0$. Therefore, $\nabla H(u)^{\mathsf{T}}$ is nonsingular.

(c) The proof of part(c) is similar to that of part(b), in which the only option is to replace $\nabla P_\mu(w)$ with $V \in \partial P_0(w)$.

(d) According to the definition of $\Psi_\mu(u(t))$ and Eq. (5.48), it is clear that

$$\frac{d\Psi_\mu(u(t))}{dt} = \nabla\Psi_\mu(u(t))\,\frac{du(t)}{dt} = -\rho\,\left\|\nabla\Psi_\mu(u(t))\right\|^2 \le 0.$$

Consequently, $\Psi_\mu(u(t))$ is nonincreasing with respect to t. □

Proposition 5.25 *Assume that $\nabla H(u)$ is nonsingular for any $u \in \mathbb{R}_+ \times \mathbb{R}^n \times \mathbb{R}^m \times \mathbb{R}^l$. Then,*

(a) (x^*, y^*, z^*) satisfies the KKT conditions (5.44) if and only if $(0, x^*, y^*, z^*)$ is an equilibrium point of the neural network (5.48);
(b) under Slater's condition, x^* is a solution to the problem (5.43) if and only if $(0, x^*, y^*, z^*)$ is an equilibrium point of the neural network (5.48).

Proof (a) Because $\phi_0 = \phi_{\text{NR}}$ when $\mu = 0$, it follows that (x^*, y^*, z^*) satisfies the KKT conditions (5.44) if and only if $H(u^*) = 0$, where $u^* = (0, x^*, y^*, z^*)^{\mathsf{T}}$. Since $\nabla H(u)$ is nonsingular, we have that $H(u^*) = 0$ if and only if $\nabla\Psi_\mu(u^*) = \nabla H(u^*)^{\mathsf{T}} H(u^*) = 0$. Thus, the desired result follows.

(b) Under Slater's condition, it is well known that x^* is a solution to the problem (5.43) if and only if there exist y^* and z^* such that (x^*, y^*, z^*) satisfies the KKT conditions (5.44). Consequently, by part (a), it follows that $(0, x^*, y^*, z^*)$ is an equilibrium point of the neural network (5.48). □

Proposition 5.26 *(a) For any initial point $u_0 = u(t_0)$, there exists a unique continuously maximal solution $u(t)$ with $t \in [t_0, \tau)$ for the neural network (5.48), where $[t_0, \tau)$ is the maximal interval of existence.*
(b) If the level set $\mathcal{L}(u_0) := \left\{u \mid \Psi_\mu(u) \le \Psi_\mu(u_0)\right\}$ is bounded, then τ can be extended to $+\infty$.

Proof The proof follows exactly the same reasoning as that of Proposition 5.3, and is therefore omitted here. □

Proposition 5.27 *Assume that $\nabla H(u)$ is nonsingular and that u^* is an isolated equilibrium point of the neural network (5.48). Then, the solution of the neural network (5.48) with any initial point u_0 is Lyapunov stable.*

Proof From Lemma 5.3, we only need to argue that there exists a Lyapunov function over some neighborhood Ω of u^*. Now, we consider the smoothed merit function

$$\Psi_{\mu}(u) = \frac{1}{2}\|H(u)\|^2.$$

Since u^* is an isolated equilibrium point of (5.48), there is a neighborhood Ω of u^* such that

$$\nabla\Psi_{\mu}(u^*) = 0 \text{ and } \nabla\Psi_{\mu}(u(t)) \neq 0, \ \forall u(t) \in \Omega\backslash\{u^*\}.$$

By the nonsingularity of $\nabla H(u)$ and the definition of Ψ_{μ}, it is easy to obtain that $\Psi_{\mu}(u^*) = 0$. From the definition of Ψ_{μ}, we claim that $\Psi_{\mu}(u(t)) > 0$ for any $u(t) \in \Omega\backslash\{u^*\}$, where Ω is a neighborhood of u^*. Suppose not, namely, $\Psi_{\mu}(u(t)) = 0$. It follows that $H(u(t)) = 0$. Then, we have $\nabla\Psi_{\mu}(u(t)) = 0$ which contradicts with the assumption that u^* is an isolated equilibrium point of (5.48). Thus, $\Psi_{\mu}(u(t)) > 0$ for any $u(t) \in \Omega\backslash\{u^*\}$. Furthermore, by the proof of Lemma 5.9(d), we know that for any $u(t) \in \Omega$

$$\frac{d\Psi_{\mu}(u(t))}{dt} = \nabla\Psi_{\mu}(u(t))\frac{du(t)}{dt} = -\rho\left\|\nabla\Psi_{\mu}(u(t))\right\|^2 \leq 0. \tag{5.53}$$

Consequently, the function Ψ_{μ} is a Lyapunov function over Ω. This implies that u^* is Lyapunov stable for the neural network (5.48). □

Proposition 5.28 *Assume that $\nabla H(u)$ is nonsingular and that u^* is an isolated equilibrium point of the neural network (5.48). Then, u^* is asymptotically stable for neural network (5.48).*

Proof From the proof of Proposition 5.27, we consider again the Lyapunov function Ψ_{μ}. By Lemma 5.3 again, we only need to verify that the Lyapunov function Ψ_{μ} over some neighborhood Ω of u^* satisfies

$$\frac{d\Psi_{\mu}(u(t))}{dt} < 0, \ \forall u(t) \in \Omega\backslash\{u^*\}. \tag{5.54}$$

In fact, by using (5.53) and the definition of the isolated equilibrium point, it is not hard to check that the equation (5.54) is true. Hence, u^* is asymptotically stable. □

Proposition 5.29 *Assume that u^* is an isolated equilibrium point of the neural network (5.48). If $\nabla H(u)^{\mathsf{T}}$ is nonsingular for any $u = (\mu, x, y, z) \in \mathbb{R}_+ \times \mathbb{R}^n \times \mathbb{R}^m \times \mathbb{R}^l$, then u^* is exponentially stable for the neural network (5.48).*

Proof From the definition of $H(u)$, we know that H is semismooth. Hence, by Lemma 5.9, we have

$$H(u) = H(u^*) + \nabla H(u(t))^{\mathsf{T}}(u - u^*) + o(\|u - u^*\|), \ \forall u \in \Omega\backslash\{u^*\}, \tag{5.55}$$

where $\nabla H(u(t))^{\mathsf{T}} \in \partial H(u(t))$ and Ω is a neighborhood of u^*. Now, we let

$$g(u(t)) = \|u(t) - u^*\|^2, \quad t \in [t_0, \infty).$$

Then, we have

$$\begin{aligned}
\frac{dg(u(t))}{dt} &= 2(u(t) - u^*)^{\mathsf{T}} \frac{du(t)}{dt} \\
&= -2\rho(u(t) - u^*)^{\mathsf{T}} \nabla \Psi_\mu(u(t)) \\
&= -2\rho(u(t) - u^*)^{\mathsf{T}} \nabla H(u) H(u).
\end{aligned} \tag{5.56}$$

Substituting Eq. (5.55) into Eq. (5.56) yields

$$\begin{aligned}
&\frac{dg(u(t))}{dt} \\
&= -2\rho(u(t) - u^*)^{\mathsf{T}} \nabla H(u(t))(H(u^*) + \nabla H(u(t))^T (u(t) - u^*) + o(\|u(t) - u^*\|)) \\
&= -2\rho(u(t) - u^*)^{\mathsf{T}} \nabla H(u(t)) \nabla H(u(t))^{\mathsf{T}} (u(t) - u^*) + o(\|u(t) - u^*\|^2).
\end{aligned}$$

Because $\nabla H(u)$ and $\nabla H(u)^{\mathsf{T}}$ are nonsingular, we claim that there exists an $\kappa > 0$ such that

$$(u(t) - u^*)^{\mathsf{T}} \nabla H(u) \nabla H(u)^{\mathsf{T}} (u(t) - u^*) \geq \kappa \|u(t) - u^*\|^2. \tag{5.57}$$

Otherwise, if $(u(t) - u^*)^{\mathsf{T}} \nabla H(u(t)) \nabla H(u(t))^{\mathsf{T}} (u(t) - u^*) = 0$, it implies that

$$\nabla H(u(t))^{\mathsf{T}} (u(t) - u^*) = 0.$$

Indeed, from the nonsingularity of $H(u)$, we have $u(t) - u^* = 0$, i.e., $u(t) = u^*$ which contradicts with the assumption of u^* being an isolated equilibrium point. Consequently, there exists an $\kappa > 0$ such that (5.57) holds. Moreover, for $o(\|u(t) - u^*\|^2)$, there is $\varepsilon > 0$ such that $o(\|u(t) - u^*\|^2) \leq \varepsilon \|u(t) - u^*\|^2$. Hence,

$$\frac{dg(u(t))}{dt} \leq (-2\rho\kappa + \varepsilon)\|u(t) - u^*\|^2 = (-2\rho\kappa + \varepsilon) g(u(t)).$$

This implies

$$g(u(t)) \leq e^{(-2\rho\kappa + \varepsilon)t} g(u(t_0))$$

which means

$$\|u(t) - u^*\| \leq e^{-\rho\kappa + \frac{\varepsilon}{2}} \|u(t_0) - u^*\|.$$

Thus, u^* is exponentially stable for the neural network (5.48). □

Next, we consider a neural network by using ϕ^p_{FB} for solving (5.43). Recall that the generalized FB merit function $\phi^p_{\text{FB}} : \mathbb{R}^n \times \mathbb{R}^n \to \mathbb{R}^n$ associated with second-order cone is defined by

$$\phi^p_{\rm FB}(x, y) := \sqrt[p]{|x|^p + |y|^p} - (x + y).$$

In view of the KKT conditions (5.44) again, we denote

$$L(x, y, z) = \nabla f(x) - A^{\mathsf{T}} y + \nabla g(x) z,$$

$$H(u) := \begin{bmatrix} Ax - b \\ L(x, y, z) \\ \phi^p_{\rm FB}(z, -g(x)) \end{bmatrix}.$$

Therefore, we consider the merit function as below

$$\begin{aligned} \Psi^p_{\rm FB}(u) &:= \frac{1}{2}\|H(u)\|^2 \\ &= \frac{1}{2}\|\phi_p(z, -g(x))\|^2 + \frac{1}{2}\|L(x, y, z)\|^2 + \frac{1}{2}\|Ax - b\|^2, \end{aligned}$$

where $u = (x, y, z)^{\mathsf{T}} \in \mathbb{R}^n \times \mathbb{R}^m \times \mathbb{R}^l$. From Proposition 3.26, we know that

$$\phi^p_{\rm FB}(z, -g(x)) = 0 \Longleftrightarrow z \in \mathcal{K},\ \ -g(x) \in \mathcal{K},\ \ -z^{\mathsf{T}} g(x) = 0.$$

Hence, the KKT conditions (5.44) are equivalent to $H(u) = 0$, i.e., $\Psi^p_{\rm FB}(u) = 0$. Then, it follows that the KKT conditions (5.44) are equivalent to the following unconstrained minimization problem with zero optimal value via the merit function approach:

$$\min \Psi^p_{\rm FB}(u) := \frac{1}{2}\|H(u)\|^2. \tag{5.58}$$

Accordingly, the neural network for solving the nonlinear SOCP (5.43) is naturally considered as below:

$$\frac{du(t)}{dt} = -\rho \nabla \Psi^p_{\rm FB}(u), \ \ u(t_0) = u_0, \tag{5.59}$$

where $\rho > 0$ is a time scaling factor. In fact, if $\tau = \rho t$, then $\frac{du(t)}{dt} = \rho \frac{du(\tau)}{d\tau}$. Hence, it follows from (5.59) that $\frac{du(\tau)}{d\tau} = -\nabla \Psi^p_{\rm FB}(u)$. For simplicity and convenience, one can set $\rho = 1$.

Lemma 5.10 *For* $z = (z_1, z_2) \in \mathbb{R} \times \mathbb{R}^{n-1}$ *and* $x = (x_1, x_2) \in \mathbb{R} \times \mathbb{R}^{n-1}$ *with* $z \succeq_{\mathcal{K}} x$*, we have* $\lambda_i(z) \geq \lambda_i(x)$ *for* $i = 1, 2$.

Proof Since $z \succeq_{\mathcal{K}} x$, we may express $z = x + y$ where $x = (x_1, x_2) \in \mathbb{R} \times \mathbb{R}^{n-1}$, $y = (y_1, y_2) \in \mathbb{R} \times \mathbb{R}^{n-1}$ and $y = z - x \succeq_{\mathcal{K}} 0$. This implies $y_1 \geq \|y_2\|$ and

$$\begin{aligned}\lambda_1(z) &= (x_1 + y_1) - \|x_2 + y_2\| \\ &\geq (x_1 + y_1) - \|x_2\| - \|y_2\| \\ &\geq x_1 - \|x_2\| \\ &= \lambda_1(x).\end{aligned}$$

Thus, we have

$$\begin{aligned}\lambda_2(z) &= (x_1 + y_1) + \|x_2 + y_2\| \geq (x_1 + y_1) + \big|\|x_2\| - \|y_2\|\big| \\ &= \begin{cases} x_1 + y_1 + \|x_2\| - \|y_2\|, & \text{if } \|x_2\| \geq \|y_2\| \\ x_1 + y_1 - \|x_2\| + \|y_2\|, & \text{if } \|x_2\| < \|y_2\| \end{cases} \\ &\geq \begin{cases} x_1 + \|x_2\|, & \text{if } \|x_2\| \geq \|y_2\| \\ x_1 + y_1, & \text{if } \|x_2\| < \|y_2\| \end{cases} \\ &\geq x_1 + \|x_2\| \\ &= \lambda_2(x)\end{aligned}$$

which is the desired result. □

Lemma 5.11 *Let $w := w(x, y) = |x|^p + |y|^p$, $t = t(x, y) := \sqrt[p]{w}$ and $g^{\text{soc}}(x) := |x|^p$. Then, the following three matrices*

$$\begin{aligned}&\nabla g^{\text{soc}}(t) - \nabla g^{\text{soc}}(x), \\ &\nabla g^{\text{soc}}(t) - \nabla g^{\text{soc}}(y), \\ &(\nabla g^{\text{soc}}(t) - \nabla g^{\text{soc}}(x))\,(\nabla g^{\text{soc}}(t) - \nabla g^{\text{soc}}(y))\end{aligned}$$

are all positive semi-definite for $p = \frac{n}{2}$ with $n \in \mathbb{N}$.

Proof From the expression of $\nabla g^{\text{soc}}(x)$ in Lemma 3.29, that is, (3.140)–(3.141), we know that the eigenvalues of $\nabla g^{\text{soc}}(x)$ for $x_2 \neq 0$ are

$$b(x) - c(x),\ a(x),\ \cdots,\ a(x),\ \text{and } b(x) + c(x).$$

Let $w := (w_1, w_2) \in \mathbb{R} \times \mathbb{R}^{n-1}$. Then, applying (3.136) gives

$$\begin{aligned}w_1 &= \tfrac{|\lambda_2(x)|^p + |\lambda_1(x)|^p}{2} + \tfrac{|\lambda_2(y)|^p + |\lambda_1(y)|^p}{2} \\ w_2 &= \tfrac{|\lambda_2(x)|^p - |\lambda_1(x)|^p}{2}\bar{x}_2 + \tfrac{|\lambda_2(y)|^p - |\lambda_1(y)|^p}{2}\bar{y}_2,\end{aligned}$$

where $\bar{x}_2 = \frac{x_2}{\|x_2\|}$ if $x_2 \neq 0$, and otherwise $\bar{x}_2$ is an arbitrary vector in $\mathbb{R}^{n-1}$ satisfying $\|\bar{x}_2\| = 1$. Similar situation applies for $\bar{y}_2$. Thus, we will proceed the proof by discussing two cases: $w_2 = 0$ or $w_2 \neq 0$.

Case 1. For $w_2 = 0$, we have $\nabla g^{\mathrm{soc}}(t) = p\sqrt[q]{w_1}I$ where

$$w_1 = \frac{|\lambda_2(x)|^p + |\lambda_1(x)|^p}{2} + \frac{|\lambda_2(y)|^p + |\lambda_1(y)|^p}{2}. \tag{5.60}$$

Under the condition of $w_2 = 0$, there are the following two subcases.

(i) If $x_2 = 0$, then $w_1 = |x_1|^p + \frac{|\lambda_2(y)|^p+|\lambda_1(y)|^p}{2}$, which implies that $p\sqrt[q]{w_1} \geq p\,\mathrm{sgn}(x_1)|x_1|^{p-1}$. Hence, we see that the matrix $\nabla g^{\mathrm{soc}}(t) - \nabla g^{\mathrm{soc}}(x)$ is positive semi-definite. Indeed, if $x \neq 0$, $\nabla g^{\mathrm{soc}}(t) - \nabla g^{\mathrm{soc}}(x)$ is positive definite.

(ii) If $x_2 \neq 0$, it follows from $w_2 = 0$ that

$$\left|\frac{|\lambda_2(x)|^p - |\lambda_1(x)|^p}{2}\right| = \left|\frac{|\lambda_2(y)|^p - |\lambda_1(y)|^p}{2}\right|. \tag{5.61}$$

We want to prove that the matrix $\nabla g^{\mathrm{soc}}(t) - \nabla g^{\mathrm{soc}}(x)$ is positive semi-definite. It is sufficient to show that

$$p\sqrt[q]{w_1} \geq \max\{b(x) - c(x), a(x), b(x) + c(x)\}.$$

It is obvious that $p\sqrt[q]{w_1} - (b(x) - c(x)) > 0$ when $\lambda_1(x) < 0$. When $\lambda_1(x) \geq 0$, using (5.60) and $\lambda_2(x) \geq \lambda_1(x)$, we have

$$\begin{aligned} & p\sqrt[q]{w_1} - (b(x) - c(x)) \\ & \geq p\sqrt[q]{|\lambda_1(x)|^p} - p\,\mathrm{sgn}(\lambda_1(x))|\lambda_1(x)|^{p-1} \\ & \geq 0. \end{aligned}$$

Next, we verify that $p\sqrt[q]{w_1} - a(x) \geq 0$. For $|\lambda_1(x)| \geq |\lambda_2(x)|$, it is clear that $p\sqrt[q]{w_1} - a(x) \geq 0$. For $|\lambda_1(x)| < |\lambda_2(x)|$, it follows from $\lambda_2(x) \geq \lambda_1(x)$ that $x_1 > 0$, which yields

$$\frac{|\lambda_2(x)|^p - |\lambda_1(x)|^p}{\lambda_2(x) - \lambda_1(x)} \leq \frac{\lambda_2(x)^p - |\lambda_1(x)|^p}{\lambda_2(x) - |\lambda_1(x)|}.$$

Let $p = \frac{n}{m}$ $(n, m \in \mathbb{N})$, $a = \lambda_2(x)^{\frac{1}{m}}$ and $b = |\lambda_1(x)|^{\frac{1}{m}}$. From $p > 1$, it follows that $n > m$. Then, we have $0 \leq b < a$ and

$$a(x) = \frac{a^n - b^n}{a^m - b^m} = \frac{a^{n-1} + a^{n-2}b + \cdots + ab^{n-2} + b^{n-1}}{a^{m-1} + a^{m-2}b + \cdots + ab^{m-2} + b^{m-1}}.$$

Now, letting $f(v) = \frac{a^n - v^n}{a^m - v^m}$ with $v \in [0, a]$, we obtain

$$f'(v) = \frac{-nv^{n-1}(a^m - v^m) + mv^{m-1}(a^n - v^n)}{(a^m - v^m)^2}.$$

In addition, it follows from $f'(v) = 0$ that

$$\frac{a^n - v^n}{a^m - v^m} = \frac{n}{m} v^{n-m}.$$

Since $f(0) = \frac{a^n}{a^m} = a^{n-m}$ with $v = 0$ and $f(a) = \frac{n}{m} a^{n-m}$ with $v = a$, it is easy to verify that $f(b) \leq \frac{n}{m} a^{n-m}$ for $0 \leq b < a$, i.e.,

$$\frac{|\lambda_2(x)|^p - |\lambda_1(x)|^p}{\lambda_2(x) - \lambda_1(x)} \leq p\, |\lambda_2(x)|^{p-1}.$$

Hence, we have

$$\begin{aligned}
& p\sqrt[q]{w_1} - a(x) \\
\geq\, & p\sqrt[q]{\max\{|\lambda_2(x)|^p, |\lambda_1(x)|^p\} + \min\{|\lambda_2(y)|^p, |\lambda_1(y)|^p\}} \\
& - \frac{|\lambda_2(x)|^p - |\lambda_1(x)|^p}{\lambda_2(x) - \lambda_1(x)} \\
\geq\, & p\sqrt[q]{\lambda_2(x)^p} - p|\lambda_2(x)|^{p-1} \\
\geq\, & 0,
\end{aligned}$$

where the first inequality holds due to (5.61). Lastly, we also see that

$$\begin{aligned}
& p\sqrt[q]{w_1} - (b(x) + c(x)) \\
\geq\, & p\sqrt[q]{\max\{|\lambda_2(x)|^p, |\lambda_1(x)|^p\} + \min\{|\lambda_2(y)|^p, |\lambda_1(y)|^p\}} \\
& - p\, \mathrm{sgn}(\lambda_2(x))|\lambda_2(x)|^{p-1} \\
\geq\, & p\sqrt[q]{\max\{|\lambda_2(x)|^p, |\lambda_1(x)|^p\}} - p\, \mathrm{sgn}(\lambda_2(x))|\lambda_2(x)|^{p-1} \\
\geq\, & 0.
\end{aligned}$$

To sum up, under this case $x_2 \neq 0$, we prove that the matrix $\nabla g^{\text{soc}}(t) - \nabla g^{\text{soc}}(x)$ is positive semi-definite.

Case 2. For $w_2 \neq 0$, from the expression of $t(x, y)$ and the properties of the spectral values of the vector-valued function $|x|^p$ with $p = \frac{n}{2}$ for $n \in \mathbb{N}$, all the eigenvalues of the matrix $\nabla g^{\text{soc}}(t)$ are

$$b(t) - c(t) \leq a(t) \leq b(t) + c(t). \tag{5.62}$$

When $x_2 = 0$, we note that

$$\begin{aligned}
& b(t) - c(t) - p\, \mathrm{sgn}(x_1)|x_1|^{p-1} \\
=\, & p\left[\sqrt[p]{\lambda_1(w)}\right]^{p-1} - p\, \mathrm{sgn}(x_1)|x_1|^{p-1}
\end{aligned}$$

$$
\begin{aligned}
&= p\left[\frac{|\lambda_2(x)|^p + |\lambda_1(x)|^p}{2} + \frac{|\lambda_2(y)|^p + |\lambda_1(y)|^p}{2}\right.\\
&\qquad \left. - \left\|\frac{|\lambda_2(x)|^p - |\lambda_1(x)|^p}{2}\bar{x}_2 + \frac{|\lambda_2(y)|^p - |\lambda_1(y)|^p}{2}\bar{y}_2\right\|\right]^{\frac{p-1}{p}}\\
&\quad - p\,\mathrm{sgn}(x_1)|x_1|^{p-1}\\
&\geq p\,|x_1|^{p-1} - p\,\mathrm{sgn}(x_1)|x_1|^{p-1}\\
&\geq 0,
\end{aligned}
$$

where $\bar{y}_2$ denotes $\bar{y}_2 = \frac{y_2}{\|y_2\|}$ when $y_2 \neq 0$, and otherwise $\bar{y}_2$ is an arbitrary vector in $\mathbb{R}^{n-1}$ satisfying $\|\bar{y}_2\| = 1$. Now, applying the relation of the eigenvalues in (5.62), we have

$$
b(t) + c(t) \geq a(t) \geq p\,\mathrm{sgn}(x_1)|x_1|^{p-1},
$$

which implies that the matrix $\nabla g^{\mathrm{soc}}(t) - \nabla g^{\mathrm{soc}}(x)$ is positive semi-definite.

When $x_2 \neq 0$, we also note that

$$
b(t) - c(t) - (b(x) - c(x)) = p\left[\sqrt[p]{\lambda_1(w)}\right]^{p-1} - p\,\mathrm{sgn}(\lambda_1(x_1))\,|\lambda_1(x_1)|^{p-1}.
$$

For $\lambda_1(x) < 0$, it is clear that $b(t) - c(t) - (b(x) - c(x)) \geq 0$. For $\lambda_1(x) \geq 0$, we have $\lambda_2(x) \geq \lambda_1(x) \geq 0$, which leads to

$$
\begin{aligned}
&\lambda_1(w)\\
&= \frac{|\lambda_2(x)|^p + |\lambda_1(x)|^p}{2} + \frac{|\lambda_2(y)|^p + |\lambda_1(y)|^p}{2}\\
&\quad - \left\|\frac{|\lambda_2(x)|^p - |\lambda_1(x)|^p}{2}\bar{x}_2 + \frac{|\lambda_2(y)|^p - |\lambda_1(y)|^p}{2}\bar{y}_2\right\|\\
&\geq \frac{|\lambda_2(x)|^p + |\lambda_1(x)|^p}{2} - \frac{|\lambda_2(x)|^p - |\lambda_1(x)|^p}{2}\\
&\quad + \frac{|\lambda_2(y)|^p + |\lambda_1(y)|^p}{2} - \left|\frac{|\lambda_2(y)|^p - |\lambda_1(y)|^p}{2}\right|\\
&\geq |\lambda_1(x)|^p.
\end{aligned}
$$

Thus, it follows that $b(t) - c(t) - (b(x) - c(x)) \geq 0$. Moreover, since $t \succeq_{\mathcal{K}} |x|$, by Lemma 5.10 and the eigenvalue of $|x|$ being $|\lambda_1(x)|$ and $|\lambda_2(x)|$, we have

$$
\lambda_2(t) \geq \max\{|\lambda_1(x)|, |\lambda_2(x)|\} \quad \text{and} \quad \lambda_1(t) \geq \min\{|\lambda_1(x)|, |\lambda_2(x)|\}. \tag{5.63}
$$

When $p = \frac{n}{2}$ with $n \in \mathbb{N}$, then, we have

$$
a(t) - a(x) = \frac{\lambda_2(t)^{\frac{n}{2}} - \lambda_1(t)^{\frac{n}{2}}}{\lambda_2(t) - \lambda_1(t)} - \frac{|\lambda_2(x)|^{\frac{n}{2}} - |\lambda_1(x)|^{\frac{n}{2}}}{\lambda_2(x) - \lambda_1(x)}.
$$

If $|\lambda_2(x)| < |\lambda_1(x)|$, it is obvious that $a(t) - a(x) \geq 0$. If $|\lambda_2(x)| \geq |\lambda_1(x)|$, in light of $\lambda_2(x) \geq \lambda_1(x)$, we obtain that $x_1 \geq 0$ and $\lambda_2(x) \geq 0$. Now, let

$$a := \lambda_2(t)^{\frac{1}{2}},\ b := \lambda_1(t)^{\frac{1}{2}},\ c := \lambda_2(x)^{\frac{1}{2}} \text{ and } d := |\lambda_1(t)|^{\frac{1}{2}}.$$

Then, we obtain that

$$\begin{aligned}
& a(t) - a(x) \\
&= \frac{a^n - b^n}{a^2 - b^2} - \frac{c^n - d^n}{c^2 - d^2} \\
&= \frac{(a^{n-1} + a^{n-2}b + \cdots + ab^{n-2} + b^{n-1})(c+d)}{(a+b)(c+d)} \\
&\quad - \frac{(a+b)(c^{n-1} + c^{n-2}d + \cdots + cd^{n-2} + d^{n-1})}{(a+b)(c+d)} \\
&= \frac{a^{n-1}c + bc(a^{n-2} + a^{n-3}b + \cdots + ab^{n-3} + b^{n-2})}{(a+b)(c+d)} \\
&\quad + \frac{ad(a^{n-2} + a^{n-3}b + \cdots + ab^{n-3} + b^{n-2}) + b^{n-1}d}{(a+b)(c+d)} \\
&\quad - \frac{ac^{n-1} + ad(c^{n-2} + c^{n-3}d + \cdots + cd^{n-3} + d^{n-2})}{(a+b)(c+d)} \\
&\quad - \frac{bc(c^{n-2} + c^{n-3}d + \cdots + cd^{n-3} + d^{n-2}) + bd^{n-1}}{(a+b)(c+d)},
\end{aligned}$$

which together with (5.63) implies that

$$a \geq c,\ \ b \geq d \geq 0 \text{ and } a(t) - a(x) \geq 0.$$

In addition , we also verity that

$$b(t) + c(t) - (b(x) + c(x)) = p(\lambda_2(t))^{p-1} - p\,\mathrm{sgn}(\lambda_2(x))|\lambda_2(x)|^{p-1} \geq 0.$$

Therefore, for any $x \in \mathbb{R}^n$, we have

$$\begin{aligned}
& x^\mathsf{T}(\nabla g^{\text{soc}}(t) - \nabla g^{\text{soc}}(x))x \\
&= x^\mathsf{T}\nabla g^{soc}(t)x - x^\mathsf{T}\nabla g^{\text{soc}}(x)x \\
&= [b(t) - c(t) + (n-2)a(t) + b(t) + c(t)]\,x^\mathsf{T}x \\
&\quad - [b(x) - c(x) + (n-2)a(x) + b(x) + c(x)]\,x^\mathsf{T}x \\
&\geq 0,
\end{aligned}$$

which shows that the matrix $\nabla g^{\text{soc}}(t) - \nabla g^{\text{soc}}(x)$ is positive semi-definite.

With the same arguments, we can verify that the matrix $\nabla g^{\mathrm{soc}}(t) - \nabla g^{\mathrm{soc}}(y)$ is also positive semi-definite.

Finally, using the properties of eigenvalues of symmetric matrix product, i.e.,

$$\lambda_i(AB) \geq \lambda_i(A)\lambda_{\min}(B), \quad i = 1, \cdots, n, \quad \forall A, B \in \mathbb{S}^{n\times n},$$

where $\mathbb{S}^{n\times n}$ denotes n order symmetric matrix, we easily achieve that the matrix $(\nabla g^{\mathrm{soc}}(t) - \nabla g^{\mathrm{soc}}(x))(\nabla g^{\mathrm{soc}}(t) - \nabla g^{\mathrm{soc}}(y))$ is also positive semi-definite. □

Remark 5.3 From the above proof of Lemma 5.11, when $x \neq 0$ and $y \neq 0$, we have the matrices

$$\nabla g^{\mathrm{soc}}(t) - \nabla g^{\mathrm{soc}}(x),\ \nabla g^{\mathrm{soc}}(t) - \nabla g^{\mathrm{soc}}(y),\ \left(\nabla g^{\mathrm{soc}}(t) - \nabla g^{\mathrm{soc}}(x)\right)\left(\nabla g^{\mathrm{soc}}(t) - \nabla g^{\mathrm{soc}}(y)\right)$$

are all positive definite.

Proposition 5.30 *Let Ψ^p_{FB} be defined as in (5.58).*

(a) The matrix $\nabla g^{\mathrm{soc}}(x)$ is positive definite for all $0 \neq x \in \mathcal{K}$.
(b) The function Ψ^p_{FB} for $p \in (1, 4)$ is continuously differentiable everywhere. Moreover, $\nabla \Psi^p_{\mathrm{FB}}(u) = \nabla H(u) H(u)$ where

$$\nabla H(u) = \begin{bmatrix} A^{\mathsf{T}} & \nabla_x L(x, y, z) & -\nabla g(x) V_1 \\ 0 & -A & 0 \\ 0 & \nabla g(x)^{\mathsf{T}} & V_2 \end{bmatrix} \tag{5.64}$$

with

$$V_1 = \begin{cases} 0 & \text{if } w(z, -g(x)) = |z|^p + |-g(x)|^p = 0, \\ \nabla g^{\mathrm{soc}}(x)\nabla g^{\mathrm{soc}}(t)^{-1} - I & \text{if } w(z, -g(x)) \in \mathrm{int}(\mathcal{K}), \\ \frac{\mathrm{sgn}(-g_1(x))|-g_1(x)|^{p-1}}{\sqrt[q]{|-g_1(x)|^p + |z_1|^p}} - 1 & \text{if } w(z, -g(x)) \in \partial\mathcal{K}\backslash\{0\}, \end{cases}$$

and

$$V_2 = \begin{cases} 0 & \text{if } w(z, -g(x)) = |z|^p + |-g(x)|^p = 0, \\ \nabla g^{\mathrm{soc}}(z)\nabla g^{\mathrm{soc}}(t)^{-1} - I & \text{if } w(z, -g(x)) \in \mathrm{int}(\mathcal{K}), \\ \frac{\mathrm{sgn}(z_1)|z_1|^{p-1}}{\sqrt[q]{|-g_1(x)|^p + |z_1|^p}} - 1 & \text{if } w(z, -g(x)) \in \partial\mathcal{K}\backslash\{0\}, \end{cases}$$

with $t := \sqrt[p]{w(z, -g(x))}$.

Proof (a) For all $0 \neq x \in \mathcal{K}$, if $x_2 = 0$, it is obvious that the matrix $\nabla g^{\mathrm{soc}}(x) = p\,\mathrm{sgn}(x_1)|x_1|^{p-1} I$ is positive definite. If $x \neq 0$, from the expression of $\nabla g^{\mathrm{soc}}(x)$ in Lemma 3.29 and $x \in \mathcal{K}$, we have $b(x) > 0$. In order to prove that the matrix $\nabla g^{\mathrm{soc}}(x)$ is positive definite, it suffices to show that the Schur complement of $b(x)$

in the matrix $\nabla g^{\text{soc}}(x)$ is positive definite. In fact, from the expression of $\nabla g^{\text{soc}}(x)$, the Schur complement has the form

$$\begin{aligned}& a(x)I + (b(x) - a(x))\bar{x}_2\bar{x}_2^{\mathsf{T}} - \frac{c^2(x)}{b(x)}\bar{x}_2\bar{x}_2^{\mathsf{T}} \\ &= a(x)\left(I - \bar{x}_2\,\bar{x}_2^{\mathsf{T}}\right) + b(x)\left(1 - \frac{c^2(x)}{b(x)}\right)\bar{x}_2\bar{x}_2^{\mathsf{T}}.\end{aligned}$$

Since $x \in \mathcal{K}$, we have $\lambda_2(x) \geq \lambda_1(x) \geq 0$, which implies that $a(x) > 0$ and $b(x) > c(x) \geq 0$. Note that the matrices $I - \bar{x}_2\bar{x}_2^{\mathsf{T}}$ and $\bar{x}_2\bar{x}_2^{\mathsf{T}}$ are positive semi-definite. Thus, the Schur complement is positive definite. Further, we obtain that $\nabla g^{\text{soc}}(x)$ is positive definite for all $0 \neq x \in \mathcal{K}$.

(b) From Proposition 3.28, we know that the function ψ_{FB}^p for $p \in (1, 4)$ is continuously differentiable everywhere. Hence, in view of the definition of the function Ψ_{FB}^p and the chain rule, the expression of $\nabla \Psi_{\text{FB}}^p(u)$ is obtained. □

Assumption 5.2 (a) The SOCP problem (5.43) satisfies Slater's condition.
(b) The matrix $\left[A^{\mathsf{T}} \;\; \nabla g(x)\right]$ is full column rank, and the matrix $\nabla_x L(x, y, z)$ is positive definite on the null space $\{u \mid Au = 0\}$ of A.

We also briefly comment on Assumption 5.2(a) and (b). Assumption 5.2(a) corresponds to Slater's condition, a standard and widely used regularity condition in the field of optimization. When the constraint function g is linear, Assumption 5.2(b) is equivalent to the commonly used condition that $\nabla^2 f(x)$ is positive definite.

Proposition 5.31 *Let $p = \frac{n}{2} \in (1, 4)$ with $n \in \mathbb{N}$. Then, the following hold.*

(a) Under the condition of Assumption 5.2, $\nabla H(u)$ is nonsingular for $u = (x, y, z) \in \mathbb{R}^n \times \mathbb{R}^m \times \mathbb{R}^l$ with $(z, -g(x)) \neq 0$.
(b) Every stationary point of Ψ_p is a global minimizer of problem (5.58) for $(z, -g(x)) \neq 0$.
(c) $\Psi_p(u(t))$ is nonincreasing with respect to t.

Proof (a) Suppose $\xi = (s, t, v) \in \mathbb{R}^n \times \mathbb{R}^m \times \mathbb{R}^l$. From the expression (5.64) of $\nabla H(u)$ in Proposition 5.30, to show the nonsingularity of $\nabla H(u)$, it is enough to prove that

$$\nabla H(u)\xi = 0 \quad \Longrightarrow \quad s = 0, \;\; t = 0 \;\text{ and }\; v = 0.$$

Indeed, by $\nabla H(u)\xi = 0$, we have

$$-At = 0, \quad A^{\mathsf{T}}s + \nabla_x L(x, y, z)t - \nabla g(x)V_1 v = 0 \tag{5.65}$$

and

$$\nabla g(x)^{\mathsf{T}} t + V_2 v = 0. \tag{5.66}$$

From (5.65), it follows that

$$t^{\mathsf{T}} \nabla_x L(x, y, z)\, t - t^{\mathsf{T}} \nabla g(x) V_1 v = 0. \tag{5.67}$$

Moveover, by Eq. (5.66), we obtain

$$t^{\mathsf{T}} \nabla g(x) = -v^{\mathsf{T}} V_2^{\mathsf{T}}. \tag{5.68}$$

Then, combining (5.67) and (5.68), this yields that

$$t^{\mathsf{T}} \nabla_x L(x, y, z)\, t + v^{\mathsf{T}} V_2^{\mathsf{T}} V_1 v = 0.$$

By Lemma 5.11 and Assumption 5.2(b), it is not hard to see that $t = 0$. In addition, from (5.65) and (5.66), we have

$$A^{\mathsf{T}} s - \nabla g(x) V_1 v = 0 \quad \text{and} \quad V_2 v = 0.$$

Due to Assumption 5.2(b) again, we also obtain

$$s = 0 \quad \text{and} \quad V_1 v = 0.$$

Thus, combining Lemma 5.11 with the expression V_1 and V_2 in Proposition 5.30, we have $v = 0$. Therefore, $\nabla H(u)^{\mathsf{T}}$ is nonsingular.

(b) Suppose that u^* is a stationary point of Ψ^p_{FB}. This says $\nabla \Psi^p_{\text{FB}}(u^*) = 0$, and from Proposition 5.30, we have $\nabla H(u^*) H(u^*) = 0$. According to part(a), $\nabla H(u)$ is nonsingular. Hence, it follows that $H(u^*) = 0$, i.e., $\Psi^p_{\text{FB}}(u^*) = 0$, which says u^* is a global minimizer of (5.58).

(c) By the definition of $\Psi^p_{\text{FB}}(u(t))$ and (5.59), it is clear that

$$\frac{d\Psi^p_{\text{FB}}(u(t))}{dt} = \nabla \Psi^p_{\text{FB}}(u(t)) \frac{du(t)}{dt} = -\rho \left\| \nabla \Psi^p_{\text{FB}}(u(t)) \right\|^2 \le 0.$$

Therefore, $\Psi^p_{\text{FB}}(u(t))$ is nonincreasing with respect to t. □

Proposition 5.32 *Assume that $\nabla H(u)$ is nonsingular for any $u \in \mathbb{R}^n \times \mathbb{R}^m \times \mathbb{R}^l$ and $p = \frac{n}{2} \in (1, 4)$ with $n \in \mathbb{N}$. Then,*

(a) (x^*, y^*, z^*) satisfies the KKT conditions (5.44) if and only if (x^*, y^*, z^*) is an equilibrium point of the neural network (5.59);
(b) under Slater's condition, x^* is a solution of the problem (5.43) if and only if (x^*, y^*, z^*) is an equilibrium point of the neural network (5.59).

Proof (a) It is easy to prove that (x^*, y^*, z^*) satisfies the KKT conditions (5.44) if and only if $H(u^*) = 0$ where $u^* = (x^*, y^*, z^*)^{\mathsf{T}}$. According to the condition that $\nabla H(u)$

is nonsingular, we have that $H(u^*) = 0$ if and only if $\nabla \Psi^p_{\text{FB}}(u^*) = \nabla H(u^*)^\top H(u^*) = 0$. Then, the desired result follows.

(b) Under Slater's condition, it is well established that x^* is a solution to the problem (5.43) if and only if there exist y^* and z^* such that (x^*, y^*, z^*) satisfies the KKT conditions (5.44). Therefore, by part (a), it follows that (x^*, y^*, z^*) is an equilibrium point of the neural network (5.59). □

Proposition 5.33 *For any fixed $p = \frac{n}{2} \in (1, 4)$ with $n \in \mathbb{N}$, the following hold.*

(a) For any initial point $u_0 = u(t_0)$, there exists a unique continuously maximal solution $u(t)$ with $t \in [t_0, \tau)$ for the neural network (5.59), where $[t_0, \tau)$ is the maximal interval of existence.
(b) If the level set $\mathcal{L}(u_0) := \left\{u \mid \Psi^p_{\text{FB}}(u) \le \Psi^p_{\text{FB}}(u_0)\right\}$ is bounded, then τ can be extended to $+\infty$.

Proof Again, the proof is exactly the same as the one for Proposition 5.3, and therefore omitted here. □

Proposition 5.34 *Assume that $\nabla H(u)$ is nonsingular and u^* is an isolated equilibrium point of the neural network (5.59). Then, the solution of the neural network (5.59) with any initial point u_0 is Lyapunov stable.*

Proof From Lemma 5.3, we only need to argue that there exists a Lyapunov function over some neighborhood Ω of u^*. To this end, we consider the smoothed merit function for $p = \frac{n}{2} \in (1, 4)$ with $n \in \mathbb{N}$

$$\Psi^p_{\text{FB}}(u) = \frac{1}{2}\|H(u)\|^2.$$

Since u^* is an isolated equilibrium point of (5.59), there is a neighborhood Ω of u^* such that

$$\nabla \Psi^p_{\text{FB}}(u^*) = 0 \quad \text{and} \quad \nabla \Psi^p_{\text{FB}}(u(t)) \neq 0, \;\; \forall u(t) \in \Omega \backslash \{u^*\}.$$

By the nonsingularity of $\nabla H(u)$ and the definition of Ψ^p_{FB}, it is easy to see that $\Psi^p_{\text{FB}}(u^*) = 0$. In view of the definition of Ψ^p_{FB}, we claim that $\Psi^p_{\text{FB}}(u(t)) > 0$ for any $u(t) \in \Omega\backslash\{u^*\}$, where Ω is a neighborhood of u^*. If not, that is, $\Psi_p(u(t)) = 0$, it follows that $H(u(t)) = 0$. Then, we have $\nabla \Psi^p_{\text{FB}}(u(t)) = 0$, which contradicts with the assumption that u^* is an isolated equilibrium point of (5.59). Thus, $\Psi^p_{\text{FB}}(u(t)) > 0$ for any $u(t) \in \Omega\backslash\{u^*\}$. Moreover, by the proof of Lemma 5.31(c), we know that for any $u(t) \in \Omega$

$$\frac{d\Psi^p_{\text{FB}}(u(t))}{dt} = \nabla \Psi^p_{\text{FB}}(u(t)) \frac{du(t)}{dt} = -\rho \left\|\nabla \Psi^p_{\text{FB}}(u(t))\right\|^2 \le 0. \tag{5.69}$$

Therefore, the function Ψ^p_{FB} is a Lyapunov function over Ω. This implies that u^* is Lyapunov stable for the neural network (5.59). □

Proposition 5.35 *Assume that $\nabla H(u)$ is nonsingular and u^* is an isolated equilibrium point of the neural network (5.59). Then, u^* is asymptotically stable for neural network (5.59).*

Proof From the proof of Proposition 5.34, we consider again the Lyapunov function Ψ^p_{FB} for $p = \frac{n}{2} \in (1, 4)$ with $n \in \mathbb{N}$. By Lemma 5.3 again, we only need to verify that the Lyapunov function Ψ^p_{FB} over some neighborhood Ω of u^* satisfies

$$\frac{d\Psi^p_{\text{FB}}(u(t))}{dt} < 0, \quad \forall u(t) \in \Omega \backslash \{u^*\}. \tag{5.70}$$

In fact, by using (5.69) and the definition of the isolated equilibrium point, it is not hard to check that the equation (5.70) is true. Hence, u^* is asymptotically stable. □

Proposition 5.36 *Assume that u^* is an isolated equilibrium point of the neural network (5.59). If $\nabla H(u)^{\mathsf{T}}$ is nonsingular for any $u = (x, y, z) \in \mathbb{R}^n \times \mathbb{R}^m \times \mathbb{R}^l$, then u^* is exponentially stable for the neural network (5.59).*

Proof From the definition of $H(u)$, Propositions 3.27 and 3.28, we have

$$H(u) = H(u^*) + \nabla H(u(t))^{\mathsf{T}}(u - u^*) + o(\|u - u^*\|), \quad \forall u \in \Omega \backslash \{u^*\}, \tag{5.71}$$

where $\nabla H(u(t))^{\mathsf{T}} \in \partial H(u(t))$ and Ω is the neighborhood of u^*. Now, letting

$$g(u(t)) = \|u(t) - u^*\|^2, \quad t \in [t_0, \infty),$$

we have

$$\begin{aligned}
\frac{dg(u(t))}{dt} &= 2(u(t) - u^*)^{\mathsf{T}} \frac{du(t)}{dt} \\
&= -2\rho(u(t) - u^*)^{\mathsf{T}} \nabla \Psi^p_{\text{FB}}(u(t)) \\
&= -2\rho(u(t) - u^*)^{\mathsf{T}} \nabla H(u) H(u).
\end{aligned} \tag{5.72}$$

Substituting (5.71) into (5.72) yields

$$\begin{aligned}
\frac{dg(u(t))}{dt} &= -2\rho(u(t) - u^*)^{\mathsf{T}} \nabla H(u(t))(H(u^*) \\
&\quad + \nabla H(u(t))^{\mathsf{T}}(u(t) - u^*) + o(\|u(t) - u^*\|)) \\
&= -2\rho(u(t) - u^*)^{\mathsf{T}} \nabla H(u(t)) \nabla H(u(t))^{\mathsf{T}}(u(t) - u^*) \\
&\quad + o(\|u(t) - u^*\|^2).
\end{aligned}$$

Since $\nabla H(u)$ and $\nabla H(u)^{\mathsf{T}}$ are nonsingular, we claim that there exists an $\kappa > 0$ such that

$$(u(t) - u^*)^\mathsf{T} \nabla H(u) \nabla H(u)^\mathsf{T} (u(t) - u^*) \geq \kappa \|u(t) - u^*\|^2. \tag{5.73}$$

Otherwise, if $(u(t) - u^*)^\mathsf{T} \nabla H(u(t)) \nabla H(u(t))^\mathsf{T} (u(t) - u^*) = 0$, it implies that

$$\nabla H(u(t))^\mathsf{T} (u(t) - u^*) = 0.$$

Indeed, from the nonsingularity of $H(u)$, we have $u(t) - u^* = 0$, i.e., $u(t) = u^*$, which contradicts with the assumption of u^* that is an isolated equilibrium point. Therefore, there exists an $\kappa > 0$ such that (5.73) holds. Moreover, for $o(\|u(t) - u^*\|^2)$, there is $\varepsilon > 0$ such that $o(\|u(t) - u^*\|^2) \leq \varepsilon \|u(t) - u^*\|^2$. Hence,

$$\frac{dg(u(t))}{dt} \leq (-2\rho\kappa + \varepsilon)\|u(t) - u^*\|^2 = (-2\rho\kappa + \varepsilon) g(u(t)).$$

This implies

$$g(u(t)) \leq e^{(-2\rho\kappa + \varepsilon)t} g(u(t_0)),$$

which means

$$\|u(t) - u^*\| \leq e^{-\rho\kappa + \frac{\varepsilon}{2}} \|u(t_0) - u^*\|.$$

Thus, u^* is exponentially stable for the neural network (5.59). □

Proposition 5.32 suggests that the parameter p, typically set to $p = \frac{n}{2} \in (1, 4)$, must be chosen within this interval due to the theoretical smoothness of Ψ^p_{FB} being established only for $p \in (1, 4)$ in the SOC setting. This raises a natural question: can the results be extended to more general values of p? In other words, is it possible to relax the condition $p = \frac{n}{2} \in (1, 4)$ to include a broader range of real values? This remains an open question and warrants further investigation.

For simulation results of the two neural networks considered in this section for solving the general convex SOCP (5.43), we refer the reader to [46, 47]. In the next section, we broaden our focus to the second-order cone constrained variational inequality (SOCCVI) problem, which generalizes both SOCP formulations (5.27) and (5.43) as special cases.

5.2.3 *Neural Networks for SOCCVI*

The variational inequality (VI) problem, originally introduced by Stampacchia and collaborators [41, 48–51], has garnered significant attention from researchers across diverse fields, including engineering, mathematics, optimization, transportation science, and economics; see, for example, [52–54]. It is well known that VIs encompass a wide array of mathematical problems, such as systems of equations, complementarity problems, and certain classes of fixed-point problems. For comprehensive discussions on finite-dimensional VI problems and their associated solution methods, we refer the interested reader to the authoritative survey by Facchinei and Pang [30], the monograph by Patriksson [55], the survey article by Harker and Pang [56], and the Ph.D. thesis of Hammond [57], along with the references therein.

In this section, we focus on solving the second-order cone constrained variational inequality (SOCCVI) problem, in which the feasible set is defined by a Cartesian product of second-order cones (SOCs). Specifically, the SOCCVI problem seeks a point $x \in C$ such that

$$\langle F(x), y - x \rangle \geq 0, \quad \forall y \in C, \tag{5.74}$$

where the feasible set C is finitely representable as

$$C = \{x \in \mathbb{R}^n \mid h(x) = 0, \ -g(x) \in \mathcal{K}\}. \tag{5.75}$$

Here $\langle \cdot, \cdot \rangle$ denotes the Euclidean inner product, $F : \mathbb{R}^n \to \mathbb{R}^n$, $h : \mathbb{R}^n \to \mathcal{R}^l$ and $g : \mathbb{R}^n \to \mathbb{R}^m$ are continuously differentiable functions and $\mathcal{K}$ is a Cartesian product of second-order cones, expressed as

$$\mathcal{K} = \mathcal{K}^{m_1} \times \mathcal{K}^{m_2} \times \cdots \times \mathcal{K}^{m_p}, \tag{5.76}$$

where $l \geq 0$, $m_1, m_2, \cdots, m_p \geq 1$, $m_1 + m_2 + \cdots + m_p = m$, and

$$\mathcal{K}^{m_i} := \left\{ (x_{i1}, x_{i2}, \cdots, x_{im_i})^{\mathsf{T}} \in \mathbb{R}^{m_i} \mid \|(x_{i2}, \cdots, x_{im_i})\| \leq x_{i1} \right\}$$

with $\|\cdot\|$ denoting the Euclidean norm and $\mathcal{K}^1$ the set of nonnegative reals $\mathbb{R}_+$. A special case of equation (5.76) is $\mathcal{K} = \mathbb{R}^n_+$, namely the nonnegative orthant in $\mathbb{R}^n$, which corresponds to $p = n$ and $m_1 = \cdots = m_p = 1$. When h is affine, an important special case of the SOCCVI problem corresponds to the KKT conditions of the convex second-order cone program (CSOCP):

$$\begin{aligned} &\min \ f(x) \\ &\text{s.t.} \ \ Ax = b, \ -g(x) \in \mathcal{K}, \end{aligned} \tag{5.77}$$

where $A \in \mathbb{R}^{l \times n}$ has full row rank, $b \in \mathbb{R}^l$, $g : \mathbb{R}^n \to \mathbb{R}^m$ and $f : \mathbb{R}^n \to \mathbb{R}$. Furthermore, when f is a convex twice continuously differentiable function, problem (5.77) is equivalent to the following SOCCVI problem: Find $x \in C$ satisfying

$$\langle \nabla f(x), y - x \rangle \geq 0, \quad \forall y \in C,$$

where

$$C = \{x \in \mathbb{R}^n \mid Ax - b = 0, \ -g(x) \in \mathcal{K}\}.$$

Analogous to other optimization problems, the SOCCVI problem (5.74)–(5.75) can be solved by analyzing its KKT conditions:

$$\begin{cases} L(x,\mu,\lambda) = 0, \\ \langle g(x), \lambda \rangle = 0, \ -g(x) \in \mathcal{K}, \ \lambda \in \mathcal{K}, \\ h(x) = 0, \end{cases} \tag{5.78}$$

where

$$L(x,\mu,\lambda) = F(x) + \nabla h(x)\mu + \nabla g(x)\lambda \tag{5.79}$$

is the variational inequality Lagrangian function, $\mu \in \mathbb{R}^l$ and $\lambda \in \mathbb{R}^m$.

Recall that the Fischer-Burmeister function associated with SOC, which is semismooth and defined by

$$\phi_{\text{FB}}(a,b) = \left(a^2 + b^2\right)^{1/2} - (a+b).$$

Accordingly, we consider the smoothed Fischer-Burmeister function given by

$$\phi_{\text{FB}}^{\varepsilon}(a,b) = \left(a^2 + b^2 + \varepsilon^2 e\right)^{1/2} - (a+b) \tag{5.80}$$

with $\varepsilon \in \mathbb{R}_+$ and $e = (1, 0, \cdots, 0)^{\mathsf{T}} \in \mathbb{R}^n$.

Lemma 5.12 *Let $\phi_{\text{FB}}^{\varepsilon}$ be defined as in (5.80) and $\varepsilon \neq 0$. Then, $\phi_{\text{FB}}^{\varepsilon}$ is continuously differentiable everywhere and*

$$\nabla_{\varepsilon}\phi_{\text{FB}}^{\varepsilon}(a,b) = e^{\mathsf{T}} L_z^{-1} L_{\varepsilon e}, \quad \nabla_a \phi_{\text{FB}}^{\varepsilon}(a,b) = L_z^{-1} L_a - I, \quad \nabla_b \phi_{\text{FB}}^{\varepsilon}(a,b) = L_z^{-1} L_b - I,$$

where $z = \left(a^2 + b^2 + \varepsilon^2 e\right)^{1/2}$, I is identity mapping and $L_a = \begin{bmatrix} a_1 & a_2^{\mathsf{T}} \\ a_2 & a_1 I_{n-1} \end{bmatrix}$ for $a = (a_1; a_2) \in \mathbb{R} \times \mathbb{R}^{n-1}$.

Proof The proof follows a similar argument to that of Lemma 3.10(a), which computes the gradient of $\phi_{\text{FB}}^{\varepsilon}$, and is therefore omitted here. □

By employing the smoothed Fischer-Burmeister function defined in (5.80), the KKT system (5.78) can be reformulated as the following unconstrained smooth minimization problem:

$$\min \Psi(w) := \frac{1}{2}\|S(w)\|^2. \tag{5.81}$$

Here $\Psi(w)$, $w = (\varepsilon, x, \mu, \lambda) \in \mathbb{R}^{1+n+l+m}$, is a merit function, and $S(w)$ is defined by

$$S(w) = \begin{pmatrix} \varepsilon \\ L(x,\mu,\lambda) \\ -h(x) \\ \phi_{\text{FB}}^{\varepsilon}(-g_{m_1}(x), \lambda_{m_1}) \\ \vdots \\ \phi_{\text{FB}}^{\varepsilon}(-g_{m_p}(x), \lambda_{m_p}) \end{pmatrix},$$

with $g_{m_i}(x), \lambda_{m_i} \in \mathbb{R}^{m_i}$. In other words, $\Psi(w)$ given in (5.81) is a smooth merit function for the KKT system (5.78). Based on the smooth minimization problem (5.81), it is natural to propose the following neural network model for solving the KKT system (5.78) associated with the SOCCVI problem:

$$\frac{dw(t)}{dt} = -\rho \nabla \Psi(w(t)), \quad w(t_0) = w_0, \tag{5.82}$$

where $\rho > 0$ is a scaling factor. In fact, we can also adopt another merit function which is based on the FB function without the element ε. In other words, we can define

$$S(x, \mu, \lambda) = \begin{pmatrix} L(x, \mu, \lambda) \\ -h(x) \\ \phi_{\text{FB}}(-g_{m_1}(x), \lambda_{m_1}) \\ \vdots \\ \phi_{\text{FB}}(-g_{m_p}(x), \lambda_{m_p}) \end{pmatrix}. \tag{5.83}$$

The neural network model (5.82) can also be derived directly, owing to the smoothness of the squared Fischer-Burmeister function $\|\phi_{\text{FB}}\|^2$. However, it is observed that the gradient mapping $\nabla\Psi$ involves more intricate expressions, particularly because the term $(-g_{m_i}(x))^2 + \lambda_{m_i}^2$ may lie either on the boundary or in the interior of the second-order cone, as discussed in Proposition 3.4. This leads to increased computational cost in practical implementations. Therefore, introducing the one-dimensional smoothing parameter ε not only leaves the theoretical results unaffected, but also simplifies the numerical computations.

To facilitate the analysis of the properties of the neural network model (5.82), we impose the following assumption, which is employed to avoid the singularity of $\nabla S(w)$; see [58].

Assumption 5.3 (a) the gradients $\{\nabla h_j(x) | j = 1, \cdots, l\} \cup \{\nabla g_i(x) | i = 1, \cdots, m\}$ are linear independent.

(b) $\nabla_x L(x, \mu, \lambda)$ is positive definite on the null space of the gradients $\{\nabla h_j(x) \mid j = 1, \cdots, l\}$.

When the SOCCVI problem (5.74)–(5.75) arises as the KKT system of a convex second-order cone program (CSOCP) of the form (5.77), where both h and g are linear functions, Assumption 5.3(b) becomes equivalent to the commonly used condition that $\nabla^2 f(x)$ is positive definite; see, for example, [13, Corollary 1].

Proposition 5.37 *Let $\Psi : \mathbb{R}^{1+n+l+m} \to \mathbb{R}_+$ be defined as in (5.81). Then, $\Psi(w) \geq 0$ for $w = (\varepsilon, x, \mu, \lambda) \in \mathbb{R}^{1+n+l+m}$ and $\Psi(w) = 0$ if and only if (x, μ, λ) solves the KKT system (5.78).*

Proof The proof is straightforward. □

Proposition 5.38 *Let $\Psi : \mathbb{R}^{1+n+l+m} \to \mathbb{R}_+$ be defined as in (5.81). Then, the following results hold.*

(a) The function Ψ is continuously differentiable everywhere with

$$\nabla \Psi(w) = \nabla S(w) S(w),$$

where

$$\nabla S(w) = \begin{bmatrix} 1 & 0 & 0 & \operatorname{diag}\left\{\nabla_{\varepsilon}\phi^{\varepsilon}_{\mathrm{FB}}(-g_{m_i}(x), \lambda_{m_i})\right\}_{i=1}^{p} \\ 0 & \nabla_x L(x,\mu,\lambda)^{\mathsf{T}} & -\nabla h(x) & -\nabla g(x)\operatorname{diag}\left\{\nabla_{g_{m_i}}\phi^{\varepsilon}_{\mathrm{FB}}(-g_{m_i}(x), \lambda_{m_i})\right\}_{i=1}^{p} \\ 0 & \nabla h(x)^{\mathsf{T}} & 0 & 0 \\ 0 & \nabla g(x)^{\mathsf{T}} & 0 & \operatorname{diag}\left\{\nabla_{\lambda_{m_i}}\phi^{\varepsilon}_{\mathrm{FB}}(-g_{m_i}(x), \lambda_{m_i})\right\}_{i=1}^{p} \end{bmatrix}$$

.

(b) Suppose that assumption 5.3 holds. Then, $\nabla S(w)$ is nonsingular for any $w \in \mathbb{R}^{1+n+l+m}$. Moreover, if $(0, x, \mu, \lambda) \in \mathbb{R}^{1+n+l+m}$ is a stationary point of Ψ, then $(x, \mu, \lambda) \in \mathbb{R}^{n+l+m}$ is a KKT triple of the SOCCVI problem.

(c) $\Psi(w(t))$ is nonincreasing with respect to t.

Proof Part(a) follows from the chain rule. For part(b), we know that $\nabla S(w)$ is nonsingular if and only if the following matrix

$$\begin{bmatrix} \nabla_x L(x,\mu,\lambda)^{\mathsf{T}} & -\nabla h(x) & -\nabla g(x)\operatorname{diag}\left\{\nabla_{g_{m_i}}\phi^{\varepsilon}_{\mathrm{FB}}(-g_{m_i}(x), \lambda_{m_i})\right\}_{i=1}^{p} \\ \nabla h(x)^{\mathsf{T}} & 0 & 0 \\ \nabla g(x)^{\mathsf{T}} & 0 & \operatorname{diag}\left\{\nabla_{\lambda_{m_i}}\phi^{\varepsilon}_{\mathrm{FB}}(-g_{m_i}(x), \lambda_{m_i})\right\}_{i=1}^{p} \end{bmatrix}$$

is nonsingular. In fact, from [58, Theorem 3.1] and [58, Proposition 4.1], the above matrix is nonsingular and $(x, \mu, \lambda) \in \mathbb{R}^{n+l+m}$ is a KKT triple of the SOCCVI problem if $(0, x, \mu, \lambda) \in \mathbb{R}^{1+n+l+m}$ is a stationary point of Ψ. It remains to show part(c). By the definition of $\Psi(w)$ and (5.82), it is not difficult to compute

$$\frac{d\Psi(w(t))}{dt} = \nabla \Psi(w(t))^{\mathsf{T}} \frac{dw(t)}{dt} = -\rho \left\|\nabla \Psi(w(t))\right\|^2 \le 0. \tag{5.84}$$

Therefore, $\Psi(w(t))$ is a monotonically decreasing function with respect to t. □

We are now prepared to analyze the behavior of the solution trajectory of (5.82) and to establish the properties corresponding to three types of stability for an isolated equilibrium point.

Proposition 5.39 *(a) If $(x, \mu, \lambda) \in \mathbb{R}^{n+l+m}$ is a KKT triple of SOCCVI problem, then $(0, x, \mu, \lambda) \in \mathbb{R}^{1+n+l+m}$ is an equilibrium point of (5.82).*

(b) If Assumption 5.3 holds and $(0, x, \mu, \lambda) \in \mathbb{R}^{1+n+l+m}$ is an equilibrium point of (5.82), then $(x, \mu, \lambda) \in \mathbb{R}^{n+l+m}$ is a KKT triple of SOCCVI problem.

Proof (a) From Proposition 5.37 and $(x, \mu, \lambda) \in \mathbb{R}^{n+l+m}$ being a KKT triple of SOCCVI problem, it is clear that $S(0, x, \mu, \lambda) = 0$. Hence, $\nabla\Psi(0, x, \mu, \lambda) = 0$. Besides, by Proposition 5.38, we know that if $\varepsilon \neq 0$, then $\nabla\Psi(\varepsilon, x, \mu, \lambda) \neq 0$. This shows that $(0, x, \mu, \lambda)$ is an equilibrium point of (5.82).

(b) It follows from $(0, x, \mu, \lambda) \in \mathbb{R}^{1+n+l+m}$ being an equilibrium point of (5.82) that $\nabla\Psi(0, x, \mu, \lambda) = 0$. In other words, $(0, x, \mu, \lambda)$ is the stationary point of Ψ. Then, the result is a direct consequence of Proposition 5.38(b). □

Proposition 5.40 *(a) For any initial state $w_0 = w(t_0)$, there exists exactly one maximal solution $w(t)$ with $t \in [t_0, \tau(w_0))$ for the neural network (5.82).*
(b) If the level set $\mathcal{L}(w_0) = \{w \in \mathbb{R}^{1+n+l+m} \mid \Psi(w) \leq \Psi(w_0)\}$ is bounded, then $\tau(w_0) = +\infty$.

Proof (a) Since S is continuous differentiable, ∇S is continuous, and therefore, ∇S is bounded on a local compact neighborhood of w. That means $\nabla\Psi(w)$ is locally Lipschitz continuous. Thus, applying Lemma 5.1 leads to the desired result.

(b) This proof is similar to the one of Case(i) in Proposition 5.3(b). □

Remark 5.4 We point out that whether the level sets

$$\mathcal{L}(\Psi, \gamma) := \{w \in \mathbb{R}^{1+n+l+m} \mid \Psi(w) \leq \gamma\}$$

are bounded for all $\gamma \in \mathbb{R}$ is still open. It seems that there needs more subtle properties of F, h and g to finish it.

Proposition 5.41 *(a) Let $w(t)$ with $t \in [t_0, \tau(w_0))$ be the unique maximal solution to (5.82). If $\tau(w_0) = +\infty$ and $\{w(t)\}$ is bounded, then $\lim_{t\to\infty} \nabla\Psi(w(t)) = 0$.*
(b) If Assumption 5.3 holds and $(\varepsilon, x, \mu, \lambda) \in \mathbb{R}^{1+n+l+m}$ is the accumulation point of the trajectory $w(t)$, then $(x, \mu, \lambda) \in \mathbb{R}^{n+l+m}$ is a KKT triple of SOCCVI problem.

Proof With Proposition 5.38(b) and (c) and Proposition 5.39, the arguments are exactly the same as those for [10, Corollary 4.3]. Thus, we omit them. □

Proposition 5.42 *Let w^* be an isolated equilibrium point of the neural network (5.82). Then, the following results hold.*

(a) w^* is asymptotically stable.
(b) If Assumption 5.3 holds, then it is exponentially stable.

Proof Since w^* is an isolated equilibrium point of (5.82), there exists a neighborhood $\Omega^* \subseteq \mathbb{R}^{1+n+l+m}$ of w^* such that

$$\nabla\Psi(w^*) = 0 \quad \text{and} \quad \nabla\Psi(w) \neq 0 \quad \forall w \in \Omega^* \setminus \{w^*\}.$$

Next, we argue that $\Psi(w)$ is indeed a Lyapunov function at x^* over the set Ω^* for (5.82) by showing that the conditions in (5.2) are satisfied. First, notice that $\Psi(w) \geq 0$. Suppose that there is an $\bar{w} \in \Omega^* \setminus \{w^*\}$ such that $\Psi(\bar{w}) = 0$. Then, we can easily obtain that $\nabla\Psi(\bar{w}) = 0$, i.e., $\bar{w}$ is also an equilibrium point of (5.82), which clearly contradicts the assumption that w^* is an isolated equilibrium point in Ω^*. Thus, we prove that $\Psi(w) > 0$ for any $w \in \Omega^* \setminus \{w^*\}$. This together with (5.84) shows that the condition in (5.2) are satisfied. Because w^* is isolated, from (5.84), we have

$$\frac{d\Psi(w(t))}{dt} < 0, \quad \forall w(t) \in \Omega^* \setminus \{w^*\}.$$

This implies that w^* is asymptotically stable. Furthermore, if Assumption 5.3 holds, we can obtain that ∇S is nonsingular. In addition, we have

$$S(w) = S(w^*) + \nabla S(w^*)(w - w^*) + o(\|w - w^*\|), \quad \forall w \in \Omega^* \setminus \{w^*\}. \tag{5.85}$$

From $\|S(w(t))\|$ being a monotonically decreasing function with respect to t and (5.85), we can deduce that

$$\begin{aligned}\|w(t) - w^*\| &\leq \|(\nabla S(w^*))^{-1}\| \|S(w(t)) - S(w^*)\| + o(\|w(t) - w^*\|)\\ &\leq \|(\nabla S(w^*))^{-1}\| \|S(w(t_0)) - S(w^*)\| + o(\|w(t) - w^*\|)\\ &\leq \|(\nabla S(w^*))^{-1}\| \left[\|(\nabla S(w^*))\| \|w(t_0) - w^*\| + o(\|w(t_0) - w^*\|)\right]\\ &\quad + o(\|w(t) - w^*\|).\end{aligned}$$

That is,

$$\begin{aligned}&\|w(t) - w^*\| - o(\|w(t) - w^*\|)\\ &\leq \|(\nabla S(w^*))^{-1}\| \left[\|(\nabla S(w^*))\| \|w(t_0) - w^*\| + o(\|w(t_0) - w^*\|)\right].\end{aligned}$$

The above inequality implies that the neural network (5.82) is also exponentially stable. □

As demonstrated in Sect. 5.2.1, the cone projection function can also be utilized to construct a neural network for solving the SOCCVI problem (5.74)–(5.75). To this end, we begin by introducing some notation. Specifically, we define the function $U : \mathbb{R}^{n+l+m} \to \mathbb{R}^{n+l+m}$ and the vector w as follows:

$$U(w) = \begin{pmatrix} L(x, \mu, \lambda) \\ -h(x) \\ -g(x) \end{pmatrix}, \quad w = \begin{pmatrix} x \\ \mu \\ \lambda \end{pmatrix}, \tag{5.86}$$

where $L(x, \mu, \lambda) = F(x) + \nabla h(x)\mu + \nabla g(x)\lambda$ is the Lagrange function. To avoid confusion, we emphasize that, for any $w \in \mathbb{R}^{n+l+m}$, we have

$$w_i \in \mathbb{R}, \quad \text{if} \quad 1 \le i \le n+l,$$
$$w_i \in \mathbb{R}^{m_{i-(n+l)}}, \quad \text{if} \quad n+l+1 \le i \le n+l+p.$$

Then, we may write (5.86) as

$$U_i = (U(w))_i = (L(x,\mu,\lambda))_i,\ w_i = x_i,\ i = 1,\ldots,n,$$
$$U_{n+j} = (U(w))_{n+j} = -h_j(x),\ w_{n+j} = \mu_j,\ j = 1,\ldots,l,$$
$$U_{n+l+k} = (U(w))_{n+l+k} = -g_k(x) \in \mathbb{R}^{m_k},\ w_{n+l+k} = \lambda_k \in \mathbb{R}^{m_k},\ k = 1,\ldots,p,\ \sum_{k=1}^{p} m_k = m.$$

With this, the KKT conditions (5.78) can be further recast as

$$\begin{aligned} &U_i = 0, \quad i = 1, 2, \ldots, n, n+1, \ldots, n+l, \\ &\langle U_J, w_J \rangle = 0,\ U_J = (U_{n+l+1}, U_{n+l+2}, \cdots, U_{n+l+p})^{\mathsf{T}} \in \mathcal{K}, \\ &w_J = (w_{n+l+1}, w_{n+l+2}, \cdots, w_{n+l+p})^{\mathsf{T}} \in \mathcal{K}. \end{aligned} \tag{5.87}$$

Thus, (x^*, μ^*, λ^*) is a KKT triple for (5.74) if and only if (x^*, μ^*, λ^*) is a solution to (5.87).

It is well known that the nonlinear complementarity problem, which is denoted by NCP(F, K) and to find an $x \in \mathbb{R}^n$ such that

$$x \in K, \quad F(x) \in K \quad \text{and} \quad \langle F(x), x \rangle = 0$$

where K is a closed convex set of $\mathbb{R}^n$, is equivalent to the following VI(F, K) problem: finding an $x \in K$ such that

$$\langle F(x), y - x \rangle \ge 0 \quad \forall y \in K.$$

Furthermore, if $K = \mathbb{R}^n$, then NCP(F, K) becomes the system of nonlinear equations

$$F(x) = 0.$$

Based on the above, solution of (5.87) is equivalent to solution of the following VI problem: find $w \in \mathbb{K}$ such that

$$\langle U(w), v - w \rangle \ge 0, \quad \forall v \in \mathbb{K}, \tag{5.88}$$

where $\mathbb{K} = \mathbb{R}^{n+l} \times \mathcal{K}$. In addition, by applying the Lemma 1.1(d), its solution is equivalent to solution of below projection formulation

$$\Pi_{\mathbb{K}}(w - U(w)) = w \quad \text{with} \quad \mathbb{K} = \mathbb{R}^{n+l} \times \mathcal{K}, \tag{5.89}$$

where function U and vector w are defined in (5.86). Now, according to (5.89), we give the following neural network:

$$\frac{dw}{dt} = \rho \left\{ \Pi_{\mathbb{K}}(w - U(w)) - w \right\}, \tag{5.90}$$

where $\rho > 0$. Note that $\mathbb{K}$ is a closed and convex set. For any $w \in \mathbb{R}^{n+l+m}$, $\Pi_{\mathbb{K}}$ means

$$\Pi_{\mathbb{K}}(w) = \left[\Pi_{\mathbb{K}}(w_1), \Pi_{\mathbb{K}}(w_2), \cdots, \Pi_{\mathbb{K}}(w_{n+l}), \Pi_{\mathbb{K}}(w_{n+l+1}), \Pi_{\mathbb{K}}(w_{n+l+2}), \cdots, \Pi_{\mathbb{K}}(w_{n+l+p})\right],$$

where

$$\begin{aligned}
&\Pi_{\mathbb{K}}(w_i) = w_i, \ i = 1, \cdots, n+l, \\
&\Pi_{\mathbb{K}}(w_{n+l+j}) = [\lambda_1(w_{n+l+j})]_+ \cdot u^{(1)}_{w_{n+l+j}} + [\lambda_2(w_{n+l+j})]_+ \cdot u^{(2)}_{w_{n+l+j}}, \ j = 1, \cdots, p.
\end{aligned}$$

Here, for the sake of simplicity, we denote the vector w_{n+l+j} by v for the moment, and $[\cdot]_+$ is the scalar projection, $\lambda_1(v)$, $\lambda_2(v)$ and $u_v^{(1)}$, $u_v^{(2)}$ are the spectral values and the associated spectral vectors of $v = (v_1; v_2) \in \mathbb{R} \times \mathbb{R}^{m_j-1}$, respectively, given by

$$\begin{cases} \lambda_i(v) = v_1 + (-1)^i \|v_2\|, \\ u_v^{(i)} = \frac{1}{2}\left(1, (-1)^i \frac{v_2}{\|v_2\|}\right), \end{cases}$$

for $i = 1, 2$, see [32, 59].

The dynamical system described by (5.90) can be interpreted as a recurrent neural network with a single-layer structure. To analyze the stability properties of (5.90), we begin by introducing the following lemmas and proposition, which will form the foundation of our subsequent analysis.

Lemma 5.13 *Let $L(x, \mu, \lambda)$ be the Lagrangian function defined as in* (5.79). *If the gradient of $L(x, \mu, \lambda)$ is positive semi-definite (respectively, positive definite), then the gradient of U in (5.86) is positive semi-definite (respectively, positive definite).*

Proof Since we have

$$\nabla U(x, \mu, \lambda) = \begin{bmatrix} \nabla_x L(x, \mu, \lambda)^\mathsf{T} & -\nabla h(x) & -\nabla g(x) \\ \nabla h(x)^\mathsf{T} & 0 & 0 \\ \nabla g(x)^\mathsf{T} & 0 & 0 \end{bmatrix},$$

for any nonzero vector $d = (p^\mathsf{T}, q^\mathsf{T}, r^\mathsf{T})^\mathsf{T} \in \mathbb{R}^{n+l+m}$, we obtain that

$$\begin{aligned}
d^\mathsf{T} \nabla U(x, \mu, \lambda) d &= \left(p^\mathsf{T}\ q^\mathsf{T}\ r^\mathsf{T}\right) \begin{bmatrix} \nabla_x L(x, \mu, \lambda)^\mathsf{T} & -\nabla h(x) & -\nabla g(x) \\ \nabla h(x)^\mathsf{T} & 0 & 0 \\ \nabla g(x)^\mathsf{T} & 0 & 0 \end{bmatrix} \begin{pmatrix} p \\ q \\ r \end{pmatrix} \\
&= p^\mathsf{T} \nabla_x L(x, \mu, \lambda) p.
\end{aligned}$$

This leads to the desired results. □

Proposition 5.43 *For any initial point* $w_0 = (x_0, \mu_0, \lambda_0)$ *with* $\lambda_0 := \lambda(t_0) \in \mathcal{K}$, *there exist a unique solution* $w(t) = (x(t), \mu(t), \lambda(t))$ *for neural network (5.90), Moreover,* $\lambda(t) \in \mathcal{K}$.

Proof For simplicity, we assume $\mathcal{K} = \mathcal{K}^m$. The analysis can be carried over to the general case where $\mathcal{K}$ is the Cartesian product of second-order cones. Since F, h, g are continuous differentiable, the function

$$F(w) := \Pi_{\mathbb{K}}(w - U(w)) - w \quad \text{with} \quad \mathbb{K} = \mathbb{R}^{n+l} \times \mathcal{K}^m \tag{5.91}$$

is semi-smooth and Lipschitz continuous. Thus, there exists a unique solution $w(t) = (x(t), \mu(t), \lambda(t))$ for neural network (5.90). Therefore, it remains to show that $\lambda(t) \in \mathcal{K}^m$. For convenience, we denote $\lambda(t) := (\lambda_1(t), \lambda_2(t)) \in \mathbb{R} \times \mathbb{R}^{m-1}$. To complete the proof, we need to verify two things: (i) $\lambda_1(t) \geq 0$ and (ii) $\|\lambda_2(t)\| \leq \lambda_1(t)$. First, from (5.90), we have

$$\frac{d\lambda}{dt} + \rho\lambda(t) = \rho\,\Pi_{\mathcal{K}^m}(\lambda + g(x)).$$

The solution of the above first-order ordinary differential equation is

$$\lambda(t) = e^{-\rho(t-t_0)}\lambda(t_0) + \rho e^{-\rho t}\int_{t_0}^{t} \rho e^{\rho s}\,\Pi_{\mathcal{K}^m}(\lambda + g(x))ds. \tag{5.92}$$

If we let $\lambda(t_0) := (\lambda_1(t_0), \lambda_2(t_0)) \in \mathbb{R} \times \mathbb{R}^{m-1}$ and denote $\Pi_{\mathcal{K}^m}(\lambda + g(x))$ as $z(t_0) := (z_1(t_0), z_2(t_0))$, then (5.92) leads to

$$\lambda_1(t) = e^{-\rho(t-t_0)}\lambda_1(t_0) + \rho e^{-\rho t}\int_{t_0}^{t} \rho e^{\rho s} z_1(s)ds, \tag{5.93}$$

$$\lambda_2(t) = e^{-\rho(t-t_0)}\lambda_2(t_0) + \rho e^{-\rho t}\int_{t_0}^{t} \rho e^{\rho s} z_2(s)ds. \tag{5.94}$$

Due to both $\lambda(t_0)$ and $z(t)$ belong to $\mathcal{K}^m$, there have $\lambda_1(t_0) \geq 0$, $\|\lambda_2(t_0)\| \leq \lambda_1(t_0)$ and $\|z_2(t)\| \leq z_1(t)$. Therefore, $\lambda_1(t) \geq 0$ since both terms in the right-hand side of (5.93) are nonnegative. In addition, from (5.94), it can be verified

$$\begin{aligned}
\|\lambda_2(t)\| &\leq e^{-\rho(t-t_0)}\|\lambda_2(t_0)\| + \rho e^{-\rho t}\textstyle\int_{t_0}^{t} \rho e^{\rho s}\|z_2(s)\|ds \\
&\leq e^{-\rho(t-t_0)}\lambda_1(t_0) + \rho e^{-\rho t}\textstyle\int_{t_0}^{t} \rho e^{\rho s} z_1(s)ds \\
&= \lambda_1(t),
\end{aligned}$$

which implies that $\lambda(t) \in \mathcal{K}^m$ □

Lemma 5.14 *Let* $U(w)$, $F(w)$ *be defined as in (5.86) and (5.91), respectively. Suppose* $w^* = (x^*, \mu^*, \lambda^*)$ *is an equilibrium point of neural network (5.90) with*

(x^*, μ^*, λ^*) being an KKT triple of SOCCVI problem. Then, the following inequality holds:

$$\left(F(w) + w - w^*\right)^T (-F(w) - U(w)) \geq 0. \tag{5.95}$$

Proof Notice that

$$\begin{aligned}
&\left(F(w) + w - w^*\right)^T (-F(w) - U(w)) \\
&= \left[-w + \Pi_{\mathbb{K}}(w - U(w)) + w - w^*\right]^T [w - \Pi_{\mathbb{K}}(w - U(w)) - U(w)] \\
&= \left[-w^* + \Pi_{\mathbb{K}}(w - U(w))\right]^T [w - \Pi_{\mathbb{K}}(w - U(w)) - U(w)] \\
&= -\left[w^* - \Pi_{\mathbb{K}}(w - U(w))\right]^T [w - U(w) - \Pi_{\mathbb{K}}(w - U(w))].
\end{aligned}$$

Since $w^* \in \mathbb{K}$, applying Lemma 1.1(d) gives

$$\left[w^* - \Pi_{\mathbb{K}}(w - U(w))\right]^T [w - U(w) - \Pi_{\mathbb{K}}(w - U(w))] \leq 0.$$

Thus, we have

$$\left(F(w) + w - w^*\right)^T (-F(w) - U(w)) \geq 0,$$

which is the desired result. □

Proposition 5.44 *Let $L(x, \mu, \lambda)$ be the Lagrangian function defined as in (5.79) and $w(t) := (x(t), \mu(t), \lambda(t))$. If $\nabla_x L(w)$ is positive semi-definite (respectively, positive definite), the the solution of neural network (5.90) with initial point $w_0 = (x_0, \mu_0, \lambda_0)$ where $\lambda_0 \in \mathcal{K}$ is Lyapunov stable (respectively, asymptotically stable). Moreover, the solution trajectory of neural network (5.90) is extendable to the global existence.*

Proof Again, for simplicity, we assume $\mathcal{K} = \mathcal{K}^m$. From Proposition 5.43, there exists a unique solution $w(t) = (x(t), \mu(t), \lambda(t))$ for neural network (5.90) and $\lambda(t) \in \mathcal{K}^m$. Let $w^* = (x^*, \mu^*, \lambda^*)$ be an equilibrium point of neural network (5.90). We define a Lyapunov function as below:

$$V(w) := V(x, \mu, \lambda) := -U(w)^T F(w) - \frac{1}{2}\|F(w)\|^2 + \frac{1}{2}\|w - w^*\|^2. \tag{5.96}$$

From [35, Theorem 3.2], we know that V is continuously differentiable with

$$\nabla V(w) = U(w) - \left[\nabla U(w) - I\right] F(w) + (w - w^*).$$

It is also trivial that $V(w^*) = 0$. Then, we have

$$\begin{aligned}
\frac{dV(w(t))}{dt} &= \nabla V(w(t))^T \frac{dw}{dt} \\
&= \left\{U(w) - [\nabla U(w) - I] F(w) + (w - w^*)\right\}^T \rho F(w) \\
&= \rho \left\{[U(w) + (w - w^*)]^T F(w) + \|F(w)\|^2 - F(w)^T \nabla U(w) F(w)\right\}.
\end{aligned}$$

Inequality (5.95) in Lemma 5.14 implies

$$[U(w) + (w - w^*)]^\mathsf{T} F(w) \le -U(w)^\mathsf{T}(w - w^*) - \|F(w)\|^2,$$

which yields

$$\begin{aligned}
&\frac{dV(w(t))}{dt} \\
&\le \rho\left\{-U(w)^\mathsf{T}(w - w^*) - F(w)^\mathsf{T}\nabla U(w)F(w)\right\} \\
&= \rho\left\{-U(w^*)^\mathsf{T}(w - w^*) - (U(w) - U(w^*))^\mathsf{T}(w - w^*) - F(w)^\mathsf{T}\nabla U(w)F(w)\right\}.
\end{aligned}$$

Note that w^* is the solution of the variational inequality (5.88). Since $w \in \mathbb{K}$, we therefore obtain $-U(w^*)^\mathsf{T}(w - w^*) \le 0$. Because $U(w)$ is continuous differentiable and $\nabla U(w)$ is positive semi-definite, by [34, Theorem 5.4.3], we obtain that $U(w)$ is monotone. Hence, we have $-(U(w) - U(w^*))^\mathsf{T}(w - w^*) \le 0$ and $-F(w)^\mathsf{T}\nabla U(w)F(w) \le 0$. The above discussions lead to $\frac{dV(w(t))}{dt} \le 0$. Also, by [34, Theorem 5.4.3], we know that if $\nabla U(w)$ is positive definite, then $U(w)$ is strictly monotone, which implies $\frac{dV(w(t))}{dt} < 0$ in this case.

In order to obtain $V(w)$ is a Lyapunov function and w^* is Lyapunov stable, we will show the following inequality:

$$-U(w)^\mathsf{T} F(w) \ge \|F(w)\|^2. \tag{5.97}$$

To see this, we first observe that

$$\|F(w)\|^2 + U(w)^\mathsf{T} F(w) = \left[w - \Pi_{\mathbb{K}}(w - U(w))\right]^\mathsf{T}\left[w - U(w) - \Pi_{\mathbb{K}}(w - U(w))\right].$$

Since $w \in \mathbb{K}$, applying Lemma 1.1(b) again, there holds

$$[w - \Pi_{\mathbb{K}}(w - U(w))]^\mathsf{T}[w - U(w) - \Pi_{\mathbb{K}}(w - U(w))] \le 0,$$

which yields the desired inequality (5.97). By combining equation (5.96) and (5.97), we have

$$V(w) \ge \frac{1}{2}\|F(w)\|^2 + \frac{1}{2}\|w - w^*\|^2,$$

which says $V(w) > 0$ if $w \ne w^*$. Hence $V(w)$ is indeed a Lyapunov function and w^* is Lyapunov stable. Furthermore, if $\nabla_x L(w)$ is positive definite, we have w^* is asymptotically stable. Moreover, it holds that

$$V(w_0) \ge V(w) \ge \frac{1}{2}\|w - w^*\|^2 \ \text{ for } t \ge t_0, \tag{5.98}$$

which tells us the solution trajectory $w(t)$ is bounded. Hence, it can be extended to global existence. □

Proposition 5.45 *Let $w^* = (x^*, \mu^*, \lambda^*)$ be an equilibrium point of (5.90). If $\nabla_x L(w)$ is positive definite, the solution of neural network (5.90) with initial point $w_0 = (x_0, \mu_0, \lambda_0)$ where $\lambda_0 \in \mathcal{K}$ is globally convergent to w^* and has finite convergence time.*

Proof From (5.98), the level set

$$\mathcal{L}(w_0) := \{w \mid V(w) \leq V(w_0)\}$$

is bounded. Then, the Invariant Set Theorem [36] implies the solution trajectory $w(t)$ converges to θ as $t \to +\infty$ where θ is the largest invariant set in

$$\Lambda = \left\{ w \in \mathcal{L}(w_0) \,\middle|\, \frac{dV(w(t))}{dt} = 0 \right\}.$$

We will show that $dw/dt = 0$ if and only if $dV(w(t))/dt = 0$ which yields that $w(t)$ converges globally to the equilibrium point $w^* = (x^*, \mu^*, \lambda^*)$. Suppose $dw/dt = 0$, then it is clear that $dV(w(t))/dt = \nabla V(w)^{\mathsf{T}}(dw/dt) = 0$. Let $\hat{w} = (\hat{x}, \hat{\mu}, \hat{\lambda}) \in \Lambda$ which says $dV(\hat{w}(t))/dt = 0$. From (5.95), we know that

$$dV(\hat{w}(t))/dt \leq \rho\left\{(-U(\hat{w}) - U(w^*))^{\mathsf{T}}(\hat{w} - w^*) - F(\hat{w})^{\mathsf{T}}\nabla U(\hat{w})F(\hat{w})\right\}.$$

Both terms inside the big parenthesis are nonpositive as shown in Proposition 5.44, so $(U(\hat{w}) - U(w^*))^{\mathsf{T}}(\hat{w} - w^*) = 0$, $F(\hat{w})^{\mathsf{T}}\nabla U(\hat{w})F(\hat{w}) = 0$. The condition $\nabla_x L(w)$ being positive definite leads to $\nabla U(\hat{w})$ being positive definite. Hence,

$$F(\hat{w}) = -\hat{w} + \Pi_{\mathbb{K}}(\hat{w} - U(\hat{w})) = 0,$$

which is equivalent to $d\hat{w}/dt = 0$. From the above, $w(t)$ converges globally to the equilibrium point $w^* = (x^*, \mu^*, \lambda^*)$. Moreover, with Proposition 5.44 and following the same argument as in [13, Theorem 2], the neural network (5.90) has finite convergence time. □

Simulations and numerical reports can be found in [38]. In general, both neural networks have merits of their own. In addition, using other discrete types of C-functions, ϕ^p_{NR} given by

$$\phi^p_{\text{NR}}(x, y) = x^p - [(x - y)_+]^p, \quad p > 1 \text{ being an odd integer,}$$

and $\phi^p_{\text{D-FB}}$ given by

$$\phi^p_{\text{D-FB}}(x, y) = \left(\sqrt{x^2 + y^2}\right)^p - (x + y)^p, \quad p > 1 \text{ being an odd integer,}$$

for the same dynamic model (5.82) and the same SOCCVI problem (5.74)–(5.75) were studied in [60]. Similar results—namely, Lyapunov stability, asymptotic stability, and exponential stability—can be established for this neural network model. Since the arguments closely mirror those presented earlier, we omit the details here. For comprehensive simulation results and further discussions, we refer the reader to [60].

Next, we introduce another efficient neural network for solving the SOCCVI problem (5.74)–(5.75), this time based on the smoothed cone projection mapping. The approach is motivated by revisiting the KKT conditions (5.78), in a manner analogous to the neural networks (5.82) and (5.90). However, before presenting the smoothed cone projection mapping, we must first review a few notations and foundational concepts. Recall from (5.33) that the projection mapping onto the second-order cone $\mathcal{K}^m$ is defined by

$$\Pi_{\mathcal{K}^m}(x) = [\lambda_1(x)]_+ C_1(x) + [\lambda_2(x)]_+ c_2(x),$$

where $[\,\cdot\,]_+$ means the scalar projection, $\lambda_1(x), \lambda_2(x)$ and $c_1(x), c_2(x)$ are the spectral values and the spectral vectors of $x = (x_0, \bar{x}) \in \mathbb{R} \times \mathbb{R}^{m-1}$, respectively. Indeed, plugging in $\lambda_i(x)$ as below

$$\lambda_i(x) = x_0 + (-1)^i \, \|\bar{x}\| \quad (i = 1, 2)$$

and $c_i(x)$ given by

$$c_i(x) = \begin{cases} \frac{1}{2}(1, (-1)^i \frac{\bar{x}}{\|\bar{x}\|}), & \text{if } \bar{x} \neq 0, \\ \frac{1}{2}(1, (-1)^i w), & \text{if } \bar{x} = 0, \end{cases} \quad (i = 1, 2)$$

with w being an arbitrary unit vector in $\mathbb{R}^{m-1}$, there is another expression for projection mapping:

$$\Pi_{\mathcal{K}^m}(x) = \begin{cases} \frac{1}{2}(1 + \frac{x_0}{\|\bar{x}\|})(\|\bar{x}\|\,, \bar{x}), & \text{if } |x_0| < \|\bar{x}\|\,, \\ (x_0, \bar{x}), & \text{if } \|\bar{x}\| \leqslant x_0, \\ 0, & \text{if } \|\bar{x}\| \leqslant -x_0. \end{cases}$$

The following lemma provides a formula for the directional derivative of the cone projection mapping $\Pi_{\mathcal{K}}$, as defined in (5.33). Throughout the discussion, we use $\text{int}(K)$, $\text{bd}(K)$, and $\text{cl}(K)$ to denote the interior, boundary, and closure of a set $K \subset \mathbb{R}^n$, respectively.

Lemma 5.15 *[61, Lemma 2] The projection mapping $\Pi_{\mathcal{K}^m}(\cdot)$ is directionally differentiable at x for any $d \in \mathbb{R}^m$. Moreover, the directional derivative is described by*

$$\Pi'_{\mathcal{K}^m}(x; d) = \begin{cases} J\Pi_{\mathcal{K}^m}(x)d & \text{if } x \in \mathbb{R}^m \backslash (\mathcal{K}^m \cup -\mathcal{K}^m), \\ d & \text{if } x \in \text{int}(\mathcal{K}^m), \\ d - 2\left[c_(x)^{\mathsf{T}} d\right]_- c_1(x) & \text{if } x \in \text{bd}(\mathcal{K}^m)\backslash\{0\}, \\ 0 & \text{if } x \in -\text{int}(\mathcal{K}^m), \\ 2\left[c_2(x)^{\mathsf{T}} d\right]_+ c_2(x) & \text{if } x \in -\text{bd}(\mathcal{K}^m)\backslash\{0\}, \\ \Pi_{\mathcal{K}^m}(d) & \text{if } x = 0, \end{cases}$$

where

$$J\Pi_{\mathcal{K}^m}(x) = \frac{1}{2}\begin{bmatrix} 1 & \frac{\bar{x}^{\mathsf{T}}}{\|\bar{x}\|} \\ \frac{\bar{x}}{\|\bar{x}\|} & I + \frac{x_0}{\|\bar{x}\|} I - \frac{x_0}{\|\bar{x}\|} \cdot \frac{\bar{x}\bar{x}^{\mathsf{T}}}{\|\bar{x}\|^2} \end{bmatrix},$$

$$\left[c_1(x)^{\mathsf{T}} d\right]_- := \min\left\{0, c_1(x)^{\mathsf{T}} d\right\},$$
$$\left[c_2(x)^{\mathsf{T}} d\right]_+ := \max\left\{0, c_2(x)^{\mathsf{T}} d\right\}.$$

For convenience in the subsequent discussions, we recall the definitions of the tangent cone, regular (Fréchet) normal cone, and limiting (Mordukhovich) normal cone of a closed set at a given point. These foundational concepts are well established and can be found in [62].

Definition 5.6 For a closed set $K \subseteq \mathbb{R}^n$ and a point $\bar{x} \in K$, we define the following sets:

(a) the tangent (Bouligand) cone

$$T_K(\bar{x}) := \limsup_{t \downarrow 0} \frac{K - \bar{x}}{t};$$

(b) the regular (Fréchet) normal cone

$$\hat{N}_K(\bar{x}) := \left\{v \in \mathbb{R}^n \mid \langle v, y - \bar{x}\rangle \leq o(\|y - \bar{x}\|),\ \forall y \in K\right\};$$

(c) the limiting (in the sense of Mordukhovich) normal cone

$$N_K(\bar{x}) := \limsup_{x \xrightarrow{K} \bar{x}} \hat{N}_K(x).$$

When K is a closed convex set, it is known that $T_K(\bar{x}) = \text{cl}(K + \mathbb{R}\bar{x})$ and

$$\hat{N}_K(\bar{x}) = N_K(\bar{x}) = T_K(\bar{x})^\circ = \{v \in K^\circ \mid \langle v, x\rangle \leq 0\},$$

where K° denotes the polar of K. The tangent cone and second-order tangent cone can be explicitly characterized, as stated in the following result.

Lemma 5.16 *[63, Lemma 2.5] The tangent and second-order tangent cones of $\mathcal{K}^m$ at $x \in \mathcal{K}^m$ are described, respectively, by*

$$T_{\mathcal{K}^m}(x) = \begin{cases} \mathbb{R}^m & \text{if } x \in \text{int}(\mathcal{K}^m), \\ \mathcal{K}^m & \text{if } x = 0, \\ \{d = (d_0, \bar{d}) \in \mathbb{R} \times \mathbb{R}^{m-1} \mid \langle \bar{d}, \bar{x} \rangle - x_0 d_0 \leq 0\} & \text{if } x \in \text{bd}(\mathcal{K}^m) \backslash \{0\}, \end{cases}$$

and

$$T^2_{\mathcal{K}^m}(x, d) = \begin{cases} \mathbb{R}^m & \text{if } x \in \text{int}(T_{\mathcal{K}^m}(x)), \\ T_{\mathcal{K}^m}(d) & \text{if } x = 0, \\ \{w = (w_0, \bar{w}) \in \mathbb{R} \times \mathbb{R}^{m-1} \mid \langle \bar{w}, \bar{s} \rangle - w_0 x_0 \leq d_0^2 - \|\bar{d}\|^2\} & \text{otherwise.} \end{cases}$$

We also review several notations that will be used throughout the remainder of this section. Given a sequence $\{t_n\} \in \mathbb{R}$, we write $t_n \downarrow 0$ to mean that $\{t_n\}$ is monotone decreasing and converges to zero. The distance from a point x to a set $K \subset \mathbb{R}^n$, denoted by $\text{dist}(x, K)$ is given by

$$\text{dist}(\bar{x}, K) := \inf\{\|\bar{x} - \bar{y}\| \mid \forall \bar{y} \in K\}.$$

By $\text{lin}K$, we mean the linear subspace generated by K. Given $x, y \in \mathbb{R}^n$, we write $x \perp y$ if and only if $\langle x, y \rangle = 0$.

We now introduce the smoothed natural residual function as a foundation for designing a neural network model. To begin, we define a smoothing metric projector function $\Phi : \mathbb{R}_+ \times \mathbb{R}^m \to \mathbb{R}^m$ as follows:

$$\Phi(\varepsilon, u) := \frac{1}{2}\left(u + \sqrt{\varepsilon^2 e + u^2}\right), \quad \forall (\varepsilon, u) \in \mathbb{R}_+ \times \mathbb{R}^m. \tag{5.99}$$

Observe that $\Phi(0, u) = \Pi_{\mathcal{K}^m}(u)$, which means it is an extension of cone projection mapping. Moreover, Φ is continuously differentiable on any neighborhood of $(\varepsilon, u) \in \mathbb{R} \times \mathbb{R}^m$ provided that $(\varepsilon^2 e + u^2)_0 \neq \left\|\varepsilon^2 e + u^2\right\|$. From [8], it is known that Φ is globally Lipschitz continuous and is strongly semismooth for all $(0, u) \in \mathbb{R} \times \mathbb{R}^m$. Furthermore, applying the concept of SOC-functions in [64], it can be verified that the function $\Phi(\varepsilon, u)$ given in (5.99) can alternatively be expressed as

$$\Phi(\varepsilon, u) = \phi(\varepsilon, \lambda_1)c_1 + \phi(\varepsilon, \lambda_2)c_2, \tag{5.100}$$

where $\phi(\varepsilon, t) := \frac{1}{2}(t + \sqrt{\varepsilon^2 + t^2})$, λ_i and c_i are the spectral values and the spectral vectors, respectively. Hence, we can write out the function Φ as

$$\Phi(\varepsilon,u)=\begin{cases}\frac{1}{2}u+\frac{1}{4}\begin{pmatrix}\sqrt{\varepsilon^2+\lambda_1^2}+\sqrt{\varepsilon^2+\lambda_2^2}\\ \left(\sqrt{\varepsilon^2+\lambda_1^2}+\sqrt{\varepsilon^2-\lambda_2^2}\right)\frac{\bar{u}}{\|\bar{u}\|}\end{pmatrix}, & \text{if } \bar{u}\neq 0,\\ \frac{1}{2}\begin{pmatrix}u_0+\sqrt{\varepsilon^2+u_0^2}\\ 0\end{pmatrix}, & \text{if } \bar{u}=0.\end{cases} \tag{5.101}$$

For $(\varepsilon^2 e+u^2)_0\neq\left\|\varepsilon^2 e+u^2\right\|$, we calculate the derivative of Φ with respect to ε as below:

$$\begin{aligned}\nabla_\varepsilon\Phi(\varepsilon,u)&=\frac{1}{2}\left(\frac{\partial}{\partial\varepsilon}\phi(\varepsilon,\lambda_1)c_1^\top+\frac{\partial}{\partial\varepsilon}\phi(\varepsilon,\lambda_2)c_2^\top\right)\\&=\frac{1}{2}\left(\frac{\varepsilon c_1^\top}{\sqrt{\varepsilon^2+\lambda_1^2}}+\frac{\varepsilon c_2^\top}{\sqrt{\varepsilon^2+\lambda_2^2}}\right)\end{aligned}$$

As for the differentiability of Φ with respect to u, we have two cases:

Case(i): For $u\neq 0$,

$$\nabla_u\Phi(\varepsilon,u)=\frac{1}{2}\begin{bmatrix}1+\frac{1}{2}\left(\frac{\lambda_1}{\sqrt{\varepsilon^2+\lambda_1^2}}+\frac{\lambda_2}{\sqrt{\varepsilon^2+\lambda_2^2}}\right) & Y^\top\\ Y & Z\end{bmatrix}, \tag{5.102}$$

where

$$Y=\frac{1}{2}\left(\frac{\lambda_2}{\sqrt{\varepsilon^2+\lambda_2^2}}-\frac{\lambda_1}{\sqrt{\varepsilon^2+\lambda_1^2}}\right)\frac{\bar{u}}{\|\bar{u}\|}$$

and

$$\begin{aligned}Z=&\left[1+\frac{\sqrt{\varepsilon^2+\lambda_2^2}-\sqrt{\varepsilon^2+\lambda_1^2}}{\lambda_2-\lambda_1}\right]I_{m-1}\\&+\left[\frac{1}{2}\left(\frac{\lambda_1}{\sqrt{\varepsilon^2+\lambda_1^2}}+\frac{\lambda_2}{\sqrt{\varepsilon^2+\lambda_2^2}}\right)-\frac{\sqrt{\varepsilon^2+\lambda_2^2}-\sqrt{\varepsilon^2+\lambda_1^2}}{\lambda_2-\lambda_1}\right]\frac{\bar{u}\bar{u}^\top}{\|\bar{u}\|^2}.\end{aligned}$$

Case(ii): For $\bar{u}=0$,

$$\nabla_u\Phi(\varepsilon,u)=\frac{1}{2}\left(1+\frac{u_0}{\sqrt{\varepsilon^2+u_0^2}}\right)I_m.$$

For $(\varepsilon^2 e + u^2)_0 = \left\| \varepsilon^2 e + u^2 \right\|$, Φ is nonsmooth at (ε, u), but its B-subdifferential can nevertheless be computed.

In light of the above $\Phi(\varepsilon, u)$ given in (5.99), (5.100) or (5.101), we are ready to present the smoothing NR function, which is given by

$$\phi^{\varepsilon}_{\mathrm{NR}}(x, y) := x - \Phi(\varepsilon, x - y). \tag{5.103}$$

It will be the basis of our neural network. More specifically, we define $S : \mathbb{R} \times \mathbb{R}^n \times \mathbb{R}^l \times \mathbb{R}^m \to \mathbb{R} \times \mathbb{R}^n \times \mathbb{R}^l \times \mathbb{R}^m$ by

$$S(z) = \begin{bmatrix} \varepsilon \\ L(x, \mu, \lambda) \\ h(x) \\ \phi^{\varepsilon}_{\mathrm{NR}}\left(-g_{m_1}(x), \lambda_{m_1}\right) \\ \vdots \\ \phi^{\varepsilon}_{\mathrm{NR}}\left(-g_{m_p}(x), \lambda_{m_p}\right) \end{bmatrix},$$

where $z = (\varepsilon, x, \mu, \lambda) \in \mathbb{R} \times \mathbb{R}^n \times \mathbb{R}^l \times \mathbb{R}^m$. Then, it is clear to see that solving the KKT system (5.78) is equivalent to solving the problem

$$\min \Psi(z) := \frac{1}{2} \|S(z)\|^2 . \tag{5.104}$$

Clearly, Ψ defined as in (5.104) is a merit function for (5.78) and in turn, we consider the dynamical system given by

$$\begin{cases} \dfrac{dz(t)}{dt} = -\rho \nabla \Psi(z(t)) = -\rho \nabla S(z(t)) S(z(t)), \\ z(t_0) = z_0, \end{cases} \tag{5.105}$$

where $\rho > 0$ is a scaling factor, for solving the SOCCVI. We refer to the above as "the smoothed NR neural network". The block diagram of the above neural network is presented in Fig. 5.6. The circuit for (5.105) requires $n + l + m + 1$ integrators, n processors for $F(x)$, m processors for $g(x)$, mn processors for $\nabla g(x)$, l processors for $h(x)$, ln processors for $\nabla h(x)$, $(1 + m + l)n^2$ processors for $\nabla_x L(x, \mu, \lambda)$, $2m + 2\sum_{i=1}^{p} m_i^2$ processors for Φ and its derivatives, and some analog multipliers and summers.

Let $u_{m_i} = -g_{m_i}(x) - \lambda_{m_i}$. For subsequent use in the numerical simulations, we note that

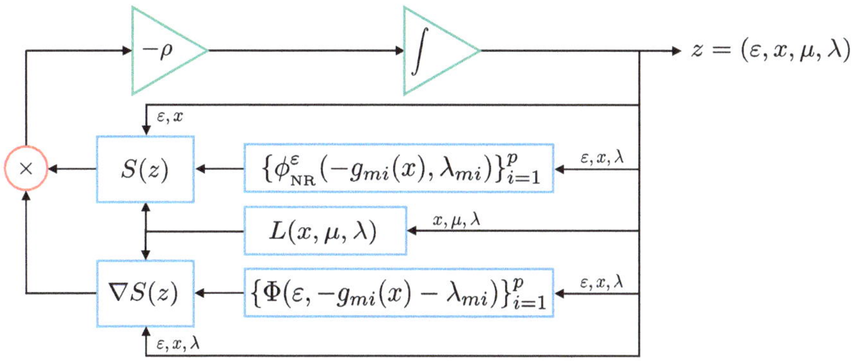

Fig. 5.6 Block diagram of the neural network with $\phi^{\varepsilon}_{\mathrm{NR}}$

$$\begin{aligned}
&\nabla S(z)\\
&= \begin{bmatrix} 1 & 0 & 0 & \{-\nabla_{\varepsilon}\Phi(\varepsilon, u_{m_i})\}_{i=1}^{p} \\ 0 & \nabla_x L(x,\mu,\lambda)^{\mathsf{T}} & \nabla h(x) & -\nabla g(x)\left(I - \mathrm{diag}\{\nabla_{u_{m_i}}\Phi(\varepsilon, u_{m_i})\}_{i=1}^{p}\right) \\ 0 & \nabla h(x)^{\mathsf{T}} & 0 & 0 \\ 0 & \nabla g(x)^{\mathsf{T}} & 0 & \mathrm{diag}\{\nabla_{u_{m_i}}\Phi(\varepsilon, u_{m_i})\}_{i=1}^{p} \end{bmatrix}\\
&= \begin{bmatrix} 1 & 0 & 0 & \{-\nabla_{\varepsilon}\Phi(\varepsilon, -g_{m_i}(x) - \lambda_{m_i})\}_{i=1}^{p} \\ 0 & \nabla_x L(x,\mu,\lambda)^{\mathsf{T}} & \nabla h(x) & -\nabla g(x)\left(I + \mathrm{diag}\{\nabla_{g_{m_i}}\Phi(\varepsilon, -g_{m_i}(x) - \lambda_{m_i})\}_{i=1}^{p}\right) \\ 0 & \nabla h(x)^{\mathsf{T}} & 0 & 0 \\ 0 & \nabla g(x)^{\mathsf{T}} & 0 & -\mathrm{diag}\{\nabla_{\lambda_{m_i}}\Phi(\varepsilon, -g_{m_i}(x) - \lambda_{m_i})\}_{i=1}^{p} \end{bmatrix}
\end{aligned}$$

It is evident that Ψ is a nonnegative function, attaining the value zero at a point $z = (\varepsilon, x, \mu, \lambda)$ if and only if (x, μ, λ) is a KKT point. Moreover, every KKT point corresponds to an equilibrium point of the system (5.105), and the converse holds under the condition that $\nabla S(z)$ is nonsingular. The stability analysis of the system (5.105) follows standard techniques and is analogous to the analysis of the smoothed Fischer-Burmeister neural network discussed earlier.

Nevertheless, our primary contributions, highlighted in the forthcoming analysis, are twofold: (i) we investigate second-order sufficient conditions that ensure the nonsingularity of $\nabla S(z)$; and (ii) we demonstrate that the proposed network exhibits superior numerical performance compared to existing neural network models for SOCCVI problems. For completeness, we present below a fundamental stability result, the proof of which is similar to earlier arguments and is therefore omitted.

Proposition 5.46 *Isolated equilibrium points of (5.105) are asymptotically stable. Moreover, we obtain exponentially stability if $\nabla S(z)$ is nonsingular.*

Proposition 5.46 highlights the critical role of the nonsingularity of the transposed Jacobian $\nabla S(z)$. In what follows, we investigate sufficient conditions that ensure this property holds. To this end, we write out the first-order optimality conditions for the SOCCVI problem (5.74)–(5.75). Let $L(x, \mu, \lambda)$ be given by (5.79) and let $(\mu, \lambda) = (\mu, \lambda_{m_1}, \cdots, \lambda_{m_p}) \in \mathbb{R}^l \times \mathbb{R}^{m_1} \times \cdots \times \mathbb{R}^{m_p} = \mathbb{R}^l \times \mathbb{R}^m$. Suppose that x^* is a solution of the SOCCVI problem (5.74)–(5.75), and Robinson's constraint qualification

$$\begin{pmatrix} \nabla h(x^*)^{\mathsf{T}} \\ -\nabla g(x^*)^{\mathsf{T}} \end{pmatrix} \mathbb{R}^n + T_{\{0_l\}\times\mathcal{K}}\left(h(x^*), -g(x^*)\right) = \mathbb{R}^l \times \mathbb{R}^m$$

holds at x^*. The first-order optimality condition is

$$\left\langle F(x^*), d\right\rangle \geq 0, \quad \forall d \in T_C(x^*), \tag{5.106}$$

where

$$T_C(x^*) = \left\{d \mid \nabla h(x^*)^{\mathsf{T}} d = 0,\ -\nabla g(x^*)^{\mathsf{T}} d \in T_{\mathcal{K}}(-g(x^*))\right\}.$$

It is known that $T_C(x^*)$ is convex and

$$N_C(x^*) = \nabla h(x^*)\mathbb{R}^l + \left\{\nabla g(x^*)\lambda \mid -\lambda \in N_{\mathcal{K}}(-g(x^*))\right\},$$

where $N_{\mathcal{K}}(y) := N_{\mathcal{K}^{m_1}}(y_{m_1}) \times N_{\mathcal{K}^{m_2}}(y_{m_2}) \times \cdots \times N_{\mathcal{K}^{m_p}}(y_{m_p})$ for $y = (y_{m_1}, \cdots, y_{m_p}) \in \mathbb{R}^m$, and

$$N_{\mathcal{K}^{m_i}}(y_{m_i}) := \left\{u_{m_i} \in \mathbb{R}^{m_i} \mid \langle u_{m_i}, v - y_{m_i}\rangle \leq 0, \quad \forall v \in \mathcal{K}^{m_i}\right\}$$

is the normal cone of $\mathcal{K}_{m_i}$ at y_{m_i}. Note that (5.106) holds if and only if $0 \in F(x^*) + N_C(x^*)$ which is equivalent to $\exists\, \mu \in \mathbb{R}^l$, $\lambda \in \mathbb{R}^m$ such that

$$L(x^*, \mu, \lambda) = 0, \quad -\lambda \in N_{\mathcal{K}}(-g(x^*))$$

and the set of multipliers (μ, λ) denoted by $\Lambda(x^*)$ is nonempty compact. Therefore, x^* satisfies the following Karush-Kuhn-Tucker condition,

$$\begin{cases} L(x^*, \mu, \lambda) = 0, \\ h(x^*) = 0, \\ -\lambda \in N_{\mathcal{K}}(-g(x^*)). \end{cases}$$

Using the metric projector and the definition of the normal cone, the KKT conditions can be expressed as

$$S(x, \mu, \lambda) = \begin{pmatrix} L(x, \mu, \lambda) \\ h(x) \\ -g(x) - \Pi_{\mathcal{K}}(-g(x) - \lambda) \end{pmatrix} = 0,$$

where

$$\Pi_{\mathcal{K}}(-g(x)-\lambda) := \left(\Pi_{\mathcal{K}^{m_1}}(-g_{m_1}(x)-\lambda_{m_1})^{\mathsf{T}}, \cdots, \Pi_{\mathcal{K}^{m_p}}(-g_{m_p}(x)-\lambda_{m_p})^{\mathsf{T}}\right)^{\mathsf{T}}.$$

It is particularly emphasized that

$$\Pi'_{\mathcal{K}}(-g(x)-\lambda; d) := \operatorname{diag}\{\Pi'_{\mathcal{K}^{m_1}}(-g_{m_i}(x)-\lambda_{m_i}; d_{m_i})\}_{i=1}^{p},$$

for $d \in \mathbb{R}^m$.

Definition 5.7 ([63]) The critical cone at x^* is defined by

$$\mathcal{C}(x^*) = \left\{d \mid d \in T_C(x^*),\ d \perp F(x^*)\right\}.$$

Proposition 5.47 *Suppose that x^* is a feasible point of the SOCCVI problem* (5.74)–(5.75) *such that $\Lambda(x^*) = \{(\mu, \lambda)\}$ is nonempty and compact. If $JF(x^*)$ is positive semidefinite and Robinson's CQ holds at x^*, then*

$$\sup_{(\mu,\lambda)\in\Lambda(x^*)} \left\{\left\langle J_x L(x^*,\mu,\lambda)d, d\right\rangle - \delta^*(\lambda \mid T^2_{\mathcal{K}}(-g(x^*), -\nabla g(x^*)^{\mathsf{T}} d))\right\} > 0, \quad \forall d \in \mathcal{C}(x^*)\backslash\{0\} \tag{5.107}$$

is the second-order sufficient condition of the SOCCVI problem (5.74)–(5.75), *where*

$$\begin{aligned}&\delta^*\left(\lambda \mid T^2_{\mathcal{K}}(-g(x^*), -\nabla g(x^*)^{\mathsf{T}} d)\right)\\ &= \begin{cases} 0 & \text{if } \lambda \in N_{\mathcal{K}}(-g(x^*)) \text{ and } \langle \lambda, -\nabla g(x^*)^{\mathsf{T}} d\rangle = 0, \\ +\infty & \text{otherwise.}\end{cases}\end{aligned}$$

Proof Let x^* be a solution of the SOCCVI problem (5.74)–(5.75). Since $JF(x^*)$ is positive semidefinite, we see that for some small $\varepsilon > 0$,

$$\left\langle F(x^*), x - x^*\right\rangle \geqslant 0, \quad \forall x \in \mathcal{B}_\varepsilon(x^*) \cap C,$$

where $\mathcal{B}_\varepsilon(x^*)$ denotes the ε-neighborhood of x^*. Equivalently,

$$x^* \in \arg\min\left\{\left\langle F(x^*), x - x^*\right\rangle \mid x \in \mathcal{B}_\varepsilon(x^*) \cap C\right\} \tag{5.108}$$

Again, due to $JF(x^*)$ being positive semidefinite, it is clear that (5.108) holds if and only if

$$x^* \in \arg\min\left\{\left\langle F(x^*), x - x^*\right\rangle + \left\langle JF(x^*)(x-x^*), x - x^*\right\rangle \mid x \in \mathcal{B}_\varepsilon(x^*) \cap C\right\}. \tag{5.109}$$

Therefore, we turn to deduce the second-order sufficient condition of (5.109). To this end, we consider the optimization problem

$$\begin{aligned} &\min\ \langle F(x^*), x - x^*\rangle + \tfrac{1}{2}\langle JF(x^*)(x - x^*), x - x^*\rangle \\ &\text{s.t. } x \in \mathcal{B}_\varepsilon(x^*) \cap C. \end{aligned} \tag{5.110}$$

First, it is known that x^* is the stationary point of problem (5.110) if and only if

$$0 \in F(x^*) + JF(x^*)(x - x^*) + N_{\mathcal{B}_\varepsilon(x^*)\cap C}(x^*) \tag{5.111}$$

where

$$N_{\mathcal{B}_\varepsilon(x^*)\cap C}(x^*) = N_{\mathcal{B}_\varepsilon(x^*)}(x^*) + N_C(x^*) = N_C(x^*) \tag{5.112}$$

On the other hand, (5.111) and (5.112) imply that $0 \in F(x^*) + N_C(x^*)$. Hence, if x^* is a solution of the SOCCVI problem (5.74)–(5.75), we conclude that x^* is the stationary point of problem (5.110).

Now, we prove that the critical cones $\mathcal{C}_p(x^*)$ and $\mathcal{C}(x^*)$ of (5.110) and the SOCCVI problem (5.74)–(5.75), respectively, are equal. Indeed,

$$\begin{aligned} \mathcal{C}_p(x^*) = \Bigg\{ d \in \mathbb{R}^n \ \Bigg|\ & \begin{pmatrix} \nabla h(x^*)^\top d \\ -\nabla g(x^*)^\top d \\ d \end{pmatrix} \in T_{\{0\}\times\mathcal{K}\times\mathcal{B}_\varepsilon(x^*)}(h(x^*), -g(x^*), x^*), \\ & \text{and } \big\langle d, F(x^*) + JF(x^*)(x - x^*)\big\rangle = 0 \Bigg\}. \end{aligned}$$

Notice that

$$\begin{aligned} & T_{\{0\}\times\mathcal{K}\times\mathcal{B}_\varepsilon(x^*)}(h(x^*), -g(x^*), x^*) \\ &= T_{\{0\}\times\mathcal{K}}(h(x^*), -g(x^*)) \times T_{\mathcal{B}_\varepsilon(x^*)}(x^*) \\ &= T_{\{0\}\times\mathcal{K}}(h(x^*), -g(x^*)) \times \mathbb{R}^n. \end{aligned}$$

This yields that

$$\begin{aligned} \mathcal{C}_p(x^*) &= \left\{ d \in \mathbb{R}^n \ \middle|\ \begin{pmatrix} \nabla h(x^*)^\top d \\ -\nabla g(x^*)^\top d \end{pmatrix} \in T_{\{0\}\times\mathcal{K}}(h(x^*), -g(x^*)),\ \big\langle d, F(x^*)\big\rangle = 0 \right\} \\ &= \mathcal{C}(x^*). \end{aligned}$$

Next, the Lagrange function of problem (5.110) is

$$\begin{aligned} \mathcal{L}(x^*, \lambda, \mu, \nu) = &\big\langle F(x^*), (x - x^*)\big\rangle + \frac{1}{2}\big\langle JF(x^*)(x - x^*), x - x^*\big\rangle \\ &+ \langle h(x), \mu\rangle + \langle g(x), \lambda\rangle + \langle x, \nu\rangle \end{aligned}$$

which gives

$$\nabla_x \mathcal{L}(x^*, \lambda, \mu, \nu) = F(x^*) + JF(x^*)(x - x^*) + \nabla h(x)\mu + \nu + \nabla g(x)\lambda,$$
$$\nabla^2_{xx} \mathcal{L}(x^*, \lambda, \mu, \nu) = JF(x^*) + \sum_{i=1}^{l} \mu_i \nabla^2 h_i(x^*) + \sum_{i=1}^{m} \lambda_i \nabla^2 g_i(x^*).$$

Here, we note that $\nabla^2_{xx} \mathcal{L}(x^*, \lambda, \mu, \nu) = J_x L(x^*, \lambda, \mu)$.

On the other hand, in light of [65, Proposition 3.269], we can check that $\{0\} \times \mathcal{K}$ is second order regular at $(h(x^*), -g(x^*))$ along the direction $(\nabla h(x^*)^\top d, -\nabla g(x^*)^\top d)$ with respect to the mapping $\begin{pmatrix} \nabla h(x^*)^\top \\ -\nabla g(x^*)^\top \end{pmatrix}$ for all $d \in \mathcal{C}(x^*)$. Then, using the definition of the second-order regularity (see [65, Definition 3.85]) yields

$$y_n = \begin{pmatrix} h(x^*) \\ -g(x^*) \end{pmatrix} + t_n \begin{pmatrix} \nabla h(x^*)^\top d \\ -\nabla g(x^*)^\top d \end{pmatrix} + \frac{1}{2} t_n{}^2 r_n, \quad \forall y_n \in \{0\} \times \mathcal{K},$$

where $t_n \downarrow 0$, $r_n = \begin{pmatrix} \nabla h(x^*)^\top w_n \\ -\nabla g(x^*)^\top w_n \end{pmatrix} + a_n$ with a_n being a convergent sequence and $t_n w_n \to 0$, $(n \to +\infty)$ such that

$$\lim_{n\to\infty} \text{dist}\left(r_n, T^2{}_{\{0\}\times\mathcal{K}}((h(x^*), -g(x^*)), (\nabla h(x^*)^\top d, -\nabla g(x^*)^\top d))\right) = 0.$$

According to the above result, for all $P_n \in \{0\} \times \mathcal{K} \times \mathcal{B}_\varepsilon(x^*)$, we have

$$P_n = \begin{pmatrix} h(x^*) \\ -g(x^*) \\ x^* \end{pmatrix} + t_n \begin{pmatrix} \nabla h(x^*)^\top d \\ -\nabla g(x^*)^\top d \\ d \end{pmatrix} + \frac{1}{2} t_n{}^2 \begin{pmatrix} r_n \\ q_n \end{pmatrix},$$

where, $t_n \downarrow 0$, $\begin{pmatrix} r_n \\ q_n \end{pmatrix} = \begin{pmatrix} \nabla h(x^*)^\top w_n \\ -\nabla g(x^*)^\top w_n \\ w_n \end{pmatrix} + \begin{pmatrix} a_n \\ b_n \end{pmatrix}$ with $\begin{pmatrix} a_n \\ b_n \end{pmatrix}$ being a convergent sequence and $t_n w_n \to 0$, $(n \to +\infty)$. Therefore, we obtain

$$\lim_{n\to\infty} \text{dist}\left(r_n, T^2{}_{\{0\}\times\mathcal{K}}\big((h(x^*), -g(x^*)), (\nabla h(x^*)^\top d, -\nabla g(x^*)^\top d)\big)\right) = 0$$

and

$$\begin{aligned}
&\lim_{n\to\infty} \text{dist}\left\{\begin{pmatrix} r_n \\ q_n \end{pmatrix}, T^2{}_{\{0\}\times\mathcal{K}\times\mathcal{B}_\varepsilon(x^*)}((h(x^*), -g(x^*), x^*), (\nabla h(x^*)^\top d, -\nabla g(x^*)^\top d, d))\right\} \\
&= \lim_{n\to\infty} \text{dist}\left\{\begin{pmatrix} r_n \\ q_n \end{pmatrix}, T^2{}_{\{0\}\times\mathcal{K}}((h(x^*), -g(x^*)), (\nabla h(x^*)^\top d, -\nabla g(x^*)^\top d)) \times T^2{}_{\mathcal{B}_\varepsilon(x^*)}(x^*, d)\right\} \\
&= \lim_{n\to\infty} \text{dist}\left\{r_n, T^2{}_{\{0\}\times\mathcal{K}}\big((h(x^*), -g(x^*)), (\nabla h(x^*)^\top d, -\nabla g(x^*)^\top d)\big)\right\} \\
&= 0,
\end{aligned}$$

and thus, $\{0\} \times \mathcal{K} \times \mathcal{B}_\varepsilon(x^*)$ is second-order regular at the point $(h(x^*), -g(x^*), x^*)$ along $\big(\nabla h(x^*)^\mathsf{T} d, -\nabla g(x^*)^\mathsf{T} d, d\big)$ with respect to the mapping $\begin{pmatrix} \nabla h(x^*)^\mathsf{T} \\ -\nabla g(x^*)^\mathsf{T} \\ I \end{pmatrix}$ for all $d \in \mathcal{C}(x^*)$, with I as the identity map.

This together with [65, Theorem 3.86] indicates that for (5.110), the second-order sufficient condition is

$$\sup_{(\lambda,\mu,\nu)\in\bar{\Lambda}(x^*)} \Big\{\nabla^2{}_{xx}\mathcal{L}(x^*, \lambda, \mu, \nu) - \delta^*\big((\mu, \lambda, \nu), T^2{}_{\{0\}\times\mathcal{K}\times\mathcal{B}_\varepsilon(x^*)}((h(x^*), -g(x^*), x^*), (\nabla h(x^*)^\mathsf{T} d, -\nabla g(x^*)^\mathsf{T} d, d))\Big\} > 0, \quad \forall d \in \mathcal{C}_p(x^*)\backslash\{0\}.$$

We can further simplify it as

$$\begin{aligned}
&\sup_{(\lambda,\mu,\nu)\in\bar{\Lambda}(x^*)} \Big\{\nabla^2{}_{xx}\mathcal{L}(x^*, \lambda, \mu, \nu)(d, d) - \delta^*\big((\mu, \lambda, \nu), T^2{}_{\{0\}\times\mathcal{K}\times\mathcal{B}_\varepsilon(x^*)}((h(x^*), -g(x^*), x^*), \\
&\qquad (\nabla h(x^*)^\mathsf{T} d, -\nabla g(x^*)^\mathsf{T} d, d))\Big\} \\
=&\sup_{(\lambda,\mu,\nu)\in\bar{\Lambda}(x^*)} \Big\{\nabla^2{}_{xx}\mathcal{L}(x^*, \lambda, \mu, \nu)(d, d) - \delta^*((\mu, \lambda, \nu), T^2{}_{\{0\}}(h(x^*), \nabla h(x^*)^\mathsf{T} d) \\
&\qquad \times T^2{}_{\mathcal{K}}(-g(x^*), -\nabla g(x^*)^\mathsf{T} d) \times T^2{}_{\mathcal{B}_\varepsilon(x^*)}(x^*, d))\Big\} \\
=&\sup_{(\lambda,\mu,\nu)\in\bar{\Lambda}(x^*)} \Big\{\nabla^2{}_{xx}\mathcal{L}(x^*, \lambda, \mu, \nu)(d, d) - \delta^*\Big((\mu, \lambda, \nu), \{0\} \times T^2{}_{\mathcal{K}}(-g(x^*), -\nabla g(x^*)^\mathsf{T} d \times \mathbb{R}^n\Big)\Big\} \\
=&\sup_{(\mu,\lambda)\in\Lambda(x^*)} \Big\{J_x L(x^*, \lambda, \mu)(d, d) - \delta^*\Big(\lambda \,|\, T^2{}_{\mathcal{K}}(-g(x^*), -\nabla g(x^*)^\mathsf{T} d)\Big)\Big\}.
\end{aligned}$$

To sum up, the second-order sufficient condition of the SOCCVI problem (5.74)–(5.75) is described by

$$\sup_{(\mu,\lambda)\in\Lambda(x^*)} \Big\{\langle J_x L(x^*, \lambda, \mu)d, d\rangle - \delta^*\Big(\lambda \,|\, T^2{}_{\mathcal{K}}(-g(x^*), -\nabla g(x^*)^\mathsf{T} d)\Big)\Big\} > 0 \quad \forall d \in \mathcal{C}(x^*)\backslash\{0\},$$

as desired. □

As shown in Proposition 5.46, the nonsingularity of $\nabla S(0, x^*, \mu^*, \lambda^*)$ is essential for ensuring that the equilibrium point of the neural network corresponds to a solution of the SOCCVI problem (5.74)–(5.75), and that it exhibits exponential stability. We now present several conditions under which the nonsingularity of $\nabla S(0, x^*, \mu^*, \lambda^*)$ can be guaranteed.

Proposition 5.48 *Suppose (x^*, μ^*, λ^*) is a KKT point of the SOCCVI problem (5.74)–(5.75). Then, $\nabla S(0, x^*, \mu^*, \lambda^*)$ is nonsingular if*

(i) $\Lambda(x^*) = \{(\mu, \lambda)\} \neq \emptyset$;
(ii) the second-order sufficient condition (5.107) holds;

(iii) $-\lambda^* \in \mathrm{int} N_{\mathcal{K}}(-g(x^*))$ *holds; and*
(iv) *the following constraint nondegeneracy holds:*

$$\begin{pmatrix} \nabla h(x^*)^\mathsf{T} \\ -\nabla g(x^*)^\mathsf{T} \end{pmatrix} \mathbb{R}^n + \mathrm{lin} T_{\{0_l\}\times\mathcal{K}}(h(x^*), -g(x^*)) = \mathbb{R}^l \times \mathbb{R}^m.$$

Proof It is enough to verify that M given by

$$M = \begin{bmatrix} \nabla_x L(x^*, \mu^*, \lambda^*)^\mathsf{T} & \nabla h(x^*) & -\nabla g(x^*)\left(I - \mathrm{diag}\{\lim_{\varepsilon\to 0} \nabla_{u^*_{m_i}} \Phi(\varepsilon, u^*_{m_i})\}_{i=1}^p\right) \\ \nabla h(x^*)^\mathsf{T} & 0 & 0 \\ \nabla g(x^*)^\mathsf{T} & 0 & \mathrm{diag}\{\lim_{\varepsilon\to 0} \nabla_{u^*_{m_i}} \Phi(\varepsilon, u^*_{m_i})\}_{i=1}^p \end{bmatrix}^\mathsf{T}$$

is nonsingular, where $u^*_{m_i} = -g_{m_i}(x^*) - \lambda^*_{m_i}$. From Lemma 5.15 and (5.102), we can deduce

$$\lim_{\varepsilon\to 0} \left[\nabla_u \Phi(\varepsilon, u)\right]^\mathsf{T} d = \Pi'_{\mathcal{K}^m}(u; d)$$

for $d \in \mathbb{R}^m$ and $u \in \mathbb{R}^m \backslash (\mathcal{K}^m \cup -\mathcal{K}^m)$ or $u \in \mathrm{int}\mathcal{K}^m$. Then, for $\varepsilon \to 0$ and $(dx, d\mu, d\lambda) \in \mathbb{R}^n \times \mathbb{R}^l \times \mathbb{R}^m$, we have

$$M \begin{pmatrix} dx \\ d\mu \\ d\lambda \end{pmatrix} = \begin{pmatrix} J_x L(x^*, \mu^*, \lambda^*)dx + \nabla h(x^*)d\mu + \nabla g(x^*)d\lambda \\ \nabla h(x^*)^\mathsf{T} dx \\ H \end{pmatrix}$$

where

$$\begin{aligned} H &= \left(I - \mathrm{diag}\{\lim_{\varepsilon\to 0} \nabla_{u^*_{m_i}} \Phi(\varepsilon, u^*_{m_i})\}_{i=1}^p\right)^\mathsf{T} (-\nabla g(x^*))^\mathsf{T} dx \\ &\quad + \left(\mathrm{diag}\{\lim_{\varepsilon\to 0} \nabla_{u^*_{m_i}} \Phi(\varepsilon, u^*_{m_i})\}_{i=1}^p\right)^\mathsf{T} d\lambda \\ &= -\nabla g(x^*)^\mathsf{T} dx - \left(\mathrm{diag}\{\lim_{\varepsilon\to 0} \nabla_{u^*_{m_i}} \Phi(\varepsilon, u^*_{m_i})\}_{i=1}^p\right)^\mathsf{T} \left[-\nabla g(x^*)^\mathsf{T} dx - d\lambda\right] \\ &= -\nabla g(x^*)^\mathsf{T} dx - \Pi'_{\mathcal{K}}\left(-g(x^*) - \lambda^*; -\nabla g(x^*)^\mathsf{T} dx - d\lambda\right). \end{aligned}$$

Therefore, we have

$$M \begin{pmatrix} dx \\ d\mu \\ d\lambda \end{pmatrix} = \begin{pmatrix} J_x L(x^*, \mu^*, \lambda^*)dx + \nabla h(x^*)d\mu + \nabla g(x^*)d\lambda \\ \nabla h(x^*)^\mathsf{T} dx \\ -\nabla g(x^*)^\mathsf{T} dx - \Pi'_{\mathcal{K}}\left(-g(x^*) - \lambda^*; -\nabla g(x^*)^\mathsf{T} dx - d\lambda\right) \end{pmatrix} \tag{5.113}$$

Suppose that $M\begin{pmatrix} dx \\ d\mu \\ d\lambda \end{pmatrix} = 0$. We need to show that $dx = 0$, $d\mu = 0$, $d\lambda = 0$. First, from the 2nd and 3rd expressions of (5.113), we obtain

$$\begin{cases} \nabla h(x^*)^\mathsf{T} dx \ = 0 \\ -\nabla g(x^*)^T dx = \Pi'_\mathcal{K}\left(-g(x^*) - \lambda^*; -\nabla g(x^*)^\mathsf{T} dx - d\lambda\right) \end{cases} \tag{5.114}$$

which implies that $dx \in \mathcal{C}(x^*)$. In addition, from the first expression of (5.113), we obtain

$$\left\langle J_x L(x^*, \mu^*, \lambda^*)dx, dx\right\rangle + \left\langle\nabla g(x^*)^\mathsf{T} dx, d\lambda\right\rangle = 0. \tag{5.115}$$

To proceed, we consider the following sets:

$$\begin{aligned} I^* &= \left\{i \mid -g_{m_i}(x^*) \in \mathrm{int}(\mathcal{K}^{m_i}), i = 1, \cdots p\right\}; \\ B^* &= \left\{i \mid -g_{m_i}(x^*) \in \mathrm{bd}(\mathcal{K}^{m_i}), g_{m_i}(x^*) \neq 0\right\}; \\ Z^* &= \left\{i \mid g_{m_i}(x^*) = 0\right\}. \end{aligned}$$

Note that

$$\mathcal{C}_\mathcal{K}(-g(x^*)) = \left\{d \in \mathbb{R}^n \mid -\nabla g(x^*)^\mathsf{T} d \in T_\mathcal{K}(-g(x^*))\right\}$$

and

$$T_\mathcal{K}(-g(x^*)) = \left\{ d \;\middle|\; \begin{array}{l} -\nabla g_0^i(x^*)^\mathsf{T} d - \frac{\nabla \bar{g}^i(x^*)^\mathsf{T} d}{g_0^i(x^*)} \geq 0, \ i \in B^* \\ -\nabla g_0^i(x^*) d + \nabla \bar{g}^i(x^*)^\mathsf{T} d \geq 0, \ i \in Z^* \end{array} \right\}$$

Since $-\lambda \perp -g(x)$, we see that

$$\lambda = \left\{ \lambda \;\middle|\; \begin{array}{l} \lambda_{m_i} = 0, \ i \in I^* \\ \lambda_{m_i} = \sigma(-g_0^i(x^*), \bar{g}^i(x^*)), \ \sigma > 0, \ i \in B^* \\ \lambda_{m_i} \in \mathrm{int}(\mathcal{K}^{m_i}), \ i \in Z^* \end{array} \right\}$$

which further yields

$$\left[-g(x^*) - \lambda^*\right]_{m_i} = \begin{cases} -g_{m_i}(x^*) \in \mathrm{int}(\mathcal{K}^{m_i}), \ i \in I^* \\ ((1 - \sigma)(-g_0^i(x^*)), \ (1 + \sigma)(-\bar{g}^i(x^*))), \ i \in B^* \\ \lambda_{m_i} \in \mathrm{int}(\mathcal{K}^{m_i}), \ i \in Z^* \end{cases}$$

On the other hand, Condition (iii) implies

$$\mathcal{C}(x^*) = \left\{ d \;\middle|\; \begin{array}{l} \nabla h(x^*)^\mathsf{T} d = 0, \ -\nabla g_{m_i}(x^*)^\mathsf{T} d = 0, \ i \in Z^* \\ -\nabla g_{m_i}(x^*)^\mathsf{T} d \in T_\mathcal{K}(-g_{m_i}(x^*)), \ \left\langle\lambda_{m_i}, -\nabla g_{m_i}(x^*)^\mathsf{T} d\right\rangle = 0, \ i \in B^* \end{array} \right\}$$

and $\mathcal{C}(x^*)$ is a linear space. Therefore, we have

$$\delta^*\left(\lambda \mid T_{\mathcal{K}}^2(-g(x^*), -\nabla g(x^*)^\top d)\right) = \sum_{i \in B^*} \frac{\lambda_0^i}{-g_0^i(x^*)} \left[\left\| \nabla g_0^i(x^*)^\top dx \right\|^2 - \left\| \nabla \bar{g}^i(x^*)^\top dx \right\|^2 \right]$$

with $\lambda_{m_i} = (\lambda_0^i, \bar{\lambda}^i)$.

Case (1). If $i \in B^*$, we have $\lambda_{m_i}^* = \left(-\sigma g_0^i(x^*), \sigma \bar{g}^i(x^*)\right)$. Then, applying Lemma 5.15 and (5.114), we obtain

$$\begin{aligned}
&\Pi'_{\mathcal{K}^{m_i}}\left(-g_{m_i}(x^*) - \lambda_{m_i}^*; -Jg_{m_i}(x^*)dx - d\lambda_{m_i}\right) \\
&= \frac{1}{2}\begin{bmatrix} 1 & w_i^\top \\ w_i & \frac{2}{1+\sigma}I - \frac{1-\sigma}{1+\sigma}w_i w_i^\top \end{bmatrix} \left(-\nabla g_{m_i}(x^*)^\top dx - d\lambda_{m_i}\right) \qquad (5.116)\\
&= A_i\left(-\nabla g_{m_i}(x^*)^\top dx - d\lambda_{m_i}\right) \\
&= -\nabla g_{m_i}(x^*)^\top dx,
\end{aligned}$$

where

$$A_i = \frac{1}{2}\begin{bmatrix} 1 & w_i^\top \\ w_i & \frac{2}{1+\sigma}I - \frac{1-\sigma}{1+\sigma}w_i w_i^\top \end{bmatrix}$$

and $w_i = \frac{-\bar{g}^i(x^*)}{\|\bar{g}^i(x^*)\|}$. Now we need to prove that $dx \in T_C(x^*)$ and

$$-\nabla g_0^i(x^*)^\top dx \geq \frac{\bar{g}^i(x^*)^\top \nabla \bar{g}^i(x^*)^\top dx}{\left\|\bar{g}^i(x^*)\right\|} \qquad (5.117)$$

From $-g_0^i(x^*) = \left\|\bar{g}^i(x^*)\right\|$, we know

$$\lambda_{m_i}^* = \begin{pmatrix} -\sigma g_0^i(x^*) \\ +\sigma \bar{g}^i(x^*) \end{pmatrix} = -\sigma g_0^i(x^*) \begin{pmatrix} 1 \\ -w_i \end{pmatrix},$$

where $\|w_i\| = 1$ and $w_i = \frac{\bar{g}^i(x^*)}{g_0^i(x^*)} = \frac{-\bar{g}^i(x^*)}{\|\bar{g}^i(x^*)\|}$ for $i \in B^*$. Hence, we achieve

$$\lambda_{m_i}^* A_i = \left(1 - \|w_i\|^2, w_i^\top - \frac{2}{1+\sigma}w_i^\top + \frac{1-\sigma}{1+\sigma}w_i^\top \|w_i\|^2\right) = (0, 0). \qquad (5.118)$$

Combining (5.116) and (5.118) yields

$$\left\langle \lambda_{m_i}^*, -\nabla g_{m_i}(x^*)^\top dx \right\rangle = 0$$

which means $dx \in \mathcal{C}(x^*)$. Then, it follows from (5.116) that

$$A_i\left(-\nabla g_{m_i}(x^*)^\top dx - d\lambda_{m_i}\right) = -\nabla g_{m_i}(x^*)^\top dx$$

$$\Longleftrightarrow (A_i - I)(-\nabla g_{m_i}(x^*)^\mathsf{T} dx) = A_i d\lambda_{m_i} \tag{5.119}$$

$$\Longleftrightarrow (1, -w_i^\mathsf{T}) \begin{bmatrix} -\frac{1}{2} & \frac{1}{2} w_i^\mathsf{T} \\ \frac{1}{2} w_i & \frac{-\sigma}{1+\sigma} I - \frac{1}{2}\frac{1-\sigma}{1+\sigma} w_i w_i^\mathsf{T} \end{bmatrix} \begin{pmatrix} -\nabla g_0^i(x^*)^\mathsf{T} dx \\ -\nabla \bar{g}^i(x^*) dx \end{pmatrix}$$

$$= (1, -w_i^\mathsf{T}) \begin{bmatrix} \frac{1}{2} & \frac{1}{2} w_i^\mathsf{T} \\ \frac{1}{2} w_i & \frac{1}{1+\sigma} I - \frac{1}{2}\frac{1-\sigma}{1+\sigma} w_i w_i^\mathsf{T} \end{bmatrix} \begin{pmatrix} d\lambda_0^i \\ d\bar{\lambda}^i \end{pmatrix}$$

In summary, we deduce that

$$\left(-1, \frac{1}{2} w_i^\mathsf{T} + \frac{\sigma}{1+\sigma} w_i^\mathsf{T} + \frac{1}{2} \cdot \frac{1-\sigma}{1+\sigma} w_i^\mathsf{T}\right) \begin{pmatrix} -\nabla g_0^i(x^*)^\mathsf{T} dx \\ -\nabla \bar{g}^i(x^*)^\mathsf{T} dx \end{pmatrix} = 0$$

which is equivalent to

$$(-1, w_i^\mathsf{T}) \begin{pmatrix} -\nabla g_0^i(x^*)^\mathsf{T} dx \\ -\nabla \bar{g}^i(x^*)^\mathsf{T} dx \end{pmatrix} = 0$$

This indicates

$$-\nabla g_0^i(x^*)^\mathsf{T} dx = \frac{\bar{g}^i(x^*)^\mathsf{T} \nabla \bar{g}^i(x^*)^\mathsf{T} dx}{\|\bar{g}^i(x^*)\|}, \tag{5.120}$$

and hence (5.117) holds.

Case (2). Let $i \in Z^*$. From the second equation of (5.114), we have

$$\Pi'_{\mathcal{K}_{m_i}} \left(0 - \lambda_{m_i}; -\nabla g_{m_i}(x^*)^\mathsf{T} dx - d\lambda_{m_i}\right) = -\nabla g_{m_i}(x^*)^\mathsf{T} dx$$

Hence, $-\nabla g_{m_i}(x^*)^\mathsf{T} dx = 0$.

Case (3). Let $i \in I^*$. Again, from the second equation of (5.114), we have

$$\Pi'_{\mathcal{K}_{m_i}} \left(-g_{m_i}(x^*), -\nabla g_{m_i}(x^*)^\mathsf{T} dx - d\lambda_{m_i}\right) = -\nabla g_{m_i}(x^*)^\mathsf{T} dx - d\lambda_{m_i} = -\nabla g_{m_i}(x^*)^\mathsf{T} dx$$

which says $d\lambda_{m_i} = 0$.

From all the above, we conclude that $dx \in \mathcal{C}(x^*)$ implies

$$\begin{cases} \nabla g_{m_i}(x^*)^\mathsf{T} dx = 0, \ i \in Z^* \\ g_0^i(x^*) \nabla g_0^i(x^*)^\mathsf{T} dx = \bar{g}^i(x^*)^\mathsf{T} \nabla \bar{g}^i(x^*)^\mathsf{T} dx, \ i \in B^*. \end{cases}$$

Applying (5.113) and (5.114), we have the following three identities.

$$J_x L(x^*, \mu^*, \lambda^*) dx + \nabla h(x^*) d\mu + \nabla g(x^*) d\lambda = 0 \tag{5.121}$$

$$\nabla h(x^*)^\mathsf{T} dx = 0 \tag{5.122}$$

$$- \nabla g(x^*)^{\mathsf{T}} dx - \Pi'_{\mathcal{K}}(-g(x^*) - \lambda^*; -\nabla g(x^*)^{\mathsf{T}} dx - d\lambda) = 0 \tag{5.123}$$

Using (5.121) and (5.122) gives

$$\begin{aligned} 0 &= \langle dx, J_x L(x^*, \mu^*, \lambda^*)dx + \nabla h(x^*)d\mu + \nabla g(x^*)d\lambda \rangle \\ &= \langle dx, J_x L(x^*, \mu^*, \lambda^*)dx \rangle - \sum_{i \in B^*} \langle -\nabla g_{m_i}(x^*)dx, d\lambda_{m_i} \rangle. \end{aligned}$$

Thus, for $i \in B^*$,

$$\begin{aligned} &\langle -\nabla g_{m_i}(x^*)^{\mathsf{T}} dx, d\lambda_{m_i} \rangle \\ &= -\nabla g_0^i(x^*)dx\, d\lambda_0^i + \langle -\nabla \bar{g}^i(x^*)dx, d\bar{\lambda}^i \rangle \\ &= \nabla g_0^i(x^*)^{\mathsf{T}} dx \cdot \frac{\bar{g}^i(x^*)}{\|\bar{g}^i(x^*)\|} d\bar{\lambda}^i - \langle \nabla \bar{g}^i(x^*)dx, d\bar{\lambda}^i \rangle \\ &= \frac{\bar{g}^i(x^*)^{\mathsf{T}} \nabla \bar{g}^i(x^*)dx}{\|\bar{g}^i(x^*)\|^2} \cdot \bar{g}^i(x^*)^{\mathsf{T}} d\bar{\lambda}^i - \langle \nabla \bar{g}^i(x^*)dx, d\bar{\lambda}^i \rangle \\ &= \left\langle (-\nabla \bar{g}^i(x^*)^{\mathsf{T}} dx)^{\mathsf{T}} \left[I - \frac{\bar{g}^i(x^*)\bar{g}^i(x^*)^{\mathsf{T}}}{\|\bar{g}^i(x^*)\|^2} \right], d\bar{\lambda}^i \right\rangle \end{aligned} \tag{5.124}$$

On the other hand, from (5.119), we have

$$\begin{aligned} &\begin{pmatrix} \frac{1}{2}\nabla g_0^i(x^*)^{\mathsf{T}} dx + \frac{1}{2} \cdot \frac{\bar{g}^i(x^*)^{\mathsf{T}}}{\|\bar{g}^i(x^*)\|} J\bar{g}^i(x^*)dx \\ \frac{1}{2}w_i \left(-\nabla g_0^i(x^*)^{\mathsf{T}} dx - w_i^{\mathsf{T}} \nabla \bar{g}^i(x^*)^{\mathsf{T}} dx \cdot \frac{1-\sigma}{1+\sigma}\right) - \frac{\sigma}{1+\sigma}(-\nabla \bar{g}^i(x^*)^{\mathsf{T}} dx) \end{pmatrix} \\ &= \begin{pmatrix} \frac{1}{2}d\lambda_0^i + \frac{1}{2}{w_i}^{\mathsf{T}} d\bar{\lambda}^i \\ \frac{1}{2}w_i(d\lambda_0^i - \frac{1-\sigma}{1+\sigma}w_i^{\mathsf{T}} d\bar{\lambda}^i) + \frac{1}{1+\sigma}d\bar{\lambda}^i \end{pmatrix} \end{aligned} \tag{5.125}$$

From (5.120), we can deduce that

$$\begin{aligned} &\frac{1}{2}w_i \left(-\nabla g_0^i(x^*)^{\mathsf{T}} dx - w_i^{\mathsf{T}} \nabla \bar{g}^i(x^*)^{\mathsf{T}} dx \cdot \frac{1-\sigma}{1+\sigma}\right) + \frac{\sigma}{1+\sigma}\nabla \bar{g}^i(x^*)^{\mathsf{T}} dx \\ &= \frac{1}{2}w_i \left(-\nabla g_0^i(x^*)^{\mathsf{T}} dx - \frac{1-\sigma}{1+\sigma} w_i^{\mathsf{T}} \nabla \bar{g}^i(x^*)^{\mathsf{T}} dx\right) + \frac{\sigma}{1+\sigma}\nabla \bar{g}^i(x^*)^{\mathsf{T}} dx \\ &= \frac{1}{2}w_i \left(-\nabla g_0^i(x^*)^{\mathsf{T}} dx + \frac{1-\sigma}{1+\sigma} \nabla g_0^i(x^*)^{\mathsf{T}} dx\right) + \frac{\sigma}{1+\sigma}\nabla \bar{g}^i(x^*)^{\mathsf{T}} dx \\ &= \frac{\sigma}{1+\sigma}\left(w_i(-\nabla g_0^i(x^*)^{\mathsf{T}} dx) + \nabla \bar{g}^i(x^*)^{\mathsf{T}} dx\right) \end{aligned} \tag{5.126}$$

and

$$\frac{1}{2}w_i \left(d\lambda_0^i - \frac{1-\sigma}{1+\sigma} w_i^{\mathsf{T}} d\bar{\lambda}^i\right) + \frac{1}{1+\sigma}d\bar{\lambda}^i$$

$$= \frac{1}{2} w_i \left(d\lambda_0^i + \frac{1-\sigma}{1+\sigma} d\lambda_0^i \right) + \frac{1}{1+\sigma} d\bar{\lambda}^i \tag{5.127}$$
$$= \frac{1}{1+\sigma} w_i d\lambda_0^i + \frac{1}{1+\sigma} d\bar{\lambda}^i$$
$$= \frac{1}{1+\sigma} \left(w_i d\lambda_0^i + d\bar{\lambda}^i \right).$$

Therefore, applying (5.125), (5.126) and (5.127) implies

$$\frac{1}{1+\sigma} \left(w_i d\lambda_0^i + d\bar{\lambda}^i \right) = \frac{\sigma}{1+\sigma} \left(w_i (-\nabla g_0^i(x^*)^\mathsf{T}) dx + \nabla \bar{g}^i(x^*)^\mathsf{T} dx \right),$$

which means

$$w_i d\lambda_0^i + d\bar{\lambda}^i = -\sigma \left(w_i \nabla g_0^i(x^*)^\mathsf{T} dx - \nabla \bar{g}^i(x^*)^\mathsf{T} dx \right). \tag{5.128}$$

Note that

$$w_i d\lambda_0^i + d\bar{\lambda}^i = \left(I - w_i w_i^\mathsf{T} \right) d\bar{\lambda}^i = \left(I - \frac{\bar{g}^i(x^*)\bar{g}^i(x^*)^\mathsf{T}}{\left\| \bar{g}^i(x^*) \right\|^2} \right) d\bar{\lambda}^i \tag{5.129}$$

Then, it follows from (5.124), (5.128), and (5.129) that

$$\left\langle -\nabla g_{m_i}(x^*) dx, d\lambda_{m_i} \right\rangle$$
$$= \left\langle -\nabla \bar{g}^i(x^*)^\mathsf{T} dx, \left(I - \frac{\bar{g}^i(x^*)\bar{g}^i(x^*)^\mathsf{T}}{\left\| \bar{g}^i(x^*) \right\|^2} \right) d\bar{\lambda}^i \right\rangle$$
$$= \sigma \left(\left\langle -\nabla g^i(x^*)^\mathsf{T} dx, w_i \left(-\nabla g_0^i(x^*)^T dx \right) \right\rangle - \left\| \nabla \bar{g}^i(x^*)^\mathsf{T} dx \right\|^2 \right)$$
$$= \sum_{i \in B^*} \frac{\lambda_0^i}{-g_0^i(x^*)} \left(\left\| \nabla g_0^i(x^*)^\mathsf{T} dx \right\|^2 - \left\| \nabla \bar{g}^i(x^*)^\mathsf{T} dx \right\|^2 \right)$$
$$= \delta^* \left(\lambda \mid T_{\mathcal{K}}^2(-g(x^*); -\nabla g(x^*)^\mathsf{T} dx) \right).$$

This together with (5.115) yields

$$\left\langle J_x L(x^*, \mu^*, \lambda^*) dx, dx \right\rangle - \delta^* \left(\lambda \mid T_{\mathcal{K}}{}^2(-g(x^*); -\nabla g(x^*)^\mathsf{T} dx) \right) = 0.$$

Now using the second-order sufficient condition (condition (ii)), we reach $dx = 0$. Plugging this into (5.121) leads to

$$\nabla h(x^*) d\mu + \nabla g(x^*) d\lambda = 0.$$

Applying (5.123) together with condition (iv) yields $d\mu = 0$ and $d\lambda = 0$. Thus, the matrix M is nonsingular, and the proof is complete. □

We illustrate the effectiveness of the smoothed natural residual (NR) neural network in solving several representative SOCCVI problems defined by (5.74)–(5.75). In addition, we provide a comprehensive numerical comparison between the neural network (5.105) and other neural network models from the SOCCVI literature. Specifically, we consider six standard benchmark problems for SOCCVI and employ the MATLAB solver *ode23s* for all simulations. The stopping criterion is set to $\|\nabla\Psi(z(t))\| \leq 1 \times 10^{-6}$. For each example, the neural network (5.105) is simulated from five randomly generated initial points z_0. The solution trajectories for these test problems are displayed in Figs. 5.7, 5.8, 5.9, 5.10, 5.11 and 5.12. Notably, all trajectories successfully converge to the SOCCVI solution, denoted by x^*.

Example 5.2 Let

$$F(x) = \begin{pmatrix} 2x_1 + x_2 + 1 \\ x_1 + 6x_2 - x_3 - 2 \\ -x_2 + 3x_3 - \frac{6}{5}x_4 + 3 \\ -\frac{6}{5}x_3 + 2x_4 + \frac{1}{2}\sin x_4 \cos x_5 \sin x_6 + 6 \\ \frac{1}{2}\cos x_4 \sin x_5 \sin x_6 + 2x_5 - \frac{5}{2} \\ -\frac{1}{2}\cos x_4 \cos x_5 \cos x_6 + 2x_6 + \frac{1}{4}\cos x_6 \sin x_7 \cos x_8 + 1 \\ \frac{1}{4}\sin x_6 \cos x_7 \cos x_8 + 4x_7 - 2 \\ -\frac{1}{4}\sin x_6 \sin x_7 \sin x_8 + 2x_8 + \frac{1}{2} \end{pmatrix}$$

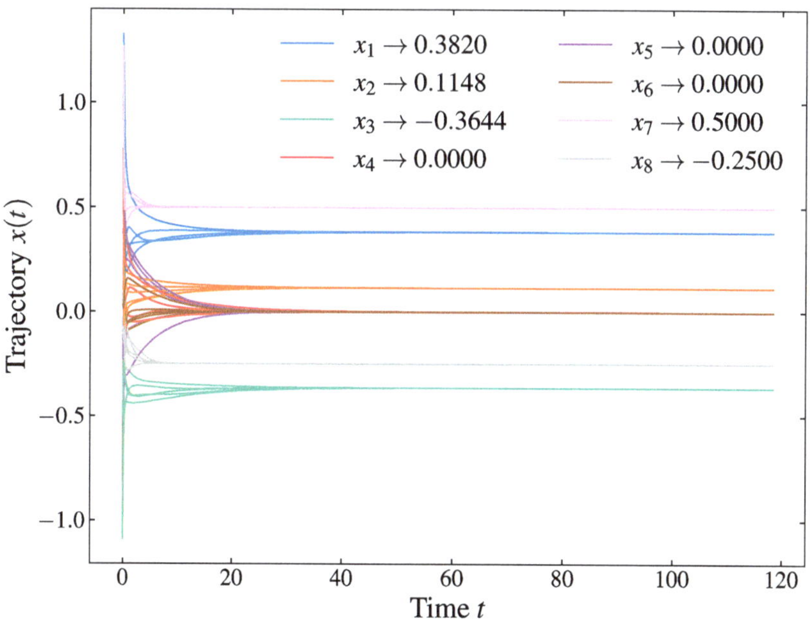

Fig. 5.7 Convergence of $x(t)$ to the SOCCVI solution in Example 5.2 using five random initial points, where $\rho = 10^3$

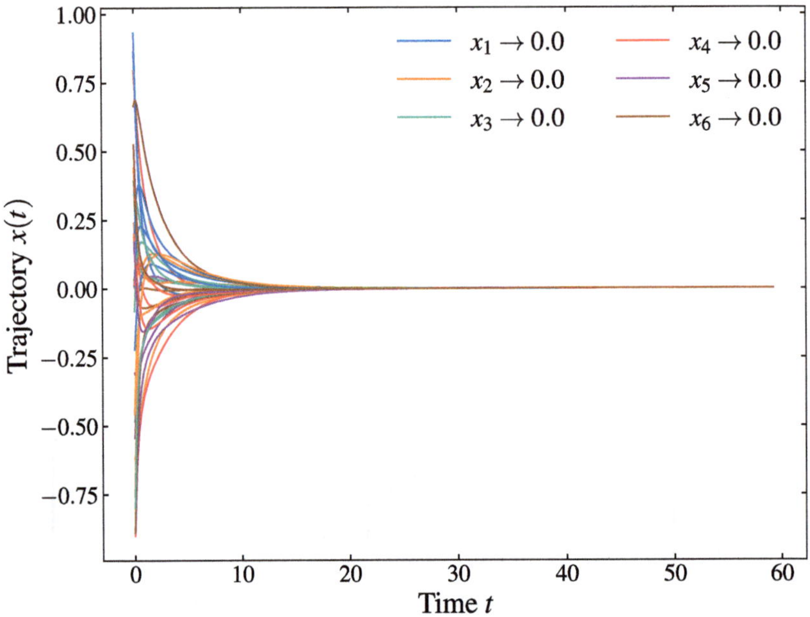

Fig. 5.8 Convergence of $x(t)$ to the SOCCVI solution in Example 5.3 using five random initial points, where $\rho = 10^3$

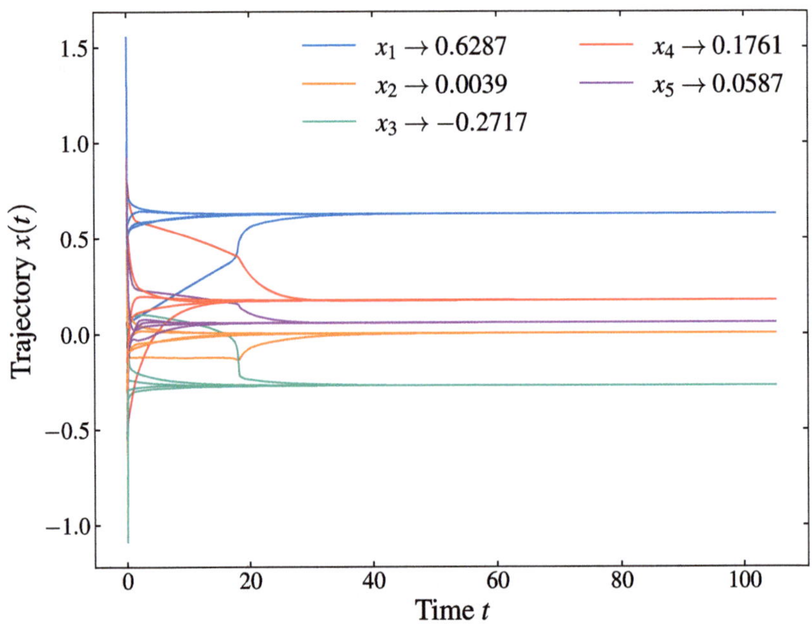

Fig. 5.9 Convergence of $x(t)$ to the SOCCVI solution in Example 5.4 using five random initial points, where $\rho = 10^3$

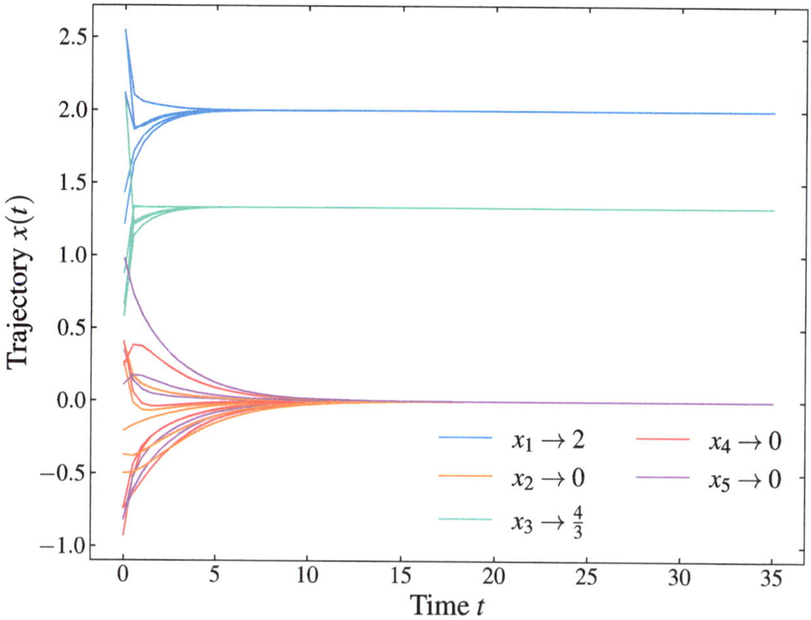

Fig. 5.10 Convergence of $x(t)$ to the SOCCVI solution in Example 5.5 using five random initial points, where $\rho = 10^3$

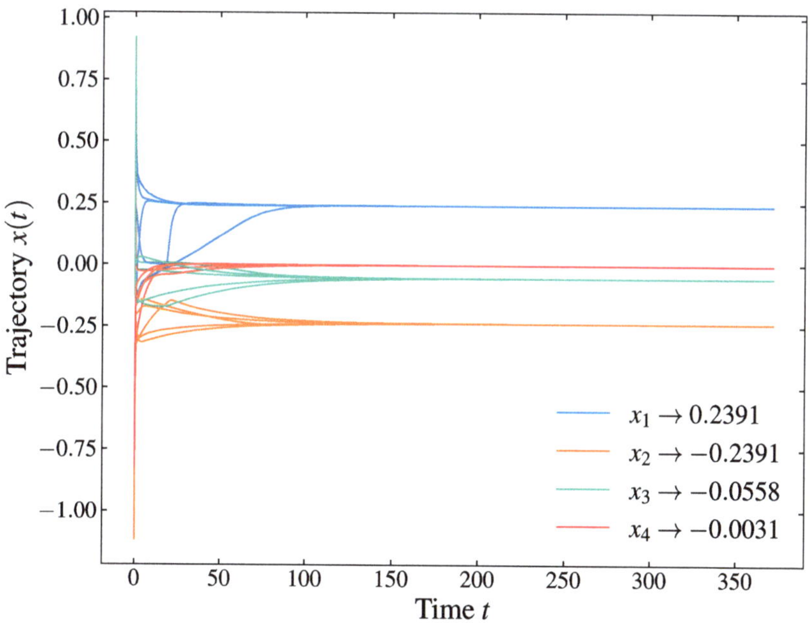

Fig. 5.11 Convergence of $x(t)$ to the SOCCVI solution in Example 5.6 using five random initial points, where $\rho = 10^3$

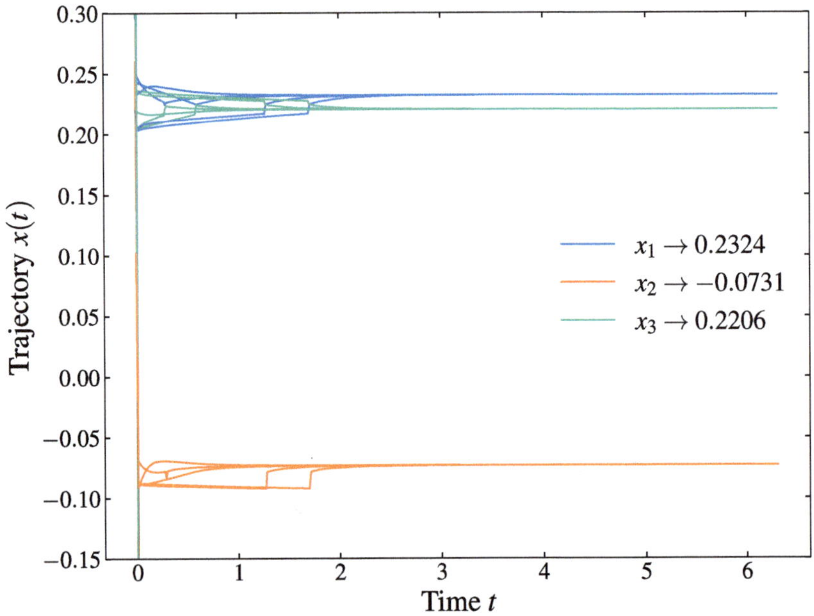

Fig. 5.12 Convergence of $x(t)$ to the SOCCVI solution in Example 5.7 using five random initial points, where $\rho = 10^3$

$$C = \{x \in \mathbb{R}^4 \mid -g(x) = x \in \mathcal{K}^3 \times \mathcal{K}^3 \times \mathcal{K}^2\}.$$

Here, $x^* = (0.3820, 0.1148, -0.3644, 0.0000, 0.0000, 0.0000, 0.5000, -0.2500)$ is the SOCCVI solution.

Example 5.3 Let

$$\left\langle \frac{1}{2}Dx, y - x \right\rangle \geq 0, \quad \forall y \in C$$

where

$$C = \{x \in \mathbb{R}^n \mid Ax - a = 0, \; Bx - b \preceq 0\},$$

$A \in \mathbb{R}^{l\times n}$, $B \in \mathbb{R}^{m\times n}$, $D \in \mathbb{R}^{n\times n}$ is symmetric, $d \in \mathbb{R}^n$, $a \in \mathbb{R}^l$ and $b \in \mathbb{R}^m$, with $l + m \leq n$.

As in [58, Example 5.1], we let

$$D = (D_{ij})_{n\times n}, \text{ where } D_{ij} = \begin{cases} 2, & i = j \\ 1, & |i - j| = 1 \\ 0, & otherwise \end{cases},$$

$$A = \begin{bmatrix} I_{l\times l} & 0_{l\times(n-l)} \end{bmatrix}_{l\times n}, B = \begin{bmatrix} 0_{m\times(n-m)} & I_{m\times m} \end{bmatrix}_{m\times n}, a = 0_{l\times 1}, b = (e_{m_1}, \ldots, e_{m_p}),$$

where $e_{m_i} = (1, 0, \cdots, 0)^T \in \mathbb{R}^{m_i}$. We set $l = m = 3$ and $n = 6$ for the simulations, and the SOCCVI has $x^* = (0, 0, 0, 0, 0, 0)$ as its solution.

Example 5.4 Let

$$F(x) = \begin{pmatrix} x_3\exp(x_1x_3) + 6(x_1 + x_2) \\ 6(x_1 + x_2) + \dfrac{2(2x_2 - x_3)}{\sqrt{1 + (2x_2 - x_3)^2}} \\ x_1\exp(x_1x_3) - \dfrac{2x_2 - x_3}{\sqrt{1 + (2x_2 - x_3)^2}} \\ x_4 \\ x_5 \end{pmatrix}$$

and

$$C = \{x \in \mathcal{R}^5 \mid h(x) = 0, -g(x) \in \mathcal{K}^3 \times \mathcal{K}^2\},$$

with

$$h(x) = -62x_1^3 + 58x_2 + 167x_3^3 - 29x_3 - x_4 - 3x_5 + 11,$$

$$g(x) = \begin{pmatrix} -3x_1^3 - 2x_2 + x_3 - 5x_3^3 \\ 5x_1^3 - 4x_2 + 2x_3 - 10x_3^3 \\ -x_3 \\ -x_4 \\ -3x_5 \end{pmatrix}.$$

Here, $x^* = (0.6287, 0.0039, -0.2717, 0.1761, 0.0587)$.

Example 5.5 Let

$$F(x) = \begin{pmatrix} 2x_1 - 4 \\ e^{x_2} - 1 \\ 3x_3 - 4, \\ -\sin(x_4) \\ x_5 \end{pmatrix}$$

and

$$C = \left\{x \in \mathbb{R}^5 \mid -g(x) = x \in \mathcal{K}^5\right\}.$$

Here, $x^* = \left(2, 0, 1.3333, 0, 0\right)$.

Example 5.6 Consider the variational inequality

$$F(x) = \begin{pmatrix} 4x_1 - \sin x_1 \cos x_2 + 1 \\ -\cos x_1 \sin x_2 + 6x_2 + \frac{9}{5}x_3 + 2 \\ \frac{9}{5}x_2 + 8x_3 + 3 \\ 2x_4 + 1 \end{pmatrix}$$

and

$$C = \left\{x \in \mathbb{R}^4 \mid h(x) = 0, -g(x) \in \mathcal{K}^2\right\},$$

with

$$h(x) = \begin{pmatrix} x_1^2 - \frac{1}{10}x_2x_3 + x_3 \\ x_3^2 + x_4 \end{pmatrix} \quad \text{and} \quad g(x) = \begin{pmatrix} -x_1 \\ -x_2 \end{pmatrix}.$$

Here, $x^* = (0.2391, -0.2391, -0.0558, -0.0031)$.

Example 5.7 Consider the CSOCP [33]

$$\begin{aligned} &\min \exp(x_1 - x_3) + 3(2x_1 - x_2)^4 + \sqrt{1 + (3x_2 + 5x_3)^2} \\ &\text{s.t.} - g(x) = \begin{pmatrix} 4x_1 + 6x_2 + 3x_3 - 1 \\ -x_1 + 7x_2 - 5x_3 + 2 \\ x_1 \\ x_2 \\ x_3 \end{pmatrix} \in \mathcal{K}^2 \times \mathcal{K}^3. \end{aligned}$$

For this CSOCP, $x^* = (0.2324, -0.07309, 0.2206)$ is the approximate solution. This problem can be recast as an SOCCVI problem as discussed in the beginning of Sect. 5.2.3.

Example 5.8 We consider the grasping-force optimization problem for multifingered robotic hands [66, 67], which involves determining the minimum force that each finger must exert on an object so as to maintain the finger's grasp. In particular, we consider the problem in [67] involving a three-fingered robotic hand with fingers positioned at $(0, 1, 0)$, $(1, 0.5, 0)$ and $(0, -1, 0)$. The optimization problem is given by

$$\begin{aligned} &\min \tfrac{1}{2}y^{\mathsf{T}}y \\ &\text{s.t. } Gy = -f_{ext} \\ &\quad\sqrt{y_{i1}^2 + y_{i2}^2} \leq \mu_i y_{i3} \quad (i = 1, \ldots, m) \end{aligned}$$

where $y = (y_{11}, y_{12}, y_{13}, \ldots, y_{m3}) \in \mathbb{R}^{3m}$, $G \in \mathbb{R}^{6\times 3m}$ is the grasping transformation matrix, f_{ext} is the (time-varying) external wrench, and μ_i is the friction coefficient at finger i. As in [67], we let $\mu_i = \mu = 0.6$ for all i,

$$G = \begin{bmatrix} 0 & 1 & 0 & 0 & 0 & -1 & 1 & 0 & 0 \\ 0 & 0 & -1 & 0 & -1 & 0 & 0 & 0 & 1 \\ -1 & 0 & 0 & -1 & 0 & 0 & 0 & -1 & 0 \\ -1 & 0 & 0 & -0.5 & 0 & 0 & 0 & 1 & 0 \\ 0 & 0 & 0 & 1 & 0 & 0 & 0 & 0 & 0 \\ 0 & -1 & 0 & 0 & -1 & 0.5 & 1 & 0 & 0 \end{bmatrix},$$

and $f_{ext} = (0, f_c \sin\theta(t), -Mg + f_c \cos\theta(t), 0, 0, 0)^{\mathsf{T}}$, where $g = 9.8$ m/s^2, M is the mass of the object (assumed to be 0.1 kg), $f_c = Mv^2/r$ and $\theta(t) = vt/r$. The hand moves along a circular path of radius $r = 0.5$ m and constant velocity $v = 0.4\pi$ m/s.

In order to use our neural network, we recast the above problem as an SOCCVI. First, we let $(x_{i1}, x_{i2}, x_{i3}) = (\mu f_{i3}, f_{i1}, f_{i2})$. By this transformation, it can be shown that the problem corresponds to the SOCCVI with F, g, and h given as below:

$$F(x) = \operatorname{diag}\left(1/\mu^2, 1, 1, 1/\mu^2, 1, 1, 1/\mu^2, 1, 1\right) x$$

$$-\, g(x) = x \in \mathcal{K}^3 \times \mathcal{K}^3 \times \mathcal{K}^3$$

$$h(x) = \bar{G}x + f_{ext}$$

where

$$\bar{G} = \begin{pmatrix} 0 & 0 & 1 & -1/\mu & 0 & 0 & 0 & 1 & 0 \\ -1/\mu & 0 & 0 & 0 & 0 & -1 & 1/\mu & 0 & 0 \\ 0 & -1 & 0 & 0 & -1 & 0 & 0 & 0 & -1 \\ 0 & -1 & 0 & 0 & -0.5 & 0 & 0 & 0 & -1 \\ 0 & 0 & 0 & 0 & 1 & 0 & 0 & 0 & 0 \\ 0 & 0 & -1 & 0.5/\mu & 0 & -1 & 0 & 1 & 0 \end{pmatrix}.$$

We note that external wrench f_{ext} applied varies over time. In Fig. 5.13, we show the optimal force required as time varies from 0 to 1 s.

We now present a comparative analysis of the five neural network models discussed in this section for solving the SOCCVI problem. The first model is based on the smoothed Fischer-Burmeister (FB) function $\phi^{\varepsilon}_{\mathrm{FB}}$, where the smoothing parameter ε is gradually reduced to zero. The second model, given by (5.90), is formulated using a projection-based approach derived from an equivalent transformation of the KKT conditions. In addition, two neural networks were proposed in [60], each constructed from discrete generalizations of the FB and NR functions, namely, $\phi^{p}_{\mathrm{D-FB}}$ and ϕ^{p}_{NR}, as defined in (3.177) and (3.176), respectively. For clarity, we refer to these two networks as "DFB-NN" and "DNR-NN".

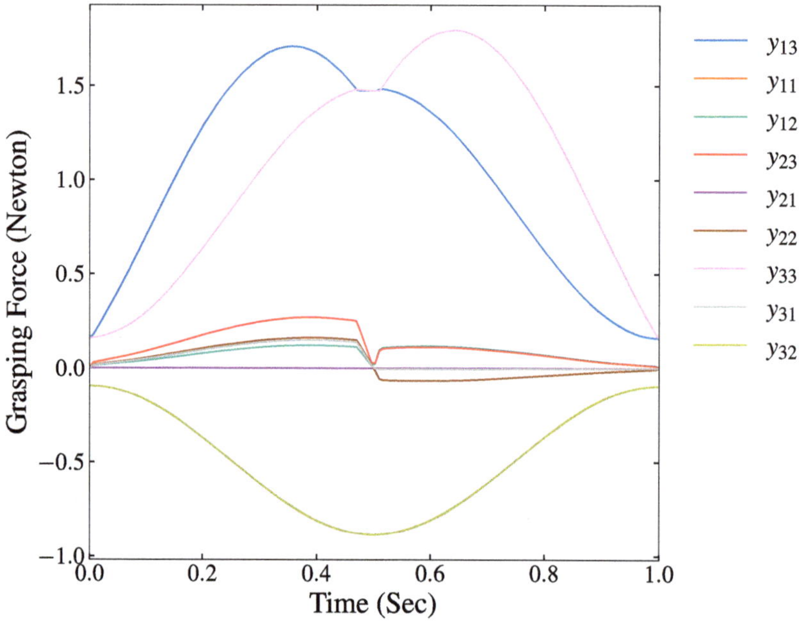

Fig. 5.13 Time-varying optimal grasping force for the three-fingered robotic hand

We begin by summarizing our findings based on several numerical experiments that assess the performance of the four previously studied neural networks, as reported in [38] and [60]. Among these models, the smoothed FB neural network consistently demonstrated the most robust numerical performance. It proved to be more reliable in solving SOCCVI problems across a range of instances and exhibited less sensitivity to variations in initial conditions. The projection-based neural network also performed well numerically. In cases where both the smoothed FB and projection-based models converged to the SOCCVI solution, the latter generally achieved faster convergence times. However, it was also observed that the projection-based neural network failed to solve certain problems that the smoothed FB neural network could handle successfully. Lastly, the DFB-NN and DNR-NN models, constructed from discrete generalizations of the FB and NR functions, were found to be considerably more sensitive to initial conditions, which often hindered their convergence reliability in practice.

We now compare the performance of our proposed smoothed NR neural network (5.105) with the four neural network models previously discussed. Overall, our model demonstrates superior stability and convergence properties. To support this claim, we conduct simulations on a suite of benchmark SOCCVI problems. Table 5.1 presents a summary of the simulation results for our model and the four existing neural networks from the literature. A check mark ("✓") indicates successful convergence to a solution of the SOCCVI, while a cross ("✗") denotes failure to do so. To assess and compare

Table 5.1 Summary of successful and unsuccessful simulation results for the five neural networks

	Ex. 1	Ex. 2	Ex. 3	Ex. 4	Ex. 5	Ex. 6
NN (5.105) by $\phi^{\varepsilon}_{\text{NR}}$	✓	✓	✓	✓	✓	✓
NN (5.82) by $\phi^{\varepsilon}_{\text{FB}}$	✓	✓	✓	✓	✓	✓
NN (5.90) by $\Pi_{\mathcal{K}}$	✓	✓	✗	✗	✗	✓
DFB-NN by $\phi^{p}_{\text{D-FB}}$	✗	✓	✗	✗	✓	✗
DNR-NN by ϕ^{p}_{NR}	✗	✓	✗	✗	✗	✗

convergence rates, we simulate the solution trajectories $z(t) = (x(t), \mu(t), \lambda(t))$ and compute the error $\|x(t) - x^*\|$, where x^* denotes the known SOCCVI solution. The error trajectories for each problem instance are displayed in Figs. 5.14, 5.16, 5.17, 5.18 and 5.19. Our key findings are summarized as follows:

- As shown in Table 5.1, only the neural network (5.105) based on the smoothed natural residual function $\phi^{\varepsilon}_{\text{NR}}$ and the smoothed FB neural network (5.82) based on $\phi^{\varepsilon}_{\text{FB}}$, successfully solved all the tested SOCCVI problems. The projection-based neural network achieved moderate performance, solving approximately half of the problems. In contrast, the two discrete-based neural networks, DFB-NN and DNR-NN, demonstrated the lowest success rate across the test suite.

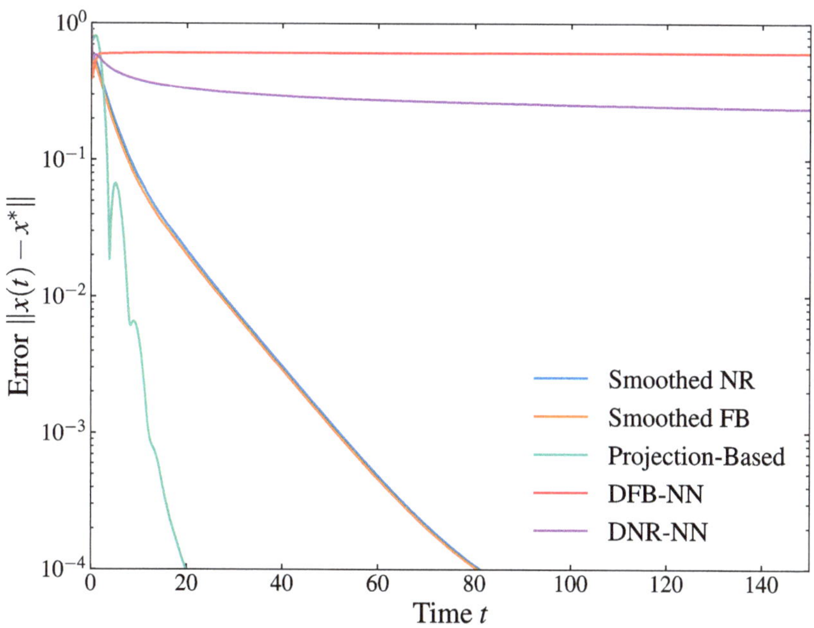

Fig. 5.14 Comparison of decay rates of $\|x(t) - x^*\|$ for the five neural networks for Example 5.2

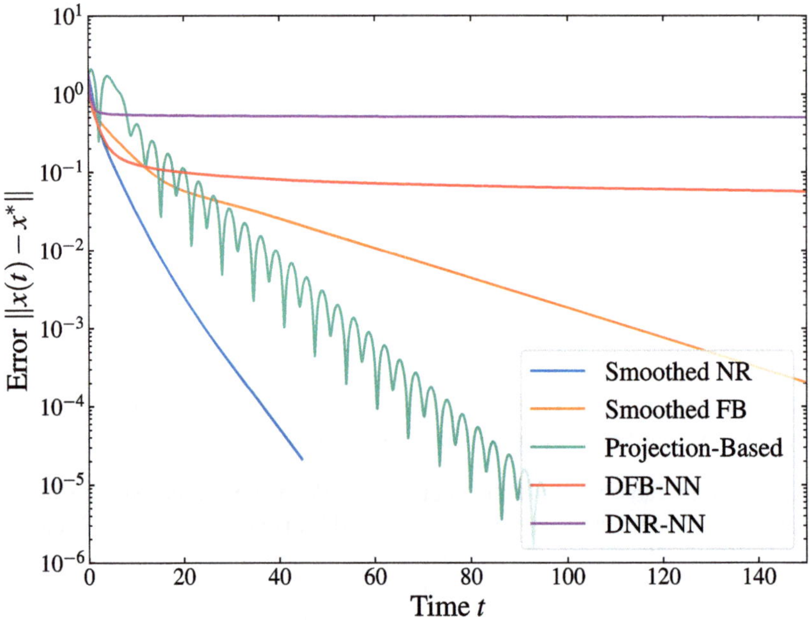

Fig. 5.15 Comparison of decay rates of $\|x(t) - x^*\|$ for the five neural networks for Example 5.3

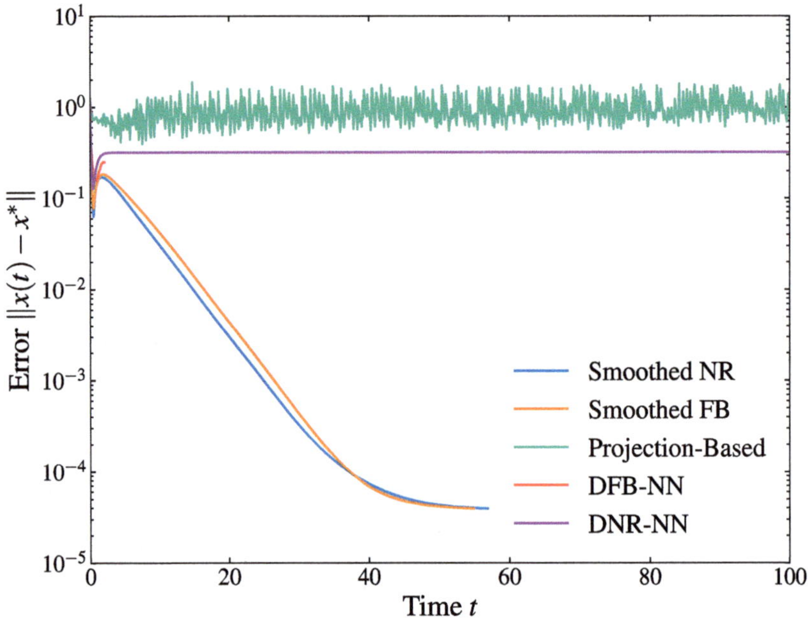

Fig. 5.16 Comparison of decay rates of $\|x(t) - x^*\|$ for the five neural networks for Example 5.4

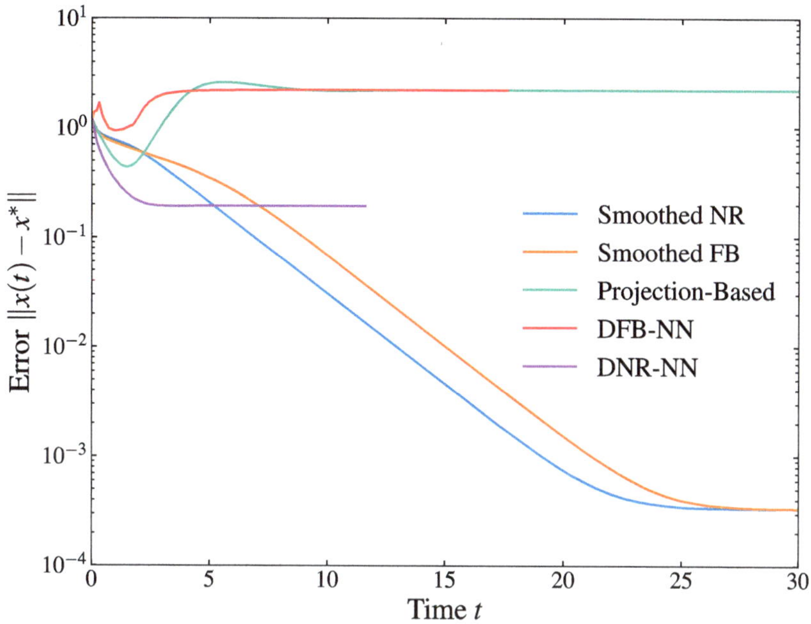

Fig. 5.17 Comparison of decay rates of $\|x(t) - x^*\|$ for the five neural networks for Example 5.5

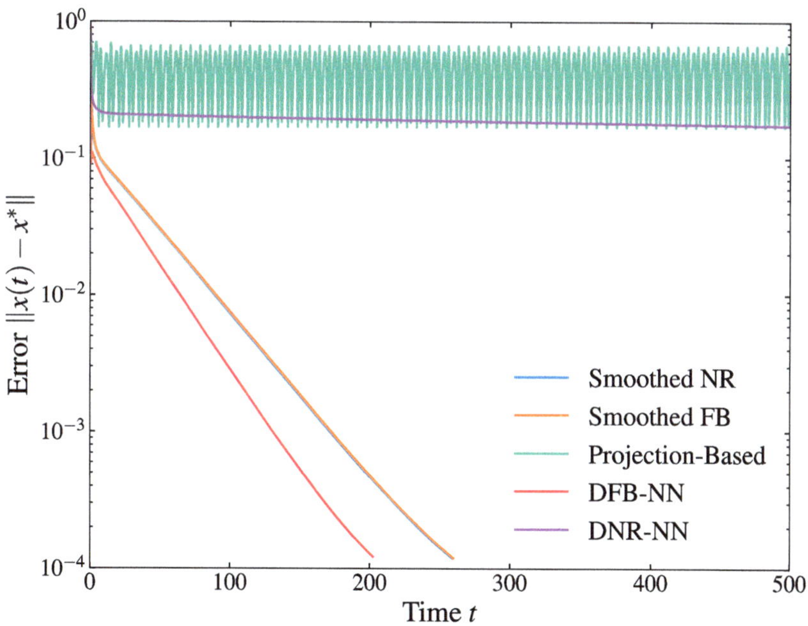

Fig. 5.18 Comparison of decay rates of $\|x(t) - x^*\|$ for the five neural networks for Example 5.6

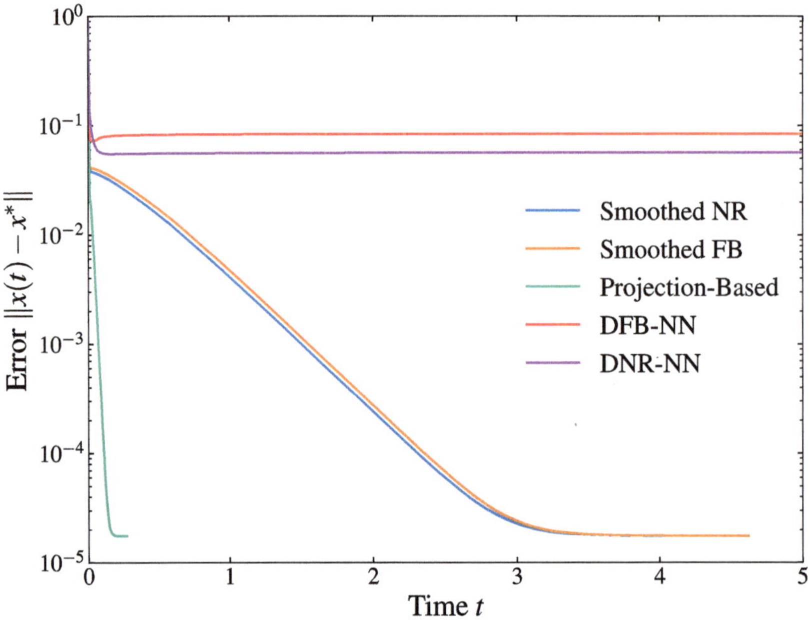

Fig. 5.19 Comparison of decay rates of $\|x(t) - x^*\|$ for the five neural networks for Example 5.7

- The projection-based neural network (5.90), constructed using the cone projection operator $\Pi_{\mathcal{K}}$, exhibits a notably fast convergence rate when it successfully approaches the SOCCVI solution, as observed in Examples 5.2 and 5.7 (see Fig. 5.14 and Fig. 5.19, respectively). However, in certain instances, specifically, Example 5.4 and Example 5.6, its trajectories display oscillatory behavior, ultimately failing to converge to the solution.
- Despite the rapid convergence exhibited by the projection-based neural network (5.90) using $\Pi_{\mathcal{K}}$, the smoothed NR neural network (5.105) based on $\phi^{\varepsilon}_{\text{NR}}$ still outperforms it in certain cases, as illustrated in Fig. 5.15. It is also worth noting that while both DFB-NN (using $\phi^{p}_{\text{D-FB}}$) and DNR-NN (using ϕ^{p}_{NR}) are able to solve Example 5.3, their convergence rates are exceedingly slow.
- The error plots presented in Figs. 5.14, 5.15, 5.16, 5.17, 5.18 and 5.19 indicate that the smoothed NR and smoothed FB neural networks exhibit nearly identical convergence rates. Moreover, both models demonstrate a notable insensitivity to variations in initial conditions. These observations suggest that the smoothed NR and smoothed FB neural networks are particularly well-suited for designing robust and reliable neural network frameworks for solving SOCCVI problems.
- Our numerical experiments indicate that the smoothed NR neural network (5.105), based on $\phi^{\varepsilon}_{\text{NR}}$, is less sensitive to initial conditions than the smoothed FB neural network (5.82), based on $\phi^{\varepsilon}_{\text{FB}}$. For example, in Example 5.2, it can be readily verified that the smoothed FB neural network fails to converge when initialized at $z_0 = (0, \ldots, 0)^{\mathsf{T}}$ or $z_0 = (1, \ldots, 1)^{\mathsf{T}}$.

Based on the above observations, we conclude that the smoothed NR neural network (5.105), based on $\phi^{\varepsilon}_{_{\rm NR}}$, and the smoothed FB neural network (5.82), based on $\phi^{\varepsilon}_{_{\rm FB}}$, demonstrate the best overall performance in solving SOCCVI problems. However, as previously noted, the smoothed FB neural network exhibits greater sensitivity to initial conditions, with divergence more likely to occur from certain starting points compared to the smoothed NR model.

Furthermore, a notable advantage of our smoothed NR neural network lies in its computational efficiency. In contrast, the smoothed FB neural network incurs significantly higher computational cost, primarily due to the complexity involved in evaluating the derivatives of the smoothed FB function $\phi^{\varepsilon}_{_{\rm FB}}$, as defined in (5.80). To see this, recall from [38, Lemma 3.1] that for any $\varepsilon \neq 0$,

$$\begin{aligned}
\nabla_{\varepsilon}\phi^{\varepsilon}_{_{\rm FB}}(a,b) &= e^{\mathsf{T}} L_z^{-1} L_{\varepsilon e},\\
\nabla_{a}\phi^{\varepsilon}_{_{\rm FB}}(a,b) &= L_z^{-1} L_a - I,\\
\nabla_{b}\phi^{\varepsilon}_{_{\rm FB}}(a,b) &= L_z^{-1} L_b - I,
\end{aligned}$$

where $z = (a^2 + b^2 + \varepsilon^2 e)^{1/2}$ and $L_a = \begin{bmatrix} a_1 & a_2^{\mathsf{T}} \\ a_2 & a_1 I_{n-1} \end{bmatrix}$ for $a = (a_1, a_2)^T \in \mathbb{R} \times \mathbb{R}^{n-1}$. It is important to note that all of the formulas discussed above involve the computation of matrix inverses, which can be computationally intensive. In particular, the smoothed FB neural network requires more extensive calculations during implementation, primarily due to the complexity of its underlying function and associated derivatives.

Finally, we offer a few remarks regarding the computational complexity of the five neural network models considered. While the architecture of the smoothed FB neural network is structurally similar to that of the smoothed NR neural network (see [38]), our findings indicate that the smoothed NR network offers superior convergence behavior. In contrast, the DFB-NN and DNR-NN models, though slightly less complex (see [60]), exhibit limited robustness and slower convergence. The projection-based neural network has the simplest architecture and the lowest computational complexity. However, as previously discussed, it suffers from stability issues and is not as reliable across problem instances.

To conclude, we summarize the neural network models discussed in this section in Table 5.2.

Table 5.2 Stability comparisons of neural networks considered in this section

	Section 5.2.1	Section 5.2.2	Section 5.2.3
Problem	$\min\ f(x)$ s.t. $Ax = b$ $x \in \mathcal{K}$	$\min\ f(x)$ s.t. $Ax = b$ $-g(x) \in \mathcal{K}$	$\langle F(x), y - x\rangle \geq 0, \quad \forall y \in C$ $C = \{x \mid h(x) = 0, -g(x) \in \mathcal{K}\}$
ODE	Using $\Pi_{\mathcal{K}}$		Using $\Pi_{\mathcal{K}}$
	Using $\phi_{\rm FB}$	Using $\phi^p_{\rm FB}$	Using $\phi^{\varepsilon}_{\rm FB}$
		Using ϕ_{μ}	Using $\phi^{\varepsilon}_{\rm NR}$
			Using $\phi^p_{\rm D-FB}$
			Using $\phi^p_{\rm NR}$
Stability	$\Pi_{\mathcal{K}}$ (Lyapunov)		$\Pi_{\mathcal{K}}$ (Lyapunov, asymptotical)
	$\phi_{\rm FB}$ (Lyapunov, asymptotical)	$\phi^p_{\rm FB}$ (Lyapunov, asymptotical, exponential)	$\phi^{\varepsilon}_{\rm FB}$ (Lyapunov, asymptotical, exponential)
		ϕ_{μ} (Lyapunov, asymptotical, exponential)	$\phi^{\varepsilon}_{\rm NR}$ (Lyapunov, asymptotical, exponential)
			$\phi^p_{\rm D-FB}$ (Lyapunov, asymptotical, exponential)
			$\phi^p_{\rm NR}$ (Lyapunov, asymptotical, exponential)

References

1. J.J. Hopfield, D.W. Tank, Neural computation of decision in optimization problems. Biol. Cybern. **52**, 141–152 (1985)
2. D.W. Tank, J.J. Hopfield, Simple neural optimization networks: an A/D converter, signal decision circuit, and a linear programming circuit. IEEE Trans. Circuits Syst. **33**, 533–541 (1986)
3. C. Dang, Y. Leung, X. Gao, K. Chen, Neural networks for nonlinear and mixed complementarity problems and their applications. Neural Netw. **17**, 271–283 (2004)
4. S. Effati, A.R. Nazemi, Neural network and its application for solving linear and quadratic programming problems. Appl. Math. Comput. **172**, 305–331 (2006)
5. S. Effati, A. Ghomashi, A.R. Nazemi, Application of projection neural network in solving convex programming problems. Appl. Math. Comput. **188**, 1103–1114 (2007)
6. Q. Han, L.-Z. Liao, H. Qi, L. Qi, Stability analysis of gradient-based neural networks for optimization problems. J. Global Optim. **19**, 363–381 (2001)
7. X. Hu, J. Wang, Solving pseudomonotone variational inequalities and pseudoconvex optimization problems using the projection neural network. IEEE Trans. Neural Netw. **17**, 1487–1499 (2006)
8. X. Hu, J. Wang, A recurrent neural network for solving a class of general variational inequalities. IEEE Trans. Syst. Man Cybern.-B **37**, 528–539 (2007)
9. M.P. Kennedy, L.O. Chua, Neural network for nonlinear programming. IEEE Tans. Circuits Syst. **35**, 554–562 (1998)
10. L.-Z. Liao, H. Qi, L. Qi, Solving nonlinear complementarity problems with neural networks: a reformulation method approach. J. Comput. Appl. Math. **131**, 342–359 (2001)
11. Y. Xia, H. Leung, J. Wang, A projection neural network and its application to constrained optimization problems. IEEE Trans. Circuits Syst.-I **49**, 447–458 (2002)
12. Y. Xia, H. Leung, J. Wang, A genarl projection neural network for solving monotone variational inequalities and related optimization problems. IEEE Trans. Neural Netw. **15**, 318–328 (2004)

13. Y. Xia, H. Leung, J. Wang, A recurrent neural network for solving nonlinear convex programs subject to linear constraints. IEEE Trans. Neural Netw. **16**, 379–386 (2005)
14. M. Yashtini, A. Malek, Solving complementarity and variational inequalities problems using neural networks. Appl. Math. Comput. **190**, 216–230 (2007)
15. S.H. Zak, V. Upatising, S. Hui, Solving linear programming problems with neural networks: a comparative study. IEEE Trans. Neural Netw. **6**, 94–104 (1995)
16. S.-K. Oh, W. Pedrycz, S.-B. Roh, Genetically optimized fuzzy polynomial neural networks with fuzzy set-based polynomial neurons. Inf. Sci. **176**, 3490–3519 (2006)
17. A. Shortt, J. Keating, L. Monlinier, C. Pannell, Optical implementation of the Kak neural network. Inf. Sci. **171**, 273–287 (2005)
18. G. Zhang, A neural network ensemble method with jittered training data for time series forecasting. Inf. Sci. **177**, 5329–5340 (2007)
19. R.K. Miller, A.N. Michel, *Ordinary Differential Equations* (Academic Press, 1982)
20. J.-S. Chen, H.-T. Gao, S.-H. Pan, An R-linearly convergent derivative-free algorithm for the NCPs based on the generalized Fischer-Burmeister merit function. J. Comput. Appl. Math. **232**(2), 455–471 (2009)
21. C. Kanzow, M. Fukushima, Equivalence of the generalized complementarity problem to differentiable unconstrained minimization. J. Optim. Theory Appl. **90**, 581–603 (1996)
22. F. Facchinei, J. Soares, A new merit function for nonlinear complementarity problems and a related algorithm. SIAM J. Optim. **7**, 225–247 (1997)
23. J.-S. Chen, C.-H. Ko, S.-H. Pan, A neural network based on generalized Fischer-Burmeister function for nonlinear complementarity problems. Inf. Sci. **180**, 697–711 (2010)
24. L. Qi, J. Sun, A nonsmooth version of Newton's method. Math. Program. **58**, 353–367 (1993)
25. S. Wiggins, *Introduction to Applied and Nonlinear Dynamical Systems and Chaos* (Springer-Verlag, New York Inc, 2003)
26. C. Geiger, C. Kanzow, On the resolution of monotone complementarity problems. *Computational Optimization and Applications*, vol. 5, pp. 155–173
27. J.-S. Chen, S.-H. Pan, A family of NCP functions and a descent method for the nonlinear complementarity problem. Comput. Optim. Appl. **40**, 389–404 (2008)
28. J.P. LaSalle, Stability theory for ordinary differential equations. J. Diff. Equ. **4**, 57–65 (1968)
29. J.H. Alcantara, J.-S. Chen, Neural networks based on three classes of NCP-functions for solving nonlinear complementarity problems. Neurocomputing **359**, 102–113 (2019)
30. F. Facchinei, J.-S. Pang, *Finite-Dimensional Variational Inequalities and Complementarity Problems* (Springer, New York, 2003), pp.1–2
31. J.H. Alcantara, J.-S. Chen, A novel generalization of the natural residual function and a neural network approach for the NCP. Neurocomputing **413**, 368–382 (2020)
32. J.-S. Chen, P. Tseng, An unconstrained smooth minimization reformulation of the second-order cone complementarity problem. Math. Program. **104**, 293–327 (2005)
33. C. Kanzow, I. Ferenczi, M. Fukushima, On the local convergence of semismooth Newton methods for linear and nonlinear second-order cone programs without strict complementarity. SIAM J. Optim. **20**, 297–320 (2009)
34. J.M. Ortega, W. Rheinboldt, *Iterative Solution of Nonlinear Equations in Several Variables* (SIAM, Philadelphia, 2000)
35. M. Fukushima, Equivalent differentiable optimization problems and descent methods for asymmetric variational inequality problems. Math. Program. **53**(1), 99–110 (1992)
36. R. Golden, *Mathematical Methods for Neural Network Analysis and Design, Cambridge* (The MIT Press, MA, 1996)
37. C.-H. Ko, J.-S. Chen, C.-Y. Yang, Recurrent neural networks for solving second-order cone programs. Neurocomputing **74**(17), 3646–3653 (2011)
38. J.-H. Sun, J.-S. Chen, C.-H. Ko, Neural networks for solving second-order cone constrained variational inequality problem. Comput. Optim. Appl. **51**(2), 623–648 (2012)
39. F. Alizadeh, D. Goldfarb, Seond-order cone programming. Math. Program. **95**, 3–51 (2003)
40. N. Kalouptisidis, *Sigunal Processing Systems* (Theory and design, New York, Wiley, 1997)

41. O. Mancino, G. Stampacchia, Convex programming and variational inequalities. J. Optim. Theory Appl. **9**, 3–23 (1972)
42. M. Fukushima, Z.-Q. Luo, P. Tseng, Smoothing functions for second-order cone complementarity problems. Comput. Optim. Appl. **17**, 159–182 (2000)
43. S. Hayashi, N. Yamashita, M. Fukushima, A combined smoothing and regularization method for monotone second-order cone complementarity problems. SIAM J. Optim. **15**, 593–615 (2005)
44. A. Cichocki, R. Unbehauen, *Neural Networks for Optimization and Signal Processing* (Wiley, New York, 1993)
45. J.-S. Pang, L. Qi, Nonsmooth equation: motivation and algorithms. SIAM J. Optim. **3**, 443–465 (1993)
46. X.-H. Miao, J.-S. Chen, C.-H. Ko, A smoothed NR neural network for solving nonlinear convex programs with second-order cone constraints. Inf. Sci. **268**(June), 255–270 (2014)
47. X.-H. Miao, J.-S. Chen, C.-H. Ko, A neural network based on the generalized FB function for nonlinear convex programs with second-order cone constraints. Neurocomputing **203**(August), 62–72 (2016)
48. P. Hartman, G. Stampacchia, On some nonlinear elliptic differential functional equations. Acta Math. **115**, 271–310 (1966)
49. J. Lions, G. Stampacchia, Variational inequalities. Commun. Pure Appl. Math. **20**, 493–519 (1967)
50. G. Stampacchia, Formes bilineares coercives sur les ensembles convexes. Comptes Rendus Academie Sci. Paris **258**, 4413–4416 (1964)
51. G. Stampacchia, Variational inequalities, Theory and Applications of Monotone Operators, in *Proceedings of the NATO Advanced Study Institute*, Venice, pp. 101–192 (1968)
52. M. Aghassi, D. Bertsimas, G. Perakis, Solving asymmetric variational inequalities via convex optimization. Oper. Res. Lett. **34**, 481–490 (2006)
53. Y. Kanno, J.A.C. Martins, A. Pinto da Costa, Three-dimensional quasi-static frictional contact by using second-order cone linear complementarity problem. Int. J. Numer. Methods Eng. **65**, 62–83 (2006)
54. D. Kinderlehrer, G. Stampacchia, *An Introduction to Variational Inequalities and Their Applications* (Academic Press, New York, 1980)
55. M. Patriksson, *Nonlinear Programming and Variational Inequality Problems: A Unified Approach, Applied Optimization, 23* (Kluwer, Dordrecht, 1998)
56. P. Harker, J.-S. Pang, Finite-dimensional variational inequality and nonlinear complementarity problems: a survey of theory, algorithms and applications. Math. Program. **48**, 161–220 (1990)
57. J. Hammond, *Solving asymmetric variational inequality problems and systems of equations with generalized nonlinear programming algorithms*, Ph.D. Dissertation, Department of Mathematics, MIT, Cambridge, 1984
58. J.-H. Sun and L.-W. Zhang, *A globally convergent method based on Fischer-Burmeister operators for solving second-order-cone constrained variational inequality problems*, Computers and Mathematics with Applications, vol. 58), pp. 1936–1946, 2009
59. S.-H. Pan, J.-S. Chen, Interior proximal methods and central paths for convex second-order cone programming. Nonlinear Anal. Theory Methods Appl. **73**(9), 3083–3100 (2010)
60. J.-H. Sun, X.-R. Wu, B. Saheya, J.-S. Chen, and C.-H. Ko, *Neural network for solving SOCQP and SOCCVI based on two discrete-type classes of SOC complementarity functions*, Mathematical Problems in Engineering, vol. 2019, Article ID 4545064, 18 pages, 2019
61. J.V. Outrata, D.F. Sun, On the coderivative of the projection operator onto the second order cone. Set-Valued Anal. **16**, 999–1014 (2008)
62. R.T. Rockafellar, R.J.-B. Wets, *Variational Analysis* (Springer, Berlin, 1998)
63. J.-F. Bonnans, H.C. Ramírez, Perturbation analysis of second-order cone programming problems. Math. Program. **104**, 205–227 (2005)
64. J.-S. Chen, *SOC Functions and Their Applications, Springer Optimization and Its Applications 143* (Springer, Singapore, 2019)

65. J.F. Bonnans, A. Shapiro, *Perturbation analysis of optimization problems* (Spring, New York, 2000)
66. M.S. Lobo, L. Vandenberghe, S. Boyd, H. Lebret, Application of second-order cone programming. Linear Algebra Appl. **284**, 193–228 (1998)
67. Y. Xia, J. Wang, L.-M. Fok, Grasping-foce optimization for multifingered robotic hands using a recurrent neural network. IEEE Trans. Robot. Autom. **20**, 549–554 (2004)

Bibliography

1. R.T. Rockafellar, Monotone operators and the proximal point algorithm. SIAM J. Control. Optim. **14**, 877–898 (1976)

J. Chen, *Complementarity Functions in Optimization*, Springer Optimization and Its Applications 228, https://doi.org/10.1007/978-3-032-06924-5

Index

J. Chen, *Complementarity Functions in Optimization*, Springer Optimization and Its Applications 228, https://doi.org/10.1007/978-3-032-06924-5

MIX
Papier aus verantwortungsvollen Quellen
Paper from responsible sources
FSC® C105338

If you have any concerns about our products,
you can contact us on
ProductSafety@springernature.com

In case Publisher is established outside the EU,
the EU authorized representative is:
Springer Nature Customer Service Center GmbH
Europaplatz 3, 69115 Heidelberg, Germany

Printed by Libri Plureos GmbH
in Hamburg, Germany